T0180511

Advances in Intelligent Systems and Computing

Volume 930

Series Editor

Janusz Kacprzyk, Systems Research Institute, Polish Academy of Sciences,
Warsaw, Poland

Advisory Editors

Nikhil R. Pal, Indian Statistical Institute, Kolkata, India
Rafael Bello Perez, Faculty of Mathematics, Physics and Computing,
Universidad Central de Las Villas, Santa Clara, Cuba
Emilio S. Corchado, University of Salamanca, Salamanca, Spain
Hani Hagras, Electronic Engineering, University of Essex, Colchester, UK
László T. Kóczy, Department of Automation, Széchenyi István University,
Gyor, Hungary
Vladik Kreinovich, Department of Computer Science, University of Texas
at El Paso, El Paso, TX, USA
Chin-Teng Lin, Department of Electrical Engineering, National Chiao
Tung University, Hsinchu, Taiwan
Jie Lu, Faculty of Engineering and Information Technology,
University of Technology Sydney, Sydney, NSW, Australia
Patricia Melin, Graduate Program of Computer Science, Tijuana Institute
of Technology, Tijuana, Mexico
Nadia Nedjah, Department of Electronics Engineering, University of Rio de Janeiro,
Rio de Janeiro, Brazil
Ngoc Thanh Nguyen, Faculty of Computer Science and Management,
Wrocław University of Technology, Wrocław, Poland
Jun Wang, Department of Mechanical and Automation Engineering,
The Chinese University of Hong Kong, Shatin, Hong Kong

The series "Advances in Intelligent Systems and Computing" contains publications on theory, applications, and design methods of Intelligent Systems and Intelligent Computing. Virtually all disciplines such as engineering, natural sciences, computer and information science, ICT, economics, business, e-commerce, environment, healthcare, life science are covered. The list of topics spans all the areas of modern intelligent systems and computing such as: computational intelligence, soft computing including neural networks, fuzzy systems, evolutionary computing and the fusion of these paradigms, social intelligence, ambient intelligence, computational neuroscience, artificial life, virtual worlds and society, cognitive science and systems, Perception and Vision, DNA and immune based systems, self-organizing and adaptive systems, e-Learning and teaching, human-centered and human-centric computing, recommender systems, intelligent control, robotics and mechatronics including human-machine teaming, knowledge-based paradigms, learning paradigms, machine ethics, intelligent data analysis, knowledge management, intelligent agents, intelligent decision making and support, intelligent network security, trust management, interactive entertainment, Web intelligence and multimedia.

The publications within "Advances in Intelligent Systems and Computing" are primarily proceedings of important conferences, symposia and congresses. They cover significant recent developments in the field, both of a foundational and applicable character. An important characteristic feature of the series is the short publication time and world-wide distribution. This permits a rapid and broad dissemination of research results.

** Indexing: The books of this series are submitted to ISI Proceedings, EI-Compendex, DBLP, SCOPUS, Google Scholar and Springerlink **

More information about this series at http://www.springer.com/series/11156

Álvaro Rocha · Hojjat Adeli ·
Luís Paulo Reis · Sandra Costanzo
Editors

New Knowledge in Information Systems and Technologies

Volume 1

 Springer

Editors
Álvaro Rocha
Departamento de Engenharia Informática
Universidade de Coimbra
Coimbra, Portugal

Luís Paulo Reis
Faculdade de Engenharia/LIACC
Universidade do Porto
Porto, Portugal

Hojjat Adeli
The Ohio State University
Columbus, OH, USA

Sandra Costanzo
DIMES
Università della Calabria
Arcavacata di Rende, Italy

ISSN 2194-5357   ISSN 2194-5365   (electronic)
Advances in Intelligent Systems and Computing
ISBN 978-3-030-16180-4   ISBN 978-3-030-16181-1   (eBook)
https://doi.org/10.1007/978-3-030-16181-1

Library of Congress Control Number: 2019934961

© Springer Nature Switzerland AG 2019
This work is subject to copyright. All rights are reserved by the Publisher, whether the whole or part of the material is concerned, specifically the rights of translation, reprinting, reuse of illustrations, recitation, broadcasting, reproduction on microfilms or in any other physical way, and transmission or information storage and retrieval, electronic adaptation, computer software, or by similar or dissimilar methodology now known or hereafter developed.
The use of general descriptive names, registered names, trademarks, service marks, etc. in this publication does not imply, even in the absence of a specific statement, that such names are exempt from the relevant protective laws and regulations and therefore free for general use.
The publisher, the authors and the editors are safe to assume that the advice and information in this book are believed to be true and accurate at the date of publication. Neither the publisher nor the authors or the editors give a warranty, expressed or implied, with respect to the material contained herein or for any errors or omissions that may have been made. The publisher remains neutral with regard to jurisdictional claims in published maps and institutional affiliations.

This Springer imprint is published by the registered company Springer Nature Switzerland AG
The registered company address is: Gewerbestrasse 11, 6330 Cham, Switzerland

Preface

This book contains a selection of papers accepted for presentation and discussion at The 2019 World Conference on Information Systems and Technologies (WorldCIST'19). This Conference had the support of IEEE SMC (IEEE Systems, Man, and Cybernetics Society), AISTI (Iberian Association for Information Systems and Technologies/Associação Ibérica de Sistemas e Tecnologias de Informação), GIIM (Global Institute for IT Management), and University of Vigo. It took place at La Toja, Galicia, Spain, April 16–19, 2019.

The World Conference on Information Systems and Technologies (WorldCIST) is a global forum for researchers and practitioners to present and discuss recent results and innovations, current trends, professional experiences and challenges of modern Information Systems and Technologies research, technological development and applications. One of its main aims is to strengthen the drive toward a holistic symbiosis between academy, society, and industry. WorldCIST'19 built on the successes of WorldCIST'13 held at Olhão, Algarve, Portugal; WorldCIST'14 held at Funchal, Madeira, Portugal; WorldCIST'15 held at São Miguel, Azores, Portugal; WorldCIST'16 held at Recife, Pernambuco, Brazil; WorldCIST'17 held at Porto Santo, Madeira, Portugal; and WorldCIST'18 took place at Naples, Italy.

The Program Committee of WorldCIST'19 was composed of a multidisciplinary group of more than 200 experts and those who are intimately concerned with Information Systems and Technologies. They have had the responsibility for evaluating, in a 'blind review' process, the papers received for each of the main themes proposed for the Conference: (A) Information and Knowledge Management; (B) Organizational Models and Information Systems; (C) Software and Systems Modeling; (D) Software Systems, Architectures, Applications and Tools; (E) Multimedia Systems and Applications; (F) Computer Networks, Mobility and Pervasive Systems; (G) Intelligent and Decision Support Systems; (H) Big Data Analytics and Applications; (I) Human–Computer Interaction; (J) Ethics, Computers and Security; (K) Health Informatics; (L) Information Technologies in Education; (M) Information Technologies in Radiocommunications; and (N) Technologies for Biomedical Applications.

The Conference also included workshop sessions taking place in parallel with the conference ones. Workshop sessions covered themes such as: (i) Air Quality and Open Data: Challenges for Data Science, HCI and AI; (ii) Digital Transformation; (iii) Empirical Studies in the Domain of Social Network Computing; (iv) Health Technology Innovation: Emerging Trends and Future Challenges; (v) Healthcare Information Systems Interoperability, Security and Efficiency; (vi) New Pedagogical Approaches with Technologies; (vii) Pervasive Information Systems.

WorldCIST'19 received about 400 contributions from 61 countries around the world. The papers accepted for presentation and discussion at the Conference are published by Springer (this book) in three volumes and will be submitted for indexing by ISI, Ei Compendex, Scopus, DBLP, and/or Google Scholar, among others. Extended versions of selected best papers will be published in special or regular issues of relevant journals, mainly SCI/SSCI and Scopus/Ei Compendex indexed journals.

We acknowledge all of those that contributed to the staging of WorldCIST'19 (authors, committees, workshop organizers, and sponsors). We deeply appreciate their involvement and support that was crucial for the success of WorldCIST'19.

April 2019   Álvaro Rocha
                                                                                          Hojjat Adeli
                                                                                        Luís Paulo Reis
                                                                                       Sandra Costanzo

Organization

Conference

General Chair

Álvaro Rocha   University of Coimbra, Portugal

Co-chairs

Hojjat Adeli   The Ohio State University, USA
Luis Paulo Reis   University of Porto, Portugal
Sandra Costanzo   University of Calabria, Italy

Local Chair

Manuel Pérez Cota   University of Vigo, Spain

Advisory Committee

Ana Maria Correia (Chair)   University of Sheffield, UK
Andrew W. H. Ip   Hong Kong Polytechnic University, China
Cihan Cobanoglu   University of South Florida, USA
Chris Kimble   KEDGE Business School and MRM, UM2,
           Montpellier, France
Erik Bohlin   Chalmers University of Technology, Sweden
Eva Onaindia   Universidad Politecnica de Valencia, Spain
Eugene H. Spafford   Purdue University, USA
Gintautas Dzemyda   Vilnius University, Lithuania
Gregory Kersten   Concordia University, Canada
Janusz Kacprzyk   Polish Academy of Sciences, Poland

João Tavares	University of Porto, Portugal
Jon Hall	The Open University, UK
Karl Stroetmann	Empirica Communication & Technology Research, Germany
Kathleen Carley	Carnegie Mellon University, USA
Keng Siau	Missouri University of Science and Technology, USA
Salim Hariri	University of Arizona, USA
Marjan Mernik	University of Maribor, Slovenia
Michael Koenig	Long Island University, USA
Miguel-Angel Sicilia	Alcalá University, Spain
Peter Sloot	University of Amsterdam, the Netherlands
Reza Langari	Texas A&M University, USA
Robert J. Kauffman	Singapore Management University, Singapore
Wim Van Grembergen	University of Antwerp, Belgium

Program Committee

Abdul Rauf	RISE SICS, Sweden
Adnan Mahmood	Waterford Institute of Technology, Ireland
Adriana Peña Pérez Negrón	Universidad de Guadalajara, Mexico
Adriani Besimi	South East European University, Macedonia
Agostinho Sousa Pinto	Polytecnic of Porto, Portugal
Ahmed El Oualkadi	Abdelmalek Essaadi University, Morocco
Alan Ramirez-Noriega	Universidad Autónoma de Sinaloa, Mexico
Alberto Freitas	FMUP, University of Porto, Portugal
Aleksandra Labus	University of Belgrade, Serbia
Alexandru Vulpe	Politehnica University of Bucharest, Romania
Ali Alsoufi	University of Bahrain, Bahrain
Ali Idri	ENSIAS, Mohammed V University, Morocco
Almir Souza Silva Neto	IFMA, Brazil
Amit Shelef	Sapir Academic College, Israel
Ana Isabel Martins	University of Aveiro, Portugal
Ana Luis	University of Coimbra, Portugal
Anabela Tereso	University of Minho, Portugal
Anacleto Correia	CINAV, Portugal
Anca Alexandra Purcarea	Politehnica University of Bucharest, Romania
André Marcos Silva	Centro Universitário Adventista de São Paulo (UNASP), Brazil
Aneta Poniszewska-Maranda	Lodz University of Technology, Poland
Angeles Quezada	Instituto Tecnologico de Tijuana, Mexico
Ankur Singh Bist	KIET, India
Antoni Oliver	University of the Balearic Islands, Spain
Antonio Borgia	University of Calabria, Italy

Antonio Jiménez-Martín	Universidad Politécnica de Madrid, Spain
Antonio Pereira	Polytechnic of Leiria, Portugal
Armando Toda	University of São Paulo, Brazil
Arslan Enikeev	Kazan Federal University, Russia
Benedita Malheiro	Polytechnic of Porto, ISEP, Portugal
Borja Bordel	Universidad Politécnica de Madrid, Spain
Branko Perisic	Faculty of Technical Sciences, Serbia
Carla Pinto	Polytechnic of Porto, ISEP, Portugal
Carla Santos Pereira	Universidade Portucalense, Portugal
Catarina Reis	Polytechnic of Leiria, Portugal
Cédric Gaspoz	University of Applied Sciences Western Switzerland (HES-SO), Switzerland
Cengiz Acarturk	Middle East Technical University, Turkey
Cesar Collazos	Universidad del Cauca, Colombia
Christophe Feltus	LIST, Luxembourg
Christophe Soares	University Fernando Pessoa, Portugal
Christos Bouras	University of Patras, Greece
Ciro Martins	University of Aveiro, Portugal
Claudio Sapateiro	Polytechnic Institute of Setúbal, Portugal
Cristian García Bauza	PLADEMA-UNICEN-CONICET, Argentina
Cristian Mateos	ISISTAN-CONICET, UNICEN, Argentina
Daniel Lübke	Leibniz Universität Hannover, Germany
Dante Carrizo	Universidad de Atacama, Chile
David Cortés-Polo	Fundación COMPUTAEX, Spain
Edita Butrime	Lithuanian University of Health Sciences, Lithuania
Edna Dias Canedo	University of Brasilia, Brazil
Eduardo Albuquerque	Federal University of Goiás, Brazil
Eduardo Santos	Pontifical Catholic University of Paraná, Brazil
Egils Ginters	Riga Technical University, Latvia
Eliana Leite	University of Minho, Portugal
Emiliano Reynares	CONICET-CIDISI UTN FRSF, Argentina
Evandro Costa	Federal University of Alagoas, Brazil
Fatima Azzahra Amazal	Ibn Zohr University, Morocco
Fernando Bobillo	University of Zaragoza, Spain
Fernando Moreira	Portucalense University, Portugal
Fernando Ribeiro	Polytechnic Castelo Branco, Portugal
Filipe Portela	University of Minho, Portugal
Filippo Neri	University of Naples, Italy
Fionn Murtagh	University of Huddersfield, UK
Firat Bestepe	Republic of Turkey, Ministry of Development, Turkey
Fouzia Idrees	Shaheed Benazir Bhutto Women University, Pakistan
Francesca Venneri	University of Calabria, Italy

Francesco Bianconi — Università degli Studi di Perugia, Italy
Francisco García-Peñalvo — University of Salamanca, Spain
Francisco Valverde — Universidad Central del Ecuador, Ecuador
Frederico Branco — University of Trás-os-Montes e Alto Douro, Portugal
Gabriel Pestana — Universidade Europeia, Portugal
Galim Vakhitov — Kazan Federal University, Russia
George Suciu — BEIA, Romania
Ghani Albaali — Princess Sumaya University for Technology, Jordan
Gian Piero Zarri — University Paris-Sorbonne, France
Giuseppe Di Massa — University of Calabria, Italy
Gonçalo Paiva Dias — University of Aveiro, Portugal
Goreti Marreiros — ISEP/GECAD, Portugal
Graciela Lara López — University of Guadalajara, Mexico
Habiba Drias — University of Science and Technology Houari Boumediene, Algeria
Hafed Zarzour — University of Souk Ahras, Algeria
Hamid Alasadi — University of Basra, Iraq
Hatem Ben Sta — University of Tunis at El Manar, Tunisia
Hector Fernando Gomez Alvarado — Universidad Tecnica de Ambato, Ecuador
Hélder Gomes — University of Aveiro, Portugal
Helia Guerra — University of the Azores, Portugal
Henrique da Mota Silveira — University of Campinas (UNICAMP), Brazil
Henrique S. Mamede — University Aberta, Portugal
Hing Kai Chan — University of Nottingham Ningbo China, China
Hugo Paredes — INESC TEC and Universidade de Trás-os-Montes e Alto Douro, Portugal
Ibtissam Abnane — Mohammed V University in Rabat, Morocco
Imen Ben Said — Université de Sfax, Tunisia
Ina Schiering — Ostfalia University of Applied Sciences, Germany
Inês Domingues — University of Coimbra, Portugal
Isabel Lopes — Instituto Politécnico de Bragança, Portugal
Isabel Pedrosa — Coimbra Business School ISCAC, Portugal
Isaías Martins — University of Leon, Spain
Ivan Lukovic — University of Novi Sad, Serbia
Jan Kubicek — Technical University of Ostrava, Czech Republic
Jean Robert Kala Kamdjoug — Catholic University of Central Africa, Cameroon
Jesús Gallardo Casero — University of Zaragoza, Spain
Jezreel Mejia — CIMAT Unidad Zacatecas, Mexico
Jikai Li — The College of New Jersey, USA
Jinzhi Lu — KTH Royal Institute of Technology, Sweden
Joao Carlos Silva — IPCA, Portugal

João Manuel R. S. Tavares	University of Porto, FEUP, Portugal
João Reis	University of Lisbon, Portugal
João Rodrigues	University of Algarve, Portugal
Jorge Barbosa	Polytecnic Institute of Coimbra, Portugal
Jorge Buele	Technical University of Ambato, Ecuador
Jorge Esparteiro Garcia	Polytechnic Institute of Viana do Castelo, Portugal
Jorge Gomes	University of Lisbon, Portugal
Jorge Oliveira e Sá	University of Minho, Portugal
José Álvarez-García	University of Extremadura, Spain
José Braga de Vasconcelos	Universidade New Atlântica, Portugal
Jose Luis Herrero Agustin	University of Extremadura, Spain
José Luís Reis	ISMAI, Portugal
Jose Luis Sierra	Complutense University of Madrid, Spain
Jose M. Parente de Oliveira	Aeronautics Institute of Technology, Brazil
José Machado	University of Minho, Portugal
José Martins	Universidade de Trás-os-Montes e Alto Douro, Portugal
Jose Torres	University Fernando Pessoa, Portugal
José-Luís Pereira	Universidade do Minho, Portugal
Juan Jesus Ojeda-Castelo	University of Almeria, Spain
Juan M. Santos	University of Vigo, Spain
Juan Pablo Damato	UNCPBA-CONICET, Argentina
Juncal Gutiérrez-Artacho	University of Granada, Spain
Justyna Trojanowska	Poznan University of Technology, Poland
Katsuyuki Umezawa	Shonan Institute of Technology, Japan
Khalid Benali	LORIA, University of Lorraine, France
Korhan Gunel	Adnan Menderes University, Turkey
Krzysztof Wolk	Polish-Japanese Academy of Information Technology, Poland
Kuan Yew Wong	Universiti Teknologi Malaysia (UTM), Malaysia
Laila Cheikhi	Mohammed V University, Rabat, Morocco
Laura Varela-Candamio	Universidade da Coruña, Spain
Laurentiu Boicescu	E.T.T.I. U.P.B., Romania
Leonardo Botega	University Centre Eurípides of Marília (UNIVEM), Brazil
Leonid Leonidovich Khoroshko	Moscow Aviation Institute (National Research University), Russia
Letícia Helena Januário	Universidade Federal de São João del-Rei, Brazil
Lila Rao-Graham	University of the West Indies, Jamaica
Luis Alvarez Sabucedo	University of Vigo, Spain
Luis Mendes Gomes	University of the Azores, Portugal
Luiz Rafael Andrade	Tiradentes University, Brazil
Luis Silva Rodrigues	Polytencic of Porto, Portugal

Luz Sussy Bayona Oré	Universidad Nacional Mayor de San Marcos, Peru
Magdalena Diering	Poznan University of Technology, Poland
Manuel Antonio Fernández-Villacañas Marín	Technical University of Madrid, Spain
Manuel Pérez Cota	University of Vigo, Spain
Manuel Silva	Polytechnic of Porto and INESC TEC, Portugal
Manuel Tupia	Pontifical Catholic University of Peru, Peru
Marco Ronchetti	Università di Trento, Italy
Mareca María PIlar	Universidad Politécnica de Madrid, Spain
Marek Kvet	Zilinska Univerzita v Ziline, Slovakia
María de la Cruz del Río-Rama	University of Vigo, Spain
Maria João Ferreira	Universidade Portucalense, Portugal
Maria João Varanda Pereira	Polytechnic Institute of Bragança, Portugal
Maria José Sousa	University of Coimbra, Portugal
María Teresa García-Álvarez	University of A Coruna, Spain
Marijana Despotovic-Zrakic	Faculty Organizational Science, Serbia
Mário Antunes	Polytecnic of Leiria and CRACS INESC TEC, Portugal
Marisa Maximiano	Polytechnic of Leiria, Portugal
Marisol Garcia-Valls	Universidad Carlos III de Madrid, Spain
Maristela Holanda	University of Brasilia, Brazil
Marius Vochin	E.T.T.I. U.P.B., Romania
Marlene Goncalves da Silva	Universidad Simón Bolívar, Venezuela
Maroi Agrebi	University of Polytechnique Hauts-de-France, France
Martin Henkel	Stockholm University, Sweden
Martín López Nores	University of Vigo, Spain
Martin Zelm	INTEROP-VLab, Belgium
Mawloud Mosbah	University 20 Août 1955 of Skikda, Algeria
Michal Adamczak	Poznan School of Logistics, Poland
Michal Kvet	University of Zilina, Slovakia
Miguel António Sovierzoski	Federal University of Technology - Paraná, Brazil
Mihai Lungu	Craiova University, Romania
Milton Miranda	Federal University of Uberlândia, Brazil
Mircea Georgescu	Al. I. Cuza University of Iasi, Romania
Mirna Muñoz	Centro de Investigación en Matemáticas A.C., Mexico
Mohamed Hosni	ENSIAS, Morocco
Mokhtar Amami	Royal Military College of Canada, Canada
Monica Leba	University of Petrosani, Romania

Muhammad Nawaz	Institute of Management Sciences, Peshawar, Pakistan
Mu-Song Chen	Dayeh University, China
Nastaran Hajiheydari	York St John University, UK
Natalia Grafeeva	Saint Petersburg State University, Russia
Natalia Miloslavskaya	National Research Nuclear University MEPhI, Russia
Naveed Ahmed	University of Sharjah, United Arab Emirates
Nelson Rocha	University of Aveiro, Portugal
Nelson Salgado	Pontifical Catholic University of Ecuador, Ecuador
Nikolai Prokopyev	Kazan Federal University, Russia
Niranjan S. K.	JSS Science and Technology University, India
Noemi Emanuela Cazzaniga	Politecnico di Milano, Italy
Noor Ahmed	AFRL/RI, USA
Noureddine Kerzazi	Polytechnique Montréal, Canada
Nuno Melão	Polytechnic of Viseu, Portugal
Nuno Octávio Fernandes	Polytechnic Institute of Castelo Branco, Portugal
Paôla Souza	Aeronautics Institute of Technology, Brazil
Patricia Zachman	Universidad Nacional del Chaco Austral, Argentina
Paula Alexandra Rego	Polytechnic Institute of Viana do Castelo and LIACC, Portugal
Paula Viana	Polytechnic of Porto and INESC TEC, Portugal
Paulo Maio	Polytechnic of Porto, ISEP, Portugal
Paulo Novais	University of Minho, Portugal
Paweł Karczmarek	The John Paul II Catholic University of Lublin, Poland
Pedro Henriques Abreu	University of Coimbra, Portugal
Pedro Rangel Henriques	University of Minho, Portugal
Pedro Sobral	University Fernando Pessoa, Portugal
Pedro Sousa	University of Minho, Portugal
Philipp Brune	University of Applied Sciences Neu-Ulm, Germany
Piotr Kulczycki	Systems Research Institute, Polish Academy of Sciences, Poland
Prabhat Mahanti	University of New Brunswick, Canada
Radu-Emil Precup	Politehnica University of Timisoara, Romania
Rafael M. Luque Baena	University of Malaga, Spain
Rahim Rahmani	Stockholm University, Sweden
Raiani Ali	Bournemouth University, UK
Ramayah T.	Universiti Sains Malaysia, Malaysia
Ramiro Delgado	Universidad de las Fuerzas Armadas ESPE, Ecuador

Ramiro Gonçalves	University of Trás-os-Montes e Alto Douro and INESC TEC, Portugal
Ramon Alcarria	Universidad Politécnica de Madrid, Spain
Ramon Fabregat Gesa	University of Girona, Spain
Reyes Juárez Ramírez	Universidad Autonoma de Baja California, Mexico
Rui Jose	University of Minho, Portugal
Rui Pitarma	Polytechnic Institute of Guarda, Portugal
Rui S. Moreira	UFP & INESC TEC & LIACC, Portugal
Rustam Burnashev	Kazan Federal University, Russia
Saeed Salah	Al-Quds University, Palestine
Said Achchab	Mohammed V University in Rabat, Morocco
Sajid Anwar	Institute of Management Sciences, Peshawar, Pakistan
Salama Mostafa	Universiti Tun Hussein Onn Malaysia, Malaysia
Sami Habib	Kuwait University, Kuwait
Samuel Fosso Wamba	Toulouse Business School, France
Sanaz Kavianpour	University of Technology, Malaysia
Sandra Costanzo	University of Calabria, Italy
Sandra Patricia Cano Mazuera	University of San Buenaventura Cali, Colombia
Sergio Albiol-Pérez	University of Zaragoza, Spain
Shahnawaz Talpur	Mehran University of Engineering and Technology, Jamshoro, Pakistan
Silviu Vert	Politehnica University of Timisoara, Romania
Simona Mirela Riurean	University of Petrosani, Romania
Slawomir Zolkiewski	Silesian University of Technology, Poland
Solange N. Alves-Souza	University of São Paulo, Brazil
Solange Rito Lima	University of Minho, Portugal
Sorin Zoican	Polytechnica University of Bucharest, Romania
Souraya Hamida	University of Batna 2, Algeria
Stefan Pickl	UBw München COMTESSA, Germany
Sümeyya Ilkin	Kocaeli University, Turkey
Syed Asim Ali	University of Karachi, Pakistan
Taoufik Rachad	Mohammed V University, Morocco
Tatiana Antipova	Institute of certified Specialists, Russia
The Thanh Van	HCMC University of Food Industry, Vietnam
Thomas Weber	EPFL, Switzerland
Timothy Asiedu	TIM Technology Services Ltd., Ghana
Tom Sander	New College of Humanities, Germany
Tomaž Klobučar	Jozef Stefan Institute, Slovenia
Toshihiko Kato	University of Electro-Communications, Japan
Tzung-Pei Hong	National University of Kaohsiung, Taiwan

Valentina Colla	Scuola Superiore Sant'Anna, Italy
Veronica Segarra Faggioni	Private Technical University of Loja, Ecuador
Victor Alves	University of Minho, Portugal
Victor Georgiev	Kazan Federal University, Russia
Victor Hugo Medina Garcia	Universidad Distrital Francisco José de Caldas, Colombia
Vincenza Carchiolo	University of Catania, Italy
Vitalyi Igorevich Talanin	Zaporozhye Institute of Economics and Information Technologies, Ukraine
Wolf Zimmermann	Martin Luther University Halle-Wittenberg, Germany
Yadira Quiñonez	Autonomous University of Sinaloa, Mexico
Yair Wiseman	Bar-Ilan University, Israel
Yuhua Li	Cardiff University, UK
Yuwei Lin	University of Roehampton, UK
Yves Rybarczyk	Universidad de Las Américas, Ecuador
Zorica Bogdanovic	University of Belgrade, Serbia

Workshops

First Workshop on Air Quality and Open Data: Challenges for Data Science, HCI and AI

Organizing Committee

Kai v. Luck	Creative Space for Technical Innovation, HAW Hamburg, Germany
Susanne Draheim	Creative Space for Technical Innovation, HAW Hamburg, Germany
Jessica Broscheit	Artist, Hamburg, Germany
Martin Kohler	HafenCity University Hamburg, Germany

Program Committee

Ingo Börsch	Technische Hochschule Brandenburg, Brandenburg University of Applied Sciences, Germany
Susanne Draheim	Hamburg University of Applied Sciences, Germany
Stefan Wölwer	HAWK University of Applied Sciences and Arts Hildesheim/Holzminden/Goettingen, Germany
Kai v. Luck	Creative Space for Technical Innovation, HAW Hamburg, Germany

Tim Tiedemann   Hamburg University of Applied Sciences,
                                     Germany
Marcelo Tramontano   University of São Paulo, Brazil

Second Workshop on Digital Transformation

Organizing Committee

Fernando Moreira   Universidade Portucalense, Portugal
Ramiro Gonçalves   Universidade de Trás-os-Montes e Alto Douro,
                                     Portugal
Manuel Au-Yong Oliveira   Universidade de Aveiro, Portugal
José Martins   Universidade de Trás-os-Montes e Alto Douro,
                                     Portugal
Frederico Branco   Universidade de Trás-os-Montes e Alto Douro,
                                     Portugal

Program Committee

Alex Sandro Gomes   Universidade Federal de Pernambuco, Brazil
Arnaldo Martins   Universidade de Aveiro, Portugal
César Collazos   Universidad del Cauca, Colombia
Jezreel Mejia   Centro de Investigación en Matemáticas A.C.,
                                     Mexico
Jörg Thomaschewski   University of Applied Sciences, Germany
Lorna Uden   Staffordshire University, UK
Manuel Ortega   Universidad de Castilla–La Mancha, Spain
Manuel Peréz Cota   Universidade de Vigo, Spain
Martin Schrepp   SAP SE, Germany
Philippe Palanque   Université Toulouse III, France
Rosa Vicardi   Universidade Federal do Rio Grande do Sul,
                                     Brazil
Vitor Santos   NOVA IMS Information Management School,
                                     Portugal

First Workshop on Empirical Studies in the Domain of Social Network Computing

Organizing Committee

Shahid Hussain   COMSATS Institute of Information Technology,
                                     Islamabad, Pakistan
Arif Ali Khan   Nanjing University of Aeronautics
                                     and Astronautics, China
Nafees Ur Rehman   University of Konstanz, Germany

Program Committee

Abdul Mateen	Federal Urdu University of Arts, Science & Technology, Islamabad, Pakistan
Aoutif Amine	ENSA, Ibn Tofail University, Morocco
Gwanggil Jeon	Incheon National University, Korea
Hanna Hachimi	ENSA of Kenitra, Ibn Tofail University, Morocco
Jacky Keung	City University of Hong Kong, Hong Kong
Kifayat Alizai	National University of Computer and Emerging Sciences (FAST-NUCES), Islamabad, Pakistan
Kwabena Bennin Ebo	City University of Hong Kong, Hong Kong
Mansoor Ahmad	COMSATS University Islamabad, Pakistan
Manzoor Ilahi	COMSATS University Islamabad, Pakistan
Mariam Akbar	COMSATS University Islamabad, Pakistan
Muhammad Khalid Sohail	COMSATS University Islamabad, Pakistan
Muhammad Shahid	Gomal University, DIK, Pakistan
Salima Banqdara	University of Benghazi, Libya
Siti Salwa Salim	University of Malaya, Malaysia
Wiem Khlif	University of Sfax, Tunisia

First Workshop on Health Technology Innovation: Emerging Trends and Future Challenges

Organizing Committee

Eliana Silva	University of Minho & Optimizer, Portugal
Joyce Aguiar	University of Minho & Optimizer, Portugal
Victor Carvalho	Optimizer, Portugal
Joaquim Gonçalves	Instituto Politécnico do Cávado e do Ave & Optimizer, Portugal

Program Committee

Eliana Silva	University of Minho & Optimizer, Portugal
Joyce Aguiar	University of Minho & Optimizer, Portugal
Victor Carvalho	Optimizer, Portugal
Joaquim Gonçalves	Instituto Politécnico do Cávado e do Ave & Optimizer, Portugal

Fifth Workshop on Healthcare Information Systems Interoperability, Security and Efficiency

Organizing Committee

José Machado	University of Minho, Portugal
António Abelha	University of Minho, Portugal

| Luis Mendes Gomes | University of Azores, Portugal |
| Anastasius Mooumtzoglou | European Society for Quality in Healthcare, Greece |

Program Committee

Alberto Freitas	University of Porto, Portugal
Ana Azevedo	ISCAP/IPP, Portugal
Ângelo Costa	University of Minho, Portugal
Armando B. Mendes	University of Azores, Portugal
Cesar Analide	University of Minho, Portugal
Davide Carneiro	University of Minho, Portugal
Filipe Portela	University of Minho, Portugal
Goreti Marreiros	Polytechnic Institute of Porto, Portugal
Helia Guerra	University of Azores, Portugal
Henrique Vicente	University of Évora, Portugal
Hugo Peixoto	University of Minho, Portugal
Jason Jung	Chung-Ang University, Korea
Joao Ramos	University of Minho, Portugal
José Martins	UTAD, Portugal
Jose Neves	University of Minho, Portugal
Júlio Duarte	University of Minho, Portugal
Luis Mendes Gomes	University of Azores, Portugal
Manuel Filipe Santos	University of Minho, Portugal
Paulo Moura Oliveira	UTAD, Portugal
Paulo Novais	University of Minho, Portugal
Teresa Guarda	Universidad Estatal da Península de Santa Elena, Ecuador
Victor Alves	University of Minho, Portugal

Fourth Workshop on New Pedagogical Approaches with Technologies

Organizing Committee

Anabela Mesquita	ISCAP/P.Porto and Algoritmi Centre, Portugal
Paula Peres	ISCAP/P.Porto and Unit for e-Learning and Pedagogical Innovation, Portugal
Fernando Moreira	IJP and REMIT – Univ Portucalense & IEETA – Univ Aveiro, Portugal

Program Committee

| Alex Gomes | Universidade Federal de Pernambuco, Brazil |
| Ana R. Luís | Universidade de Coimbra, Portugal |

Armando Silva	ESE/IPP, Portugal
César Collazos	Universidad del Cauca, Colombia
Chia-Wen Tsai	Ming Chuan University, Taiwan
João Batista	CICE/ISCA, UA, Portugal
Lino Oliveira	ESMAD/IPP, Portugal
Luisa M. Romero Moreno	Universidade de Sevilha, Espanha
Manuel Pérez Cota	Universidade de Vigo, Espanha
Paulino Silva	CICE & CECEJ-ISCAP/IPP, Portugal
Ramiro Gonçalves	UTAD, Vila Real, Portugal
Rosa Vicari	Universidade de Rio Grande do Sul, Porto Alegre, Brazil
Stefania Manca	Instituto per le Tecnologie Didattiche, Italy

Fifth Workshop on Pervasive Information Systems

Organizing Committee

Carlos Filipe Portela	Department of Information Systems, University of Minho, Portugal
Manuel Filipe Santos	Department of Information Systems, University of Minho, Portugal
Kostas Kolomvatsos	Department of Informatics and Telecommunications, National and Kapodistrian University of Athens, Greece

Program Committee

Andre Aquino	Federal University of Alagoas, Brazil
Carlo Giannelli	University of Ferrara, Italy
Cristina Alcaraz	University of Malaga, Spain
Daniele Riboni	University of Milan, Italy
Fabio A. Schreiber	Politecnico Milano, Italy
Filipe Mota Pinto	Polytechnic of Leiria, Portugal
Hugo Peixoto	University of Minho, Portugal
Gabriel Pedraza Ferreira	Universidad Industrial de Santander, Colombia
Jarosław Jankowski	West Pomeranian University of Technology, Szczecin, Poland
José Machado	University of Minho, Portugal
Juan-Carlos Cano	Universitat Politècnica de València, Spain
Karolina Baras	University of Madeira, Portugal
Muhammad Younas	Oxford Brookes University, UK
Nuno Marques	New University of Lisboa, Portugal
Rajeev Kumar Kanth	Turku Centre for Computer Science, University of Turku, Finland

Ricardo Queirós   ESMAD- P.PORTO & CRACS - INESC TEC,
                               Portugal
Sergio Ilarri   University of Zaragoza, Spain
Spyros Panagiotakis   Technological Educational Institute of Crete,
                               Greece

Contents

Information and Knowledge Management

My Employer's Prestige, My Prestige

Tom Sander[1](✉) and Phoey Lee Teh[2]

[1] New College of the Humanities, London, UK
tom.sander@nchlondon.ac.uk
[2] Department of Computing and Information Systems,
Faculty of Science and Technology, Sunway University, Subang Jaya, Malaysia
Phoeyleet@sunway.edu.my

Abstract. Employer branding is an essential component that attracts potential candidates to companies. Social media, particularly employer rating platforms, provide many opportunities to present a company's employer brand. Individuals use these platforms to collect information and evaluations about potential employers and companies could utilise these platforms to present themselves favourably. Based on social capital theory, this study examined the variables of support and benefit as reasons why individuals share information about their employers on employer rating platforms. The influence of demographic factors on the use of these platforms was also investigated. Data was collected from 309 respondents via an online survey, and analysed using the t-test, Spearman's correlation, and analysis of variance (ANOVA) with the least significant difference (LSD) method. Only descriptive statistics, distribution of responses, and statistically significant results are presented.

Keywords: Employer branding · Rating platforms · Social capital ·
Social media · Recruitment

1 Introduction

The labour market is currently changing to a candidate's market where there is demand for skilled employees but the number of potential candidates to meet this demand is decreasing. Labour market is refers to the availability of employment and labour, in terms of supply and demand. The market becomes more competitive and companies are finding it difficult to identify suitable employees from the pool of candidates (Sander 2013). Moreover, the importance of employer branding—how companies project themselves as favourable employers—is increasing. One way that companies can brand themselves positively is through the Internet.

The Internet is a powerful word-of-mouth marketing platform (Mukherjee and Banerjee 2017). The exchange of information, experience and knowledge in digital media creates value for a company. The value of information available allow individuals to aid in their decision-making. For example, TripAdvisor serves as a hotel-rating platform for travellers and as a way for hotels to improve their service based on ratings. The consistency in ratings, either good or bad, influences a traveller's decision when selecting a hotel (Khoo et al. 2017).

© Springer Nature Switzerland AG 2019
Á. Rocha et al. (Eds.): WorldCIST'19 2019, AISC 930, pp. 3–11, 2019.
https://doi.org/10.1007/978-3-030-16181-1_1

Social media provide new channels for job seekers to collect and share information about potential employers. These channels can be used by anyone with access to the Internet, making information easily accessible and available worldwide. It has become increasingly common for employees to share information with each other about their employers on social media. Job seekers could use such exchange of information when considering which company to apply to (Balaji et al. 2016; Luarn et al. 2015). Companies' reputations are also at stakes as a negative report about them can be read by anyone anytime (Vergeer 2014). Job seekers might not apply to a company that has been rated negatively. As such, companies could use employer rating platforms to brand themselves positively and therefore, reduce destructive criticism.

Employer rating platforms are online software-based tools that employees used to evaluate companies (Dabirian et al. 2017). Job seekers turn to these platforms to seek realistic and authentic information about potential employers (Li and Bernoff 2011; Sander et al. 2017). Individuals are able to anonymously share information such as company culture, benefits, and leadership behaviour as well as describe their working experience on these platforms (Teh et al. 2014; Wasko 2005). Due to the anonymity of these accounts, the information is deemed to be genuine and hence, trusted by job seekers (Bakir and McStay 2017). Individuals also use this channel to support each other, such as by encouraging (or discouraging) job seekers in applying to a particular company. Some companies use employer rating platforms to advertise themselves to potential candidates. They would use their employees as company ambassadors by asking them to rate the companies favourably. In short, these platforms provide an opportunity for employees to evaluate their companies and also for companies to react to these evaluations and communicate with potential candidates.

Previously, research has conducted on the reason of how to attract potential candidates to apply to their company through rating platform. The rating of the employees supports the success of the recruiting process which candidates have information to find for as a decision (Sander et al. 2015). The question of "Why would you evaluate your employer on an employer rating platform (e.g. Kununu or Glassdoor)?" has resulted in several feedbacks such as "I like to support my employer to be recognised as a good employer", "I like to inform other people about my employer" and "I like to motivate potential candidates to apply". These comments have exhibited the support of the employer to pursue suitable candidates for the company and the ratings or comments of the employees inform other people about an employer. To reiterate, information on employer rating platforms is deemed authentic as it came from the employees themselves and not from the company's branding or marketing department or communications officer. Individuals place more trust in information from "normal" employees as compared to information given by companies media centre or communication department (Klein et al. 2012). This is an advantage of employer rating platforms and the power of word-of-mouth.

The current study aimed to identify the reasons that motivate individuals to share information about their employers on employer rating platforms. To explore the benefit of employer rating platform is to present the prestige of the employer (Sander et al. 2017). Could a positive image of the employer increase the prestige of their employees? Could the prestige of an employer motivate suitable candidates to apply to the company? Employees would want experienced and suitable colleagues, which are essential for companies to be successful. Hence, employees evaluate their company on rating platforms to ensure that the company show one's true colour. Successful and competitive companies are a motivation for employees to work there as company success secures the employment and survival of the company.

2 Motivators to Share Information on Employer Rating Platforms

Social capital theory explains the exchange of information in social networks (Finkbeiner 2013). The Internet, particularly social media, has become an accepted channel or opinion-mining platform to share and obtain information. With the Internet, individuals have easy access to resources and information without any cost (Fussell et al. 2006). As the Internet creates large networks that connect people and is an infinite source of information, it has become a place to exchange knowledge and experience. Individuals can refer to these platforms to aid in their own decision-making or influence the decisions of others.

For employees, employer rating platforms are a place to share and exchange information as well as obtain needed information on companies (Hampton et al. 2011). In a way, individuals support one another by providing information or resources to each other. Such information or resources might not be obtainable if individuals did not receive such support. Individuals tend to receive information or resources from a third party, sometimes anonymous, via the Internet. This form of support is from an external source such as a machine, a rating platform or a person (Moon 2004). Hence, this leads to the first statement of the study examining the variable of support:

> "Individuals use employer rating platforms to exchange information about their employers to support other individuals"

Benefits are a valuable resource that improves the situation of an individual (e.g. prestige increases prosperity). Benefits have a positive influence on the individual and are hence desirable. A typical online benefit is the reputation and trust of information (Daigremont et al. 2008; Ellison et al. 2007), which are important for online users to obtain power and influence over others. Online users expect individuals that use their provided information to be obliged to them (Hampton and Wellman 2003; Hlebec et al. 2006). Thus, this leads to the second statement of the study examining the variable of benefit:

> "Individuals provide information on employer rating platforms to seek benefits"

The current study also investigated if demographic factors would influence the use of employer rating platforms. Demographic factors are important for the labour market and decision to select employees. The demographic factors tested in this study consisted of age, work experience, gender, and education level. This leads to the third statement of the study:

> "The use of employer rating platforms is influenced by demographic factors"

3 Methodology

The study was carried out in cooperation with a project at University of Ludwigshafen, Germany. As employer rating platforms are online tools (Evans and Mathur 2005; Wright 2005), the study recruited individuals with access to the Internet. An online survey conducted in the German language was forwarded randomly to over 900 individuals between November and December 2017, but only 309 individuals responded.

In terms of age, 2.5% of respondents were below 21 years, 57.6% were between 21 and 30 years, 6% were between 31 and 40 years, and 18.5% were above 40 years. Young individuals aged between 21 and 30 years formed the majority of respondents. These are individuals who have familiarised themselves with social media and rating platforms, using them in their daily life to evaluate products and services online (Miguéns et al. 2008). These platforms are an important marketing tool that young individuals trust and use when it comes to making a decision.

In terms of gender, 38.1% of respondents were male while the remaining 61.9% were female. As for education level, 30% of respondents have a school degree, 40.6% have an apprenticeship, 18.7% have a three-year university degree, 9.5% have a university degree of more than three years, and 1.2% have a doctorate degree or higher.

The survey comprised five items related to employment, rated on a 6-point Likert scale ranging from stages 1 (full agreement) to 6 (full disagreement). Survey's responses were analysed using the t-test, Spearman's correlation, and analysis of variance (ANOVA) with the least significant difference (LSD) method. Descriptive statistics, distribution of responses, and only statistically significant results are presented in the paper.

4 Results

The descriptive statistics presented in Table 1 show differences between the two variables, support and benefit. All items measuring support have a mode of 2, two items have a median of 2, and one item has a median of 3. In other words, the opinion to use employer rating platforms to transfer and share information is confirmed. On the other hand, the two items measuring benefit have a mode of 3 and 6 and a median of 3 and 4 respectively, indicating responses closer to full disagreement in the Likert scale.

Table 1. Descriptive statistics of survey responses

Item	N	Mean	Median	Mode	Standard deviation
I like to support my employer to be recognised as a good employer	308	2.67	2	2	1.391
I like to inform other people about my employer	309	2.65	2	2	1.379
I like to motivate potential candidates to apply	308	2.79	3	2	1.471
I like to provide feedback to my employer on an anonymous channel	308	2.97	3	3	1.540
Because the positive prestige of my employer has a positive influence on my prestige	307	4.12	4	6	1.610

The distribution of responses presented in Table 2 confirms the results and tendencies for all items. The first four items have a weak tendency to the stages 1 to 3. Surprisingly, the fifth item is the only item that exhibits a percentage of over 63.8% for stages 4 to 6. The fourth items averagely constitute 67.9% for stages 1 to 3. The results indicate a clear tendency of individuals using employer rating platforms to support other individuals. However, the results are not clear on whether or not individuals use employer rating platforms to seek benefits.

Table 2. Distribution of survey responses, Results presented in percentage (%); N = 307 – 309

Item	1 (full agreement)	2	3	4	5	6 (full disagreement)	Stage 1–3	Stage 4–6
I like to motivate potential candidates to apply	22.7	25.6	22.4	15.3	7.5	6.5	70.7	29.3
I like to inform other people about my employer	20.7	33.0	23.3	12.3	4.5	6.1	77.0	23.0
I like to support my employer to be recognized as a good employer	20.1	34.7	20.8	12.7	5.8	5.8	75.6	24.4
I like to provide feedback to my employer on an anonymous channel	19.5	24.0	24.4	12.7	10.6	8.8	67.9	32.1
Because the positive prestige of my employer has a positive influence on my prestige	7.2	12.1	16.9	16.6	19.5	27.7	36.2	63.8

In terms of the influence of demographic factors, age and gender were not found to be statistically significant factors in the use of employer rating platforms. This finding on gender is consistent with previous work that reported no statistically significant gender differences in the use of social media platforms (Sander et al. 2016). Spearman's correlation showed that only work experience was negatively correlated to the item "Because the positive prestige of my employer has a positive influence on my prestige", with a correlation coefficient of -0.180 ($p = 0.003$, n = 272). This indicates that individuals with more work experience are less involved in the use of employer rating platforms as compared to individuals with less work experienced.

To analyse the influence of education level, ANOVA with the LSD method was used. Statistically significant mean differences among the education levels were found for three of the five survey items. For the item "I like to inform other people about my employer", statistically significant differences between different education levels were found. The results are presented in Table 3.

Table 3. The mean difference between education levels for the item "I like to inform other people about my employer"

Education (I)	Education (J)	Mean difference (I–J)	Standard error	Sig.
University degree of more than three years (i.e. Master's degree)	Apprenticeship	0.600	0.300	0.047
	Three-year university degree (i.e. Bachelor's degree)	0.900	0.330	0.007
Doctorate degree or higher	School degree	1.663	0.799	0.038
	Apprenticeship	1.693	0.795	0.034
	Three-year university degree (i.e. Bachelor's degree)	1.994	0.807	0.014

For the item "I like to motivate potential candidates to apply", statistically significant differences were found among the education levels of doctorate degree or higher, school degree, apprenticeship, and three-year university degree. The mean difference between doctorate degree or higher and university degree of more than three years was not statistically significant. The results are presented in Table 4.

Table 4. The mean difference between education levels for the item "I like to motivate potential candidates to apply"

Education (I)	Education (J)	Mean difference (I–J)	Standard error	Sig.
Doctorate degree or higher	School degree	1.831	0.864	0.035
	Apprenticeship	1.886	0.860	0.029
	Three-year university degree (i.e. Bachelor's degree)	2.119	0.873	0.016

For the item "I like to provide feedback to my employer on an anonymous channel", statistically significant differences were found among the education levels of university degree of more than three years, school degree, and three-year university degree. The results are presented in Table 5.

Table 5. The mean difference between education levels for the item "I like to provide feedback to my employer on an anonymous channel".

Education (I)	Education (J)	Mean difference (I–J)	Standard error	Sig.
University degree of more than three years (i.e. Master's degree)	School degree	0.835	0.350	0.018
	Three-year university degree (i.e. Bachelor's degree)	0.789	0.374	0.036

These results do not indicate a stable continuous influence of education level on the different survey items as statistically significant differences were only reported for some survey items. The results do not present a general difference among the demographic factors.

5 Discussion and Practical Implications

Overall, the results confirm the importance and use of employer rating platforms in exchanging information. The first statement "Individuals use employer rating platforms to exchange information about their employers to support other individuals" was positively rated in the survey and hence, confirmed. Individuals like to use employer rating platforms to support others by providing information about their companies that can encourage or discourage job seekers in applying to their companies. This want to support others seems altruistic and intrinsic. Companies could utilise this intrinsic motivator by having their employees be ambassadors and promote the companies on employer rating platforms.

However, the second statement "Individuals provide information on employer rating platforms to seek benefits" was not confirmed. Based on the results, it does not seem that individuals use employer rating platforms to obtain benefits for themselves. Individuals are neither keen to improve their prestige nor make others feel obligated to them for having provided information. They would not use the chance to notify their employers through anonymous feedback on employer rating platforms to improve their working conditions. Further analysis is required to examine this statement.

In terms of demographic factors, the results have provided an interesting insight regarding education level which was found to influence the use of employer rating platforms. Individuals with higher education levels were more motivated to use employer rating platforms. Work experience was found to have a negative influence on only one item "Because the positive prestige of my employer has a positive influence on my prestige". In terms of gender, no statistically significant difference between male and female respondents were found; this is consistent with past research that examined gender differences in other social media platforms. Age was also not found to be a demographic factor that influences the use of employer rating platforms.

The current study only looked at the variables of support and benefit. Future research could examine further the motivation of individuals towards the discussion about the prestige of their company, and how it affects them proceeding to their next move of employment. Future research could also adopt a multicultural perspective by examining companies in different companies or multinational companies that have branches in different locations.

The results of the study highlight the importance of employer rating platforms and why individuals share information on these platforms. This study is important for companies to understand employee behaviour and why there is a need to motivate their employees to publish information on these platforms. Companies need to utilise employer rating platforms as a way to present their company favourably to potential candidates and job seekers.

References

Bakir, V., McStay, A.: Fake news and the economy of emotions: problems, causes, solutions. Digit. Journal. **0811**, 1–22 (2017)

Balaji, M.S., Khong, K.W., Chong, A.Y.L.: Determinants of negative word-of-mouth communication using social networking sites. Inf. Manag. **53**, 528–540 (2016)

Dabirian, A., Kietzmann, J., Diba, H.: A great place to work!? Understanding crowdsourced employer branding. Bus. Horiz. **60**(2), 197–205 (2017)

Daigremont, J., Skraba, R., Legrand, P., Hiribarren, V., Beauvais, M.: Social communications: applications that benefit from your real social network. In: International Conference on Intelligence in the Next Generation Networks (2008)

Ellison, N.B., Steinfield, C., Lampe, C.: The benefits of Facebook "friends:" social capital and college students' use of online social network sites. J. Comput. Mediat. Commun. **12**, 1143–1168 (2007)

Evans, J.R., Mathur, A.: The value of online surveys. Internet Res. **15**(2), 195–219 (2005)

Finkbeiner, P.: Social media and social capital: a literature review in the field of knowledge management. Int. J. Manag. Cases **15**, 6–20 (2013)

Khoo, F.S., Teh, P.L., Ooi, P.B.: Consistency of online consumers' perceptions of posted comments: an analysis of TripAdvisor reviews. J. Inf. Commun. Technol. **2**(2), 374–393 (2017)

Fussell, H., Harrison-Rexrode, J., Kennan, W.R., Hazleton, V.: The relationship between social capital, transaction costs, and organizational outcomes: a case study. Corp. Commun.: Int. J. **11**(2), 148–161 (2006)

Hampton, K.N., Lee, C., Her, E.J.: How new media affords network diversity: direct and mediated access to social capital through participation in local social settings. New Media Soc. **13**(7), 1031–1049 (2011)

Hampton, K., Wellman, B.: Neighboring in Netville: how the internet supports community and social capital in a wired suburb. City Community **2**(4), 277–311 (2003)

Hlebec, V., Manfreda, K.L., Vehovar, V.: The social support networks of internet users. New Media Soc. **1**, 9–32 (2006)

Klein, S., Bertschek, I., Falck, O., Mang, C.: Neue Informations-und Kommunikationstechnik: Lösung für den Arbeitsmarkt der Zukunft? Wirtschaftsdienst **92**(11), 723–736 (2012)

Li, C., Bernoff, J.: Groundswell, Winning in a World Transformed by Social Technologies. Harvard Business Review Press, Boston (2011)

Luarn, P., Yang, J.-C., Chiu, Y.-P.: Why people check in to social network sites. Int. J. Electron. Commer. **19**(4), 21–46 (2015)

Miguéns, J., Baggio, R., Costa, C.: Social media and tourism destinations: TripAdvisor case study. In: IASK ATR 2008, vol. 2008, pp. 1–6 (2008)

Moon, B.-J.: Consumer adoption of the internet as an information search and product purchase channel: some research hypotheses. Int. J. Internet Mark. Advert. **1**(1), 104–118 (2004)

Mukherjee, K., Banerjee, N.: Effect of social networking advertisements on shaping consumers' attitude. Glob. Bus. Rev. **18**(5), 1291–1306 (2017)

Sander, T.: New circumstances for the labor market under the consideration of social media. Commun. Glob. Inf. Technol. **5**, 41–52 (2013)

Sander, T., Sloka, B., Kalkis, H.: The trust of the information from employer rating platforms. Int. J. Web Portals **9**(1), 13–28 (2017)

Sander, T., Sloka, B., Teh, P.L.: Gender difference in the use of social network sites. In: Project Management Development – Practice and Perspectives, Riga, pp. 324–332 (2016)

Sander, T., Teh, P.L., Majláth, M.: User preference and channels use in the employment seeking process. In: Michelberrger, P. (ed.) Management, Enterprise and Benchmarking in the 21st Century II, pp. 239–248. Obuda University, Budapest (2015)

Teh, P.-L., Huah, L.P., Si, Y.-W.: The intention to share and re-shared among the young adults towards a posting at social networking sites. In: Rocha, Á., Correia, A.M., Tan, F.B., Stroetmann, K.A. (eds.) New Perspectives in Information Systems and Technologies, Volume 1. AISC, vol. 275, pp. 13–21. Springer, Cham (2014). https://doi.org/10.1007/978-3-319-05951-8_2

Vergeer, M.: Peers and sources as social capital in the production of news: online social networks as communities of journalists. Soc. Sci. Comput. Rev. **33**(3), 277–297 (2014)

Wasko, M.: Why should I share? Examining social capital and knowledge contribution in electronic networks of practice. MIS Q. **29**(1), 35–57 (2005)

Wright, K.B.: Researching internet-based populations: advantages and disadvantages of online survey research, online questionnaire authoring software packages, and web survey services. J. Comput.-Mediat. Commun. **10**(3), 1–15 (2005)

The Four Major Factors Impacting on the Future of Work

Michal Beno[(✉)]

VSM/City University of Seattle, Panonska cesta 17, 851 04 Bratislava, Slovakia
michal.beno@vsm-student.sk

Abstract. Automation, as manifested in modern technological innovation, is expected to lead to fundamental changes in the future of work. The impact on jobs, however, is uncertain. Several people expect to lose their jobs, while others are placing their faith in history and trust that new and better jobs will be created. To understand the future of work, we believe it is essential to explore four major factors that will impact on the future of work: (1) Technological progress, IT platforms, the sharing and knowledge economy; (2) Demographic, social and environmental changes; (3) Globalisation and glocalisation; and (4) Labour flexibility. Our aim is to gain a better understanding of the dynamics of the future of work by examining these four key factors that influence today's labour market, because this market is agile, since people can work anywhere at any time. In summary, seeing automation as synonymous with job losses is not correct. We contend that it is a mistake to believe that globalisation and technological advances lead to a reduction in the demand for human employees. However, it is possible that the opposing viewpoints of those who agree and those who disagree with this opinion are causing a polarisation of the workforce. Changes in our society, such as the constantly evolving demography, as well as environmental issues and ICT, have an influence on the way we work, and when, how and where we work.

Keywords: Future of work · Four major impacting factors · Labour market

1 Introduction

Clearly we are living in a world of fundamental transformation in the way that employers and employees work. Work is not only an integral part of our lives, but also one of the most important topics of human concern throughout the world [1–4]. Today's labour markets are undergoing a primary change and are facing less of a job crisis than a work revolution [5].

Looking back over history, the first evidence of a human work revolution was what was called the Agrarian Revolution, which involved domestic animals, new farming technologies, food production and population growth [6, 7]. It brought a radical change that was needed at the time. Subsequently four main industrial changes occurred in human history. The First Industrial Revolution from 1760 to 1840 was characterised by an era of mechanisation, including industries mechanised by steam energy, and textile, metallurgy and metal works. The era was symbolised by the steam engine and the train.

© Springer Nature Switzerland AG 2019
Á. Rocha et al. (Eds.): WorldCIST'19 2019, AISC 930, pp. 12–24, 2019.
https://doi.org/10.1007/978-3-030-16181-1_2

The Second Industrial Revolution from 1870 to 1914 was represented by innovations in communication, transport and manufacturing – the telegraph, telephone, Edison's light bulb, first flight, Henry Ford's Model T and others. This period is symbolised by steel. The Third Industrial Revolution from 1969 to 2000 includes new technologies, such as the Internet, computers and different forms of renewable energy, developments that have changed history. Production by modern methods using computers and new designs of machinery symbolises this era. The Fourth Industrial Revolution, also called the digital revolution, has been with us since the middle of the last century. Schwab [7] emphasises that this period is characterised by a much more ubiquitous and mobile Internet, by smaller and more powerful sensors that have also become cheaper, and by artificial intelligence and machine learning. He adds that the difference between the other three revolutions is the fusion of technologies and their interaction across the physical, digital and biological domains.

In our view, automation and thinking avatars are partly replacing human's tasks and jobs through the creation of modern skills needed for future organisations. Various highly disruptive technologies, such as artificial intelligence (AI), robotics, blockchain, the Internet of things (IoT), autonomous vehicles, 3D printing, nano- and biotechnology, materials science, energy storage and others [6, 7] are transforming social, economic and political systems in often unpredictable ways. Jack Ma, founder of Alibaba, a Chinese e-commerce giant, noted that "in the next 30 years, the world will see much more pain than happiness because social conflicts will have an impact on all sorts of industries and walks of life" [8, 9]. He stresses further that "machines will do what human beings are incapable of doing and will partner and cooperate with humans, rather than become mankind's biggest enemy" [10].

In this paper, we will outline four very different factors that will have implications for the world of work: (1) Technological progress and the sharing economy; (2) Demographic, social and environmental changes; (3) Globalisation and glocalisation; and (4) Labour flexibility. The impact of these on the future of work cannot be ignored by governments, organisations or individuals, as we are living through an essential transformation in the way we work. Automation and thinking machines are replacing human tasks and jobs. The resultant changed demographic trends and consumption have been causing environmental degradation. Continuing globalisation has a major influence on work, working conditions and work security all over our connected world. Thus the nature of work has changed, and this includes more flexible work anywhere and at any time.

This paper is based on a robust, evidence-based approach including key elements such as a comprehensive literature review and full analysis of trends and disruptions. In our study, we carried out a detailed literature analysis of publications related to the future of work. Drawing on this analysis, the four factors were explored systematically. Our aim is to gain a better understanding of the dynamics of future work and to illustrate the role of these four key factors that influence today's labour market, because this market is agile and people can work anywhere at any time.

2   The Four Major Impacting Factors

The greatest challenge in understanding the future of work lies in identifying the underlying implications for the four chosen factors and in understanding the roles played by these factors. Under ideal circumstances, all the parties involved (the individual, businesses and other employers, social and governmental institutions) will recognise the fundamental evolution in the nature of work. In this section, we review the major components that collectively constitute the future of work. On the basis of our experience and the literature review, we have identified four major impacting factors that shape and influence the nature of future work and the future workforce. These momentous changes raise huge organisational and social challenges, pose problems for the requisite human talent, and will result in risk, disruption and political and social revolution. Our preview describes the main factors that will influence thinking about many possible scenarios that could develop for the future of work.

2.1   Technological Progress, IT Platforms, the Sharing and Knowledge Economy

Technological advances like automation, robotics, AI, sensors and data have created entirely new ways of getting work done. This technology transformation of the nature of the workforce and also of organisations has led to a redesigning of most jobs.

Initially, automation focused on routine tasks, such as clerical work, bookkeeping, basic paralegal work and reporting. With modern technological advances and ever-increasing computing power (in other words, the digital revolution), it is now also increasingly likely that non-routine tasks will become automated [11]. Mandl et al. [12] emphasise that "cooperation among self-employed workers and SMEs is a traditional way of doing business to overcome the limitations".

Frey and Osborne [13] forecast that within a few decades many occupations will disappear in the face of new means of production. Gruen [14] states that "automation tends to take jobs but the invention of new complex tasks creates new jobs". As far as we know, technology is leading to the disappearance of jobs, but only technology can save them. This confirms the report by the McKinsey Global Institute [15], which predicts that by 2030 as many as 800 million jobs could be lost to automation worldwide. The report also states that, as in the past, technology will not have a purely negative impact on jobs, but new jobs will also be created. Nübler [16] recognises that technological change is non-linear, uncertain and a complex process that comes in waves, thereby driving job-destruction and job-creation at the same time. These phases do not happen automatically, but are brought about by a variety of forces at the economic, societal and political levels. Furthermore, the study says that advances in AI and robotics will have a drastic effect on everyday working lives [15]. We believe that jobs will not disappear entirely, but many will be redefined. Considering automation to be synonymous with job losses is not correct. Ford [17] explains that the job positions facing the most risk are those which "are on some level routine, repetitive and predictable". There is a high probability of automation for jobs like cashiers, counter and rental clerks, and telemarketers [13]. According to the report by Deloitte [18], there is a high likelihood that over 100 000 jobs in the legal sector could be automated.

Occupations at least risk of automation include recreational therapists, first-line supervisors of mechanics, installers, repairers, occupational therapists and healthcare social workers [13], in other words those where there is a close relationships with clients.

Part of the technological change in work is linked to IT platforms. Sharing-economy platforms have fostered competition and redefined industry boundaries in a range of businesses [19] and workplaces. We divide the evolution of the sharing economy into three phases: (1) communication (e.g. Yahoo, Aol); (2) Web 2.0 and social media (e.g. Google, Facebook, Twitter); and (3) the sharing economy. Sharing-economy workers drive you home, deliver groceries and office lunches or rent your home.

According to Stephany [20], sharing economy can be a confusing term. In practice, various definitions of this term are found, e.g. gig economy or on-demand economy, collaborative economy, access economy, peer-to-peer economy [21, 22]. Stephany [23] has provided a concise definition with five main limbs: "The sharing economy is the **value** in taking **underutilised assets** and making them **accessible online** to a community, leading to a **reduced need for ownership** of those assets". We believe that the sharing economy will experience significant growth in the future.

Employment in the knowledge-based economy is characterised by an increasing demand for more highly-skilled workers [24]. Powell and Snellman [25] define the knowledge economy as "production and services based on knowledge-intensive activities that contribute to an accelerated pace of technical and scientific advance, as well as rapid obsolescence. The key component of a knowledge economy is a greater reliance on intellectual capabilities than on physical inputs or natural resources".

2.2 Demographic, Social and Environmental Changes

Needs and demands for goods and services have undergone a transformation worldwide. The total world population in August 2018 was in excess of 7.63 billion people [26], and the trend is upward [27]. According to the available data, China is the most populous country, with more than 1.4 billion people, followed by India with 1.355 billion [26]. With figures like these to contend with, it is essential for consumption by the population and natural resources to be monitored.

As stated by the Technology at Work v2.0 report [28], the working age population in industrial countries peaked at 388 million in 2011, with the expectation of a decline by 30 million by 2030. Growth in working age populations for industrial countries has slowed from 1% per year for the period 1970–90 to 0.4% per year for 1990–2010, and the figure is expected to fall by 0.3% per year over the next 20 years. Growth in human capital may also slow. These authors believe that automation could pose more risks to jobs than demographic changes. But we assume that demographic change, imploding birth rates and the ageing society have rapidly moved to the centre of public debate worldwide. Buck et al. [29] highlight the following national/international factors relating to work development: *birth rates, the timing of births, family and household composition, age structure of the total population and of the labour force, labour force participation and employment rates among various groups of people, definition of working age population, and normal and early retirement.*

Movement of social values and behaviour creates new market spaces. Changes in the way we work have led to the expansion of various business ideas, in combination with social and health issues. Grantham [30] stresses the change in the approach to workplace design and the move from cost-based measures of building performance to more human-centric behavioural measures of wellness and well-being. This is confirmed by the growing global interest in this topic. For example, the United Nations [31] has placed Good Health and Well-being in the third place of the target list of 17 development goals by 2030. Maintaining a healthy and productive workforce remains a current priority, and is more demanding nowadays in the context of changes in organisations and the environment [32].

In the literature, it is stated that humans or human capital can offer companies a competitive advantage [33–35]. Schleicher [36] suggests that "knowledge is no longer stacked up in silos. What is required is the capacity to think across disciplines, connect ideas and 'construct information': these 'global competencies' will shape our world and the way we work and live together". Diversity of the workforce brings a strategic marketplace advantage to the workplace, as stated by Hunt et al. [37]. Managing populations of people in different age, gender, ethnicity and cultural groups brings a greater understanding of how different people collaborate to work together.

One of the major challenges facing organisations at present is to transform the culture of the organisation to fulfil new business realities and imbue employees with a sense of changing values [38]. Therefore those organisations that are able to complete this cultural transformation successfully and utilise the knowledge, skills and abilities of the employees will have a great advantage over those that fail to do so. Yet many organisations that try to carry out this transformation of culture are often unsuccessful. One of the possible reasons for this is that the motivations, needs and interests of employees differ according to the generation they are born in [39]. For this reason, Blattner and Walter [40] highlight the need to create a deep, engaging and highly efficient organisational culture that integrates the power of several generations and offers employees the opportunity to use their own talents and skills. These generations include those of **veterans** (born in 1922–1945), **baby boomers** (1946–1964), **generation X** (1965–1980), **generation Y** (millennials, 1981–2000) and, lastly, **generation Z** (1995–2012).

The effects of climate change are not highly apparent in the everyday life of the current generation, but its consequences are increasingly noticeable across continents. There is a strong consensus among the scientific community that the earth is warming and that this is caused by human CO_2 emissions. From our perspective, changing weather patterns significantly affect economic activity. All sectors should therefore prepare to face a range of new threats and opportunities, including co-operation. As reported by the OECD [41], the world economy is projected to nearly quadruple by 2050, with a growing demand for energy and natural resources as a result. Global greenhouse gas (GHG) emissions are expected to increase by 50%, primarily due to a 70% growth in energy-related CO_2.

2.3   Globalisation and Glocalisation

Huws et al. [42] emphasise that there is a growing awareness that the globalisation process is playing an increasingly important role in shaping work patterns. In the era of globalisation, most economists and policymakers have asserted that trade liberalisation has a strong potential to contribute to growth and that those effects will be beneficial to employment. This belief has strongly influenced the liberalisation policies of the past 25 years in multilateral, regional and bilateral settings. Yet evidence from surveys illustrates that negative perceptions of the labour-market effects of trade are frequent and persistent among the population, particularly in the industrialised world, but increasingly also in developing countries. Recent surveys show a rising concern about income and job security [43], although there are no indications that the interviewees had negative perceptions of globalisation. We feel that most people believe in the positive growth effects of globalisation, as is also shown by McKeon's [44] claims that this process has obliterated trade barriers, narrowed income gaps and bolstered economic co-operation among nations.

The definition of globalisation is a source of contention among scholars across the social sciences. This term became popular in the last decade of 20th century, although the phenomenon itself is actually much older, with periods of globalisation in the 16th and 19th centuries. Current globalisation is marked by several phenomena occurring at the same time: *new markets linked globally operating 24 hours a day; new technological tools; new actors, such as multinational corporations, global networks of non-governmental organisations (NGOs)* and *other groups transcending national boundaries; and new international rules increasingly binding national governments and reducing the scope for national policy.* Globalisation is not so much driven by technological progress, but is rather the outcome of political and ideological change. It is a human-led (or rather state-led) process [45].

OECD [46] defines globalisation as "generally used to describe an increasing internationalisation of markets for goods and services, the means of production, financial systems, competition, corporations, technology and industries. Amongst other things this gives rise to increased mobility of capital, faster propagation of technological innovations and an increasing interdependency and uniformity of national markets". We understand this term generally as a process of creating a worldwide network of business and markets. This causes greater mobility of goods, services, capital and workforces around the world. The process is made possible through the rapidly decreasing cost of ICT.

It is argued that globalisation has created the modern social, cultural, and political problems of the world. For example, it is blamed for increased competition. It ignores the fact that people in poorer countries need jobs. It encourages the production of goods for the world market, rather than being self-sufficient, and global jobs reduce unrest and increase stability. And, finally, interdependent economies mean less war. But the main problem is that the World Trade Organisation (WTO) makes the rules for globalisation, and globalisation is accelerating the exodus of manufacturing and white-collar jobs from countries.

Employment in the informal economy has generally increased around the world, and various forms of non-standard employment have emerged in most regions of the developed world. Global trade and investment patterns tend to benefit capital, especially companies that can move quickly and easily across borders, and to handicap labour, especially lower-skilled workers who cannot migrate easily, if at all [47]. To increase their global competitiveness, more and more investors are moving to countries that have low labour costs or shifting to informal employment arrangements. Globalisation also tends to benefit large companies that can capture new markets quickly and easily, to the detriment of small and micro entrepreneurs who face difficulties gaining knowledge of, let alone access to, emerging markets. In sum, globalisation puts pressure on low-skilled workers and small producers by weakening their bargaining power and subjecting them to increasing competition. But globalisation can also lead to new opportunities for those who work in the economy (formal or informal) in the form of new jobs in telecentres, homeworking, teleworking, e-working, mobile working, etc.

A few salient trends depict the globalisation of work: *declining labour force participation ratios; a shift from employment in industry to employment in services; more precarious work; continuing or increasing youth unemployment; a decreasing labour share of national income; increasing wage differentiation; enterprises become transnational; production processes change; and migration* [45].

Cowen [48] labelled 1990 as the decade of globalisation, but the year 2000 is considered to be glocalisation [49]. This term is defined by Friedman [50] as the ability of a culture to encounter other powerful cultures, to absorb influences that naturally fit into and can enrich that culture, to resist foreign affairs. We understand this term as a link between global and local in growth and diversity, without overwhelming them. Some authors regard glocalisation as the real path of globalisation [51]. Globalisation has increased the social space, leading to borderless economic, ecological, financial, social, political and cultural dimensions for traditional societies, driven by a world changing at an unprecedented rate. Stueckelberger [52] emphasises that "the technological and economic speed of globalization has to slow down a bit (decelerate) and the ethical, cultural and political globalization has to speed up substantially (accelerate)".

2.4 Work Flexibility

The Industrial Revolution brought employees from their homes to the factories. With ICT, the reverse is possible, with employees now able to move back to their homes [53]. At present, workplace innovation refers to a number of specific actions, such as teleworking, telecommuting, remote-work, networking, digital nomadic work, flexi-place, networking and many other variants [54, 55], alternative payment schemes, employee empowerment and autonomy, task rotation, multi-skilling, teamwork and team autonomy. We believe that technology is an important enabler of workplace innovation. Increasing traffic congestion and the rise in the cost of petrol also makes flexible working conditions more appealing and less stressful to employees.

The flexibility for jobholders to be able to work at any time in any place is technically feasible for many employees and has been for many years. In the literature, the following seems to be an accepted category now: subject, working anywhere, for over forty years [56–58]. Workplace flexibility has been defined as an opportunity for workers to choose, with their choice influenced by when, where and for how long they will engage in work-related tasks [59] in a variety of forms.

Using technological innovations, more and more organisations have started to redesign their approach to work. We feel strongly that central to this new approach is the fact that employees are asked to organise their work flexibly. The ICT technologies provide the possibility for workers to be monitored; technology forms only one part of the management process. Managing these workers involves a significant amount of trust. This means that flexible working is an essential part of the future of work. The era when the man went to the office and the woman stayed at home has now long gone. As stated in the World Development Report 2012, women make up nearly 51.8% of participation in labour markets, while the male rate fell slightly from 82.0% to 77.7% [60].

The latest data from the Flexible working survey demonstrate that work flexibility is still a huge priority in our working life: 67% wish they had this opportunity, for 47% it is not encouraged, and 40% of people would choose it over a pay rise. Although 58% of workers are offered flexible working, 24% do not make use of it. A massive 70% of workers strongly believe that flexible working would make a job more attractive, and if employers plan to hire top talent, flexible working is an imperative. Among those in the survey, 58% believe that working away from the office would help them to become more motivated, and 53% think they would be more productive [61].

3 Discussion

As we see it, the key attribute of the future of work is the time horizon. The time scale generally informs our vision of the future, and out of that the challenges, threats and opportunities arise. Relationships between individuals, employers, organisations and policymakers also develop over time; this means that work and society as a whole have changed over time and are continuing to do so. This raises a number of questions: If we believe that modern technology increases productivity, is the spread of flexibility also increasing? The next question is the relationship between modern work forms and health, and its impact on society. It seems likely that these will probably be renewed through a decrease of permanent employment contracts or an increase in the modern forms of flexible work.

Manyika [62] reviewed three aspects related to this discussion as follows: (1) the impact of artificial intelligence and automation on work and jobs, and whether we will have enough work and jobs left after that has run its course; (2) the changing models for work and work structure; and (3) whether any of those kinds of evolved work models will become the future, and whether people can work effectively and sustainably and earn sufficient living wages with enough support.

4 Conclusions

It is hard to predict the future. But this paper is an attempt to demonstrate four major factors impacting on the future of work. The recent wave of innovation and technological progress certainly has an influence on future work. Some believe that automation will destroy jobs on a massive scale, and they forecast a jobless future, but, as we know, technology can save these jobs. This is supported by historical experience, which demonstrates that such destruction is followed by job creation.

Each of the factors has distinct implications for the future of work. But the implications described require that action must be taken. To prepare employees for tomorrow's world of work, our study indicates four key areas for consideration by individuals, employees, employers, organisations and policymakers. The above discussion shows that all the following factors play an essential part in impacting on future work, namely technology, innovation, changes in society, globalisation and work flexibility. Therefore, we may assume that each of them has an important part to play in the overall impact.

Globalisation is a major contributing factor in today's world and structural change in the labour market. It has enabled individuals to become independent of a particular organisation. These independent workers enjoy flexibility in their work. To optimise the positive effects and minimise the negative ones, we must find a balance between flexibility, security and well-being.

The future of jobs needs to be shaped in line with social and political consensus. The beauty of the modern "office" technology is that it gives people the freedom to shape the work environment to fit individual needs and work styles. Work flexibility helps both companies and professionals embrace the workforce of the future.

Since the recruitment of employees is no longer confined to national borders, it can be seen that the concept of diversity of the workforce is emerging.

This example highlights the difficulties involved in forecasting change and shows the need to take a broad view when examining a question of this nature. This paper is an overview of a proposal to initiate further debate on the existence of other factors in this topic and could lead to further research into the analysis of other factors related to the future of work.

In summary, to regard automation as synonymous with job losses is not correct. We believe it is a mistake to believe that globalisation and technological advances lead to a reduction in the demand for human employees. However, the opposing viewpoints of those who agree and those who disagree with this opinion are probably causing a polarisation of the workforce. We believe that changes in society, such as our constantly evolving demography, as well as environmental issues and ICT, have a significant impact on the way we work, and on when, how and where we work. In our changing world of work, there will be winners and losers. The digitisation of work is creating new governance issues relating to the utilisation of information systems. Global labour instruments and the implementation of policies with the potential of integration into global markets will be crucial for a stable future of work. Getting the right information to the right people all over the globe at the right time 24 x 7 is the primary aim of the information system. This impacts on our mutual communication using modern ICT. Through the implementation of information systems by means of information, knowledge, communication and relationships, we conquer geographical and sometimes cultural boundaries. These systems are available anywhere and at any time, thereby creating modern and interesting jobs. Today, devices that we can hold in one hand are more powerful than any computers we have ever had in the past. The Internet, wireless local area networks, cellular networks, Wi-Fi, mobile Internet and apps have made the entire world accessible, allowing us to communicate and collaborate. As the world of information technology moves forward, we will constantly be challenged by new capabilities and innovations that will both amaze and disturb us. This will be a test to us and will require a new way of thinking about the world. Businesses and individuals alike must inculcate an awareness of these approaching changes and prepare for them.

References

1. Burgmann, V.: The Strange Death of Labour History, Bede Nairn and Labor History, pp. 69–81. Pluto Press, Sydney (1991)
2. Irving, T.: Challenges to Labour History. University of New South Wales Press, Sydney (1994)
3. Moody, C.J., Kessler-Harris, A.: Perspectives on American Labor History – The Problems of Synthesis. Northern Illinois University Press, DeKalb (1989)
4. Van der Linden, M.: The End of Labour History? Cambridge: Cambridge University Press (1993)
5. World Employment Confederation: The Future of Work (2016). https://www.wecglobal.org/fileadmin/templates/ciett/docs/WEC___The_Future_of_Work_-_What_role_for_the_employment_industry.pdf. Accessed 24 Sept 2018
6. Asian Development Bank: ASEAN 4.0: What does the Fourth Industrial Revolution mean for regional economic integration? (2017). https://www.adb.org/sites/default/files/publication/379401/asean-fourth-industrial-revolution-rci.pdf. Accessed 24 Sept 2018
7. Schwab, K.: The Fourth Industrial Revolution. World Economic Forum, Cologny/Geneva (2016)
8. Price, R.: ALIBABA'S JACK MA: 'In the next 30 years, the world will see much more pain than happiness' (2017). https://www.businessinsider.de/alibaba-jack-ma-automation-decades-of-pain-2017-4?r=UK&IR=T. Accessed 25 Sept 2018
9. Solon, O: Alibaba founder Jack Ma: AI will cause people 'more pain than happiness' (2017). https://www.theguardian.com/technology/2017/apr/24/alibaba-jack-ma-artificial-intelligence-more-pain-than-happiness. Accessed 25 Sept 2018
10. Langlois, S.: A grim warning from the usually upbeat Jack Ma: there will be pain (2017). https://www.marketwatch.com/story/the-usually-upbeat-jack-ma-there-will-be-pain-2017-04-24. Accessed 25 Sept 2018
11. OECD: Future of Work and Skills (2017). https://www.oecd.org/els/emp/wcms_556984.pdf. Accessed 25 Sept 2018
12. Mandl, I., et al.: New Forms of Employment. Publications Office of the European Union, Eurofond, Luxembourg (2015)
13. Frey, C.B., Osborne, M.A.: The future of employment. How susceptible are jobs to computerisation? Technol. Forecast. Soc. Change **114**(C), 254–280 (2017)
14. Gruen, D.: Technological Change and the Future of Work (2017). https://www.pmc.gov.au/news-centre/domestic-policy/technological-change-and-future-work. Accessed 25 Sept 2018
15. McKinsey Global Institute: Jobs Lost, Jobs Gained: Workforce transitions in a time of Automation (2017). https://www.mckinsey.com/~/media/mckinsey/featured%20insights/future%20of%20organizations/what%20the%20future%20of%20work%20will%20mean%20for%20jobs%20skills%20and%20wages/mgi%20jobs%20lost-jobs%20gained_report_december%202017.ashx. Accessed 25 Sept 2018

16. Nübler, I.: New technologies: a jobless future or golden age of job creation? International Labour Office, Research Department, Research Department working paper no. 13. ILO, Geneva (2016)

17. Ford, M.: Attention White-Collar Workers: The Robots Are Coming For Your Jobs (2015). https://www.npr.org/sections/alltechconsidered/2015/05/18/407648886/attention-white-collar-workers-the-robots-are-coming-for-your-jobs. Accessed 25 Sept 2018

18. Deloitte: Deloitte Insight: Over 100,000 legal roles to be automated (2016). https://www.legaltechnology.com/latest-news/deloitte-insight-100000-legal-roles-to-be-automated/. Accessed 25 Sept 2018

19. Constantiou, I., Marton, A., Tuunainen, V.K.: Four models of sharing economy platforms. MIS Q. Exec. **16**(4), 231–251 (2017)

20. Stephany, A.: Alex Stephany: how to understand the sharing economy (2014). https://www.lsnglobal.com/opinion/article/16302/alex-stephany-how-to-understand-the-sharing-econom. Accessed 25 Sept 2018

21. Steinmetz, K.: Exclusive: See How Big the Gig Economy Really Is (2016). http://time.com/4169532/sharing-economy-poll/. Accessed 25 Sept 2018

22. Selloni, D.: New Forms of Economies: Sharing Economy, Collaborative Consumption, Peer-to-Peer Economy. Springer International Publishing AG, CoDesign for Public-Interest Services, Research for Development (2017). https://doi.org/10.1007/978-3-319-53243-1_2

23. Stephany, A: The Business Sharing: Making it in the New Sharing Economy. Palgrave Macmillan, Hampshire (2015)

24. OECD: The Knowledge-Based Economy (1996). https://www.oecd.org/sti/sci-tech/1913021.pdf. Accessed 25 Sept 2018

25. Powell, W.W., Snellman, K.: The knowledge economy. Ann. Rev. Sociol. **30**, 199–220 (2004)

26. World Population Review (2018). http://worldpopulationreview.com/. Accessed 25 Sept 2018

27. Worldometers (2018). http://www.worldometers.info/world-population/. Accessed 25 Sept 2018

28. CITI GPS. Global Perspectives & Solutions: Technology at Work v2.0, The Future Is Not What It Used to Be (2016). https://www.oxfordmartin.ox.ac.uk/downloads/reports/Citi_GPS_Technology_Work_2.pdf. Accessed 26 Sept 2018

29. Buck, H., Kistler, E., Mendius, H.G.: Demographic Change in the World of Work. Bundesministerium für Bildung und Forschung, Stuttgart (2002)

30. Grantham, C.: Shifting the Paradigm From Wellness to Well-Being (2018). https://workdesign.com/2018/02/shifting-the-paradigm-from-wellness-to-wellbeing/. Accessed 26 Sept 2018

31. United Nations: About the Sustainable Development Goals (2015). https://www.un.org/sustainabledevelopment/sustainable-development-goals/. Accessed 26 Sept 2018

32. Kowalski, T.H.P., Loretto, W., Redman, T.: Call for papers: special issue of international journal of human resource management: well-being and HRM in the changing workplace. Int. J. Hum. Resour. Manag. **26**, 123–126 (2017)

33. Luthans, F., Youssef, C.M.: Human, social, and now positive psychological capital management: investing in people for competitive advantage. Org. Dyn. **33**, 143–160 (2004)

34. Pasban, M., Nojedeh, S.H.: A review of the role of human capital in the organization. Procedia Soc. Behav. Sci. **230**, 249–253 (2016)

35. Richard, O.: Racial diversity, business strategy, and firm performance: a resource-based view. Acad. Manag. J. **43**, 164–177 (2001)

36. OECD: 21st Century Skills: Learning for the Digital Age (2017). https://www.oecd-forum. org/users/50593-oecd/posts/20442-21st-century-skills-learning-for-the-digital-age. Accessed 26 Sept 2018

37. Hunt, V., Layton, D., Prince, S.: Why diversity matters (2015). https://www.mckinsey.com/ business-functions/organization/our-insights/why-diversity-matters. Accessed 26 Sept 2018

38. Graen, G., Grace, M.: New Talent Strategy: Attract, Process, Educate, Empower, Engage and Retain the Best (2015). https://www.shrm.org/hr-today/trends-and-forecasting/special-reports-and-expert-views/Documents/SHRM-SIOP%20New%20Talent%20Strategy.pdf. Accessed 26 Sept 2018

39. Twenge, J.M., Campbell, S.M., Hoffman, B.J., Lance, C.E.: Generational differences in work values: leisure and extrinsic values increasing, social and intrinsic values decreasing. J. Manag. **36**, 117–142 (2010)

40. Blattner, J., Walter, T.J.: Creating and sustaining a highly engaged company culture in a multigenerational workplace. Strateg. HR Rev. **14**, 124–130 (2015)

41. OECD: OECD Environmental Outlook to 2050: The Consequences of Inaction (2012). http://www.oecd.org/environment/indicators-modelling-outlooks/49846090.pdf. Accessed 26 Sept 2018

42. Huws, U, Jagger, N., O'Regan, S.: Teleworking and Globalization. IES Report 358, Brighton (1999)

43. ILO: Trade and Employment from Myths to Facts (2011). http://www.ilo.org/wcmsp5/ groups/public/—ed_emp/documents/publication/wcms_162297.pdf/. Accessed 26 Sept 2018

44. McKeon, A.: How the Globalization Cycle Could Impact HR and Your Organization (2017). https://www.adp.com/spark/articles/2017/10/how-the-globalization-cycle-could-impact-hr-and-your-organization.aspx. Accessed 26 Sept 2018

45. Van der Hoeven, R.: Globalization of Work, 2015 UNDP Human Development Report Office (2015). http://hdr.undp.org/sites/default/files/van_der_hoeven_hdr_2015_final.pdf. Accessed 26 Sept 2018

46. OECD: Globalization (2013). https://stats.oecd.org/glossary/detail.asp?ID=1121. Accessed 26 Sept 2018

47. Rodrik, D.: Has Globalization Gone Too Far? Institute for International Economics, Washington, D.C. (1997)

48. Cowen, T.: Creative Destruction: How Globalization is Changing the World's Cultures. Princeton University Press, Princeton (2002)

49. Westover, J.H.: Globalization, Labor, and the Transformation of Work: Readings for Seeking a Competitive Advantage in an Increasingly Global Economy. Common Ground Research Networks, Champaign (2010)

50. Friedman, T.L.: The Lexus and the Olive Tree. Farrar, Straus Giroux, New York (1999)

51. Robertson, R.: Globalization: Social Theory and Global Culture. Sage Publications, London (1992)

52. Stueckelberger, C.: Global Trade Ethics. An Illustrated Overview. WCC Publications, Geneva (2002)

53. Simitis, S.: The Juridification of labour relations. Comp. Labor Law **93**, 93–142 (1986)

54. Bates, P., Bertin, I., Huws, U.: E-Work in Ireland. IES Report 394, Brighton, 72 p. (2002)

55. Huws, U.: Statistical indicators of eWork, A Discussion paper. IES, Report 385, Brighton, 30 p. (2001)

56. Nilles, J.M.: Telecommunications and organizational decentralization. IEEE Trans. Commun. **23**(10), 1142–1147 (1975)

57. Wilkes, R.B., Frolick, M.N.: Critical issues in developing successful tele-work programs. J. Syst. Manag. **45**(7), 30 (1994)

58. Hunton, J.E.N., Strand, C.: The impact of alternative telework arrangements on organizational commitment: insights from a longitudinal field experiment. J. Inf. Syst. **24**(1), 67–90 (2010)
59. Bal, P.M., De Lange, A.H.: From flexibility human resource management to employee engagement and perceived job performance across the lifespan: a multi-sample study. J. Occup. Organisational Psychol. **88**(1), 126–154 (2014)
60. The World Bank: World Development Report 2012, Gender Equality and Development (2012). http://siteresources.worldbank.org/INTWDR2012/Resources/7778105-1299699968583/7786210-1315936222006/Complete-Report.pdf. Accessed 26 Sept 2018
61. Powwownow: Flexible Working in 2017. https://www.powwownow.co.uk/smarter-working/flexible-working-statistics-2017. Accessed 26 Sept 2018
62. Manyika, J.: What is the future of work? (2017). https://www.mckinsey.com/featured-insights/future-of-work/what-is-the-future-of-work. Accessed 28 Sept 2018

The Future of the Digital Workforce: Current and Future Challenges for Executive and Administrative Assistants

Anabela Mesquita[1,2(✉)], Luciana Oliveira[2], and Arminda Sequeira[2]

[1] Polytechnic of Porto and Algoritmi RC, Porto, Portugal
[2] Polytechnic of Porto/ISCAP, Porto, Portugal
{sarmento,lgo,arminda}@iscap.ipp.pt

Abstract. Changes brought by the 4th Industrial Revolution and digitalisation impact directly in the way we live, shape the organisations and change the way we work. This article explores and anticipates the scope and depth of the impacts that current and emerging technologies are imposing to the jobs of administrative and executive assistants. Under the scope of the global digital transformation, we present sets of tasks that currently are or will be depreciated and automated by technologies, impacting the role of these professionals. In this scenario, we discuss which new competencies, comprising new knowledge, skills and abilities, are increasingly required and point out strategies for repositioning the profession in the digital transformation era.

Keywords: Digital transformation · Workforce · Competencies · Executive assistants · Administrative assistants

1 Introduction

The 4th Industrial Revolution and the digitalisation are transforming the way we live, socialise and work. Besides the changes being visible in the business models, in communication and collaboration, in the relationship between our working and private life, in the structure and organisational hierarchies, technologies are also impacting the employment itself. According to a study carried out by the Oxford University [1], 47% of the job occupations in the United States are at risk of being replaced by technologies or automatisation and work becomes more digital, virtual and remote. Moreover, according to the report of the WEF 2018 [2], "the wave of technological advancement is set to reduce the number of workers required for certain tasks" (p. v). It is also stated that there will be an increase in the "demand for new roles" and a "decrease demand for others". Of course, these changes will not be the same all over the world and in all industries, but the trend is set globally.

This scenario is enabled by some factors such as: (1) technological advancement and adoption (mobile internet, artificial intelligence, big data analytics and cloud computing); (2) trends in robotisation; (3) changing geography or the way we produce and distribute value in the value chain); (4) changing employment types; (5) a net

© Springer Nature Switzerland AG 2019
Á. Rocha et al. (Eds.): WorldCIST'19 2019, AISC 930, pp. 25–38, 2019.
https://doi.org/10.1007/978-3-030-16181-1_3

positive outlook for jobs; (6) a decrease of some others; (7) emerging in-demand roles; (8) growing skills instability and (9) reskilling is and will be imperative [2]. Finally, the WEF 2018 report [2] also points out the strategies for addressing the skills gap in the organisations: (1) companies expect to "hire wholly new permanent staff already possessing skills relevant to new technologies", (2) "seek to automate the work tasks concerned completely", and (3) "retrain existing employees" [2].

One of the professions where this impact will be significant in the next years is the one related to administrative tasks (administrative assistants, executive assistants, executive secretaries, etc.) [3]. Being aware of the challenging scenario, promoted by the digital transformation, it is necessary to understand what are the tasks that may be carried out by the applications/software/robots available and those that can (still) only be performed by humans. In parallel, as there are several technological possibilities in the market, how can the existing human assistants make the most out of the technologies and thus making their job easier and use the spare time in tasks requiring more cognition? Of course, this will imply the development of (new) competencies, namely those related to communication, teamwork, problem-solving, critical and digital skills, just to name a few. Moreover, it is necessary to understand if the educational institutions that are presently preparing those professionals are ready to shift and/or upgrade the content and the pedagogical approaches needed to prepare such (new) professionals. As such, the purpose of this study is to contribute with a reflection about the future of the administrative occupations, including the identification of possible tasks to be replaced by or complemented with the technologies as well as those that can still only be performed by humans (if any). Results will allow us to think about the future of work and the implications of digitalisation for the current and future workforce.

In order to achieve the objective of this research, first we do a literature review concerning the digital transformation in the workforce by covering the technologies enabling this digital transformation followed by the analysis of the functions and tasks of the administrative assistant and the use of the technologies in their daily profession. Finally, we present an analysis of the tasks carried out by these professionals, and we identify technologies already in the market that may replace the professional or help him/her to perform his/her task. The results can be used to reflect upon the present/future curricula that HEI are offering programs in this field. To conclude, we reflect on the results and identify some concerns about this profession and point directions for future research.

2 Digital Transformation in the Workforce

Technologies are enabling several changes, not only in the organization of the companies, in the business models but also in the way people work, in the professions and in the skills needed to perform tasks. In this section, first we present some of the changes related to the companies, then the technologies enabling these changes and finally, the changes in the profession of administrative assistants.

2.1   Changes in the Organization of Work and People

According to the literature, the main changes occurring in the organisation of work and people are visible in the following areas [4, 5]:

- Organisation of work – these changes concern the way people work and how companies organise themselves. While more traditional companies try to adapt to the technologies introducing telework, mobile work, the digital native companies present a more agile organisation of work, such as projects, as well as collaborative spaces and crowdsourcing. This flexibility is visible in the place of work (the employee can work anywhere), the working hours (any time), the relationship between employers and employee and even in the approach (this kind of flexibility allows companies to find in the market the right person they need for a specific task). This led to the emergence of digital platforms that incorporate mechanisms to validate the identity of the service provider and the reliability of the participants. E. g. TaskRabbits (https://www.taskrabbit.com/), Airbnb (https://www.airbnb.pt/), UberEats (https://www.ubereats.com/pt-PT/stores/), Etsy (https://www.etsy.com/).
- The relationship between private and working life – due to the changes in the organization of work, the activities are dissociated from a physical space, a building and a schedule. The working hours are no longer relevant as what matters are the results achieved. And if on the one hand, it seems to contribute to balance private and professional life, as tasks are performed according to the convenience of the professional, on the other hand it means that people are always connected and accessible, eliminating barriers between the work and family life, which might constitute a problem if the professional is not able to set boundaries.
- The format of work, communication and collaboration – Digital transformation also affects the way people work, communicate and collaborate. Teamwork gains importance to the extent that knowledge as a resource can only be developed and advanced within a group of people [4]. New ways of teamwork and collaboration will emerge, enhanced by the technologies (virtual teams). The organisational structures will be replaced by project work and self-organised teams as traditional hierarchies are too rigid and slow to constitute an answer to the challenges
- Performance and talent management – the increase in the use of technologies at work leads to an increase in the need for digital competencies. The essential competencies related to the use of the computer will be necessary for all occupations. Besides, routine tasks will be automatised, increasing the demand for cognitive competencies and those related to problem solving and creativity in order to handle successfully those tasks that are not automatised. Moreover, the markets will be more dynamic, demanding people to continuously adapt to new situations and forcing them to be more agile. This will lead to investing in lifelong learning as the lifetime of knowledge is permanently shorter. A certain level of resilience is also crucial.
- Organisational hierarchies – the digital transformation is leading to the increase of the influence of workers. Technologies allow workers to participate more in decision making, involving them in real-time in the different subjects and decisions. The increased responsibility in one's work is facilitated by increased access to

information and increased transparency of information, making it easier to find contacts, people for specific topics/tasks, and direct contact with several interlocutors (inside and outside the organization), which seems to flatten the hierarchical levels.

This overall organizational scenario has sound impacts specially in the scope of the typical job description of AAs and EAs as these professionals usually perform tasks in a traditional 9 to 5 schedule, being close to the management teams and smoothing the work for them, acting as a facilitator but simultaneously as a strainer and a connection link between management and the other departments within a company. Given the current digital transformation, it is necessary to discuss its impacts on the role of these professionals.

2.2   Technologies in the Digital Transformation

In the context mentioned above, it is relevant to consider which technologies are enabling this digital transformation, being aware that these might change every day.

One of the technologies concerns the mobile internet and the cloud as they allow to offer more efficient services and the opportunity to increase productivity. Moreover, the cloud allows the quick dissemination of the service models, based on the internet. Big Data allows the retrieval, storage and analysis of large amounts of structured and non-structured data. This analysis is and will constitute the basis of the intelligent services, requiring complex knowledge. The Internet of Things is and will generate large amounts of information, patterns and knowledge as never seen before. Crowdsourcing, collaborative consumption and peer-to-peer interaction constitute technology-based processes that allow companies to have access to talents, promote mass production and enable the emergence of small/family companies. Robotics allow that production tasks that are being developed by robots with high dexterity and intelligence to be more convenient, precise and cheaper. Additionally, the transportation sector is witnessing changes due to the introduction of autonomous vehicles. Artificial intelligence and machine learning allow the automatization of tasks within a cognitive dimension, augmenting human intelligence. Finally, 3D and 4D printing allow to increase productivity to levels never seen before as 3D printing permits an adjusted production to real-time demand, creating and improving the supply chain and global nets [2, 6, 7].

As expressed in the previous section, technology lies at the core of most of the changes that we may witness in professions [8]. As for Susskind and Susskind [8], "traditionally, practical expertise has been held in people's heads, textbooks and filing cabinets". Nowadays, it is being stored and represented in digital form, in a variety of machines, systems and tools. Nevertheless, no matter how complex the systems, the impact of technology on the professions can be categorized under two categories - automation (sustaining technologies - those that support and enhance traditional ways of operating in an organisation or an industry) and innovation (disruptive technologies - those that fundamentally challenge and change working practices) [8].

In automation, technologies are used to suppress some inefficient activities. The focus is on streamlining manual or administrative work. However, the old ways of operating are not ignored or discarded. Actually, this automation can be transformative, i.e., people can use technology to transform the way a task is done (e.g. the use of Skype in a meeting - in this situation, the interaction is still real-time and face-to-face but at distance). In innovation, technologies enable ways of making practical expertise available that just was not possible without the system in question. Innovative systems provide services at a lower cost, or at a higher quality or in a more convenient way than in the past [8].

2.3   The Career of Administrative Assistants and the Use of Technologies

In this section, we briefly present the profession of Executive/Administrative Assistants by identifying the main tasks that they are expected to perform. In this career we identified two levels: the basic level - Administrative Assistants (AA) - whose primary focus is to perform operational tasks, and the top level - Executive Assistant or Executive Secretary (EA) - whose tasks are more related with communication, problem solving, negotiation, and support to the top-level executives of a company.

At the basic level, the professional has instrumental knowledge or knowledge competence [9] consisting mainly of clerical, computer and electronics in the user perspective.

At the top level – Executive Assistant or Executive Secretary – further knowledge, skills and abilities[1] [10] are required, focusing primarily on customer and personal service, information and data literacy; communication and collaboration, digital content creation, safety, problem solving [5] as well as mastering foreign languages and understanding and negotiating with partners form foreign cultures individuals, administration and management knowledge. The daily operations require additional skills such as service orientation – actively looking for ways to help people; coordination – adjusting actions in relation to others' actions; social perceptiveness – being aware of others' reactions and understanding the reasoning behind the behaviour; monitoring/assessing performances (self-assessment and other individuals and organizations to make corrections and/or improvements) [10].

The core tasks of these job positions, Administrative Assistants and Executive Assistants, tend to focus mainly on aid/support tasks that are provided to executives of all specialization areas: management, economy, accounting, finance, etc. They consist, most of the times, in extensions of a specific job occupation in a company, providing support to the implementation of a large set of diversified tasks. These job positions are, therefore, typically not domain specific in order to be flexible enough to accommodate the needs and demands of the broad range of jobs that they support. By not being domain specific, AAs and AEs are not, however, unspecialised jobs. Specialisation occurs at the technical level, and on the multi-domain of personal and technological skills. However, it is precisely on the job of 'assisting others' that these jobs collide directly with the role of the current and future technological developments that have been made available to industry.

---

[1] Competence, in this context, is understood as the proficiency developed through training or experience and is the sum of skills, knowledge and abilities. Skill define specific learned activity; Ability is the capacity of performing a task regardless of the proficiency; ability can be innate and not learned;

The professional intervention of AAs and EAs is centred mainly on four different domains: organising tasks; connection and business communication tasks; data generation and data management; representation tasks [11]. Many of these tasks will suffer the impact of AI and automation either because the tasks it selves became automated or because each person takes care of the task as it becomes intuitive and/or simpler, without the need of an assistant.

Typically organising tasks consists of: database storage and maintenance; schedule management and contact selection; work trip arrangements; set up meetings (either face-to-face or remote) taking care of physical conditions of the place; data collection to minute's elaboration; data collection to support decisions; office supply management; applying rules of organisational and institutional protocol; integrated systems networks supervision; work teams coordination and concrete projects; administrative procedures implementation;

The business communication and connection tasks consist mainly of: establishing contacts with business partners, both through digital or conventional means; elaborate different messages fit to each receiver; keep up and master technological tools; develop interpersonal internal and external contacts and acts as intermediary in solving conflicts; master foreign languages and establish contacts with partners from different countries/markets.

Data generation and data management consist mainly of: preservation and grant access, through established rules, to the archives that constitutes the organisational historical memory; relevant data collection about the world, markets and competition; take decisions on delegated matters; extract relevant information from available internal data making suggestions and course of actions, aiming organisational efficiency.

The representation tasks consist specially in "acting in the name of" a person or even an organisation: hosting foreign, multicultural visitors and implementing and supervising of business meetings and contacts taking into consideration multicultural peculiarities; support and supervise the integration and socialisation in the organisational culture of foreign employees; support and provide relevant information about other cultures to employees going abroad; establish connection among geographically disperse organisational units through distant communication technologies.

3 Research Design

As said, the purpose of this exploratory study is to contribute with a reflection about the future of the administrative occupations, including the identification of possible tasks to be replaced by or complemented with the technologies as well as those that can still only be performed by humans (if any). In order to achieve these goals, as this is an exploratory study, we first identified the main tasks that these professionals are expected to perform and considered those scoring at least 50% on the importance scale for the job position, as those are identified as core tasks. For that, we took into consideration the description of the jobs under the codes Executive Secretaries and Executive Administrative Assistants (code 43-6011.00) and Secretaries and Administrative Assistants, Except Legal, Medical and Executive (code 43-6014.00) [10, 12]. We used the information available on the O*Net OnLine (www.onetonline.org),

Effortless HR (www.effortlesshr.com) and Career Planner (job-descriptions. careerplanner.com), as information sources. As a second step, we did a thorough analysis of tools available/publicised on the internet with the potential to replace the identified job position's tasks. Finally, we did a match between the technologies and the tasks, identifying those tasks that have the potential to replace the person and those that will remain performed by humans. Results are presented in the next section.

4 Presentation and Discussion of Results

Table 1 presents the summary of results concerning the core tasks conducted by the job occupation of the Administrative Assistant. In the first column, we present the main tasks being performed by those professionals by order of importance to the job position. In the second column, we present sets of technologies found that might replace or complement those tasks.

Table 1. Tasks performed by the administrative assistant and the diagnose.

Task	Diagnose
Use computers for various applications, such as database management or word processing	Partially deprecated. Database management will be performed by machines, using technologies such as AI. Tasks related to word processing will partially be replaced by dashboards for the generation of performance reports, for instance. Some word processing operations might still be conducted by these professionals for addressing specific purposes
Answer telephones and give information to callers, take messages, or transfer calls to appropriate individuals	Soon to be replaced by automatic call distribution (ACD) with applications as: Capterra (https://www.capterra.com/call-center-software) and applications for Customer Relationship Management (CMR) and Enterprise Resource Planning (ERP): most important suppliers of the CRM systems include Salesforce.com, Microsoft, SAP and Oracle, com Salesforce.com representing 18,4% of the market, Microsoft representing 6,2%, SAP representing 12,1% and Oracle representing 9,1% of the market in 2015 [13]
Create, maintain, and enter information into databases	Soon to be deprecated, as digital information is growingly available, particularly concerning the creation of databases and entering information. Current challenges in handling and maintaining large databases of information are associated with data analysis tasks

(continued)

Table 1. (*continued*)

Task	Diagnose
Set up and manage paper or electronic filing systems, recording information, updating paperwork, or maintaining documents, such as attendance records, correspondence, or other material	Soon to be replaced by ERP. Setting up and managing systems is necessary, but not as a core task for this job position. Digitising documents or paperwork is soon to be deprecated, as information in digital format is growingly available
Operate office equipment, such as fax machines, copiers, or phone systems and arrange for repairs when equipment malfunctions	All information will be digital. Repairs will require specialised staff with skills from other job positions
Greet visitors or callers and handle their inquiries or direct them to the appropriate persons according to their needs	Soon to be replaced by automatic call distribution (ACD) and chatbots on social media channels and websites. As for personal contact, may be maintained or will be absorbed by other jobs, such as Public Relations, operating at non-operation levels
Maintain scheduling and event calendars	Soon to be deprecated. Replaced by AI digital assistants such as Siri (Apple), Cortana (Microsoft), Alexa (Amazon), Bixby (Samsung) and Google Assistant
Complete forms in accordance with company procedures	Procedures will be enforced by SAP or other similar systems. Completing forms will be a task for every domain specialist in the organisation
Schedule and confirm appointments for clients, customers, or supervisors	Soon to be replaced by AI digital assistants such as Siri (Apple), Cortana (Microsoft), Alexa (Amazon), Bixby (Samsung) and Google Assistant or absorbed
Make copies of correspondence or other printed material	Soon to be deprecated. Replaced by ERP and SAP systems
Locate and attach appropriate files to incoming correspondence requiring replies	Soon to be deprecated. Replaced by ERP and SAP systems, namely those with built-in natural language processing (NLP) for document scanning
Operate electronic mail systems and coordinate the flow of information, internally or with other organisations	Soon to be deprecated. To be replaced by SAP and enforcement of decentralization of information flows and definition of levels of access to documents

(*continued*)

Table 1. (*continued*)

Task	Diagnose
Compose, type, and distribute meeting notes, routine correspondence, or reports, such as presentations or expense, statistical, or monthly reports	Composing presentations may still be required as it may require cognitive operations related to support one's speech visually. Current automation relies mainly on the layout provision, with AI systems, such as https://www.beautiful.ai/. Several services worldwide are offering customer service though, for both. All types of distribution tasks will be deprecated. Statistics reports will be automatically generated from digital Data-warehouse, through dashboards (models) created in applications such as Power BI (Microsoft)
Open, read, route, and distribute incoming mail or other materials and answer routine letters	Soon to be deprecated. Replaced by ERP and SAP systems, namely those with built-in natural language processing (NLP) for document scanning
Provide services to customers, such as order placement or account information	Soon to be deprecated. Depending on the platform, will be replaced by chatbots, ERP and SAP systems
Review work done by others to check for correct spelling and grammar, ensure that company format policies are followed, and recommend revisions	Technologies will conduct these operations first-hand, with no need for another person to revise. Tools such as Grammarly or the new machine learning-based grammar checker to be launched by Google. Regarding the format, policies will be enforced by automated verification systems
Conduct searches to find needed information, using such sources as the Internet	Critical information gathering and organising will remain relevant, but not the ability to find information as an end in itself. AI-powered digital assistants can aid professionals in internet search and navigation
Manage projects or contribute to committee or teamwork	As it consists of critical thinking and knowledge management, tasks will remain, but not specifically in this job description. It may be moved to higher levels of the hierarchy because of its non-operational, cognitive-intellectual nature
Mail newsletters, promotional material, or other information	Soon to be deprecated. Replaced by CRM operations and already being presented to customers through ACD systems. Social media and website chatbot also provide alternative sustainable options to humans in this task

(*continued*)

Table 1. (*continued*)

Task	Diagnose
Order and dispense supplies	Soon to be deprecated. Replaced by CRM and ERP and SAP operations and assigned to each organizational member
Learn to operate new office technologies as they are developed and implemented	This task will remain, but not exclusively for this job occupation. It will include not only office technologies, but productivity, team management, and information management The need for technological update will remain as a requirement for most jobs around the globe. If most of the routine operations of Administrative Assistants are deprecated, this training task will most likely be absorbed by specialized on-demand services

As it is possible to observe in Table 1, the vast majority of tasks performed by Administrative Assistants, whose occupancy is mainly at the operational level, will be deprecated soon, as they can be automated by technology, leading to the extinction of the job position. We also believe that the remaining tasks, those that cannot be fully automated by technology, will be absorbed by other job positions in companies, such as middle-managers, since technology can provide greater autonomy in performing them.

In Table 2 we present the analysis and results concerning the Executive Assistant/Secretary job position.

Table 2. Tasks performed by the executive assistant/secretary and the diagnose.

Task	Diagnose
Manage and maintain executives' schedules	Soon to be deprecated. Replaced by AI digital assistants such as Siri (Apple), Cortana (Microsoft), Alexa (Amazon), Bixby (Samsung) and Google Assistant
Make travel arrangements for executives	Soon to be deprecated. Replaced by AI digital assistants such as Siri (Apple), Cortana (Microsoft), Alexa (Amazon), Bixby (Samsung) and Google Assistant
Prepare invoices, reports, memos, letters, financial statements, and other documents, using word processing, spreadsheet, database, or presentation software	Most of the documents needed will be automatically generated by information systems based on ERP, and big data supported dashboards. Documents requiring critical thinking, creativity and knowledge management may still be required and, according to the time and complexity they require, may still be conducted by EAs

(*continued*)

Table 2. (*continued*)

Task	Diagnose
Coordinate and direct office services, such as records, departmental finances, budget preparation, personnel issues, and housekeeping, to aid executives	If coordination tasks are part of a routine process, they will be automated through ERP systems, just like any other routine tasks. Technology is propelling executives to include technological systems as their primary aid
Answer phone calls and direct calls to appropriate parties or take messages	Soon to be deprecated. Replaced by automatic call distribution (ACD)
Prepare responses to correspondence containing routine inquiries	Soon to be deprecated. Routine process will be automated through ERP systems, namely those with built-in natural language processing (NLP) for document scanning
Open, sort, and distribute incoming correspondence, including faxes and email	Soon to be deprecated. Routine process will be automated through ERP systems, namely those with built-in natural language processing (NLP) for document scanning
Greet visitors and determine whether they should be given access to specific individuals	Personal contact may be maintained as a core task of EA or will be absorbed by other job positions, depending on how human resource costs are held at the organisation. Personal contact (as in Public Relations), will be, however, one of the core competencies of EA in terms of representing executives in meetings and other teamwork activities
Prepare agendas and make arrangements, such as coordinating catering for luncheons, for committee, board, and other meetings	Coordination tasks that cannot be performed by AI digital assistants will remain as competencies of EA
Conduct research, compile data, and prepare papers for consideration and presentation by executives, committees, and boards of directors	Knowledge management and tasks requiring cognitive-intellectual function and critical thinking such as research will remain as core competencies
Perform general office duties, such as ordering supplies, maintaining records management database systems, and performing basic bookkeeping work	Soon to be deprecated. Routine process will be automated through ERP systems
File and retrieve corporate documents, records, and reports	Routine process will be automated through ERP systems. Digital information will be widely available and access to it will be managed by systems which will provide distinct access levels inside and outside the company

(*continued*)

Table 2. (*continued*)

Task	Diagnose
Read and analyse incoming memos, submissions, and reports to determine their significance and plan their distribution	Routine process will be automated through ERP systems. However, knowledge management and tasks requiring critical thinking (cognitive-intellectual function), such as determining significance, will remain as a competence of the job position
Provide clerical support to other departments	Knowledge management, critical thinking and teamwork will remain as competencies of the job position. If support is merely clerical, the task may be deprecated
Attend meetings to record minutes	Soon to be deprecated. Replaced by AI digital assistants such as Siri (Apple), Cortana (Microsoft), Alexa (Amazon), Bixby (Samsung) and Google Assistant. Alternatively, meeting managers can make use of specialized transcription software and meetings management, such as: – Dragon: https://www.nuance.com/dragon.html (transcription software with dictation) – Software to manage meetings: https://www.meetingbooster.com/ and https://www.capterra.com/meeting-software/

From the analysis of Table 2, concerning the job position of Executive Assistants/Secretaries, it is possible to verify that some tasks are expected to be automated, namely de ones related to routine processes. There are, however, other tasks, namely the ones that require cognitive-intellectual operations, such as critical thinking, knowledge management and teamwork that are expected to remain in the job description and in the tasks performed. We consider these to be higher order tasks that will still require human intervention, despite the advances in artificial intelligence.

It is also important to notice that, either professionals will be replaced by technology, or they will be required to undertake continuous, advanced, cross-functional high-tech training to respond to the digitalisation of tasks. As technology grows in applicability and complexity, more time is required to learn how to handle it and to keep up with new features and new releases. Time, as a determinant factor of every learning curve, will be key, either in striping tasks from these jobs or in continuously adding tasks to them, in which technology handling will be core. Considering the business field, top-level white collars, such as CEOs, senior managers, middle managers, etc. are not always available, and it is not even considered in their job description to devote large amounts of time to technological or computer training. They are expected to absorb additional tasks and functions as technology offers more convenience and autonomy for operations such as: booking a meeting, managing their own schedules, make travel arrangements with the aid of a PDA, for instance. Some of the tasks performed by AS and ES will be performed entirely by digital systems, such as answering and transferring calls, typing data, manage documents, etc. Operations related to handling complex technology that requires hard training together with higher cognitive skills will rapidly become the core for AA/EA.

The use of technologies in the occupation of AA may imply that the current entry level of this career might disappear, as almost all tasks can be performed or automated by technology. However, on a best-case scenario, the existing professionals will find their professional tasks facilitated as they can use the technologies as a facilitator or a complement. Nevertheless, in the most challenging scenario, these professionals might need to evolve to the next stage EA. This implies performing new/other tasks and work in a more autonomous and independent way, as well as the development of new competencies such as creativity, negotiation, communication, leadership, problem-solving, creative thinking, just to name a few. However, this is only possible if the professionals continue to improve its personal and technological competencies throughout life, as digital transformation implies a constant update of knowledge and skills.

Additionally, the AA might also specialise in one domain and become the right arm or even occupy a position in the middle of the hierarchy (e.g. human resources, marketing, financial department), as a specialisation strategy that guarantees the relevance of the tasks performed, supported by field-specific knowledge. Finally, according to the literature, these professionals might opt for a complete restructure of their careers and become virtual/remote assistants, creating their own companies and providing their services at distance to a set of persons/companies in the most diverse areas.

5 Conclusions and Future Work

As we have seen, digital transformation is changing the way we work and even changing careers as is the case in the administrative area. In this situation, the professional might be replaced by the technologies as these can perform their tasks or, in the more optimistic scenario, technologies might help to perform tasks more efficiently. Anyway, schools, educators and governments should be aware of how jobs are rapidly changing in order to address the current and future demands of companies, concerning the roles that these jobs will have in industry/companies [3]. There is a clear need to adjust and adapt current training curricula to the alterations that have been introduced by technology in the market, and there is also the need to foresee and prepare for future changes and impacts on the jobs that current students will occupy.

Taking into consideration that this is an area where the impact of digital transformation will be evident in the future, several possibilities for future research emerge. Future research should be conducted to analyze the type of cognitive work behind each of the identified tasks and map it to automation technologies (artificial intelligence, machine learning, predictive analytics, etc.). Results could show which tasks are more susceptible to being replaced. We also identify the need to replicate this study by region/country as well as by domain area as those challenges might change from one country/area to another. Moreover, there is the need to go deeper in this analysis and to study how these changes are happening and what are the challenges that these professionals are facing as well as to identify how they are reacting to this digital transformation. It is already possible to identify some movements in order to create groups and platforms for virtual/remote assistants managed by humans, and it is necessary to understand how this is being introduced in the business market.

References

1. Frey, C.B., Osborne, M.A.: The future of employment: how susceptible are jobs to computerisation? Technol. Forecast. Soc. Change **114**, 254–280 (2017)
2. WEF: The Future of Jobs Report 2018, Geneva, pp. vii–xix (2018)
3. Wike, R., Stokes, B.: Pew research center. In: Advanced and Emerging Economies Alike, Worries about Job Automation (2018). http://www.pewglobal.org/2018/09/13/in-advanced-and-emerging-economies-alike-worries-about-job-automation/?fbclid=
IwAR02CjIGbpQ1PNYepFmL6gQaK87w4lAm66EcNMsFDwnXb_dTLJBHRMR6uLY
4. Schwarzmüller, T., et al.: How does the digital transformation affect organizations? Key themes of change in work design and leadership. Manag. Rev. **29**(2), 114–138 (2018)
5. Vuorikari, R., et al.: DigComp 2.0: the digital competence framework for citizens. Update phase 1: the conceptual reference model. Joint Research Centre (2016)
6. Boneva, M.: Challenges related to the digital transformation of business companies. In: Innovation Management, Entrepreneurship and Sustainability (IMES 2018), pp. 101–114. Vysoká škola ekonomická v Praze (2018)
7. INTUI: Twenty Trends that Will Shape the Next Decade (2010)
8. Susskind, R.E., Susskind, D.: The Future of the Professions: How Technology Will Transform the Work of Human Experts. Oxford University Press, Cambridge (2015)
9. Lyotard, J.-F.: The Postmodern Condition: A Report on Knowledge, vol. 10. University of Minnesota Press, Minneapolis (1984)
10. O*NET, N.C.f.O.N.D. Executive Secretaries and Executive Administrative Assistants. 43-6011.00, 15 September 2018. https://www.onetonline.org/link/details/43-6011.00
11. Sequeira, A., Santana, C.: O Trabalho Especializado do Secretariado/Assessoria: a comunicação assertiva como competência diferenciadora (2016). https://issuu.com/anavieira34/docs/anais_cisa2016
12. O*NET, N.C.f.O.N.D. Secretaries and Administrative Assistants, Except Legal, Medical, and Executive. 43-6014.00, 15 September 2018. https://www.onetonline.org/link/details/43-6014.00
13. Columbus, L.: Gartner CRM market share update: 41% Of CRM systems are SaaS-based, Salesforce dominating market growth. Julkaistu **6** (2014). https://www.forbes.com/sites/louiscolumbus/2014/05/06/gartners-crm-market-share-update-shows-41-of-crm-systems-are-saas-based-with-salesforce-dominating-market-growth/

Project Management Practices
at Portuguese Startups

Anabela Tereso$^{(\boxtimes)}$, Celina P. Leão, and Tobias Ribeiro

Production and Systems Department/Centre ALGORITMI,
University of Minho, Campus de Azurém, 4804-533 Guimarães, Portugal
{anabelat, cpl}@dps.uminho.pt,
tobias.m.ribeiro@gmail.com

Abstract. Nowadays, due to the continuous changes in technology and markets, companies need to use the best project management practices to effectively manage their projects. The use of these practices however will not have the same results in different types of projects and in different organizations. The practices used by larger companies may have different results if used by smaller companies. With this in mind and focused on project management practices used by Portuguese startups, this study explores the value that project management represents for this type of companies. The study also focused on scale-ups, which are startups that have already undergone an initial phase of maturation and have more complex structures and processes, thus allowing us to see how the project management practices change with the evolution of the organization. This way, it was possible to observe that project management is seen as an essential factor within this type of organizations. These organizations seem to opt for a more agile approach to project management, thus taking advantage of the flexibility typically offered by such approaches. This type of approach seems to continue as startups mature, where there is only an increase in the formalization and complexity of project management practices and tools.

Keywords: Project management practices · Startups · Scale-ups

1 Introduction

Organizations are increasingly focused on using Project Management (PM) methods and practices to increase the success of their projects.

According to the results presented by the Project Management Institute (PMI) Pulse of the Profession Report, the success rate of corporate projects is increasing and the average amount of money wasted on pour project performance is reducing over years. This success is due to the correct application of a professional PM system. In 2018, the results of this study demonstrated that regardless of whether using a predictive, iterative, incremental or agile approach, organizations that use some form of formal PM approach are successfully meeting their goals [1].

These benefits have already been widely discussed in the literature, but almost always in the context of PM practices associated with mature and large companies. For this study, the main focus was on small companies, with special attention to companies

© Springer Nature Switzerland AG 2019
Á. Rocha et al. (Eds.): WorldCIST'19 2019, AISC 930, pp. 39–49, 2019.
https://doi.org/10.1007/978-3-030-16181-1_4

called startups. These companies are usually characterized by small activity time and they are usually looking for a position in the market [2].

To better understand this knowledge gap, the literature closest to this field is related to Small and Medium Enterprises (SMEs), and can give a perspective to better understand how the projects are treated within the universe of startups. SMEs and startups share similar characteristics and challenges. According to Murphy and Lewith [3], SMEs can be described as more flexible and closer to customers than large companies, also having a greater capacity to respond to threats and better internal communication. The differences between SMEs and large companies have profound implications, and it is possible to highlight four main differences: processes (simpler and more informal); procedures (with less standardization); structure (employees with more multi-tasking); and people (more risk-averse due to fear of failure) [4].

With the constant technological evolution and the increase in customer demand levels, organizations are forced to discover new strategies/methods of PM that can deal with these problems. The lean and agile methods are among the different methods that are able to solve these types of problems [5].

The choice of the startup type of organizations for this study is pertinent, because according to a study by Informa D&B, entrepreneurship in Portugal has evolved considerably over the last years. The creation of startups is increasing, influencing job creation and the Portuguese business environment [6].

Taking into account the above, this paper reports on a study of the context of PM in Portuguese startups, which in turn, has as aspiration to be a contribution for this type of companies to be able to understand the importance of PM practices and what their competitors use and also to be a base for future research on these topics. The research was done trying to find answers to the following research questions: Which approaches to PM are most appropriate for Portuguese startups? What is the level of interest of Portuguese startups in the use of PM? What are the PM practices and tools used in Portuguese startups? What are the biggest limitations and challenges of Portuguese startups in PM?

2 Literature Review

Nowadays, due to the constant technology and markets evolution, companies no longer choose to use a classic and closed management form. According to Kerzner [7], in order to respond to the changing environment, it is necessary to abandon this inefficient traditional structure and replace it with PM. Unlike traditional management, PM allows better interaction, organization, coordination and communication between different departments of a company. The result of this type of modern management provides better efficiency, effectiveness and productivity. PM helps companies to reduce time to market, to better use limited resources, to manage technology complexity, to respond to stakeholder satisfaction, and to increase their competitive advantage [8].

According to the PMI [9], PM is the application of knowledge, tools, techniques and skills to project activities, in order to meet their requirements. Besner and Hobbs [10] suggest that PM tools and techniques are practices, that if well implemented, can be considered as an "added value" to the organization. These practices can increase project success rates and bring competitive advantages. In another perspective, Thomas and Mullaly [11] state that the direct influence of the benefits of using project management on the profits and revenues of organizations has not yet been proven, and that this use may lead to higher costs in the short term. Most of the benefits offered by project management are associated with improvement in less tangible aspects such as stakeholder satisfaction, taking into account time, cost, quality and project processes. These benefits can help prepare organizations for future activities.

There are several standards that can help with the implementation of PM practices. Most of these standards emerged from the early 1970s, through associations dedicated to project management professionals. Some examples are: PMBOK, Prince2, ICB, P2M and APMBOK [12]. There are also agile methodologies that can be used to manage project. They emerged from software development projects, but are now used in a wide variety of project types. Generally, these methodologies are used in projects that involve a certain degree of uncertainty, risk and complexity such as new product development. Some examples of these methodologies are: Extreme Programming (XP); Scrum; Crystal; Feature Driven Development; Dynamic Systems Development Method; and Kanban [9]. Agile methodologies give more value to people, interactions, and collaboration with the client than to processes, tools, contracts, and plans. They are based on several principles: continuous innovation; product adaptation; shorter delivery time; adapting people to reliable processes and results [13].

Startups can be characterized as young enterprises, with a simple organizational structure, and informal communication. Normally, decision-making is centered on the founder or Chief Executive Officer (CEO) and is done in a style of trial and error with regard to daily business choices [14]. According to Blank and Dorf [15], a startup is not only a smallest version of a large company. It can be defined as a temporary organization in search of a scalable, repeatable and profitable business model. The European Startup Monitor (ESM) [16] considers that startups are companies under the age of ten and have high aspirations in terms of employee growth and sales. From another point of view, Ries [17] defines startup as a human institution created to design new products and services in situations of extreme uncertainty. When a startup exceeds the first initial barriers and achieves a scalable and repeatable business model, it can get a new denomination of scale-up. These companies were the focus of this study, namely on the use and relevance of project management practices.

There are not many studies on project management at startups so we searched for similar studies in SMEs. According to the studies of Turner et al. [18, 19], SMEs use simpler practices and tools and not complex standards or methodologies. They presented the following practices as the most used by SMEs: Client Requirements; Road Map; Milestones; Agile; Scrum; Work Breakdown/Activity list; Scope/Resource Schedule; Burn Down; PM Software; Risk Management; Project Office; Domain knowledge; Responsibility Assignment Matrix. These list was compared to the one collected in our study.

3   Research Methodology

Survey research was the strategy used to collect information. This strategy allows to question individuals on a topic and then describe their responses [20]. The researchers choose to study different startups from different areas, in order to have a more general set of results. In order to make data collection more dynamic and less standardized, the researchers found more appropriate to use semi-structured interviews. In semi-structured interviews researchers draw up a list of topics and issues to be developed, but these may vary from interview to interview. This means that the researcher may omit some questions in particular interviews, taking into account the specific organizational context that is found in relation to the research topic. The order of the questions may also vary depending on the flow of the conversation [21]. Content analysis was used to analyze the interviews [22]. The final text produced after the transcriptions and compilation of the semi-structured interviews, carried out with the support of the script developed for this purpose, was analyzed in order to obtain meanings and answers to the research questions. The categories considered coincide with the different parts defined in the interview script.

Each interview started with a brief introduction of the researcher, the topic of study, and the concise description of the purpose of the research. The script contained the main objectives of the research, starting with a first section dedicated to ascertaining the professional profile of the interviewees, being this technique used to break the ice and make them feel more comfortable. Before starting the interviews, the researcher asked permission to record the contents of the conversation, in order to later facilitate its transcription. Each interview lasted 40 min on average.

The script of the semi-structured interview is divided into 5 parts. The first part, as mentioned above, is dedicated to the interviewee' profile characterization in general terms, such as sex, qualifications, occupation and level of experience in PM. The second part was designed to give a brief description of organizations and identify their characteristics in terms of products, structure, risk, innovation and competitive strategy. Part three refers to the company's past and current strategy for PM and its implications for the organization's progress. Part four is intended to provide more in-depth information, observing the practices and PM tools used by the organizations, inquiring in what context they are used. The last part seeks to identify the main challenges and limitations of organizations in the area of PM, emphasizing the challenges from the point of view of flexibility and scalability.

The sample was obtained by the technique of non-probabilistic sampling by convenience and snowball [23]. In order to make the data collection richer, the researchers tried to select startups from different sectors, with different dimensions and in different contexts (e.g. 4 startups are included in incubators). In order to obtain a future perspective of PM in a startup, two of the participating organizations are organizations that have already gone through the initial startup phase and are currently in an advanced phase of scale-up. Of the 21 organizations contacted, only 9 agreed to participate in the research, resulting in an acceptance rate of approximately 43%. All interviews were conducted in person and in place. All organizations received an email before the interviews were held, to know in advance the topics to be addressed. During the course of the interviews, the results obtained began to be similar, so the researchers felt that it would not be necessary to extend the sample of 9 organizations. Table 1 shows a brief description of the organizations that participated in the study, and Table 2 the details of the 11 interviewees of the organizations that participated in the study.

Table 1. Characterization of organizations.

Startup	Years of existence	No of employees	Company description	Code
Startup 1	3	6	Treatment of statistical data; Health sector	ST1
Startup 2	1	4	Artificial intelligence; Customer service	ST2
Startup 3	3	6	Freight transport service	ST3
Startup 4	2	8	Entertainment; Digital Marketing and Event Management	ST4
Startup 5	5	11	Creative services industry; Digital and multimedia	ST5
Startup 6	3	8	Monitoring software; Occupational health	ST6
Startup 7	1	2	Consulting; Construction; Investments	ST7
Scaleup 1	10	32	Digital products and services; Web design	SCA1
Scaleup 2	9	>100	Applications and computer software	SCA2

Table 2. Characterization of interviewees

Organization/interviewee	Role	Experience
ST1/1	Chief Operating Officer (COO)	Low
ST1/2	Engineer	Medium
ST2/1	Chief Executive Officer (CEO)	High
ST3/1	Chief Executive Officer (CEO)	High
ST4/1	Founder	Medium
ST5/1	Chief Executive Officer (CEO)	High
ST6/1	Project Manager/Engineer	Medium
ST7/1	Founder	Low
SCA1/1	Project Manager	High
SCA1/2	Project and Product Manager	High
SCA2/1	Engineer	High

4 Results and Analysis

For the characterization of interviewees' profile, gender, level of education and experience in PM, more specifically related to certification in this area, were considered. A curious result is that only 9% of respondents are female. Regarding the level of education of the interviewees, it can be considered quite high: 84% with a master's degree; 9% with a degree and 9% with secondary education. Although the high level of education, almost none of the interviewees has any type of training in the area of PM. Of the few academic contacts with this area, most was directed to PM in software development. Only two of the respondents had some kind of contact with more traditional tools and techniques, such as the ones described in PMBOK. Only 18% of respondents had some type of certification in this area, which in turn was related to agile PM methodologies, more specifically in Scrum.

A description of the characteristics of the organizations was then required in relation to their organizational structure and decision making style, risk positioning, innovation, growth strategy, and their source of funding. A summary of the characteristics of the organizations is presented in Table 3.

Table 3. Summary of the characteristics of the organizations (♠ startups ✈ scale-ups)

Criteria	Startups	Scale-ups
Organizational structure	Simple, flexible and informal ♠♠♠♠♠♠♠	Formal ✈ Informal ✈
Decision making style	Performed informally; centralized in founders or CEO. ♠♠♠♠♠♠ Decentralized ♠	Decentralized ✈ Centralized ✈
Risk positioning	High risk ♠♠♠♠♠♠♠	Low risk; comfortable situation ✈ ✈
Innovation	Something crucial ♠♠♠♠♠♠♠	Something important ✈ ✈
Growth strategy	In terms of market penetration through internationalization, or by increasing the customer base in the respective markets ♠♠♠♠♠	Related to the business model ✈ Increase number of employees ✈
Source of funding	Internal financing ♠♠♠ External financing ♠♠♠♠	Sustainable ✈ ✈ Still need some external financing ✈

A simple structure, an informal environment, and a centralized decision making are characteristics that are shared by all startup organizations, regardless of the industry to which they belong. In the case of scale-ups, it was noticed that with the increase of the organization it is natural that the decision making becomes more decentralized and the environment more formal.

All startups, regardless of their context, have confirmed that the risk factor is something that is always present in their daily life. For scale-ups, as the organization gets a more scalable business model, the risk decreases.

Confirming the characteristics presented by Ries [17], innovation was seen by all startups as an essential element of the organization, evidenced in the characteristics of the product/service or in the business model. The importance of this factor seemed to be maintained in the scale-ups represented in the study, demonstrating that innovation is a value that seems to be transversal throughout the maturation of the startup.

In terms of growth, not all startups have shown a huge ambition to increase the number of employees and turnover as referred by the European Startup Monitor [16]. Most startups were just looking to penetrate their respective markets through internationalization or by increasing the customer base in the respective markets.

Lastly, one surprising fact that has emerged in this sample was the fact that some startups are using a type of internal financing, that is, the founders are willing to take high risks. The scale-ups show to be more sustainable on this matter.

Regarding the strategy of companies in relation to PM, we asked about: the type of projects they had (internal/external); what was their official PM structure; the culture of the organization (formal/informal PM); vision of the utility of PM; lean startup concepts application; and staff certified with PM. This is summarized in Table 4.

Table 4. Summary of the strategies in relation to PM

Criteria	Startups	Scale-ups
Projects	Internal 🚀 External 🚀🚀🚀🚀 Internal and External 🚀🚀	Internal and External 🚀 🚀
Official PM system	Agile or Scrum Methodologies 🚀🚀🚀🚀 Not using any methodology or standard 🚀🚀🚀	Scrum 🚀 🚀
Culture of the organization (formal / informal PM)	Informal 🚀🚀🚀🚀🚀🚀 Formal 🚀	Formal 🚀 🚀
PM, competitive strategy / competitive advantage	Useful in the organization and its value as a competitive tool 🚀🚀🚀🚀🚀🚀🚀	An indispensable tool for the organization 🚀 🚀
Lean Startup	Concepts related to Lean Startup identified and consciously applied 🚀🚀🚀 Unconscious application 🚀	
Staff certified with PM	Certification is not an essential element for organization 🚀🚀🚀🚀🚀🚀🚀	Certification in PM as an important element for the organization 🚀 🚀

These results show that PM is important to startups and scale-ups, being more informal in the startups and tending to be more formal for scale-ups, and seen as useful for the startups and indispensable for the scale-ups, where PM certification is also considered relevant. Lean startup is applied in some of the organizations studied.

The PM practices most used in the studied organizations are shown in Table 5. In this results we can highlight the difference between the use of less formalized agile methodologies for the companies with less maturity and formal Scrum or Kanban methodologies used by the companies with more maturity. The same applies to risk management and the use of PM software.

Compared to the study of Turner et al. [18, 19] on practices in SMEs the difference was that in SMEs are used two practices not referred by the organizations under our study, namely Project Office and Responsibility Assignment Matrix. But in our study Kanban was referred as a practice not present in the SMEs study.

Finally, Table 6 shows a summary of the limitations and challenges in the PM strategy felt by the studied organizations. It is visible the limitations in the application of PM practices for startups and how it decreases when the organization grows, but all considered important to improve PM practices in their organizations.

Table 5. PM practices most used in the studied organizations

Practices	ST1	ST2	ST3	ST4	ST5	ST6	ST7	SCA1	SCA2
Client Requirement	X	X	X	X	X	X	X	X	X
Road Map	X		X		X	X		X	X
Milestones				X			X		
WBS/Activity list	X		X	X	X	X		X	X
Agile	X		X		X	X		X	X
Scrum						X		X	X
Kanban	X	X			X			X	X
Scope/Resource Schedule		X		X			X		
Burndown						X		X	X
Domain Knowledge				X			X		X
Risk Management						X		X	X
PM Software	X				X	X		X	X

Table 6. Summary of the limitations and challenges in the PM strategy

Criteria	Startups	Scale-ups
Limitations in application of PM	It has drawbacks 🚀🚀🚀🚀🚀🚀🚀	There is a decrease in the limitations of PM as the organization grows 🚀 🚀
Need to improve PM in the organization	Yes 🚀🚀🚀🚀🚀🚀🚀	There is always room for improving PM in the organization (continuous process) 🚀 🚀
Scalability of PM strategy	Changes are needed in the future 🚀🚀🚀🚀🚀🚀	Scalability has to be planned 🚀 🚀

5 Conclusions and Future Research

The concepts of innovation, risk and flexibility are present in the definition of what it means to be a startup. These young companies need to have the innovative ability to generate value for customers and to penetrate their respective markets. The search for a business model that is scalable and profitable carries risks, thus creating an environment of uncertainty that involves the company in this early phase of the organization's life cycle. Startups seek to receive constant feedback from the market, which in turn will have an influence on the way PM is developed in the organization. Startups require a lighter approach to project management, which promotes flexibility and iterations, rather than more structured and sequential approaches, characteristic of more traditional approaches to project management. In this way, it is not surprising to find in this research, and based on data gathered during semi-structured interviews to 7 startups and 2 scale-ups, that startups opt more for agile strategies for the development of their projects, since it is the approach most compatible with the cycle of experimentation characteristic of this type of companies.

In the universe of Portuguese startups, the value of the scientific area of PM seems to be well founded. Portuguese startups are managed through the use of projects, and PM is seen as essential to the organizations, although their low level of maturity. There is not much academic knowledge about some concepts of this scientific area and certification of employees is seen as not fundamental. However, it is our believe that PM can represent a competitive advantage for the young companies like startups and scale-ups in Portugal, since it provides tools and techniques capable of managing projects scope, time and cost, besides other areas such as quality, stakeholders and risk, so important for this type of organizations.

Most Portuguese startups seem to opt for project management practices and tools with more agile features like Kanban and Scrum. The degree of formalization of the practices and tools of project management seemed to vary with the maturity of the organizations. The organizations under study revealed to use the following practices: Client Requirement; Road Map; Milestones; WBS/Activity list; Agile; Scrum; Kanban; Scope/Resource Schedule; Burndown; Domain Knowledge; Risk Management and PM Software. Compared to a previous study on SMEs [18, 19], there are similar tools being used. Organizations just did not refer the use of Project Office and Responsibility Assignment Matrix as SMEs seem to use. But Kanban was a practice referred in our study that is not mentioned in the SMEs study.

The main limitations observed in the context of the Portuguese startups in the application of PM are related to the rigidity, cost and time associated with some PM practices, which are the main reasons for their more informal use in an initial phase of the startup. The lack of resources is also a problem of Portuguese startups, which makes them reluctant to invest on the area of PM, at an early stage. The more mature Portuguese startups present in the study are a proof that investing in PM can be a crucial part of an organization's development.

A limitation of this study was the size and the geographic location of the sample, so future studies on this matter are important to consolidate conclusion and information gathered on the subject that may be used as a reference for startups. In future work it might also be interesting to isolate circumstances and study only startups belonging to an incubator or a particular industry.

Acknowledgements. This work has been supported by FCT – Fundação para a Ciência e Tecnologia within the Project Scope: UID/CEC/00319/2019.

References

1. PMI: Success in disruptive times. In: Pulse of the Profession 2018. Project Management Institute (2018)
2. Racolta-Paina, N.-D., Mone, S.-D.: Start-up marketing: how to become a player on the B2B services market in Romania. Manag. Mark. **4**, 63–78 (2009)
3. Murphy, A., Ledwith, A.: Project management tools and techniques in high-tech SMEs in Ireland. In: High Technology Small Firms Conference (2006)
4. Ghobadian, A., Gallear, D.: TQM and organization size. Int. J. Oper. Prod. Manag. **17**, 121–163 (1997)
5. Rodríguez, P., Mäntylä, M., Oivo, M., Lwakatare, L.E., Seppänen, P., Kuvaja, P.: Advances in using agile and lean processes for software development. Adv. Comput. **113**, 135–224 (2018)
6. Informa D&B: O Empreendedorismo em Portugal (2017)
7. Kerzner, H.: Project Management Best Practices: Achieving Global Excellence. Wiley, New York (2018)
8. Patanakul, P., Iewwongcharoen, B.: An empirical study on the use of project management tools and techniques across project life-cycle and their impact on project success. J. Gen. Manag. **3**, 41–66 (2010)
9. PMI: A Guide to the Project Management Body of Knowledge (PMBOK GUIDE). Project Management Institute, Pennsylvania (2017)
10. Besner, C., Hobbs, B.: The perceived value and potencial contribution of project management practices to project success. Proj. Manag. J. **37**, 37–48 (2006)
11. Thomas, J., Mullaly, M.: Understanding the value of project management: first steps on an international investigation in search of value. Proj. Manag. J. **38**, 74–89 (2007)
12. Sanjuan, A.G., Froese, T.: The application of project management standards and success factors to the development of a project management assessment tool. Procedia - Soc. Behav. Sci. **74**, 91–100 (2013)
13. Highsmith, J.: Agile Project Management. Addison-Wesley Professional, Boston (2004)
14. Lester, D.L., Parnell, J.A., Carraher, S.: Organizational life cycle: a five-stage empirical scale. Int. J. Organ. Anal. **11**, 339–354 (2003)
15. Blank, S., Dorf, B.: The Startup Owner's Manual: The Step-By-Step Guide for Building a Great Company. K&S Ranch, Inc., Pescadero (2012)
16. Kollmann, T., Stöckmann, C., Hensellek, S., Kensbock, J.: European startup monitor (2016)
17. Ries, E.: The Lean Startup: How Today's Entrepreneurs Use Continuous Innovation to Create Radically Successful Businesses. Crown Publishing Group, New York (2011)
18. Turner, R., Ledwith, A., Kelly, J.: Project management in small to medium-sized enterprises: matching processes to the nature of the firm. Int. J. Project Manag. **28**, 744–755 (2010)

19. Turner, R., Ledwith, A., Kelly, J.: Project management in small to medium-sized enterprises: tailoring the practices to the size of company. Manag. Decis. **50**, 942–957 (2012)
20. Jackson, S.L.: Research Methods and Statistics: A Critical Thinking Approach. Cengage Learning, Boston (2015)
21. Saunders, M., Lewis, P., Thornhill, A.: Research Methods for Business Students. Pearson Education, London (2009). ISBN 978-0-273-71686-0
22. Huberman, A.M., Miles, M.B.: Qualitative Data Analysis: An Expanded Sourcebook. Sage Publications, Thousand Oaks (1994)
23. Robinson, O.C.: Sampling in interview-based qualitative research: a theoretical and practical guide. Qual. Res. Psychol. **11**, 25–41 (2014)

The Application of Clustering Techniques to Group Archaeological Artifacts

N. Mikhailova[1], E. Mikhailova[1,2(✉)], and N. Grafeeva[1,2]

[1] Saint-Petersburg State University, Saint Petersburg, Russia
e.mikhaylova@spbu.ru
[2] ITMO University, Saint Petersburg, Russia

Abstract. Modern methods of data analysis are rarely used in archaeology. Meanwhile, it is archaeology that opens up impressive opportunities for various interdisciplinary studies at the junction of archaeology, chemistry, physics and mathematics. XRF analysis, which has long been used to determine the qualitative and quantitative composition of discovered archaeological artifacts, among other things, provides arrays of digital information that can be used by machine learning methods for more accurate clustering or classification of artifacts. This is especially true for artifacts that are presented in the form of fragments of ancient ceramic amphorae or glass vessels. Such fragments, as a rule, represent the mass of the fragments mixed among themselves. There is a need to divide them into groups and then restore them as a single artifact from the detected fragments of one group. This paper presents a comparative analysis of the application of different clustering methods to combine artifacts into groups with similar properties.

Keywords: X-ray fluorescence analysis (XRF analysis) ·
Chemical composition · Archaeological artifacts · Clustering · Ceramics · Glass

1 Introduction

Data analysis methods have obtained a wide application in various branches of science, those methods allow to solve problems that previously required human actions. One of such scientific areas is archeology.

A set of quantitative and qualitative indicators, such as shape, material, color, has been examined by scientists during the analysis of the artifact for many years. However, evolution of physical and chemical methods of analysis had made it possible to move from general characteristics of materials to more precise ones. In particular, it became possible to determine the chemical composition of archaeological samples with high accuracy. The chemical composition is an important characteristic of archaeological artifacts, as it reflects raw material composition. So analysis of the chemical composition of artifacts that were made of ceramics [1] or glass [2] for example, is important to determine relations and trade links between different regions or settlements within the same region.

In this paper, we will consider approaches to solve the problem of determination the similarities and differences between archaeological artifacts. In particular, the possibility of using unsupervised machine learning methods in the context of the problem described above.

© Springer Nature Switzerland AG 2019
Á. Rocha et al. (Eds.): WorldCIST'19 2019, AISC 930, pp. 50–57, 2019.
https://doi.org/10.1007/978-3-030-16181-1_5

2 Data Description

Provided data is a result of X-ray fluorescence (XRF) analysis of archaeological samples. These samples are aluminosilicates – ceramics or glass. The result of XRF analysis is their qualitative and quantitative (%) elemental composition. It is worth drawing attention that some cells are empty. Empty cell means that atoms of the chemical element was not detected in the sample by XRF. Table 1 provides an example of initial data.

Table 1. An example of initial data.

	Al	Si	Fe	Ca	Mg	Ti	...	Zn	Ba
1	20.197	53.627	15.542	4.751	2.437	1.619	...		0.640
2	21.533	41.598	11.505	17.191	2.954	0.878	...	0.032	0.368
3	18.943	43.561	9.683	21.393	2.479	0.914	...	0.026	
...	...	...	...	...	...	...	...	...	...

3 Overview

Cluster analysis [3] allows us to split a set of objects into disjoint subsets. These subsets are called clusters, each cluster should contain similar objects, while objects from different classes should have significant differences. This is one of the tasks solved by the unsupervised learning methods.

One of the most common tasks is dimensionality reduction. It may help to visualize data or prepare it for further processing. Principal Component Analysis (PCA) [4] is one of the methods for linear dimensionality reduction. The main idea is that some features are possibly correlated with some others. So several correlated features can be replaced by new one. As a result, the dimensionality of the space will be reduced and the loss of information will be minimized.

K-Means [5] is one of the most popular clustering algorithms. It takes random points as centroids and moves it to the center of attributed set iteratively. The main problem is selection of number of clusters. Here can be used the following criterion.

$$J(C) = \sum_{k=1}^{K} \sum_{i \in C_k} \|x_i - \mu_k\|^2 \to \min_C \tag{1}$$

where C is a set of clusters with cardinality K and μ is a centroid of a cluster. However, the optimum is reached in case when each cluster contains exactly one object. So the following formula is used to choose number of clusters.

$$D(k) = \frac{|J(C_k) - J(C_{k+1})|}{|J(C_{k-1}) - J(C_k)|} \to \min_k \tag{2}$$

Mean-shift [6] clustering is another clustering method, it is based on the idea that centroids of clusters should be near of its local density maxima. The candidate points (possible centroids) are iteratively updated in accordance with the following equation.

$$x_i^{t+1} = x_i^t + \frac{\sum_{x_j \in N(x_i)} K(x_j - x_i) x_j}{\sum_{x_j \in N(x_i)} K(x_j - x_i)} \tag{3}$$

where x_i^t is centroid candidate i on iteration t, $N(x_i)$ are sets of samples in the neighborhood of x_i, K is kernel function. So the method moves a centroid in direction of the maximum increase in the density in its neighborhood.

DBSCAN [7] is another one clustering method. It supposes that clusters are high density areas separated by areas of low density. This method is able to detect clusters of any shape and does not require to select the number of clusters. It requires to define two parameters, which define 'dense' formally. As a result, we get clusters and possibly 'noise' – objects that could not be attributed to any cluster.

The approaches described above were iterative, in addition to them there are hierarchical clustering methods [8]. These methods build a complete tree of nested clusters. Result can be represented as a dendrogram or tree. The root of tree is cluster containing all objects, the leaves are clusters with only one object. So algorithm builds tree by merging or splitting clusters step by step. There are different metrics for the merge strategy, Ward is one of them. It minimizes within clusters variance.

We should be able to evaluate the quality of results from clustering. There are external and internal metrics. External metrics cannot be used in this paper since these metrics use the true labels for data. Internal metrics are based only on the initial data and results of clustering.

The Silhouette coefficient [9] is one way to evaluate the quality of results from clustering. It allows us to estimate the quality of the clustering using only the unlabeled sample and the clustering result. First, we have to determine the coefficient for each sample, it can be calculated according to the following equation

$$s_{x_i} = \frac{b - a}{\max(a, b)} \tag{4}$$

where a is the mean of distances between object x_i and other objects within cluster, b is the mean of the distance between object x_i and objects from the nearest cluster (different from the one the object belongs to). The Silhouette coefficient is the mean of values of the coefficients for each object. More formally

$$S = \frac{\sum_{i=1}^{N} s_{x_i}}{N} \tag{5}$$

where N is the number of objects. The Silhouette coefficient values are in the $[-1, 1]$ range. The value close to -1 corresponds to incorrect clustering, $+1$ to clustering with well separated and dense clusters.

The Calinski-Harabaz index [10] is another way to evaluate the quality of results from clustering using only the unlabelled sample and the clustering result. This index is a ratio of the between-clusters mean variance and within-cluster mean variance:

$$CH = \frac{Tr(B_K)/(K-1)}{Tr(W_K)/(N-K)} \tag{6}$$

where B_K is the between-clusters variance matrix, W_K is the within-cluster variance matrix defined by

$$W_K = \sum_{q=1}^{K} \sum_{x \in C_q} (x - c_q)(x - c_q)^T \tag{7}$$

$$B_K = \sum_q n_q (c_q - c)(c_q - c)^T \tag{8}$$

where K is the number of clusters, N is the number of samples in our data, C_q is the set of samples in cluster q, c_q is its centroid and n_q is its cardinality. The Calinski-Harabaz index is higher when clusters are well separated and quite dense.

4 Experiments

All experiments were performed using Python 3.

The initial data required pre-processing, because it contained empty cells. Empty cell means that chemical element was not detected in the sample, so each empty cell has been filled with zero before further processing.

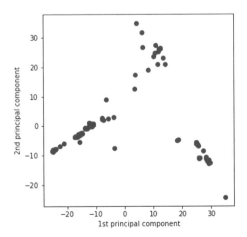

Fig. 1. Initial data.

As was described above PCA designed to reduce dimensionality that may help with data visualization. Figure 1 shows initial data with dimensionality reduced to 2 dimensions by PCA.

A number of experiments have been performed. Different parameters of methods and different approaches to pre-processing of initial data were used to determine more accurate clustering.

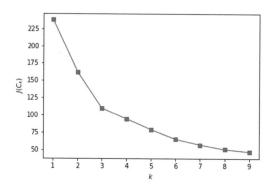

Fig. 2. $J(C)$ values

Figure 2 shows the graph with $J(C)$ values that have been calculated to determine the optimal number of clusters for K-Means method. The values of $J(C)$ decreases significantly when the number of clusters is less than 3 and not so significantly after 3. So 3 was selected as an optimal number of clusters. Figure 3 shows K-Means result with dimensionality reduced to 2 dimensions by PCA.

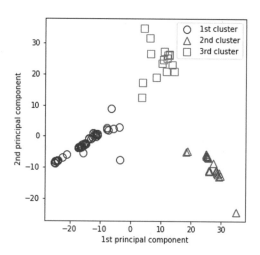

Fig. 3. K-Means result.

Mean-shift method does not require to set number of clusters manually. We have used various parameters values for this method. Each clustering result was evaluated and result with the highest score is shown on Fig. 4 with dimensionality reduced to 2 dimensions by PCA. As can be seen in this case data were split into 4 clusters.

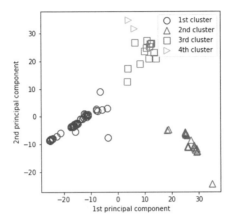

Fig. 4. Mean-shift result.

DBSCAN method also does not require to set number of clusters manually. Various parameters values were used in the method, each clustering result was evaluated. Figure 5 shows clustering by DBSCAN method result with the highest score, where we can see that data have been split into 3 clusters and 4 objects have been determined as noise. Objects marked as noise have not been used during evaluation.

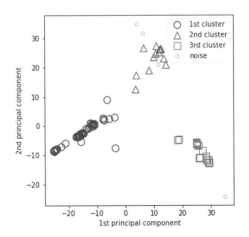

Fig. 5. DBSCAN result.

The dendrogram built by Hierarchical clustering with Ward method is shown on Fig. 6, here we can see 3 well separated clusters. Also it provides opportunity to see how clusters were built step by step.

The best clustering result by each method has been selected to compare different clustering methods. Table 2 shows values of Silhouette coefficient and Calinski-Harabaz index for each method. Selected methods have shown slightly different results. Using these values, we can assume that DBSCAN method has produced the best splitting on our data. However, all selected methods are appropriate for our problem.

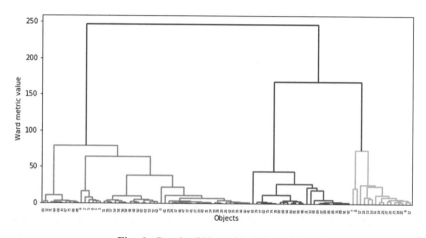

Fig. 6. Result of hierarchical clustering.

Table 2. Comparison of the best results of different clustering methods.

Metrics	Optimal value	Clustering method			
		K-Means	Mean-shift	DBSCAN	Hierarchical (Ward)
Silhouette	Max	0.6899	0.695	0.7146	0.6899
Calinski-Harabaz	Max	173.643	156.043	233.22	173.643

5   Conclusion

During the research provided data were successfully split into several groups (clusters) based on the qualitative and quantitative chemical composition of the archaeological samples. Principal Component Analysis method was used to reduce dimensionality before clustering and for visualization of data in 2 dimensions. K-Means, Mean-shift, DBSCAN and Hierarchical (Ward) clustering methods were used with different parameters. All results were evaluated using Silhouette coefficient and Calinski-Harabaz index. The best result of each method has been selected to compare it with other methods results. The values of quality indicators showed that all selected clustering methods can be successfully applied to the elemental composition of ceramic or glass archaeological samples, but nevertheless DBSCAN algorithm proved to be the best.

We believe that XRF-analysis and clustering algorithms can successfully solve the problems of grouping or separating archaeological artifacts, especially if these artifacts are fragments of ancient ceramic amphorae or glass vessels. This will allow to group fragments of artifacts that belong to the same object or group complete artifacts of similar origin. We hope to attract archaeological museums to cooperation and believe that we are waiting for interesting discoveries.

Acknowledgment. The research for this paper was financially supported by the Russian Federal Ministry for Education and Science (Grant No. 16-57-48001 IND_omi).

References

1. Bishop, L.R., Holley, R.G., Rands, R.L.: Ceramic compositional analysis in archaeological perspective. In: Advances in Archaeological Method and Theory, vol. 5, pp. 275–330 (1982)
2. Petit-Dominguez, D.M., Gimenez, R.G.: Chemical and statistical analysis of Roman glass from several Northwestern Iberian archaeological sites. Mediterr. Archaeol. Archaeom. **14**, 221–235 (2014)
3. Tryon, R.C.: Cluster Analysis: Correlation Profile and Orthometric (factor) Analysis for the Isolation of Unities in Mind and Personality. Edwards Brother, Inc., Ann Arbor (1939)
4. Pearson, K.: On lines and planes of closest fit to systems of points in space. Philos. Mag. **2**, 559–572 (1901)
5. Hartigan, J.A., Wong, M.A.: A k-means clustering algorithm. J. R. Stat. Soc. Ser. C (Appl. Stat.) **28**, 101–108 (1979)
6. Comaniciu, D., Meer, P.: Mean shift: a robust approach toward feature space analysis. IEEE Trans. Pattern Anal. Mach. **24**, 603–619 (2002)
7. Ester, M., Kriegel, H.P., Sander, J., Xu, X.: Density-based algorithm for discovering clusters in large spatial databases with noise. In: Proceedings of the 2nd International Conference on Knowledge Discovery and Data Mining, pp. 226–231 (1996)
8. Joe, H., Ward, J.: Hierarchical grouping to optimize an objective function. J. Am. Stat. Assoc. **58**, 236–244 (1963)
9. Rousseeuw, P.J.: Silhouettes: a graphical aid to the interpretation and validation of cluster analysis. J. Comput. Appl. **20**, 53–65 (1987)
10. Calinski, T., Harabasz, J.: A dendrite method for cluster analysis. Commun. Stat. **3**, 1–27 (1974)

Segmentation of Magnetic Anomalies in the Conduct of Archaeological Excavations

Sofya Ilinykh[1], Natalia Grafeeva[1,2], Elena Mikhailova[1,2(✉)],
and Olga Egorova[2]

[1] Saint-Petersburg State University, Saint Petersburg, Russia
e.mikhaylova@spbu.ru
[2] ITMO University, Saint Petersburg, Russia

Abstract. There are different ways to identify sites for future excavation in archaeology. One of these methods is based on magnetic properties of hidden underground objects and is carried out by means of magnetometric survey. In the pre-selected points the magnetic induction module of geomagnetic field is measured. Then the data is preprocessed (for example, changes in the Earth's total magnetic field are taken into account). After that, the researchers build a map of the area and estimate the probability that there are objects of archaeological value in the selected area. Based on this map the decision is taken whether it is worth to conduct excavations. However, both useful anomalies caused by the presence of archaeological sites and anomalies of a random nature can be simultaneously presented on the map. Unfortunately, sometimes it is quite difficult for a human to detect with one's eyes what anomalies are useful in the picture. In this article a process of automatic allocation of segments of archaeological objects is described. For this purpose specific smoothing and adaptive threshold binarization methods as well as clustering algorithms are considered.

Keywords: Magnetometric survey · Image segmentation · Archeology

1 Introduction

Archaeology is the science that studies the past of mankind based on real physical sources. Excavations necessary for the search of historically important objects are always very resource-intensive, so it is important to correctly determine the location of future excavations. From a scientific point of view, archaeological objects are physical bodies with physical properties. Magnetic properties reflect the state of the object and the result of changes that have occurred to it. So one of the ways to determine the location of future excavations is based on the magnetic properties of the objects hidden under the ground and is carried out using the so-called magnetometric survey. Magnetometric survey is usually performed on a square of surface with dimensions from 50 * 50 m to 100 * 100 m at the proposed excavation site. This square is called a tablet. The module of magnetic induction of the geomagnetic field is measured at pre-selected points.

The data are then processed for the first time (for example, changes in the Earth's total magnetic field are taken into account). After that, researchers build a map of the tablet and detect the probability that there are objects of archaeological value in it, then they take a

© Springer Nature Switzerland AG 2019
Á. Rocha et al. (Eds.): WorldCIST'19 2019, AISC 930, pp. 58–67, 2019.
https://doi.org/10.1007/978-3-030-16181-1_6

decision whether to conduct excavations. In the map built according to the magnetometric survey, anomalies can be caused by human activity in the past (deviations from the average value of the magnetic field). For example, such anomalies can be caused by foundations of old buildings, wells, tombs, roads, etc. Magnetometric survey is a non-destructive way of a primary exploration and does not require enormous expenditure of resources, as it allows to study the objects located under the ground without excavations. Moreover, this method is worth to use in cases when excavation is difficult or even impossible. Thus, the method described above would be convenient to use, but there are some difficulties encountered at the stage of processing the results. First of all, the search for anomalies that may be of scientific interest is hampered by the presence of soils with heterogeneous physical properties, metal debris and a high level of noise. In addition, a noise can be caused, for example, by the errors of the devices. As a result, it is sometimes quite difficult for a human to find fragments of a magnetometric image, indicating the presence of valuable archaeological artifacts. Therefore, when carrying out magneto-metric researches it is necessary to consider that magnetometric data contain the fol-lowing components:

- regional background data in which smooth data changes are related to geological features of the Earth's surface;
- noise in which data changes are related to the magnetic in homogeneity of the soil and instrumental errors during measurement;
- local anomalies in which data changes are generated by some objects.

Our work was aimed at finding local anomalies. However, it was necessary to take into account that not all such anomalies will be interesting: some of them could have been created by metal objects, or small archaeological artifacts. We were interested in large-scale archaeological artifacts - the remains of building foundations, wells, roads, etc. In the context of this article the anomalies created by large-scale archaeological objects will be called useful anomalies.

2   Related Work

More information on usage of magnetometry in archeology and on processing of measurement results of magnetometric devices can be found in [10]. According to [6], there are two main groups of methods to work with magnetometric data: in one group there are methods focused on the magnetometric nature of the data and in the other group there are methods for working with magnetometric images as simple images.

Indeed, if we take into account the nature of the data, we can use the methods developed for the interpretation of the magnetometric survey results. The input data at point x is represented as a sum:

$$B(x) = R(x) + P(x) + A(x) \tag{1}$$

where R - the regional background level, P - the interference anomaly, A - the local anomaly at point x.

In this case the researchers usually use certain algorithms (Shepherd filter [2]; Malovichenko method [11]; optimal energy filter [12], etc.) to isolate and subtract noise from regional background, and then to find local anomalies. The main description of these methods is in [8]. An attempt to use some of these methods is described in [3, 4, 6].

Another approach is based on the visualization of data obtained from magneto-metric measurements. For example, the value at each point can be reflected in shades from black to white. Then the researcher has the opportunity to analyse the data as an image and, accordingly to apply such methods as various filters, Fourier transform and so on. Two different approaches for image processing are considered within the framework of the task [9]:

- methods based on working directly with image pixels. For example, logarithmic and power transformations, application of Sobel [7], Gauss [5] filters and various computer vision methods;
- methods modifying the signal generated by the Fourier transform [6].

In our work, we tried to take into account some features of the analyzed data, but generally adhered to methods focused on image processing.

3 Data Description

Data sets for this work were provided by the Laboratory of magnetic resonance in the weak fields of the Department of nuclear physics of the Faculty of physics, St. Petersburg University. The data is the result of magnetometric surveys in the places of possible archaeological excavations. Each dataset contains about 15 thousands of three-dimensional points: the first two components are coordinates, and the third is the magnetic field value at that point minus the average field value rounded to integer. The coordinates of the points are measured in increments of 0.5. Such input data sets can be represented in two-dimensional images, where the third component is indicated by color. Examples of visualization of such datasets are presented in Fig. 1.

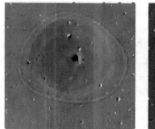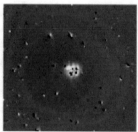

Fig. 1. Examples of input datasets with point anomalies.

In Fig. 1 black dots surrounded by white areas are clearly visible (in numerical data they are represented by large deviations from average value of the quadrangle). These are the point anomalies, caused by metal junk or measurement error. Usually, they cover no more than 1–2% of the whole area. They are not always objects of archaeological research, but they can make analysis of significant objects more difficult. Thus, our first idea was to smooth the values in such points. Then some small meaningful deviations will be more noticeable when visualized. For example, this smoothing can be achieved by a simple method of arithmetic mean, but instead of time series, space surrounding the interference points can be used. During the data analysis it was noticed that large deviations occur in places where metal junk is found, both positive and negative. At the same time, there are not many such points in the quadrangle where such deviations occur. There are two approaches for finding point anomalies based on these ideas.

4 Magnetometric Data Smoothing

Our initial data was extremely noisy. This was due to the fact that magnetometric measurements were carried out in places that have been used by humans not only for everyday lives for centuries, but that have been also the places for numerous battles. That is why the direct use of filters (Laplace, Prewitt, Sobel, Roberts, Schar, etc.), which are usually chosen to select segments of images [13–17] was almost impossible. Thus in our work we had to pay special attention to getting rid from point anomalies in the original data that were out of our interest in this study, as well as of measuring instruments errors and that is was the reason why we had to offer special methods of data smoothing.

- *Frequency-based approach.* First approach is based on sorting magnetic value by frequency of its appearance on a set area. Afterwards points that appear rarely enough (i.e. make up no more than M percent of the area) are selected. Table 1 shows a table of magnetic values and their corresponding frequencies, values that needed to be smoothed are highlighted. In this example M equals 2%. Figure 2 (left) shows values of the examined area with selected values to be smoothed marked in dark grey.
- *Deviation-based approach.* In the second approach points are considered to be interfered when their value deviates from the average by more than parameter δ. Figure 2 (right) shows selected points with $\delta = 9$. The average value is -3.5.

Table 1. Magnetic values and a percent of area covered by each value

Value	−14	−13	−9	3	4	−16	−12	−10	−6	−3	5	6	−8	2	7	−1	−4	−5	0	−2
Area (%)	2	2	2	2	2	3	3	3	3	3	3	5	5	5	6	6	8	9	13	16

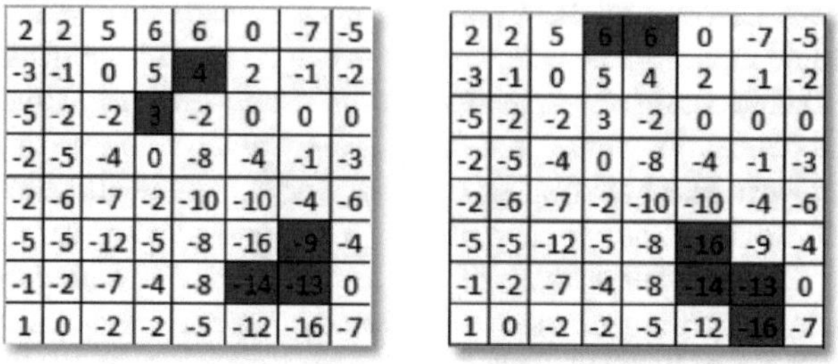

2	2	5	6	6	0	-7	-5
-3	-1	0	5	4	2	-1	-2
-5	-2	-2	3	-2	0	0	0
-2	-5	-4	0	-8	-4	-1	-3
-2	-6	-7	-2	-10	-10	-4	-6
-5	-5	-12	-5	-8	-16	-9	-4
-1	-2	-7	-4	-8	-14	-13	0
1	0	-2	-2	-5	-12	-16	-7

2	2	5	6	6	0	-7	-5
-3	-1	0	5	4	2	-1	-2
-5	-2	-2	3	-2	0	0	0
-2	-5	-4	0	-8	-4	-1	-3
-2	-6	-7	-2	-10	-10	-4	-6
-5	-5	-12	-5	-8	-16	-9	-4
-1	-2	-7	-4	-8	-14	-13	0
1	0	-2	-2	-5	-12	-16	-7

Fig. 2. Detecting point anomalies using frequency (left) and deviation (right) methods.

After the point anomalies have been found, their values should be smoothed. Following solution has been considered: points are assigned an average value of some neighborhood that does not include point anomalies. During the work it was found that in most cases 8 is an optimal number of points in a neighborhood. Figure 3 shows interference points (marked in gray) and their corresponding neighbors (marked in light gray). The interference points were found using the deviation method, where parameter δ is equal to 9.

2	2	5	6	6	0	-7	-5	-5	-5	-6
-3	-1	0	5	4	2	-1	-2	-5	-3	-4
-5	-2	-2	3	-2	0	0	0	-2	-1	-3
-2	-5	-4	0	-8	-4	-1	-3	-3	-1	2
-2	-6	-7	-2	-10	-10	-4	-6	-3	-3	3
-5	-5	-12	-5	-8	-16	-9	-4	1	-1	2
-1	-2	-7	-4	-8	-14	-13	0	1	-2	2
1	0	-2	-2	-5	-12	-16	-7	-5	-3	-1
1	-1	-2	-2	-3	-7	-16	-12	-11	-12	-9
-1	-2	-5	1	-3	-4	-13	-13	-11	-11	-6
-1	-3	-1	0	1	-2	-9	-16	-12	-7	-7
-1	-3	-2	2	3	0	-4	-11	-12	-4	3
-2	-2	2	1	-1	-2	2	-2	-9	9	2
-2	-2	0	3	3	0	0	-1	-12	-13	1

Fig. 3. Example of localization point anomalies and their neighborhoods

It should be noted that the smoothing efficiency is achieved by applying the smoothing algorithm to small squares. In all our subsequent experiments, we divided the original squares into 25 parts and used the frequency method.

5 Adaptive Threshold Binarization

As mentioned earlier, the use of different filters is one of the most well-known ways to segment images. We carried out many experiments with different filters and parameters on smoothed and non-smoothed data. However, the result in all cases turned out to be far from the desired, since instead of separating the noise from the useful anomalies, the filters visually strengthened both the useful anomalies and the noise. That is why it was decided to test the threshold binarization algorithm. Threshold binarization is used in the field of computer vision mainly to search for segments of objects in black and white images. All source data points are divided into two categories, depending on whether their threshold value is higher or lower, and are colored black and white, respectively. The threshold value is usually constant and is set by the researcher. For example, the average value of the tablet could be taken as a threshold value. However, in magnetometric images, it is often possible to observe sharp changes in the values of magnetic induction within the same tablet, which means that it is irrational to choose one threshold value for the entire data set. This problem can be solved by splitting the tablet into several smaller squares or by applying a modified threshold binarization called adaptive threshold binarization. The threshold value in adaptive threshold binarization is calculated separately for each pixel based on the neighborhood of a predetermined radius and can be calculated in several ways:

- arithmetic mean of the neighborhood;
- arithmetic mean of the maximum and minimum neighboring values;
- weighted sum minus constant C, where weights are pixel values from the neighborhood of a given size after applying the Gauss filter [5], and C is a constant that is manually selected by the researcher.

The third method is most suitable for the selection of segments in magnetometric data, since it is considered the most successful in the detection of segments, while the first two are used mainly for rather rough image smoothing, e.g. they are used in the field of photo processing. After all pixels are divided into two categories, primary filtering is performed, where the first category of pixels is colored black and the second category is colored white.

6   Clustering Using the DBScan Algorithm

Since adaptive threshold binarization allocates in addition to the necessary useful anomalies many small "pieces" that do not belong to large anomalies, it is necessary to learn how to interpret its results correctly. One option is to try to cluster the results of adaptive binarization. Clustering is the task of dividing objects into groups, and each of the groups obtained must be as homogeneous as possible. In images representing adaptive threshold binarization results, clusters are objects represented by a group of connected white pixels. It is appropriate to use density-based clustering algorithms, such as DBScan, to isolate such objects. This algorithm divides objects into clusters based on the proximity of the points within those objects. According to this algorithm, points that have a sufficient number of neighbors in a certain neighborhood are written to the same cluster. Accordingly, to start the method, you need to choose two parameters:

- eps-neighborhood in which the neighbors of the point are found;
- minPts-the minimum number of neighbors that a point must have.

The paper [1] described attempts to apply DBScan to magnetometric images, but a lot of noise and heterogeneity of the initial data did not allow to obtain a qualitative result. In this paper, the clustering algorithm was applied to preprocessed images whose points after binarization can take only two values: 0 or 1. Selecting the parameters, you can get results a little better or a little worse, but in general, the result is the same: the main part of the anomalies-interference is clearly visible and is a small cluster, and larger clusters correspond to parts of the useful anomalies.

7 Merging of Close Clusters

After the image has been divided into clusters, it is necessary to formally determine which of the clusters are noises and which are the desired useful anomalies. During the experiments, it was found out that in most cases the parts of the desired useful anomalies are close to each other. For example, Fig. 4 (left image) shows that a thin contour, which is a useful anomaly, consists of several separate arcs that are almost close to each other. This observation brought an idea to consider the distance between clusters and combine two clusters into one if the distance between clusters is less than the specified *cluster_distance* parameter. The distance between clusters S1 and S2 can be calculated as the distance of "nearest neighbors":

$$cluster\_distance(S_1, S_2) = min_{x_1 \in S_1, x_2 \in S_2} d(x_1, x_2) \tag{2}$$

where d – Euclidean distance.

8 Secondary Filtration

After merging close clusters, attention should be payed to the clusters of small size (with a small number of pixels), which have not found a suitable neighbors to join with. Such clusters are unlikely to indicate the presence of large archaeological artifacts and can be removed. One can get rid of them by simple filtering: for example, you can delete clusters that are smaller than some specified value of the *cluster_size* parameter.

9 Experiments

In the previous sections the algorithm stages for the allocation of magnetic anomalies segments were described, which data sets went through during analysis. Let us briefly recall these stages:

- data preprocessing (smoothing);
- primary filtration (detecting threshold) and adaptive threshold binarization;
- application of the DBScan algorithm;
- merging close clusters;
- secondary filtering (getting rid of small clusters).

It is worth noting that the peculiarity of magnetometric images and their interpretation complexity was also caused by their diversity. There are visually similar initial data, but there is no single "template" for them, so there were some difficulties at the point of selection of algorithm parameters. In our work we have considered several types of initial data that have different noise levels and contain different amounts of useful anomalies. We have selected and described the parameters of the main methods used in the algorithm. However, you should remember that the selected parameters will not be universal even for similar data sets. Although the parameters from the examples below can be taken as initial parameters in the study of any magnetometric image, each data set requires an individual selection of parameters.

Fig. 4. The results of the algorithm for the input datasets from Fig. 1

9.1   A Dataset with a Noticeable Noise and Useful Anomalies

The left dataset in Fig. 1 shows the input data, first of all, with an interesting contour, which is a very much curved ellipse. It can be assumed that the noises, which in the initial data are visible as vertical bands, make the selection of parameters very complicated, and may not allow the algorithm to highlight the contours of useful anomalies. However, even if this happens, such noise is rare, as it is a result of measurement errors in the magnetometric data. Thus, if the algorithm detects "ellipse" together with vertical stripes, it will be a good result. We used the following parameters: minPts = 3, eps = 2, cluster_distance = 5, cluster_size = 100 and got the result shown in Fig. 4 (left image). Note that vertical noiseis partially removed and other noise almost completely disappears. Thus, the algorithm works quite well on the original data even with strong noise.

9.2   A Dataset with Useful Anomalies

On the middle dataset shown in Fig. 1, there are a lot of point anomalies and it is difficult to see the segments of the useful anomalies. The input data was preprocessed by means of the smoothing algorithm for magnetometric data, afterwards more segments became visible. Smoothed data contains much less noise and, in addition, the image itself becomes more contrast. Then the adaptive threshold binarization algorithm and the DBScan clustering algorithm were applied. We used parameters: minPts = 3, eps = 2, cluster_distance = 5, cluster_size = 100 and got the result, which shown in Fig. 4 (middle image). There are clearly visible outlines in the inner circle, in the outer arc and roads. Moreover, now the anomaly in the lower left corner becomes visible, which is very difficult to see with the naked eye even on the smoothed data. Unfortunately, the internal useful anomaly (apparently, the road) is determined only partially.

9.3   A Dataset Without Useful Anomalies

As even a naked eye can see, any segments of useful anomalies in the right dataset in Fig. 1 do not exist. There are only noises created by metal debris that look like black dots surrounded by white. It is obvious that a really good algorithm that can find useful anomalies in this data set should show their absence. If we take the parameters that are close to the parameters in the previous examples, we get the following result: there is one cluster of small size, formed by a cluster of closely spaced noise (Fig. 4, right image).

Summing up, we can say that the developed algorithm really achieves the goal: the segments are detected on the basis of input data or smoothed input data that contain useful anomalies, while on magnetometric images containing only noise, the segments of useful anomalies are not found.

10   Conclusion

In the framework of the analysis, we investigated various methods for working with magnetometer data and algorithms of segment extraction in magnetic images. An algorithm was developed to determine the segments of magnetic anomalies representing large archaeological sites. The developed algorithm was tested on the available data, the paper describes the selection of algorithm parameters on various initial data. In the process of testing the algorithm, the following conclusions were made:

- The developed algorithm really highlights the segments of useful magnetic anomalies.
- If there are no archaeological objects and large area noise in the data, the algorithm shows the absence of useful anomalies.
- In case of highly noisy initial data, preprocessing with a special smoothing algorithm should be used.

The main difficulty of the analysis of magnetometric images is the problem of separation of useful anomalies from point anomalies. Despite the positive results achieved, the current algorithm is not able to solve this problem perfectly. Since the task of automated extraction of segments on magnetometric images is quite relevant, further researchers are invited to work in the following areas:

- Search for a formal criterion to evaluate the results of the algorithm.
- Improvement of the described algorithm and its efficiency.
- Development of algorithms to identify the type of segments.

References

1. Volzhina, E., Chudin, A., Novikov, B., Grafeeva, N., Mikhailova, E.: Discovering geo-magnetic anomalies: a clustering-based approach. In: 2016 International FRUCT Conference on Intelligence, Social Media and Web (ISMW FRUCT), pp. 1–7 (2016)
2. Gonzalez, R., Wood, R.: Digital Image Processing. Prentice Hall, Englewood Cliffs (2008)
3. Grafeeva, N., Mikhailov, E., Pashkova, M., Ilinykh, S.: Magnetometry data processing to detect archeological sites. In: 17th International Multidisciplinary Scientific GeoConference SGEM 2017, pp. 393–400 (2017)
4. Grafeeva, N., Paskova, M., Ilinykh, S.: Smoothing algorithm for magnetometric data. In: Second Conference on Software Engineering and Information Management (SEIM-2017), pp. 31–36 (2017)
5. Liu, J.-L., Feng, D.-Z.: Two-dimensional multi-pixel anisotropic Gaussian filter for edge-line segment. Image Vis. Comput. **32**, 37–53 (2014)
6. Mikhailova, V., Grafeeva, N., Mikhailova, E., Chudin, A.: Magnetometry data processing to detect archaeological sites. Pattern Recognit. Image Anal. **26**(4), 789–799 (2016)
7. Bui, T., Spicin, V.: Analysis of methods for edge detection on digital images. In: Proceedings of TUSUR, no. 2, pp. 221–223 (2010)
8. Dudkin, V., Koshelev, I.: Methods of complex interpretation of the results of magnetometric survey of archaeological sites, Kiev, Thought tree (1999)
9. Kazhdan, A., Guskov, O.: Digital Image Processing. Technosphere (2006)
10. Kutaisov, V., Smekailov, T.: Materials for the archaeological map of crimea: remote and geophysical studies of ancient settlements in the North-Western Crimea, Dolia (2014)
11. Malovichenko, A.: Methods of Analytical Continuation of Gravity Anomalies and Their Application to the Problems of Gravity Exploration, Technique (1956)
12. Nikitin, A., Petrov, A.: Theoretical basis of Geophysical Data Processing. RGGU, Moscow (2008)
13. Lakshmi, S., Sankaranarayanan, V.: A study of edge detection techniques for segmentation computing approaches. Comput. Aided Soft Comput. Tech. Imaging Biomed. Appl. **20**, 35–41 (2010)
14. Thakare, P.: A study of image segmentation and edge detection techniques. Int. J. Comput. Sci. Eng. **3**(2), 899–904 (2011)
15. Ramadevi, Y.: Segmentation and object recognition using edge detection techniques. Int. J. Comput. Sci. Inf. Technol. **2**(6), 153–161 (2010)
16. Senthilkumaran, N., Rajesh, R.: Edge detection techniques for image segmentation—a survey of soft computing approaches. Int. J. Recent Trends Eng. **1**(2), 250–254 (2009)
17. Sehgal, U.: Edge detection techniques in digital image processing using fuzzy logic. Int. J. Res. IT Manag. **1**(3), 61–66 (2011)

Automatic Document Annotation with Data Mining Algorithms

Alda Canito[1](✉) [iD], Goreti Marreiros[1] [iD],
and Juan Manuel Corchado[2] [iD]

[1] GECAD - Research Group on Intelligent Engineering
and Computing for Advanced Innovation and Development,
Polytechnic of Porto, Porto, Portugal
alrfc@isep.ipp.pt
[2] Department of Computer Science, University of Salamanca, Salamanca, Spain

Abstract. By combining both semantically annotated documents and semantically annotated services, it is possible for digital solutions to automatically retrieve and assign documents not only to their own services but also to those provided by others, thus improving and optimizing the experience of its users. Most of the information exchanged in and between services is still either in paper form or over email and is mostly unstructured and in lack of any form of annotation. Manual and semi-automatic approaches are not suitable to deal with the huge amounts of heterogeneous and constantly flowing data existent in this scenario, thus raising the issue of automatic annotation. In this paper, three data mining algorithms are used to annotate a set of documents and their results compared to manually provided annotations.

Keywords: Semantic annotation · Data mining · Information Extraction

1 Introduction

Implementation in large scale of digital solutions faces many challenges, mainly because we are facing a heterogeneous and highly distributed environment, with many services being dispersed geographically and/or supplied concurrently through different channels. Additionally, a large amount of services is still made available in a purely 'traditional' fashion, i.e., in a non-electronic, in-person basis. These conditions make it particularly hard for a user to navigate existing options and understand what documents and information they must supply and where in order to achieve their goals.

Finding and deploying the correct services requires a common knowledge formalization such that different entities can facilitate/consume each other's services to provide the citizen a simple, straightforward and transparent experience when dealing with digital services. The semantic annotation of these services allows them to be described by virtue of their purposes, of the problems they pertain to solve and of the information they use or facilitate [1, 2]. This new layer of information makes it possible to describe, among other things: (i) the contents of the exchanged messages; (ii) the sequence in which the messages must be exchanged in order for them to make sense; (ii) the state changes resulting of such exchanges. Such descriptions allow a higher

© Springer Nature Switzerland AG 2019
Á. Rocha et al. (Eds.): WorldCIST'19 2019, AISC 930, pp. 68–76, 2019.
https://doi.org/10.1007/978-3-030-16181-1_7

degree of automation in the tasks of discovery, invocation, composition and compensation of available services [1].

In order to automatically assign documents to services, it is imperative to properly manipulate the information they contain. Information Extraction (IE) approaches can be applied to these documents, with semantic technologies, such as ontologies, used to describe the output of these approaches and often to guide the extraction process [2, 3]. This metadata adds a semantic layer to the documents, in a manner that is commonly described as semantic annotation. Because ontologies are a formal and explicit specification of a shared conceptualization, they make the documents understandable and processable by both human and machine, improving system interoperability, document querying and easing further information retrieval tasks [3]. However, in most cases, these documents are either manually annotated or not annotated at all. While manual and semi-automatic annotation approaches exist, their reliance on human intervention does not make them suitable for all scenarios: especially when large amounts of constantly-flowing information are involved [3]. Data mining algorithms can be used to automate the task of identifying entities and patterns in (often large) volumes of data and can be combined with the domain knowledge provided by the ontologies to generate annotations with a larger confidence value. In this paper, a set of documents is annotated by different data mining algorithms, and the quality of these annotations is assessed by comparing them to a manually supplied set.

This document is structured as follows: (1) Introduction, (2) Semantic Annotation, which presents the concepts, technologies and challenges of this area and how data mining algorithms can be applied, (3) Experiments and Results, showing the experimental setup, the configurations employed, the experimentations' results and respective discussion and, finally, (4) Conclusions.

2 Semantic Annotation

Information contained in documents is most commonly provided in the form of natural language texts. This information can be retrieved and made useful using Information Extraction (IE) techniques [4]. Such information, i.e. the classes of objects and their relationships, must be specified in some way. In Ontology-Based Information Extraction systems, as the name implies, ontologies are used to specify which entities and attributes to look for in natural text and to describe the results [4]. The goal of the semantic annotation process is to assess which parts of the document correspond the concepts described in the ontology, and thus the result is a set of mappings between document fragments and ontology concepts [5].

There are three main types of annotators described in literature: (i) manual annotators, which, as the name indicates, require human intervention. For the best results to be obtained, a user with a deep understanding of the modelled knowledge is desired. This, however, is often not the case, which can result in syntactic errors or incorrect references, issues that are exacerbated when dealing with complex domain ontologies. Nevertheless, the resulting annotations are generally of good quality, even though this process is often time-consuming and unsuited for large corpora [4]. The second type of annotation, (ii) automatic annotation, aims to generate the annotations without any human intervention. Many of the existing tools use IE techniques, pattern matching or rule-based algorithms, among other approaches, with different levels of success [6]. Completely automated processes are, however, still rare, as most of the existing tools expect some human intervention before or after the annotation processes [3]. As such, they are actually used for (iii) semi-automatic annotation. This type of annotation requires human input in terms of rules, template annotations and structures which will then be used by algorithms to make new annotations. Additionally, the generated annotations can be further reviewed by domain experts, correcting any wrong or missed annotations generated by the algorithms, with this information serving as new inputs in an iterative fashion until a certain threshold is achieved [4].

Even though the digitalization of services is expanding, many entities hold a large share of their information either in paper or over email, without any sort of annotations beyond those performed manually. This information is virtually unqueryable and searches must be done in person. Additionally, new documents in the same situation are created daily, rendering manual annotation extremely ineffective. Many different approaches have been used when it comes to the automatization of the annotation processes over unstructured data [7, 8]. Some of these rely on data mining algorithms, which can be combined with the semantic information embedded in domain ontologies for adding constraints or to present results, among other enhancements, in what is commonly described as Semantic Data Mining [8]. As an input, the ontology can supply a number of descriptions and individuals which can be found in text, which ultimately means that the annotation problem can be framed as a multi-label classification problem. For our experiment, we will focus on three different data mining approaches: Classification, Clustering and Named-Entity Recognition.

Classification approaches attempt to assign fragments of text into predefined categories and can be done either manually or automatically [8]. For automatic approaches, this process is divided in two steps: an initial training step, where specific text fragments are properly separated into different disjoint categories (or classes), and a testing step, in which new text fragments must be assigned to these categories.

Clustering approaches attempt to generate groups (or clusters) of samples by calculating the distance between existing observations [9]. In the case of document analysis, the clusters can be comprised of words, sentences or sections of text which are considered similar according to the distance function in use.

Named-Entity Recognition [10], as the name indicates, is the task of finding entities mentioned in unstructured text. In order to do so, these entities must either be previously identified and supplied in through dictionary, or the patterns they follow be properly described (e.g. by using a regular expression).

3   Experiments and Results

As stated in [8], the quality and cover of the domain ontology is highly influential on the quality of the resulting annotations. A good ontology is therefore described as one which covers the categorization domains, whose classes and properties include descriptive labels and has a good measure of named instances properly categorized. The information supplied by the ontology can be used to calculate the semantic similarity between its instances, class declarations and the information in the documents [11]. Since this means algorithm implementations are independent of the domain ontology in use, an ontology with a large set of descriptive labels was selected for the preliminary experiments.

MartineTLO [12] is a top-level ontology for describing the marine domain. This ontology can be used, among other things, to taxonomically classify marine life, having its classes and properties extensively labeled [13]. As a top-level ontology, it does not include instances; it was manually instantiated with 450 instances of different species extracted from its warehouse [14], which also include descriptive labels (such as common names and scientific names). As for the documents to annotate, a total of 20 Wikipedia pages pertaining to the same domain were selected and manually annotated with the instances, resulting in a total of 1325 annotations. These manually annotated texts will serve as our golden standard, with which the automatically generated annotations will be compared. The results are evaluated in terms of *precision*, *recall* and their harmonic mean, the *F1 measure,* as they are the measures most commonly used for evaluating Information Extraction performance [4].

Precision (1) reflects the relevance of the generated annotations, by diving the number of correct annotations by total annotations found:

$$precision = \frac{True\,Positives}{True\,Positives - False\,Positives} \tag{1}$$

Recall (2) is the number of correct annotations effectively made divided by the total number of annotations that should have been found:

$$recall = \frac{True\,Positives}{True\,Positives + False\,Negatives} \tag{2}$$

Finally, F1 measure (3) is the harmonic mean of the previous measures:

$$F1 = 2\frac{precision * recall}{precision + recall} \tag{3}$$

As previously mentioned, in our experiment three different approaches were considered, namely: Named Entity Recognition, Classification and Clustering, the specific implementations of which were provided by the LingPipe [15] toolkit. The labels of classes, properties and individuals were used to train the algorithms.

3.1 Classification

The classification algorithm applied follows the prediction by partial match (PPM) approach [16]. It works by using the set of descriptions and labels as the training set, which will generate the decision trees. This is a multi-label classification problem, with one class per each ontological entity which effectively work as the algorithm's language model. The corpus is consumed one character at a time and, if a word has been found, its joint classification is computed, attributing the word the best category match. Additionally, the cross-entropy is calculated and taken in account when generating models from training data [16]. In our experiment, we established that the generated model should have a probability level over 90 and a cross-entropy below 2,5. The algorithm was run several times with different parameters in order to establish which generated the best annotations. Table 1 shows the parameters used in each run and their respective results in terms of precision, recall and F1 measure.

Table 1. Parameter variation and results for the Classification algorithm

Prob. level	Cross-entropy	True positives	False positives	False negatives	Precision	Recall	F1
90	0,5	1325	564	0	0,701	1,000	0,825
90	1	1325	564	0	0,701	1,000	0,825
90	2	1325	564	0	0,701	1,000	0,825
90	2,5	1293	138	32	0,904	0,976	0,938
95	0,5	1293	138	32	0,904	0,976	0,938
95	1	1293	138	32	0,904	0,976	0,938
95	2	1293	138	32	0,904	0,976	0,938
95	2,5	1293	138	32	0,904	0,976	0,938
99	0,5	1293	138	32	0,904	0,976	0,938
99	1	1293	138	32	0,904	0,976	0,938
99	2	1293	138	32	0,904	0,976	0,938
99	2,5	1293	138	32	0,904	0,976	0,938

It is possible to understand that lower Probability Levels together with low Cross-Entropy generate poor results: although the recall is high, they have a very low precision, meaning the number of false positives is high and thus the algorithm is generating annotations in text fragments that shouldn't have been annotated. However, a Probability Level of 90 combined with the highest Cross-Entropy (2,5) shows a slight decrease in recall but a considerable increase in precision. This is an interesting situation, since cross-entropy can be used to assess the inaccuracy of the model: with lower cross-entropy values relating to a better accuracy, and higher relating in lower accuracy. This effect can be seen in the rise of false positives obtained with a higher cross-entropy. However, because the model is less strict, it also allowed for a higher number of true positives, thus providing a more balanced F1 measure. When raising the probability level over 95, however, it seems that the rigidness of the model is at its lowest possible, and neither raising the probability level or the cross-entropy has any effect on the number of true or false positives.

3.2 Clustering

The Clustering algorithm applied is a modified version of agglomerative clustering (single-link), where the labels supplied by the ontology are the initial clusters [17]. It consumes segments of text and computes the edit distance between these and the known clusters. If this distance is below a user-defined threshold, the segment is annotated, and the process continues. The calculation of this distance relies on the number of permutations, additions and removals that are required to obtain the label from the text fragment and this value comes in the range [0–99] (0 being an exact match and 100 being a complete mismatch). Table 2 shows the results obtained with different threshold values:

Table 2. Threshold variation for the Clustering algorithm

Threshold	True positives	False positives	False negatives	Precision	Recall	F1
2	1228	25	97	0,980	0,927	0,953
5	1291	61	34	0,955	0,974	0,965
10	1291	140	34	0,902	0,974	0,937
20	1295	257	30	0,834	0,977	0,900
30	1325	564	0	0,701	1,000	0,825

As expected, increasing the threshold does result in a higher number of false positives. However, the initial run, with the lowest threshold, had a low number of false positives, but also a high number of false negatives – thus implying that a lot more annotations were being generated than expected. Slowly increasing the threshold results in some reevaluations, with the number of false positives rising but the number of false negatives lowering considerably, resulting in a slightly more balanced F1 measure. The number of true positives also shows a slight improvement. Raising the threshold further results in further reevaluations, which manifest in improvements on the number of true positives and false negatives but comes with the cost of inflating the number of false positives, resulting in worse precision and F1 measure.

3.3 Named Entity Recognition

The idea of a Named Entity refers to a unit of textual information expressing an entity (e.g. a person, a location or a company, but also numerical expressions such as dates or prices) and are frequently cases of individuals in text [18]. The Named Entity Recognition algorithm aims to find and classify exact mentions of entities in text [19]. In order to do so, it needs a source ontology that defines the classes and samples of text or patterns that it must find, which are once more provided by the labels and properties in the ontology. Then, the algorithm consumes the corpora not in sequences of characters but in sequences of chunks, which can be provided through the use of different tokenization techniques. In this scenario, the *IndoEuropeanTokenizer* of the LingPipe toolkit has been applied. The results can be seen in Table 3, below:

Table 3. Named Entity Recognition results

True positives	False positives	False negatives	Precision	Recall	F1
1228	25	97	0,980	0,927	0,953

While these results show a good F1 measure, the comparatively low recall is the result of a particularly high number of false negatives. This a symptom of low flexibility in the annotation process, meaning that the classifier was leaving out a significant number of annotations. This is understandable if we consider that the way entities are mentioned in the text is not always identical to the way they are described in the dictionary.

3.4 Comparative Analysis

Now that the best parameter configurations for each algorithm are known, a comparative analysis can be made. Table 4 shows the best result for each algorithm:

Table 4. Best results for Classification, Clustering and Named Entity Recognition

Algorithm	True positives	False positives	False negatives	Precision	Recall	F1
Classification	1293	138	32	0,904	0,976	0,938
Clustering	1291	61	34	0,955	0,974	0,965
NER	1228	25	97	0,980	0,927	0,953

We can now compare the behavior of the different algorithms. The Classification algorithm apparently tends to generate more annotations, with the downside of also generating more incorrect annotations. However, because of this, the probability of missing annotations is also lower. As seen in Table 2, the Clustering algorithm is the most versatile, with a wider range of results. It does show that the bigger the edit distance allowed, more annotations are found, but with a higher error. It is possible to fine tune the parameters to reach a good balance of false positives and false negatives, in a combination that attains the best F1 measure between all three algorithms. As the NER attempts to find exact mentions of the entities in the text, its low number of true positives was to be expected. This happens because, as mentioned above, the entities in text are not always mentioned in the exact same way as they appear in the dictionary. Ultimately, the algorithms ability to identify the entries in text will depend on how the chucks are processed, which leaves some room for the documented miss-annotations. While this algorithm does not appear to be particularly suited for this kind of scenario, it does have a much lower rate of false positives as a result of its more conservative approach. Nonetheless, this approach may produce better results over strongly typed data, whose entities may follow a structure closer to that presented in the dictionary.

4 Conclusions

Even though a growing number of electronic means are available, a significant portion of communications between users and digital services is made in the form of paper or email. These organizations are thus faced with an ever-growing volume of data which they must manually associate with their services. Assigning a layer of semantic information to these documents is essential in order to automatically associate them with the correct services; and because more frequently then not these data are unstructured, Data Mining algorithms become useful tools for the task of semantic annotation.

In this paper, three different approaches were selected to find ontological entities in text: Classification, Clustering and Named-Entity Recognition, and the implementations supplied by the LingPipe Toolkit were used. While these implementations may be simple, by fine-tuning their parameters relatively good results were obtained.

Through a comparative analysis of these results, it was possible to notice that the weak points of one algorithm could potentially be overcome by the strong points of others. This suggests that some combination of these algorithms may produce better results, which will be the subject of further study by the authors.

Acknowledgements. The present work has been developed under the EUREKA - ITEA2 Project INVALUE (ITEA-13015), INVALUE Project (ANI|P2020 17990), and has received funding from FEDER Funds through NORTE2020 program and from National Funds through FCT under the project UID/EEA/00760/2013.

References

1. McIlraith, S.A., Son, T.C., Zeng, H.: Semantic web services. IEEE Intell. Syst. **16**(2), 46–53 (2001)
2. Uren, V., et al.: Semantic annotation for knowledge management: requirements and a survey of the state of the art. In: Web Semantics: science, services and agents on the World Wide Web 4.1, pp. 14–28 (2006)
3. Abioui, H., et al.: Semantic annotation of documents: a comparative study. Int. J. Adv. En. Manage. Sci. **2**(11)
4. Wimalasuriya, D.C., Dou, D.: Ontology-based information extraction: an introduction and a survey of current approaches. J. Inf. Sci **36**, 306–323 (2010)
5. Pech, F., et al.: Semantic annotation of unstructured documents using concepts similarity. Sci. Program. **2017**, 10 (2017)
6. Oliveira, P., Rocha, J.: Semantic annotation tools survey. In: 2013 IEEE Symposium on Computational Intelligence and Data Mining (CIDM). IEEE (2013)
7. Corcho, O.: Ontology based document annotation: trends and open research problems. Int. J. Metadata Semant. Ontol. **1**(1), 47–57 (2006)
8. Dou, D., Wang, H., Liu, H.: Semantic data mining: a survey of ontology-based approaches. In: Semantic Computing (ICSC). IEEE (2015)
9. Jain, A.K.: Data clustering: 50 years beyond k-means. Pattern Recogn. Lett. **31**(8), 651–666 (2010)
10. Nadeau, D., Satoshi, S.: A survey of named entity recognition and classification. Lingvisticae Investig. **30**, 3–26 (2007)
11. Allahyari, M., Kochut, K.J., Janik, M.: Ontology-based text classification into dynamically defined topics. In: 2014 IEEE International Conference on Semantic Computing (ICSC). IEEE (2014)
12. Martine TLO. https://www.ics.forth.gr/isl/MarineTLO/. Accessed 20 Nov 2018
13. Martine Top Level Ontology Specification. https://www.ics.forth.gr/isl/ontology/content-MTLO/html/index.html. Accessed 20 Nov 2018
14. Martine TLO Warehouse. https://www.ics.forth.gr/isl/MarineTLO/#warehouse. Accessed 20 Nov 2018

15. LingPipe. http://alias-i.com/lingpipe. Accessed 30 Nov 2018
16. Teahan, W.J.: Text classification and segmentation using minimum cross-entropy. In: Content-Based Multimedia Information Access, vol. 2 (2000)
17. Murtagh, F., Contreras, P.: Algorithms for hierarchical clustering: an overview. Wiley Interdisc. Rev.: Data Min. Knowl. Discov. **2**(1), 86–97 (2012)
18. Cimiano, P., Völker, J.: Towards large-scale, open-domain and ontology-based named entity classification. In: Proceedings of the International Conference on Recent Advances in Natural Language Processing (RANLP) (2005)
19. Alvarado, A.B.R., Arevalo, I.L., Leal, E.T.: The acquisition of axioms for ontology learning using named entities. IEEE Lat. Am. Trans. **14**(5), 2498–2503 (2016)

Generation Z and the Technology Use During a Trip

Pedro Liberato[1], Cátia Aires[1], Dália Liberato[1(✉)], and Álvaro Rocha[2]

[1] School of Hospitality and Tourism, Polytechnic Institute of Porto,
Porto, Portugal
`{pedrolib,dalialib}@esht.ipp.pt`, `aires.ca@hotmail.com`
[2] Department of Informatics Engineering, University of Coimbra,
Coimbra, Portugal
`amrocha@dei.uc.pt`

Abstract. This article aims to contribute to the deepening of scientific knowledge about the specificities of the Generation Z behavior regarding the tourism sector and the use of Information Technologies, during a trip. The objective of this research is to evaluate the destination choice by Generation Z and conclude if the reasons for the choice of destination (city of Porto - Portugal) positively influence the use of technological resources during the trip. To achieve the defined objectives, 400 validated questionnaires were gathered from tourist belonging to generation Z at Porto on the main streets of tourist attraction. The results obtained highlight tourist services importance focused on accommodation, transport, catering, and attractions in general, that must always be available to tourists, since the need to obtain clarification on the tourist destinations before and during the trip, is essential and positively influences the satisfaction of the consumer experience.

Keywords: E-tourism · Generation Z · Technology · Tourism destination

1 Introduction

Porto is one of the oldest cities in Europe, with a historic center classified by UNESCO as World Cultural Heritage since 1996 [1]. As a tourist destination, it has increased its reputation internationally due to the prizes for Best European Destination in 2012, 2014 and for the third time in 2017 [2]. Also, in 2017, the municipality of Porto obtained the first place in the dimension living in the Portugal City Brand Ranking [3].

Technology has significantly changed the way tourism is perceived, encompassing individuals, space, products, services, has a dynamic character itself, and increasingly consumer involvement, and co-creation, reflecting greater relevance for the promotion of the destination [4]. Experience is characterized as a subjective and cognitive activity of an individual, where it acquires knowledge and skills in the involvement or exposure to a specific event, linking the emotions, feelings and sensations [5, 6]. One of the consequences of technological advancement has been the digitization of words, audio, video and images, which means that content can be viewed and consumed in different digital formats and in several or various mobile devices [7]. The tourism sector uses the

© Springer Nature Switzerland AG 2019
Á. Rocha et al. (Eds.): WorldCIST'19 2019, AISC 930, pp. 77–90, 2019.
https://doi.org/10.1007/978-3-030-16181-1_8

image to promote destinations and attract new tourists, and today, the sharing of these images is done mostly through social networks by tourists/visitors themselves who take photographs of their experiences and share in the digital world (social networks, tourist sites, among others), influencing the decision of other potential tourists regarding the destination choice [8].

Technologies have played a decisive role not only for the competitiveness of tourism organizations, but also for the tourists' experience. Most of the research about travel information, reservations and payments during the preparation phase of the trip is done through the internet. The same occurs during travel. The internet, mobile phones and other technologies provide travelers with relevant information sometimes essential to travel, most diverse and useful due to easy accessibility and connection [9, 10]. According to [8], is possible to have access to information about how the tourist perceives the destination through the use of their social networks, interpreting how they communicate with their Internet contacts regarding the destination visited, and also with the digital footprint of the destinations they visit, such as hotel check-in, tourists attractions, credit card use (registration), among others. Facebook profile or a search in Google, leaves a footprint that indicates the tourist's journey from the decision to purchase a trip or share a destination with their group of friends or family.

While tourists travel, it demonstrates their thoughts/perception regarding the destination, stay, during their publications in real time, sharing reviews online through their comments and hashtags [8]. Technology can be used in many ways, and one of the characteristics that most benefits tourism is based on data analysis (digital footprint) that can be made, during the tourist's Internet usage, on the consumer's journey (Fig. 1). This proposal encompasses synergies between four dimensions: technology, humanization, territoriality and context. In this way, there is a link between the sector entities, the local and regional network of service providers and the extension of services that manage the experiences in the destination, reservations and purchases of tours and activities, such as accommodation and moments inspiration/decision/planning with the availability of online content and services so the consumer can make the best decision and plan their trip and perceive, from their motivations, the multiple experiences that can be made in the destination (Porto City Council, 2018). In this investigation, the evaluation of the tourist experience satisfaction is focused on the tourists belonging to Generation Z, in the city of Porto, during the tourist experience, where the use of technology can lead to satisfaction or dissatisfaction concerning the destination.

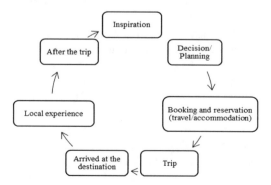

Fig. 1. Consumer's journey Source: Porto City Council (2018), adapted.

The first phase (Table 1), the one before the trip, focuses on: where the tourist collects information about the destination, to know the most interesting places to visit, how to collect information and make reservations about accommodation, safety, which public transport to use during the experience, and everything that could be useful for organizing the trip. During the trip, the tourist may want to complement the information already searched in the previous phase, or know for example, which way is the closest path go to get a desired place, the sights, museums or restaurants closer, among other useful information that can enrich the stay. Apps and the use of mobile devices creates interactivity with the destination, as well as allows the recording of moments and sharing on social networks [11–14]. According to [15], technologies in tourism have increasingly emergent fields, regarding the tourism experience, such as the virtual reality as a form of mobility interfering in the equation between tourism and physical movement. Virtual Reality (VR) is a technology that allows its users to visualize a simulated and interactive scenario (in the sense that the user can modify the virtual environment and move the virtual objects with the use of sensors and other devices), however, this technology doesn't allow interaction between the user and the real world [16]. Unlike VR, the augmented reality (AR) inserts virtual objects into real environment, in real time, where there's an interconnection between virtual elements (in 3D) and real elements easily used by users [17]. The appearance of portable devices, such as tablets and smartphones, have led to the creation of a universe of applications, so users can access a wide variety of content daily in a more simplified way in areas such as leisure, tourism, health, education and online shopping [7].

Table 1. Tourist experience phases

Before	During
Research	Access to additional information
Organization	Moment recording
Reservations	Interactivity
Payment methods	Payment methods
	Technology available

Source: Compiled by the authors

The evolution of technology is increasingly directed towards mobility, with the accession of portable devices by the population, which is increasingly demanding in terms of demand for the best offer, with more features (additional features to the device) and systems operations. Mobile applications are systems embedded in the software of electronic devices, such as smartphones and tablets. Apps are installed on devices through online stores: Google play, App Store, or Windows Phone Store [18]. For tourism purposes, mobile applications have emerged as basic tools for supplementary information to users, but with technological developments new fields have emerged within applications such as reservations, purchases, check-in of airplane tickets, among many other options such as the booking of hotels, cars, restaurants, and information applications as mobile guides to improve the tourist experience before and during their visit at a certain location/destination [18, 19].

This paper is divided in three parts. The first part, the literary review, emphasizes the important role of technology for tourism, and for the generation Z, as well as technology offer in Porto as a tourism destination. The second part describes the research methodology, the methodology of data collection, presents and discusses the results of the research. The third and final part, the conclusion, discusses the innovative perspective introduced in the analysis of the generation Z and the choice of technological resources during the trip.

2 Literature Review

Technology in tourism can be identified through four key attributes: information, accessibility, interactivity and customization, which interact with utility and the perception of the role of technology. Information is the combination of quality information and reliable information that is transmitted online, for example, on tourism websites. Accessibility is the level of access that each tourist/traveler has and their online traffic. Interactivity facilitates the immediate actions of tourists when they access to the technologies of the destination on the moment, for example, through the feedback provided by the dialog of a virtual assistant (chatbox). Finally, customization is the ability each tourist can find and customize their own travel [9]. [20] argues that the process of online access is focused on: destinations, intermediaries, transport, accommodation, restaurants, other tourist services and that tourist access begins before, during and after the trip, which creates social, followed by technological trends, pursuing innovation [20]. In this sense, technology has significantly changed the way tourism experience is perceived, experience (covering people, space, products, services or cultures) already has a dynamic character, and increasingly the involvement of the consumer, the co-creation and implementation of the technology gain relevance for the promotion of the destination. Technological resources for research during the trip allow visitors to have access to information about exhibitions/events or locations and other useful information [21], such as information on tourist attractions, accessibility, accommodation availability, prices, search of geographic information and weather [22], using smartphones, apps and other types of new technologies [16]. There may be an interconnection with place or even other tourists, discovering new information about the destination or new places to visit, all through the mobile technology available that helps search and share content anywhere, anytime [9]. Tourism destinations that make use of technology are able to influence and improve the quality of tourism experiences, because the information stored in the databases allow us to know how tourists receive the information, use the resources and evaluate the points of interest of a tourist destination, being one of the most important aspects to support new decisions in the market and for urban area managers, taking assertive and informed decisions [23].

The access and use of the Internet have become a constant, technological devices are increasingly easy to carry, and have more associated functions. From mobile phones to digital assistants with varied functions such as email usage, built-in photo machines, GPS, music downloads, reaching areas such as cars with network access and a new set of mobile broadband networks that allow wireless communication, encouraging the development of more sophisticated location-based services using GPS [24].

Seven official mobile applications (app) were created in Porto, and are refered on the platform VisitPorto, and they are available for Android, IOS or Windows mobile: Porto Card, Oporto Insight, Tales & Tours, Taggeo, VPorto, OPORTOnityCity, and Farol City Guides. The download is free, however, to use some services is necessary the purchase. They present contents and services about the city of Porto for the tourist/travelers. Regarding public transport in the city, there are also applications, such as the iMetro do Porto (Porto metro application). It provides information about schedules, planning of routes and check the status of the lines, prices, among other information. Tourists can also check a list of other free applications corresponding to bus schedules, and other transports, that help to get around the city. However, these applications aren't official (not from the DMO), such as: Moovit, Move-me.AMP, Porto.bus, AppCaris, Anda Porto, UmovePorto, Citymapper, as well as complementary services such as Uber.

According to [25], Generation Z travelers are gaining more and more relevance in the travel market for both leisure and business travel, and the traditional marketing method isn't the most effective method to influence this generation (in all sectors: accommodation, travel, food, activities, among others). Advertising must be attractive. This type of travelers interacts and perceives the information in a different way, which should be suitable for several devices and screens (various sizes) because they use multiple mobile devices. Regarding the accommodation, it's the most favorable generation to stay with family and friends if possible, being the hotel the second most favorable option, considering alternative accommodation as well [25].

It's important to emphasize three aspects of consumer behavior regarding this generation in tourism: they seek the best price-quality promotions, they claim that they only live once and that is why they have to enjoy life regardless of the danger or risks they may face; want to fulfill all the wishes, and don't mind spending all day at the beach sleeping, or relaxing in a spa. It's also important, the attraction of other activities, the exploration of destination, search of recommended places, the perspective of an adventure, sightseeing, cultural visits, among other plans [25]. In this sense, fourteen reasons for the choice of holidays were presented in order of relevance: activities carried out during the trip; unique experience; cultural experience; outdoor activities; memorable places; low prices (important but not the most important); a destination to go on a holiday with friends; feeling welcome (well-being) during the stay; promotions and discounts; gastronomic experiences; recommendations from family and friends; unique place for taking photos; recommendations online from other travelers; and recommendations of tourism professionals [25]. They seek to live a true cultural experience, interested in what is true and willing to communicate with the local community, being one of the factors most requested by this generation [26].

Regarding the technology use for the destination selection, the use of mobile phone (smartphone), laptop and mobile devices in general prevails to search for tourist destinations and make reservations. The selection of the destination is guided by the discrepancy of choice between national and international tourism (European tourist of Generation Z), since 64% of their travels are international destinations and the remaining 36% are domestic [25].

3   Methodology

A quantitative approach was considered appropriate for the research methodology. A survey was used as the data gathering technique for this investigation. For the survey application methodology, the direct or non-probabilistic sampling method was used. The data collection took place in the main tourist's streets of the city of Porto concerning Generation Z tourists. Within the methods of direct or non-probabilistic sampling, the convenience sampling method was used, in which the sample is selected according to the availability of the target population elements. Table 2 presents the sample technical data, where it's possible to identify the study population, the sample period, the data collection instrument used, the process of obtaining data and subsequent analysis.

Table 2. Sample technical data

Sample	Generation Z (18–28 years old)
Geographical scope	Oporto city
Sample size	400 Valid questionnaires
Data collection period	29th July to 19th August 2018
Data collection method	Direct or non-probabilistic
Method	Survey by questionnaire
Maximum error	4,9%
Trust level	95% z = 1,96

Source: Compiled by the authors

4   Results

The aim of this research is to evaluate the destination choice (city of Porto) by Generation Z. The hypothesis presented in this investigation is: "The reasons for the choice of destination Porto positively influence the use of technological resources during the trip". According to [27] (which analyzes data on destination demand, overnight in the municipality and the digital search for it), the reasons most often presented for choosing the city of Porto as a destination were: previous recommendation; get to know the city; proximity to the place of residence or holiday. The beauty of the city, the cultural offer, the history and heritage, the wine and gastronomy, and the value for money were also indicated [27].

For the study of the hypothesis, it was established the relations between the questions "main reason of the trip" and "what are the reasons for choosing the city of Porto" with the questions "assessing the use of the following types of applications and/or information available on the Internet during your Experience in Porto".

In Table 3, we verified that for all dimensions, there aren't statistically significant differences between the main reasons of the trip. In the sample, the assessment of the dimensions under analysis presents the differences illustrated between the main reasons of the trip, but aren't statistically significant.

Table 3. Descriptive statistics and Kruskall-Wallis tests

		N	Average	Deviation default	KW	p
Applications and/or information in the Internet during tourist experience	Leisure/Holiday	340	3,67	0,57	6,20	0,102
	Visit Family/Friends	34	3,65	0,54		
	Professional reasons	14	**3,94**	0,53		
	Other	12	3,92	0,47		
Available applications to move around	Leisure/Holiday	308	3,48	0,72	6,27	0,099
	Visit Family/Friends	31	3,35	0,80		
	Professional reasons	14	**3,74**	0,63		
	Other	11	**3,83**	0,68		
Applications during the tourist experience	Leisure/Holiday	277	**3,37**	0,61	2,03	0,566
	Visit Family/Friends	27	3,22	0,58		
	Professional reasons	11	**3,47**	0,36		
	Other	9	3,34	0,45		
Technological resources during the tourist experience	Leisure/Holiday	316	**3,43**	0,62	1,00	0,801
	Visit Family/Friends	32	3,38	0,70		
	Professional reasons	13	3,34	0,96		
	Other	11	**3,43**	0,75		
Payment services to consume on destination	Leisure/Holiday	302	3,29	0,68	7,60	0,055
	Visit Family/Friends	33	3,30	0,83		
	Professional reasons	12	**3,87**	0,77		
	Other	12	**3,56**	1,10		
Virtual and augmented reality	Leisure/Holiday	279	**3,59**	0,56	4,02	0,259
	Visit Family/Friends	31	3,47	0,83		
	Professional reasons	13	**3,50**	1,25		
	Other	11	3,19	0,83		
Use of proximity device technology	Leisure/Holiday	322	**3,63**	0,68	1,93	0,587
	Visit Family/Friends	33	**3,50**	0,85		
	Professional reasons	14	**3,71**	0,59		
	Other	11	**3,79**	1,05		

Source: Compiled by the authors

Regarding Table 4, there are statistically significant positive relationships between: "Applications and/or information on the Internet during the tourist experience" and the reasons "history and heritage" and "other"; "Applications available to move" and the reasons "the cultural offer" and "other"; "Technological resources during the tourist experience" and the reason "proximity to the place of residence or holiday"; The dimension "payment services to consume in the destination" and the reasons "the cultural offer", "the History and Heritage", "the Wine and gastronomy" and "other".

Table 4. Pearson correlation: relations between the dimensions associated with the technologies evaluation and "what are the reasons for choosing the city of Porto"

		Applications and/or information on the Internet during tourist experience	Applications available to move around	Applications during the tourist experience	Technological resources during the tourist experience	Payment services to consume on destination
Get to know the city	r	0,005	0,043	0,078	0,028	**0,105***
	p	0,917	0,417	0,160	0,589	0,046
	N	400	364	324	372	359
Have been recommended	r	−0,038	−0,064	−0,038	−0,059	−0,041
	p	0,448	0,221	0,498	0,258	0,444
	N	400	364	324	372	359
Proximity to the place of residence or holiday	r	−0,004	0,102	0,091	**0,121***	0,078
	p	0,940	0,052	0,102	0,019	0,139
	N	399	364	324	371	358
Cultural offer	r	0,084	**0,139****	0,093	−0,007	**0,126***
	p	0,093	0,008	0,094	0,893	0,017
	N	400	364	324	372	359
The beauty of the city	r	0,005	−0,042	−0,011	−0,039	0,022
	p	0,917	0,421	0,849	0,451	0,677
	N	400	364	324	372	359
History and Heritage	r	**0,114***	−0,025	0,031	-0,069	**0,133***
	p	0,023	0,642	0,575	0,185	0,012
	N	398	362	322	370	357
Wine and gastronomy	r	0,076	−0,011	−0,056	−0,063	**0,132***
	p	0,128	0,829	0,319	0,229	0,012
	N	400	364	324	372	359
Quality-price ratio	r	0,095	0,044	0,055	0,045	0,065
	p	0,058	0,405	0,326	0,391	0,223
	N	399	363	323	371	358
Other	r	**0,101***	**0,140****	0,071	0,026	**0,121***
	p	0,043	0,007	0,203	0,622	0,021
	N	400	364	324	372	359

Source: Compiled by the authors

We can see from Table 5, regarding the results of Kruskall-Wallis tests, the relationships between the "evaluation of the use of the following types of applications and/or information available on the Internet during the tourism experience in Porto" and "main reason for Trip", which for "Internet access (paid)", "Language translators" and "Webcams (places, city points, automobile traffic, beaches, tourist sites)", there are statistically significant differences between the main reasons of the trip. The information on "Internet access (paid)" is more important for the reason professional reasons and less for leisure/vacation, information about "language translators" and about "Webcams (places, city points, automobile traffic, beaches, tourist sites)" is more important for the reason professional reasons, with statistically significant differences.

Table 5. Descriptive statistics and Kruskall-Wallis tests: relations between Informations available on internet during the trip and the reasons to visit Porto

		N	Average	Standard deviation	KW	p
Tourist support/tour guides (maps, itineraries, circuits, etc.)	Leisure/vacation	333	**4,32**	0,84	0,95	0,812
	Visit family/friends	34	4,18	0,90		
	Professional reasons	13	4,31	0,85		
	Other	12	4,25	1,14		
Wi-Fi Access (free)	Leisure/vacation	330	4,34	0,96	3,54	0,316
	Visit family/friends	34	4,35	0,73		
	Professional reasons	14	**4,50**	1,16		
	Other	12	4,08	0,90		
Internet access (fee required)	Leisure/vacation	307	2,96	1,12	8,04	***0,045**
	Visit family/friends	30	3,27	1,08		
	Professional reasons	13	**3,69**	0,85		
	Other	10	3,30	1,16		
Cultural agenda (museums, exhibitions, cinemas, concerts, shows, prices, ticket reservations)	Leisure/vacation	324	4,09	0,90	0,28	0,965
	Visit family/friends	32	**4,13**	0,61		
	Professional reasons	13	4,08	1,04		
	Other	11	4,09	0,54		

(continued)

Table 5. (*continued*)

		N	Average	Standard deviation	KW	p
Weather forecast	Leisure/vacation	333	4,37	0,85	5,09	0,165
	Visit family/friends	33	4,33	0,92		
	Professional reasons	14	**4,79**	0,43		
	Other	10	**4,70**	0,67		
Accommodation, catering and similar (prices, availability, contacts, bookings, etc.)	Leisure/vacation	336	4,38	0,91	2,12	0,548
	Visit family/friends	34	4,47	0,75		
	Professional reasons	14	**4,64**	0,84		
	Other	12	4,42	1,08		
Transport management (air, land, timetables, check-in, prices, reservations, etc.)	Leisure/vacation	330	4,30	0,92	2,91	0,406
	Visit family/friends	34	4,09	0,97		
	Professional reasons	14	**4,36**	1,22		
	Other	12	**4,50**	0,67		
Car rental (companies, prices, promotions, reservation, etc.)	Leisure/vacation	287	3,39	0,77	5,70	0,127
	Visit family/friends	29	3,52	0,78		
	Professional reasons	12	**3,92**	0,79		
	Other	9	3,44	1,01		
Travel agency (programs, promotions, reservations tourist packages, …)	Leisure/vacation	287	3,45	0,71	0,96	0,811
	Visit family/friends	30	3,40	0,72		
	Professional reasons	10	**3,50**	0,53		
	Other	9	**3,67**	0,50		
Language translators	Leisure/vacation	314	2,34	1,05	10,44	***0,015**
	Visit family/friends	32	2,31	0,93		
	Professional reasons	12	2,50	1,09		
	Other	11	**3,36**	0,81		
Webcams (local, parts of the city, car traffic, beaches, places of interest)	Leisure/vacation	295	2,19	1,06	8,18	***0,042**
	Visit family/friends	31	2,06	0,85		
	Professional reasons	11	2,18	0,87		
	Other	8	**3,13**	0,35		

*p < 0,05 **p < 0,01 Source: Compiled by the authors

Table 6. Descriptive statistics and Kruskall-Wallis tests: Relations between "evaluation of the use of the following applications available to travel in Oporto" and "main reason of the trip"

		N	Average	Standard deviation	KW	p
Moovit	Leisure/vacation	233	3,25	1,05	1,31	0,727
	Visit family/friends	28	3,21	1,23		
	Professional reasons	12	3,25	1,36		
	Other	9	**3,67**	0,50		
Move-me.AMP	Leisure/vacation	243	3,94	1,19	1,67	0,643
	Visit family/friends	27	3,78	1,22		
	Professional reasons	12	**4,08**	1,16		
	Other	9	**4,33**	1,00		
Porto.bus	Leisure/vacation	259	3,22	0,82	4,31	0,229
	Visit family/friends	28	3,25	0,80		
	Professional reasons	12	**3,58**	0,79		
	Other	11	3,45	0,69		
iMetro do Porto	Leisure/vacation	269	3,49	0,91	4,78	0,188
	Visit family/friends	28	3,54	1,14		
	Professional reasons	14	**3,93**	1,00		
	Other	11	**4,00**	1,10		
AppCaris	Leisure/vacation	240	3,09	0,85	4,46	0,216
	Visit family/friends	28	3,00	0,90		
	Professional reasons	12	**3,50**	1,00		
	Other	10	3,40	0,70		
Uber	Leisure/vacation	276	4,25	1,06	2,96	0,398
	Visit family/friends	30	4,03	1,13		
	Professional reasons	12	**4,50**	0,90		
	Other	11	**4,55**	0,69		
Cabify	Leisure/vacation	234	3,71	1,00	5,96	0,113
	Visit family/friends	27	3,52	1,12		
	Professional reasons	12	**3,75**	1,06		
	Other	10	**4,30**	0,95		
Anda Porto!	Leisure/vacation	256	**4,12**	1,03	6,49	0,090
	Visit family/friends	29	3,66	1,14		
	Professional reasons	14	3,71	1,14		
	Other	9	**4,00**	1,12		
UmovePorto	Leisure/vacation	231	3,23	0,69	3,93	0,269
	Visit family/friends	26	**3,38**	0,85		
	Professional reasons	12	3,08	0,79		
	Other	9	3,44	0,88		
Citymapper	Leisure/vacation	253	2,25	1,33	8,27	***0,041**
	Visit family/friends	28	2,46	1,50		
	Professional reasons	12	2,92	1,56		
	Other	9	**3,56**	1,59		

*p < 0,05 **p < 0,01 Source: Compiled by the authors

For "Citymapper", there are statistically significant differences between the main reasons of the trip. For the remaining items, there aren't statistically significant differences between the main reasons of the trip. The use of "Citymapper" is more important for the other reason, followed by professional reasons and less for leisure/vacation. Differences were statistically significant (Table 6).

5 Conclusions

The technology is part of the Generation Z trip, and as we can verify by the results, it's observed the use of the technology in the pre-trip phase. Therefore, overall, we can conclude that the hypothesis "The reasons for choosing the destination Porto positively influence the use of technological resources during the trip" is verified, concerning the meaningful relationships found. One of the advantages of using mobile applications in tourism is the power to enrich the tourist experience in the destination, resulting in a change of value in the way information is perceived, creating an interaction between tourists and the surrounding environment. This solution can respond to companies, entities and professionals in the tourism sector as a way to reach a greater number of tourists, providing additional information and services, including tailored to the profile of each user [16]. Nowadays, tourists don't travel without mobile phone and internet access, because their use has become a daily practice, which can potentialize and transform the tourist experience. In this sense, the local DMO's must be conscious to this new reality [19]. The use of technology can transform a conventional tourism experience into a better and more appealing experience [5, 6]. The availability of mobile applications for tourist use in the destination makes possible the satisfaction of the tourist experience [19]. This way it's necessary to appeal the tourist entities for the constant contents updating of online information and updating the technologies involved in the destination. These updates should, in particular, focus on tourist services (in all sectors: accommodation, transport, catering, and attractions in general), should always be available to tourists, since the need to obtain clarification on the tourist destinations before and during the trip, is essential and positively influences the satisfaction of the consumer experience [9].

References

1. UNESCO Portugal (2017). https://www.unescoportugal.mne.pt/pt/temas/proteger-o-nosso-patrimonio-e-promover-a-criatividade/patrimonio-mundial-em-portugal/centro-historico-do-porto
2. European Best Destination (2017). https://www.europeanbestdestinations.com/best-of-europe/european-best-destinations-2017/
3. Blom Consulting City Brand Ranking. Portugal City Branding (2018). https://www.bloom-consulting.com/pdf/rankings/Bloom_Consulting_City_Brand_Ranking_Portugal.pdf
4. Liberato, P., Alén, E., Liberato, D.: Digital technology in a smart destination: the case of Porto. J. Urban Technol. **25**(1), 75–97 (2018). https://doi.org/10.1080/10630732.2017.1413228

5. Neuhofer, B., Buhalis, D., Ladkin, A.: A typology of technology-enhanced tourism experiences. Int. J. Tourism Res. **16**, 340–350 (2014). https://doi.org/10.1002/jtr.1958

6. Beliatskaya, I.: Understanding enhanced tourist experiences through technology: a brief approach to the Vilnius case. Rev. Investig. Tur. **7**, 17–27 (2017)

7. Martínez, M., Ugarte, T., Lorenzo, F.: The smart city apps as the core of place branding strategy: a comparative analysis of innovation cases. Rev. Estud. Comun. **22**(42), 119–135 (2017). https://doi.org/10.1387/zer.17813

8. Santos, G., Cabral, B., Gosling, M., Christino, J.: As redes sociais e o turismo: uma análise do compartilhamento no Instagram do Festival Cultura e Gastronomia de Tiradentes. Rev. Iberoam. Tur.- RITUR, Penedo **7**(2), 60–85 (2017)

9. Huang, C., Goo, J., Nam, K., Yoo, C.: Smart tourism technologies in travel planning: the role of exploration and exploitation. Inf. Manag. **54**(6), 757–770 (2017). https://doi.org/10.1016/j.im.2016.11.010

10. Kim, H., Xiang, Z., Fesenmaier, D.: Use of the internet for trip planning: a generational analysis. J. Travel Tour. Mark. **32**(3), 276–289 (2015). https://doi.org/10.1080/10548408.2014.896765

11. Cutler, S., Carmichael, B.: The dimensions of the tourist experience. In: Morgan, M., Lugosi, P., Ritchie, B. (eds.) The Tourism and Leisure Experience: Consumer and Managerial Perspectives. Channel View Publications, Bristol (2010)

12. Cutler, S.: Exploring the moments and memory of tourist experiences in Peru. Theses and Dissertations (Comprehensive) (2015)

13. Díaz-Meneses, G.: A multiphase trip diversified digital and varied background approach to analysing and segmenting holidaymakers and their use of social media. J. Destin. Mark. Manag. (2017). http://dx.doi.org/10.1016/j.jdmm.2017.07.005

14. Marujo, N.: Turismo, turistas e experiências: abordagens teóricas. Revista Turydes: Turismo y Desarrollo, no. 20, junio 2016 (2016). http://www.eumed.net/rev/turydes/20/turistas.html

15. Monaco, S.: Tourism and the new generations: emerging trends and social implications in Italy. J. Tour. Futur. **1**, 7–15 (2018). https://doi.org/10.1108/JTF-12-2017-0053

16. Martins, M., Malta, C., Costa, V.: Viseu mobile: um guia turístico para dispositivos móveis com recurso à Realidade Aumentada. Dos Algarves: Multidiscip. J. **26**(1), 8–26 (2015). https://doi.org/10.18089/DAMeJ.2015.26.1.1

17. Oliveira, R., Correa, C.: Virtual Reality como estratégia para o marketing turístico. Revista Turydes: Turismo y Desarrollo (2017). http://www.eumed.net/rev/turydes/23/virtual-reality.html

18. Biz, A., Azzolim, R., Neves, A.: Estudo dos Aplicativos para Dispositivos Móveis com Foco em Atrativos Turísticos da Cidade de Curitiba (PR). Anuais do Seminário da ANPTUR (2016). ISSN 2359-6805

19. Florido-Benítez, L., Martínez, B., Robles, E.: El beneficio de la gestión de relación entre las empresas y turistas a través de las aplicaciones móviles. ARA: J. Tour. Res./Rev. Investig. Tur. **5**, 57–69 (2015). ISSN 2014-4458

20. Dexeus, R.: Innovación en el sector turístico. Segittur. In: I Congreso Mundial de la OMT de Destinos Turísticos Inteligentes (2017). http://www.segittur.es/es/sala-de-prensa/detalle-documento/Innovacin-en-el-sector-turstico-/#.WyGi0qdKjIU

21. Kuflik, T., Wecker, A., Lanir, J., Stock, O.: An integrative framework for extending the boundaries of the museum visit experience: linking the pre, during and post visit phases. Inf. Technol. Tour. **15**, 17–47 (2014). https://doi.org/10.1007/s40558-014-0018-4

22. Ramos, C.: Os sistemas de informação para a gestão turística. Rev. Encontros Cient. Tour. Manag. Stud. (2010). http://www.scielo.mec.pt/scielo.php?script=sci_arttext&pid=S1646-24082010000100011&lng=pt&nrm=iso

23. D'Aniello, G., Gaetta, M., Reformat, M.: Collective perception in smart tourism destinations with rough sets. In: 2017 3rd IEEE International Conference on Cybernetics, p. 1 (2017). https://doi.org/10.1109/cybconf.2017.7985765
24. Mackay, K., Vogt, C.: Information technology in everyday and vacation contexts. Ann. Tour. Res. 3, 1380–1401 (2012). https://doi.org/10.1016/j.annals.2012.02.001
25. Expedia (2017). https://info.advertising.expedia.com/european-travel-and-tourism-trends-for-german-british-french-travellers
26. Haddouche, H., Salomone, C.: Generation Z and the tourist experience: tourist stories and use of socia networks. J. Tour. Futures 1, 69–79 (2018). https://doi.org/10.1108/JTF-12-2017-0059
27. Almobaideen, W., Krayshan, R., Allan, M., Saadeh, M.: Internet of Things: geographical routing based on healthcare centers vicinity for mobile smart tourism destination. Technol. Forecast. Soc. Chang. 123, 342–350 (2017). https://doi.org/10.1016/j.techfore.2017.04.016

Information and Communication Technologies in Creative and Sustainable Tourism

Ana Ferreira[1], Pedro Liberato[1], Dália Liberato[1(✉)], and Álvaro Rocha[2]

[1] School of Hospitality and Tourism, Polytechnic Institute of Porto, Porto, Portugal
{anaferreira, pedrolib, dalialib}@esht.ipp.pt
[2] Department of Informatics Engineering, University Coimbra, Coimbra, Portugal
amrocha@dei.uc.pt

Abstract. In this article we intend to show the importance of information and communication technologies (ICT) in the election and diffusion of creative and sustainable tourism.

Creative tourism, by focusing its action on the way of life and the identity of local communities, will be a tourism that proposes activities that are based on local identity and memory and consider the preservation of available resources. In this sense, we ask: What is the role of ICT for the valuation of creative tourism? Can ICT define the course of tourism in each territory?

Firstly, we intend to clarify the pertinence of this interdisciplinary research - sustainable creative tourism and information and communication technologies. Next, we emphasize that the diffusion of information is a fundamental asset for tourism development, since the tourist value of a destination depends largely on its ability to affirm itself as a national and international brand.

The methodology followed will be based on the revision of the available bibliography leading to the consolidation of the idea that ICT are an indispensable tool for the promotion of creative tourism.

Keywords: Information and communication technologies · ICT · Creative tourism · Sustainable tourism

1 Introduction

The ongoing technological development in the 21st century, also known as the Digital Information Age, allows us to reflect on the importance of new communication and information technologies (ICT). In fact, the increasing use of these technologies has revolutionized the business world by changing the paradigms in people and organizations' way of thinking and acting. Tourism, as an economic activity, has not been unaware of this new reality [1].

The distribution of information about tourism products via Internet is the area in which technological innovations had the greatest impact on companies associated with tourism [1]. ICT can be effective for the development and growth of the tourism

© Springer Nature Switzerland AG 2019
Á. Rocha et al. (Eds.): WorldCIST'19 2019, AISC 930, pp. 91–100, 2019.
https://doi.org/10.1007/978-3-030-16181-1_9

market. The encouragement of e-commerce practices with the use of ICT facilitates the organization of tourism industry agents and can stimulate cooperation between the several actors [1].

Based on this new scenario, this article aims to reflect on the influence of the use of the new ICT in the tourism activity. The operationalization of all the activities of the tourism chain has been increasingly marked by a wide range of choices presented online. The widespread use of ICT in people and organizations' daily life, the computer applications, which allow and foster individual initiative and increasing environmental awareness of tourists create new tourist practices with different impacts on the territory. Creative tourism, which is an example of a different approach to this economic activity, opposes the traditional mass vision by focusing on unique proposals. These intend to promote the sense of respect for nature, that is, we can associate creative tourism with concerns about the sustainability of tourism so that we can benefit from ICT regarding their activities and subsequent diffusion. In fact, creative tourism is opposite to traditional mass tourism by focusing on the presentation of unique proposals.

2 Creative Tourism

Tourism is an intensive information industry. One of the reasons for a high exchange of information between the various actors is the nature of the tourism product. Compared to other products sold online the tourism product is immaterial, heterogeneous and non-durable. A trip involves several elements [2]: transportation, accommodation, insurance, financial services, guide services, tours, among others. Considering the diversity of the product, travel agents or tourists, planning a trip, need to access several different sources of information.

In addition, the product is immaterial since the tourist cannot touch or view the product before going on a trip. Consequently, it is extremely important to have reliable information about all the elements that make up the product. The product cannot be stored either. If a hotel room or a seat on an airplane is empty, that means loss of revenue. For this reason, distribution and cash management is a key factor in this industry. Thus, the possibility of establishing mechanisms to increase this interoperability will provide better conditions for market growth, especially for small and medium-sized enterprises.

Nowadays, the increase in tourism flow is a reality with a growing trend despite the objectives of tourists becoming increasingly diversified. In fact, according to [1, 3], the new tourist - who seeks experience - is a more experienced and sophisticated consumer who can structure his or her own tourism by rejecting standardized and less structured proposals. In recent years, we have witnessed the evolution of a tourist with greater social and environmental awareness who wants to have access to the "spirit of the place", which becomes a co-creator and not a mere consumer. This new tourist pressures the market in order to force the tourism activity to meet their needs and expectations, thus emerging creative tourism.

[4] understands creative tourism as "one that offers visitors the opportunity to develop their creative potential through active participation in courses, workshops and learning experiences, characteristic of the culture of the destinations visited" [5]; this

type of tourism promotes the experience of the visitors, bringing them closer to the residents' cultural reality, which will give them a unique and memorable experience of intangible value. Doing things instead of just seeing things allows the visitor to experience feelings and emotions, creating opportunities for self-development, that is, building one's identity. In order to achieve this type of experience associated with a certain place, creative tourism depends on an integrated action. According to [6], "a tourism system consists of a complex set of interrelations between different institutions and organizations that work in an integrated way, in order to offer tourists a unique experience in their stay in a certain destination".

In fact [7] reinforces this idea when defines creative tourism as "travel directed toward an engaged and authentic experience, with participative learning in the arts, heritage, or special character of a place, and it provides a connection with those who live in this place, creating this living culture".

According to [3], technology is an interface with the culture that is being visited, as a window to the cultures that are intended to visit.

Nowadays, the modern tourist accesses information daily, often using digital media. Through social networks tourists share their experiences and exchange information on holiday destinations, sightseeing, giving and getting travel tips in the blink of an eye [1]. New friends are made online and we receive constantly new offers, appealing destinations and challenging holiday activities.

[8] argue that recent research [4] indicates that "the development of creative activities also requires a holistic, multi-sectoral approach that integrates experiential programming, the use of Internet communication technologies and visitor engagement." According to [5] "New technologies and new media now play an important role in the way people communicate and live a touristic experience. This can both enhance the experience and make it easier for visitors to interact and communicate with local people while visiting a destination".

We must therefore consider ICT as an excellent tool to improve these interrelations between different institutions and organizations that work together to offer tourists a unique holiday experience.

The main attraction of creative tourism offers tourists the opportunity to experience new emotions and take active part in courses and learning activities or programs where tourists can embrace the spirit and culture of the chosen destination. However, we must not forget that tourists have become more demanding and make more informed decisions. Due to the development of social networks and the availability of gadgets such as translators, navigators and local SIM-cards, new communication formats between tourists and the locals have emerged. Youth tourism organizations are highly active in this field, as evidenced by the existence of educational programs, language learning programs, and work and travel programs [9].

As a result, tourism professionals will have to satisfy their new requests and respond in an increasingly fast and personalized way using adequate tools.

3   21st Century – The Digital Information Age

Nowadays the development of new technologies has an impact on the way people live, appreciate life and the way they investigate creative businesses. When network technology, digital imaging technology, interactive technology and many other technical means become 'materials' of artistic creation, people are no longer satisfied to appreciate art passively, "they want to participate in the dialogue and exchanges with the works" [10]. People demand not only reliable but also fast information. We can search the Internet in less than a second and travel agencies as well as other institutions (…) must use ICT to its utmost performance – they must speed up the delivery of data and analysis, particularly in the context of providing fast, relevant and detailed information on creative offers.

ICT have promoted access and linkage of information to all forms of communication all over the world. In fact, the Internet, has accelerated business and the exchange of data and information around the world. The widespread use of ICT in the tourism industry has transformed the role played by each of the various intervenient: travel agencies, tour operators, providers of a wide range of services and products [2]. On the one hand, information and communication systems provide consumers detailed and up-to-date information on the availability of services, prices, which leads to increased sales. On the other hand, ICT enable efficient and fast communication between consumers and suppliers. According to [11] "social networks are changing the way tourists plan their trips. These websites allow users to interact and share their opinions with others about touristic attractions, hotels and restaurants. Probably the largest community of travel/tourism presented online is Tripadvisor".

It is important to bear in mind that there has been a recent growth in the use of mobile phones and devices worldwide. The strongest reason given by customers who use the internet to purchase tourism products and destinations is the idea that the price is lower, and it is easier to access information.

With the emergence of e-commerce, a new type of consumer has appeared. A more demanding e-consumer who often uses the network to exchange online information on specific holiday destinations [12]. ICT, in addition to supporting tourism service providers, play an increasingly important role in satisfying tourists needs.

In fact, e-commerce, not only sells services, but also helps the development of activities such as market research [12]. The internet has been steadily considered a multimedia tool that can satisfy the various needs of tourism, since its values technologies that can transmit sensations such as the use of sound, appealing images and movements/videos. In short, online tourism represents one of the most powerful toolkits to promote a tourism product. Moreover, new forms of interaction between suppliers and customers facilitate the knowledge of the profile of the consumer/visitor and allow the use of the information to treat different customers in different ways. Furthermore, the internet is also a potentially significant means of promotion and destination marketing for the different world destinations [2].

In fact, the internet has become a people-oriented network, being a valuable source of information and an easy and fast means of communication. [12] lists some of the benefits of e-commerce for consumers:

- it enables them to check various sales sites, 24-hours a day, all year round;
- it offers consumers more purchase options, giving them access to more sellers and products;
- it enables consumers to make rapid comparisons between prices and products, as they can access several sites simultaneously;
- in some cases, where the product is digital, it enables rapid delivery by means of the internet itself;
- it facilitates and promotes competition, which can result in discounts for the consumer.

It is also important to mention that the common use of mobile phones and technological devices has allowed tourists to change the way they plan their trip, and the way they get involved in the touristic experience. [5] state that "mobile applications in tourism are also increasingly successful – building on their ability to offer support to tourists within a destination, giving them access to information anytime and anywhere" as well as giving them the possibility to be connected to the world and share their experiences with their friends [13].

[5] refers that an example of the use of new technology in a tourism experience is within the concept of "transmedia storytelling", which uses different platforms to tell stories and allows tourists to have a personalized visit, tailored to their tastes and participation.

Overall, we can say that digital technologies such as new media, mobile applications, social networks (Facebook, Blogger, Twitter, Youtube), and videos uploaded online, besides providing tourists with information about the destination and activities they might attend while visiting, also enable them to share the touristic experience [1, 2].

4   The Role of ICT in the Development of Sustainable Creative Tourism

Nowadays tourism is one of the most important economic sectors, and it's undergoing a significant transformation with the use of Information Technology as a business tool. According to [1, 2, 14], prior to the use of the Internet for tourism business, the main electronic means of tourism distribution and trading are Global Distribution Systems (GDS). GDS is a system used by travel agencies, which enables transactions in real time and offers functions such as: timetable information, availability, price quotes for tourism services throughout the world, reservations of seats and special dietary requirements, sale and issue of airline tickets and other client services [1, 2, 14].

Currently, regarding Internet access, customers can plan and program the entire journey, to any part of the world, without leaving their own home. They can book airline and hotel tickets, make payments, rent cars, check the weather conditions at the destination, find out the exchange rate, learn about the culture of the destination, check the necessary documentation [1, 2].

The use of ICT can generate gains in all tourism activity in general and in the creative tourism segment especially since, as a small market segment, it is often more difficult to maintain itself as a profitable economic activity.

Furthermore, with widespread use of digital technologies travelling to foreign countries has become less and less exotic, increasingly prosaic, as they only retain a semblance of the exotic in the world of the publicity or the returning tourist. The intercultural experience of the commonplaces is summarized by internet mailing lists and social networks; transnational media and consumer goods; real-time and low-cost travel [5, 12, 15].

Freedom and flexibility acquired by the consumer through direct and individual access to tourism services has significantly altered the structure of the tourism industry. In fact, ICT can provide cost advantages, by offering an innovative, agile, safe and quality service, which will allow customers to respond more quickly to their needs in a shorter time, increasing the range of individual products in any country or location will have equivalent opportunities in access to potential customers. With the increasing demand for unique experiences that can meet the needs of the tourist, ICT can become important tools allowing professionals to satisfy the customers' personal demands and, therefore, offer a wider variety of creative and personalized choices [2, 5, 6].

It's the moment for tourism companies to reevaluate their strategic positions, considering the competitive advantage arising from the use of the Internet, as the increased competition and the growing fragmentation of the markets means that they must find new forms of reaching consumers [16].

The internet forces those involved in the tourism business to re-think their business models and forms of relationship with commercial partners and clients in order to adapt to the current technology. In their relationship with the client, the travel agents become consultants. So, they will need to obtain large amounts of quality information on the innumerable factors that are involved in tourism business, such as: national and international legislation, packages, services, suppliers, catering, destinations, prices and transport [2, 5, 6, 17]. This reinforces digital technologies role because the agents of the future will need electronic libraries of information to research and serve their clients better.

The active participation and interaction of the tourist with the local community allows to develop the feeling of greater respect for the preservation of the essential characteristics of each community, that is, it can contribute to strengthen the community's ties with the territory - the sustainable local development. In fact, according to [5] additionally the literature review has shown that both participation and interaction are relevant because tourism preferences for destination activities have been evolving towards more participative behavior.

Each tourist is a different person, carrying a unique blend of experiences, motivations and desires. Tourists are increasingly informed and have more language and technological skills and are increasingly concerned with environmental issues. Thus, the demand for creative and sustainable tourism will increase [2, 15].

Nowadays, to make the journey more exciting, internet allows the tourist to feel the place. The experience becomes real by the digital technologies that present the narrative of places linking sounds, colors, images. All of us agree on the important role played by the audio and visual resources to create an adequate atmosphere, improving the feeling of immersion in the place to go [1, 2].

Creative tourism can also be a way of developing very specific relational links related to the interests of the individuals involved. This is also interesting because it often represents a physical manifestation of virtual networks – people travel to meet people who they encounter in online communities and come together because the embedded skills and practices in many creative activities cannot be exchanged without physical co-presence.

[5] indicates that the development of creative activities also requires a holistic, multi-sectoral approach that integrates experiential programming, the use of Internet communication technologies (ICT) and visitor engagement. The authors add that "success requires partnering with people working in the creative industries to foster innovative approaches that provide credibility and quality service in terms of new visitor experiences and products" [5].

ICT empower consumers to identify, customize and purchase tourism products and support the globalization of the industry by providing tools for developing, managing and distributing offers worldwide. ICT are becoming a determinant of territorial competitiveness while provide a powerful tool that can bring advantages in promoting and strengthening the sustainable tourism's strategy and operations [2].

New technologies and new media now play an important role in the way people communicate and live a touristic experience. This can both enhance the experience and make it easier for visitors to interact and communicate with local people while visiting a destination [1].

[5] affirms the existence of new forms of interaction and of new opportunities for intercultural communication while traveling have enhanced the development of creative tourism in the past twenty years.

[5] argues that with the development of digital media technology, interactive technology and enough integration of art and technology, the form for the public to participate and experience public art has basically changed. Public art can make artists, works and the audience interact by the image, voice, and act identification, and realize of two-way exchange of art creation and art experience, breaking the one-way mode of art activities [5].

In addition, we can also state as [18] says, that creative tourism (as an unprecedented economic activity), when promoted dynamically and connected with the local population, can fulfill the articulating and inductive role of sustainable development, even more so in an integrated way with other economic activities. This interaction can be greatly enhanced by ICT. In fact, the use of ICT can encourage an environmentally more integrated demand for products and services.

According to [19], sustainable tourism must, among other actions, "respect the socio-cultural authenticity of host communities, conserve their built and living cultural heritage and traditional values, and contribute to inter-cultural understanding and tolerance" [19]. At the same time [17] add that, "for tourism to become more sustainable requires constant education, monitoring and collaboration, which can be achieved based on ICT". These authors add that ICT helps to distribute and make accessible information, to create environmental and cultural awareness, to monitor environmental resources and to reduce energy use. In this contribution, all interrelationships of tourism and IT are seen to have positive outcomes for sustainability [17]. This will help the diversified stakeholders to better understand their responsibility in the Sustainable Tourism process and make them more aware of appropriate and ethical behaviors. As a matter of fact, destinations need to adopt ICT and become technology experts, eco-efficient and environmentally innovative in their activities. Not doing this may lead to economic and environmental deterioration of destinations. The identification of a collection of ICT-based tools/applications and their respective uses in destination management can have wide-ranging uses for Sustainable Tourism Development as well as it can be used to progress the already its existing approaches such as visitor management techniques and indicator development [15, 19].

5 Final Considerations

Nowadays, the tourism market is divided into two major models of commercialization, the traditional and the electronic, with clients for both models, the space occupied by each being undefined. Therefore, ICT are inseparable from tourism, having contributed decisively to its transformation. Tourism is a sector that requires an intensive use of information and one of the reasons for such a high exchange of information has to do with the nature of the tourism product. This one is different because the tourist cannot see or touch it before making the trip. As stated, [2] Information Communication Technologies are modifying the industry structure and developing a whole range of opportunities and threats. ICT empowers consumers to identify, customize and purchase tourism products and supports the globalization of the industry by providing tools for developing, managing and distributing offerings worldwide [2].

Basically, companies seek information for decision-making in the tourism market, following the wishes of customers. ICT doesn't create tourism services or products. It facilitates access to information for the client and stimulates curiosity, but destinations create services [1]. Although the tourist knows the destination through the internet with photos, films, 3D images, tourists need to experience, witness the place to feel the experience. Thus, the creation of new products and services is due to the desire of the clients, since we can only measure the quality of services through customer satisfaction. The use of ICT is irreversible. New digital technologies challenges tourism professionals and they should take advantage of the offered benefits, developing new features: being flexible to change, entrepreneurial, creative, critical and capable of leveraging their business with new technologies.

Overall, as mentioned in [15] report, new technologies create opportunities for innovative crossovers between tourism and the creative industries. These technologies can facilitate new tourist experiences, as well as offer new ways to develop and disseminate these experiences through social networks. Creative input can also add value and increase the accessibility of technology by providing effective consumer interfaces and attractive design for tourism experiences. Technology has enabled new creative intermediaries, such as bloggers and creative content producers, to become involved in the creation and distribution of creative tourism experiences. These new intermediaries largely operate outside conventional tourism distribution systems and are likely to have a growing influence on the tourism choices of consumers in the future [12].

All things considered, ICT ensure new tourists through the distribution of secure, simple, accessible information, once travelers became more demanding.

References

1. Liberato, P., Liberato, D., Abreu, A., Alén, E., Rocha, Á.: The information technologies in the competitiveness of the tourism sector. In: Proceedings of the International Conference on Information Technology and Systems (ICITS 2018). Advances in Intelligent Systems and Computing, vol. 721 (2018). https://doi.org/10.1007/978-3-319-73450-7_9
2. Buhalis, D., O'Connor, P.: Information communication technology revolutionizing tourism. Tourism Recreat. Res. **30**(3), 7–16 (2005). https://doi.org/10.1080/02508281.2005.11081482
3. Richards, G.: Creativity and tourism: the state of the art. Ann. Tourism Res. **38**(4), 1225–1253 (2011). https://doi.org/10.1016/j.annals.2011.07.008
4. Richards, G.: Trajetórias do desenvolvimento turístico - da cultura à criatividade? Encontros Cient. **6**, 9–15 (2010)
5. Castro, J.: New Technology and Creative Tourism – A case study for the city of Porto, Dissertação de Mestrado em Gestão de Indústrias Criativas. Escola das Artes da Universidade Católica Portuguesa, Porto (2012)
6. Campos, A., Mendes, J., Oom do Vale, P., Scott, N.: Co-creation of tourist experience: a literature review. Curr. Issues Tourism **21**(4), 369–400 (2018). https://doi.org/10.1080/13683500.2015.1081158
7. UNESCO: Towards Sustainable Strategies for Creative Tourism. Discussion Report of the Planning Meeting for 2008 International Conference on Creative Tourism, Santa Fe, New Mexico, USA, October, pp. 25–27 (2006)
8. Hull, J., Sassenberg, U.: Creating new cultural visitor experiences on islands: challenges and opportunities. J. Tourism Consum. Pract. **4**(2), 91–110 (2012). http://www.tourismconsumption.org/JTCPVOL4NO2HULLSASSENBERG.pdf. Accessed 22 June 2012
9. Gordin, V., Matetskaya, M.: Creative tourism in Saint Petersburg: the state of the art. J. Tourism Consum. Pract. **4**(2), 55–68 (2012). http://www.tourismconsumption.org/JTCPVOL4NO2GORDINMATETSKYA.pdf. Accessed 22 June 2012
10. Feng, W., Xu, Z.: Research on interaction of pubic art in the context of new technologies. In: Proceedings IEEE, 10th International Conference on Computer-Aided Industrial Design and Conceptual Design, Wenzhou (2009). https://doi.org/10.1109/caidcd.2009.5375068
11. Abrantes, J.L., Kastenholz, E., Lopes, R.: Web 2.0 and impacts in tourism. J. Tourism Dev. **4**(17/18), 91–93 (2012). http://www.ua.pt/event/invtur/PageText.aspx?id=15503. Accessed 22 May 2012
12. Turban, E.: Eletronic Commerce – A managerial Perspective. Prentice-Hall, New Jersey-EUA (2000)
13. Alves, A.P., Ferreira, S., Quico, C.: Location based transmedia storytelling: the travelplot Porto experience design. J. Tourism Dev. **4**(17/18), 95–99 (2012). http://www.ua.pt/event/invtur/PageText.aspx?id=15503. Accessed 22 May 2012
14. Bissoli, M.: Planejamento turístico municipal com suporte em Sistemas de Informação. Futura, São Paulo (1999)
15. OECD: Tourism and the Creative Economy, OECD Studies on Tourism, OECD Publishing (2014). http://dx.doi.org/10.1787/9789264207875-en

16. Abreu, N.R., Costa, E.B.: Um estudo sobre a viabilidade da utilização de Marketing na Internet no Setor Hoteleiro de Maceió. 24º Encontro Anual da Associação Nacional dos Programas de Pós-graduação em Administração, Florianópolis, Anais. ANPAD, Rio de Janeiro (2000)

17. Gossling, S.: Tourism, information technologies and sustainability: an exploratory review. J. Sustain. Tourism **25**(7), 1024–1041 (2017). https://doi.org/10.1080/09669582.2015.1122017

18. Junqueira, L.D.M.: Cadeia produtiva da industria cultural criativa possíveis conexões com o turismo criativo. Rosa dos Ventos Turismo e Hospitalidade **10**(3), 517–537 (2018). https://doi.org/10.18226/21789061.v10i3p517

19. UNEP & UNWTO: Making Tourism More Sustainable - A Guide for Policy Makers (2005). http://www.unep.fr/shared/publications/pdf/dtix0592xpa-tourismpolicyen.pdf

The Importance of Project Management Competences: A Case Study in Public Administration

Eliane Gonzales Meirelles, Anabela Tereso[(✉)], and Cláudio Santos

Production and Systems Department/Centre ALGORITMI, University of Minho, Campus de Azurém, 4804-533 Guimarães, Portugal
gonzaleseli2011@hotmail.com,
{anabelat,claudio.santos}@dps.uminho.pt

Abstract. Public administration is increasingly implementing new manage-ment practices borrowed from the private sector with the aim of improving quality service. Public funded projects have specific characteristics, namely regulatory, in resources usage and accountability, that may be related in different ways to existing competence models. However, it is generally understood that project management practices can improve quality and effectiveness in public funded projects. This paper presents a study on the evaluation of competences of project team members of public funded projects in the public administration in Brazil. The research methodology employed document analysis and a survey with team members directly involved in programs and projects in Metropolitan Region of Manaus and in the Social and Environmental Program PROSAMIM. The conclusions indicate that, although project teams of public funded projects strive to achieve project goals, there are still several competence gaps in ele-mentary project management practices. Future work will focus on the devel-opment of a competency model that aims to improve project success in the public sector.

Keywords: Project management · Skills · Public administration

1 Introduction

Project management is a field that has been applied in organizations since ancient times, with the aim of getting the best results from projects through a more efficient and effective management. Project management practices in the public sector are quite recent, however, the increasing demand for quality and efficiency is driving further implementations in many public sector divisions [1].

The efficient management of project is benefiting organizations in the way it increases the probabilities of achieving their objectives, since it enables the control of many factors, such as costs, schedule and risks, thus favoring decision-making on the necessary actions that take projects in the right directions [2]. One of the greatest issues concerning the public sector is the achievement of organizational excellence. Literature suggests that there are great possibilities to develop changes in the mentality, behavior and in the rules towards achieving improvements in the public sector [3].

© Springer Nature Switzerland AG 2019
Á. Rocha et al. (Eds.): WorldCIST'19 2019, AISC 930, pp. 101–111, 2019.
https://doi.org/10.1007/978-3-030-16181-1_10

This paper presents an analysis on the competences of project team members in the public sector of Brazil, and their importance in the management of public projects and their results.

The objective of this research is to identify the project management competences that need to be nurtured in the public sector to improve project success, so the research question is: what project management competences need to be nurtured in public administration to improve project success?

The researchers, based on document analysis and a survey, identified individual competences deemed necessary for project managers that deal with specific circumstances of public funded projects. The results may contribute to the development of training programs that are more adapted to the needs of the public sector.

2 Literature Review

Knowledge is one of the main pillars of modern society. Recent changes have made personal competences even more important in achieving professional success, in a global scenario where organizational competiveness is critical [4].

The main causes for project failure reported in the literature include: the lack of objectives definition from the early stages of projects, the complexity in team management, top management commitment, absence of planning and control, resistance to changes and inadequate communication [5].

Projects have been gaining increasing importance to organizations since they can contribute directly to the achievement of competitive advantages. As such, institutions have been promoting project management practices in increasingly professional manner. However, project success has not been the usual practice [6].

Public funded projects are required to deliver benefits to society that justify their investments. Stakeholders, particularly customers and subcontractors, are among the most interested in the implementation of project management practices [7].

The need to improve public administration in Brazil requires delivering public services with greater quality. The modernization of public administration is aimed at achieving its social purpose, thus driving improvements in its performance, transparency and battle against corruption and valuation of public services [8].

Organizations aim to attract people that are committed to their professional activities and that are able to aggregate value through their competences development. This is particularly noted in project oriented organizations due to their inherent nature that demands goals achievements [9].

The competences of project managers play a crucial role in delivering successful projects. These competences require specific personal, knowledge and skills traits in many areas, being the interpersonal, technical and cognitive aptitude among the most important competences [10].

ICB4 (Individual Competence Baseline 4) is a framework that describes three competences domains – People, Practice and Perspective - for project, programs and portfolio management. The "People" domain focuses on individuals' personal and social traits. The "Practice" domain portray technical aspects, and the "Perspective" domain refers to the contextual competences that must be mastered within and in a wider environment [11].

Qualified professionals will be increasingly more important, namely the skills that enable professionals to address dynamic and uncertain environments. Competence can be defined as the proper application of knowledge, skills and abilities towards the achievements of goals and results [11]. In that sense, individuals' competences play a critical role in the management of projects, programs and portfolios, and have the potential of realizing benefits not only to organizations but to society as a whole. The nurturing of individual competences requires constant learning in environments characterized by frequent changes, that in turn may have an effect on behavioral traits of an individual [12].

A project team typically includes staff from various disciplines, that embody knowledge and skills from several areas, working together to achieve projects goals [13]. The importance of a particular competence may vary from project to project, but their impact may be categorized in the following areas: project management success factors, stakeholders' management, goals and project requirements, risks and opportunities, quality, teamwork and communication, among others [11].

Projects should contribute to organizations' strategic plans. Therefore, a project manager must be capable of making decisions according to the organization strategy, demonstrating to top management that the project is aligned with the organization mission and to the execution of business strategies [11].

3   Research Methods

This study was carried out in the Metropolitan Region of Manaus, composed of 13 municipalities. The Social and Environmental Program PROSAMIM was also studied, being a relevant program for this region, contributing to give relevance to the environmental, urban and social issues of the city. This program seeks to revitalize the Manaus creeks that present housing and environmental problems [14].

The research strategy used in this study was a survey, with a deductive approach, and as such can be categorized as an exploratory and descriptive type of research. The data collection was done through document analysis and a questionnaire. Data was collected in the area of infrastructure and basic sanitation. The participants were selected through non-probability sampling, using convenience sampling. The selection was done from people working directly with projects/programs in several public administration departments in the above mentioned region. Data collection began in August and was completed in September 2017. During the application of the questionnaires it was possible to have an initial idea of the current state in project management and to clarify the doubts that arose.

The questionnaire focused on the survey of project success factors and project management competences. The data was analyzed quantitatively, taking into account the answers obtained, allowing to draw some conclusions relevant to this research.

Secondary data was collected through document analysis of the models previously used by government agencies and of project management literature available in books, scientific papers, internet, and documents provided by the entities involved in the Metropolitan Region of Manaus and PROSAMIM Program.

The questionnaire was divided into the following categories to facilitate data analysis:

1. Characterization of the respondent: the first part of the questionnaire includes information related to the profile of the respondent, namely: Gender; Age; Education; Current position in the organization; and Experience in the function.
2. Projects success factors: in the second part, it is questioned which percentage of projects was canceled before completion. Respondents are then asked to rank, by degree of importance, the factors they consider influencing project success, initiatives that could be taken to improve project management practices and critical factors that may influence projects failure.
3. Competences of project team members: the third part of the survey is intended to identify the skills that respondents feel most important for the performance of their work activities, where each respondent indicates if he or she uses the competence more or less frequently when managing their projects, according to a numerical scale of 1 to 5, where 1 means "No Knowledge/Experience" and 5 means "Great Knowledge/Experience".

In the questionnaire developed, the project management competences found in ICB4 are used: Perspective (5 elements), People (10 elements), Practice (14 elements), being a total of 29 competences. These questions helped to determine whether project management practices, with an emphasis on competences, are applied in the public sector.

4 Results and Discussion

A research was carried out with project team members of Brazilian public organizations in the Metropolitan Region of Manaus and PROSAMIM program, with the objective of mapping the competences of public project team members, and to answer the research question proposed.

The organizations surveyed employ 250 people on average and have projects with different characteristics and team sizes, depending on their complexity and duration. The organizations analyzed confirmed informally that project management is very important for them and that they have the support of top management for its accomplishment. Data collection comprised a distribution of 52 questionnaires to people working on projects in the organizations mentioned above, of which 36 were valid and 16 were discarded because they were not complete.

With regards to the characterization of the respondents, it was observed that the sample has the following gender distribution: male (47%) and female (53%), with half of the respondents aged over 40 years, and the other half aged below 40 years. The most representative age range is from 41 to 45 years (22% of the sample). The highest percentage of respondents have between 1 and 3 years of project management experience (36%), but there is also a high percentage of people with more than 10 years of experience in the function (33%). In the item referring to academic training, the respondents had high-school degree (8%), bachelor degree (39%) and master degree (53%). None of the respondents have a PhD degree. In terms of areas occupied professionally, the areas were represented as: Functional manager (8%); Project Manager (11%); Team member (31%); Other (50%). This final value represents people with a role not defined in the questionnaire.

In order to gather information about the rate of successful projects, the percentage of projects canceled before completion was questioned, and only 14% of the respondents answered that no project was canceled. Cancel rates between 1%–25%, 26%–50%, 51%–75% and 76%–100% were reported by 42%, 22%, 8% and 14% of the participants, respectively. These values indicate that the success rate needs to be improved.

Regarding the factors that influence project success, the results are presented in Table 1. Respondents were asked to sort from 1 to 10 the 10 factors in Table 1, where 1 represents the most important and 10 the least important. For each factor a total score was calculated, multiplying by 10 the number of answers that placed it in 1st place, by 9 the number of answers that placed it in 2nd place, and so on. The total score was then normalized into a scale of 1 to 10, but where 1 represents the least importance and 10 represents the most important, to be easier to relate the values.

Table 1. Factors influencing project success.

Factors – Project success	Total score	Score (1..10)
Data analysis	232	9.85
Good communication	220	8.97
Experienced project managers	221	9.04
Monitoring and control	171	5.35
Stakeholder participation	112	1.00
Competent team	234	10.00
Motivation of the project team	181	6.09
Competence of the project managers	210	8.23
Quality management	189	6.68
Clear vision of objectives	200	7.49

Analyzing the results, it was verified that the factor considered most relevant for project success was "Competent team" (10 points), so it is recognized that the teamwork competences are very relevant to project success, as also mentioned by PMBOK Guide [15]. With more than 9 points, "Data analysis" appears, confirming the need of information to improve project management, and "Experienced project managers", also relate to competence issues. Next, there are two factors with more than 8 points: "Good communication" and "Competence of the project managers". Again, competences emerge as relevant. The other factors, although also relevant, were considered less relevant by the participants of this study.

Then, participants were asked to rank, by degree of importance, the 14 initiatives that should be taken to improve project management practices, where 1 represents the most important and 14 the least important. The calculations of the scores were done in the same way as explained in the previous case. The results are shown in Table 2.

Table 2. Initiatives to improve project management practices.

Initiatives – Improve PM practices	Total score	Score (1..10)
Improve project planning	376	10.00
Qualify the team	359	9.44
Improve communication	307	7.72
Improve conflict resolution	328	8.41
Elaborate costs and deadlines with adequate forecasts	348	9.07
Approve the project on time	263	6.26
Improve human resources management	240	5.50
Improve leadership	263	6.26
Favor delivery on time	206	4.38
Have appropriate material	259	6.13
Anticipate requests made by customers	201	4.21
Have an adequate preliminary study	297	7.39
To promote the cultural values defined	109	1.17
Improve and favor decision making	104	1.00

The initiative indicated as most relevant is "Improve project planning", followed by "Qualify the team" and "Elaborate costs and deadlines with adequate forecasts". It can be understood that planning and qualification are critical aspects of project management that need further improvement.

In order to complete this type of questions, participants were asked to sort the 8 critical factors that influence project failure, from 1 to 8. The results obtained are shown in Table 3, with the calculations of the scores having been made as above.

Table 3. Critical factors that influence project failures.

Critical factors - Project failures	Total score	Score (1..10)
Negative influence of the political will of stakeholders	188	10.00
Inadequate organizational structure	169	7.29
Incompetent team	168	7.14
Project goal bad defined	175	8.14
Lack of stakeholder involvement	129	1.57
Lack of top management support	125	1.00
Poor planning	182	9.14
Lack of competence of the project manager	162	6.29

Respondents consider that the most critical factors that influence project failures are "Negative influence of the political will of stakeholders", "Poor planning", reinforcing previous results, and "Project goal bad defined". The last two factors may be related to the competence of those responsible for initial planning.

For a better understanding of the status of the participants in the study, in terms of their competences as defined by ICB4 (perspective, people and practice), the questions were divided into two parts, one including the degree of knowledge and another the experience. The scale used was 1 to 5, with 1 meaning no knowledge/experience and 5 meaning a lot of knowledge/experience.

The results for perspective competences are presented in Table 4.

Table 4. Knowledge and experience for perspective competences.

Perspective competences	Knowledge (1..5)	Experience (1..5)
Strategy	2.20	4.43
Governance, structures and processes	1.00	1.00
Compliance, standards and regulations	1.00	1.00
Power and interest	4.67	5.00
Culture and values	5.00	2.14

The competences that respondents seem to have more knowledge about are "Power and interest" and "Culture and values". In terms of experience, "Power and interest" remains, but "Strategy" is also ranked high. There seems to be more weaknesses in competences related to "Government, structures and processes" and "Compliance, standards and regulations", which may indicate the need for training in these areas.

The results for people competences are presented in Table 5.

Table 5. Knowledge and experience for people competences.

People competences	Knowledge (1..5)	Experience (1..5)
Self-reflection and self-management	2.07	1.44
Personal integrity and reliability	1.00	1.00
Personal communication	4.47	3.89
Relations and engagement	5.00	5.00
Leadership	3.93	3.67
Teamwork	4.73	3.67
Conflict and crisis	1.27	2.78
Resourcefulness	2.60	1.44
Negotiation	1.80	1.67
Result orientation	3.40	4.11

The competence that respondents seem to have more knowledge and experience is "Relationships and engagement". Then "Teamwork" and "Personal communication". Still, with more than 3.5 points, "Leadership" emerges. All other competences have lower values of importance, except "Results orientation" that appears with high experience value (4.11), but less than 3.5 for knowledge.

Finally, the results for practice competences are presented in Table 6.

Table 6. Knowledge and experience for practice competences.

Practice competences	Knowledge (1..5)	Experience (1..5)
Project design	5.00	2.80
Requirements and objectives	2.56	1.00
Scope	1.89	2.80
Time	3.44	4.00
Organization and information	3.89	5.00
Quality	3.00	1.60
Finance	1.00	1.00
Resources	3.00	3.60
Procurement	2.11	2.80
Plan and control	4.11	4.00
Risk and opportunity	4.11	4.40
Stakeholders	3.44	3.40
Change and transformation	3.67	4.20

The competence that respondents seem to have more knowledge is "Project design", being "Organization and information" the one with higher experience value. "Plan and control" and "Risk and opportunity" also have values greater than 4 in both parts. Next "Change and transformation" and "Time" are also relatively well-quoted. Table 7 presents a summary of the results of this part of the study with the worst competences.

Table 7. Summary of worst competences in the three areas.

Competence	Perspective	People	Practice
Less knowledge	Governance, structures and processes/Compliance, standards and regulations	Personal integrity and reliability	Finance
Less experience	Governance, structures and processes/Compliance, standards and regulations	Personal integrity and reliability	Finance/Requirements and objectives

The "Governance, structures and processes" competence is related to the understanding of and the alignment with the established structures, systems and processes of the organization that provide support for projects and influence the way they are organized, implemented and managed. "Compliance, standards and regulations" describes how to interpret and balance the external and internal restrictions in a given area such as country, company or industry. These seem to be the perspective competences that most need to be improved.

On the part of the people it was verified that the competence "Personal integrity and reliability" is pointed out as the weakest one in terms of knowledge and experience. Individuals must demonstrate personal integrity and reliability because a lack of these qualities may lead to a failure of the intended results. This also needs to be improved.

In the practice competences, respondents revealed to have less knowledge/experience of "Finance" and "Requirements and objectives". Finance includes all activities required for estimating, planning, gaining, spending and controlling financial resources. Requirements and objectives allows to establish the relationship between what stakeholders want to achieve and what the project is going to accomplish. These two practice competences also need to be improved.

5 Conclusions and Future Research

Public organizations face great challenges to respond to increasingly complex societal problems. Some have fragile structures due to insufficient resources, both financial and human, and knowledge weaknesses in general and also in the specific area of project management. The main objective of this study was to verify the positioning of the organizations that provide public services, namely municipalities, in relation to project management. It was carried out in the Metropolitan Region of Manaus, composed of 13 municipalities. The Social and Environmental Program PROSAMIM was also studied, being a relevant program for this region. This program seeks to revitalize the Manaus creeks that present housing and environmental problems [14].

Managing projects involves the need for multiple skills and this becomes a constant challenge for project managers who need to have knowledge and experience in many areas. The research was focused on answering the following research question: what project management competences need to be nurtured in public administration to improve project success?

The strategy used for the research was survey, using document analysis and a questionnaire to collect data. The questionnaire was distributed to 52 people working on projects of the entities mentioned above and 36 valid responses were received.

The questionnaire focused on the survey of project success factors and project management competences. The competencies analyzed where the 29 competencies proposed in ICB4. The data was analyzed quantitatively.

The results allowed to conclude that the success of managed projects can be improved, taking into account the rates reported. Only 14% of respondents reported that no project had been canceled.

Regarding the factors that influence the success of the projects, the most relevant factor was the competence of the team, so it is recognized that competences are very relevant in project success.

The opinion of respondents was also asked on the importance of initiatives that could be taken to improve project management practices. The results allow us to conclude that are several issues related to the qualification of the personnel involved in project management.

As for the critical factors that could most influence project failures, the results indicate lack of competence of those responsible for setting goals and planning.

Finally, the competences the respondents considered better knowing and having more experience on were collected, so as to try to define the gaps, based on the competences defined in ICB4, divided into the areas of Perspective, People and Practice.

A limitation of this paper is the reduced sample size. In future work, the questionnaire applied should be refined and applied to a large sample, from which the results obtained in this study will be complemented with, for example, semi-structured interviews with decision-makers in this area, to formulate a model of improvement actions aimed at increasing the success of public projects, not only in Brazil but also in other similar organizations in the World.

Acknowledgements. This work has been supported by FCT – Fundação para a Ciência e Tecnologia within the Project Scope: UID/CEC/00319/2019.

References

1. Pestana, C.V.S., Valente, G.V.P.: Gerenciamento de projetos na administração pública: da implantação do escritório de projetos à gestão de portfólio na secretaria de estado de gestão e recursos humanos do Espírito Santo. In: III Congresso Consad de Gestão Pública. Consad, Brasília (2010)
2. Quadros, A.S., Carvalho, H.G.: O gerenciamento da comunicação de projetos públicos: como adaptar os processos do PMBOK/PMI à realidade da administração pública. Rev. Bras. Planej. Desenvolv. **1**, 52–60 (2012)
3. Amaral, H.K.D.: Desenvolvimento de competências de servidores na administração pública brasileira. Rev. Serv. Público Brasília. **57**, 549–563 (2006)
4. Pant, I., Baroudi, B.: Project management education: the human skills imperative. Int. J. Project Manag. **26**, 124–128 (2008)
5. Young, T.L.: Successful Project Management. Kogan Page Publishers, London (2013)
6. Rabechini Jr., R., Pessoa, M.S.P.: Um modelo estruturado de competências e maturidade em gerenciamento de projetos. Rev. Prod. **15**, 34–43 (2005)
7. Forsythe, P.J.: In pursuit of value on large public projects using "spatially related value-metrics" and "virtually integrated precinct information modeling". Procedia-Soc. Behav. Sci. **119**, 124–133 (2014)
8. Matias-Pereira, J.: Manual de gestão pública contemporânea. Atlas, São Paulo (2009)
9. de Oliveira Albergaria Lopes, R., Sbragia, R., Linhares, E.: The psychological contract and project management as a core competence of the organization. Procedia - Soc. Behav. Sci. **226**, 148–155 (2016)
10. Vlahov, R.D., Mišić, S., Radujković, M.: The influence of cultural diversity on project management competence development–the Mediterranean experience. Procedia-Soc. Behav. Sci. **226**, 463–469 (2016)
11. IPMA: Individual Competence Baseline for Project, Programme & Portfolio Management, International Project Management Association (2015)
12. Tavares, E.S.: Uma contribuição para os processos da gerência de projetos através da gerência do conhecimento. Universidade de São Paulo (2004)
13. Liu, J.Y.C., Chen, H.H.G., Jiang, J.J., Klein, G.: Task completion competency and project management performance: the influence of control and user contribution. Int. J. Project Manag. **28**, 220–227 (2010)

14. Segundo, R.F.: Impactos do Programa PROSAMIM para os ribeirinho do Igarape do Quarenta na cidade de Manaus. Universidade Federal do Pará (2014)
15. PMI: A Guide to the Project Management Body of Knowledge (PMBOK GUIDE). Project Management Institute, Pennsylvania (2017)

Improvement of Industrialization Projects Management: An Automotive Industry Case Study

Diana Fernandes[✉], Anabela Tereso, and Gabriela Fernandes

Production and Systems Department/Centre ALGORITMI, University of Minho,
Campus de Azurém, 4804-533 Guimarães, Portugal
pg31496@alunos.uminho.pt,
{anabelat,g.fernandes}@dps.uminho.pt

Abstract. Nowadays, in response to the technological evolution of the markets, organizations need to effectively improve their product development process. The product development can be seen as a global project where the industrialization project is a subproject of the product development project. The use of an integrated project management process helps to ensure projects completion according to the schedule plan and to keep the cost inside the budget and the quality with acceptable standards. The main objective of this research was the improvement of the project management practices in industrialization projects in a company of the automotive sector. An integrated project management process was developed and key project management tools and techniques were proposed, considering the project organizational context. The stakeholders who participated in this research study perceived that improvements proposed will be an added value to the company, however, due to the time constraints, it was not possible to test and quantify the real benefits of the proposed improvements on the management of industrialization projects.

Keywords: Project management · Industrialization projects ·
Integrated process of project management · Automotive industry

1 Introduction

The constant search for new and better products leads to the creation of new processes of product development, these being crucial to pay special attention to the aspects of quality and safety of the final product. Due to the high competitiveness of the automotive industry, time to market must be reduced, therefore requiring tailored and standardized processes [1, 2]. Data from a survey carried out by the Economist Intelligence Unit [3] shows that well-defined methodologies in organizations give more confidence to project managers helping to deliver their projects with success. Faced with these facts, project management (PM) evolved from a set of recommended processes to a mandatory methodology, being crucial to the survival of the companies. However, it remains a challenge because there are still many projects that exceed budget, delay or fail to meet their objectives, as evidenced in several studies [4, 5]. According to the Gartner Group's worldwide survey, the percentage of unsuccessful

© Springer Nature Switzerland AG 2019
Á. Rocha et al. (Eds.): WorldCIST'19 2019, AISC 930, pp. 112–121, 2019.
https://doi.org/10.1007/978-3-030-16181-1_11

projects (failed and con-tested) as of 2012 was 61% [6]. However, a more recent study from PMI, in 2016, found that projects using proven project management practices were 2.5 times more successful [7]. Besner and Hobbs [8] study reveals significant differences in project management practices in different contexts and types of projects. The process of choosing the project management model to be developed in an organization is therefore very complex because many dimensions must be considered simultaneously, the effects of which may be contradictory [9]. According to Besner and Hobbs [8] these dimensions may vary depending on project size (in terms of budget and duration), project typology, degree of complexity, project similarity, and degree of project innovation. However, despite these variations in specific practices, Besner and Hobbs [10] confirm the position that project management is a generic discipline with a wide range of applicability.

The objective of this research was, through a study in a company in the automotive industry, to answer the following research question: How to improve project management practices in automotive industrialization projects? The goal was to determine which project management practices could be improved, and how, in order to improve the success of industrialization projects. In order to answer this central research question, the following specific objectives were defined: (1) Identification of the project management practices and difficulties at the company (As-Is Model); (2) Definition of an integrated project management process and identification of best project management practices to mitigate the difficulties and respond to the needs of the industrialization projects (To-Be Model).

This paper is organized as follows. In Sect. 2 the literature review that contributed to this research is presented. Section 3 presents the research methodology used. Section 4 presents the case study analysis and Sect. 5 the proposal of an integrated project management process to improve PM practices in the company. Finally, Section, 6 presents the conclusions.

2   Literature Review

This section presents a literature review on project management of industrialization projects and project management practices.

2.1   Project Management of Industrialization Projects

PM refers to the individual management of each project. It is the combination of people, techniques and systems needed to manage the resources necessary to successfully reach the end of the project [11]. It means doing what is necessary to complete the project within the established objectives. PMI [12] defines project management as being "the application of knowledge, skills, tools, and techniques to project activities to meet the project requirements".

The success of a project is usually measured at the end of the project. The project management success usually measured at project end, being a way to help the project to meet the budget, schedule and quality criteria. Each of these criteria is based on the comparison of the project objectives with what was actually achieved. For example, quality can be measured through the amount of rework or level of customer satisfaction.

Project management success is then linked to project success, and the two are inseparable [13]. However, success always varies from project to project taking into account the organizational context and the different stakeholders [14].

In industrialization projects specific context, the role of project management may not be as evident as it should be [15]. According to Perrotta, Araújo, Fernandes, Tereso and Faria [16] the process of industrialization involves the analysis/understanding of all product requirements, the development of prototypes to increase product maturity before moving to the design of the manufacturing line. Typically, clients (internal or external, depending on the origin of the request) participate in this process, receive some prototypes to make their own evaluation and then provide feedback to the industrializing company. This feedback can cause changes in the product, resulting in new requirements to be met by the product. New prototypes are then built and the process is repeated until all customer requirements are considered, allowing the production line development process to be started. Throughout the process of industrialization, the quality of the product must always be evaluated.

2.2   Project Management Practices

There are several studies about project management practices, including tools and techniques, which can help project manager professionals, in different contexts.

According to a study by White and Fortune [17], the most used tools in project management were: "off the shelf" software (77%); Gantt charts (64%); and cost-benefit analysis (37%). This study is in line with a more recent survey by Besner and Hobbs [18], which was based on a questionnaire to 753 experienced project management participants. The results were a ranking of 70 techniques and tools by their degree of use. However, this study could not be applied directly to our study without first analyzing the context of the organization. A more recent study by Fernandes, Ward, and Araújo [19] also aimed to identify the most useful project management practices and the results present some similarities with the Besner and Hobbs study.

Perrotta, Fernandes, Araújo, Tereso and Faria [15] carried out a study, on the automotive industry, where the best practices for industrialization projects were identified. A survey was carried out with 17 industrialization project managers. Project planning is the practice at the top of the list, due to the importance of respecting the customer's deadlines, being a guide to develop the industrialization schedule and milestones. Second is lessons learned. According to the authors, knowledge gained through the development of new products can be useful for future projects, either in relation to new projects using similar technology or projects of the same client with specific needs. Third is ECR (Engineering Change Request). Due to the fast-paced environment and the innovative development of new manufacturing lines, it is very common to have process-related changes and, as such, ECR must be completed to formalize, approve and record these changes [15].

A study from Tereso, Ribeiro, Fernandes, Loureiro and Ferreira [20] allowed to evaluate the 20 tools and practices most used for each group of project management processes in private companies. The top 20 cover the overall project management life cycle from initiation to project closing, but particular relevance is given to tools and techniques from planning, and curiously, to tools and techniques from closing.

3   Research Methodology

The methodology adopted presupposes a deductive approach that begins with the literature review to identify theories and studies, and then to apply it to practice. An analysis of the company's processes and procedures, regarding project management practices, was also made, in order to identify the main problems and difficulties. Subsequently, in order to answer the research question, semi-structured interviews were applied to the four project managers and the functional process manager of the organization under study. These interviews aimed to understand how the projects were managed and collect ideas to improve the process. After analyzing the data from these interviews, a 2-hour focus group was conducted for a preliminary validation of the integrated project management process proposed by the researchers. It was an unstructured focus group, allowing the presentation of the integrated process and providing openness for improvement suggestions. The methodology used, inspired in Perrotta, Araújo, Fernandes, Tereso and Faria [16], allowed to identify and decompose the existent project management practices, defining the As-Is model. Then this model was used as a starting point for the development of the To-Be model.

4   Case Study Analysis

The industrialization in the company under study begins with the study of the production process, still during the development of the product, extending through the approval phase. This includes the validation of processes and tools to be implemented for the pre-series and series production, using tests that are articulated and updated simultaneously with the evolutions of the product. Included in industrialization is the logistics part, which defines the packaging and transportation of the product.

The process development life cycle is divided into the following phases: (1) Study: consists of the design of the injection tool, plating jig, control gauge, packaging and other tools needed for the process; (2) Construction: construction phase of the injection tool and all the tools necessary to obtain the final product, according to customer's requirements (plating jig, control gauge, packaging and others); (3) Testing and validation of processes and tools: preparation phase of injection tool tests, tests to the plating jig and general test with all the tools necessary for the product. After this process and closed tools, the final validation is carried out through the production of initial samples.

With the nomination of the project, the client provides the last 2D plan and 3D of the part ordered, in order to be analyzed in detail and thus to define the necessary tools. The industrialization projects aim to develop tools that allow the subsequent production in series of a chrome part that meets the requirements of the customer.

With the approval of the tools study, it is possible to proceed to the construction phase and then to the tests, in order to validate all the outputs. Upon completion of the project, the company begin the serial production of the parts.

A common process of industrialization usually lasts 24 months, however there are factors that can affect the duration of this process, such as: (1) Complexity of the part; (2) Workload, especially of the tool maker; (3) Any modifications to the part to ensure conformity in the assembly; (4) Deadlines agreed with the client.

With regard to the number of people directly involved in a project, this can range from eight to ten people, depending on the project complexity.

With regard to the project costs, these vary according to the typology of the product (size, complexity, client requirements). However, the tool that requires greater investment is the injection tool. It is not possible to quantify the average cost of this tool due to the fact that it varies with several factors, such as the number of cavities, their location and how to inject the part.

In the course of describing project management practice at the company, and in the interviews conducted, several difficulties were encountered that conditioned the development of a project. The main objective of these interviews was to survey all the difficulties and problems experienced, as well as to identify the practices and tools already used in the organization (Table 1).

5   Proposal of PM Initiatives to Improve PM Practice

During this section, a proposal for an integrated project process will be presented, thus describing the life cycles of the industrialization process development and project management. Proposals for improvements to the tools and techniques used in project management will also be presented. The proposed new practices/tools were based on literature review and interviews.

To validate the improvements, a focus group was performed. The stakeholders who participated have accepted the improvement proposals willingly, however there are proposals that involve a restructuring of the organization, so it is necessary to perform another focus group with the top management. Yet, due to time limitations and incompatibility of agendas, this was postponed to future work.

5.1   Project Initiation

This group of processes occurs after receipt of the client award, which can be done by sending the order of the tools, or with an official nomination letter. This phase includes the realization of the project charter, which corresponds to the definition of the project's characteristics, namely in terms of time, scope and cost. It contains a small description of the main deliverables and also appoints the project team that will follow up. Subsequent to the realization of the project charter, the contract revision meeting (Kick off) is made. Initiation is the stage at which the project is described in detail, so that its main purpose and objectives are understood.

Regarding D1 difficulty (Table 1), it was possible to verify that there are projects awarded that are not always feasible in the first phase, and that there is a need to make changes to the product at the beginning of the project, which may contribute to eventual delays. The proposal presented by the researchers involve the creation of a new function in the company, namely the function of pre-project. The objective is to create viable alternatives to start the project phase. The person who would hold this position would be the link between the client, the budgeting department and the commercial department and would always intervene on the feasibility phase. This function would act in two main moments: (1) when the complexity of the project at the risk and technological level justifies it; (2) when the project is considered strategic (e.g. strategic customer, billing volume above normal, extension of the operation area).

Table 1. Difficulties found in the project management process (As-Is model).

ID	Project phases	Difficulties (D)	Causes according to the interview analysis
D1	Project initiation	Existence of awarded non-feasible projects	Necessary change the part geometry in the middle of the project due to the lack of a depth feasibility analysis
D2		Limited autonomy of the project manager due to the company's organizational structure	Currently the structure of the organization is functional, where each team member of the project has a hierarchical superior from a functional department
D3		The company's management system does not include two important standards, the development and innovation management standard NP 4457, and the project management standard ISO 21500	Implementation of these two norms by the importance assigned to these standards in the management system of an organization
D4	Initial planning	Responsibility of each project stakeholder is unclear	Project responsibilities matrix does not existent
D5		Lack of detail of the work to be executed	There is no detail of the work to be done which contributes to a PM non-standardized methodology
D6	Project follow-up	Lack of some technical knowledge in meetings with the client	The project manager is the main interlocutor with the client, however he does not have the same technical knowledge as for example the tool technician to respond assertively and immediately
D7		Loss of information during project follow-up	The existence of several projects simultaneously, in different phases and with different teams, leads to the loss of information
D8		Difficulty in establishing contact with some of the elements of the team, which causes an increase of time in the execution of the project activities, being the client is more and more demanding	Production and logistics departments' physical distant from the projects department
D9	Closing	Incomplete project closure documentation	Lack of a lessons learned template and project closure report with little information

After awarding the project, this person with this new function, would inform the project manager of the status so that the PM activities could begin.

Another proposal of the researchers in the project initiation phase would be to change the structure of the organization. The organization has a functional structure, where each project team member has a hierarchical superior. This type of structure tends to cause conflicts within the team. In order to avoid this situation, a structure by projects is proposed, thus giving full authority to the project manager regarding its team. He would only report to one person (project sponsor), reducing the communication channels and facilitating the communication. This would allow to overcome the difficulty D2. Still in this group of processes, more precisely in the integration management knowledge area, it was possible to identify that the Enterprise Management System does not include two important norms, both for the automotive industry and for project management, namely the NP 4457, standard for research, development and innovation management and the project management standard, ISO 21500 (D3). The researchers proposed that the company invest on the implementation of these two standards, as they will enhance the credibility and image of the organization, which will contribute to a potential increase in projects to be awarded. The acquisition of new customers and continuous improvement of their processes are other benefits that are expected with the implementation of these two standards. It is important to emphasize that the implementation of these standards naturally involves costs, however, the organization must take into account that the costs initially invested can bring a positive return in the medium/long term, increasing efficiency and productivity.

5.2 Project Initial Planning

After the project charter is formalized, the project initial planning will start, where all the subsidiary plans of the different areas of knowledge will be defined and integrated into the final project management plan.

In this phase, the researchers propose two new tools for project management: (1) the construction of a responsibility matrix that did not exist and will allow a clearer definition of the responsibility of each project stakeholder; (2) The construction of a WBS that will allow to have a more precise detail of the work to be done throughout the project.

5.3 Project Follow-Up

The project follow-up phase concerns all work carried out in the development of the process, namely the study of the tools, the monitoring and control of the construction of the tools in the supplier and, finally, test and validation. This phase is responsible for monitoring, analyzing and recording the project progress, in order to meet the objectives that were defined. It also provides a clear view of the project performance.

During this process, corrective actions may be necessary, namely changes in the product, which may come from both the client and the organization that develops the project. To allow these changes it is important to add a re-planning process, which aims to reformulate the already existing planning, in order to avoid possible inaccuracies in time and cost estimates.

In the organization under study, it was possible to perceive some lack of technical knowledge in the meetings between the client and the project manager, both in the injection process and molds, and in the electroplating process (D6). To solve this situation, it was proposed the integration of a technical team member in the meetings where technical aspects are discussed. Meeting minutes are also proposed to avoid loss of information.

Increasingly, the client requires closer project monitoring, periodically requesting information and updating the project status. Thus, project team weekly meetings were proposed, whose objective is to improve the communication between the team members and to reduce the time of execution of the project activities, caused by the lack of response of these members. As an auxiliary tool for these meetings, the researchers created a template called Open Point List. The use of meeting minutes were also proposed.

5.4   Project Closing

Project closing phase includes the finalization of all project management activities. It integrates, among other things, customer acceptance, store of information and project handover to serial production, which is usually done after three months from the date of star of production.

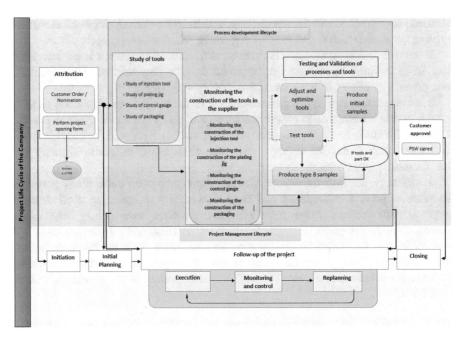

Fig. 1. Proposed integrated project management process between project management lifecycle and process development management lifecycle.

Generally, at this stage, an analysis and validation of the lessons learned is performed, which in the studied organization did not happen. This is an important tool in this industry context, such as Perrotta, Fernandes, Araújo, Tereso and Faria [15] show in their study. The researchers proposed a template so that throughout the project all lessons learned are recorded. Additionally, the project closure report, which aims to formalize the closure of project management activities, was also improved.

At various points in the project it is possible to verify the interconnection between project management and the industrialization process development. In this way, and as shown in Fig. 1, the integration of the processes of the both life cycles are presented.

6   Conclusions

This paper presents a study about improving project management in industrialization projects, developed in the context of the automotive industry. The work was developed in the form of a case study which may be of interest to other similar companies of the same area. However, as evidenced in the literature review, the organizational context must always be considered, so that the project management tools and techniques proposed are context-adjusted [8–10].

It was possible to verify that the group of processes that the company needs to focus more on are from the project closing phase. According to the project managers, the reason why this group of processes is still a weak point in the company is because the awarded projects are always increasing, resulting in lack of time to focus on closure of ongoing projects, which contributes to neglecting this important project phase.

In industrialization projects, it is essential to document the lessons learned acquired throughout the project, as these can be useful for new projects that may use similar technology, or projects of the same client with specific needs [15].

It is important to emphasize the contribution of the studied organization in order to provide viable alternatives to ensure the success of the project management process. Although this study has a more applied aspect, it can be considered for possible future studies that want to develop an approach to the management of industrialization projects in similar organizations.

In summary, the researchers consider that the organization under study perceives the value of project management and has the necessary resources to implement the proposed improvements. Some tools and good practices have already been implemented, such as the project charter, the kick off meeting and the Gantt Chart. It is noteworthy that the company showed a strategic orientation for the future coincident with the proposed improvements. These proposals are understood as fundamental for the organization to create value, through the satisfaction of its customers and its employees.

Acknowledgements. This work has been supported by FCT – Fundação para a Ciência e Tecnologia within the Project Scope: UID/CEC/00319/2019.

References

1. Gobetto, M.: From Industrial Strategies to Production Resources Management, Through the Industrialization Process and Supply Chain to Pursue Value Creation. Operations Management in Automotive Industries. Springer, Netherlands (2014)
2. Fernandes, G., Ward, S., Araújo, M.: Developing a framework for embedding useful project management improvement initiatives in organizations. Proj. Manag. J. **45**(4), 81–108 (2014)
3. EIU: Industrial manufacturing: Managing for success. Economist Intelligence Unit Limited, pp. 1–18 (2010)
4. Pinto, R., Dominguez, C.: Characterization of the practice of project management in 30 portuguese metalworking companies. Procedia Technol. **5**, 83–92 (2012)
5. The Standish Group: The Standish group: the chaos report. Retrieved from Project Smart (2014)
6. Spalek, S.: Success factors in project management. Literature review. In: Education and Development Conference INTED2014, Valencia, Spain (2014)
7. Costa, I., Fernandes, G., Tereso, A.: Integration of project management with NPD process: a metalworking company case study. In: 23rd ICE/IEEE International Conference on Engineering, Technology and Innovation, Madeira Island, Portugal, 27–29 June 2017 (2017)
8. Besner, C., Hobbs, B.: Contextualized project management practice: a cluster analysis of practices and best practices. Proj. Manag. J. **17**(1), 17–34 (2013)
9. Besner, C., Hobbs, B.: An empirical identification of project management toolsets and a comparison among project types. Proj. Manag. J. **43**, 24–46 (2012)
10. Besner, C., Hobbs, B.: Project management practice, generic or contextual: a reality check. Proj. Manag. J. **39**, 16–33 (2008)
11. Norat, S., et al.: Manual de Gestão de Projetos, 1st edn. Brasil (2011)
12. PMI: A Guide to the Project Management Body of Knowledge, Sixth edn. Project Management Institute, Newtown Square (2017)
13. Munns, A., Bjeirmi, B.: The role of project management in achieving project success. Int. J. Proj. Manag. **14**(2), 81–87 (1996)
14. Milosevic, D., Patanakul, P.: Project standardized project management may increase development projects success. Int. J. Proj. Manag. **23**, 181–192 (2005)
15. Perrotta, D., Fernandes, G., Araújo, M., Tereso, A., Faria, J.: Usefulness of project management practices in industrialization projects - a case study. In: 23rd ICE/IEEE International Technology Management Conference, 27–29 June 2017 (2017)
16. Perrotta, D., Araújo, M., Fernandes, G., Tereso, A., Faria, J.: Towards the development of a methodology for managing industrialization projects. Procedia Comput. Sci. **121**, 874–882 (2017)
17. White, D., Fortune, J.: Current practice in project management—an empirical study. Int. J. Proj. Manag. **20**(1), 1–11 (2002)
18. Besner, C., Hobbs, B.: The perceived value and potential contribution of project management practices to project success. Proj. Manag. J. **37**, 37–48 (2006)
19. Fernandes, G., Ward, S., Araújo, M.: Identifying useful project management practices: a mixed methodology approach. Int. J. Inf. Syst. Proj. Manag. **1**(4), 5–21 (2013)
20. Tereso, A., Ribeiro, P., Fernandes, G., Loureiro, I., Ferreira, M.: Project Management Practices in Private Organizations. Proj. Manag. J. **50**(1), 1–17 (2018)

A Digital Strategy for SMEs in the Retail Business that Allows the Increase of Sales

Alesandro Anthony Huayllas Iriarte[✉] [iD],
Betsy Andrea Reinaltt Higa [iD], Alfredo Barrientos Padilla,
and Rosario Villalta Riega

Universidad Peruana de Ciencias Aplicadas, Lima, Peru
{u201310460, u201311630, pcsiabar,
rosario.villalta}@upc.edu.pe

Abstract. This document shows the benefits of implementing a digital strategy for SMEs that is aimed at increasing sales. The proposed strategy is based on the analysis of the needs of the company and the alignment of its business objectives with social media objectives. The main objectives of the de-signed digital strategy are the attraction of potential customers and, subsequently, the increase of online sales.

Keywords: Digital strategy · Social Media · ROI · Metrics · Web 2.0 · Analytics

1 Introduction

Social networks are an innovative tool to promote different businesses. In Latin America, there is an average usage of 6 h per month per visitor, which makes it the region with the greatest involvement in social networks worldwide [1]. According to the report "Estado de Social Media en América Latina 2018", made by comScore in 2017, 76% of the communication between brands and users occurs on Facebook; in second place, Instagram with 20%; and lastly, Twitter with 4% [2].

However, not all opportunities are pleasant to corporations because consumers freely publish their opinions about any product, which may have a positive or negative impact on the reputation of companies and their profits [3]. Therefore, these platforms require special attention and management from companies [4].

An alternative to improve the management of social media information is by raising awareness among companies about the importance of designing a good marketing plan according to their needs, regardless of the size, sector, and resources available to the company [5]. This helps direct the marketing plan which mainly addresses contents to be implemented in different media [6]. This paper will introduce a digital strategy based on the Social Media Measurement Framework by Altimeter, consisting of four steps that allow the company to design a strategy aligned with its objectives and to measure its results [7].

© Springer Nature Switzerland AG 2019
Á. Rocha et al. (Eds.): WorldCIST'19 2019, AISC 930, pp. 122–131, 2019.
https://doi.org/10.1007/978-3-030-16181-1_12

2   Literature Review

To implement a digital strategy for a SME, two important aspects of the Social Media Measurement Framework have been considered: Building Guidelines and Measuring Guidelines.

2.1   Building Guidelines

According to Nieves Fernandez, building an appropriate digital strategy for the company is a key process to achieve a wider reach [8]. Likewise, the Assisting Attraction Classification reference framework, made by Junge Shen, details that for the design of a digital strategy, the company must define how its performance will be measured, since based on this, decisions will be made for its optimization [3]. For this purpose, the following steps are considered:

Strategy. The main objectives of the company are aligned with those intended to achieve under the digital strategy, in order to define the projects and activities that will help fulfill it. The content to be used will depend on the approach selected by the company [9].

Metrics. The metrics help to learn the results of the strategy. Every metric must have a meaning or context for an analysis process since, without it, the data does not add value to the business [10].

Organization. The resources, analytical knowledge, and tools owned and needed by the company must be evaluated for the development and measurement of the digital strategy [11].

Tools. According to Javier Serrano, the company should choose the tools that will help the business to interpret properly data obtained from the digital strategy [12]. This is supported by the reference framework made by Evelyn Ruiz, who mentions the importance of using digital tools to improve the analysis obtained and the website positioning.

2.2   Measuring Guidelines

According to the reference framework made by Dan Puiu, the importance of analyzing the data comprehensively to generate information of greater value for the company is considered [4]. For this reason, indicators, metrics, and KPIs (Key Performance Indicators) are established according to the approach of the digital strategy, which is chosen according to the company's objectives. The Social Media Measurement Framework suggests 6 approaches which are grouped according to their characteristics:

- **Qualitative**

Brand Health. This approach allows the building of a powerful brand that conveys trust to users [13].

Customer Experience. This approach allows the development of customer loyalty. According to Pitre, this is achieved by learning the needs of the customers and adapting the offer to their particular situation [14].

Innovation. This approach allows companies to get feedback from users in social networks and measures the impact of ideas to identify new opportunities [15].

- **Quantitative**

Marketing Optimization. This approach consists in generating content for social media that allows users to access a web portal to generate a conversion (purchases, successful registration, and subscription, among other objectives) [16].

Income Generation. This approach is aimed at increasing sales and conversions through customers. For this purpose, the business environment must be constantly evaluated in order to establish a reference model [17].

Operational Efficiency. This approach is intended to involve social networks in the resolution of doubts and problems to reduce costs [18].

3 Contribution

3.1 Basis

An SME implements a digital strategy for the purpose of contributing to the fulfillment of its objectives. In order to validate this, an adequate strategy has been created for a Peruvian SME in the retail business based on the four steps of the Social Media Measurement Framework (strategy, metrics, organization, and tools).

3.2 Development

Strategy. After analyzing the SME and understanding the context in which it operates, we aligned its objectives with social media objectives in order to choose one or more approaches for the digital strategy [19] (Table 1).

Table 1. Defined objectives

Objectives	
Increase its followers by 15% on the Instagram fan page and 10% on Facebook within two months for more brand presence	Brand health
Create a website so that it serves as a web catalog in the first month	Marketing optimization
Get traffic of at least 50 users to the website in the first month	Marketing optimization
Increase sales by 15% compared to the last month (March 2018)	Marketing optimization

Once the objectives are identified, these are linked to the proposed approaches that will be incorporated into the digital strategy, which are Brand Health and Marketing Optimization.

In addition, the projects to be executed within the strategy are listed. These are (Table 2):

Table 2. Projects to be executed within the strategy

Projects
Organic and paid content in the company's social networks
Integration of the new website and social networks with digital analytics tools

Metrics. When the SME has selected the approach, they must define the metrics and indicators (KPIs) that help verify that the amount of time and money invested produces an adequate return on investment [20]. The issues and metrics proposed by the Social Media Measurement Framework and used for the validation—which are Brand Health and Marketing Optimization—will be detailed below.

- **Brand Health Metrics**

  The Social Media Measurement Framework suggests five issues related to the company's reputation according to the company's needs. In this case, the selected topic was Conversation and Sentiment Drivers.

Conversation and Sentiment Drivers. Sentiment is how users feel in relation to the brand. It is the task of the brand to be able to define which elements should be considered as positive or negative sentiments. On the other hand, a conversation occurs when the brand is mentioned in a post in which two or more users share their opinion [21].

Metric 01 - Sentiment through time. It consists in monitoring how the sentiment of users has changed over a certain period of time.
Metric 02 - Source of sentiment. It consists in measuring the sentiment for each social network of the company and identifying where they get a better perception.
Metric 03 - Number of followers. It is a basic indicator that helps learn if the audience has increased. However, the brand must know that a high number of followers does not ensure a high number of conversions [23].
Metric 04 - Interactions. It consists in adding the number of positive interactions by social network and evaluating how it has evolved over time.
Metric 05 - Mentions of the brand. It consists in knowing and analyzing what is said about the brand when it is mentioned by users on sites that are external to the page.

- **Marketing Optimization Metrics**

  The Social Media Framework suggests analyzing marketing campaigns in the following order: content, campaign, and channel; and analyzing moments of greatest impact of the campaign and which one (if there is more than one) was the most active.

For this analysis, conversions are taken into consideration; this is when the user performs an activity that has been previously defined by the company. This can range from an interaction, to a message, or to an activity such as a web purchase or data entry [24].

Content Performance.

> *Metric 1 - Conversions of contents. It consists in calculating the amount of conversions that the contents received as well as the sentiment that was generated.*
> *CTR: It helps identify if the image or video shown had the expected impact on users.*
> *Engagement: It helps identify which content generated the greatest impact on the target audience.*
> *Bounce Rate: It helps to identify contents that were not related to the messages and the addressed landing page.*

Campaign Performance.

> *Metric 1 - Campaign conversions. It consists in calculating the number of conversions generated on the web for each campaign.*

$$Conversion\ Rate = \frac{Objectives\ achieved}{Total\ visits} * 100$$

> *Metric 2 - Return on investment. It consists in calculating the profitability of the campaign.*

$$ROI = \frac{Profit - Investment}{Investment} * 100$$

Channel Performance.

> *Metric 1 - Conversions per channel. It consists in calculating the number of conversions per channel and finding the rate of each one to measure its performance.*

Organization. At this stage, the SME must organize its resources to be able to develop the digital strategy. There are various options to manage itself, considering that new mechanisms for the development of better strategies have been emerging as marketing has been evolving. The organization is divided into two components: the workforce, which involves all the SME staff that will be part of the strategy implementation, and all social media resources, which will be used to fulfill the objectives [23] (Tables 3 and 4).

Tools. The SME selects the tools that will help the creation of contents for campaigns, and the creation of the website and tools that will allow the obtaining of data for a continuous digital strategy analysis.

There is not a single tool that helps the analytical process, as there are multiple platforms that have facilitated the measurement of results after the implementation of a digital strategy [24]. All these tools help enable the analysis of data obtained from the strategy, as the data is not useful for the company without an analytical process [25] (Table 5).

Table 3. Workforce component - SME staff

Resources: workforce
Content planning staff
Content creation staff
Implementation staff
Web analytics staff

Table 4. Workforce component - social media resources

Resources: social media
Facebook page - Facebook business
Instagram
Website
Digital analytics tools

Table 5. Digital tools

Tools	
Magento	Website building
Google analytics	Monitoring marketing optimization metrics
Google tag manager	Capture events to understanding behavior of clients in the website
Facebook business	Publicity in Facebook and Instagram fan pages and monitoring of brand health metrics
Adobe photoshop	Edition and creation of visual content for the digital channels
UTM builder	Tracking of traffic from social networks to the website

4   Validation

The digital strategy was validated for a Peruvian SME in the retail business. After working with the company's Marketing and Design Team, it was possible to create an adequate strategy. The following results were obtained at the end of April 2018 (Table 6):

It was possible to increase the number of followers on social networks; the posts had a wider reach, and no brand mentions were identified (Tables 7, 8 and 9).

The Facebook campaign to get traffic to the web has not been compared with the previous month as it was the first time that the company had an active website. Furthermore, it was possible to receive sales through this channel, even though the objective was related to the web catalog, and to create a retail business opportunity.

Table 6. 2018 digital strategy implementation results

	Brand health			
Facebook	Metric	March	April	Variation
	Followers	10205	10594	3.8%
	Likes on fan page	10095	10446	3.5%
	Organic reach	6539	7534	15.2%
	Paid reach	19581	25104	28.2%
Instagram	Instagram followers	258	553	114.3%
	Average of likes per post	13.8	16.3	18.1%

Table 7. Marketing optimization and interaction

Marketing optimization			
Facebook campaign - objective: interaction			
Metric	March	April	Variation
Reach	19581	25104	28.2%
Interactions	3389	5026	48.3%
Investment	50 PEN	65 PEN	30.0%
CPI	0.01 PEN	0.01 PEN	-
Clicks	2132	5493	157.6%
Page likes	-	88	-

Table 8. Marketing optimization and traffic to web

Marketing optimization	
Facebook campaign - objective: traffic to web	
Metric	April
Reach	10892
Impressions	20492
Clicks on reach	614
CTR	3.70%
Investment	57 PEN
CPI	0.09 PEN

Table 9. Marketing optimization

Marketing optimization	
Metric	April
Visits	959
Conversions	17
Conversion rate	1.77%

The website results will be found below (Table 10):

Table 10. Website results

Results from website	
Metric	April
Visits	959
Conversions	17
Conversion rate	1.77%
Pages per session	15.48
Average time	4:06 min
Users	674

Based on this, it was possible to make an analysis based on the needs of the company and thus provide recommendations which ranged from the type of content and the manner it was presented to the target audience. Given that it is the first time that the company obtains tangible data, there will be upcoming opportunities where it will be able to compare results and analyze moments in which the public is willing to purchase more, the media or format in which it must invest more, and which products have greater preference.

5   Conclusions

Networks and social media significantly help SMEs to become known, obtain potential customers, communicate with users, and thus generate more income. However, since they do not have tangible results from their campaigns and they do not know how much these results influence the fulfillment of their objectives, many SMEs do not know how profitable the digital media can be.

For this reason, it is essential to design a digital strategy that has one or more approaches aligned to the business objectives and to have a preview of the business strategy in order to – through digital analytics tools – compare results and see how the performance of the strategy went. This is how the business will monitor the performance of digital media, which will influence the fulfillment of its objectives.

SMEs have the opportunity to use social media and networks in their digital campaigns so that users fulfill a particular objective; however, thanks to an additional work of digital analytics, the company will recognize the resources that contribute more to the fulfillment of its objectives, and which other needs must be fulfilled in order to obtain the most optimal results.

References

1. Marchan, I., Castro, A.: El Estado de Social Media en América Latina, 27 February 2017. https://www.comscore.com/lat/Prensa-y-Eventos/Presentaciones-y-libros-blancos/2017/El-Estado-de-Social-Media-en-America-Latina. Accessed 1 June 2018
2. Vega, F., Castro, A.: El Estado de Social Media en América Latina, 13 April 2018. https://www.comscore.com/lat/Insights/Presentations-and-Whitepapers/2018/Estado-de-Social-Media-en-America-Latina-2018. Accessed 1 June 2018
3. Kaplan, A.M., Haenlein, M.: Users of the world, unite! The challenges and opportunities of social media. Bus. Horiz. **53**(1), 59–68 (2017)
4. Li, D., Li, Y., Ji, W.: Gender identification via reposting behaviors in social media. IEEE Access **6**, 2879–2888 (2017)
5. Yejas, D.A.A.: Estrategias de marketing digital en la promoción de Marca Ciudad. EAN Mag. (80), 59–72 (2016)
6. Morales, M.D.O., Aguilar, L.J., Marín, L.M.G.: Los desafíos del Marketing en la era del big data. e-Ciencias de la Información **6**(1), 1–31 (2016)
7. Etlinger, S., Li, C.: A Framework for Social Analytics. Altimeter Group, pp. 1–40, 10 August 2011
8. Fernández-Villavicencio, N.G., Novoa, J.L.M., García, C.S., Fernández, M.E.S.M.: Revisión y propuesta de indicadores (KPI) de la Biblioteca en los medios sociales. Res. Stud. **36**(1), 1–14 (2014)
9. Prieto, N.J.: La creciente importancia de las redes sociales en la estrategia de marketing de la empresa. El caso GAM. Universidad de Valladolid, pp. 50–110, 12 July 2014
10. Marjani, M., Nasaruddin, F., Gani, A., Karim, A., Hashem, I.A.T., Siddiqa, A., Yaqoob, I.: Big IoT data analytics: architecture, opportunities, and open research challenges. IEEE Access **5**, 5247–5261 (2017)
11. Fernández-Villavicencio, N.G.: Bibliotecas, medios y métricas de la web social. Anales de Documentación **19**(1), 1–13 (2016)
12. Serrano-Puche, J.: Herramientas web para la medición de la influencia digital: Análisis de Klout y Peerindex. El profesional de la información **21**(3), 298–303 (2014)
13. Ananda, A.S., Hernández-García, Á., Lamberti, L.: RENL: a framework for social media marketing strategy. Universidad Politécnica de Madrid, pp. 1–12, 15 June 2014
14. Montero Pitre, L.K.: Marketing Digital Como Mecanismo Para Optimizar Las Ventas En pymes. Universidad Militar Nueva Granada, pp. 1–20, 24 November 2015
15. Peters, K., Chen, Y., Kaplan, A.M., Ognibeni, B., Pauwels, K.: Social media metrics - a framework and guidelines for managing social media. J. Interact. Mark. **27**(4), 281–298 (2014)
16. Taladriz-Mas, M.: Los servicios de información y el retorno de la inversión: cómo llegar a conocerlo. El profesional de la información **22**(4), 281–285 (2014)
17. Gascón, J.F.F.I., et al.: Reputational factor and social media: comparative metric proposals. Red de Revistas Científicas de América Latina, el Caribe, España y Portugal **32**(7), 615–629 (2016)
18. Ko, N., Jeong, B., Choi, S., Yoon, J.: Identifying product opportunities using social media mining: application of topic modeling and chance discovery theory. IEEE Access **6**, 1680–1693 (2017)
19. Medranda, A.V.: Influencia del marketing digital en el proceso de decisión de compra. Universidad Estatal Península de Santa Elena, pp. 1–5, 5 May 2015

20. Chávez Yépez, H.F.: Social media marketing como herramienta estratégica en el posicionamiento de marca en la rama de la confección. Universidad Técnica de Ambato, pp. 25–30, 28 October 2017
21. Lai, E.L.: Topic time series analysis of microblogs. IMA J. Appl. Math. **81**, 409–431 (2016)
22. Castello-Martinez, A.: El estudio del retorno de la inversión y el impacto en la relación de la comunicación empresarial y publicitaria en plataformas sociales: herramientas disponibles en el mercado. In: Proceedings of the 2nd National Congress on Methodology for Communication Research, vol. 2, pp. 411–428, 14 November 2014
23. Anandhan, A.: Social media recommender systems: review and open research issues. IEEE Access **6**, 15608–15628 (2018)
24. González Fernández Villavicencio, N.: Qué entendemos por usuario como centro del servicio. Estrategia y táctica en marketing. El profesional de la información **24**(1), 5–13 (2015)
25. Marjani, M.: Big IoT data analytics: architecture, opportunities, and open research challenges. IEEE Access **5**, 5247–5261 (2017)

Multi-split HDFS Technique for Improving Data Confidentiality in Big Data Replication

Mostafa R. Kaseb[1(✉)], Mohamed H. Khafagy[1], Ihab A. Ali[2,3],
and ElSayed M. Saad[2]

[1] Faculty of Computers and Information, Fayoum University, Fayoum, Egypt
{mrk00,mhk00}@fayoum.edu.eg
[2] Faculty of Engineering, Helwan University, Helwan, Egypt
EHAB_ALI02@h-eng.helwan.edu.eg, ihabali@qec.edu.sa,
elsayed012@gmail.com
[3] Qassim University, Burayda, Saudi Arabia

Abstract. In this paper, the Secure Distributed Redundant Independent Files (SDRIF) approach addresses some issues found with the Redundant Independent Files (RIF) approach and it mainly introduced data confidentiality that the RIF approach lacks to offer. It works similar to RIF, but the generated parity is not stored in one separate file. The generated parity blocks are distributed among all four data parts.

The CPSDRIF, which is the model produced when combining SDRIF with cloud providers (CP), introduces data security through the multi-split HDFS technique by distributing the parity blocks and reducing the size of the SDRIF block to the HDFS block.

According to the experimental results to the CPSDRIF system using the TeraGen benchmark, it is found that the data confidentiality using CPSDRIF have been improved as compared to CPRIF. Also, the storage space is reduced by 33.3% with CPSDRIF system compared to other models and improved the data writing by 32% and reading by about 31%.

Keywords: Big data · Cloud storage providers · HDFS ·
Redundant Independent Files (RIF) · Data confidentiality · Data security

1 Introduction

Cloud computing is a model for enabling convenient, on-demand network access to a shared pool of configurable computing resources that can be rapidly provisioned and released with minimal management effort or cloud provider interaction [1].

Big Data [2] and security [3] are a major challenge for the performance of the cloud storage systems. Hadoop Distributed File Systems (HDFS) [4], Google File System (GFS) [5] and others are widely utilized to store big data [6]. The HDFS replicates and saves data as multiple copies to achieve availability and reliability, but HDFS increases storage and resources consumption. Users of these cloud storage systems usually store sensitive and important information. Therefore the trustworthiness of the providers is so important. Among the various security threats, three main issues should be given special attention. These are data Confidentiality, Integrity and Availability [3, 7].

© Springer Nature Switzerland AG 2019
Á. Rocha et al. (Eds.): WorldCIST'19 2019, AISC 930, pp. 132–142, 2019.
https://doi.org/10.1007/978-3-030-16181-1_13

HDFS is the most famous because it is an open source and it is designed to store big data sets and to stream them at a high bandwidth suitable for the client applications. In storage, HDFS separates file system into file data and metadata. NameNode stores the metadata of the file system. The file data is stored in the DataNodes [4].

A comparative survey of the most common HDFS replication models was provided by [8]; Pipeline [4, 8] designs to replicate each data block with multiple copies to provide double fault tolerance. But, pipeline increases storage, resources consumption and take a lot of time to replicate. The work in [9] presented a technique for efficient replica placement on HDFS that can improve data transfer rate and throughput by replicating data blocks in parallel. The main drawback of the parallel replication is the one failure problem. Reconfigured Lazy [10] reduces the execution time and increases the writing throughput. The Enhancement Lazy replication [10] improves availability, reduces the execution time and increases the writing throughput by introducing an extra DataNode. But, it results in a significant increase in storage space and resource consumption in large-scale storage systems. The redundancy overhead becomes a scalability bottleneck, significantly increases the cost [11]. In the data replication, the original data is set to W bytes, so the total data storage is 3*W bytes in replication models if double fault tolerance. The time of writing a data block is assumed to be T_W and reading a data block is assumed to be T_R. Therefore, the total time spent for these replication model is $T_W+T_R+\delta t$ where δt depends on the type of replication models [12].

In [13], RIF approach reduced storage space, resources consumption, operational costs. Also, it improved the reading and writing performance in big data replication compared to CPHDFS [13] with two replicas. But, it suffers from limited reliability, availability, increased data recovery time that they are solved by High Availability Redundant Independent Files (HARIF) approach [14].

2   The Proposed Model

The proposed system model is a secure decentralized hybrid model. That model is called CPSDRIF (Fig. 1). It is produced when combining the SDRIF with CP. The main objective of the proposed CPSDRIF model is to overcome the problems of data confidentiality in addition to data integrity and data availability. Also, it reduces storage, resources consumption and operational costs in the cloud storage providers. The CP provide HDFS without replica, and the SDRIF acts as a service layer. The SDRIF is independent of HDFS because it is built above CP. The SDRIF interacts with HDFS without changing HDFS properties.

2.1   The Proposed System Architecture Composed Five Components

1. **The Data Uploader is the First Component.** The client is responsible for uploading his big data to suggested service (SDRIF) before writing it to CP in parallel.

2. **The Multi-split SDRIF is the Second Component,** which is responsible for splitting data before it is stored to CP in parallel.

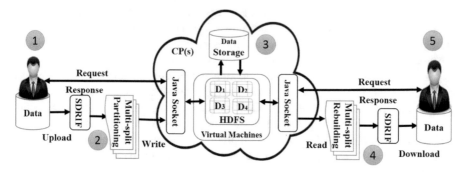

Fig. 1. Proposed system architecture (CPSDRIF).

The SDRIF approach is a technique for reducing storage, resources consumption and increasing data security by offering data confidentiality, integrity, and availability of big data replication. The SDRIF is similar to the RIF, but differs in the way data are distributed. The SDRIF is responsible for distributed storage by dividing the original data (D) (set on W bytes) into three parts (set to D1, D2, D3). The generated parity part (Dp) is created by XORing the three parts (D1: D3). The Dp is distributed on all four data parts (D1, D2, D3, D4) as shown in Fig. 2. Each of the four parts is 0.33 W bytes and the four parts are stored into four HDFS files as separate files on CP.

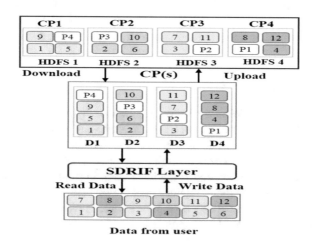

Fig. 2. Multi-split SDRIF Layer and Write/Read data to/from CP.

Simply, the first block of data is stored in the first file, the second block is stored in the second file, the third block is stored in the third file and the parity block (P1) of the first three blocks is stored in the fourth file. The fourth block is stored in the fourth file, the fifth block is stored in the first file, the sixth block is stored in the second file and the parity block (P2) of the second three blocks is stored in the third file. The seventh block is stored in the third file, the eighth block is stored in the fourth file, the ninth block is stored in the first file and the parity block (P3) of the third three blocks is stored in the second file and so on. Therefore, each of HDFS files contains the collection of blocks with load balance [15] between four files (Fig. 2). Where D1 consists of blocks (1, 5, 9, P4), D2 consists of blocks (2, 6, P3, 10), D3 consists of blocks (3, P2, 7, 11) and D4 consists of blocks (P1, 4, 8, 12). Which: $P1 = 1 \oplus 2 \oplus 3$, $P2 = 4 \oplus 5 \oplus 6$, $P3 = 7 \oplus 8 \oplus 9$, $P4 = 10 \oplus 11 \oplus 12$ and so on. Figure 2 illustrates that the size of the HDFS block (Bc) is equal to the size of the SDRIF block after splitting the data (Bs = 128 MB).

(A) The SDRIF (Multi-split the HDFS Block (Bc = 128 MB))

SDRIF introduces data confidentiality in big data replication by multi-split the HDFS block (Bc = 128 MB) in the CP by reducing the size of the SDRIF block (Bs) to the HDFS block and distributing parity blocks. The size of the data block (Bs) is split into 1 MB or 2 MB up to 128 MB. This means that Bs multiple splits are based on the information that is important to the client according to Eq (1).

$$B_S = 2^i \, MB \qquad\qquad 0 \le i \le 7 \qquad\qquad (1)$$

The writing process preserves some data security by distributing data and splitting data into small block size (Bs). When the block size (Bs) after splitting by SDRIF is 1 MB and is stored in CPSDRIF; each 1 MB of data will be collected from a different location of the original data. The data in these blocks (Bc) is not concatenated, 96 MB plain text from 96 location of the original data and 32 MB encrypted by the XOR as shown Fig. 3. When the block size (B_S) is increased to 2 MB, each 2 MB of data will be collected from a different location of the original data. The data in these blocks (BC) is 96 MB plain text from 48 location of the original data and 32 MB encrypted by the XOR process and so on. Due to, when the block size (B_S) increases the data security decreases. In CPDRIF, the client can increase or decrease (B_S) depending on his needs.

Fig. 3. The B_C (128 MB) on CPSDRIF and the B_S (1 MB) from SDRIF.

3. **The CP(s) is the Third Component,** Each of a cloud provider is composed of Java socket, Virtual Machine (VM), Hadoop distributed file system and data storage.

4. **The SDRIF Downloader is the Fourth Component,** which is responsible for reconstructing files to original data from CPSDRIF in parallel as separate files.

5. **The Data Downloader is the Fifth Component,** The client is responsible for reading data from CPSDRIF in parallel as independent files.

2.2 Storage Space on CPSDRIF Model

Suggested Service SDRIF is implemented on the original data. Therefore, the data storage space in each of the four files in the CPSDRIF model is one-third of the original data, which is 0.33 W bytes. Thus, the total data storage is only 1.33 times of the original data which is the same as the CPRIF model.

2.3 Multi-split Data Writing in CPSDRIF Model: (WCPSDRIF)

When the client writes data into the secure decentralized hybrid model (CPSDRIF) needs to one-third the amount of the original data to each file from four files as independent files. So, the writing time is 0.33Tw+ts into HDFS file on CPSDRIF and it is similar to CPRIF. ts is the time of splitting data and running the XOR process. Figure 4 illustrates the flowchart of multi-split WCPSDRIF algorithm to write data.

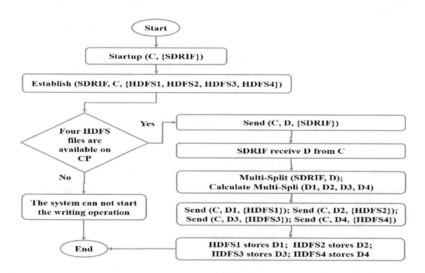

Fig. 4. Flowchart of multi-split WCPSDRIF algorithm.

2.4 Reconstruct Data Reading from the CPSDRIF Model: RCPSDRIF)

When the client reconstructs data from a CPSDRIF model, the time to read four HDFS files from CPSDRIF is 0.33Tr + tcom and it is similar to CPRIF. Because there are four separate files and read them in parallel, tcom is the time to combine the four files.

Figure 5 illustrates the flowchart of multi-split reconstructing RCPSDRIF algorithm to read data from CPSDRIF.

3   Experimental Results and Discussion

The CPSDRIF model proposed in this paper will evaluate the performance of storage, resources consumption, writing and reading operation regarding the performance of CPRIF model [13], CP (Pipeline) [4], CP (parallel) [9] and CP (Reconfigured-Lazy) [10] models. That called CPHDFS, because it is based on multiple copies replication techniques for HDFS. The suggested models were implemented on servers (Dell T320) running multiple virtual nodes and with the TeraGen [16] benchmark tool, different performance aspects were tested.

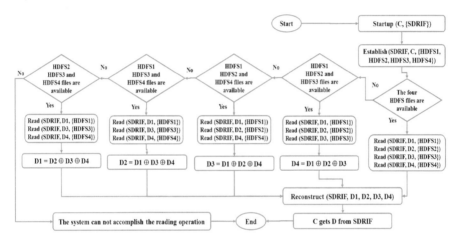

Fig. 5.   Flowchart of multi-split reconstructing RCPSDRIF algorithm.

3.1   Comparison Between the SDRIF and RIF with Respect to the Bc Block Format

From Fig. 6, the block (Bc) on CPRIF contains sequential 128 MB data from original data. Because of the block size (B_S = 128 MB) is equal to the block size (B_C = 128 MB). Therefore, CPRIF has not provided data confidentiality for the user. On the other hand, the CPSDRIF has introduced data confidentiality for the user. Because of the block (Bc) on CPSDRIF does not contain sequential data from original data and is injected with 32 MB encrypted data.

138   M. R. Kaseb et al.

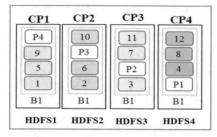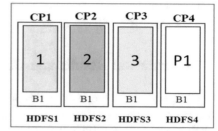

Bc block On CP(s) after SDRIF (Bs=32)   Bc block On CP(s) after RIF (Bs=128)

Fig. 6. Comparison between DRIF and RIF to the Bc (128 MB) block format on CP(s).

If block size (Bs) is equal 1 MB (Fig. 3), every block size (Bc) is injected by 32 MB is encrypted on CPSDRIF and each 1 MB from 32 MB has distributed on Bc with 96 MB is not sequential from original data. Therefore, when a hacker reads block from CPSDRIF, he doesn't know where the original data and encrypted data. So, he doesn't get any information from data. Thus, the CPSDRIF offers confidential data for users.

3.2 Comparison Among the CPSDRIF Versus the CPSDRIF and Traditional CPHDFS Replication Models

We will calculate the probability that hackers will get information if the client uses HDFS replication, RIF approach and SDRIF with variable block size (Bs).

The data is stored in Bc (128 MB) sequentially

For CPHDFS Model: If the hackers read all the blocks (Bc), the information is likely to be available to them. Because the data is stored in Bc (128 MB) sequentially. So, the percentage to get the information is 100% after reading all the blocks.

For CPRIF Model: if the hackers read all the blocks (Bc), the percentage to get the information is 75% (Fig. 7). Because of every four blocks (Bs = 128 MB), there are three blocks that have sequential data and the fourth block has encrypted data.

For CPSDRIF Model: If the hackers read all the blocks (Bc), the percentage to get the information is very small. Because of every block (Bc) of CPSDRIF includes a part of encrypted data and other parts are not sequential. Also, the sequential data depends on the size of the SDRIF block. So, the CPSDRIF provides more data confidentiality.

From Fig. 7. It is clear that encrypted data are about 0% on CPHDFS. The encrypted data are improved about 25% on CPRIF. The encrypted data are improved about 99.5% on CPSDRIF. In CHSDRIF; the encrypted data are reduced to about 25% with increased Bs to 128 MB.

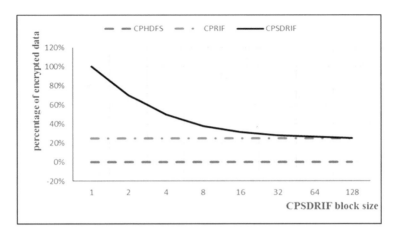

Fig. 7. Comparison of CPSDRIF, CPRIF and CPHDFS to provide encrypted data

3.3 Resources Consumption

Reducing the number of replicas the system performance will increase. Because the number of available machines (DataNodes) can be used for other operation. Assuming that the CP has 30 machines. From Fig. 8. It is clear that the number of machines in CPSDRIF is less than CPHDFS with the same data blocks for the writing operation. For example, when CPHDFS writes 12 blocks in HDFS, their 12 DataNodes are busy to write data blocks and also there are 12 busy DataNodes for writing replication blocks at the same time and only 6 DataNodes are free. On the other hand, in the CPSDRIF, CPSDRIF replication model will generate only 4 extra blocks for replication. So, there are 16 busy DataNodes and 14 free machines which in turn increase system performance and decrease power consummation. Also, the number of free and busy DataNodes in CPSDRIF is the same number compared to CPRIF.

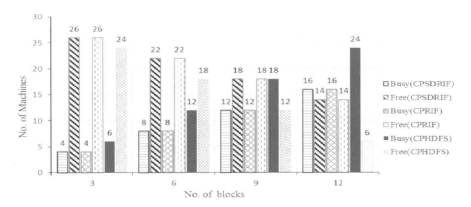

Fig. 8. Comparing a number of busy and free DataNodes in CPSDRIF, CPRIF and CPHDFS.

3.4   Data Storage

It is clear from Fig. 9. That the proposed model CPSDRIF consumes the data storage space by 66% compared to CPHDFS model. Also, it is the same size compared to CPRIF model. This is because CPHDFS model uses multiple copies technology (R = 2) while CPSDRIF and CPRIF use the XORing process for replication of the original data.

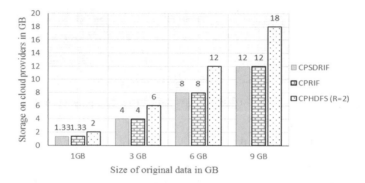

Fig. 9. Comparison of CPSDRIF, CPRIF and CPHDFS at storage.

3.5   The Writing Performance

From Fig. 10. It is clear that CPSDRIF improves the average of execution time for writing data by 32% compared to CPHDFS models. Also, it is approximately equal to the average execution time compared to CPRIF model. Using two connection channels only. Using more than two connection channel the performance should increase.

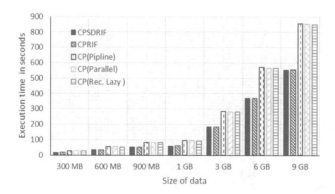

Fig. 10. Comparison among CPSDRIF, CPRIF and CPHDFS at an average time of writing data.

3.6 The Reading Performance

From Fig. 11, it is clear the average of the execution time for reading is improved by 31% for CPSDRIF model compared to other models. But, it is increased by 3.5% compared to CPRIF model. Using two connection channels only.

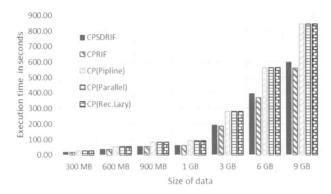

Fig. 11. Comparison among CPSDRIF, CPRIF and CPHDFS at an average time of reading data.

4 Conclusion and Future Work

This paper provided an improved model for offering data confidentiality in big data replication. That called CPSDRIF. Also, it reduced storage space, operational costs, resources consumption, improved data security, the reading and the writing performance compared to other CPHDFS models. The SDRIF overcame the lacks of the RIF approach by introducing data confidentiality through multi-split HDFS block by distributing the parity blocks and reducing the size of the SDRIF block to the HDFS block.

Experimental results show that the CPSDRIF model saved the storage space by 33% compared to other CPHDFS replication models, improved the cost time for writing data by 32% and improved the cost time for reading data by 31%. But, it is increased by 3.5% compared to CPRIF. In the future, we are going to increase the number of replication that will achieve data confidentiality, integrity, availability and reliability.

Acknowledgments. This work is funded by Academy of Scientific Research & Technology, Egypt through a contract with Fayoum University, Egypt.

References

1. Mell, P., Grance, T.: The NIST definition of cloud computing (2011)
2. Dittrich, J., Quiané-Ruiz, J.-A.: Efficient big data processing in Hadoop MapReduce. Proc. VLDB Endow. **5**(12), 2014–2015 (2012)

3. Maddineni, V.S.K., Ragi, S.: Security techniques for protecting data in cloud computing (2012)

4. Shvachko, K., Kuang, H., Radia, S., Chansler, R.: The hadoop distributed file system. In: 2010 IEEE 26th Symposium on Mass Storage Systems and Technologies (MSST), pp. 1–10 (2010)

5. Ghemawat, S., Gobioff, H., Leung, S.-T.: The Google file system, vol. 37, no. 5. ACM (2003)

6. Sahal, R., Khafagy, M.H., Omara, F.A.: Comparative study of multi-query optimization techniques using shared predicate-based for big data. Int. J. Grid Distrib. Comput. **9**, 229–240 (2016)

7. Mahmoud, H., Hegazy, A., Khafagy, M.H.: An approach for big data security based on Hadoop distributed file system. In: 2018 International Conference on Innovative Trends in Computer Engineering (ITCE), pp. 109–114 (2018)

8. Abead, E.S., Khafagy, M.H., Omara, F.A.: A comparative study of HDFS replication approaches. Int. J. IT Eng. **3**(8), 5–11 (2015)

9. Patel, N.M., Patel, N.M., Hasan, M.I., Patel, M.M.: Improving data transfer rate and throughput of HDFS using efficient replica placement. Int. J. Comput. Appl. **86**(2) (2014)

10. Abead, E.S., Khafagy, M.H., Omara, F.A.: An efficient replication technique for hadoop distributed file system. Int. J. Sci. Eng. Res. **7**(1), 254–261 (2016)

11. Wu, S., Zhu, W., Mao, B., Li, K.-C.: PP: popularity-based proactive data recovery for HDFS RAID systems. Futur. Gener. Comput. Syst. **86**, 1146–1153 (2018)

12. Li, J., Zhang, P., Li, Y., Chen, W., Liu, Y., Wang, L.: A data-check based distributed storage model for storing hot temporary data. Futur. Gener. Comput. Syst. **73**, 13–21 (2017)

13. Kaseb, M.R., Khafagy, M.H., Ali, I.A., Saad, E.M.: Redundant Independent Files (RIF): a technique for reducing storage and resources in big data replication. In: WorldCIST 2018 Advances in Intelligent Systems and Computing, vol. 745, pp. 182–193. Springer, Cham (2018)

14. Kaseb, M.R., Khafagy, M.H., Ali, I.A., Saad, E.M.: An improved technique for increasing availability in big data replication. Futur. Gener. Comput. Syst. **91**, 493–505 (2019)

15. Sarhan, E., Ghalwash, A., Khafagy, M.: Queue weighting load-balancing technique for database replication in dynamic content web sites. In: Proceedings of the 9th WSEAS International Conference on Applied Computer Science, pp. 50–55 (2009)

16. TeraGen benchmark. https://hadoop.apache.org/docs/r1.0.4/api/org/apache/hadoop/examples/terasort/TeraGen.html. Accessed 10 Aug 2018

Social Media, Evolutionary Psychology, and ISIS: A Literature Review and Future Research Directions

Sylvie Borau[(⊠)] and Samuel Fosso Wamba

Toulouse Business School, 1 Place Alphonse Jourdain, 31068 Toulouse, France
{s.borau, s.fosso-wamba}@tbs-education.fr

Abstract. The paper aims at conducting a robust literature review of the emerging literature on social media use by Islamic State (ISIS) to promote their ideology and recruit future members and supporters. Then, the study uses evolutionary psychology as a theoretical framework to analyze these papers in order to underscore the mechanisms underlying young occidental males' willingness to become terrorists. Our key findings, implications for research and practice are discussed.

Keywords: Social media · Terrorism · ISIS · Evolutionary psychology

1 Introduction

The use of social media platforms such as Facebook, Twitter, Youtube, or Snapchat is growing exponentially worldwide. For example, on average 350,000 tweets are sent per minute which corresponds to over 500 million tweets per day [1], whilst Facebook remains the largest social media network with 2.23 billion monthly active users [2]. These social media platforms were primarily developed to facilitate pictures, information sharing and ideas among virtual communities and networks. However, they are now among key preferred communications tools used by terrorists' organizations including the Islamic State (ISIS) to promote their ideology and recruit future members and supporters. For example, in 2018, 60% of internet-related counter-terrorism prosecutions were related to Islamist content material. These Islamist related contents were spread on a variety of platforms: 33% on encryption (coded messages), 17% on Youtube, 17% on Google, 17% on Twitter, 8% on Instagram, and 8% on Facebook [3], and thus allowing ISIS to disseminate terrorist material, to communicate between members, and, most importantly, to recruit new members, particularly among the young generation.

Terrorism and the recruitment of new terrorist members have been extensively studied in psychology, using different theoretical frameworks such as social identity and social power [4] or attachment theory [5]. Evolutionary psychology inspired a complementary approach to the understanding of terrorism [6, 7]. However, very few studies have used evolutionary psychology as a theoretical framework to underscore the mechanisms underlying young occidental males' willingness to become terrorists, and the role of social media in this process. Therefore, this study intends to fill the

© Springer Nature Switzerland AG 2019
Á. Rocha et al. (Eds.): WorldCIST'19 2019, AISC 930, pp. 143–154, 2019.
https://doi.org/10.1007/978-3-030-16181-1_14

existing knowledge gap identified in the literature. More specifically, this study aims at conducting a robust literature review of the emerging literature on social media use by ISIS. Then, the study uses evolutionary psychology theories to analyze these papers in order to underscore the mechanisms underlying young occidental males' willingness to become terrorists.

The rest of the paper is organized as follows: after the introduction, the theoretical issues are presented, followed by a section on our research methodology, and another section describing our findings and the discussion. The paper ends with a conclusion (Table 1).

Table 1. Social media technologies adapted from [8, 9].

Platform types	Examples
Social networking	
Friends, groups, events	Facebook
Followers	Twitter
Circles	Google+
Connections	Linkedin
Friends, crush lists, communities	Orkut, VK "VKontakte"
Media platform	
Video	Youtube, SocialCam, FacebookLive
Photo	Flickr
Mobile Photos	Instagram, Justpaste.it
Location-based platform	
Check-in	Foursquare
Location of circles	Google lattitude
Location of friends	Find myfriends
Local reviews and check-ins	Yelp
Check-ins, local deals	GroupOn
Check-ins, friends locations	Facebook places
Crowdsourcing platforms	
Translation	Amara
Labor	Amazon's M-Turk
Geo-location labor, verification	Crisismappers
Labor	Crowdflower, people per hour
Combination platform	
Social networking, media and crowdsourcing	Reddit, Pinterest
Social networking, media, location-based and crowdsourcing	Facebook
Social networking, media and location-based	Pair, Meebo
Device types	
Desktop computers, laptops, gaming devices, smart televisions, tablets, smartphones, "dumb phones", vehicles and augmented reality devices	

2   Theoretical Issues

2.1   Evolutionary Psychology and Men's Competition

According to evolutionary psychology, individuals' behavior is driven by two main motivational forces: survival, and reproduction. Sexual selection is the evolutionary strategy that favors the reproductive success of an organism, and this strategy that favors reproduction can sometimes be stronger than those that favor survival [10, 11]. Kin selection is the evolutionary strategy that favors the reproductive success of an organism's relatives, and this strategy can sometimes be stronger than those that favor the organism's own survival and reproduction [12].

Based on the sexual selection theory, because males commit fewer resources to parental investment and exhibit greater reproductive variance than females [13, 14], they engage in more intrasexual competition for mating access to females. As a result of these asymmetrical parental investment and reproductive variance, the following patterns of intrasexual competition have been regularly observed across cultures [15–17]: men compete more than women for social status and material resources; and men are more likely than women to engage in physical contests in the course of this competition. This highest contest competition among men translates into greater body size and strength than females, higher rates of aggression, higher use of weapons and threat displays [18, 19]. Male higher competitiveness and violence can potentially lead to higher status and greater resources, which in turn result in greater reproductive success. Many researches have shown that men's dominance and social status translate into higher reproductive success in both traditional and industrial societies [15, 39]. Men who succeed in acquiring - sometimes violently - status and dominance are preferred by women (as long-term mates), because these men have access to resources that can contribute to offspring survival [20]. While acting violently is a way to compete and get access to these resources, showing devotion and prosociality can also help to improve individuals one's status among the community. Religion and religious rituals have been shown to act as costly signals of devotion, prosociality and even physical strength [21, 22], and these traits are generally preferred by women when selecting a mate. Based on the kin selection theory, some scholars argue that religion fosters strong relationships within the group that can be as strong as, or even stronger than family relationships, sometimes even substituting the family and triggering the same mechanism as kinship altruism [23].

To sum up, high status - obtained through violence and/or devotion - has historically translated into access to resources (e.g. food, territory, social support), and men with resources are more likely to be chosen by women as a mate because they are more able to contribute to offspring survival. High status can also be obtained through self-sacrificial altruism, this time to contribute to relative's survival.

2.2   Evolutionary Psychology and Terrorism

Terrorism is an example of an extremely costly behavior that can lower the chances of survival, but that can potentially increase the chances of reproduction, either for the individual or for their relatives.

When men fail to obtain dominance and high status as well as to attract mates, they might become more violent, as violence is a strategy often used by people who do not succeed in obtaining resources and mates in a peaceful way [24]. More specifically, when men's economic perspectives are uncertain (because they belong to a low-income group and/or because they have not achieved a high educational level), they seem to engage in more intrasexual competition. That is, these men may engage in more risky and violent behaviors, in order to increase their resources and their reproductive success [7]. For example, marginal men with low status and resources are more likely to perform rape [25]. Also, in societies with a low ratio of males compared to females, the occurrences of war tend to be higher [26]. The same is true in traditional polygamous societies that leave some men with no access to sexual relationships and marriages. In case of war within such traditional societies, some men even engage in the capture of wives from other villages [27, 28].

Moreover, studies demonstrated that, across cultures and societies, most murderers and murder victims are men, especially young men [29–33], and younger males are more likely to perform suicide bombings than older ones. Extreme behaviors such as suicide terrorism, which can appear maladaptive as it annihilates one's reproductive success [7], can be - at least partially - explained by sexual selection: some young men who lack decent status can go as far as choosing to become suicidal terrorists because they believe that, as a martyr, they will become heroes and get access to 72 virgins once in heaven.

Kin selection can also help understand such extreme behaviors as families of suicide terrorists usually benefit from an increased status and are awarded large sums of money [6, 34]. Kin selection also favors strong relationships between members of the same religion and can increase the degree of violence towards other groups. Because forming cohesive male coalitions relies on a distinction between ingroup and outgroup [35], perceptions of outgroup tend to facilitate intergroup aggression [7]. This desire to build strong male coalitions is more acute among young men who are in a period of transition and who are trying to find a meaning for life and a sense of belonging [36, 37]. Religion provides these men a sense of collective and distinctive identity that can increase violence towards other groups.

2.3   Evolutionary Psychology, Terrorism and Communication on Social Media

ISIS communication strategy to recruit young occidental men on social media seems to strongly rely on incentives related to violence and competition for status, resources, mates and a group identity that these men are potentially lacking. Male fighters joining the Islamic state are given a job and a house, and they are promised a wife —sometimes more than one. The group even ranks the women and rewards the most valuable ones— usually foreign women—to the best recruits [38]. Some male fighters can also have access to Yazidi and Shia sex slaves [38]. The Islamic states have also instituted a payment system for every child born. Finally, even after their death, martyrs become heroes and obtain intensive media coverage. ISIS also strongly activates in-group identity—by creating an environment of trust as well as "secrecy and exclusion of other friends and family members" [38] while triggering out-group competition through conspiracy theories against western countries (see Table 2).

Table 2. ISIS recruitment techniques through the evolutionary psychology lens

Dimensions		ISIS recruitment techniques
Competition	Process of gaining or winning something by defeating or establishing superiority over others: men tend to be more competitive than women to acquire resources [15]	Activation of competition among men: In-group competition (competition among fighters: ranks, army etc.) & out-group competition (competition with the "enemies")
Violence/physical con-test	Behavior involving physical force intended to hurt, damage, or kill someone or something: men have a greater likelihood of engaging in physical contests in the course of the resource competition [39]	ISIS encourages and rewards violence
Status	Relative social & professional position, prestige, recognition: men tend to compete more than women to achieve a high status [16, 17]	Fighters become heroes. They become martyrs if they die. In Western countries, they also get intensive media coverage
Material resources	A supply from which benefit is produced (money, housing, cars…). Men tend to compete more than women to obtain material resources [16, 17]	Fighters get a job and a house. ISIS has also instituted a payment system for every child you have in the Islamic State
Mating	Pairing of opposite-sex usually for the purposes of sexual reproduction: men give greater importance to the physical attractiveness and youth of prospective mates and are more likely to perform polyamory [17]	Promise of a wife — and sometimes more than one. Some also have access to sex slaves [38]. They are also promised 72 virgins as a reward for jihadists/martyrs
Signaling morality through religion	Signaling is the idea that one party credibly conveys some information about itself to another party.	Joining ISIS = signaling morality/religion. ISIS communicates that fighting against Bashar El Assad, Russia and the West is a humanitarian act to save the Syrian population
Kinship Ingroup identity (Bonding, friendship, love)	High amount of attention, common interests and concerns among a group [5]	ISIS starts by building trust with a single contact (regular emails, phone calls etc.). They build strong relationships within the group
(Fear and hate of the outside world)	Sense of threat: fear of the outside world. [5] Membership in a lower-status group triggers outgroup competition	ISIS creates "an environment of secrecy and exclusion of other friends and family members". Conspiracy theories against western countries. ISIS convinces men that they are humiliated (by Israel and the West)

3 Methodology

As a reminder, the main objective of this study is to assess the knowledge development level of social media use by ISIS in order to disseminate terrorist material, communicate between members, and, most importantly, recruit new members, using an evolutionary psychology perspective. Therefore, this study uses a comprehensive review, a research approach that has been already used to analyze big data [40], the internet of things and big data [41], as well as electronic commerce [42]. More precisely, the study develops a classification framework (Table 2), conducts a comprehensive review using the following keywords: "Social Media" or Facebook or Twitter or YouTube and Terrorism and ISIS, and eventually achieve a final validation of the proposed classification framework. The search was conducted within the world's largest abstract and citation database of peer-reviewed literature called SCOPUS on October 25, 2018. The output of the search query provided 18 relevant journal articles written in English [43–60].

4 Results and Discussion

In the section below, all key findings will be presented and discussed.

Table 3 displays the distribution of papers by year of publication, and clearly shows that the first journal articles stored in the SCOPUS database were published in 2015 (two articles, accounting for 11%). The following year, 5 articles (28%) were published on the topic, followed by a decrease in publications on the subject area (only 2 articles (11%)) in 2017, and then, by another increase to 9 relevant articles (50%) by October 2018. This high number of papers published on the topic in 2018 is probably linked to the worldwide threat posed by the ISIS organization.

Table 3. Distribution of papers per year of publication

Year of publication	Number of papers	Percentage
2018	9	50%
2017	2	11%
2016	5	28%
2015	2	11%
Total	18	100%

Table 4 shows the distribution of papers by subject area, and as expected, "Social Sciences" is the most significant subject area discussed, with 14 articles (43.8%). This is followed by "Computer Science" with 4 articles (12.5%), "Arts and Humanities", "Decision Sciences" and "Engineering" with 3 articles each (9.4%). The last group is made of "Biochemistry, Genetics and Molecular Biology", "Chemical Engineering", "Environmental Science", "Mathematics" and "Multidisciplinary" with only 1 article each (3.1%).

Table 4. Distribution of papers by subject area

Subject area	Number of papers	Percentage
Social Sciences	14	43.8%
Computer Science	4	12.5%
Arts and Humanities	3	9.4%
Decision Sciences	3	9.4%
Engineering	3	9.4%
Biochemistry, Genetics and Molecular Biology	1	3.1%
Chemical Engineering	1	3.1%
Environmental Science	1	3.1%
Mathematics	1	3.1%
Multidisciplinary	1	3.1%
Total*	32	100.0%

*Some articles are calculated more than once since they cover more than one subject area.

Prior studies have reported a good number media platforms used by ISIS for terrorism propaganda, including a new social media network called "Diaspora" [51], and well-known platforms such as Twitter, Facebook and Youtube (Table 5). Also, it has been reported that ISIS uses other communication channels such as video game, hip hop social media, testimonials, documentaries, magazine [51], and archive.org, Tumblr, Telegram, e-mail [51] to achieve its propaganda objectives. That notwithstanding, "YouTube, Facebook and Twitter are the most important instruments of the communication that are valued by the representatives of the ISIS today", according to a study by [50] (p. 636).

Table 5. social media technologies discussed in the papers (used by terrorists to prepare/report or by a third-party to report them).

Platform types	Examples
Social networking	
Friends, groups, events	Facebook: [43, 45, 48, 50, 53, 55, 58, 60]
Followers	Twitter: [43, 44, 48–51, 53, 54, 57, 58, 60]
Friends, crush lists, communities	VK "VKontakte": [55]
Media platform	
Video	Youtube: [43–45, 50, 51, 53, 58, 60]

Table 6 shows the results of a content analysis of the papers analyzed in this research, using evolutionary psychology as a theoretical framework. The results of this analysis show that Isis triggers evolutionary mechanisms to communicate and recruit young occidental males on social media. More specifically, ISIS online communication strategy mainly refers to male violence and competition for status, resources, and mates, while reassuring men about their morality through religion. ISIS online communication strategy also activates in-group identity and triggers out-group competition. Our proposed classification framework is validated.

Table 6. ISIS recruitment techniques presented in the papers

Dimensions	ISIS com	Articles
Competition	Activating male competition	[51, 52, 59, 60]: inter-confessional competition, between Sunnis and Shiites [46, 48]: humiliation of targeted communities [60]: destruction of the West [48]: sex of perpetrators: 90% males
Violence/physical con-test	Promoting male violence	[43, 44, 51–53, 59]: use of harsh and violent language and narrative [45, 46, 50, 54, 57, 58]: images of violent acts, atrocities, violence porn [59]: inter-confessional violence [51]: revenge seekers
Status	Granting of hero status	[43, 53]: glamourizing fighters [46]: life of glory/self-achievement/status [54]: strength/pride/confidence. Isis offers a distinct position, making the fighters heroes [59]: honor and glory [51]: status seekers/fantasy of power & fame [57]: new rock stars, high media coverage [58]: power & control/military success
Material resources	Giving access to resources	[43]: promise of prosperity [51]: economic control [54]: promoting people welfare, images of normal and abundant life, gaining financial resources
Mating	Promising a wife & giving access to sex slaves	[43]: wife market [51]: beards are sexy. The hippest group [54]: "the sexiest Jihadi group on the block" "Islam is punk rock" [54]: sexual conquests/encourages sexual slavery and rape [58]: marriage and rape
Signaling morality through religion	Signaling morality and a devoted life	[44, 46, 58]: religious duty [51]: shows charitable side [53]: moral crusaders [54]: divine righteousness [57]: serving the greater good [59]: high devotion

(*continued*)

Table 6. (*continued*)

Dimensions	ISIS com	Articles
Kinship/Ingroup identity -Bonding, friendship, love	Building trust and friendship within the group	[46]: friendship among peer networks [51]: helping the ingroup [53]: strong group identity based on self-esteem & self-belonging [54]: powerful unified group. Images of men having fun [58]: defender of Muslims against oppression [59]: ice-cream eating contest for children
Kinship/Ingroup identity -Fear and hate of the outside world	Spreading fear of the outside world & conspiracy theories	[52]: hate and fear against other confessions [51, 56]: fearful messages against the infidels, executions [56]: terror social media engagement [53, 58]: Muslim persecution abroad [53]: them vs. us war type mentality [58]: conspiracy theory [60]: Zionist crusader, against the West

5   Conclusion

Understanding why men can decide to become radicalized and violent by joining ISIS is crucial to develop efficient awareness campaigns among the most vulnerable individuals. This research uses evolutionary psychology as a theoretical framework to underscore the mechanisms used by Isis on social media to motivate men to engage in violent extremism. More precisely, we developed a classification framework, we conducted a comprehensive review using relevant keywords, and we validated our proposed classification framework. We showed that Isis uses evolutionary drivers (i.e. survival, reproduction, and kin selection) in their communication on social media to motivate men to join ISIS: they communicate about violence, competition, and morality to gain access to status, resources, and mates, while triggering in-group identity and fear of the out-group. Future research should investigate whether increased awareness of these drivers and communication techniques would change men's attitude towards Isis communication campaigns on social media. Also, it will be interesting to develop strategies and tools to ensure that social media tools are not used to promote terrorism. Likewise, further research needs to investigate the use of social media to detect hate speeches.

References

1. Twitter (2018). http://www.internetlivestats.com/twitter-statistics/
2. Facebook   (2018).   https://www.statista.com/statistics/264810/number-of-monthly-active-facebook-users-worldwide/

3. Malik, N.: The fight against terrorism online: here's the verdict (2018). https://www.forbes. com/sites/nikitamalik/2018/09/20/the-fight-against-terrorism-online-heres-the-verdict/ #4ff996904dc5. Accessed 11 Nov 2018
4. Wright, J.: A social identity and social power perspective on terrorism. J. Terror. Res. **6**(3) (2015)
5. Stein, A.: Terror, Love & Brainwashing (2017)
6. Liddle, J.R., Bush, L.S., Shackelford, T.K.: An introduction to evolutionary psychology and its application to suicide terrorism. Behav. Sci. Terror. Polit. Aggress. **3**(3), 176–197 (2011)
7. Liddle, J.R., Shackelford, T.K., Weekes-Shackelford, V.A.: Why can't we all just get along? Evolutionary perspectives on violence, homicide, and war. Rev. Gen. Psychol. **16**(1), 24 (2012)
8. Akter, S., Wamba, S.F.: Big data and disaster management: a systematic review and agenda for future research. Ann. Oper. Res. (2017)
9. Gupta, R., Brooks, H.: Using Social Media for Global Security. Wiley, Hoboken (2013)
10. Buss, D.M.: Human social motivation in evolutionary perspective: grounding terror management theory. Psychol. Inq. **8**(1), 22–26 (1997)
11. Clutton-Brocke, T.: Sexual selection in males and females. Science **318**, 1882–1885 (2007)
12. Hamilton, W.D.: The genetical evolution of social behaviour. II. J. Theor. Biol. **7**(1), 17–52 (1964)
13. Buss, D.M., Shackelford, T.K.: Human aggression in evolutionary psychological perspective. Clin. Psychol. Rev. **17**(6), 605–619 (1997)
14. Geary, D.C.: Evolution and proximate expression of human parental investment. Psychol. Bull. **126**, 55–77 (2000)
15. Puts, D.A., Bailey, D.H., Reno, P.L.: Contest competition in men. In: Buss, D.M. (ed.) The Handbook of Evolutionary Psychology. Wiley, Hoboken (2015)
16. Shackelford, T.K., Schmitt, D.P., Buss, D.M.: Universal dimensions of human mate preferences. Pers. Individ. Differ. **39**, 447–458 (2005)
17. Schmitt, D.P.: Fundamentals of human mating strategies. In: Buss, D.M. (ed.) The Handbook of Evolutionary Psychology. Wiley, Hoboken (2015)
18. Andresson, M.: Sexual Selection. Princeton University Press, Princeton (1994)
19. Hill, A.K., Bailey, D.H., Puts, D.A.: Gorillas in our midst? Human sexual dimorphism and contest competition in men. In: On Human Nature: Biology, Psychology, Ethics, Politics, and Religion. Academic Press, New York (2017)
20. Puts, D.: Human sexual selection. Curr. Opin. Psychol. **7**, 28–32 (2016)
21. Sosis, R., Alcorta, C.: Signaling, solidarity, and the sacred: the evolution of religious behavior. Evol. Anthropol.: Issues News Rev. **12**(6), 264–274 (2003)
22. Power, E.A.: Discerning devotion: testing the signaling theory of religion. Evol. Hum. Behav. **38**(1), 82–91 (2016)
23. Whitehouse, H., Lanman, J.A., Downey, G., Fredman, L.A., Swann Jr., W.B., Lende, D.H., McCauley, R.N., Shankland, D., Stausberg, M., Xygalatas, D., Whitehouse, H.: The ties that bind us: ritual, fusion, and identification. Curr. Anthropol. **55**(6), 674–695 (2014)
24. Wilson, M.: Conflict and homicide in evolutionary perspective. In: Bell, R., Bell, N. (eds.) Sociobiology and the Social Sciences, pp. 45–62. Texas Tech University Press, Lubbock (1989)
25. Thornhill, N.W., Thornhill, R.: An evolutionary analysis of psychological pain following rape. The effects of victim's age and marital status. Ethol. Sociobiol. **11**, 177–193 (1990)
26. Divale, W., Harris, M.: Population, warfare, and the male supremacist complex. Am. Anthropol. **78**, 521–538 (1976)
27. Ayres, B.: Bride theft and raiding for wifes in cross-cultural perspective. Anthropol. Q. **47**, 238–252 (1974)

28. Barnes, R.H.: Marriage par capture. J. R. Anthropol. Inst. **5**, 57–73 (1999)
29. Archer, J.: Sex differences in aggression in real-world settings: a meta-analytic review. Rev. Gen. Psychol. **8**(4), 291 (2004)
30. Archer, J.: Does sexual selection explain human sex differences in aggression? Behav. Brain Sci. **32**, 249–266 (2009)
31. Daly, M., Wilson, M.: Killing the competition. Hum. Nat. **1**(1), 81–107 (1990)
32. Lester, D.: Questions and Answers about Murder. Charles Press, Philadelphia (1991)
33. Walker, R.S., Bailey, D.H.: Body counts in lowland South American violence. Evol. Hum. Behav. **34**(1), 29–34 (2013)
34. Blackwell, A.D.: Terrorism, heroism, and altruism: kin selection and socioreligious cost–benefit scaling in Palestinian suicide attack. Poster Session Presented at the 17th Annual Human Behavior and Evolution Society Conference, Austin, TX (2005)
35. Durant, R.: Collective violence: an evolutionary perspective. Aggress. Violent Behav. **16**, 428–436 (2011)
36. Atran, S.: The devoted actor: unconditional commitment and intractable conflict across cultures. Curr. Anthropol. **57**(S13), S192–S203 (2016)
37. Atran, S., Sheikh, H., Gomez, A.: Devoted actors sacrifice for close comrades and sacred cause. Proc. Natl. Acad. Sci. **111**(50), 17702–17703 (2014)
38. Bloom   M.:   (2017).   http://www.huffingtonpost.com/mia-bloom/isis-marriage-trap_b_6773576.html
39. Glowacki, L., Wrangham, R.: Warfare and reproductive success in a tribal population. Proc. Natl. Acad. Sci. **112**(2), 348–353 (2015)
40. Wamba, S.F., Akter, S., Edwards, A., Chopin, G., Gnanzou, D.: How 'big data' can make big impact: findings from a systematic review and a longitudinal case study. Int. J. Prod. Econ. **165**, 234–246 (2015)
41. Riggins, F.J., Wamba, S.F: Research directions on the adoption, usage, and impact of the internet of things through the use of big data analytics. In: Book Research Directions on the Adoption, Usage, and Impact of the Internet of Things through the Use of Big Data Analytics, pp. 1531–1540 (2015)
42. Ngai, E.W.T., Wat, F.K.T.: A literature review and classification of electronic commerce research. Inf. Manag. **39**(5), 415–429 (2002)
43. Speckhard, A., Shajkovci, A., Wooster, C., Izadi, N.: Mounting a facebook brand awareness and safety ad campaign to break the ISIS brand in Iraq. Perspect. Terror. **12**(3), 50–66 (2018)
44. Shehabat, A., Mitew, T.: Black-boxing the black flag: anonymous sharing platforms and ISIS content distribution tactics. Perspect. Terror. **12**(1), 81–99 (2018)
45. Murrell, C.: The global television news agencies and their handling of user generated content video from Syria. Media War Confl. **11**(3), 289–308 (2018)
46. McElreath, D., Doss, D.A., McElreath, L., Lindsley, A., Lusk, G., Skinner, J., Wellman, A.: The communicating and marketing of radicalism: a case study of isis and cyber recruitment. Int. J. Cyber Warf. Terror. **8**(3), 26–45 (2018)
47. Klausen, J., Marks, C.E., Zaman, T.: Finding extremists in online social networks. Oper. Res. **66**(4), 957–976 (2018)
48. Cunliffe, E., Curini, L.: ISIS and heritage destruction: a sentiment analysis. Antiquity **92**(364), 1094–1111 (2018)
49. Benigni, M., Joseph, K., Carley, K.M.: Mining online communities to inform strategic messaging: practical methods to identify community-level insights. Comput. Math. Organ. Theory **24**(2), 224–242 (2018)
50. Alyousef, Y., Zanuddin, H.: Saudi Arabian government crisis management and prevention strategies: has it been effective to curb the presence of radical groups in the social media? Int. J. Eng. Technol. (UAE) **7**(2.29 Special Issue 29), 633–638 (2018)

51. Al-Rawi, A.: Video games, terrorism, and ISIS's jihad 3.0. Terror. Polit. Violence **30**(4), 740–760 (2018)
52. Botz-Bornstein, T.: The "futurist" aesthetics of ISIS. J. Aesthet. Cult. **9**(1), 1271528 (2017)
53. Awan, I.: Cyber-extremism: ISIS and the power of social media. Society, **54**(2), 138–149 (2017)
54. Melki, J., Jabado, M.: Mediated public diplomacy of the Islamic state in Iraq and Syria: the synergistic use of terrorism, social media and branding. Media Commun. **4**(2A), 92–103 (2016)
55. Johnson, N.F., Zheng, M., Vorobyeva, Y., Gabriel, A., Qi, H., Velasquez, N., Manrique, P., Johnson, D., Restrepo, E., Song, C., Wuchty, S.: New online ecology of adversarial aggregates: ISIS and beyond. Science **352**(6292), 1459–1463 (2016)
56. Golan, G.J., Lim, J.S.: Third person effect of ISIS's recruitment propaganda: online political self-efficacy and social media activism. Int. J. Commun. **10**, 4681–4701 (2016)
57. Galily, Y., Yarchi, M., Tamir, I., Samuel-Azran, T.: The Boston game and the ISIS match: terrorism, media, and sport. Am. Behav. Sci. **60**(9), 1057–1067 (2016)
58. Aistrope, T.: Social media and counterterrorism strategy. Aust. J. Int. Aff. **70**(2), 121–138 (2016)
59. Celso, A.N.: Zarqawi's legacy: Al Qaeda's ISIS "renegade". Mediterr. Q. **26**(2), 21–41 (2015)
60. Celso, A.: The "caliphate" in the digital age: the Islamic state's challenge to the global liberal order. Int. J. Interdiscip. Glob. Stud. **10**(3), 1–14 (2015)

Improvement of the Applicability of the General Data Protection Regulation in Health Clinics

Isabel Maria Lopes[1,2,3](✉), Teresa Guarda[1,4], and Pedro Oliveira[3]

[1] Centro ALGORITMI, Guimarães, Portugal
[2] UNIAG (Applied Management Research Unit),
Polytechnic Institute of Bragança, Bragança, Portugal
[3] School of Technology and Management,
Polytechnic Institute of Bragança, Bragança, Portugal
{isalopes,pedrooli}@ipb.pt
[4] Universidad Estatal Peninsula de Santa Elena – UPSE, La Libertad, Ecuador
tguarda@gmail.com

Abstract. The General Data Protection Regulation (full name: Regulation on the protection of natural persons with regard to the processing of personal data and on the free movement of such data, and repealing Directive) entered into force on May 25 2018, but was approved on April 27 2016. The General Data Protection Regulation (GDPR) aims to ensure the coherence of natural persons' protection within the European Union (EU), comprising very important innovative rules that will be applied across the EU and will directly affect every Member State. The GDPR considers a 'special category of personal data', which includes data regarding health, since this is sensitive data and is therefore subject to special conditions regarding treatment and access by third parties. This premise provides the focus of this research work, where the applicability of the GDPR in health clinics in Portugal is analysed. Such analysis is based on a study by [1], who presents the results of a survey regarding the GDPR applicability in health clinics six months before its enforcement. This work aims to present the evolution of the regulation applicability six months after its enforcement. The results are discussed in light of the data collected from a survey and possible future works are identified.

Keywords: Regulation (EU) 2016/679 · General data protection regulation · Personal data · Health clinics

1 Introduction

The EU established a two-year transitional period for companies to implement the necessary changes until May 25 2018 in order to ensure the full compliance of their data treatment with the rules imposed by the GDPR.

The GDPR comes at a time when businesses are increasingly data driven. The volume of personal information that they are collecting and keeping is forever increasing with the information becoming, in many cases, a key business asset. Also,

© Springer Nature Switzerland AG 2019
Á. Rocha et al. (Eds.): WorldCIST'19 2019, AISC 930, pp. 155–165, 2019.
https://doi.org/10.1007/978-3-030-16181-1_15

some data is considered to be more 'sensitive' than other, namely data regarding health, children, among others.

Businesses that understand their data protection obligations and seek to meet them in an intelligent way will be best placed to unlock the benefits of the personal data that they hold. Getting data protection right is not just a matter of legal compliance. It also makes sound business sense. So, whether evolutionary or revolutionary, the GDPR requires a step change for businesses in their management and delivery of personal data and privacy. Planning is required, priorities need to be set and resources allocated, but no responsible business can afford to turn a blind eye to the GDPR's many requirements [2].

Since we are over the deadline imposed to companies for implementing the regulation, it is relevant to assess the companies' level of preparation to comply with the GDPR demands by comparing the present results with those of studies conducted before the end of the deadline. Many industry sectors could have been chosen, but this research work focused on the health sector, through a survey conducted in health clinics in Portugal. The aim was to determine the point to which these companies are in compliance with the new personal data regulation.

The structure of the present work consists of an introduction, followed by a desk review on the general data protection regulation and its implementation. The following section focuses on the research methodology, identifying the target population and the structure of the survey. The results of the study are discussed in Sect. 4, followed by the conclusions drawn from the study. Finally, the limitations of this research work are identified and possible future studies are proposed.

2   General Data Protection Regulation

The enforcement of the GDPR on natural persons' protection regarding personal data treatment and movement, which repeals the Directive 95/46/CE of October 24 1995, poses innumerable challenges to both public and private entities as well as to all the agents whose activities involve the treatment of personal data.

Although the full application of the new GDPR has been set for May 25 2018, date from which the directive 95/46/CE will be effectively repealed, its enforcement on May 25 2016 dictated the need for an adaptation to all the aspects changed or introduced by the regulation. Such adaptation of the present systems and models as well as of best practices regarding personal data treatment and protection by companies is now an imperative stemming from the regulation in order to safeguard its full applicability from May 25 2018. In Fig. 1, we can see all the stages which the GDPR has undergone.

However, before focusing directly on the new regulation, it is important to clarify exactly how the document defines 'personal data' since its protection is the focus of the act.

The GDPR defines personal data in a broad sense so as to include any information related to an individual which can lead to their identification, either directly, indirectly or by reference to an identifier. Identifiers include [3]:

- Names.
- Online identifiers such as social media accounts.

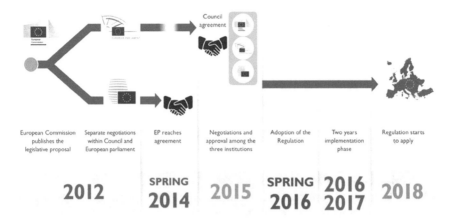

Fig. 1. Stages of the GDPR [4].

- Identification numbers (e.g., passport numbers).
- Data regarding location (e.g., physical addresses).
- Any data that can be linked to the physical, physiological, genetic, mental, economic, cultural or social identity of a person.

Companies collecting, transferring and processing data should be aware that personal data is contained in any email and also consider that third parties mentioned in emails also count as personal data and, as such, would be subject to the requirements of the GDPR [5].

The GDPR requirements apply to each member state of the European Union, aiming to create more consistent protection of consumer and personal data across EU nations. The GDPR mandates a baseline set of standards for companies that handle EU citizens' data to better safeguard the processing and movement of citizens' personal data.

The Regulation is based on a lot of familiar legal practice from the existing legislation, and many articles have the same content. However, a number of new initiatives are introduced.

Among the new initiatives, the following are to be highlighted [6]:

- Partial harmonisation of rules as well as interpretation of rules and case-law across the European countries
- Removal of the obligation to notify the supervisory authority when it comes to processing of personal data
- Some degree of one-stop-shop where the main establishment of the company interacts with only one European supervisory authority
- New rights for data subjects
- New obligations for controllers and processors
- More cooperation between the European supervisory authorities
- Introduction of significant penalties including administrative fines for failing to comply with the Regulation.

According to another author [7], the main innovations of the General Data Protection Regulation are:

1. New rights for citizens: the right to be forgotten and the right to a user's data portability from one electronic system to another.
2. The creation of the post of Data Protection Officer (DPO).
3. Obligation to carry out Risk Analyses and Impact Assessments to determine compliance with the regulation.
4. Obligation of the Data Controller and Data Processor to document the processing operations.
5. New notifications to the Supervisory Authority: security breaches and prior authorisation for certain kinds of processing.
6. New obligations to inform the data subject by means of a system of icons that are harmonised across all the countries of the EU.
7. An increase in the size of sanctions.
8. Application of the concept 'One-stop-shop' so that data subjects can carry out procedures even though this affects authorities in other member states.
9. Establishment of obligations for new special categories of data.
10. New principles in the obligations over data: transparency and minimisation of data.

Among these points representing the main innovations imposed by the new legislation, we highlight point nine, in which the regulation recognises that health data integrates the 'special categories of data' considering that such data is sensitive and therefore subjected to special limitations regarding access and treatment by third parties.

Health data may reveal information on a citizen's health condition as well as genetic data such as personal data regarding hereditary or acquired genetic characteristics which may disclose unique information on the physiology or health condition of that person. The protection of such health data imposes particular duties and obligations to the companies operating in this sector.

All organisations, including small to medium-sized companies and large enterprises, must be aware of all the GDPR requirements and be prepared to comply.

3 Research Methodology

The choice of an appropriate data collection to characterise the implementation of the GDPR in medical clinics fell on the survey technique, since it enables a clear, direct and objective answer to the questions asked to the respondents. Also, the universe under study comprises thousands of clinics, among which 190 were surveyed, numbers which make the adoption of alternative research techniques not recommendable if not impossible.

The aim of the survey was to characterise the current state of health clinics with regards to the implementation of the GDPR, in other words, determine their level of knowledge and preparation regarding the issue of personal data protection and privacy, as well as their evolution since the conduction of the last survey.

3.1 Population

The first survey was sent to 190 clinics, but only 57 gave an effective reply, which corresponds to a response rate of 30%. The sample subjects were selected randomly based on the kind of clinic and their location distributed throughout the 18 inland Portuguese districts as well as Madeira and the Azores.

Among the 190 contacts established, 35 replied via telephone and 22 via email after a first telephone contact.

In as many cases as possible, the respondent to the survey was the person in charge of the clinic's IT department. When there was no such person, the respondent was the person in charge of the clinic.

In order to compare the two surveys, one conducted before and another after the enforcement of the GDPR, the same clinics were contacted. The first study was conducted between October and December 2017. The second study was carried out between October and November 2018.

3.2 Structure

The structure of the survey resulted from a desk review on personal data protection and the study of the legal framework Regulation (EU) 2016/679 of the European Parliament and of the Council of April 27 – General Regulation on Data Protection.

The questions of the survey, of individual and confidential response, were organised in three groups.

The first group aimed to obtain a brief characterisation of the clinic as well as of the respondent. The two following groups contained questions concerning the GDPR applicability, preceded by the paramount core question: 'has the clinic implemented the measures imposed by the GDPR yet?'

After responding to this central question and when the answer was negative, respondents were asked whether they intended to implement such measures, since they were not in compliance with the regulation, and if so, whether the implementation process was already in motion. When the respondents did not intend to adopt any measure, they were asked about whether or not they were aware of the fines they may have to pay for the non-compliance with the regulation and why they did not intend to adopt such measures.

A positive answer to the central question would lead to the group of questions targeted at the companies which are already in compliance with the regulation or which are implementing the measures imposed. Some of the questions asked within this group were as follows: Are you aware of the GDPR? What impacts and challenges will clinics face in the compliance with the regulation? What stage of the implementation of the GDPR are you in? Have you identified or designated anyone for the post of Data Protection Officer? Has any training or awareness raising session been held about the new rules? Is the protection of personal data a priority in this clinic?

The survey was quite extensive. However, this study focuses particularly on the analysis of the core questions so as to assess the evolution between two different time windows.

4   Results

The first group of questions concerned the characterisation of the clinics as well as that of the respondent to the survey. Since such data is confidential, Table 1 shows the characterisation of the clinics involved in the study according to their type. The second survey targeted the same clinics involved in the first study. All the 57 clinics responded, with the slight change being that in three of them, the person who answered the survey was not the same since they were not available.

Table 1. Clinics surveyed

Type of clinic	Clinics surveyed
Nursing	12
Dental	16
Ophthalmology	4
Medical and Diagnostics	15
Orthopaedics	3
Physiotherapy	7

For the question about whether or not the companies had started or completed the process of implementation of the measures imposed by the GDPR, the results were as follows: in the first study (2017), 43 (75%) answered no and 14 (25%) answered yes; in the second study (2018), 39 (69%) replied no and 18 (31%) said that they had already started or completed the adoption of such measures (Fig. 2).

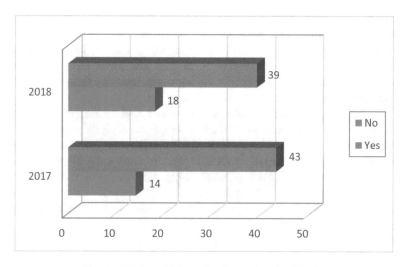

Fig. 2. Clinics which are implementing the GDPR.

In the first study, only 4 (28%) of the 14 clinics which gave a positive answer consider to be in compliance with the legislation. The remaining 10 clinics (72%) are still implementing the measures. In the second study, there was a small evolution since from the 18 clinics in the implementation stage, 7 (39%) consider that the process is completed, and the remaining 11 (61%) still have some measures to implement.

These numbers seem residual when we observe that from the 57 clinics surveyed, only 7% (2017) and 12% (2018) actually have implemented the GDPR. Over one year, which was the lapse of time between the two studies, the evolution observed was of 5%, which corresponds to 3 clinics more to have completed the process. Such improvement seems insufficient when taking into account that the second study was conducted almost 6 months after the enforcement of the GDPR.

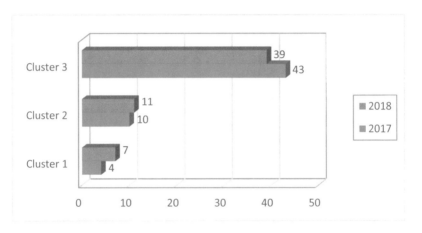

Fig. 3. Clusters according to the GDPR transitional implementation.

For a better understanding of the results, we can group the clinics into three clusters (Fig. 3):

- Cluster 1 – Clinics in compliance with the regulation;
- Cluster 2 – Clinics which are implementing the measures imposed by the regulation;
- Cluster 3 – Clinics which are not in compliance with the regulation.

Since this study focuses on the implementation of the GDPR, emphasis will be given to clusters 1 and 2 as cluster 3 comprises clinics which are not implementing the regulation.

The majority of the subjects surveyed are aware of the obligations and challenges posed by the new general data protection regulation, although this seems a contradiction since only 25% and 31% of the clinics have adopted or are adopting the measures imposed.

The implementation of the regulation requires a higher or lower level of demands depending on the size of the company as well as on whether they were already or not in compliance with the principles enshrined in the directive n. 95/46/CE.

When the clinics in cluster 2 were questioned about the implementation stage of the GDPR they were in (gather, analyse or implement), the answers in 2017 were 30%, 20% and 50%, respectively; and in 2018, the answers were 9%, 27% and 64%, respectively (Fig. 4).

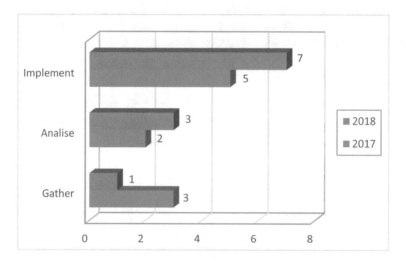

Fig. 4. Stage of the GDPR implementation.

The results show that from 2017 to 2018, the number of clinics implementing the measures imposed by the regulation increased significantly in the implementation stage. We can also highlight that there is only one clinic still in the gathering stage in 2018.

The implementation will enable the creation of conditions to make the GDPR an integrating part of the organisation's activities as well as to make it monitorable. After the conclusion of these implementation stages, a compliance assessment must be conducted periodically since the data is not immutable and even the company business and activity may undergo changes which may make the measures initially implemented inadequate to the new circumstances.

When asked how they had implemented the new measures enshrined in the regulation, the respondents gave the same answer, namely that there was nobody in the company with enough knowledge to conduct the process. They stated to have hired the services of external companies for guidance in order to be able to meet the requirements imposed by the GDPR.

Also, we determined that among the four clinics (2017) and seven clinics (2018) which said to be in compliance with the regulation, only one has identified and designated the person who will be responsible for the data treatment, the Data Protection Officer (DPO).

Overall, the respondents showed to be sensitive to the importance of both board and workers' training. However, no training or awareness raising session has been held concerning the new rules to be adopted, but such sessions were said to take place soon. It is paramount to ensure that workers are aware of the GDPR implications and such sessions are the most appropriate way to communicate the new data protection rules to collaborators.

The results obtained with regard to the acknowledgement of the sanctions and fines companies are subjected to are presented in Fig. 5.

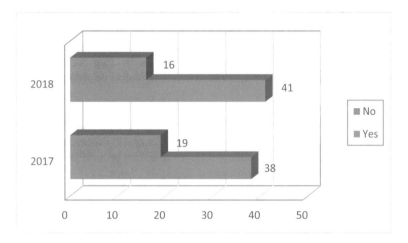

Fig. 5. Acknowledgement of the sanctions.

The GDPR reinforces the power of authorities and increases the fines. These sanctions are more burdensome and can reach the sum of 20 million euros or 4% of the overall turnover for the previous year.

Of the total number of clinics responding to the survey, most consider the stipulated two-year transitional period given to companies to adapt to the new GDPR insufficient.

The time taken to implement the GDPR will always depend on the complexity of the company's business activity, its organisational maturity, the volume and variety of the personal data used, the adequacy and flexibility of its information systems and on all its stakeholders' availability and willingness. We could say that companies only take action at the last minute. However, this study shows that even after the deadline for the implementation of the measures imposed by the regulation, only 12% actually do have such measures implemented.

One of the grounds supporting the GDPR was the reinforcement of citizens' rights regarding the way companies and organisations collect and use their personal data. All the respondents to this survey agree with this principle and consider this regulation of high relevance and importance.

It is not enough for a company to claim that they comply with the regulation, they have to make proof that the personal data they use within the scope of their activity is being protected in accordance with the regulation.

5 Conclusion

Most of the information which was previously shared in paper is currently shared digitally, thus posing new digital challenges and threats concerning security and privacy, namely regarding personal data protection in a society increasingly digital [8].

There are currently 28 laws of data protection based on the 1995 EU Data Protection Directive, which was implemented over 20 years ago and is gradually being replaced by the new GDPR [5]. Considering the advances witnessed in information and communication technologies over the last two decades, such laws can only be totally inadequate to the necessary protection of both individuals and companies' data.

The digital impact and transformation of recent years is visible in several sectors. The health sector is no exception and such transformation is an indisputable fact. Digital revolution brings along inevitable concerns regarding users' data security, privacy and protection, especially as far as health and clinical information is concerned [8–10].

The implementation of the regulation implies the definition of procedures, records and policies. Both people and technologies represent critical success factors to its implementation. Therefore, it might be relevant to carry out further research to determine to what extent this GDPR, although targeted at data protection, might not be as well a booster for the digital transformation of health clinics.

Acknowledgments. UNIAG, R&D unit funded by the FCT – Portuguese Foundation for the Development of Science and Technology, Ministry of Science, Technology and Higher Education. Project no. UID/GES/4752/2019.

This work has been supported by FCT – Fundação para a Ciência e Tecnologia within the Project Scope: UID/CEC/00319/2019.

References

1. Lopes, I.M., Oliveira, P.: Implementation of the general data protection regulation: a survey in health clinics. In: 13th Iberian Conference on Information Systems and Technologies, vol. 2018, pp. 1–6 (2018)
2. Allen & Overy: Preparing for the General Data Protection Regulation (2018)
3. European Parliament and Council: Regulation (EU) 2016/679 of the European Parliament and of the Council of 27 April 2016, Off. J. Eur. Union (2016)
4. Goubau, T.: How GDPR will change personal data control and personal data control an affect everyone in construction. https://www.aproplan.com/blog/construction-news/gdpr-changes-personal-data-control-construction. Accessed 20 July 2018
5. Ryz, L., Grest, L.: A new era in data protection. Comput. Fraud Secur. **2016**(3), 18–20 (2016)
6. Guideline General Data Protection Regulation – Implementation in Danish Companies. In: Montensen, H. (eds.) The Danish ICT and Electronics Federation, DI Digital (2016)
7. Díaz Díaz, E.: The new European union general regulation on data protection and the legal consequences for institutions. Church, Commun. Cult. **1**, 206–239 (2016)
8. SPMS – Serviços Partilhados do Ministério da Saúde, Privacidade da informação no setor da saúde (2017)

 9. Martins, J., Gonçalves, R., Branco, F., Pereira, J., Peixoto, C., Rocha, T.: How ill is online health care? An overview on the Iberia Peninsula health care institutions websites accessibility levels. In: New Advances in Information Systems and Technologies, pp. 391–400. Springer (2016)
10. Martins, J., Gonçalves, R., Oliveira, T., Cota, M., Branco, F.: Understanding the determinants of social network sites adoption at firm level: a mixed methodology approach. Electron. Commer. Res. Appl. **18**, 10–26 (2016)

The Use of LinkedIn for ICT Recruitment

Guilherme Pinho[1], João Arantes[1], Tiago Marques[1],
Frederico Branco[2], and Manuel Au-Yong-Oliveira[3(✉)]

[1] Department of Electronics, Telecommunications and Informatics,
University of Aveiro, 3810-193 Aveiro, Portugal
{guilhermepinho, arantesjoao, tiago.pm}@ua.pt
[2] INESC TEC and University of Trás-os-Montes e Alto Douro,
Vila Real, Portugal
fbranco@utad.pt
[3] GOVCOPP, Department of Economics,
Management, Industrial Engineering and Tourism,
University of Aveiro, 3810-193 Aveiro, Portugal
mao@ua.pt

Abstract. LinkedIn is undoubtedly a market leader when it comes to social networks that are aimed for professional use, demonstrating over time that it is a highly valued instrument used by recruiters, as well as a successful way to complement/replace organic recruitment. One of the areas with the most current demand and specific requirements/skills is Information and Communication Technologies (ICT) and, as in all scientific fields, there is a huge focus on recruiting the best candidate in the shortest time possible. The key goals of this article are to understand the relevance of LinkedIn use by an ICT recruiter taking into account the specific requirements/skills often desired by companies, to perceive if ICT students and workers with an updated LinkedIn profile are contacted by recruiters, and comprehend if it is true to say that recruitment via a social network is faster and does not detract from the quality of the hired candidate, compared with organic recruitment. A survey was performed focusing on students and workers in the ICT field and two semi-structured interviews were conducted in a consulting firm as well as in a software house. The results obtained suggest that LinkedIn is an essential recruitment tool for the ICT companies/consulting firms, but it is important to emphasize that most companies combine organic recruitment with LinkedIn recruitment. In our study, 89% of the respondents with an updated LinkedIn account have already been contacted by recruiters, proving that LinkedIn can certainly increase the probability of being hired.

Keywords: LinkedIn · Social network · Online recruitment · ICT

1 Introduction

As time passes, with the continuous evolution of technology, the internet has proven to be an essential tool in all professional fields, including in Information and Communication Technologies (ICT). Particularly, it has revolutionized how recruitment is done, making the process more efficient in general. The focus of this study is to explore

© Springer Nature Switzerland AG 2019
Á. Rocha et al. (Eds.): WorldCIST'19 2019, AISC 930, pp. 166–175, 2019.
https://doi.org/10.1007/978-3-030-16181-1_16

the importance of a professional social network (LinkedIn) in ICT recruitment, its advantages as well as disadvantages.

Social networking in the workplace has become a crucial communication tool in many businesses. It provides a platform for creating communities based on similar interests, hobbies or knowledge [1–3]. Therefore, social networking rapidly became a "trendy" way to find a job. With that, in 2002, a business-oriented social network called LinkedIn was founded in Mountain View, USA [4].

LinkedIn allows registered users to create a reliable network of contacts and knowledge. Currently, LinkedIn is seen as an employment opportunity, so it is essential to keep your profile and curriculum vitae updated; always looking to make new contacts to grow your personal network can also be of help in getting more job offers from employers. One of the main reasons for the higher level use of LinkedIn among the social networking sites is related to socializing with the sole purpose of building professional relationships, which is not the case at all with Facebook, for example [5].

Before LinkedIn, the difficulty to recognize passive candidates on regular online job boards was very high, making the recruitment processes conducted by companies less efficient and more protracted.

The key feature in LinkedIn is the way the profile is structured, like a curriculum, giving importance to professional experience and academic skills. This profile can be visible to almost all LinkedIn members, allowing a viable network of contacts, adding value to future opportunities [6].

LinkedIn is now the world's largest professional online service, used by more than 562 million professionals worldwide at the time of writing [7], proving that online recruitment is replacing the "old" organic recruitment.

In general, all the 10 biggest companies in their respective countries are registered on LinkedIn, and almost all companies update their profiles every week [1]. However, it is important to highlight that companies, in general, use LinkedIn recruitment alongside with organic recruitment, showing that online recruitment is undoubtedly becoming more and more used, but will hardly completely substitute the organic recruitment process – that is, it is impossible to conduct an interview via LinkedIn as strictly and professionally as a face-to-face interview.

Herein we thus discuss the recruitment via LinkedIn's world – what is the impact of having a LinkedIn profile from the point of view of ICT companies and ICT candidates, what are the advantages and disadvantages. We aim to create a basis for future research on how the recruitment market will develop in the upcoming decades. One might even conclude that the future of recruitment is to perfectly combine both online and organic recruitment, that is, to conjugate the advantages of both methods to make recruitment more efficient, faster and without compromising the quality of the selected candidates.

Four main research questions were determined, as follows:

- Is LinkedIn recruitment revolutionizing the way ICT companies/ICT candidates find a new job?
- What is the impact of LinkedIn hiring processes on ICT companies?
- What are the advantages/disadvantages of online recruitment in comparison with organic recruitment?
- How important is it to combine online recruitment with organic recruitment?

From this point on, this article is structured in five sections, which are as follows: literature review, methodology, results, discussion and conclusion/future research.

The literature review provides a general approach on recruitment in LinkedIn and establishes some of the ideas/conclusions related to the theme that have already been researched and published by other authors. The methodology was based on an online survey, of our own authorship, that was targeted at students and workers in ICT areas. Besides that, two ICT recruiters were also contacted, in order to contribute with further information to this article. In the results, an ample look at the data obtained is taken from the steps executed in the methodology. In the discussion, the authors go over the data provided, and study the possible reasons that led to those results and how they may have an impact on the rest of the ICT recruitment community, showing which situations led to advantages or disadvantages, both for the employers and the employees. Finally, the conclusion and future research sums up what can be taken from this article, by answering the questions set forth in this introduction; we also give a suggestion for future research that can be followed from this point forward.

2   Recruitment in LinkedIn: Literature Review

Recruitment is an essential part of talent management and can be defined as 'the process of searching for the right talent and stimulating them to apply for jobs in the organization' [5].

As the global economy expanded dramatically between 2002 and 2007, business leaders and human resource managers worried about the intensifying international competition for talent [8].

Adding to that, the massive growth of social media and Internet capacities and capabilities has added numerous other sourcing possibilities and activities [5]. With that massive growth appeared LinkedIn, a business-oriented social networking website founded in 2002 [7]. Unlike other social networks such as Facebook, which are often purely recreational, LinkedIn emphasizes a user's professional connections. LinkedIn allows users to further their careers by searching for jobs, finding connections (networking) at a particular company, and receiving recommendations/offers from other users [4].

It is hard to deny the impact that LinkedIn, which rightfully bills itself as "the world's largest professional network" [7], has had on the recruiting function. The network has forever changed the way organizations connect with talent, giving them unprecedented access to both active and passive candidates. A lot of companies have changed the way they look for new employees and have implemented a new strategy: hire recruiters only for the purpose of looking for possible candidates on LinkedIn, making the recruitment process more specific and taking into account the skills that the company is looking for to fill the job [9].

By scrutinizing in detail the influential works of Koch, Gerber and de Klerk [5], Pavlícek and Novák [10], Böhmová and Novák [1] and Silva, Silva and Martins [6], it is possible to conclude the following:

The impact of not having the right people in place to manage and confront business challenges led to a change of strategy in big companies, creating in-house recruitment [5].

Before, finding possible candidates on regular job boards was a difficult task, something that nowadays is rather easy while using LinkedIn [10].

A study done by Jobs2web Inc. shows that companies look through about 219 applications per job on a major job board before finding someone to hire. Now, using LinkedIn, they only look at 33 applications per hire on their own corporate career site and 32 per hire when a job seeker types the job they are looking for into a search engine [1, 11].

Dwight Scott, a recruiter with ExecuSearch in New York, says he does not spend much time studying applicants' resumes because he prefers searching LinkedIn for potential hires (65% of his placements in 2014 have been done via LinkedIn) [10].

In every three LinkedIn users, two have a college degree and in every four, one is postgraduate, and 60% of users are confident that LinkedIn is an important tool for their journey. LinkedIn works as a network of professional contacts and increases the possibility of getting proposals for employment and, with that, possible invitations to interviews and even getting hired [6].

Recruiters rely on the social network to find even the highest skilled executives. For example, Oracle CFO was recruited via LinkedIn in 2008 [10].

The most important parameters that might be interesting to recruiters are the employment experience, the schools where the user studied, the years of experience and the description of how they present themselves [5].

LinkedIn gathers so much information/data, that it can even be helpful for decision-making processes; in other words, with the data provided by LinkedIn it is possible to understand where to open new facilities (number of skilled workers potentially available) [1].

The percentage of recruiters that use LinkedIn are so significant that it is safe to say that, as time passes, LinkedIn can build a monopoly of recruitment and become the undisputed place to search for a job. On the other hand, it is also important to highlight that LinkedIn still has its limitations and does not provide all the solutions to the already existing recruitment problems, so if not used wisely can easily become one more example of the "spray and pray" approach [5, 10].

3   Methodology

After doing the literature review, it is safe to say that LinkedIn is a must-have tool for professionals in any area, and a powerful tool for recruitment, considering that a growing number of companies are choosing LinkedIn to recruit employees. Having the skills and job experience gathered in an online platform can give one a new opportunity, even if the user is not looking to change jobs at the time. However, the authors were aiming their research at ICT, trying to find out if this trend would also apply here. However, studies involving ICT as a focus group were lacking in research. To get a bigger picture of the reality of the recruitment in this area, an online survey, of our own authorship, was given to students and professionals in the ICT field, using Google

Forms (the questionnaire may be accessed here: https://docs.google.com/forms/d/
10YkUqggdMmtMmGl4fZyYgglST4F52UOFyUhGPaZdhpA/edit). It was firstly
given to a validation group, and the feedback received resulted in three
changes/additions to the original form, resulting in a final form with eleven questions
for LinkedIn users and seven questions for nonusers. It was then shared in Facebook
groups of Electronics, Telecommunications and Informatics, and via private message to
the authors' personal network, in order to acquire some information about the experiences and opportunities of the users while on LinkedIn. The survey was available
online between the 24th of October and the 1st of November 2018, and it was
anonymous. The most relevant questions are described in the "Results" section of this
article.

Many companies were also contacted in order to understand the importance of this
platform to them and to know how often they use it to recruit passive candidates.
LinkedIn provides specific and exclusive tools for recruiters, for different prices
depending on different needs. A free user cannot contact people that are not in their
personal network, they have a daily limit of profiles they can look at, and they cannot
filter many search options – which constitute many limitations. Not every business uses
this tool – only if they are willing to pay for it. The authors wanted to understand if it is
worth using it, so they have contacted eighteen ICT companies in order to obtain
responses. The process was not easy, mostly because these companies only provide one
general e-mail address for every affair. Despite the lack of responses from many of
them, the authors were able to contact, in a more "personal way", two recruiters directly
from their LinkedIn profiles, leading to a semi-structured interview (both held on the
30th of October 2018 – one via e-mail, and the other via the telephone and during a
recorded telephone call lasting for 20 min). Both responses obtained – from a recruiter
working in a consulting firm and from another working in a software house – were
confidential, so the names of the companies and the persons of contact will not be
mentioned in this article. Of note is that each interview contact was followed-up on, a
second time, to clarify some of the issues brought up during the initial interviews, and
so further questions were asked and duly answered.

The interview questions included ones about: whether the firm uses LinkedIn as a
platform for recruitment in ICT and with what frequency? The percentage of ICT
candidates they have interviewed via LinkedIn? The percentage of employees of the
firm who were recruited via LinkedIn? What results better – spontaneous candidates or
candidates responding to an advert for a vacancy or professionals recruited via LinkedIn? Does the firm use the premium functionality of LinkedIn? Is there an advantage
in using the premium functionality of LinkedIn? Do they believe that LinkedIn will be
able to replace organic recruitment?

4   Results

The discussion of the results will, firstly, take into consideration the information
provided by the survey given to the students and employees, and then will consider the
information collected by the interviews of the ICT recruiters.

A survey was performed involving students (Bachelor's, Master's and PhDs) and workers from the ICT field, leading to 93 responses. Due to the proximity of the authors with the University of Aveiro, Portugal, 92% of the responses were from students or ex-students from this academy. Most of the responses were also provided by Master's students and professionals already working (88%). 75% of the inquired have a LinkedIn profile (68% of the students and 100% of the professionals), and most of the ones who aren't registered on the platform are willing to become users (83%). Only 2% of the inquired have never heard about this platform.

Having a profile on LinkedIn does not mean much if the profile does not provide the key information meant to be on the platform, so it was also asked if the users with a LinkedIn profile have their information updated with their experience, skills, recommendations and publications, leading to a 73% "yes" response. It was also important to understand if the users were active on their profiles, so the survey included a question asking if they often add people or companies from the ICT field to their personal networks, where it was found that 70% of the inquired did, despite most of them (89%) never writing, sharing or commenting on LinkedIn publications. Now, aiming more at the recruitment part, 74% of the inquired have been contacted via LinkedIn for job opportunities. All the PhD students and the professionals that are already working have been contacted at least once, and 63% of the Bachelor's and Master's students have already been contacted at least once as well. Now, looking for the acceptance rate, most of the students replied that they didn't get the position in the company or they are still waiting for a response. Note that, in many cases, companies with a high demand of specialized employees start looking for students before they finish their degrees, and they stay in the company's database until they complete their studies, only to be contacted again afterwards. On the other hand, 53% of the ICT professionals that have been contacted via LinkedIn for job opportunities have been accepted for the position. A relation was found between being active in this network and the interest of the recruiters: 89% of the users that add ICT connections and have their profiles updated have already been contacted for job positions. The last question of the survey asked the respondents to point out some disadvantage(s) or something that LinkedIn could change to provide a better user experience. One of the answers pointed out that, at the time of writing, ICT candidates-to-be were often harassed, by recruiters, with multiple opportunities that, sometimes, wouldn't even correspond to the users' skills.

All this user information gathered on one platform becomes a must-look place for recruiters. For this reason, the authors wanted to analyse the perspective of the ICT recruiters. At the time of writing, a lot of the recruitment going on in the ICT field was done by consulting firms, so they can provide reliable information about the recruitment process. The information provided by the consulting firm showed that they use LinkedIn on a daily basis, and around 35% of the interviews held by the firm are with candidates they have found searching on LinkedIn. At the time of writing, around 30% of this company's new employees came from LinkedIn candidates.

The other recruiter, the software house, does not use LinkedIn very often for recruitment, but searches for candidates through consulting firms.

Following a semi-structured interview, it was possible to get around other topics and receive different perspectives about the recruitment process. One feedback was that LinkedIn can be really powerful when you are looking for very specific competences, usually fulfilled by a senior profile. Organic or spontaneous appliances hardly will lead to a resumé as good as the one they were specifically looking at, because the paid features of LinkedIn allow you to filter and search for exactly what you need and give results in "no time". Another topic of discussion was about the receptivity of the ICT professionals to accept to do an interview. There is a lot of demand for professionals in this field and recruiters are always looking for possible candidates, so it is not unusual that the same person might be contacted a couple of times every week by different (or even sometimes the same) company. This could lead to a not-so-pleasant experience for LinkedIn users. In one of the interviews, the recruiter pointed out that it is possible to notice a reduction in the number of profiles showed in a search, comparing to recent years, concluding that users are probably starting to turn they profiles invisible for recruiters. It was also referred that, sometimes, ICT professionals do not even respond to recruiters or respond in a not-so-polite way. It was mentioned that "LinkedIn is a powerful tool but, when used excessively, can show the other side of the coin, that can lead to a bad image/reputation of the company". The recruiter from the software house pointed out that "recommendation programs" are the best method to recruit, where the employees could recommend a person for a position in the company and, if selected, the company would give a bonus or a prize to the one who recommended the chosen candidate, making it a win-win situation for both the employee and the company. Considering the above, there always exists some "responsibility" on the part of the employee regarding their recommendations so, usually, the company tends to see these recommended candidates as more trustworthy employees.

5   Discussion

Having more options available allows employers to make better decisions when it comes to hiring new professionals. Nowadays, people have to compete not only with the active candidates but also with the passive candidates that recruiters find for a given position. That might be a problem for the active candidate if they are applying for a specific position – it means more competition for them – making this a disadvantage to the candidate but an advantage to the company. Otherwise, the demand for ICT professionals is very high [12], meaning opportunities for every professional in the field, as long as they are willing to be flexible enough to adapt to a new company's conditions. So, if the candidate doesn't get that specific position, it is probable that they will find a new opportunity that might even come via LinkedIn.

In the ICT, a lot of companies recruit new employees with no work experience. Human resources are increasingly starting to pay more attention to recently graduated students that bring new ideas and different visions to the workplace [13], something that is highly important in technological areas, since everything in this field changes at a high rhythm. LinkedIn plays its role in the recruitment of these new generations, and this is supported by the results of the survey that the authors conducted. ICT students and professionals are aware of the importance of LinkedIn, since 75% of them have a profile, and most of those who don't are willing to join the platform as well as to keep information about their professional endeavours up-to-date.

Following this way of conduct, companies are able to categorize individuals based on the information they provide on LinkedIn, and this platform allows them to look at the users' profiles in a clean, homogenous and organized way; and that translates into positive actions between company and individual, since this research points out that people active on LinkedIn have been contacted in the past (89%), which ends up being advantageous to these individuals.

Despite the interest of ICT companies in this new generation, millennials are known for frequently job-hopping in the first 5 years of their career [14, 15], making them less reliable, especially in the ICT, where usually new employees go through a significant initial training process (with associated costs). It is important to hold new employees for longer periods of time, so it might be interesting to study, in future research, if it is more likely for an employee to leave the company if he has applied organically to the current job or if he was contacted via LinkedIn.

So far, we have discussed how much of a positive factor LinkedIn can be in recruitment and how much it facilitates the contact between companies and professionals. It is clear that not everything is perfect on LinkedIn. The idea of social networks is based on free access [16], which happens on LinkedIn for a simple profile, but the main advantages for users and recruiters come with a price, literally. In the ICT, some companies find that the money paid for having a recruiter profile in LinkedIn is worth it, so they search for profiles daily and contact users frequently. When a professional sets himself as available to receive job offers on LinkedIn, a vast number of companies will have an interest in approaching that individual, especially if he possesses competences that are rare among the pool of available professionals. Applying the same routine to many other companies, this results in multiple contacts reaching out to the same user, up to a point that the whole process can start to be annoying; this leads to a lack of response, rude responses or even turning the profile private/invisible for recruiters. This turns out to be something that, at the time of writing, was not very focused on in the literature review done by the authors: a saturation and a new turning point for the use of LinkedIn in the ICT.

This is where organic recruitment steps in, because it allows for a stricter approach of the professionals by the companies, even though it will, most likely, take longer than a LinkedIn approach. The "recommendations programs" are popular among companies and don't rely on any social network but on the personal network of the individuals, and it is more likely that a recommended person will stay with the job because, usually, when a person is recommended, it is because he or she has an interest in the position and fulfils the need that the company is looking for. This can save a lot of time in the recruitment process.

Some professionals and students still don't use social media, and the same applies to LinkedIn. It is common to hear that companies are selling their clients' data to advertisement companies and the growth of social media has boosted the development of this new "data-selling business" [17, 18], so it is not so strange that some people would prefer to stay "offline" rather than fill all their information on websites they might not trust.

6   Conclusion and Suggestion for Future Research

Nowadays, when ICT companies and organizations have a big urge to find the best professionals, social networks like LinkedIn have erupted in popularity, and completely revolutionized the recruitment market, providing information about competences/ education/experience from professionals available to fill in positions in the ICT industry. The recruitment process is also more accelerated, because the recruiters have all the information they need at their disposal, even without having had the first contact with the professionals. This is good for companies, but also for the professionals, because it gives them a better perspective of the opportunities existent in the industry. This study concluded that most of the Master's and PhD students as well as professionals in the ICT field have a LinkedIn profile, and are aware of its importance. The more complete the profiles are and the more active the professionals become on the social platform, the better the chances of being approached by companies.

Even though time is saved by using LinkedIn, it can work as a double-edged sword, because being harassed by multiple companies can become exhausting and make professionals distance themselves from this method of recruitment. Therefore, organic recruitment must coexist with online recruitment. It provides a deeper vision of the individual regarding his competences and other aspects like responsibility, matureness and autonomy.

It might be interesting to study, in the future, if it is more likely for an employee to leave a company if he has applied organically for a job or if he was recruited via LinkedIn, and if there is any difference in the performance between those same two types of candidates. Do LinkedIn candidates make better employees than organic candidates? If so, how will the recruitment market develop in the upcoming decades in this field? These above-mentioned questions can become the basis for future research with the intent of predicting the evolution of the recruitment market, making recruitment in the future even more accurate, considering the very specific skills that ICT companies are always looking for.

Acknowledgements. The conclusions of this work were only possible due to the people that responded to our online survey and due to the interviews conducted with the two companies which accepted to respond, to whom we would like to give thanks.

References

1. Böhmová, L., Novák, R.: How employers use LinkedIn for hiring employees in comparison with job boards, pp. 189–194
2. Gonçalves, R., Martins, J., Rocha, Á.: Internet e redes sociais como instrumentos potenciadores de negócio. RISTI-Revista Ibérica de Sistemas e Tecnologias de Informação, pp. 09–11 (2016)
3. Martins, J., Gonçalves, R., Oliveira, T., Cota, M., Branco, F.: Understanding the determinants of social network sites adoption at firm level: a mixed methodology approach. Electron. Commer. Res. Appl. **18**, 10–26 (2016)
4. Gregersen, E.: LinkedIn. https://www.britannica.com/topic/LinkedIn. Accessed 18 Oct 2018

5. Koch, T., Gerber, C., De Klerk, J.J.: The impact of social media on recruitment: Are you LinkedIn? SA J. Hum. Resour. Manag. **16**, 1–14 (2018)
6. Silva, C., Silva, S., Martins, D.: The LinkedIn platform in human resources recruitment (2017)
7. Sobre o LinkedIn. https://about.linkedin.com/pt-br. Accessed 07 Nov 2018
8. Beechler, S., Woodward, I.C.: The global "war for talent". J. Int. Manag. **15**(3), 273–285 (2009)
9. Hanigan, M.: How LinkedIn fundamentally ruined recruitment, 4 February 2015. https://www.entrepreneur.com/article/242554. Accessed 18 Oct 2018
10. Pavlíček, A., Novák, R.: Social media and industry 4 : 0 – Human resources in the age of LinkedIn. In: IDIMT 2018, pp. 199 209 (2018)
11. Zhang, L.: 3 smart ways to attract recruiters to your LinkedIn profile, 6 October 2014. https://www.forbes.com/sites/dailymuse/2014/10/06/3-smart-ways-to-attract-recruiters-to-your-linkedin-profile/#372582823959. Accessed 18 Oct 2018
12. Calé, P.: Emprego: Procura é tanta que já são os profissionais de TI que "escolhem" salários, 26 March 2018. https://tek.sapo.pt/noticias/negocios/artigos/emprego-procura-e-tanta-que-ja-sao-os-profissionais-de-ti-que-escolhem-salarios. Accessed 04 Nov 2018
13. Bartakova, G.P., Gubiniova, K., Brtkova, J., Hitka, M.: Actual trends in the recruitment process at small and medium-sized enterprises with the use of social networking. Econ. Ann. **164**(3–4), 80–84 (2017)
14. Berger, G.: Will this year's college grads job-hop more than previous grads? April 12 2016. https://blog.linkedin.com/2016/04/12/will-this-year_s-college-grads-job-hop-more-than-previous-grads. Accessed 02 Nov 2018
15. Au-Yong-Oliveira, M., Gonçalves, R., Martins, J., Branco, F.: The social impact of technology on millennials and consequences for higher education and leadership. Telemat. Inform. **35**, 954–963 (2018)
16. Aspridis, G., Kazantz, V., Kyriakou, D.: Social networking websites and their effect in contemporary human resource management, a research approach. Mediterr. J. Soc. Sci. **4**(1), 29–46 (2013)
17. Diaby, M., Viennet, E., Launay, T.: Toward the next generation of recruitment tools. In: Proceedings of the 2013 IEEE/ACM International Conference on Advances in Social Networks Analysis and Mining, ASONAM 2013, pp. 821–828 (2013)
18. Martins, J., Gonçalves, R., Pereira, J., Cota, M.: Iberia 2.0: a way to leverage web 2.0 in organizations. In: 2012 7th Iberian Conference on Information Systems and Technologies (CISTI), pp. 1–7. IEEE (2012)

The Role of Technologies:
Creating a New Labour Market

Ana Isabel Vieira[1], Eva Oliveira[1], Francisca Silva[1], Marco Oliveira[1],
Ramiro Gonçalves[2], and Manuel Au-Yong-Oliveira[1,3(✉)]

[1] Department of Economics, Management, Industrial Engineering and Tourism,
University of Aveiro, 3810-193 Aveiro, Portugal
{anasilva23, evafcoliveira, franciscafs,
marcosilveiraoliveira, mao}@ua.pt
[2] INESC TEC and University of Trás-os-Montes e Alto Douro,
Vila Real, Portugal
ramiro@utad.pt
[3] GOVCOPP, Aveiro, Portugal

Abstract. Nowadays, workers and students preparing for a career are aware
that they must constantly update their skills in order to fulfil both the market's
and companies' demands. The environment is packed with technologies that
help us by optimizing our tasks on a daily basis. However, the majority of
opinions can agree that these tools may have a bigger impact on the labour
market than it was ever expected. What is the future of jobs as we know them,
and what impact will future technology innovations have on society? Lots of
factors enter the picture when this particular topic is discussed – including the
economy in general, productivity rates, salaries and expenses, people's stress
and satisfaction levels, among others. To achieve this study's main purpose,
questionnaires were distributed, which were the basis for the creation of a
hypothetical model that represents our understanding of the subject considering
the sample's feedback. This study also advances our view concerning the literature review and the opinions of those surveyed, aiming to provide a strong
and consistent theory of what the market labour will be; that is, which jobs are
likely to lose practice, which ones may subsist, and which activities might grow
with the circumstances' influence.

Keywords: Technology · Mutation · Adaptation · Skills · Future

1 Introduction

When it comes to the role of new technology and its influence on our daily routine,
opinions turn quite ambiguous. Some defend how much humans are irreplaceable even
with the presence of the best machinery; others already feel that robots and technology
in general can and will replace human labour - or, at least, demand its improvement. As
[6] said productivity's accelerated increase and the inverse relation between production
and the generation of jobs points to the occurrence of important modifications in the
field of human labour - which cannot be mistaken with the end of the labour argument.
In fact, this means exactly the opposite: the creation of a new world [6].

© Springer Nature Switzerland AG 2019
Á. Rocha et al. (Eds.): WorldCIST'19 2019, AISC 930, pp. 176–184, 2019.
https://doi.org/10.1007/978-3-030-16181-1_17

The motivation of the current article was people's consciousness on the current subject. Also, a lot of us can agree that the chosen topic can easily intrigue workers in general, making this study more interesting in the way that the final product will provide new information. Taking into consideration the novelty of the theme, it must be referred that the collected data was particularly recent - from 2000 to 2018. This allowed the building of a strong authors' opinion, the basis for the development of our questionnaire. Thus – as is discussed in the Methodology section - it was more intuitive to understand which concepts were already being explored in the literature and which ones would be interesting to probe in 2018.

This study intends to provide updated information sources, so that people can learn what is required of them in order to maintain their jobs, but also, to specify which jobs will, hypothetically, be more influenced by technology – according to our sample. In terms of research questions, four of them were determined:

- Will jobs, in general, not be performed by humans in the future? [secondary questions - What does our literature review tell us? How about our sample's perception?]
- Do workers feel stressed about this issue?
- Which jobs will never suffer major changes?
- Do people agree with this technological evolution?

In the next section we do a literature review that advances different authors' perspectives on the research theme. The main goal was to understand what knowledge already exists in the literature. Furthermore, the literature review provided a solid basis for our analysis and discussion. Following the literature review we present the methodology followed in our study. In the methodology section we describe how we chose to build an original questionnaire and to describe what its main purpose was. Finally, a theoretical model was created from both the literature review and the questionnaire answers.

2   Labour Market and Technologies - Literature Review

No other aspect of the human life has been through so many constant variations like work has. One decade has shown more complex modifications than all of the 20th century, so it is likely that in the next ten years extraordinary changes will take place [6, 14, 16]. Throughout centuries, work as we know it has become a process of "automatization". Hader was the first person to use this word and it basically refers to the process in which raw material is inserted and processed without any human contact [4].

"Labour would become less important and workers would be replaced by machines" was the prediction of John Maynard Keynes and Wassily Leontief that also gave the motto to Richard Works' analysis [10, p.1]. [10] has supported this in the study by Daron Acemoglu and Pascual Restrepo that related robots to jobs, and whose results showed positive and negative consequences in wages and employability levels. Thus, society in general must bear in mind that we live in a time of massive changes that have been registered, with jobs which will disappear and jobs which will be created [12]. Consequently, this will demand new qualifications and frequent updates from workers.

These changes are aligned with new levels of specialization: the cohabitation of man and intelligent machine will lead to successive role adjustments, it being up to humans to make the most out of the experience and expertise of their new working "colleagues". New levels of knowledge can be achieved, deep innovations can take place – as long as virtuous cycles between both parties can be created [12]. As a technology negative effect, there is the fact that robots can replace humans at work. On the other hand, the relation price-production constitutes the positive factor, as confirmed by the study - "increasing the number of robots decreases the cost of production" [10, p.1].

The most affected industry is manufacturing, while management is the only irreplaceable human job. In the field of education, the analysis considers that the higher the level of education, the less the impact is felt on workers [10]. However, Acemoglu and Restrepo concluded that if robots' multiplication goes on, there can be considerable losses in human employment. The only way of stopping human jobs from disappearing is by adopting public politics and developing the Third Sector, constituted by non-governmental organizations and community activities [4].

As technology shapes the way we work, it also shapes the type of work available. While the most obvious effect has been the handling of repetitive tasks by machines, one of the more surprising consequences has been the disappearance of middle-skilled roles. From 1980 to 2005, some types of middle-skilled jobs saw a decline of up to 54%, and over the same period, low-skilled service jobs and high-skilled managerial and professional work increased by 30%. This trend is likely to become even more pronounced over time, with Oxford University predicting that 47% of jobs in the US and a third of those in Europe will be replaced by technology within the next two decades [7].

According to the Boston Consulting Group, investment in robotics sees a sharp increase when a machine or platform becomes 15% cheaper than a worker. In the US, this moment has already arrived: the cost of operating a welding machine for an hour is now $8, while the cost of employing a worker for the same period is $25 [7]. What this suggests is that organizations will have to start developing a talent acquisition model that relies more heavily on seeking the right skills than the right experience. Managers and recruitment professionals will need to identify people who fit work opportunities in terms of their aptitudes rather than previous experience – this will be a major shift [7].

"Will there be any jobs left in the future?" is the starting point of [8, p.1], where only two possible answers arise: (1) jobs will subsist but in a different way and inequality will increase; and (2) human jobs will be replaced by machines except for a minority of skilled people. [8] defends the first one and starts explaining why, by considering the consequences in the labour market. In his opinion, knowing technology's reach is not enough for predicting which jobs will be automatized, as that is something feasible but with high costs [8]. After analysing a certain amount of data, the author realizes that low-skilled humans will be the less affected, since they are cheaper and sometimes can develop complex tasks. The high-skilled workers will be protected as well, because their jobs are way too difficult to be automatized; also, they will be needed for the jobs that will appear, as well as for the management area. Therefore, the most likely to be replaced are the average workers - in the author's opinion.

[11] investigated the relationship between modern technologies and employment. The author does an extensive research on the existing literature and affirms that

innovation does not necessarily mean jobs' disappearance. While it leads to rendering some tasks obsolete, it also creates the need for new jobs. It is explained that "in the past, innovations were mainly labour-friendly; the literature regards innovation and technology as the main drivers of economic growth and new employment" [11, p.4]. [11] also enlightens that the main cause for our association of "technology" to "job disappearance" is because of our lack of imagination regarding new types of jobs. Though it is correct to say that some jobs will cease to exist, others will be created, but with different required skills. Plus, tasks that require creative and social skills will always be needed.

3   Methodology

The methodology used in this study of an exploratory nature was developed within the scope of the literature review, related to science and technology. The gathering of some of the literature was done in databases like Scopus, Springer and Rcaap (Portugal's open access scientific repositories). In terms of research, it was found difficult to access updated information, taking into consideration that this article's theme is recent and even the congregated data do not reveal a common answer among the authors – which also revealed itself as a motivation and a source of interest for this study.

After being aware of the topics discussed by different authors, it was decided that a questionnaire will be designed in order to collect people's feelings and experiences in what concerns technology and their current and future labour market. The questions presented had as a main goal to find out people's perspectives on the probability of their jobs' substitution by machinery, in what amount of time would that happen and, for those who do not believe this event will ever occur, the question of why they feel that could happen was asked. Moreover, it was of interest to learn the respondents' level of stress concerning the situation. Finally - as is developed in the Theoretical Model - people were asked about jobs that will disappear, those that will "survive" and the new ones that will appear.

The reason why this tool was chosen (after being duly validated in a test run, before the "go-live") is because it was considered a straightforward way to get to a larger number of people and to be able to analyse data more precisely afterwards. Also, as the current theme is quite recent and unstudied, the questionnaire allows the congregation of real and fresh information to answer the questions mentioned in the Introduction. Thus, our work is considered qualitative since we categorized our sample's jobs in several categories – as is discriminated in the Theoretical Model – to make the results' explanation more intuitive. Consequently, respondents were, in some questions, allowed to give an open answer – a factor that allowed a deeper analysis of different points of view concerning the same topic.

The collected sample includes 113 people, of whom 73% are female and 27% male. Around 80% of them are aged 18 to 25, 8% are aged 37 to 50; 5% - 26 to 36; 4% are older than 50, while only 3% of the respondents are younger than 18. In terms of residence, 85% of our participants come from the Centre region of Portugal, 12% from the north of the country and 1% from the south. Also, nearly 2% are Portuguese emigrants. The main research questions to be answered in this context were:

- Do people consider that new jobs will appear together with technological advancements? If yes, which ones?
- Consequently, which jobs will disappear?
- Which current jobs will survive in the future labour market?

In the fourth segment of the current article, a theoretical model was developed. The questionnaire's results were explored, and, after that, a hypothesis was formulated with the aim of answering the current research questions, as can be observed in Fig. 1. This was followed by a discussion on the fieldwork, establishing a connection with the literature – where different opinions match and differ. Finally, conclusions were made, providing an opinion regarding the articles read and the sample's contribution.

4   A Theoretical Model

In order to create the hypothetical theory represented in Fig. 1, an original question-naire was developed, as was mentioned in the Methodology section.

Figure 1 is the result of an intuitive analysis of the data collected. Note how, in Fig. 1, [traditional] Communication/Marketing are in danger but will also still exist in the future, perhaps in other forms, due to new technology and to the rising in popularity (and reach) of digital influencing (especially with the younger generations and the attention and power that YouTubers command) [15].

Taking into consideration the collected statistics, people perceive that there are jobs that will disappear and that new ones will certainly appear. Thus, 17% of our respondents answered that they believe their replacement by some sort of technology is likely to happen, while overall, 83% of them do not believe at all in that substitution.

Referring to people's current jobs, 65% of them are still students – whose predicted professions will be discriminated afterwards. Then, in order to provide a clearer per-ception of our sample, the answers provided were categorized in (1) Administration, (2) Marketing and Communication (which include Journalism and Public Relations), (3) Design and Multimedia, (4) Engineering, (5) Education, (6) Geography, (7) Man-agement, (8) Industry, (9) Services, (10) Health, (11) Politics, (12) Law, (13) Arts (including Music and Spectacle arts) (14) Investigation, (15) Information Technologies and (16) Not answered. Considering the current 65% comprised of students, the pre-vious categories were also used to predict their future jobs. In a more advanced approach, data was analysed evaluating the gathered professions as if all people were already in the labour market.

When it comes to the sample's perception on whether their job will or will not be substituted by a machine, 82.5% do not believe in that, while only the remaining 17.5% feel that is likely to happen. These 17.5% currently work in areas such as Marketing and Communication, Administration, Management, Geography, Economy and Services (bank tasks, to specify). However, there are people from the same categories who do not agree with these opinions.

People who feel their jobs will be replaced by machinery in some way were also asked when could that happen: 41.2% feel that it might happen approximately 10 years from now; 23.5% selected an interval between 11 and 15 years; 29.4% believe this could happen 16 to 20 years from now; finally, a smaller percentage chose the option "in more than 20 years".

When it comes to explaining why humans are irreplaceable, people argue that robots will never develop skills such as intuition, creativity, innovation and empathy. Also, human intervention is considered to sustain interpersonal relations, strategic planning, cultural perception, conflict management, market analysis, problem resolution and decision-making capacities. The public reinforces the importance of medical intervention and diagnosis as well as its' human side. Education, for instance, is claimed to be incomparable to any machine due to its relational and emotional component.

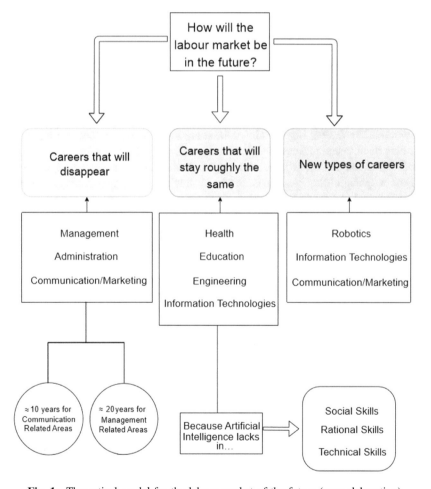

Fig. 1. Theoretical model for the labour market of the future (own elaboration)

After being aware of the general vision of our sample of society regarding the future, we aimed to understand whether participants felt any stress concerning the current issue. Therefore, the collected data shows that 37.5% of respondents are not worried at all that they might be substituted; however, with just 1.7% less, 36% feel mildly stressed over the situation; at last, 22.8% already experience intense stress, while a minority of 3.5% experiences high-level stress over the matter. Based both on the literature review as well as on the questionnaires, there are reasons to believe that *Medium-Skill tasks* have a higher chance to disappear, since they are not complex nor cheap enough to maintain. Repetitive and simple tasks also have a greater probability to be substituted by machinery. In the bigger picture, it can be said that if the costs of production are lower when automatization enters the scenario, then those jobs will probably be cut. *Communication and Marketing* related jobs are visibly assigned to two opposite perspectives. This happens due to the fact that the respondents' opinions are not consensual.

5   Discussion on the Field Work

Taking into consideration the reviewed literature and the information collected from the questionnaire, it can be concluded that the idea of [7, 8] about the low- and high-skilled workers not being replaced by robots is shared by people in general – specifically, with the current questionnaire respondents. Consequently, why only keep the low- and high-skill workers? In the big organizations, these are the ones who would not be replaced, since they do not need professional knowledge and are cheaper – in the case of the low-skilled – and, on the other hand, the high-skilled ones are needed to take on management responsibilities. As an example of low-skilled workers, there are maids, hairdressers and fast food servers [8]. If these are not replaced, then the medium-skilled workers will be the first ones to disappear. But have we considered that no one needs to be replaced in the first place? Why hasn't work just kept to its origins? This is a very good question. The explanation is evolution. The world evolves every day. Every day, a new technology that improves products and services' needs is launched. It's the same with companies: it is cheaper and, in some cases, more productive to have a machine instead of employees doing the work. Moreover, people are problematic and complain; they want to build a family and that also implies other complications. To organizations, this traduces in less production and breaks up the product development. If they rely on machines instead of humans, all these problems are solved. To humans - who need a job and, in consequence, a salary to survive, that is prejudicial. However, when it comes to a CEO's perspective, machines will be preferred over workers, in order to generate more profit.

Therefore, once we talk about profit it is imperative to understand how profitable this will be and when can companies reap the rewards. Our enquiries revealed some concerns about how industries will get the money to cover a massive production of robots, knowing – or expecting – that everything related to robotics and artificial intelligence is expensive. The Boston Consulting Group [7] has concluded that the investment in robotics is something worthwhile – not immediately, but in the future – since organizations will save 15% more than by employing human workers. This means that hiring robots or creating them could be more expensive at first, but it has to be seen as an investment in the long run. Afterwards, the robot is ready to work tirelessly, only needing maintenance once in a while.

Finally, it is also relevant that a big amount of information given by the literature review is denied or unknown by the people who answered the questionnaires. Which can lead to another question: are people really aware of technology's impact on jobs? Probably not. Proof of this is that 82.5% claimed not to believe that their position can be replaced by a machine, while several authors defend the opposite theory – even if not all agree on which types of human skills are more predictably to be affected.

6   Conclusions and Suggestions for Future Research

After observing all the parameters, it is safe to say that we are now living in an era where technology is taking over the world – and its people. Nowadays, it is almost impossible to get through the day without using a mobile phone, while during the 1950's this was unthinkable. Over the years, emerging technologies have become an addiction. The process of job-automatization is just a consequence. Note that even the mobile phone came to replace the role of the postman. Now, postmen are rarely seen in the streets due to the usefulness of email – through which we can receive news as well as bank extracts.

This is a topic far from having a "sure" answer; even with all the information that has been collected, there is still a huge amount of material to explore. For now, according to the presented model, jobs in areas like marketing, robotics and digital influencing stood out as the ones that already exist, to a certain extent, but which are likely to grow exponentially.

Today, every organization which wants to be known at a national or international level must be present on online platforms and social media. Additionally, to achieve their targets, to analyse the market (competitors, threats, opportunities, customers' needs) and to establish personal relations with clients, marketing professionals are needed. Digital influencing as a job came up as a consequence of these new environments and as a way of promoting products. Which leads to a third job theory – influencers need constant management and analysis of their work and public, so personal social network managers will be needed – as some respondents referred. Finally, the robotics field is definitely an area that still has a lot to show.

To conclude, it is considered that – for future studies - it might be interesting to analyse, in more depth, new emerging jobs. Also, it would be enriching to perform interviews to understand what people really think about the issue at hand. Throughout the study, the survey participants stated that humans are unique regarding creativity and empathy, but this can generate the question: humans are also the main source of conflict; so if machines take humans' place, will there be no conflict? Moreover, to humans - who need a job and, in consequence, a salary to survive, the massive development of robotics may be prejudicial. However, when it comes to a CEO's perspective, machines will be preferred over workers to generate more profit. This easily leads to concerns regarding social implications: if people are not occupied with work, which activities will emerge? And what will generate income? Furthermore, a question that it would be important to answer, in the future, is whether robots can combine stored information with a creative process to be able to idealize new products.

Finally, the fact that 65% of the sample were students may be considered a limitation of the study – as they are not currently in the job market. However, on the other hand, students from top universities are very aware of the challenges that await them in the job market, which they are about to enter, and being a topic much focused upon in the media, and may thus provide important insights.

References

1. Carvalho, A.M.: O impacto da tecnologia no mercado de trabalho e as mudanças no ambiente de produção. Revista Evidência **6**(6), 153–172 (2010)
2. Cattani, A., Holzmann, L.: Dicionário de Trabalho e Tecnologia. Editora Zouk, Porto Alegre (2011)
3. BrightHR.: A future that works. a report by BrightHR – the futures agency (2015)
4. Graetz, G.: The Impact of Technology on the Labor Market. Employment Advisory Council, Paris (2016)
5. Works, R.: The impact of technology on labor markets. Mon. Labour Rev. **140**, 1–2 (2017). Bureau of Labor Statistics
6. Dachs, B.: The impact of new technologies on the labour market and the social economy. Science and Technology Options Assessment (STOA), European Parliament, 64 p. (2018). https://doi.org/10.2861/68448
7. Liz, C., Romao, R., Filipe, I., Portela, P., Carlos, F.: Work 4.0. COTEC Europe Summit 2018: "Work 4.0 - Rethinking the Human-Technology Alliance" (2018)
8. Gonçalves, R., Martins, J., Pereira, J., Cota, M., Branco, F.: Promoting e-commerce software platforms adoption as a means to overcome domestic crises: the cases of Portugal and Spain approached from a focus-group perspective. In: Trends and Applications in Software Engineering, pp. 259–269. Springer (2016)
9. Martins, J., Gonçalves, R., Oliveira, T., Cota, M., Branco, F.: Understanding the determinants of social network sites adoption at firm level: a mixed methodology approach. Electron. Commer. Res. Appl. **18**, 10–26 (2016)
10. Gonçalves, R., Martins, J., Branco, F., Perez-Cota, M., Au-Yong-Oliveira, M.: Increasing the reach of enterprises through electronic commerce: a focus group study aimed at the cases of Portugal and Spain. Comput. Sci. Inf. Syst. **13**, 927–955 (2016)

Hybrid Machine Translation Oriented to Cross-Language Information Retrieval: English-Spanish Error Analysis

Juncal Gutiérrez-Artacho[1] , María-Dolores Olvera-Lobo[2,3] ,
and Irene Rivera-Trigueros[1(✉)]

[1] Department of Translation and Interpreting,
Faculty of Translation and Interpreting, University of Granada,
C/Buensuceso, 11, 18003 Granada, Spain
{juncalgutierrez, irenerivera}@ugr.es
[2] Department of Information and Communication, Colegio Máximo de Cartuja,
University of Granada, Campus Cartuja s/n, 18071 Granada, Spain
molvera@ugr.es
[3] CSIC, Unidad Asociada Grupo SCImago, Madrid, Spain

Abstract. The main objective of this study focuses on analysing the automatic translation of questions (intended as query inputs to a Cross-Language Information Retrieval System) and on the creation of a taxonomy of translation errors present in hybrid machine translation (HMT) systems.

An analysis of translations by HMT systems was carried out. From these, there is a proposal of a type 1, 2 or 3 error taxonomy weighted according to their level of importance. Results indicate that post-editing is an essential task in the automatic translation process.

Keywords: Cross-language information retrieval ·
Hybrid machine translation systems · Translation errors · Post-editing

1 Introduction

Cross-language information retrieval (CLIR) is centred upon the search for documents, reconciling queries and documents which are written in different languages [1]. In CLIR systems, translating queries is the most frequent option since they are shorter texts than the documents and their translations have limited computational costs [2–4]. One of the most used resources for undertaking translation processes in the field of CLIR is machine translation (MT) [5–8]. Although the automatic translations offered by these systems lack the level of excellence of human translations [9, 10], they are useful within the translation process. In this regard, it is interesting to analyse the functioning of MT and the errors detected in the translation of questions from the point of view of their potential as CLIR tools.

The translation market can be defined as global, decentralised, specialised, dynamic, virtual and demanding [11]. Globalisation and the eagerness of businesses to expand into international markets has meant an increase in MT, as in many cases it is

© Springer Nature Switzerland AG 2019
Á. Rocha et al. (Eds.): WorldCIST'19 2019, AISC 930, pp. 185–194, 2019.
https://doi.org/10.1007/978-3-030-16181-1_18

impossible to satisfy the demand for translations with human translators alone. In addition, what is sought is the maximum cost reduction possible [12]. This situation has caused the profile of the translator to change, as ever more companies in the linguistic sphere are expanding their traditional offer of translation services to include services related to MT and post-editing [13, 14].

For the evaluation of MT systems and contribute to their improvement, error classification plays a key role. The existence of taxonomies including inaccuracies and the most common errors will facilitate their identification at the post-editing stage [15, 16]. The aim of this study focuses on analysing the automatic translation of questions (intended as query inputs to a CLIR system) and in the creation of a taxonomy of translation errors present in MT systems for Spanish (SP) and English (EN).

2 Machine Translation and Post-editing

One of the current trends in MT−along with Neural and Adaptive MT−is the combination of different types of architectures, giving rise to hybrid technologies [17, 18]. These systems combine the advantages of rule-based MT (RBMT) and statistical MT (SMT). Hybrid machine translation (HMT) systems attempt to solve the problems detected in these two technologies with the objective of producing better quality translations [19–21].

The identification and classification of translation errors is essential for the assessment of the effectiveness of MT systems. A number of different proposals have been implemented according to the motivation of the research, the languages used, or the fact that human or machine translations are being assessed [15].

Laurian [22] distinguished three main types of error: (a) errors in isolated words, (b) errors in the expression of relationships and (c) errors in the structure or presentation of the information. Furthermore, other typologies [23] are based on the quantitative distribution of the errors found during research undertaking and include categories related to morphology, syntax, lexicon, punctuation, style, textual coherence, textual pragmatism and literal translations of the original text. A more detailed proposal [24] establishes a typology organised around four categories of main errors-lexicon, syntax, grammar and errors due to deficiencies in the original text-which, are divided into subcategories. Some classifications [15, 25–27] develop exhaustive hierarchies structured into different levels and depending on the linguistic elements affected.

An interesting aspect in the post-editing of MT systems is the cognitive effort required to correct lexical, grammatical or style errors [28]. Indeed, post-editing work is accelerated and can be carried out much more efficiently when there is guidance to facilitate the task. The guidelines depend on different factors and vary according to the desired quality or post-editing type employed [16]. In 1985, Wagner [29] offered some post-editing recommendations, which constitutes, still today, a reference as they can be applied to different types of post-editing. These guidelines are still in force and have laid the groundwork for other authors to expand on them [16, 30–32].

3   Methodology

The MT systems used in the study needed to be free of charge, contain Spanish and English amongst the languages available and apply hybrid technology. Systran and ProMT were the only two HMT systems that fulfilled all the requirements[1]. In 2009, Systran introduced the first MT hybrid engine onto the market. For its part, ProMT presented in 2012 the ProMT DeepHybrid system.

The corpus used is a collection of questions proposed by CLEF (Cross-Language Evaluation Forum). These collections are used in this type of forum to carry out the assessment of IRS (Information Retrieval Systems) and their techniques, allowing for comparative studies [33–38]. Two collections of questions about European legislation from the ResPubliQA (2009 and 2010) track, related to the Europarl corpus were used, which includes European Parliament acts in various languages [39]. The corpus, comprised, of a sample of 100 questions, was translated from EN-SP and vice-versa by ProMT and Systran. This gave the result of 400 translations. There was an analysis of all errors produced by the HMT systems. The proposed error taxonomy takes other existing classifications as a base [7, 15, 22–27]. To establish the weighting assigned to each error a sample of 200 translations was taken. There was an identification of type 1 errors (minor); type 2 (medium) and type 3 (serious). This process was well defined in order to avoid ambiguity when performing the evaluation. Finally, the most frequent errors were determined in order to assess HMT systems.

4   Results

4.1   Error Taxonomy

The taxonomy covers five large groups: (a) orthography, (b) lexicon, (c) grammar, (d) semantics and (e) discourse. Each of these groups presents various levels (Fig. 1):

Orthography
This section includes punctuation errors, confusion between upper and lower case and spelling errors. For example, when translating *registered designations of origin* Systran translates it as *denominaciones de origen registradoas* (spelling error), instead of *registradas*, which would be the correct word.

Lexicon
This group includes errors referring both to omissions and additions, which can be unnecessary, if they affect functional words like prepositions or articles, or essential ones, if they affect the content of the translation. Errors have also been detected in the translation of abbreviations, initials and proper or institutionalized nouns. In this category are also those words or expressions that the MT systems has not translated.

---

[1] Nowadays, Systran has already implemented Neural Machine Translation in its MT systems.

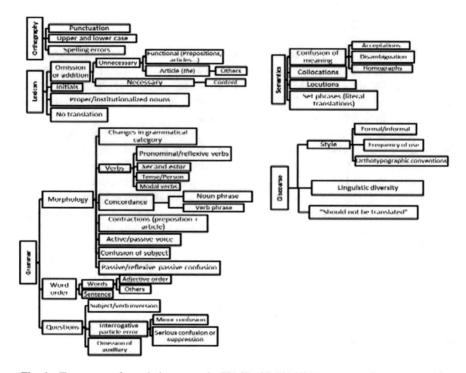

Fig. 1. Taxonomy of translation errors in EN-SP, SP-EN HMT systems (Own authorship)

Grammar

This group includes morphology errors, which affect word order and errors specific to interrogative sentences. The errors related to morphology are those connected to (i) changes in the grammar category; (ii) errors in verbs, either tense or person; (iii) confusion between the verbs *ser* and *estar*; (iv) errors in the translation of English modal verbs, reflexive verbs or pronoun verbs; (v) concordance errors, both in the verbal and noun phrase; (vi) errors in the contraction of the article; (vii) errors in the use of the passive voice, and (viii) confusion when determining the subject of the sentence.

Semantics

This category includes errors of meaning, either because there has been a confusion of acceptance with a homographic word, or due to a disambiguation problem. This group also contains errors in the translation of collocations, locutions and set phrases, as some of these are not identified as such and are literally translated.

Discourse

At the level of discourse, there has been an identification of style errors produced either by the use of an inappropriate register or due to orthotypographic conventions. There is also the inclusion of the errors related to linguistic diversity, in this case, between British and American English, and errors created by the translation of words or expressions that should be conserved in their original language.

Once this taxonomy was established, errors were grouped according to their level of importance (Table 1). Type 1 errors are minor, they correspond to errors that do not alter the meaning of the question. These errors are usually minor syntax errors which are not related to the content in itself. Type 2 errors are considered of medium importance, as they are errors, of either syntax or content, which modify the meaning of the question, although without making it unintelligible. In this case, errors in collocations are frequent. Type 3 errors include syntax or content errors that modify the meaning of the question in a way that makes it unintelligible.

Table 1. Types of translation errors in English-Spanish, Spanish-English HMT systems

Type 1 errors	Orthography	• Upper and lower case
	Lexicon	• Omissions or unnecessary additions
	Grammar	• Modal verb error • Confusion of passive and reflexive voice • Contraction of preposition or article • Error in word order • Minor confusion in interrogative particle
	Discourse	• Errors in linguistic diversity • Style errors
Type 2 errors	Orthography	• Punctuation errors
	Grammar	• Changes in grammatical category • Confusion of pronominal and reflexive verbs • Errors in verb tense and/or person • Verb or noun phrase concordance errors • Serious confusion or suppression of interrogative particle
	Semantics	• Collocation errors • Expression errors
Type 3 errors	Orthography	• Spelling errors
	Lexicon	• Omissions or unnecessary additions • Error in the translation of proper or institutionalized nouns • Error in the translation of initials • No translation
	Grammar	• Confusion of verbs *ser* and *estar* • Confusion of active and passive voice • Confusion of subject • Error in sentence order • Subject/verb inversion • Omission of auxiliary verb
	Semantics	• Confusion of meaning errors (disambiguation, acceptation or homographs) • Literal translation of set phrases
	Discourse	• "Should not be translated"

The most common type 3 errors were those related to the order of the elements in the sentence (20.97%) and those errors caused by the confusion of meaning, either due to an acceptation error or confusion with a homograph (31.18%), or due to a disambiguation error (18.82%) (Table 2).

Table 2. Type 3 errors in the analysed HMT

			Systran	ProMT	Total
Type 3 errors	Orthography	Spelling errors	3	1	4 (2.15%)
	Lexicon	Omission	1	1	2 (1.08%)
		Proper/institutionalized nouns	5	11	16 (8.60%)
		Initials	2	3	5 (2.69%)
		No translation	3	5	8 (4.30%)
		Addition	1	2	3 (1.61%)
	Grammar	Subject/verb inversion	7	1	8 (4.30%)
		Verb *ser/estar*	3	1	4 (2.15%)
		Sentence order	29	10	39 (20.97%)
		Omission of auxiliary verb	1	-	1 (0.54%)
	Semantics	Acceptation/homographs	23	35	58 (31.18%)
		Disambiguation	19	16	35 (18.82%)
	Discourse	"Should not be translated"	1	2	3 (1.61%)
Total			98	88	186

For medium errors (Table 3), there is a prominence of collocation errors (37.36%) and concordance errors, both in the verb phrase (18.68%) and the noun phrase (10.99%).

Table 3. Type 2 errors in the analysed HTM

			Systran	ProMT	Total
Type 2 errors	Grammar	Changes in grammatical category	3	5	8 (8.79%)
		Pronominal/reflexive verb	2	3	5 (5.49%)
		Verb tense	-	6	6 (6.59%)
		NP concordance	4	6	10 (10.99%)
		VP concordance	9	8	17 (18.68%)
		Serious/omission interrogative particle	4	2	6 (6.59%)
	Semantics	Collocation	19	15	34 (37.36%)
		Locution	4	1	5 (5.49%)
Total			45	46	91

In the minor errors (Table 4) there is a prominence of functional additions or omissions. (17.11% and 7.24%, respectively), errors in word order (15.13%), errors in the interrogative particle (9.21%) and errors in the use of upper case (14.47%). However, regarding the latter, it must be pointed out that in a corpus of questions on European legislation there were numerous names of bodies, institutions, committees, etc., therefore, this type of error was expected.

Table 4. Type 1 errors in the analysed HMT

			Systran	ProMT	Total
Type 1 errors	Orthography	Upper or lower case	15	7	22 (14.47%)
	Lexicon	Functional omission	6	5	11 (7.24%)
		Functional addition	16	10	26 (17.11%)
	Grammar	Word order (others)	15	8	23 (15.13%)
		Passive/reflexive passive	12	1	13 (8.55%)
		Modal verb	2	6	8 (5.26%)
		Interrogative particle	11	3	14 (9.21%)
		Preposition + article contraction	2	2	4 (2.63%)
	Discourse	Linguistic diversity	1	2	3 (1.97%)
		Style errors	21	7	28 (18.42%)
Total			101	51	152

A total of 422 errors were identified according to Table 5, of which 34.4% were minor (type 1), 20.1% were medium (type 2) and 45.5% were considered as serious (type 3). ProMT is the MT system that obtained the best results, although without large differences, returning 43.3% of the total errors compared to the 56.6% of Systran.

Table 5. Types of translation errors in the analysed HMTs

Type 1 ProMT	50 (11.9%)	Type 1 Systran	95 (22.5%)	Total Type 1	145 (34.4%)
Type 2 ProMT	44 (10.4%)	Type 2 Systran	41 (9.7%)	Total Type 2	85 (20.1%)
Type 3 ProMT	89 (21.1%)	Type 3 Systran	103 (24.4%)	Total Type 3	192 (45.5%)
Total ProMT	**183 (43.4%)**	**Total Systran**	**239 (56.6%)**	**Total**	**422**

If the languages involved are considered according to Table 6, when translating SP-EN it is observed how most errors are type 3, followed by type 1 errors, whereas there is a reduced percentage of Type 2. In SP-ES translation, there is a predominance of Type 3 errors and there is also an increase in type 2 errors, this could be since translating to Spanish can produce concordance errors (type 2 errors). In terms of total errors, the HMTs perform better overall when translating SP-EN.

Table 6. Errors according to translation language pair

Total errors	Total SP-EN	184 (43.60%)	Type 1 SP-EN	76 (41.3%)	Type 2 SP-EN	16 (8.7%)	Type 3 SP-EN	92 (50%)
	Total EN-SP	238 (56.39%)	Type 1 EN-SP	69 (29%)	Type 2 EN-SP	69 (29%)	Type 3 EN-SP	100 (42%)

5   Conclusions

In machine translations numerous types of errors appear that depend on the grammar of the languages involved, the topic of the translations or their complexity, amongst other factors. MT systems, although constituting support tools, require post-editing, as a human task linked to professional translators, which plays a fundamental role in the translation process.

The main contribution of this study, put forward from the perspective of CLIR, focuses on the establishment of a taxonomy of errors specific to the MT of questions, a type of input frequently employed in their CLIR queries. In addition, there has also been an in-depth analysis regarding existing classifications, concerning errors related to style, register, frequency of use and errors related to orthotypographic conventions. Regarding errors related to verbs, there has also been an identification of new cases such as those caused by the confusion of pronoun or reflexive verbs, the verbs *ser* – used to talk about permanent or lasting attributes – and *estar* – used to indicate temporary locations and states, and in the translation of modal verbs. Referring to errors caused by functional omissions or additions, already identified in other typologies, there is a distinction of those errors produced by the incorrect translation of the article, *the*. Finally, in order-related errors, with regards to the already existing classifications, those related to adjective order are added.

Lastly, it is worth mentioning that, although the quality of MT is still deficient, as proved in previous work [21], the demand for this type of translation tool is generalised and growing, especially in the multilingual context of the Internet. Therefore, we should focus our efforts in their improvement.

References

1. Zhou, D., Truran, M., Brailsford, T., Wade, V., Ashman, H.: Translation techniques in cross-language information retrieval. ACM Comput. Surv. **45**, 1:1–1:44 (2012)
2. Banchs, R.E., Costa-Jussà, M.R.: Cross-language document retrieval by using nonlinear semantic mapping. Appl. Artif. Intell. **27**, 781–802 (2013)
3. Sharma, V.K., Mittal, N.: Cross lingual information retrieval (CLIR): review of tools, challenges and translation approaches. In: Satapathy, S., Mandal, J., Udgata, S., Bhateja, V. (eds.) Information Systems Design and Intelligent Applications. Advances in Intelligent Systems and Computing, vol. 433. Springer, New Delhi (2016)
4. Hull, D.A., Grefenstette, G.: Querying across languages: a dictionary-based approach to multilingual information retrieval. In: Proceedings of the 19th Annual International ACM SIGIR Conference on Research and Development in Information Retrieval - SIGIR 1996, pp. 49–57. ACM Press, New York (1996)

5. Olvera-Lobo, M., Gutierrez-Artacho, J.: Language resources used in multi-lingual question-answering systems. Online Inf. Rev. **35**, 543–557 (2011)
6. García-Santiago, L., Olvera-Lobo, M.D.: Analysis of automatic translation of questions for question-answering systems. Inf. Res. **15**(4), paper 450 (2010). http://InformationR.net/ir/15-4/paper450.html
7. Olvera-Lobo, M.D., Garcia-Santiago, L.: Analysis of errors in the automatic translation of questions for translingual QA systems. J. Doc. **66**, 434–455 (2010)
8. Madankar, M., Chandak, M.B., Chavhan, N.: Information retrieval system and machine translation: a review. Procedia Comput. Sci. **78**, 845–850 (2016)
9. Allen, J.: Post-editing. In: Somers, H.L. (ed.) Computers and Translation: A Translators Guide, pp. 297–317. John Benjamins, Amsterdam/Philadelphia (2003)
10. Koponen, M.: Is machine translation post-editing worth the effort? A survey of research into post-editing and effort. J. Spec. Transl. **25**, 131–148 (2016)
11. Olvera-Lobo, M.D., Castro-Prieto, M.R., Quero-Gervilla, E., Munoz-Martin, R., Munoz-Raya, E., Murillo-Melero, M., Robinson, B., Senso-Ruiz, A., Vargas-Quesada, B., Dominguez-Lopez, C.: Translator training and modern market demands. Perspect. Transl. **13**, 132–142 (2005)
12. Lagarda, A.L., Ortiz-Martinez, D., Alabau, V., Casacuberta, F.: Translating without in-domain corpus: machine translation post-editing with online learning techniques. Comput. Speech Lang. **32**, 109–134 (2015)
13. Temizöz, Ö.: Postediting machine translation output: subject-matter experts versus professional translators. Perspectives (Montclair) **24**, 646–665 (2016)
14. Torres-Hostench, O., Cid-Leal, P., Presas, M. (coords.) El uso de traducción automática y posedición en las empresas de servicios lingüísticos españolas: Informe de investigación ProjecTA 2015, Bellaterra (2016)
15. Costa, Â., Ling, W., Luís, T., Correia, R., Coheur, L.: A linguistically motivated taxonomy for machine translation error analysis. Mach. Transl. **29**, 127–161 (2015)
16. Mesa-Lao, B.: Introduction to post-editing – the CasMaCat GUI 1. Introduction: Why post-editing MT outputs? (2013)
17. Costa-Jussà, M.R., Fonollosa, J.A.R.: Latest trends in hybrid machine translation and its applications. Comput. Speech Lang. **32**, 3–10 (2015)
18. Labaka, G., España-Bonet, C., Màrquez, L., Sarasola, K.: A hybrid machine translation architecture guided by syntax. Mach. Transl. **28**, 91–125 (2014)
19. Hunsicker, S., Yu, C., Federmann, C.: Machine learning for hybrid machine translation. In: Proceedings of the Seventh Workshop on Statistical Machine Translation, Montreal, pp. 312–316 (2012)
20. Tambouratzis, G., Athena, I., Centre, R., Amaroussiou, P.: Comparing CRF and template-matching in phrasing tasks within a Hybrid MT system. In: Proceedings of the 3rd Workshop on Hybrid Approaches to Translation (HyTra), pp. 7–14 (2014)
21. Gutiérrez-Artacho, J., Olvera-Lobo, M.-D., Rivera-Trigueros, I.: Human post-editing in hybrid machine translation systems: automatic and manual analysis and evaluation. In: Rocha, A., Adeli, H., Reis, L., Costanzo, S. (eds.) Trends and Advances in Information Systems and Technologies, WorldCIST8 2018. Advances in Intelligent Systems and Computing, vol. 745, pp. 254–263. Springer, Cham (2018)
22. Laurian, A.M.: Machine translation: what type of post-editing on what type of documents for what type of users. In: Proceedings of the 10th International Conference on Computational Linguistics and 22nd Annual Meeting on Association for Computational Linguistics, pp. 236–238 (1984)
23. Krings, H.: Repairing Texts: Empirical Investigations of Machine Translation Post-editing Processes. Kent State University Press, Kent (2001)

24. Schäfer, F.: MT post-editing: how to shed light on the "unknown task" - Experices made at SAP. In: Joint Conference on 8th International Workshop European Association for Machine Translation, 4th Controlled Language Application Workshop, pp. 133–140 (2003)
25. Farreús, M., Costa-Jussà, M.R., Morse, M.P.: Study and correlation analysis of linguistic, perceptual, and automatic machine translation evaluations. J. Am. Soc. Inf. Sci. Technol. **63**, 174–184 (2012)
26. Sivakama, S., Prema, V., Savitha, G.: A comparative study of occurrence of errors in machine translation in a multilingual environment. Eng. Sci. Int. J. (ESIJ) **3**, 1–4 (2016)
27. Vilar, D., Xu, J., D'Haro, L., Ney, H.: Error analysis of statistical machine translation output. In: Proceedings of LREC (2006)
28. Vieira, L.N.: Cognitive Effort in Post-Editing of Machine Translation: Evidence from Eye Movements, Subjective Ratings, and Think-Aloud Protocols (2016). http://hdl.handle.net/10443/3130
29. Wagner, E.: Post-editing systran - a challenge for commission translators. Terminol. Trad. **3**, 1–7 (1985)
30. O' Brien, S.: Researching and Teaching Post-Editing (2009). http://www.mt-archive.info/MTS-2009-OBrien-ppt.pdf
31. TAUS: MT Post-editing Guidelines. https://www.taus.net/academy/best-practices/postedit-best-practices/machine-translation-post-editing-guidelines
32. Hu, K., Cadwell, P.: A comparative study of post-editing guidelines. Balt. J. Mod. Comput. **4**, 346–353 (2016)
33. Olvera-Lobo, M.-D., Gutiérrez-Artacho, J.: Evaluación de los sistemas QA de dominio abierto frente a los de dominio especializado en el ámbito biomédico. In: I Congreso Español de Recuperación de Información (CERI 2010), Madrid, pp. 161–169 (2010)
34. Olvera-Lobo, M.D., Gutiérrez-Artacho, J.: Question-answering systems as efficient sources of terminological information: an evaluation. Heal. Inf. Libr. J. **27**, 268–276 (2010)
35. Olvera-Lobo, M.D., Gutiérrez-Artacho, J.: Multilingual question-answering system in biomedical domain on the web: an evaluation. Lecture Notes in Computer Science, vol. 6941, pp. 83–88 (2011)
36. Olvera-Lobo, M.-D., Gutierrez-Artacho, J.: Performance analysis in web-based question answering systems. Rev. Esp. Doc. Cient. **36**(2), e009 (2013)
37. Olvera-Lobo, M.-D., Gutierrez-Artacho, J.: Question answering track evaluation in TREC, CLEF and NTCIR. In: Rocha, A., Correia, A.M., Costanzo, S., Reis, L.P. (eds.) New Contributions in Information Systems and Technologies, pp. 13–22 (2015)
38. Gutiérrez Artacho, J.: Recursos y herramientas lingüísticos para los sistemas de búsqueda de respuestas monolingües y multilingües (2015)
39. Koehn, P.: Europarl: a parallel corpus for statistical machine translation. MT Summit **11**, 79–86 (2005)

Fake News and Social Networks: How Users Interact with Fake Content

Manuel Au-Yong-Oliveira[1(✉)], Carlota P. A. Carlos[2], Hugo Pintor[3],
João Caires[3], and Julia Zanoni[2]

[1] GOVCOPP, Department of Economics, Management,
Industrial Engineering and Tourism, University of Aveiro,
3810-193 Aveiro, Portugal
mao@ua.pt
[2] Department of Physics, University of Aveiro, 3810-193 Aveiro, Portugal
{carlotapereiracarlos,julia.ines}@ua.pt
[3] Department of Electronics, Telecomunications and Informatics,
University of Aveiro, 3810-193 Aveiro, Portugal
{hrcpintor,higinocaires}@ua.pt

Abstract. Fake news, fabricated stories, rumours, misleading headlines (the so-called clickbaits) are nothing new. The difference of the current context is how fast and wide they spread online, mainly through social network websites, reaching sometimes millions of users. Recently, this problem gained visibility due to its influence on political systems, especially electoral processes and political polarization. The current study addresses these matters, studying how users interact with news in social media platforms, with a deeper insight concerning Facebook and Twitter. The findings not only provide evidence of the importance of social networks as gateways for news, but our data also shows that consumer trust in news is worryingly low, combined with high levels of concern about false news stories.

Keywords: Fake news · Social networks · Facebook · Twitter

1 Introduction

Throughout history there have been several changes in media technology. In recent years, the spotlight has shifted to social media. Platforms such as Facebook have a radically different structure from their predecessors – content can be spread between users without significant third-party filtering, fact checking or editorial review. An individual user with no track record or reputation is sometimes capable of reaching as many readers as CNN or the *New York Times*. Additionally, digitisation and recent advances in technology, such as the acceptance and diffusion of the smartphone, propelled most newspapers to establish an online presence [1, 2]. The ever-growing popularity of social network sites (SNSs) made newspapers extend their online experiments further, including platforms such as Facebook and Twitter (the former being the biggest worldwide and the latter emphasizing the rapid diffusion of information, thus our interest in them). A clear motivator is the amount of traffic these

© Springer Nature Switzerland AG 2019
Á. Rocha et al. (Eds.): WorldCIST'19 2019, AISC 930, pp. 195–205, 2019.
https://doi.org/10.1007/978-3-030-16181-1_19

platforms have. YouTube, Facebook, Twitter, Instagram and Reddit, for example, are all within the top 15 of the world's most visited websites in terms of pages viewed, time spent, percentage of traffic from search and total sites linking in [3].

Following the 2016 US election, a specific concern has arisen about the effect of false stories – "fake news" as they have been referred to – circulating on social media platforms. We define fake news as news articles that are deliberately and verifiably false, and that could delude readers. This type of content imposes social costs by making it difficult for consumers to infer the true state of the world. There appear to exist two main motivations for providing fake news. The first is monetary: news articles that go viral on social media can draw significant advertising revenue when users click to the original site. The second motivation is ideological, with providers seeking to grant an advantage to candidates, policies, parties, etc. that they favour [4]. The spread of false news stories has become such a tremendous problem in today's society that multiple platforms (sites, apps, plugins, etc.) that allow people to fact check news, news sites and twitter profiles and identify social media "bots" have been created [5, 6].

In Portugal, false news stories have been less of a concern when compared to countries such as the US, UK and Brazil, due to the more stable political situation – the latter are all polarised countries where recent or ongoing elections or referendum campaigns have been affected by disinformation and misinformation. Nonetheless, more and more cases of fake news in Portuguese media are being reported. In the summer of 2017, for example, during one of the deadliest wildfires in Portugal's history, national and international media reported the crash of a Canadair plane tackling the fires. Later, this was confirmed to be false – there was no record of any aircraft crash whatsoever [7, 8]. Recently, one of Portugal's main brands of newspaper reportedly discovered several Canadian-based websites that hosted false news stories about Portuguese politicians. These were then spread by several Facebook groups with thousands of members [9].

Taking into consideration the impact of fake news in other countries and in international affairs, and considering the next few months' foreseen volatile political environment, in Portugal, due to next year's legislative elections, we believe it to be important to study how social network users interact with false content in news. Our goal with this paper is to study the trust in news, particularly in Portugal, as well as the user perception of the impact of the Internet and of SNSs on quality journalism. We also investigate the perception of our sample on how common fake news is as well as its impact on individuals.

The dimension of this problem is significant. Facebook alone removed more than 1.5 billion fake accounts in the last six months [36].

2   Fake News and Social Networks – Literature Review

In the 19th century, cheap newsprint and better-quality presses permitted newspapers to expand their reach very significantly. One century later, as radio and television became key players, observers worried that these new platforms would harm policy debates, privileging charismatic candidates over those who might have more capability to lead [10]. Between 2000 and the early 2010s, the growth of online news invoked a new set of concerns [11–13]. Nowadays, the attention has shifted to the dissemination of false content in social network websites. With their growing in popularity, SNSs have become an emerging focus of scholarly research [14–16]. This study focuses on two of the most popular ones – Facebook and Twitter – since they are being adopted as news media by newspapers as alternative publishing platforms [2, 17, 18].

Facebook

Facebook is one of the largest SNSs on earth, becoming one of the most popular means of communication [19–21]. Launched in 2004, this SNS was targeted at college students – Facebook allowed university students to create and establish social connections relevant to their university experiences [22, 23]. As of June 2018, Facebook had 2.23 billion active users, who visited the site at least once a month, and 1.47 billion daily active users, who visited the site daily [24]. To access this platform, the user needs to register an account to create a profile. If a registered user adds another as a Facebook friend (they send a request and the other confirms it) their "friendship" is established. Users may post status updates, upload photos and videos and instantly message their friends. Facebook also offers a large range of applications and games.

Twitter

Twitter is a SNS with microblogging characteristics, founded in 2006 [25]. Internet users can access Twitter and see public accounts without registering, but in order to "tweet", one must register. To tweet is to post a short (280 character limit) status update, with the possibility to add media. In 2018, the number of monthly active users was 335 million [26]. Twitter users build connections with each other through "following" – a user can follow any other user (without any type of approval if it is not a private account) and can see in real time all the tweets from those she/he follows. At first glance, Twitter might appear unappealing, due to its limited functionalities, yet Twitter's success lies in its simplicity: Twitter users can "tweet" anytime, anywhere, which makes Twitter a powerful distribution channel for breaking news. For instance, when Michael Jackson died suddenly in 2009, the first tweet was reported 20 min after the initial 911 call, an hour before mainstream news media broke the news [27].

Digital news consumption has been shown to be so important that multiple organizations and institutes have issued several reports on the topic. For example, the Reuters Institute for the Study of Journalism has commissioned yearly reports, since 2012, to understand how news is being consumed in a range of countries [28–32]. They found that the growth in social media for news reached a standpoint in 2017, with the rise in the usage of messaging apps for news, as consumers look for more private (and less confrontational) spaces to communicate. A concern was also reported for fake content online. Data suggests that users feel that the lack of rules and viral algorithms encourages low quality and 'fake news' to spread quickly [33]. The same was true for 2018 [34]. The average level of trust in the news found in this year's report (across all countries) was of 44%. In contrast, only 23% claimed to trust the news they find in social media and over half to agree or strongly agree that they are concerned about what is real and fake on the Internet. This concern is highest in countries like Brazil, Spain, and the United States where intense social media use merges with extremely polarised political situations. Interestingly, Portugal is ranked (jointly) highest for trust in news across the

37-country survey. However, a low confidence in news in social media (29%) and suspicions about the legitimacy of online news content raise some concerns.

3 Methodology

A literature review was performed, which combined with the growing awareness of the consequences of false news stories, exposed the need to study how users interact with online news. It is also important to see how these false stories are affecting the users' opinion of journalism in general. Online databases, as well as available online newspapers and websites were sources of all the data collected for this part of the study. The goal of this research was then determined, as stated in the introduction.

Although several impactful articles related to these topics were found, only one presented a comprehensive and global study of digital news consumption and the global effects of fake news – most of the studies were specific to one situation or country in particular. Therefore, considering the impact of fake news in other countries and in international affairs, we believe it to be important to study further how users interact with false content in news. Research was conducted through an online questionnaire (shared on the online networks Facebook and Twitter, of the authors) from the 11th to the 17th of October 2018, and to which 803 responses were obtained. The sample was thus a convenience sample. "In the field of business and management, convenience samples are very common" [35], though it is not possible to generalize the findings. The results "could provide a springboard for further research or allow links to be forged with existing findings in the area" [35]. 30.1% had completed high school. 37.5% had a Bachelor's degree. 11.1% had a Master's degree. 2% had a doctoral degree. 57.5% of the respondents were female, while 42.5% were male. Participants were also asked to choose their age group (in years) – younger than 18 (18-), 18–25, 26–35, 35–50, 51–66, and older than 65 (65+) – and all age groups are represented in the collected data. 18–25 is the most represented age group, with 50.2% of the sample. Concerning other demographics, it is important to point out that 97% of the sample are Portuguese, 2% are Brazilian and 1.1% are of other nationalities.

4 Results and Data Analysis

Previous studies demonstrated the relevance of online platforms for the dissemination of news and our results are in accordance with the expectations, that online social networks are very highly used for this purpose. More than 95% (95.3%) of participants strongly believe the Internet has contributed greatly for the diffusion of information. Above 60% of the participants claim to mainly get their news online: 31.8% say they mainly read news on social media, while 29.8% prefer websites. Roughly 40% prefer more traditional news media: 30.8% claim TV as their main source of news, while 4.5%, 2.7% and 0.5% prefer printed newspapers, radio and blogs, respectively. Asked to identify the SNSs that they often use, Facebook, Youtube and Instagram were the clear winners, as seen in the graph of Fig. 1. Looking at these numbers, the importance of SNSs in news consumption is quite evident and again reinforced.

Percentage that uses each social network regulary

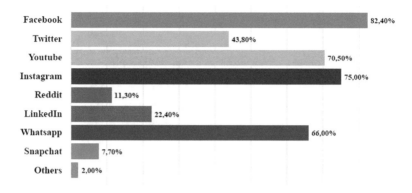

Fig. 1. SNSs often used by the respondents, with Facebook, Youtube and Instagram the clear winners. *Q11. Which social networks do you often use? Base: total sample (803).*

In agreement with other studies previously mentioned, our data also shows that consumer trust in news is worryingly low, combined with high levels of concern about so-called fake news. A large majority agree or strongly agree to be concerned about what is real and fake on the Internet – 89.1% of the sample expect the number of inaccurate news they see on social media platforms to be at least equal to the number of accurate ones and 43.5% believe the number of fake news to be actually higher. This can be seen in graph (a) of Fig. 2. More than half (58.4%) also claim the Internet has contributed to the decline in quality journalism.

Regarding the importance of the source's credibility in the news they share in SNSs, 96.35% of all of the participants in the survey agree or strongly agree it to be important. When asked who they think has the responsibility to verify news for fake and unreliable content, most believe that publishers (58.0%) have the biggest responsibility. 14.57% agree the responsibility should fall on to an external regulatory entity and 12.7% consider it to be the sole responsibility of the reader. Only 5.86% of respondents hold accountable the SNSs where news are posted. Of all the participants, 5.7% admitted to have shared a fake news story intentionally in their social media accounts.

When confronted with false content, the most common attitude is to simply ignore the post (44.6% of the sample), as shown in graph (b) of Fig. 2. The second most common is to share it with their family or their family and friends (18.1%), followed by reporting it (15.2%). Some people also point out the inaccuracy of the post, either by leaving a comment or sharing (14.4%), while others just warn the person who shared the story directly (7.7%).

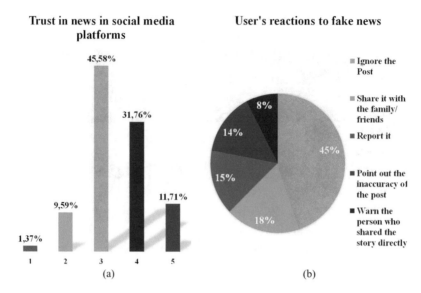

Fig. 2. (a) Level of trust in news in social media for all participants. *Q8. Do you believe social media platforms spread more fake news than trustworthy ones? (1 corresponds to "I strongly disagree" and 5 "I strongly agree") Base: total sample (803).* (b) User's reaction to fake news in social media. As we can see, when confronted with false content, the most common attitude is to simply ignore the post *Q20. When you encounter fake news in social media, which of the following do you do? Base: total sample (803).*

The Case of Facebook and Twitter

Almost 99% of the participants are Facebook and/or Twitter users. This is not, nonetheless, representative of the population as a whole, since most people came into contact with the online questionnaire through those media. Asked about how often they use these social media platforms as a source of news, 69.3% of Facebook users answered that they recur to it at least once a week and 43.6% claim to use it daily. 72.7% of Twitter users acknowledge to use it weekly as a news source, with 51.2% admitting to use it daily. Nearly three quarters (74.1%) of Facebook users claim to follow national and/or international newspaper brands and/or other credible sources[1], with 67.2% of Twitter users admitting the same.

Reflecting on the collected data for these two SNSs, it seems more users resort to Twitter as a news gateway, as opposed to Facebook. This happens even though this platform came in fourth in the graph of Fig. 1 (Facebook came in first). This seems to suggest that Facebook users are not as interested in following the news. The different characteristics of the two platforms, however, may explain this trend – while Twitter is highly focused on reading and posting tweets and media, Facebook has multiple features that users can choose from (e.g. posting/viewing photos or videos, playing games, chatting, etc.). To summarize, Facebook is a broad social network, while Twitter

---

[1] Discerning the credibility of the sources was left to the participants (no list of credible sources was supplied). Consequently, these results should be evaluated taking this into consideration.

appears to be more of social *media*, more news-oriented. It is important to point out, however, that Facebook's overall user base is much larger than Twitter's, so far more people get their news through the former.

More than a fourth (27.7%) of Facebook users agree to see news they know to be (or later find out to be) false daily, with 78.2% agreeing to find fake stories at least once a week. Only 12.9% of Twitter users, on the other hand, claim to find fake news on a day–to–day basis, while 61.7% agree on finding them at the minimum weekly. The percentages for both networks, nonetheless, are worrying. Facebook users were also asked if they had any knowledge of the reporting tool that allows people to report fake news specifically. 54.6% of the users answered affirmatively, with 15.6% claiming to use it at least once a week. However, most users, 51.8%, rarely use it. This is interesting: even though more than 78% of Facebook users agree to find fake news stories weekly and more than half seem to know of this reporting tool, most people seldom recur to it.

Age Differences

The behaviour of SNSs news consumers can look significantly different – people from different age groups, nationalities, genders, etc. interact differently. In our research, we were able to study how different age groups interact with news in general. 48.39% of the participants with ages between 18 and 50 believe the number of false news stories to be higher than those with trustworthy content. In contrast, only 22.2% of the participants with ages higher than 65 believe the same (perhaps also due to their lesser use of online social networks for news purposes – Fig. 3). On the contrary, all age groups strongly believe that the Internet has had a nefarious effect on good quality journalism and agree that the responsibility to check factual news on social media lies with the reader. They also agree on the importance of the credibility of the source – above 90% of the participants of all age groups agree or strongly agree on its importance (except for the age group 65+, where only roughly 70% are of the same opinion). However, the strongest differences were found in their chosen gateways to news, as seen in the graph of Fig. 3. The demographic push from under 35 s is towards a greater use of social platforms and less direct access, such as printed newspapers and the television, as expected. In other generations, these media still play a significant role.

Of all the participants, people under 18 were the age group with the lowest percentage of daily usage of Facebook or Twitter for news (11.1% and 40.5% respectively). In contrast, consumers between the ages of 26 and 35 and 36 and 50 have the highest percentages – 54.9% and 51.9% for Facebook, respectively, and 74.2% and 79.7% for Twitter. Younger users also worry less about following credible national/international newspapers – just 44.4% and 69.0% answered positively when asked. 26–35 and 36–50 users were the ones with the higher percentages once again: 82.9% and 84.2% for Facebook (respectively) and 90.3% and 92.2% for Twitter. Once more, the percentage in all age groups is also higher in Twitter. Twitter's daily usage as a gateway for news seems to be higher than Facebook's in all age groups.

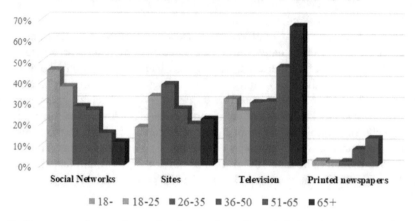

Percentage that uses each news media, by age group

Fig. 3. Percentage of respondents who chose each media as their favorite gateway to news, by age group. In the graph only the four with the highest representation in all age groups are represented, for simplicity reasons, leaving out "radio" and "blogs". The demographic push from under 35 s is towards greater use of social platforms and less direct access (printed newspapers and the television). *Q12. Which news media platform do you use more often? Base: total sample (803).*

Ethnicity Differences - Portugal and Brazil

The main nationality of the participants of our survey was Portuguese (779 of the total 803 which made up the sample), with Brazilians in second place in a much smaller number (15). The number of the latter is not enough to validate results found in quantitative studies. However, it is worth noting the following exploratory data:

- Only 70.6% of the Portuguese respondents strongly agree that the Internet has an important role in the diffusion of information, in contrast with 93.3% of the Brazilians.
- Around 60% of the Brazilians agree or strongly agree that social media platforms spread more fake news than trustworthy ones, in contrast with only 43% of Portuguese people.
- Finally, the percentage of Brazilian users that find fake news daily (in Facebook and Twitter) is much higher than for Portuguese users (47.4% and 22%, respectively).

When comparing these two ethnic groups, one finds that Brazilians seem to show a higher concern for fake news and their effect on journalism. This was to be expected, since political tensions and a highly polarised political environment marked this year's presidential elections in Brazil (in 2018), with campaigns deeply swayed by the spread of fake news in social media platforms.

5   Conclusion and Future Research

Although fabricated news is not a modern phenomenon, recently this problem gained visibility due to its influence on political systems, especially electoral processes and political polarization. Since future efforts in combating fake news and policy initiatives are likely to be conducted based upon the perceived influence of fake news, it is important that researchers study how people perceive the influence of fake news and how they interact with it. This takes on special importance as when discussing fake news regulation and media literacy intervention, one enters a complex terrain between guaranteeing freedom of speech and the limits of its exercise. The current study, therefore, addressed these matters, studying how users interact with news in social media platforms, with a deeper insight concerning Facebook and Twitter. The findings not only provided evidence of the importance of social media as gateways for news, but our data also shows that consumer trust in news is worryingly low, with high levels of concern about fake news (89.1% of our sample expect the number of inaccurate news they see on social media platforms to be at least equal to the number of accurate ones and 43.5% believe the number of fake news to be actually higher). Comparing user behaviour on Facebook and Twitter, we found that the latter appears to be more news-oriented. Age and ethnic differences in user interactions were also studied.

To reduce the spread of fake news and to address the underlying problems they have revealed in our society, it is necessary to redesign our ecosystem and culture – the way we write, consume and share news – to value and promote truth. This effort needs to be global in scope, since many countries face challenges regarding fake and trustworthy news. Our suggestion for future work is to promote intensive interdisciplinary research, similar to the study presented here, on additional social networks (e.g. WhatsApp, which is gaining popularity in countries with less freedom of speech) and in more countries, to better understand this phenomenon globally. We also intend to perform more advanced and complex statistical analyses on the data presented here.

References

1. Thurman, N., Schifferes, S.: The future of personalization at news websites. Journal. Stud. **13**(5–6), 775–790 (2012)
2. Hermida, A., Fletcher, F., Korell, D., Logan, D.: Share, like, recommend. Journal. Stud. **13**(5–6), 815–824 (2012)
3. Alexa: The top 500 sites on the web (2018). https://www.alexa.com/topsites. Accessed 17 Oct 2018
4. Allcott, H., Gentzkow, M.: Social media and fake news in the 2016 election. J. Econ. Perspect. **31**(2), 211–236 (2017)
5. OSoMe: Botometer (2018). https://botometer.iuni.iu.edu/. Accessed 22 Oct 2018
6. Bouzy, C.: Bot sentinel (2018). https://botsentinel.com/. Accessed 22 Oct 2018
7. PÚBLICO: A história da notícia de um avião que, afinal, não caiu (2017). https://bit.ly/2QjH7qX. Accessed 24 Oct 2018
8. Batchelor, T.: Portugal fires: plane crashes while fighting deadly blazes. The Independent (2017). https://ind.pn/2DpoDCq. Accessed 24 Oct 2018

9. Pena, P.: Como funciona uma rede de notícias falsas em Portugal (2018). https://bit.ly/2QdSksS. Accessed 22 Oct 2018
10. Lang, K., Lang, G.E.: Television and Politics. Transaction Publishers, Piscataway (2002)
11. Sunstein, C.R.: Republic.com. Princeton University Press, Princeton (2002)
12. Sunstein, C.R.: Republic.com 2.0. Princeton University Press, Princeton (2009)
13. Pariser, E.: The Filter Bubble. Penguin Books Ltd., London (2012)
14. Kaplan, A.M., Haenlein, M.: Users of the world, unite! The challenges and opportunities of social media. Bus. Horiz. **53**(1), 59–68 (2010)
15. Kietzmann, J.H., Hermkens, K., McCarthy, I.P., Silvestre, B.S.: Social media? Get serious! Understanding the functional building blocks of social media. Bus. Horiz. **54**(3), 241–251 (2011)
16. Steinfield, C., Ellison, N.B., Lampe, C.: Social capital, self-esteem, and use of online social network sites: a longitudinal analysis. J. Appl. Dev. Psychol. **29**(6), 434–445 (2008)
17. Barnard, S.R.: Citizens at the Gates. Springer, Cham (2018)
18. Matsa, K.E., Shearer, E.: News use across social media platforms 2018 (2018). https://pewrsr.ch/2Qee8VE. Accessed 23 Oct 2018
19. Ross, C., Orr, E.S., Sisic, M., Arseneault, J.M., Simmering, M.G., Orr, R.R.: Personality and motivations associated with Facebook use. Comput. Hum. Behav. **25**(2), 578–586 (2009). https://doi.org/10.1016/j.chb.2008.12.024
20. Martins, J., Gonçalves, R., Pereira, J., Cota, M.: Iberia 2.0: a way to leverage Web 2.0 in organizations. In: 2012 7th Iberian Conference on Information Systems and Technologies (CISTI), pp. 1–7. IEEE (2012)
21. Martins, J., Gonçalves, R., Oliveira, T., Cota, M., Branco, F.: Understanding the determinants of social network sites adoption at firm level: a mixed methodology approach. Electron. Commer. Res. Appl. **18**, 10–26 (2016)
22. Facebook: Our history. Facebook Newsroom (2018). https://newsroom.fb.com/company-info/. Accessed 23 Oct 2018
23. Ellison, N.B., Steinfield, C., Lampe, C.: The benefits of Facebook "friends:" social capital and college students' use of online social network sites. J. Comput.-Mediat. Commun. **12**(4), 1143–1168 (2007)
24. Portal, T.S.: Number of monthly active Facebook users worldwide as of 2nd quarter 2018 (in millions) (2018). https://bit.ly/2F3kqGj. Accessed 23 Oct 2018
25. Java, A., Song, X., Finin, T., Tseng, B.: Why we twitter. In: Proceedings of the 9th WebKDD and 1st SNA-KDD 2007 Workshop on Web Mining and Social Network Analysis. ACM Press (2007)
26. Portal, T.S.: Number of monthly active Twitter users worldwide from 1st quarter 2010 to 2nd quarter 2018 (in millions) (2018). https://bit.ly/2dt7OI9. Accessed 23 Oct 2018
27. Sankaranarayanan, J., Samet, H., Teitler, B.E., Lieberman, M.D., Sperling, J.: Twitter stand. In: Proceedings of the 17th International Conference on Advances in Geographic Information Systems – GIS 2009. ACM Press (2009)
28. Newman, N.: Reuters institute digital news report 2012. Reuters Institute for the Study of Journalism (2012)
29. Newman, N., Levy, D.A.L.: Reuters institute digital news report 2013. Reuters Institute for the Study of Journalism (2013)
30. Newman, N., Levy, D.A.L.: Reuters institute digital news report 2014. Reuters Institute for the Study of Journalism (2014)
31. Newman, N., Levy, D.A.L., Nielsen, R.K.: Reuters institute digital news report 2015. Reuters Institute for the Study of Journalism (2015)
32. Newman, N., Fletcher, R., Levy, D.A.L., Nielsen, R.K.: Reuters institute digital news report 2016. Reuters Institute for the Study of Journalism (2016)

33. Newman, N., Fletcher, R., Kalogeropoulos, A., Levy, D.A.L., Nielsen, R.K.: Reuters institute digital news report 2017. Reuters Institute for the Study of Journalism (2017)
34. Newman, N., Fletcher, R., Kalogeropoulos, A., Levy, D.A.L., Nielsen, R.K.: Reuters institute digital news report 2018. Reuters Institute for the Study of Journalism (2018)
35. Bryman, A., Bell, E.: Business Research Methods, 4th edn. Oxford University Press, Oxford (2015)
36. Ingraham, N.: Facebook removed over 1.5 billion fake accounts in the last six months. Engadget, 15 November 2018. https://www.engadget.com/2018/11/15/facebook-transparency-report-fake-account-removal/?guccounter=1. Accessed 07 Jan 2019

What Will the Future Bring? The Impact of Automation on Skills and (Un)employment

Manuel Au-Yong-Oliveira[1,2(✉)], Ana Carina Almeida[1],
Ana Rita Arromba[1], Cátia Fernandes[1], and Inês Cardoso[1]

[1] Department of Economics, Management, Industrial Engineering and Tourism,
University of Aveiro, 3810-193 Aveiro, Portugal
{mao,almeida.carina,ana.rita.arromba,catiafernandes,
inesteixeiracardoso}@ua.pt
[2] GOVCOPP, Aveiro, Portugal

Abstract. A big concern in seeking to understand the evolution of the future of jobs and the demand for skills with the increase of automation is shown by the wide range of existing literature regarding the subject. This study gathers relevant information about the topics of: automation, the future of employment, advantages and disadvantages of technological changes, changes in the range of skills and if people are ready for the future changes or not. Firstly, the authors give an overview on the main topics that will be discussed later on in the article and construct a theoretical framework based on the current literature. Secondly, the methodology is presented namely the field work involving 21 interviews. After that, the interview results are analyzed, discussed and compared with what was shown in the literature review. The results revealed that people in general are not aware of the impact that automation will have on their working environment. Additionally, the authors concluded that existing education systems should be reformulated since the portfolio of skills will change in the future.

Keywords: Automation · Skills · Unemployment · Technological change ·
Higher education

1 Introduction

Evolution makes the future uncertain. It scares some people but encourages others. Industries and services alike are suffering transformations. Fields like Machine Learning, Artificial Intelligence or Robotics are still not totally explored, they are actually far from being so. This is a result of the development of automation, that will certainly change the way humans face their lives.

The automation of processes consists in the manufacturing activities and material flows that are handled entirely automatically. The objectives are to free workers from dirty, dull or dangerous jobs, and to improve quality by eliminating errors and variability; and to cut manufacturing costs by replacing increasingly expensive people with ever-cheaper machines [17].

This concept brings opportunities such as increasing productivity, improving safety, ergonomics, quality, agility and flexibility [10]. To capture the value of the

© Springer Nature Switzerland AG 2019
Á. Rocha et al. (Eds.): WorldCIST'19 2019, AISC 930, pp. 206–217, 2019.
https://doi.org/10.1007/978-3-030-16181-1_20

opportunities presented, it is important that the automation strategy is aligned with the business and operations strategy, having a clear articulation of the problem and return on investment [4, 17].

Over the last century, whenever unemployment rates have risen, there have always been some who have blamed machines [14]. Fear of technology is a reaction to employment problems that happened in the past because the substitution of tasks can create a displacement of workers from the duties that are being automated [2].

Technological unemployment is a concept associated with this way of contemplating things. It has been defined as the "unemployment due to our discovery of means of economizing the use of labor outrunning the pace at which we can find new uses for labor"; and according to whom "the increase of technical efficiency has been taking place faster than we can deal with the problem of labor absorption" [9].

Workers will have to be continually retrained to work alongside machines as their jobs continue to evolve, requiring changes in skills, mind-sets and in the culture of the organization [3, 10]. The ability to reallocate tasks between robots and people helps productivity, since it allows companies to rebalance production lines as demand fluctuates [17]. Points of view like these discard technological unemployment. The regression results in [13] indicate that technological innovations have no effect on unemployment.

Bearing in mind the information considered above, this study covers two main questions:

1. What is the impact of the automation on the future of employment?
2. What is the impact of the automation on the panorama of skills?

The main contribution of this article is to complement the literature in topics such as the development of automation and the future of employment; balancing advantages and disadvantages of technological changes regarding jobs and the skills needed; and whether if people are ready for the future changes or not.

2   Literature Review

2.1   Framework on Automation and Jobs

Several Industrial Revolutions have already occurred, from the transition of some manufacturing processes to industrialization and then to the rise of new technologies. Together with the revolutions, employment has suffered changes caused by technological change.

Technological change is a complex process which comes in waves and different phases [15]. Market, social and political forces are driving the dynamics of job destruction and job creation. The author explains that those dynamics are caused by innovation in production processes, in production structures and social transformations.

Several studies have been performed to analyze and understand the relationship between automation and technological unemployment, as well as on how will the future be for them. There are two possible views: that automation contributes to unemployment or that automation boosts employment.

On the one hand, not all tasks can be completely automated [7]. Some require flexibility, making them less vulnerable to automation. At the same time, problem-solving skills are becoming relatively productive, explaining the substantial employment growth in occupations involving cognitive tasks where skilled labor has a comparative advantage.

On the other hand, the ongoing decline in manufacturing employment and the disappearance of routine jobs is causing the low rates of employment [5, 21]. The developments in robotics, artificial intelligence and machine learning enable computers to perform a range of routine physical activities better and cheaper than humans. These authors agree that much of the current debate about automation has focused on the potential for mass unemployment. But the productivity estimates assume that people displaced by automation will find other types of employment. It is expected that business processes will be transformed, as well as the workers. That will alter work-places, requiring a new degree of cooperation between workers and technology [11].

This situation has already happened in the past and the shifts that occurred in the labor force did not result in long-term mass unemployment, because they were accompanied with the creation of new types of work [11].

2.2 Impact of Automation on the Future

A lot of studies have been written focusing on this question in particular: "How is automation going to impact the future?", that is, unemployment, the labor market and society in general. A lot has been discussed about how "Robots are killing our jobs" and explanations have been given on how the savings from increased productivity are recycled back into the economy to create demand and consequently new jobs. The authors consider that some people overestimate the ability of computers to substitute humans and assume that despite the prospect that the rate of innovation will continue or even accelerate, it is unlikely that the systems resulting from that innovation will be ready for commercialization immediately [14].

Furthermore, as analyzed in [6], Germany is one of the biggest users of robots in its industry, having 7.6 robots per thousand workers, while in the meantime Europe has only 2.7 and the United States 1.6. With this research, it was discovered that people that work in industries that use and implement robots have a higher probability of keeping their jobs. A big percentage of these workers end up executing different tasks and people with higher skills in engineering and other specific subjects are favored, having higher earnings.

These two research studies had similar conclusions: robots do not affect employment itself but there will be an overall shift in the economy in the direction of higher-skill and higher-wage jobs. Consequently, it can be said that jobs are not disappearing, just shifting.

In addition, having in mind evidence from the US, it was found that between 1990 and 2007 for each robot that was implemented, 6.2 workers lost their jobs [1]. Although this is a negative scenario, the percentage of jobs affected by automation is relatively small and as such, there is no actual evidence that automating processes will make most jobs disappear.

However, there are still some individuals, like Elon Musk and Richard Branson, that argue that robotization will likely take millions of people out of work. Having this in mind, some authors discuss that this consequence can be regulated by blocking and restricting the adoption of robots [18]. Other authors have similar opinions and one of the options to stop the rise of unemployment is to slow down innovation and technological change [12].

2.3   Skills Are Changing

The impact of technological, demographic and socio-economic disruptions on business models will be felt in transformations to the employment landscape and skill requirements, resulting in challenges for recruiting, training and managing talent [20].

First, the focus will be on skills and competencies. "The Future of Jobs" is a report done as a result of an extensive survey on senior talent and strategy executives. This survey focuses on what the current and future employment trends are, what is changing both industries and business models and what are the future workforce strategies [20].

Across the industries embraced in the survey, about two thirds of the respondents (65%) report intentions to invest in the reskilling of current employees as part of their change management and future workforce planning efforts. However, companies that report both that they are confident in the adequacy of their workforce strategy and that these issues are perceived as a priority by their top management are nearly 50% more likely to plan to invest in reskilling. Also described in this report is the core set of work-relevant skills, namely Abilities, Basic Skills and Cross-functional Skills [20]:

- Abilities include Cognitive Abilities (cognitive flexibility, creativity, problem sensitivity) and Physical Abilities (physical strength and precision).
- Basic Skills encompass Content Skills (active learning, oral, reading and written expression) and Process Skills (active listening and critical thinking).
- Cross-Functional Skills cover Social Skills (emotional intelligence, negotiation, persuasion, service orientation, and teaching others), Resource Management Skills (people and time management, management of financial and material resources), Technical Skills (quality control, programming, equipment operation and control), System Skills (judgement and decision making) and Complex Problem-Solving Skills (complex problem solving).

Complex problem-solving will be required as one of the core skills. Social skills will be in higher demand than narrow Technical skills. Cognitive abilities and Process skills will be a growing part of the core skills for many industries. The skills family with the most stable demand will be Technical skills. A wide range of occupations will require a higher degree of cognitive abilities as part of their core skill set.

Among all jobs requiring physical abilities, less than one third (31%) are expected to have a growing demand for these in the future, about as many as the proportion of jobs in which physical abilities are anticipated to decline in importance (27%). Working with data and making data-based decisions will become an increasingly vital skill.

Automated occupations and sectors affected by automation are limited [20]. Occupations classified as replaceable usually are from administrative support, transport and material handling, production, construction and extraction, cleaning and maintenance of buildings and land, installation, maintenance and repair, agriculture, fishing and forestry, further food preparation, service and sales. These findings are based on the expert assessment of whether robots can perform specific activities within ten years or not.

However, the fall in employment can only manifest itself in a part of the economy for a temporary period and this can be compensated for in several ways. Historical experience was enhanced as an example where "phases of job destruction were followed by phases of job creation" [15]. A focus on new job needs with the evolution of automation and technology were highlighted [15]:

- Declining working hours and increasing income creates a wide range of product innovations, new leisure industries or services.
- Diffusion of combustion engines, electrical machines and microprocessors destroys jobs, but the new robots and learning machines require a new range of jobs once they need to be developed, designed, built or repaired.
- "The Internet of Things and the collection and commercial use of Big Data will require fundamentally new software development activities, as well as major activities in research and development".
- Diffusion of automation in service-based industries, where the customer relationship is the key to success, needs to embrace new product development to maintain and increase the long-term relationships.
- Innovative processes in sectors like energy, transport and communication require development of new infrastructure networks, leading to new jobs.

As stated in [20], despite business leaders being aware of the challenges ahead, they are not very assertive when talking about acting. Not all leaders are confident regarding the adequacy of their organizations' future workforce strategy to prepare for these shifts. It is also mentioned what the aspects are on which the companies should focus. Therefore, it is important to reinvent the human resource function, make use of data analytics, focus on talent diversity and leverage flexible working arrangements and online talent platforms. In the longer term, it is important to rethink the education systems, incentivize lifelong learning, as well as public-private collaboration.

2.4   Advantages and Disadvantages of Automation

Considering that automation is something inevitable, it is important to understand what the benefits and harms of this reality are. On the positive side, automation can bring improvements in the quality and speed of processes, thus contributing to productivity and can even allow to achieve results that go beyond human capabilities [11]. In addition, the use of machines reduces costs, complementing some tasks, increasing the performance of tasks that have not been automated and yet allowing employees to increase their productivity, since they can focus on more sophisticated responsibilities [15].

Going forward to the negative aspects, the greatest concern is undoubtedly the rising of technological unemployment, coupled with the idea that machines can replace people. However, balancing all the aspects referred to above, unemployment will not be a big deal in this possible change. Another disadvantage may be the inability to automate more complicated and non-repetitive tasks. In addition, a very high initial investment is required to acquire all the technology until it becomes operational.

Beyond the advantages and disadvantages, it is crucial to recognize some other aspects associated with automation. It is true that the social and economic capacities of each country will influence the behavior of this transition process. Thus, new jobs will be created if societies are able to adopt a forward-looking view of the future. Some authors also argue that institutions that promote social learning will be conducive to product innovation and the creation of jobs [15]. In the future it is essential to focus on the social and political dimension in order to progress with technological change.

However, some defend slightly different approaches. Industrial innovation will not only affect the less skilled workers. Thus, the governments of each country should consider the role of automation in the ethical and legal issues of the workplace, as there is always the possibility of accidents. In conclusion, governments play an important role in benefiting society in general and not negatively affecting employment [19].

3   Methodology

In this section, we present the methodology followed. An article related with the future of employment, found in a Portuguese magazine called *Visão*, motivated the research. The first step was to figure out the scientific source of the information. Then, the objective was to do a literature review on technological change, on the most wanted skills in the future, on the automation of processes and on the future of employment.

The literature review included searches in online databases, such as ScienceDirect. After the review, the authors decided that it was relevant to carry out interviews in order to understand, firsthand, the opinions of people in the job market and to see if these opinions were consistent with what existed in the literature.

The authors then held a brainstorming session to raise possible pertinent questions concerning the topic. As a result, the interview was divided into three parts:

1. **Personal data:** questions concerning the profession, gender and age of the respondents.
2. **Impact of automatization in processes:** questions concerning the people's opinion on the impact of automatization on their own job position and on jobs in general.
3. **Necessary skills and modifications:** questions concerning the skills that are required to perform a certain function, as well as if automation will change those skills. In addition, participants were asked for an advantage and a disadvantage of the automatization of processes.

The authors used their own personal networks to find people to interview and everyone approached accepted to be interviewed. In total, 21 interviews were performed and 21 was the number at which thematic saturation occurred – "in situations of "cultural competence" (i.e., well informed individuals in a homogeneous culture) […] well below 30 is a reliable sample" (Romney et al., 1986, as quoted in [16]).

To test the quality of the interview script it was first tested (with the first 2 out of the 21 interviews). The interview script was deemed to be clear and understandable and so more interviews were performed based on the interview script. The interviews lasted, on average, 15 min and were performed between the 8th and the 15th of October 2018. Four of the authors did the interviewing, separately, in order to save time. The interviews were performed at the workplaces of the interviewees – in Aveiro, in Ponte de Lima, and in Porto (in Portugal). Table 1 summarizes the profiles of the interviewees.

Considering the targets of this activity, it was interesting to interview people with different careers, and different ages and genders, in order to have a better overview of diverse mindsets and opinions. It was also relevant to understand if the mindsets concerning the automatization of processes were different depending on those factors. It should be noted that unemployed people and students were not included in the sample.

The authors decided not to record the interviews and decided to do everything anonymously except for the personal data referred to. This was done to ease the process of interviewing, the analysis of the data and to reduce problems with authorizations.

The interviews were done with 11 women and 10 men, so there was a significant balance between the quantity of opinions of each gender. All the interviewees are of Portuguese nationality, and aged between 20 and 58 years of age, at the time of the interviews (Table 1).

After performing the interviews, the information was analyzed and compared with the data gathered in the literature review to see what the pertinent conclusions were. The authors were satisfied with the information collected, though with more time more interviews could be performed on the topic, namely involving international subjects also.

4   Discussion of the Field of Work

Within the wide variety of professions, none of the interviewees said they were worried when asked about the impact of automation on their jobs (could they lose their job, or did they feel safe?). In fact, some of the answers were interesting: "automation will not step in between me and my functions", "someone has to show the machines to people" and "a human control of processes is always needed".

The authors were not expecting this because when some people were asked about "one disadvantage of automation", the answer was "unemployment". It can be a reality that some are not well informed about the topic, not being prepared for the changes that will happen.

According to [20], the overview for the future is: until 2020, thousands of employees in the categories "Office and Administrative", "Manufacturing and Production" and "Installation and Maintenance" will lose their jobs. Conversely, thousands of employees in "Business and Financial Operations", "Management" and "Sales and Related" will have wider jobs.

In general, the interviews left no doubt that the most vulnerable jobs are those which have low-skilled workers, who are easier to replace by machines. Some examples are supermarket cashier operators and industrial operators. However, others consider that "almost any occupation is replaceable".

Table 1. The profiles of the interviewees.

Occupation	Gender	Age
Teacher	F	54
Responsible for continuous improvement	M	30
Industrial director	M	58
Commercial and Marketing	M	27
Operational purchaser	F	24
Credit analyst	F	23
Process analyst	M	29
Operational procurement indirects	F	38
Accounts receivable process associate	M	26
Waitress	F	20
Horse trainer	F	46
Industrial maintenance technician	M	24
Environmental engineer	F	36
Communication officer	F	29
Sales technician	F	27
Administrative	F	43
Teacher	M	37
CEO	M	29
Civil engineering manager	F	28
Marketing	M	32
Consultant	M	31

The advantages of automation most mentioned are the "increase of productivity", "the reduction of costs", "the increase of quality" and "the speed of operations". These advantages are in agreement with some of the authors, who identified the same topics as previously mentioned in the literature review [11, 15].

Conversely, the interviewees see as disadvantages the "creation of unemployment" and the "loss of personal touch or contact" in the accomplishment of the tasks, that is, dehumanization. One of the respondents considered that "it will take me more education and acquisition of skills to perform the job" and another said that "most people are not prepared for change". The results obtained at this point require deep reflection – by academics and practitioners alike. If people are not concerned about losing their jobs, why do they mention unemployment as one of the disadvantages? This might indicate that people do not see changes coming soon or are not aware of what automation will require, because as referred to in the literature, losing jobs is not the main concern. The main concern is that the skills to perform the jobs are changing.

Another of the respondents, the horse handler, considered that "the tasks will not be done as well by robots as by people" - this might indicate that the job that this person does is more interactive, in this case with animals, or requires more communication and personal contact. Despite this, as the literature has shown, and other interviewees answered as well, this is one of the main advantages that automation will bring, since machines can operate quicker and minimize human error, leading to a cost reduction, but a quality and productivity increase.

In the third part of the interview, we asked what the most important skills for the individuals were to perform their functions. After that, when asked "Will those skills change in the future?", most answered "no". Nonetheless, one respondent had an interesting point of view: "everything is always changing, so maybe these skills will also change", which is something the authors agree with.

Most of the interviewees occupy leading and management roles, this being one of the possible justifications for the answer that the skills needed in the future are not going to change. Some of the skills mentioned are cognitive abilities like: "experience", "vision of the future", "creativity" and "flexibility" that according to the literature will be highly sought out in the future [20]. Furthermore, answers like "interpersonal relationship", "people and conflict management", "teamwork" and "leadership" which represent resource management skills will also be important in the future.

Other interviewees, by answering "no", showed that skills such as "capacity of analysis", "software programming", "technical knowledge" and "technical skills" will have a stable demand in the future, as shown in the literature [20].

Another cluster is social skills: "communication", "knowing the competitors", "negotiation", "sympathy", "customer interaction" and "being proactive", also mentioned during the interviews for industrial and non-industrial positions. This is a group which will have a higher demand across industries [20].

Finally, there were some competencies mentioned, namely "trust", "perseverance", "diversity of skills" and "dedication" that the authors think are also pertinent, but for example "availability" and "speaking other languages" seem to be replaceable and consequently less important.

Unfortunately, none of the interviewees mentioned complex problem solving nor the capacity of data analysis. Despite the diversity, this may be a consequence of the range of the interviews done (it is something that depends on the jobs focused on).

5   Conclusions and Suggestions for Future Research

Throughout the writing of this article, the authors realized that a wide range of literature on this subject already exists, which shows a concern for discussing this topic in order to better understand the future of jobs and the demand for skills with the increase of automation. As discussed above, not all is positive, and some authors have concluded that a disadvantage of automation is that it involves complex systems that need maintenance and intensive training as regards their users [8].

As concerns the impact of automation on unemployment in general, the authors agree that in fact an increase in automation will not have such a significant impact on this area but nonetheless there will be consequences. Through the research and results obtained, we have concluded that jobs may change at the structural level.

Having this subject in mind, when asked about the most important competencies to perform their function, some interviewees identified the same skills mentioned previously in the literature. However, most believe that these skills will remain the same in the future and the authors disagree and think that some of these skills will suffer changes.

As automation progresses, it will be necessary to perform other tasks that were not previously necessary or were needed on a smaller scale. With this, people are not expecting big changes and therefore are not prepared for them, making the understanding of how this gap can be addressed a necessity.

In terms of the literature there are already some "clues" on what to do regarding this issue. For example, some authors enhanced the importance of slowing down innovation and technological change [12]. However, the authors think that this is not the way to deal with the problem being analysed. Therefore, the authors would like to suggest, for future research, a follow-up of the topics studied (automation, technological change, advantages, disadvantages, what will the future bring) and how will people's perspectives change or not. The objective would be to focus on understanding why people are not prepared, what are the biggest barriers and what should be done in order to change this mindset. A bridge between the mindset of people and the future of jobs needs to be built.

Additionally, the authors conclude that adjustments in education will be needed since the main skills in the job market will be different, in the future. It becomes important to rethink education systems, adapting them to the new reality that comes with automation. The challenges linked to recruiting, training and managing talent in the future are going to be significant and warrant a concerted effort in academia to help solve them.

When the authors talk about rethinking the education system, they are thinking of creating a system which teaches students how to learn, continuously and rapidly, throughout their careers and lives, to be able to face new challenges as they appear.

As a first approach, it would be interesting to add components related to the development of non-technical competencies, such as team leading and team work, including on how to learn from one another in group settings, from the beginning of the education paths of students.

The second step would be to create an ongoing survey regarding which skills are lacking in the job market, thus creating a mechanism for receiving continuous feedback from former students – so as to understand how to adjust and improve the subjects lectured in the future – with higher education seen as a company in a continuous improvement process. Continuous improvement officers in higher education should be the norm, and not the exception, in higher education institutions.

It would also be wise to take advantage of the impact of automation in higher education institutions and create an autonomous system that would be capable of keeping track of the career progression of other former students, in a way that higher education may always be present in the development of core competencies. After formal academic training, some levels could be defined, according to the number of years of experience in the labor market, and at certain times it should be necessary to recall the students to higher education institutions to receive further training. For example, after three years of experience in the job market, a person would be recalled to do a specific training course, depending on the area, with the competencies needed for the next level of experience. This suggestion must be studied according to the market's evolution, the impacts caused by automation and the future of higher education.

The main goal of the article was achieved, which was to understand the link between automation and the skills needed in the future, and we hope with our suggestions to complement, in whatever small way, the existing literature on the topic.

References

1. Acemoğlu, D., Restrepo, P.: Robots and jobs: evidence from the US. VOX CEPR Policy Portal (2017). https://voxeu.org/article/robots-and-jobs-evidence-us. Accessed 7 Oct 2018
2. Acemoğlu, D., Restrepo, P.: Artificial Intelligence, Automation and Work. The Economics of Artificial Intelligence: An Agenda, Massachusetts Institute of Technology, Boston University (2018). https://economics.mit.edu/files/14641
3. Au-Yong-Oliveira, M., Gonçalves, R., Martins, J., Branco, F.: The social impact of technology on millennials and consequences for higher education and leadership. Telematics Inform. **35**, 954–963 (2018)
4. Branco, F., Gonçalves, R., Martins, J., Cota, M.: Decision support system for the agri-food sector–the sousacamp group case. In: New Contributions in Information Systems and Technologies, pp. 553–563. Springer, Cham (2015)
5. Charles, K., Hurst, E., Notowidigdo, M.: Manufacturing Decline, housing booms, and non-employment. NBER Working Paper No. 18949, National Bureau of Economic Research (2013). https://www.sole-jole.org/Notowidigdo.pdf
6. Dauth, W., Findeisen, S., Südekum, J., Woessner, N.: The rise of robots in the German labour market. VOX CEPR Policy Portal (2017). https://voxeu.org/article/rise-robots-german-labour-market?utm_content=buffer5120f&utm_medium=social&utm_source=twitter.com&utm_campaign=buffer. Accessed 8 Oct 2018
7. Frey, C., Osborne, M.: The future of employment: how susceptible are jobs to computerisation? Technol. Forecast. Soc. Change **114**, 254–280 (2016). https://www.sciencedirect.com/science/article/pii/S0040162516302244?via%3Dihub. https://doi.org/10.1016/j.techfore.2016.08.019
8. Frohm, J., Lindström, V., Winroth, M., Stahre, J.: The industry's view on automation in manufacturing. IFAC Proc. Volumes **39**(4), 453–458 (2006). https://www.sciencedirect.com/science/article/pii/S1474667015330925. https://doi.org/10.3182/20060522-3-fr-2904.00073
9. Keynes, J.: Economic Possibilities for our Grandchildren (1930). In: Essays in Persuasion, pp. 358–373. Harcourt Brace, New York (1932). https://assets.aspeninstitute.org/content/uploads/files/content/upload/Intro_and_Section_I.pdf
10. Manyika, J., Chui, M., Miremadi, M., Bughin, J., George, K.: Automation, robotics, and the factory of the future Human + machine: a new era of automation in manufacturing. McKinsey Global Institute (2017). https://www.mckinsey.com/business-functions/operations/our-insights/human-plus-machine-a-new-era-of-automation-in-manufacturing. Accessed 15 Oct 2018
11. Manyika, J., Chui, M., Miremadi, M., Bughin, J., George, K., Willmott, P., Dewhurst, M.: Harnessing automation for a future that works. McKinsey Global Institute (2017). https://www.mckinsey.com/featured-insights/digital-disruption/harnessing-automation-for-a-future-that-works. Accessed 12 Oct 2018
12. Marchant, G., Stevens, Y., Hennessy, J.: Technology, unemployment & policy options: navigating the transition to a better world. J. Evol. Technol. **24**(1), 26–44 (2014). https://jetpress.org/v24/marchant.htm

13. Matuzeviciute, K., Butkus, M., Karaliute, A.: Do technological innovations affect unemployment? Some empirical evidence from European countries. Economies **5**(4), 1–19 (2017). https://www.mdpi.com/2227-7099/5/4/48. https://doi.org/10.3390/economies5040048

14. Miller, B., Atkinson, R.: Are robots taking our jobs, or making them? The Information Technology and Innovation Foundation (2013). http://www2.itif.org/2013-are-robots-taking-jobs.pdf

15. Nübler, I.: New technologies: a jobless future or golden age of job creation? International Labour Office (2016). https://www.ilo.org/wcmsp5/groups/public/—dgreports/—inst/documents/publication/wcms_544189.pdf

16. Remenyi, D.: Field Methods for Academic Research – Interviews, Focus Groups & Questionnaires, p. 195. ACPI, Reading (2013)

17. Tilley, J.: Automation, robotics, and the factory of the future. McKinsey Global Institute (2017). https://www.mckinsey.com/business-functions/operations/our-insights/automation-robotics-and-the-factory-of-the-future. Accessed 15 Oct 2018

18. Vermeulen, B., Kesselhut, J., Pyka, A., Saviotti, P. The impact of automation on employment: just the usual structural change? Sustainability **10**(5), 1–27 (2018). https://www.mdpi.com/2071-1050/10/5/1661. https://doi.org/10.3390/su10051661

19. Wong, J.: How will automation affect society? World Economic Forum (2015). https://www.weforum.org/agenda/2015/01/how-will-automation-affect-society/. Accessed 10 Oct 2018

20. World Economic Forum. The Future of Jobs Report - Employment, Skills and Workforce Strategy for the Fourth Industrial Revolution (2016). https://www.weforum.org/reports/the-future-of-jobs-report-2018. Accessed 7 Oct 2018

21. Jaimovich, N., Siu, H.: The trend is the Cycle: Job Polarization and Jobless Recoveries. NBER Working Paper No. 18334, National Bureau of Economic Research (2012). https://www.nber.org/papers/w18334.pdf. https://doi.org/10.3386/w18334

Ambient Assisted Living – A Bibliometric Analysis

João Viana[1,2(✉)] ⓘ, André Ramalho[1,2] ⓘ,
José Valente[1,2] ⓘ, and Alberto Freitas[1,2] ⓘ

[1] MEDCIDS – Department of Community Medicine,
Information and Health Decision Sciences, Faculty of Medicine,
University of Porto, Porto, Portugal
joaoviana@med.up.pt
[2] CINTESIS – Centre for Health Technology and Services Research,
Porto, Portugal

Abstract. Introduction: Ambient Assisted Living technologies provide assistance and support to people with some disabilities, such as the elderly and in cases of dementia. They offer greater control and reduction of predicted economic costs as life expectancy and the incidence of dementias tend to increase. The AAL is a growing area of research and there is a need to assess the existent publications, research trends, and characteristics in this field, providing an overview of the development of AAL systems-related research. Therefore, the aim of this study is to analyze and understand the research trends and characteristics in the AAL field by means of a bibliometric analysis.

Methods: We performed a document search in Scopus with a query designed to reflect the AAL field of study, from inception up to the end of 2018. Subsequently we made several analyses to discover publication trends and patterns in this research area.

Results: There are a total of 14645 authors in the AAL field, which corresponds to a mean of 1.84 authors per document. The most productive country was Spain with 895 documents, followed by United Kingdom and Germany with 723 and 718 respectively.

Conclusion: This paper provides an overall picture of the state of the research regarding the field of ambient assisted living, identifies the most active authors and highlight relations between areas of research. It shades light on possible collaborations with research groups and across countries.

Keywords: Bibliometric analysis · Scopus · Ambient assisted living · AAL

1 Introduction

As a result of advances in the medical sciences, public health, healthcare technology and nutrition, there is a progressive increasing in the world's elderly population [1]. The childhood mortality, the declining birth rates, and the increase in population life expectancy, will continue to influence worldwide demographic changes in varying time frames and intensities [2]. It is estimated that two billion people will be 60 years of age or older by 2050, and that approximately 33 countries will have more than ten million

© Springer Nature Switzerland AG 2019
Á. Rocha et al. (Eds.): WorldCIST'19 2019, AISC 930, pp. 218–228, 2019.
https://doi.org/10.1007/978-3-030-16181-1_21

people each in this age group [3]. Over half the EU population is predicted to be over-65 by 2070 [4]. Dwyer reaffirms this result by pointing out that as for men, only a minority at or above the age of 60 lives alone [5].

Ambient Assisted Living (AAL) technologies provide assistance and support to people with some disabilities, such as the elderly and in cases of dementia [6]. They offer greater control and reduction of predicted economic costs as life expectancy and the incidence of dementias tend to increase. They also allow for an easier way to live at home for a longer period of time, with more comfort and safety, improving quality of life for older people [7, 8].

As a growing area of research [9–12], it is essential to appraise the existent publications, research trends, and characteristics in this field providing an overview of the development of AAL systems-related research. Therefore, the aim of this bibliometric analysis is to understand research trends and characteristics in the AAL field of study.

2   Methods

2.1   Bibliographic Database

Scopus is one of the world's largest abstract and citation databases of peer-reviewed research literature, with over 22,800 serial titles from more than 5000 international publishers. Furthermore, Scopus has a wide coverage, currently counting 70 million items, 70,000 main institution profiles and 12 million author profiles; it has efficient tools for retrieving and aggregating information; and it exports data in multiple formats based on previously defined output variables and provides extensive information for analysis [13].

For these reasons, Scopus was the database of peer-reviewed literature chosen for this research.

2.2   Search Strategy and Validity

The search was performed without any limitations, e.g. publication language, or year of publication.

The search query was a refinement of search queries used in Blackman et al. and Calvaresi et al. [14, 15]. Changes were made to the query to better suit our purposes, mainly reducing unwanted sensitivity; for a bibliometric analysis a systematic screening of papers is not performed. The query included terms such as "Ambient Assisted Living", "Ambient-Assisted Living", "Ambient Intelligence", "technology and aging", "telecare" and other related terms. (search query is long, but can be supplied on request).

2.3   Software and Data Analysis

All included articles were analysed by index keywords, author keywords, publication year, source title, authors, title and abstract. The index, author keywords were analysed by frequency, and a ranking of publication sources was performed by relevance.

The articles' keywords, titles and abstracts were also searched for TAALXONOMY terms [16] to determine their usage frequency. Productivity analysis was conducted by author and country rankings. Total and average number of citations were analysed by country. Authorship analysis was carried out by computing the number of single-authored articles and multi-authored publications. We also calculated the rate of articles per author, authors per article, and co-authors per articles. Overall collaboration and the frequency publications cited together i.e., co-citations, were also analysed.

VOSviewer program, a software tool designed specifically for bibliometric analysis [17], was employed for the visualization of the retrieved data. This tool provides a density map in which a colour represents clusters (e.g., countries or authors) connecting lines to indicate certain parameters (e.g., strength of collaboration between countries or authors are measured and represented by the thickness of connecting lines). The circle size or font size indicates the level of productivity or citations. On selecting the unit of analysis, thresholds were imposed for the visualization to be intelligible. The software Microsoft Excel was used to rearrange the data and perform the analysis of the exported data (i.e., comma separated values' table) from Scopus. Python was used to select keywords, authors, years and journals and perform the analysis and export data to Microsoft Excel and the software R was used to plot charts. Moreover, for network visualization of the analysis, the software VOSviewer was used.

2.4   Statistics and Ethical Considerations

This study was exempted from ethical approval, as only publicly available electronic data containing bibliographic information was used in the analysis.

3   Results

3.1   General Information

The search on Scopus was performed on the 29th of September 2018 and a total of 7949 documents were retrieved. There were no date limits imposed on the search query, therefore the documents retrieved were from inception to the date of the search.

Regarding index or author keywords, "ambient intelligence", "artificial intelligence", and "ambient assisted living" were the most frequent keywords, occurring 3168, 2016, and 1673 times respectively. The 10 most frequent index or author keywords can be seen in Table 1. The plot of 4 clusters representing the co-occurrence of all keywords is presented in Fig. 1. This representation can be analysed as a proxy for the relationship and proximity of fields of study.

Table 1. Frequency of ambient assisted living keywords found either in Index or Author's keywords

Keyword	Occurrences as index or author keywords n (%)
Ambient intelligence	3,168 (39.9)
Artificial intelligence	2,016 (25.4)
Ambient assisted living	1,673 (21.0)
Human	900 (11.3)
Ubiquitous computing	850 (10.7)
Humans	747 (9.4)
Telemedicine	732 (9.2)
Telecare	652 (8.2)
Article	617 (7.8)
Assisted living	598 (7.5)

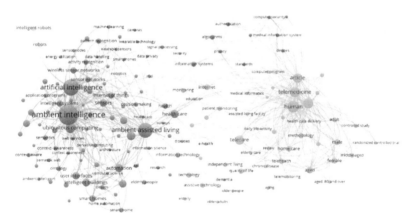

Fig. 1. Co-occurrence of index or author keywords. The colors represent the clusters created based on co-occurrence of keywords. The minimum occurrence of the words plotted was 75.

3.2   Trends and Citations

Throughout the period considered in our study, the average number of publications was 248 documents per year. An increasing trend of the number of publications was observed over the years (Table 2). The scientific production has been steadily increasing, reaching in 2017 a total of 922 publications, which comprised 11.6% of the total productivity in the studied period. Table 2 summarizes yearly scientific production and citations, while Fig. 2 shows the trends of articles (i.e. Article, Article in Press and Review), Conference Papers (i.e. Conference Paper and Conference Review) and Other (e.g. Letter, Book, Editorial) from 1987 to 2017.

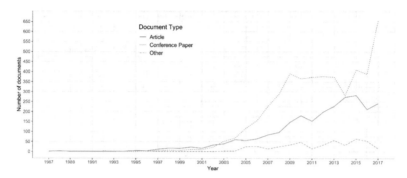

Fig. 2. AAL publication trends over the study period.

Table 2. Yearly summary of publish articles in the field of ambient assisted living. Articles—number of articles published per year and percentage from the total of articles retrieved (7949), ACLO—Articles cited at least once, ACLO%—percentage of articles cited at least once when compared to total annual publication and total citations for the articles published that year.

Year	Articles, n	%	ACLO, n	ACLO %	Total citations, n	Median	Q1	Q3
1987–2002	210	2.64						
2003	131	1.65	96	73	2123	5.0	0.0	14.0
2004	199	2.50	156	78	4395	4.0	1.0	18.5
2005	247	3.11	178	72	3897	4.0	0.0	14.0
2006	325	4.09	240	74	3391	3.0	0.0	11.0
2007	410	5.16	302	74	4345	3.0	0.0	9.0
2008	573	7.21	425	74	5243	3.0	0.0	8.0
2009	594	7.47	463	78	5507	3.0	1.0	9.0
2010	543	6.83	446	82	5287	3.0	1.0	10.0
2011	611	7.69	468	77	5752	3.0	1.0	8.0
2012	661	8.32	492	74	5261	3.0	0.0	8.0
2013	588	7.40	418	71	3973	2.0	0.0	7.0
2014	763	9.60	507	66	4647	2.0	0.0	5.5
2015	657	8.27	427	65	1904	1.0	0.0	3.0
2016	922	11.60	420	46	1175	0.0	0.0	1.0
2017	416	5.23	91	22	169	0.0	0.0	0.0
2018	5	0.06	0	0	0	0.0	0.0	0.0
2019	3	0.01	1	100	2	2.0	2.0	2.0

The article that received the highest number of citations was "Implementation of a real-time human movement classifier using a triaxial accelerometer for ambulatory monitoring", with a total of 747 citations published in the Graduate School of Biomedical Engineering, University of New South Wales, Sydney, Australia in 2006. A list with the most referenced publications by articles in the field of ambient assisted living is presented in Table 3.

Table 3. Top 10 most referenced publications in the field of ambient assisted living.

References	Title	Number of citations
Karantonis (2006)	Implementation of a real-time human movement classifier using a triaxial accelerometer for ambulatory monitoring	747
Levis (2005)	TinyOS: an operating system for sensor networks	680
Boulos (2011)	How smartphones are changing the face of mobile and participatory healthcare: an overview, with example from eCAALYX	471
Cook (2009)	Ambient intelligence: technologies, applications, and opportunities	471
Rashidi (2013)	A survey on ambient-assisted living tools for older adults	416
Dounis (2009)	Advanced control systems engineering for energy and comfort management in a building environment-a review	412
Islam (2015)	The internet of things for health care: a comprehensive survey	379
Ekeland (2010)	Effectiveness of telemedicine: a systematic review of reviews	352
Heo (2005)	Energy-efficient deployment of intelligent mobile sensor networks	321
Whitmore (2015)	The internet of things—a survey of topics and trends	296

3.3 Geographical Distribution and Journals

The most productive country was Spain with 895 documents, followed by United Kingdom and Germany with 723 and 718 respectively. In order to understand collaboration between countries a co-authorship network was created and plotted in Fig. 3.

The selected publications on ambient assisted living were published in different sources, a total of 883 and 1004, for journal and conferences respectively. Table 4 lists all preferred journals with article, article in press and review as document types, with a minimum productivity of 25 documents. Journal of Telemedicine and Telecare ranked first, with 119 (4.5% of the total) documents, followed by Journal of Ambient Intelligence and Smart Environments, with 111 publications (4.2% of the total), and Sensors (Switzerland), with 107 manuscripts (4.0% of the total).

Similarly, Table 5 lists all preferred conferences with conference paper and conference review as document types, with a minimum productivity of 25 documents. Lecture Notes in Computer Science (including subseries Lecture Notes in Artificial Intelligence and Lecture Notes in Bioinformatics) ranked first, with 1023 (21.4% of the total) documents, followed by 2017 14th International Conference on Ubiquitous Robots and Ambient Intelligence, URAI 2017, with 253 publications (5.3% of the total), and Communications in Computer and Information Science, with 155 manuscripts (3.2% of the total).

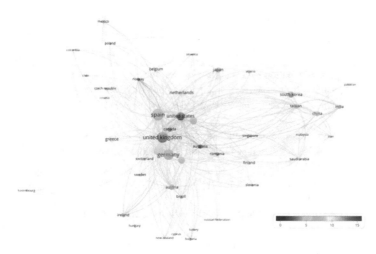

Fig. 3. Co-authorship by country. The size of the circles are proportional to the number of documents, the colours represent the countries' average citations. The countries were selected with minimum documents per country of 20.

Table 4. Ranking of journals with a minimum productivity of 50 publications in the field of ambient assisted living. IF - Impact Factor for 2017, Articles (%)—number of articles per journal and percentage from the total of articles retrieved (n = 2664 of 7949).

Journal	IF	Articles (%)
Journal of Telemedicine and Telecare	3.046	119 (4.5)
Journal of Ambient Intelligence and Smart Environments	0.878	111 (4.2)
Sensors (Switzerland)	2.475	107 (4.0)
Journal of Medical Systems	2.098	88 (3.3)
Lecture Notes in Computer Science (including subseries Lecture Notes in Artificial Intelligence and Lecture Notes in Bioinformatics)	n.a.	74 (2.8)
Journal of Ambient Intelligence and Humanized Computing	1.423	72 (2.7)
Journal of Assistive Technologies	n.a.	61 (2.3)

Table 5. Ranking of conferences with a minimum productivity of 60 publications in the field of ambient assisted living. Publications (%)—number of articles per conference and percentage from the total of articles retrieved (n = 4784 of 7949).

Conference	Articles (%)
Lecture Notes in Computer Science (including subseries Lecture Notes in Artificial Intelligence and Lecture Notes in Bioinformatics)	1,023 (21.4)
2017 14th International Conference on Ubiquitous Robots and Ambient Intelligence, URAI 2017	253 (5.3)
Communications in Computer and Information Science	155 (3.2)
CEUR Workshop Proceedings	148 (3.1)
ACM International Conference Proceeding Series	138 (2.9)
Advances in Intelligent Systems and Computing	130 (2.7)
Studies in Health Technology and Informatics	85 (1.8)

3.4 Authorship Pattern and Collaborations

There is a total of 14,645 authors found in the retrieved documents, which corresponds to a mean of 1.84 authors per document. Single-authored documents comprised a total of 1,115 (0.14% of the total) publications. Additionally, authors with a minimum productivity of 15 documents were graphically represented using an algorithm implemented by VOSviewer, as shown in Fig. 4. Table 6 lists the top 10 active authors in the field of ambient assisted living.

Table 6. Top 10 active authors in the field of ambient assisted living. SCR—Standard Competition Ranking, Articles (%)—number of authorships and percentage from the total of articles retrieved (n = 7949), % of Articles from Total—percentage of articles retrieved by the total scientific production within the study period.

Standard competition ranking (SCR)	Author	Articles, n (%)	Total articles in Scopus within the study period	% of articles from total
1st	Novais P.	74 (0.9)	324	22.8
2nd	Augusto J.C.	62 (0.8)	188	33.0
3rd	Bravo J.	60 (0.8)	144	41.7
4th	Stephanidis C.	57 (0.7)	359	15.9
5th	Hervás R.	50 (0.6)	105	47.6
6th	Nugent C.	46 (0.6)	439	10.5
7th	Neves J.	42 (0.5)	299	14.0
7th	Corchado J.M.	42 (0.5)	432	9.7
9th	Antona M.	39 (0.5)	83	47.0
10th	Bajo J.	37 (0.5)	264	14.0

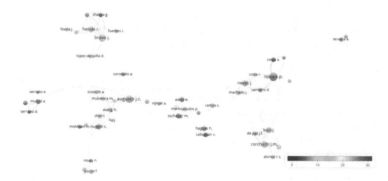

Fig. 4. Co-authorship author analysis. The size of the circles are proportional to the number of documents, the colours represent the author' average citations. The selection had the threshold of a minimum productivity per author of 15.

3.5 TALLXONOMY Usage

The most frequent TAALXONOMY terms found in field of ambient assisted living, searched in the title, abstract, index or author keywords, are the terms "environment", "care" and "communication" that appeared 2395, 1997 and 1422 times respectively. However, more than 60% of the terms had less than 10 occurrences in our retrieved documents. We also verified the terms that originated from division of the TAALX-ONOMY terms that have "and" or "&" or "," and found the most frequent terms " Information", " Telecare" and "Health Care" that appeared 2761, 1465 and 1049 times respectively.

4 Discussion

As expected, keywords "Ambient Intelligence", "Artificial Intelligence" and "Ambient Assisted Living", lead the ranking of the most used terms in the analysed period (from inception to the date of research, September 29th 2018), having weights of 40%, 25% and 21% respectively. The occurrence of terms together demonstrated in our cluster analysis the great relationship of terms related to devices, sensors and facilities, which include the terms of highest weight in publications such as "Ambient intelligence", "Artificial intelligence" and "Ambient Assisted Living" and less co-occurrences with the clusters more related to health and clinical records.

The article "Implementation of a real-time human movement classifier using a tri-axial accelerometer for ambulatory monitoring" [18] had the highest number of citations. It reflects a great interest in wearable devices capable of recording and classifying the movements of an individual in a way to determine his or her degree of functional ability and general level of activity [19]. Studies in this AAL tool show that these technologies are relevant because the elderly population benefits in reducing age-related cardiovascular risks, as well as improving muscular endurance and flexibility [20].

The increase in the number of publications in the study area started a discrete rise since 1996. This increase started initially from publications of articles, being surpassed in volume by the conference papers from 2002. An important fact occurred since 2007 when the number of conference documents arrived to be almost 300 articles more than in traditional journal publications. This scenario might be partially explained by the increased number of events related to these technologies. Regarding journal publications, our research consistently pointed out to specific journals in the subject, among them the Journal of Telemedicine and Telecare, Journal of Ambient Intelligence and Smart Environments, Sensors (Switzerland), and Journal of Medical Systems. Nowadays, AAL conference related publications more than doubled similar journal publications.

When considering co-authorship, we identified the authors' collaboration clusters. Despite the production volume, it appears that the more diverse the collaboration network, the greater is the average citations values.

5 Conclusion

This paper provides an overall picture of the state of the research regarding the field of ambient assisted living, identifying the most active authors and highlighting relations between areas of research. It shades light on possible collaborations with research groups and across countries.

Acknowledgement. ActiveAdvice – Decision Support Solutions for Independent Living using na Intelligent AAL Product and Service Cloud' (European Commission AAL Joint Programme, Ref.ª AAL-2015-2-058), financiado por fundos nacionais através da Fundação para a Ciência e Tecnologia, I.P. (FCT, Ref.ª AAL/0007/2015).

Conflicts of Interest. The authors declare no conflict of interest.

References

1. Beard, J., Biggs, S., Bloom, D.E., Fried, L.P., Hogan, P.R., Kalache, A., Olshansky, S.J.: Global population ageing: peril or promise?: Program on the global demography of aging (2012)
2. United Nations: World Population Ageing. United Nations, New York (2013)
3. Haub, C.: Fact Sheet: World Population Trends 2012 (2012)
4. Active Assisted Living Programme - Ageing Well, http://www.aal-europe.eu/. Accessed 14 Nov 2018
5. Dwyer, M., Gray, A., Renwick, M.: Factors affecting the ability of older people to live independently: a report for the international year of older persons. Ministry of Social Policy (2000)
6. Haux, R., Koch, S., Lovell, N.H., Marschollek, M., Nakashima, N., Wolf, K.H.: Health-enabling and ambient assistive technologies: past, present, future. Yearb. Med. Inf. **25** (Suppl. 1), S76–S91 (2016)

7. Ienca, M., Jotterand, F., Elger, B., Caon, M., Scoccia Pappagallo, A., Kressig, R.W., et al.: Intelligent assistive technology for Alzheimer's disease and other dementias: a systematic review. J. Alzheimer's Dis.: JAD **60**(1), 333 (2017)
8. Novitzky, P., Smeaton, A.F., Chen, C., Irving, K., Jacquemard, T., O'Brolchain, F., et al.: A review of contemporary work on the ethics of ambient assisted living technologies for people with dementia. Sci. Eng. Ethics **21**(3), 707–765 (2015)
9. Memon, M., Wagner, S.R., Pedersen, C.F., Beevi, F.H., Hansen, F.O.: Ambient assisted living healthcare frameworks, platforms, standards, and quality attributes. Sensors **14**(3), 4312–4341 (2014)
10. Al-Shaqi, R., Mourshed, M., Rezgui, Y.: Progress in ambient assisted systems for independent living by the elderly. SpringerPlus **5**, 624 (2016)
11. Geissbuhler, A., Haux, R., Kulikowski, C.A.: From ambient assisted living to global information management: IMIA-referencing publications in 2008. Yearb. Med. Inf. **18**, 143–145 (2009)
12. Costa, A., Novais, P., Simoes, R.: A caregiver support platform within the scope of an ambient assisted living ecosystem. Sensors **14**(3), 5654–5676 (2014)
13. What Content Is Included in Scopus? https://www.elsevier.com/solutions/scopus/content. Accessed 14 Nov 2018
14. Blackman, S.: Ambient assisted living technologies for aging well: a scoping review. J. Intell. Syst. **25**, 55 (2016)
15. Calvaresi, D., Cesarini, D., Sernani, P., Marinoni, M., Dragoni, A.F., Sturm, A.: Exploring the ambient assisted living domain: a systematic review. J. Ambient Intell. Humanized Comput. **8**(2), 239–257 (2017)
16. https://www.taalxonomy.eu/project/
17. van Eck, N.J., Waltman, L.: Software survey: VOSviewer, a computer program for bibliometric mapping. Scientometrics **84**(2), 523–538 (2010)
18. Karantonis, D.M., Narayanan, M.R., Mathie, M., Lovell, N.H., Celler, B.G.: Implementation of a real-time human movement classifier using a triaxial accelerometer for ambulatory monitoring. IEEE trans. Inf. Technol. Biomed. **10**(1), 156–167 (2006)
19. Biagetti, G., Crippa, P., Falaschetti, L., Orcioni, S., Turchetti, C.: Human activity monitoring system based on wearable sEMG and accelerometer wireless sensor nodes. BioMedical Eng. OnLine **17**(S1), 132 (2018)
20. Godfrey, A., Conway, R., Meagher, D.: ÓLaighin G (2008) Direct measurement of human movement by accelerometry. Med. Eng. Phys. **30**(10), 1364–1386 (2008)

A Taboo-Search Algorithm for 3D-Binpacking Problem in Containers

Paul Leon[1], Rony Cueva[1], Manuel Tupia[1(✉)],
and Gonçalo Paiva Dias[2]

[1] Department of Engineering, Pontificia Universidad Catolica del Peru,
Avenida Universitaria 1801, San Miguel, Lima, Peru
{andres.leonm, tupia.mf}@pucp.edu.pc, cueva.r@pucp.pe
[2] ESTGA/GOVCOPP, Universidade de Aveiro,
Rua Comandante Pinho e Freitas, 28, 3750-127 Águeda, Portugal
gpd@ua.pt

Abstract. One of the biggest challenges facing by logistics companies is the packing of fragile products in containers. These activities, due to the nature of the packaged products, can entail great risks for the company due to the possible losses and the cost of the transporting containers, so-called in literature as *bin-packing problem*. Being this problem of a complex computational nature (NP-hard), companies do not have exact technological solutions to establish a positional sequence of the packages that have to be arranged in containers, considering aspects such as fragility, weight, volume, among other aspects. In most cases only the volume and order of dispatch is considered. The application of bio-inspired techniques such as metaheuristics are an appropriate way to obtain approximate solutions to this kind of problems, without the difficulties of complex software implementations. In previous works of the authors, bio-inspired algorithms have been developed to solve the problem and in this paper a taboo-search algorithm is proposed to solve the 3D-bin packing variant, which has been compared with the previous algorithms obtaining approximately a waste reduction of 28%. Fundamentally, research has used real data from Peruvian ceramic industry.

Keywords: Bin packing problem · Taboo search · Genetic algorithm · Metaheuristic

1 Introduction

A key aspect to become successful in the current trading market is properly managing product's logistics and storage. Last years' trend follows the path of companies whose advantage lies in the capacity of conducting international shipments with a great number of products. Hence, the efficiency in the capacity of transportation is a mandatory requirement for sales companies to maintain or increase their share in the market [1]. Created in 1956, containers are used as a transportation and packaging cargo unit for commercialization of different products. They have evolved to such an extent that even ships have been built with the specific purpose to carry containers [2]. Companies around the world considered containers the ideal method of delivering the

© Springer Nature Switzerland AG 2019
Á. Rocha et al. (Eds.): WorldCIST'19 2019, AISC 930, pp. 229–240, 2019.
https://doi.org/10.1007/978-3-030-16181-1_22

goods to clients because of reduced time in cargo transit, more safety (in the sense of integrity and protection of the products that are sent in the containers) and above all lower costs.

Despite the benefits from the use of containers, sales companies have got into troubles in optimizing the space used within these transport units mainly due to the fact that the order and distribution of packages placed into containers is inadequate since weight, shape, volume of these items are not taken into account. The problem gets worse when personnel involved lacks the knowledge or training required to conduct the tasks at hand [3]. Therefore, proposing an order of placement of packages within the container with the aim of optimizing the space used in them, becomes the main hurdle for this research, since the solution will allow transporting more packages—in a safe way—in less numbers of containers and minimizing faults or fractures in fragile products during transport [12, 14]. Eventually, this will result in reduced transportation costs per container incurred by the company, as costs are related to the number of containers to be used. In the bibliography this problem is known as the 3D Bin Packing Problem. This is a combinatorial optimization problem focused on maximizing the space used by objects within a restricted space, e.g., a container. Objects are placed following certain boundaries to achieve the proposed objective. This problem increases its computational complexity in a non-determinist polynomial time while variables such as the number of objects or their sizes differ, that is why it is called NP-hard (a term to describe the set of problems that is at least as hard as the hardest problem that can be solved in a polynomial time [4]).

Because of this complexity, exact algorithms cannot solve 3D packaging scenarios, nor may consider many restrictions. It is necessary to use approximate algorithms to solve these situations. In this field, it is that metaheuristic algorithms such as bio-inspired ones, can yield sufficiently good and computationally economical solutions to implement in software [13].

For this research the design and construction of a metaheuristic algorithm delivering a solution to the above said problem are proposed. Likewise, the taboo-search algorithm has shown to be capable of providing very good results in combinatorial optimization problems [5]. On this premise, an algorithm previously used in the literature will be built to compare its results with those obtained by the algorithm proposed on this project, using mainly data from Peruvian companies producing ceramic for bathrooms and kitchens, which must export their products transnationally both by sea and land.

In this paper it is to find the following structure: a brief introduction to the taboo-search in Sect. 2; algorithms proposed in Sect. 3; numerical experimentation in Sect. 4; to end with the conclusions in Sect. 5.

2 Taboo Search Algorithm

The authors have developed previous works related to the variants of the bin packing problem, developing several metaheuristics (GRASP, Genetic algorithm, memetic, algorithm, among others) being the genetic algorithms those that have obtained the best results for 3D scenarios [9–11]. The main motivation of this investigation is to verify if Taboo-search algorithm can surpass the previously calculated results. This is a kind of metaheuristic algorithm that tries to solve optimization problems. Its main characteristic is that it relies on a specific heuristics to prevent areas of the search space recently visited from being visited again. This characteristic is called short-term memory and is represented by a taboo list of specific size, with this structure storing the most recent moves and marking them as moves forbidden for the following n iterations. Through this heuristic, the taboo search avoids being stuck in a cyclic state, *i.e.*, it escapes local optima while approaching global optima [6, 7].

It can be considered a better alternative to other types of heuristic and bio-inspired algorithms due to the ease of implementation and the low computational cost. In Table 1, appears the pseudo code of the TSA:

Table 1. The taboo search algorithm (TSA)

```
Input: Iterations, InitialSolutions, ListSize, NeighborhoodSize
Exit: Solution
1. BestSolution = InitialSolution
2. Iteration = 0
3. TabooList = { }
4. Whereas (StoppingCondition)
    4.1 Neighborhood = GenerateNeighborhood(BestSolution,Iteration)
    4.2 BestNeighbor = FindBestSolution(Neighborhood)
    4.3 If ( TabooListDoesNotContain(BestNeighbor) And Fitness(BestNeighbor)
> Fitness(BestSolution) )
        4.3.1 BestSolution = BestNeighbor
    End If
    4.4 If ( TabooListSize == ListSize )
        4.4.1 RemoveLast(TabooList)
    End If
    4.5 AddTabooList(BestNeighbor)
    4.6 Iteration = Iteration + 1
    End Whereas
5. Return BestSolution
```

3   Contribution

3.1   Brief Discussion About State of Art of the Problem

Three different approaches have been found in the literature for solutions to the problem [15, 16]:

- Optimization of container filling for multimodal transport
- Hybrid approach between linear programming with heuristics
- Minimization of wasted warehouse space

This allows perceiving that the resolution of the problem in 3D scenarios has been based on hybrid solutions -between the use of exact and approximate methods- that minimize the loss of spatial volume inside containers. When reviewing the most commonly used approximate methods, there are some tools that provide an alternative solution to the problem of waste of container space, but these tools only consider the space used and no other important factors such as the distribution of weight and fragility of objects.

3.2 Objective Function for Taboo-Search Algorithm

With the aim of placing the largest number of boxes in each container, the function attempts to find the exact points within these containers to place the boxes. The volume of the box to be placed is an important attribute to consider since a lot of restrictions of this problem are related to each box's own characteristic. This importance may be reflected in the following objective function formalization:

$$\text{Max} \ \frac{\sum_{i=1}^{N} \frac{1}{Ci} \sum_{j=1}^{M} Vj \, x \, kji}{N}$$

Where:

N: Number of containers used.
M: Number of boxes in total.
Vj: Volume of box j.
Kji: Indicates if box j is placed in container i (it can only have the value 0 or 1).
Ci: Capacity of container i.

To test the validity of the objective function, the following test case is presented in an extreme situation. There are two solutions, the first one has 3 containers with a box inside each one; the second one has a container with 3 boxes inside. If we consider that the boxes have a volume of 20 "cubic" units and the containers, a capacity of 100 "cubic" units, it can be raised:

$$First \ solution = \frac{\left(20 * \frac{1}{100}\right) + \left(20 * \frac{1}{100}\right) + \left(20 * \frac{1}{100}\right)}{3} = 0.2$$

$$Second \ solution = \frac{(20 + 20 + 20) * \left(\frac{1}{100}\right)}{1} = 0.6$$

3.3 Rotations and Positioning of Boxes

Here are some of the most important restrictions to consider when developing the problem. First restriction is crucial to understand how the boxes will be positioned within containers. Usually, a box can be placed in more than one manner as can be seen in the following figure [8] (Fig. 1):

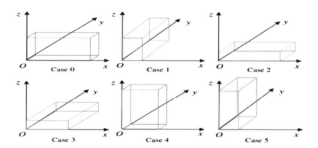

Fig. 1. Rotations of a box

Second restriction is *limit of the walls of the container*: this one indicates that none of the boxes placed can exceed the limits of the container, that is, all the boxes must be inside it. Last restriction is *weight limit*: this one is directly related to the intrinsic fragility factor of the boxes. It specifies that each box can only have a specific weight limit above.

3.4   Data Structure for Algorithms

This chapter introduces data structures mentioned in the objectives, which will be used in both algorithms. It is worth mentioning that these structures will be the same for both but will be treated in a different manner for each of the algorithms.

Container: As an international standard will be applied, all containers used will have the same dimensions: Height, Width, Length and Volume.

Box: This structure is the main part of the solution because it will not only consider boxes' dimensions but will also indicate the coordinates where the box is placed within the container and the rotation that has been used: Height, Width, Length and Volume, and Position X, Position Y and Position Z that represent the point in the X, Y or Z axis where the box is placed.

Rotation: As each box can be placed in different positions in a same point by rotating such box on its own axes, a structure that considers this concept is required.

Solution: As its own name suggests, this structure will represent the solution generated, which will contain a list of all boxes placed in each container and the extreme points present within:

- List of boxes: represents all the boxes placed in the container.
- List of extreme points: represents all the remaining extreme points after generating the solution.

3.5   Pseudocode of Genetic Algorithm

In the next sections, the pseudocodes of the proposed algorithms will be presented. It begins by presenting the genetic algorithm in Table 2.

Table 2. Genetic algorithm

Input: PopulationSize, Generations, MutationProb, CrossOverProb
Output: BestSolution
1. Population = GenerateInitialPopulation()
2. BestSolution = FindBestSolution(Population)
3. Whereas (StopCondition)
    3.1 Parents = SelectParents(Population, PopulationSize)
        3.2 Children = ReproduceParents(Parents, PopulationSize, Mutationprob, CrossOverProb)
    3.3 Children_BestSolution = FindBestSolution(Children)
    3.4 If (Fitness(Children_BestSolution) > Fitness(BestSolution.Fitness))
        3.4.1 BestSolution= Children_BestSolution
    End If
    3.5 Population = Children
End Whereas
4. Return BestSolution

3.6   Pseudocode of Taboo-Search Algorithm

Like in previous section and Tables 1, 2 and 3 shows pseudo code of TSA:

Within this pseudocode, there are four functions (line 4.1, line, 4.2, line 4.4.1 and line 4.5), but only two, "GenerateNeighborhood" and "RemoveLast", will be specified, since function "FindBestSolution" was detailed in the previous chapter and "AddTabooList" only adds a solution to a list of solutions called "TabooList".

Finally, it is worth mentioning that both "ConditionCombination" and "SizeNeighborhood", in lines 3 and 3.1, respectively, are parameters entered as global variables for all the algorithm, these parameters indicate the condition that must be met to conduct the combination and the number of neighbors to be generated by neighborhood, respectively.

Table 3. The taboo search algorithm proposed

Input: Iterations, InitialSolutions, ListSize, NeighborhoodSize
Exit: Solution
1. BestSolution = InitialSolution
2. Iteration = 0
3. TabooList = { }
4. Whereas (StoppingCondition)
    4.1 Neighborhood = GenerateNeighborhood(BestSolution,Iteration)
    4.2 BestNeighbor = FindBestSolution(Neighborhood)
    4.3 If (TabooListDoesNotContain(BestNeighbor) And Fitness(BestNeighbor)
> Fitness(BestSolution))
        4.3.1 BestSolution = BestNeighbor
    End If
    4.4 If (TabooListSize == ListSize)
        4.4.1 RemoveLast(TabooList)
    End If
    4.5 AddTabooList(BestNeighbor)
    4.6 Iteration = Iteration + 1
    End Whereas
5. Return BestSolution

Table 4. Algorithms results

Sample	Genetic algorithm results	Taboo search algorithm results
1	0.862351701350831	0.868905212716766
2	0.859659925914651	0.859659925914651
3	0.852777447283593	0.852777447283593
4	0.84968569449088	0.84968569449088
5	0.848484693392605	0.848484693392605
21	0.838743194496973	0.848570362470513
22	0.838610380146598	0.838610380146598
23	0.838360825649067	0.841123372999282
24	0.838320390885628	0.838320390885628
25	0.837857149741339	0.845669846722336
26	0.837607010878392	0.842349033764246
27	0.837453369855341	0.837453369855341
28	0.836870289488535	0.836971786283775
29	0.836569676408897	0.839018173044392
30	0.836426227161836	0.836426227161836

4   Numeric Experimentation

To obtain the results sought, both algorithms were run 55 times with different samples. It should be mentioned that as standards size containers are used, the tests were conducted with same size containers.

4.1   Algorithms Results

Below are the results obtained from the algorithms run. These results contain fitness values of each solution found, which represent the objective function mentioned in previous chapters.

Below are the results displayed within the interface created (Fig. 2):

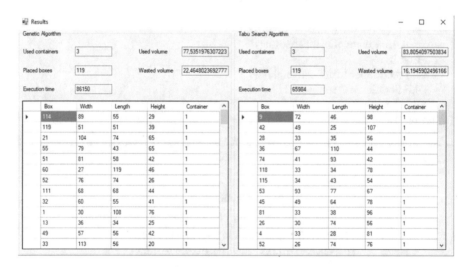

Fig. 2. Interface of algorithms results

Where volumes are represented in percentages and run time in milliseconds. Likewise, all the boxes and their respective measurements, along with the container in which they were placed, are indicated.

4.2   Kolmogorov-Smirnov Test

As mentioned in the introduction, this test will be used to prove that the results of algorithms, *i.e.* fitness of each solution found, follow a normal distribution, which is an indispensable requirement to be able to run the Z test. Usually, hypotheses for this test are that the data analyzed follow a certain distribution. For this case, the following hypotheses area considered:

- H0: algorithm results follow a normal distribution.
- H1: algorithm results do not follow a normal distribution.

These hypotheses will be advanced for both algorithms in such a way that there will be sufficient evidence to conclude that one of the two hypotheses is true.

4.2.1   Test for Genetic Algorithm

With the aim of proving that the genetic algorithm results follow a normal distribution, the following values were calculated:

- Mean of results: 0.8373238.
- Standard deviation of results: 0.008056261.

Then, with the assistance of R Studio, considering a 5% significance, the Kolmogorov-Smirnov test was obtained:

- P-value: 0.1968.

As P-value obtained is higher than 0.05, the null hypothesis is accepted, and it is concluded that genetic algorithm results follow a normal distribution (Fig. 3).

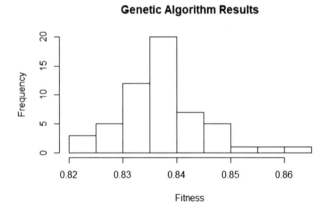

Fig. 3. Genetic algorithm results

4.2.2   Test for Taboo Search Algorithm

With the aim of proving that the genetic algorithm results follow a normal distribution, the following values were calculated:

- Mean of results: 0.8419255.
- Standard deviation of results: 0.008761028.

Then, considering a 5% significance, the Kolmogorov-Smirnov test was obtained:

- P-value: 0.4315.

As P-value obtained is higher than 0.05, the null hypothesis is accepted, and it is concluded that genetic algorithm results follow a normal distribution (Fig. 4).

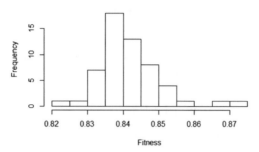

Fig. 4. Taboo algorithm results

4.2.3   Fisher F Test

This statistical test is intended to verify that variances of algorithm results are homogenous since this is required to be able to conduct the Z test. This test can be applied to different statistics; however it is most often used to check that variances of samples taken, in this case, of the algorithm results are considerably homogenous or otherwise considerably different. Therefore, the following hypotheses for both algorithms are considered:

- H0: The variances of algorithm results are considerably homogenous.
- H1: The variances of algorithm results are considerably different.

With the aim of proving that the variances of algorithm results are homogeneous the Fisher F Test was run. In this test, the following result was obtained:

- P-value: 0.5398

As the P-value obtained is higher than 0.05, the null hypothesis is accepted concluding that variances of algorithm results are considerably homogeneous.

4.2.4   Z Test

Since this test will be used to compare the means of both algorithms results, this is ideal to be able to determine which of the two delivers the best solutions and then reach a final conclusion.

In order to compare both means, it is first necessary to prove these are considerably different. After verifying this, the test may be run to determine which of the two means is the lowest and hence finally conclude which algorithm showed a better performance. For the first test (two tails) the following hypothesis are proposed (Table 5):

- H0: the means of algorithms results are considerably homogeneous.
- H1: the means of algorithms results are considerably different.

For the second test (one tail) the following hypotheses are proposed:

- H0: the mean of taboo search algorithms results is lower than the mean of genetic algorithms results.
- H1: the mean of taboo search algorithms results is higher than the mean of genetic algorithms results.

Table 5. Z test results

	Genetic algorithm	Taboo search algorithm
Mean	0,837323812	0,841925455
Variance (known)	6,49034E-05	7,67556E-05
Observations	55	55
Hypothetical difference of means	0	
Z	−2,86729349	
P(Z <= z) one tail	0,002069994	
Critical value of z (one tail)	1,644853627	
Critical value of z (two tails)	0,004139989	
Critical value of z (two tails)	1,959963985	

4.2.5 Results for the Two-Tailed Test

With the purpose of accepting or rejecting the null hypothesis of the two-tailed test, it is verified if the value of "Z" found is between −1,959963985 and +1,959963985 (values representing the interval outside the critical area). In this case, the value "Z" is found in the critical area, hence it can be stated that there is sufficient evidence to reject the null hypothesis, i.e., the means of the genetic algorithms results and the taboo search algorithms results are considerably different.

4.2.6 Results for One-Tailed Test

Thanks to the results of the two-tailed test, the one-tailed test can be continued. For this test, the value of "Z" should be compared with the critical value of one tail, in this case, −1,644853627. As the value of "Z" found is −2,86729349, this is situated in the critical area so that it may be concluded that there is enough evidence to reject the null hypothesis, that is to say, the mean of the taboo search algorithm results is higher than that of the genetic algorithm.

Finally, it might be concluded that the taboo search algorithm produces better results than the genetic algorithm to optimize the space used in the filing of containers.

5 Conclusions

First, we can conclude that the objective function proposed was easily accommodated to the development of algorithms, since the solutions provided by these contain all the components that are part of the function. Likewise, it was noted that the objective function met its purpose, which is determining if a solution is better than the other.

Second, both algorithms were correctly adapted to the optimization problem of the space used in containers. Both the genetic algorithm and the taboo search algorithm provided the expected solutions.

Third, through the numeric experimentation, it can be inferred that the taboo search algorithm delivers better results than the genetic algorithm. This is due to the fact that the working mode of the search taboo algorithm lays emphasis on the continuous search for the global optimum avoiding repetition in local optima. On the other hand, the genetic algorithm is only based on inheriting the best solutions and trying to improve them iteration after iteration, which can result in a local optimum. However, as can be seen in results shown in Table 4, the taboo search algorithm results are not very different from those of the genetic algorithm. This is because boxes' measures introduced in containers do not vary considerably in size and so the solutions found will hardly be any different.

Finally, it should be mentioned that although this research is focused on finding a solution to the problem of containers, the algorithms proposed may be applied to other type of deposits such as warehouses, shelving, railway wagons, and so on.

References

1. Liang, S., Lee, C., Huang, S.: A hybrid meta-heuristic for the container loading problem. Commun. Int. Inf. Manag. Assoc. 7(4), 73–84 (2007)
2. Tiba Group. http://www.tibagroup.com/mx/mclean-y-la-caja-que-cambio-la-historia-del-comercio. Accessed 05 Sept 2017
3. Jiménez, J., Jiménez, J.: Cubicaje: distribución a bajo costo. http://www.logisticamx.enfasis. com/articulos/72752-cubicaje-distribucion-costo. Accessed 12 Sept 2017
4. Garey, M., Johnson, D.: Computers and Intractability. A Guide to the Theory of NP-Completeness. W. H. Freeman and Company, New York (1979)
5. Werra, D., Hertz, A.: Taboo search techniques. Oper.-Res.-Spektrum 11(3), 131–141 (1989)
6. Glover, F.: Taboo search—part I. ORSA J. Comput. 1(3), 190–2016 (1989)
7. Brownlee, J.: Clever Algorithms: Nature-Inspired Programming Recipes. Lulu, Australia (2011)
8. Xuehao, F., Ilkyeong, M., Jeongho, S.: Hybrid genetic algorithms for the three-dimensional multiple container packing problem. Flex. Serv. Manuf. J. 27(2), 451–477 (2013)
9. Tupia, M., Cueva, R., Guanira, J.: A bat algorithm for job scheduling problem in ceramics production lines. In: International Conference on Infocom Technologies and Unmanned Systems (ICTUS 2017), pp. 47–53. Amity University Dubai and IEEE India, Dubai (2017)
10. Pizarro, A., Cueva, R., Tupia, M.: Implementation of a genetic algorithm to optimize the distribution of water in irrigation of peruvian farmland affected by "El Niño". In: 31st International Conference on Computers and their Applications (CATA 2016), pp. 31–38. International Society for Computers and Their Applications (ISCA), Las Vegas (2016)
11. Meneses, S., Cueva, R., Tupia, M., Guanira, J.: A genetic algorithm to solve 3D traveling salesman problem with initial population based on a GRASP algorithm. J. Comput. Methods Sci. Eng. 17(1), 1–11 (2017)
12. Viegas, J., Vieira, S., Henriques, E., Sousa, J.: A tabu search algorithm for the 3D bin packing problem in the steel industry. Lecture Notes in Electrical Engineering, vol. 321, pp. 355–364 (2015)
13. Ren, R.: Combinatorial algorithms for scheduling jobs to minimize server usage time. Doctoral thesis, Nanyang Technological University, Singapore (2018)
14. Hawa, A., Lewis, R., Thompson, J.: Heuristics for the score-constrained strip-packing problem. Comb. Optim. Appl. 11346, 449–462 (2018)
15. Crainic, T., Perboli, G., Tadei, R.: Extreme point-based heuristics for three-dimensional bin packing. INFORMS J. Comput. 20, 368–384 (2008)
16. Hifi, M., Kacem, I., Negre, S., Wu, L.: Heuristics algorithms based on a linear programming for the three-dimensional bin-packing problem. IFAC Proc. Volumes 43, 72–76 (2010)

Artificial Intelligence in Government Services: A Systematic Literature Review

João Reis[1(✉)], Paula Espírito Santo[2], and Nuno Melão[3]

[1] Institute of Social and Political Sciences (ISCSP) and GOVCOPP,
University of Lisbon, Lisbon, Portugal
jcgr@campus.ul.pt
[2] Institute of Social and Political Sciences (ISCSP) and CAPP,
University of Lisbon, Lisbon, Portugal
paulaes@iscsp.ulisboa.pt
[3] Department of Management and CI&DETS,
School of Technology and Management of Viseu,
Polytechnic Institute of Viseu, Viseu, Portugal
nmelao@estgv.ipv.pt

Abstract. The aim of this paper is to provide an overview on how artificial intelligence is shaping the digital era, in policy making and governmental terms. In doing so, it discloses new opportunities and discusses its implications to be considered by policy-makers. The research uses a systematic literature review, which includes more than one technique of data analysis in order to generate comprehensiveness and rich knowledge, we use: a bibliometric analysis and a content analysis. While artificial intelligence is identified as an extension of digital transformation, the results suggest the need to deepen scientific research in the fields of public administration, governmental law and business economics, areas where digital transformation still stands out from artificial intelligence. Although bringing together public and private sectors, to collaborate in the public service delivery, presents major advantages to policy makers, evidence has also shown the existence of negative effects of such collaboration.

Keywords: Digital transformation · Digital era · Artificial intelligence · Political science · Policy-makers · Government services · Systematic literature review · Technological development

1 Introduction

Buzzwords as digital disruption and digital transformation are predominantly used in business contexts. Although, in a much smaller scale, the digital era is also reaching new areas of study, such as the public administration [1]. In cases where governments have made huge investments in the introduction of online services to link government networks to citizens, the penetration of these services tended to be unsatisfactory and did not provide adequate returns on the investment e.g. [2]. Therefore, the low level of user acceptance of electronic government services is recognized as a huge, incumbent problem for policy makers, public administration managers and the community as a whole [3]. Moreover, according to Agarwal [4, p. 1] the "public administrators are

© Springer Nature Switzerland AG 2019
Á. Rocha et al. (Eds.): WorldCIST'19 2019, AISC 930, pp. 241–252, 2019.
https://doi.org/10.1007/978-3-030-16181-1_23

unprepared for the challenges they must face in order to cope with this non-incremental and exponential changes, as many of the existing government structures and processes that have evolved over the last few centuries will likely become irrelevant in the near future". Waves of technology, such as big data, autonomous agents and artificial intelligence (AI) have long been discussed and are reshaping government services. In most western democracies, in public administration, the real danger is represented by researchers and practitioners who are divorced from the world of public administration and are engaged in discussion and making technological decisions without understanding the implications for governance of the administrative state [5]. In line with this background, there is an emerging need for holistic understanding of the range and impact of AI-based applications and associated challenges [6].

Thus, in this article we will investigate the phenomena from two different perspectives: In a first perspective, we will analyse the digital revolution from a macro perspective, by examining that countries who have invested on artificial intelligence and how these governments have succeeded in doing so. Secondly, we intend to provide insightful recommendations on which areas artificial intelligence may be implemented at a governmental level.

Both scholars and practitioners are peremptory on the relevance of education and governmental investment to AI advancements. For instance, Mikhaylov *et al.* [7, p. 1] argues that "AI grand challenge requires collaboration between universities and the public and private sectors", while Bughin *et al.* [8] states that public education systems and workforce training programs will have to be rethought to ensure that workers have the skills to complement rather than compete with machines. The aforementioned studies consider AI challenges fragmented, given the lack of a comprehensive overview of AI challenges for the public sector. In that extent, our conceptual approach intends to analyse and compiles relevant insights and tendencies from the scientific and systematic literature review.

Motivated by the above, we have structured this article in four sections after the Introduction, as follows: We begin by a review of the basic concepts; secondly, we describe the methodological approach; then, we investigate the AI phenomenon from a theoretical perspective; in the last place, we provide the implications and suggestions for subsequent research and decisions at administrative and private levels.

2 Conceptual Background

The term "digital" comprises a pleonasm between humans and technologies. The malleability (e.g., re-programmability), homogeneity (e.g. standardized software languages) and transferability (e.g., ease of transferring digital representation of any object) is at the heart of technologies meshing digital, and often physical materiality, interwoven with human action [9]. In an attempt to define the essential elements of digital transformation, Reis *et al.* [1] carried out a systematic literature review. They have found that the most referenced elements by the majority of scholars were as follows: the *technological* element that is based on the use of new digital technologies such as social media, mobile, analytics or embedded devices; the *organizational* element that requires a change of processes or the creation of new business models; and *social* component that influences all aspects of human life, such as the goal of improving the customer experience.

Several concepts have been advanced to label digital transformation, and despite often used indistinctively in the literature, there are some differences. For example, digitization is the conversion of atoms to bits, digitalization is the transformation of all those bits into value [10]. Gobble [10] argues that digitalization may deliver some savings, most commonly, through efficiency gains and reduced error rates, but it does not change how the company does business. True digitalization, by contrast, changes everything as some authors e.g. [11] refer to it as "digital transformation". Digitization and digital transformation have been occurring in organizations since the 1950s [12]. Digital transformation may be defined as the use of technology to radically improve performance or reach of the enterprises, and it generally encompasses three key areas: customer experience, operational processes and business models [13]. The transformation of *customer experience* focus on what makes customer happier. While companies are using technology to enhance in-person sales conversation and multiple channels to enhance the customer integrated shopping experience. The transformation of *operational processes* enables companies to refocus their people on more strategic tasks as the technology gives executives deeper insights, allowing decision to be made on real-time and real data. Finally, the transformation of business *models* requires digitally modified businesses to share content across organizational silos; moreover, companies introduce digital products that complement traditional products and increasingly transform their multinational into truly global operations.

Digital technologies are bringing new fields of study to academics and innovative solutions to companies; however, established companies do not always understand their current business models well enough to know if it would suit a new opportunity or hinder it [14]. Digital talents and millennials are primary siloed in functions and academic disciplines that were designed to meet the needs of a past era – consequently, traditional academics should be encouraged to delve deeper in order to reach out more technological disciplines and transdisciplinary research agendas [15]. According to Demirkan and Spohrer [15], the new digital millennium requires new types of professionals and work practices as well as new types of citizens and social practices – therefore, the authors encourage the development of T-shaped digital professionals and citizens' future-ready innovators who uniquely combine specialization and flexibility and who also use smart machines as assistants.

As advanced by Kostin [16], three global trends within the field of digital technologies are commonly investigated: artificial intelligence, block chain and big data – we will focus on the first one. The AI was firstly initiated by McCarthy *et al.* [17], with an attempt to investigate how to make machines use language, from abstraction, to solve problems now reserved for humans. In 1995, Russell *et al.* [18] argued that several definitions and variations exists, but the concept could be defined broadly as intelligent systems with the ability to think and learn. Nowadays, it is commonly accepted that AI embodies a heterogeneous set of tools, techniques, and algorithms [19]. Jarrahi [19] reinforces that various applications and techniques fall under the broad umbrella of AI, ranging from neural networks to speech/pattern recognition, to genetic algorithms and to deep learning. Examples of these elements extend AI to include concepts as natural language processing, machine learning and machine vision. In general terms, public administration is also taking into account the AI capacities. This vision is shared by practitioners and consulting agencies, Capgemini [20] argues about the economic and societal benefits of AI to the public sector, or as they state: "AI helps us to enter a new era of sophisticated and smart public services".

3 Methodology

As the scientific production has been growing steadily, the scientific databases have played a key role in the diffusion of the scientific research. These databases (e.g., ISI, Scopus) make available statistical data, by displaying attractive and graphical knowledge, to allow researchers to understand and interpret real-life phenomenon's or scientific developments from virtually all areas of study. Due to its relevance, Pritchard [21, p. 348] has early defined the term *bibliometric* as the "application of statistical and mathematical methods set out to define the processes of written communication and the nature and development of scientific disciplines by using recounting techniques and analysis of such communication". Raan [22] also argued that bibliometric methods have been used to measure scientific progress in many disciplines of science and is a common research instrument for systematic analysis. Thus, the bibliometric method allows access to relevant knowledge about the status of scientific research in specific areas, which helps researchers to identify novel schemes among researchers [23]. In light with the above, the bibliometric analysis will use relevant quantitative data (e.g. dates) retrieved from the Web of Science database, that will support and enhance the findings of the content analysis.

The second technique is *content analysis*, which is relatively new instrument of analysis, dating back to the 18th century in Scandinavia [24] and in a scientific perspective from the second half of the 20th century on. It is identified as a systematic and replicable technique that allows compressing many words and sentences of text into fewer content categories based on explicit rules of coding, in order to allow researchers to make inferences about the author (individuals, groups, organizations, or institutions), the audience, and their culture in time [25]. Berelson [26, p. 18] defined content analysis as "a research technique for the objective, systematic and quantitative description of the manifest content of communication". However, Berelson's [26] definition did not capture the qualitative and latent perspective of the analysis [27, 28], but was later on debated by Krippendorff [29], which defined content analysis as a research technique for making replicable and valid inferences from texts (or other meaningful matter) to the contexts of their use. At first, content analysis was a time-consuming process, due to the absence of adequate data analysis technologies. For some decades, researchers have available sophisticated software programs designed for computer-assisted qualitative data analysis, as e.g., NVIVO and MaxQDA. In practical terms, content analysis comprises three stages: stating the research problem, retrieving the text or the contents, and employing sampling procedures, and inference procedures of analysis. This technique considers the presence with which certain words or particular phrases occur in the text as a means of identifying its characteristics, the sentences in the document are transformed into numbers, and the number of times in which a word occurs in the text is taken as an indicator of its significance [30]. Content analysis is different from textual analysis, as it includes words and sentences, but it also be based on a linguistic set of instruments but also includes other items as the personalities, seconds and frames [31]. However, content analysis has several advantages, but, perhaps the most important is that it can be virtually unobtrusive [32], since besides being useful for analysing in-depth interview data, it may be also used non-reactively, which means that it is adequate to similar sources of data collection (e.g., digital libraries) that allow researchers to conduct analytic studies [33].

For decades, the Science Citation Index, now the WoS (Web of Science) (owned by Clarivate Analitics) was the only large multidisciplinary citation data source world-wide. Meanwhile, Scopus, provided by Elsevier, is a second comprehensive citation database [34]. Thus, the common way to systematically review the literature is to trace publications by using the WoS, which is still one of the worldwide most-used scientific database of peer-reviewed literature. The decision to choose this database is due to reasons of transparency and easy reproduction, since the results from Scopus were not very different from the ones from ISI WoS. The online search within WoS was conducted on October 8th, 2018, by including the keywords and operators "Government" AND "Digitalization" OR "Government" AND "Artificial Intelligence" in the Title, Abstract and Keywords field of the search-engine as displayed at Table 1.

Table 1. Methodological approach

ISI web of science		Government	
		Digitalization	Artificial intelligence
Search	Title-Abstract-Keywords	126	199
Language	English	113	192
Categories	Management, Business, Public Administration, Economics, Political Science, Social Sciences Interdisciplinary, Social Issues, Operations Research Management Science, Multidisciplinary Science	37	40
Document types	Proceedings paper Articles	35	39

The review process was based on successive filters: (1) only articles written in *English* were deemed relevant for an accurate interpretation; (2) the selected *documents* were restricted to indexed scientific articles to ensure credibility, but also on conference papers with the intent of focusing on the most updated research; finally, (3) to gain a greater research focus our findings focused on, but were not limited to, areas of *management*, *social sciences* and *political science*. The WoS database has returned 74 research papers, as the exclusion rate is approximately ¾ of the initial search; this selection allowed us to gain an overall understanding of the subject under investigation and to focus on the results. This article presents the following research limitations:

the literature review is confined to its keywords and it is possible that articles indirectly related may be missing. Due to space limitations, it was not possible to list all the references (74 articles) and the content analysis (e.g., coding, categories, subcategories). The author's working material may be provided on request by contacting the first author.

4   Findings and Discussion

4.1   The Digital Scope and the Shift to Artificial Intelligence

The analysis of the digital scope and the shift to other technologies is still a vibrant topic. Recently, Kostin [16] had researched the foresight of the global digital trends. The author examines the recent developments (e.g. artificial intelligence) of the global digital trends and investigates relevant digital technologies, which are crucial to frame future strategies not only for governments and citizens, but also for international companies. In light with the above, we analyse the AI paradigm, to this end we cross-checked the ISI WoS data to understand the extent of AI divergent and convergent relationships with the concept of digitalization.

Bani and Paoli [35] refer that governments have been moving away from the digitalization of documents, processes and decision-making, within the administration, towards a new model that involves citizens in the co-production and information sharing. Grandhi *et al.* [36] corroborates this view by stating that the current business environments are more uncertain than ever before, while understanding customer/ citizen behaviour is an integral part of an organization strategic planning and execution process. Organizations which embrace digitalization are seeing the investments made in IT infrastructure, Internet of Things, machine learning and AI getting more established and aiding the decision-making process [36].

In recent years, we have also witnessed the amalgamation of government services and electronic systems, and the United Kingdom (UK) is no exception. Citizens and the state interactions have changed focus towards human centred electronic approaches, by introducing citizens with electronic services that have simplified bureaucratic mechanisms and response time [37]. In sum, the UK and the most part of EU countries are not just digitalizing government processes, there are also moving forward by involving their citizens in co-producing knowledge and information sharing, while machine learning and AI are turning the decision-making process easier. An empirical research of Russo *et al.* [38] refers that the Italian government, which is the second worldwide highest investor on governmental digitalization (Fig. 1), is considering the introduction of several online and office access points to its services to increase the penetration of provision systems not requiring a direct interaction in public administration – they argue that the digitalization of processes can save resources. The Italian investment in research and new digital technologies are due to the significant investment of e-Government services in Europe, while the diffusion of e-government services in Italy is slightly behind the European Union (EU) investment average [38]. Figure 1 illustrates the countries that are investing on the digitalization of their governments and the ones which are investing on AI.

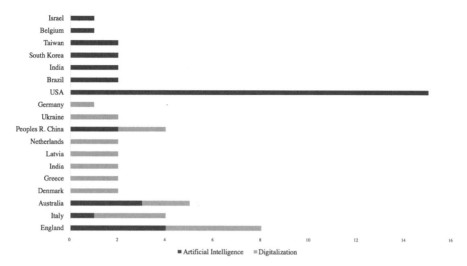

Fig. 1. Number of published articles by country concerning governmental digitalization and artificial intelligence (Source: based on ISI WoS data)

The British government had also made substantial practical contributions in the development of expert systems – the trend has been one of building from what others, particularly Americans, have done [39]. As governments cannot do the integration of AI into public service delivery on their own, the UK Government Industrial Strategy is clear that delivering on the AI grand challenge requires collaboration between universities and public private sectors [7]. Mikhaylov *et al.* [7] also argues that, despite the cross-sectorial collaborative approach is the norm in applied AI centres of excellence around the world, the popularity of this strategy entails serious management challenges that hinder their success. Therefore, the UK perspective to focus their AI investments on the synergies between the state and companies.

To the United States (US) maintain its innovative and technological competitive advantage, the US government, through its policy, is encouraging private companies to invest on new trends, such as Internet of Things, artificial intelligence, national security, and many other areas which are expected to evolve over the years ahead [40]. In sum, the US perspective is centred on governmental policies to enable and support AI technological developments. An example are the technological breakthroughs that have been sponsored by the US Government to gain military advantage – well-known technologies that challenge military traditional dominance includes autonomous vehicles, cyber technologies and artificial intelligence [41]. On the contrary, the UK and EU perspective is enhancement by collaborative synergies between companies and the state, thus this strategy somewhat enables the technological developments on AI to move forward. Although bringing together public and private sectors, collaboration in the public service delivery presents major advantages. Chou *et al.* [42] defines public-private partnership (PPP) as a strategy where governments encourage private institutions to support public construction projects by providing proper incentives based on collaboration with private institutions. It is worth noticing that Chou *et al.* [42]

acknowledges existing negative effects of such collaboration, as disputes may occur during contract management. However, a research from Cheung *et al.* [43] highlights the PPP benefits on mitigating the shortage of governmental funding and avoid public investment restriction in the UK Most EU member states and also considers the PPPs as an important tool to attract additional financial resources [44] and, overall, the years to come we will find an increasing role for PPPs in the provision of public infrastructures and services [45]. We have also noticed a lack of research in recent years concerning the governmental process digitalization, when compared with the AI that has gained strength. Figure 2 presents a timeline of published articles regarding governmental digitalization and AI.

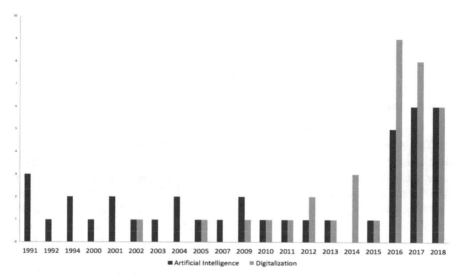

Fig. 2. Timeline of published articles regarding governmental digitalization and artificial intelligence (Source: based on ISI WoS data)

As far as digitalization is concerned, the year 2016 reached the top of scientific production, since digital technologies have had a strong impact on the governmental sector, and deserved wide scholar attention, vide e.g. digitalization of Finland's transport sector [46]. Moreover, citizens have also started to experience new technologies and applications, becoming co-producers of digital services [47]. On the other hand, recent scandals have undermined the digitalization progression vide [48], for instance the data protection showed how much sensitive we are in front of cyber-attacks and vulnerable when it comes to our personal security [49, 50]. The progress of published articles on AI is on a counter-cycle and its production is predictably increasing. Although slightly stable after 2017, the published articles on AI had increased exponentially from 2015 to 2016 which, according to Mikhaylov *et al.* [7], the observed growth is due the interest of public sector on data science and artificial intelligence capabilities to deliver policy and to generate efficiencies in high-uncertainty environments.

4.2   Will Artificial Intelligence Shape Digital Governments? if so, in Which Areas?

According to Fig. 3, a greater emphasis has been given to the digitalization of the public administration and governmental law. Thus, it is clear that digital transformation is already changing the way how government services work, although with a smaller expression when compared with the business economics perspectives. The business economics stand out (Fig. 3), and represents the relationship between governments and companies, which supports our previous findings about the E.U governments efforts to establish partnerships with the private sector to underpin technological developments – and it is at this stand that the digitalization and the artificial intelligence meet.

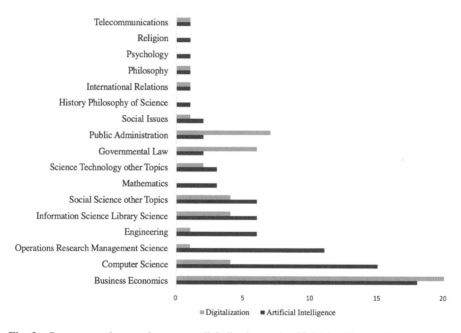

Fig. 3. Governmental research areas on digitalization and artificial intelligence (Source: based on ISI WoS data)

AI differs from digitalization at points essentially related to computer science and operations research, which are the foundation pillars for the development and implementation of the artificial intelligence at a governmental level. It is still hard to say that the AI is shaping digital governments, but it is clear at the point where partnering with private companies can improve the ways in which public administration and judicial services are delivered to its citizens.

5 Concluding Remarks

As previously evidenced, governmental digitalization is currently involving citizens to co-produce information, which is enabling the decision-making process. It is known that currently digital transformation is mainly focused on business and industry areas, with less expression to governments, although it is a predictable game-changing tendency. On the other hand, what refers to AI, we concluded that Western countries, in particular the United States, are directly supporting its introduction in governmental policies. On the contrary, the EU governments are following a different strategy, as they underpin their technological developments in collaborative synergies with leading digital companies to move AI forward. Despite the association of these synergies brings several advantages, this strategy also entails serious management challenges that may hinder their success and therefore it deserves to be studied. It is still hard to say that AI is shaping digital governments. However, it is clear that AI is playing a strong influence on governments, that is leading to an increase of public and private investments.

While AI is identified as an extension of digital transformation, the results suggest to deepen the scientific research in public administration, governmental law and business economics. We do not believe the number of published articles on digital transformation should be lower than those on AI, because the latter refers to a digital technology within many other existing ones that are used in digital transformation. However, we identify the existing margin as an AI research opportunity within the digital transformation spectrum. Further research should not only focus on areas such as business economics, computer science or operations research, as it has been done so far, but also focus on the social issues and how the public administration can effectively improve the citizens lives through the use of AI.

References

1. Reis, J., Amorim, M., Melão, N., Matos, P.: Digital transformation: a literature review and guidelines for future research. In: Trends and Advances in Information Systems and Technologies, WorldCIST, pp. 411–421. Springer, Cham (2018)
2. Hung, S., Chang, C., Yu, T.: Determinants of user acceptance of the e-government services: the case of online tax filing and payment system. Gov. Inf. Q. **23**(1), 97–122 (2006)
3. Lamberti, L., Benedetti, M., Chen, S.: Benefits sought by citizens and channel attitudes for multichannel payment services: evidence from Italy. Gov. Inf. Q. **31**(4), 596–609 (2014)
4. Agarwal, P.: Public Administration Challenges in the World of AI and Bots. Public Administration Review (2018)
5. Barth, T., Arnold, E.: Artificial intelligence and administrative discretion: implications for public administration. Am. Rev. Public Adm. **29**(4), 332–351 (1999)
6. Wirtz, B., Weyerer, J., Geyer, C.: Artificial intelligence and the public sector–applications and challenges. Int. J. Public Adm. **13**(7), 1–20 (2018)
7. Mikhaylov, S., Esteve, M., Campion, A.: Artificial intelligence for the public sector: opportunities and challenges of cross-sector collaboration. Philos. Trans. R. Soc. A **376** (2128), 20170357 (2018)

8. Bughin, J., Hazan, E., Ramaswamy, S., Chui, M., Allas, T., Dahlström, P., Henke, N., Trench, M.: Artificial intelligence–the next digital frontier. McKinsey Glob Institute. https://www.mckinsey.de/files/170620_studie_aipdf. Accessed 11 Oct 2018
9. Hinings, B., Gegenhuber, T., Greenwood, R.: Digital innovation and transformation: an institutional perspective. Inf. Organ. **28**(1), 52–61 (2018)
10. Gobble, M.: Digital strategy and digital transformation. Res.-Technol. Manag. **61**(5), 66–71 (2018)
11. Matthias, L., Juliane, K., Peter, S.: The digital future has many names – how business process management drives the digital transformation. In: 6th International Conference on Industrial Technology and Management, pp. 22–26 (2017)
12. Heavin, C., Power, D.: Challenges for digital transformation–towards a conceptual decision support guide for managers. J. Decis. Syst. **27**, 38–45 (2018)
13. Westerman, G., Bonnet, D., McAfee, A.: The nine elements of digital transformation', MIT Sloan Management Review (2014). http://sloanreview.mit.edu/article/the-nine-elements-of-digital-transformation/. Accessed 11 Oct 2018
14. Sanchez, M., Zuntini, J.: Organizational readiness for the digital transformation: a case study research. Revista Gestão & Tecnologia **18**(2), 70–99 (2018)
15. Demirkan, H., Spohrer, J.: Commentary-cultivating T-shaped professionals in the era of digital transformation. Serv. Sci. **10**(1), 98–109 (2018)
16. Kostin, K.: Foresight of the global digital trends. Strateg. Manag. **23**(1), 11–19 (2018)
17. McCarthy, J., Minsky, M.L., Rochester, N., Shannon, C.: A proposal for the dartmouth summer research project on artificial intelligence, august 31, 1955. AI Mag. **27**(4), 12 (2006)
18. Russell, S., Norvig, P.: Artificial Intelligence: A Modern Approach. Prentice-Hall, Englewood Cliffs (1995)
19. Jarrahi, M.: Artificial intelligence and the future of work: human-AI symbiosis in organizational decision making. Bus. Horiz. **61**(4), 577–586 (2018)
20. Tinholt, D., Carrara, W., Linden, N.: Unleashing the potential of artificial intelligence in the public sector. Capgemini Consulting (2017)
21. Pritchard, A.: Statistical bibliography or bibliometrics? J. Doc. **25**(4), 348–349 (1969)
22. Raan, A.: For your citations only? Hot topics in bibliometric analysis. Meas.: Interdisc. Res. Perspect. **3**(1), 50–62 (2005)
23. Zyoud, S., Fuchs-Hanusch, D., Zyoud, S., Al-Rawajfeh, A., Shaheen, H.: A bibliometric-based evaluation on environmental research in the Arab world. Int. J. Environ. Sci. Technol. **14**(4), 689–706 (2017)
24. Rosengren, K.: Advances in Scandinavia content analysis: an introduction. In: Rosengren, K.E. (ed.) Advances in Content Analysis. Beverly Hills, Sage (1981)
25. Mills, A., Durepos, G., Wiebe, E.: Encyclopedia of Case Study Research. SAGE Publications, California (2010)
26. Berelson, B.: Content Analysis in Communications Research. Free Press, New York (1952)
27. Hsieh, H., Shannon, S.: Three approaches to qualitative content analysis. Qual. Health Res. **15**(9), 1277–1288 (2005)
28. Bengtsson, M.: How to plan and perform a qualitative study using content analysis. NursingPlus Open **2**, 8–14 (2016)
29. Krippendorff, K.: Content Analysis. Sage Publications Inc., Beverly Hills (1980)
30. May, T.: Social Research: Issues, Methods and Process, 3rd edn. Open University Press, Buckingham (2001)
31. Santo, P.: Looking for social class and civil society in political discourse in Portuguese democracy (1976–2006) – content analysis approach. Comunicação Pública **10**(8), 1–11 (2015)

32. Webb, E., Campbell, D., Schwartz, R., Sechrest, L., Grove, J.: Nonreactive Measures in the Social Sciences. Houghton Mifflin, Boston (1981)
33. Berg, B.: Qualitative Research Methods for the Social Sciences, 4th edn. Allyn and Bacon, Boston (2004)
34. Raan, A.: Advances in bibliometric analysis: research performance assessment and science mapping. Bibliometrics. Use Abuse Rev. Res. Perform. **3**, 17–28 (2014)
35. Bani, M., De Paoli, S.: Ideas for a new civic reputation system for the rising of digital civics: digital badges and their role in democratic process. In: ECEG2013–13th European Conference on eGovernment: ECEG (2013)
36. Grandhi, B., Patwa, N., Saleem, K.: Data driven marketing for growth and profitability. In: 10th Annual Conference of the EuroMed Academy of Business (2017)
37. Zissis, D., Lekkas, D.: The security paradox, disclosing source code to attain secure electronic elections. In: Proceedings of the 9th European Conference on e-Government (2009)
38. Russo, C., Ghezzi, C., Fiamengo, G., Benedetti, M.: Benefits sought by citizens in multichannel e-government payment services: Evidence from Italy. Procedia-Soc. Behav. Sci. **109**, 1261–1276 (2014)
39. Bench-Capon, T., Rada, R.: Expert systems in the UK: from AI to KBS. Expert Syst. Appl. **3**(4), 397–402 (1991)
40. Rao, G., Williams, J., Walsh, M., Moore, J.: America's seed fund: how the SIBR/STTR programs help enable catalytic growth and technological advances. Technol. Innov. **18**(4), 315–318 (2017)
41. FitzGerald, B., Parziale, J.: As technology goes democratic, nations lose military control. Bull. Atomic Sci. **73**(2), 102–107 (2017)
42. Chou, J., Hsu, S., Lin, C., Chang, Y.: Classifying influential for project information to discover rule sets for project disputes and possible resolutions. Int. J. Project Manag. **34**, 1706–1716 (2016)
43. Cheung, E., Chan, A., Kajewski, S.: Reasons for implementing public private partnership projects: perspectives from Hong Kong, Australian and British practitioners. J. Prop. Invest. Financ. **27**(1), 81–95 (2009)
44. Medda, F., Carbonaro, G., Davis, S.: Public private partnerships in transportation: some insights from the European experience. IATSS Res. **36**(2), 83–87 (2013)
45. Iossa, E., Saussier, S.: Public private partnerships in Europe for building and managing public infrastructures: an economic perspective. Ann. Public Coop. Econ. **89**(1), 25–48 (2018)
46. Leviäkangas, P.: Digitalisation of Finland's transport sector. Technol. Soc. **47**, 1–15 (2016)
47. Björklund, F.: E-government and moral citizenship: the case of Estonia. Citizsh. Stud. **20**(6–7), 914–931 (2016)
48. Arun, P.: Uncertainty and insecurity in privacyless India: a despotic push towards digitalisation. Surveill. Soc. **15**(3–4), 456–464 (2016)
49. Power, D.: "Big Brother" can watch us. J. Decis. Syst. **25**(sup1), 578–588 (2016)
50. Kushzhanov, N., Aliyev, U.: Digital space: changes in society and security awareness. Bull. Nat. Acad. Sci. Repub. Kaz. **1**, 94–101 (2018)

Ontology Driven Feedforward Risk Management

Cédric Gaspoz[1], Ulysse Rosselet[1(✉)], Mathias Rossi[2],
and Mélanie Thomet[2]

[1] University of Applied Sciences Western Switzerland (HES-SO),
HEG Arc, Neuchâtel, Switzerland
ulysse.rosselet@hc-arc.ch
[2] University of Applied Sciences Western Switzerland (HES-SO),
HEG Fribourg, Fribourg, Switzerland

Abstract. Organizations are increasingly relying on projects to support their activities. With this increase, the management of these projects is becoming critical. New methods, tools and frameworks appear in order to improve the success rate of these projects. However, being complex endeavours, projects face numerous challenges that have to be addressed with proper risk management. Various risk management frameworks have been developed in order to support project managers in their task of identifying and controlling risk. Although these frameworks are effective in reducing the impact of risk, they are slow to implement. In order to more productively address these risks, we have developed an ontology to support feedforward risk management. This ontology can be implemented in project management software to provide project managers with a multi-level approach of the risks associated with their projects. The scope and breadth of the risks identified by the software will be closely related to the project data collected in the system, reducing the need for a lengthy setup. Moreover, using data collected by the project management software, the system will be able to proactively raise management awareness towards potential upcoming issues.

Keywords: Risk management · Ontology · Feedforward · OWL

1 Introduction

During the last few decades, the flattening of organizational hierarchies, the weakening of firms' boundaries in favour of networks of collaborations and the restructuring of competition between firms within and across industries has caused a large increase in project-based work [1]. Even though project management methods and tools have seen an important development over the same period, successful project execution is still a challenge for organizations. Consulting firms such as the Standish Group regularly publish alarming reports about projects success rates [2]. While the figures produced in these reports are questionable [3], studies by researchers also show that project success is still hard to achieve [4].

© Springer Nature Switzerland AG 2019
Á. Rocha et al. (Eds.): WorldCIST'19 2019, AISC 930, pp. 253–261, 2019.
https://doi.org/10.1007/978-3-030-16181-1_24

Projects are complex endeavours involving multiple actors, compressed schedules, ambitious goals, limited budget and frequently changing requirements and are therefore affected by several sources of uncertainties such as market payoff, project budget, product performance, market requirements and project schedule [5].

In this context, risk management is a critical element for successful project management. Project risks can be defined as any undesired event that may cause delays, excessive spending, unsatisfactory project results, safety or environmental hazards and total failure [6]. In order to prevent problems and deal with uncertainties, risk management techniques have been developed for project management. However, such techniques are still seldom used, as many managers still do not fully understand their value or know how to apply them in the context of project management [7].

In order to manage these risks effectively, practitioners need effective means of detecting and monitoring potential risk factors. An ontology supporting risk management and integrated to project management and ERP systems could potentially enable proactive risk management.

In this paper, we address the following research question: how can an ontology be designed and be used in project management software in order to improve preventive risk management?

This research is being carried out in collaboration with an external software company and aims to provide practitioners with an integrated tool for identifying, assessing and monitoring project risks.

2   State of the Art

First, we present the literature review for project risk management, which is the application domain of our ontology. In order to frame the concepts composing the ontology, we built a typology of project risks. This typology defines a hierarchy of ideal types of projects which have different risk profiles and are believed to ask for different approaches in terms of risk monitoring and management. Each of the resulting ideal types represents a unique combination of the organizational attributes that are believed to determine the relevant outcome(s) [8].

One important aspect of every typology is the objectives underlying its development. In our case, this objective is the identification of project types requiring different approaches in terms of risk management. With this in mind, we reviewed and combined characteristics of existing typologies of project risks from different domains in order to build our own.

2.1   Existing Typologies for Project Risk Management

A risk breakdown structure is a hierarchical structure of risk sources that is used to aid the understanding of the risks faced by a project [9]. Hillson [9] proposed a risk breakdown structure for generic projects based on the work of the Risk Management Specific Interest Group of the Project Management Institute [10]. Shown in Table 1, this risk breakdown structure comprises the general risks faced by any kind of project.

Table 1. Risk breakdown structure for generic projects [9]

Top level	Risk family	Risk factors
Project risk	Management	Corporate
		Customer & stakeholder
	External	Natural environment
		Cultural
		Economic
	Technology	Requirements
		Performance
		Application

A "risk factor" is a source of risk from the internal or external environment of the project. These factors influence the occurrence of adverse events in risk assessment and are used as independent variables. The works of Barki et al. provide us with a comprehensive review of risk factors and variables for project management that are presented in Table 2 [11, 12]. We will use these risk factors and underlying variables as the attributes characterizing the ideal types in our typology.

Table 2. Risk factors [11]

Top level	Risk factors
Project risk	Technological risk
	Project size
	Deliverable complexity
	Organizational environment
	Project complexity
	Exogenous conditions
	External agents characteristics

2.2 Ontology

In information systems (IS), ontologies are used to describe concepts and their relationships in a given domain. Their importance in the IS community is being recognized in a multiplicity of research fields and applications areas, including knowledge engineering, database design and integration and information retrieval and extraction [13]. Ontologies support the description of the entities and their interrelationships in order to structure the knowledge of a particular domain of discourse. They can be used to build a 'web of data' understandable by machines to support the Semantic Web [14], identify similarities between concepts [15] or to model a specific domain of knowledge [16]. The second property of ontologies is that they support various operations such as merging, translation, alignment, refinement of concepts, extension, specialization and inheritance.

Facing the task of ontology creation, researchers put efforts in fully or at least partially automating the generation process [16]. Ontology learning is associated with techniques supporting the extraction of content from structured or semi-structured data [17]. This presupposes the existence of data sources covering the domain of interest. In cases where this data is not available or is only partially available, a traditional way of ontology creation is recommended.

3 Ontology Driven Risk Management

3.1 Developing the Ontology

In order to support the creation of a risk management framework, we decided to create an ontology representing the domain of risk management. The idea behind the creation of an ontology is to be able to represent the whole domain, but also to be able to use ontological properties such as inferences.

To support our work, we chose to use the Methontology framework, "a methodology to build ontologies from scratch" [18]. It follows a multi-step process: specification, conceptualization, formalization, integration and implementation, along with maintenance, knowledge acquisition, documentation and evaluation.

Specification. The first step consists of establishing the purpose and scope of the ontology. In our case, the ontology will be used in enterprise risk management software, integrated in a complete enterprise resource planning (ERP) solution. Our ontology can therefore be categorized as a systems engineering one [1, 2]. End users for the risk management system and the underlying ontology will be managers and top-level executives, who will use it for preventive risk management. Automatic data collection from the ERP and self-reporting from the employees will be used to populate the system and produce the risk management indicators. The resulting scope encompasses the concepts of risk management and project management presented in the literature review above.

Based on the purpose of the ontology, we moved to define the level of formality. Uschold and Gruninger [19] classify the level of formality on a continuum from highly informal to rigorously formal, based on the language used to specify the ontology. In our case, given that the ultimate goal of the ontology is to be integrated in some sort of computerized tools, we agreed on using a semi-formal ontology. Such ontologies combine informal descriptions in plain English with some sort of formalisms to describe the concepts and relationships.

As recommended by Uschold and Gruninger [19] we followed a middle-out approach to identify the main terms for our ontology. Based on interviews of experienced project management practitioners, we identified the primary concepts of the ontology. We then confronted these initial terms to existing typologies of risks and depending on the project's need, we further specialized and generalized those terms, to make sure that the concepts were defined at an adequate level of granularity to satisfy the purpose of our ontology [18]. During this refinement process, we used intermediate classification trees of the risk-related concepts, to check the conciseness and completeness of the ontology.

Knowledge Acquisition. To have a broad understanding of the domain and to cover all of its aspects, we started by brainstorming, in order to identify potentially relevant terms. During this process, we identified more than 120 terms related to risk management and grouped them by affinity reaching 12 classes of risks. Six interviews with professionals of the fields were conducted in order to extend the list of terms and to improve our understanding of the relationship between risk management concepts. The second part of the knowledge acquisition phase consisted of an extensive literature review in order to identify risk management frameworks, risks definitions and existing categorizations of risks. Using these elements, we were able to refine our terms, formalize their relationships and agree on a classification.

The first step in this literature review was to identify the key areas of risk-related activities within projects. The distinguishing criterion was that the relevant risks are different from one sector to another. To do this, a bottom-up approach was used to ensure the reliability of the sources. This approach consisted of identifying the major risks within the projects. Afterwards, we distinguished business sectors, based on these risks.

The second phase of this literature review involved grouping the main risks for each sector of activity in the form of risk breakdown structures. Bourdeau et al. [12] reviewed the project management literature and identified general risk factors and their underlying variables which we used as attributes to characterize the ideal types in our typology. Finally, a risk framework by business area was designed using the same process, this time incorporating the main risks identified in each sector.

At a finer level of granularity, individual risk variables from the literature were listed and integrated in the risk management framework. For these individual risk variables, indicators were extracted from relevant literature. This general catalogue of risk variables and related indicators will be used by project managers to select the specific risks associated with their projects. As each project has its specificities, different risks will need to be monitored through different sets of indicators.

The flexibility of our ontology, linking multi-level risks to indicators in a per project configurable manner, will allow for preventive risk management of projects with individual risk profiles.

Conceptualization. Following the knowledge acquisition phase, we moved on to the conceptualization of the ontology. The goal of this phase is "to structure the domain knowledge in a conceptual model that describes the problem and its solution in terms of the domain vocabulary identified in the ontology specification activity" [18]. Following the methodology of Fernández-López et al. [18], we build a glossary of terms structured as a two-dimensional table organized according to high level risk factors and economic activity domains. This allowed for identifying recurring concepts, synonyms, instance candidates, verbs and typical properties for the risk management domain. We described all the concepts in our typology using a data dictionary. This dictionary consists of tables specifying the attributes for the classes of our ontology. Table 3 shows an excerpt of the data dictionary for the "indicator" and "risk variable" classes.

Integration. It is highly recommended to integrate existing ontologies during the process of creating a new ontology [18]. In this project, we chose to not follow this recommendation. There are two main reasons for doing so. First, there are no ontologies

Table 3. Data dictionary example for class attributes

Class	Attributes
Indicator	Name, Definition, Hierarchy, Measures, Frequency, Data, source, Target
Risk variable	Name, Definition, Domain, Type

covering a substantial part of our domain of interest. This would imply combining multiple ontologies having different degrees of specification into our ontology. Second, and more importantly, the goal of the project is to create an integrated framework of risk management, which implies a highly integrated definition of the concepts of the domain. Therefore, we chose to base our definitions on the literature for sound proofness, but to work on each definition by ourselves in order to guarantee the consistency of the whole ontology.

Implementation. The next step was to implement the ontology in a formal language. There are many languages supporting the codification of ontologies such as Ontolingua, LOOM, OCML, OIL, DAML + OIL and OWL. Based on the comprehensive work of Corcho et al. [20] to compare the main methodologies, tools and languages for building ontologies, we decided to use Protégé and OWL to codify our ontology. This choice was made based on our previous experience with Protégé as a tool for building and supporting ontologies. OWL was elected due its compatibility with the RDF Schema, without the limitations of RDF's expressive power [21]. OWL files and a database of instances were provided to our implementation partner to support the integration of the ontology in their project management software and ERP systems. The simplified ontology is depicted in Fig. 1.

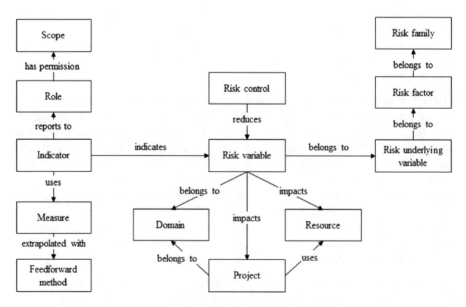

Fig. 1. Ontology for project risk monitoring and feedforward control

Evaluation. This phase consists of carrying out a technical judgement of the ontology and its implementation with respect to a frame of reference [18]. In our case the reference will be the scope and purpose of the ontology. The verification of the technical correctness of the ontology was done during its modelling in Protégé. The validation of the ontology has been possible through the configuration of a third-party decision dashboard tool that uses data from the instances to present project managers with the risk management indicators as defined in our ontology.

3.2 Application

The ontology can be implemented in a risk management module for a project management system. To support the goal of increasing project success in the long run, the ontology brings three key advantages to the risk management module.

First, the ontology allows for integration of human-related risks. The risk management module will collect and analyse data that provides information about influential contributors of project performance such as employee engagement, team members motivation state, team interrelationships, stakeholders commitment and political changes. Human-related risks will be managed at the individual, group and organizational levels.

Second, the ontology allows for automatic data collection of the vast majority of indicators. Meaningful human-related data will be collected throughout project lifecycle. Data gathering processes will rely mostly on automated collection procedures of digital data. This is a major advantage as it can be performed with no direct involvement of managers or team members, who can focus on their activities. When needed, self-reported personal data will complement data collected automatically. To motivate people to hand over personal data, proper incentives have to be defined and implemented.

Third, the ontology will support the use of preventive controls. Human-related indicators will be combined with more traditional indicators to generate alerts that will work as early warning signals. The software solution will favour feedforward controls (as opposed to feedback controls) in order to prevent risks from arising, rather than dealing with damage when observed.

4 Discussion

Most of the available approaches to dealing with risk are model-based. Modelling syntaxes vary from simple diagrams to very complex mathematical formulas. We propose an alternative approach to risk management that leverages the power of ontological conceptualization. Supported by the ontology, the risk identification, which takes place at the heart of the project management software, can be effectively presented in a multi-level project management dashboard. Compared to a classical approach, this will foster an ex-ante treatment (feedforward controls) of risks as opposed to an ex-post one (feedback controls). Detective controls come after the action (feedback) when preventive controls come before the action and are designed to prevent major problems from ever occurring [22]. This approach is expected to support project managers in effectively reducing the impact of risk by taking appropriate preventive measures.

The ontology is currently implemented and data is being collected through the ERP system, in order to assess the ontology's effectiveness in detecting risks and allowing for better risk control. The evaluation of its actual effectiveness will be the subject of further research.

5   Conclusion

Effective project management is a complex activity in which risk management plays a critical role [5]. In this paper, we have presented an ontology for risk management, based on the existing literature about project risk management and interviews of project management professionals. This ontology has been built according to the Methontology technique [18] and is integrated into the risk management module of an ERP system, forming together a new tool that provides project managers with a multi-level approach to the risks associated with their projects. Thanks to the extensive risk catalogue instantiated in the ontology, it will allow for better identification, definition, monitoring and control of the multiple risk variables that can impact a project.

Acknowledgement. We are grateful to the Swiss Commission for Technology and Innovation which provided partial funding for this work (grant number 19311.1 PFES-ES).

References

1. Whitley, R.: Project-based firms: new organizational form or variations on a theme? Ind. Corp. Change **15**(1), 77–99 (2006)
2. Clancy, T.: The Standish Group CHAOS Report, Project Smart (2014)
3. Eveleens, J.L., Verhoef, C.: The rise and fall of the chaos report figures. IEEE Softw. **27**(1), 30–36 (2009)
4. de Bakker, K., Boonstra, A., Wortmann, H.: Does risk management contribute to IT project success? A meta-analysis of empirical evidence. Int. J. Proj. Manag. **28**(5), 493–503 (2010)
5. Huchzermeier, A., Loch, C.H.: Project management under risk: using the real options approach to evaluate flexibility in R…D. Manag. Sci. **47**(1), 85–101 (2001)
6. Raz, T., Shenhar, A.J., Dvir, D.: Risk management, project success, and technological uncertainty. RD Manag. **32**(2), 101–109 (2002)
7. Todorović, M.L., Petrović, D.Č., Mihić, M.M., Obradović, V.L., Bushuyev, S.D.: Project success analysis framework: a knowledge-based approach in project management. Int. J. Proj. Manag. **33**(4), 772–783 (2015)
8. Doty, D.H., Glick, W.H.: Typologies as a unique form of theory building: toward improved understanding and modeling. Acad. Manag. Rev. **19**(2), 230–251 (1994)
9. Hillson, D.: Using a risk breakdown structure in project management. J. Facil. Manag. **2**(1), 85–97 (2003)
10. Hall, D.C., Hulett, D.T., Graves, R.: Universal Risk Project—Final report, PMI Risk SIG (2002)
11. Barki, H., Rivard, S., Talbot, J.: Toward an assessment of software development risk. J. Manag. Inf. Syst. **10**(2), 203–225 (1993)
12. Bourdeau, S., Rivard, S., Barki, H.: Evaluation du risque en gestion de projets. CIRANO Sci. Ser. **47**, 3–45 (2003)

13. Guarino, N.: Formal Ontology in Information Systems: Proceedings of the First International Conference (FOIS 1998), 6–8 June 1998, Trento, Italy, vol. 46. IOS Press (1998)
14. Berners-Lee, T., Fischetti, M.: Weaving the Web: The Original Design and Ultimate Destiny of the World Wide Web by Its Inventor. DIANE Publishing Company (2001)
15. Gómez-Pérez, A., Fernández-Lopez, M., Corcho, O.: Ontological Engineering: With Examples from the Areas of Knowledge Management, Ecommerce and the Semantic Web. Springer, London (2004)
16. Bedini, I., Nguyen, B.: Automatic ontology generation: state of the art. PRiSM Laboratory, Technical report, University of Versailles (2007)
17. Asim, M.N., Wasim, M., Khan, M.U.G., Mahmood, W., Abbasi, H.M.: A survey of ontology learning techniques and applications. Database **2018**(1), bay101 (2018)
18. Fernández-López, M., Gómez-Pérez, A., Juristo, N.: METHONTOLOGY: from ontological art towards ontological engineering. In: Proceedings of the Ontological Engineering AAAI-97 Spring Symposium Series, Stanford University, EEUU (1997)
19. Uschold, M., Gruninger, M.: Ontologies: principles, methods and applications. Knowl. Eng. Rev. **11**(2), 93–136 (1996)
20. Corcho, O., Fernández-López, M., Gómez-Pérez, A.: Methodologies, tools and languages for building ontologies. Where is their meeting point? Data Knowl. Eng. **46**(1), 41–64 (2003)
21. Antoniou, G., van Harmelen, F.: Web ontology language: OWL. In: Staab, S., Studer, R. (eds.) Handbook on Ontologies, pp. 67–92. Springer, Heidelberg (2004)
22. Hermanson, D.R., Hermanson, H.M.: The internal control paradox: what every manager should know. Rev. Bus. **16**(2), 4 (1994)

Incremental Hotel Recommendation with Inter-guest Trust and Similarity Post-filtering

Fátima Leal[1(✉)], Benedita Malheiro[3,4], and Juan Carlos Burguillo[2]

[1] Intelligent & Digital Systems, R&Di, Instituto de Soldadura e Qualidade,
Oeiras, Portugal
fdleal@isq.pt
[2] EET/UVigo – School of Telecommunication Engineering, University of Vigo,
Vigo, Spain
J.C.Burguillo@uvigo.es
[3] ISEP/IPP – School of Engineering, Polytechnic Institute of Porto, Porto, Portugal
mbm@isep.ipp.pt
[4] INESC TEC, Porto, Portugal

Abstract. Crowdsourcing has become an essential source of information for tourists and tourism industry. Every day, large volumes of data are exchanged among stakeholders in the form of searches, posts, shares, reviews or ratings. Specifically, this paper explores inter-guest trust and similarity post-filtering, using crowdsourced ratings collected from the Expedia and TripAdvisor platforms, to improve hotel recommendations generated by incremental collaborative filtering. First, the profiles of hotels and guests are created using multi-criteria ratings and inter-guest trust and similarity. Next, incremental model-based collaborative filtering is adopted to predict unknown hotel ratings based on the multi-criteria ratings and, finally, post-recommendation filtering sorts the generated predictions based on the inter-guest trust and similarity. The proposed method was tested both off-line (post-processing) and on-line (real time processing) for performance comparison. The results highlight: (*i*) the increase of the quality of recommendations with the inter-guest trust and similarity; and (*ii*) the decrease of the predictive errors with the on-line incremental collaborative filtering. Thus, this work contributes with a novel method, integrating incremental collaborative filtering and inter-guest trust and similarity post-filtering, for on-line hotel recommendation based on multi-criteria crowdsourced rating streams.

1 Introduction

Information and Communication Technologies (ICT) has changed the travel industry. Specifically, mobile devices enable users to generate large volumes of data known as crowdsourced information. Crowdsourcing is, according to Egger *et al.* [3], an outsourcing process supported by ICT and performed voluntarily by a large number of participants. Increasingly, tourists rely on crowdsourced

© Springer Nature Switzerland AG 2019
Á. Rocha et al. (Eds.): WorldCIST'19 2019, AISC 930, pp. 262–272, 2019.
https://doi.org/10.1007/978-3-030-16181-1_25

information, *e.g.*, ratings, reviews or posts of tourists, for decision making. This crowdsourced information classifies prior tourist experiences regarding tourism resources and is highly influential since it conditions the behaviour of future tourist planning and decision making [1]. However, since the crowdsourced information is shared voluntarily, it raises reliability questions. In this context, the incorporation of trust models on systems supported by crowdsourcing helps to avoid false information. Trust models quantify the trustworthiness between a pair of users, *i.e.*, trust is a one-to-one relationship.

This work addresses the problem of improving on-line hotel recommendations using a post-filtering based on the inter-guest trust and similarity. Our proposed on-line crowdsourced data processing method adopts incremental updating and relies on: (*i*) multi-criteria rating profiling based on the Personalised Weighted Rating Average (PWRA) proposed by Leal et al. [11]; (*ii*) model-based collaborative filtering via Alternating Least Squares with Weighted-λ-Regularization (ALS-WR) matrix factorisation as proposed by Leal et al. [10]; and (*iii*) *a posteriori* recommendation filtering based on inter-guest trust and similarity modelling. Finally, we compare the results of the proposed on-line and the corresponding off-line processing.

The processing of crowdsourced data streams raises data reliability questions and requires constant updating. Consequently, in on-line collaborative recommendation systems supported by crowdsourcing, it is vital to update the guest-hotel and inter-guest models incrementally. We contribute with a post-recommendation inter-guest trust and similarity filter, and with an on-line processing engine to update incrementally both models as soon as a new event occurs. Our results highlight that on-line incremental updating decreases the prediction errors, while trust and similarity post-filtering increases the quality of recommendations.

This paper is organised as follows. The Sect. 2 reviews the related work on rating-based hotel recommendation engines. The Proposed Method Section describes the algorithms used. The Experiments and Results Section includes the experiences performed and the results obtained. Finally, the Conclusions Section discusses the outcomes of this work.

2   Related Work

The crowdsourced feedback, which is volunteered by costumers typically in the form of ratings or reviews, is used by potential costumers to choose new goods or services and by businesses to suggest relevant products. However, the crowdsourced information raises trustworthiness questions. In this scenario, trust and reputation models provide solutions to minimise the effect of false information. According to Josang et al. [8], trust is based on direct experiences between stakeholders while reputation is based on third party experiences, *e.g.*, the crowd. Both are used to represent the reliability of publishers (tourists and businesses). In this scenario, trust and reputation models are particularly valuable in the case of profiles which use social information improving the quality of data.

This work applies data mining techniques to tourism crowdsourced data streams to recommend hotels using a model-based approach together with a crowd trust modelling. Specifically, we adopt the ALS-WR matrix factorisation algorithm for rating prediction. Additionally, we applied a post-recommendation filter based on crowd trust in order to improve recommendation. Therefore, this related work reviews, essentially, hotel recommendation approaches supported by crowdsourced ratings.

In the tourism domain, the crowdsourced information is growing dramatically. Particularly in hotel industry, the tourists use ratings to express their opinion regarding the different aspects of the service. The hotel recommendation has been strongly explored in terms of ratings: (*i*) Song *et al.* [13] report a rating-based hotel recommendation system which determines the similarity between the user preferences and the hotel cluster average ratings using the Euclidean Distance (ED). (*ii*) Jannach *et al.* [7] combine user and item models based on Support Vector Regression (SVR), using a weighted approach, to provide recommendations. The evaluation was performed with a data set provided by HRS.com; (*iii*) Nilashi *et al.* [12] employ Principal Component Analysis (PCA) for the selection the most representative rating (dimensionality reduction) and Expectation Maximisation (EM) and Adaptive Neuro-Fuzzy Inference System (ANFIS) as prediction techniques for hotel recommendation. The approach was tested with TripAdvisor data; and (*iv*) Farokhi *et al.* [4] explore data clustering (Fuzzy *c*-means and *k*-means) to find the nearest neighbours and, finally, predicted the unknown hotel ratings using the Pearson Correlation coefficient. The evaluation was performed with TripAdvisor data.

Although trust plays an important role in initiating interactions and building higher-quality relationships between the users [9, 18], these research works do not explore crowd trust modelling as a means to obtain more accurate tourism recommendations. The proposed crowd trust model is distributed and decentralised since trustworthiness is a one-to-one relationship established between trustor (the active guest) and trustee (another guest). Therefore, as far as we know, the processing of crowdsourced information and modelling of crowd trust for profiling and recommendation is a novel approach for hotel tourism applications whereby the related-work is sparse. Table 1 depicts a comparison of the surveyed hotel recommendation engines. We can verify that the trust modelling in hotel rating predictions has yet to be explored. While earlier work was focussed on ratings processing, here we analyse crowdsourced data in order to forecast trends and patterns in the tourism domain. Additionally, we use crowdsourced tourist hotel ratings to make personalised hotel recommendations by predicting unknown hotel ratings both on and off-line, and sorting them according to the inter-guest trustworthiness. Finally, we tested our proposed method with two different crowdsourced data sets: (*i*) TripAdvisor (TA) and; (*ii*) Expedia (E).

3   Proposed Method

Tourism crowdsourcing platforms are widely used for travel planning. They gather and share not only hotel-related data (*e.g.*, description, price, location, official star rating, amenities, *etc.*), but also tourist-related data (*e.g.*, multicriteria ratings or textual reviews). The proposed method addresses the problem

Table 1. Comparison of hotel recommendation engines

Approach	Evaluation	Prediction	Crowd trust
Song *et al.* [13]	–	SVR	
Jannach *et al.* [7]	HRS	–	–
Nilashi *et al.* [12]	TA	ANFIS	
Farokhi *et al.* [4]	TA	k-means	
Our proposal	**TA & E**	**ALS–WR**	✓

how to personalise hotel recommendations when processing on-line crowdsourced multi-criteria ratings. The hotel recommendation engine is composed of: (i) a multi-criteria profiling method proposed by Leal *et al.* [10]; (ii) a model-based collaborative filtering algorithm (ALS-WR); and (iii) a post-recommendation filter using inter-guest trust modelling and the Collinearity and Proximity Similarity (CPS) presented by Veloso *et al.* [16]. This recommendation engine was assessed using both on and off-line evaluation protocols.

Multi-criteria Profiling uses the multi-criteria personalised weighted rating average (PWRA) proposed by Leal *et al.* [10] to combine the ratings given by the guests to the different accommodation dimensions, *i.e.*, cleanliness, service, staff, *etc.*

Recommendation incorporates a Model-based Collaborative Filtering module which implements the ALS-WR matrix factorisation algorithm. This algorithm represents guests and hotels as vectors of latent factors. The rating matrix (R_{g*h}) holds for all guests and hotels the corresponding guest (g) hotel (h) PWRA rating.

For recommendation purposes, the algorithm factorises the matrix R_{g*h} into two latent matrices: (i) the user-factor matrix P; and (ii) the item-factor matrix Q. Each row of P or Q represents the relation between the corresponding latent factor with the guest g or hotel h, respectively, and the λ regularises the learned factors [5].

Finally, R is approximated by the product of P and Q. In each ALS–WR iteration, P and Q are sequentially fixed to solve the optimisation problem. Once the latent vectors converge, the algorithm terminates. The selected regularisation weight λ, dimensionality of latent feature space (k) and number of iterations (n) minimise the prediction errors. In off-line, the final matrix holds all rating predictions used for recommendation, whereas, in on-line, the model is incrementally updated whenever a user rating event occurs. In the latter case, the recommendation engine generates a list of hotel predictions for the active guest.

Post-filtering relies on the Trust & Similarity methods to sort the generated predictions according to the inter-guest trustworthiness and similarity. The proposed crowd trust model is distributed and decentralised since trustworthiness is a one to one relationship established between the trustor (active guest a) and the trustee (guest g). The trustworthiness quantifies the closeness between two guests, taking into account the set of relevant co-rated hotels. Specifically, a relevant co-rated hotel is a hotel which was identically rated by both guests, *i.e.*, $|r_{a,h} - r_{g,h}| \leqslant v_t$, where $r_{a,h}$ is the rating given by the active guest a to hotel h, $r_{g,h}$ the rating given by guest g to hotel h and v_t a pre-defined threshold value. This threshold value, which was experimentally determined by selecting the one leading to smallest prediction errors, was set to 10 % of the rating scale. Therefore, we establish the trust $T_{a,g}$ that the active guest a has on guest g using Eq. 1 where $n_{a,g}$ is the number of relevant co-rated hotels between guests a and g and N_a is the total number of hotels rated by the active guest a.

$$T_{a,g} = \frac{n_{a,g}}{N_a} \tag{1}$$

This inter-guest trust model is incrementally updated and stored in the so-called trust matrix, whenever a new event occurs. From the active guest a perspective, a trustworthy guest g has a $T_{a,g} \geqslant 40\%$. This threshold value, which was obtained experimentally, leads to the smallest prediction errors.

The CPS determines the similarity between the profiles of the active guest a and guest g, $\hat{P}_a$ and $\hat{P}_g$, by combining the Cosine Similarity (CS) and Chebyshev Distance Dissimilarity (CDD). While the CS establishes the collinearity, the CDD determines the most disparate rating between both profiles. Therefore, this metric provides a better perception concerning the real similarity between two guests. The combining parameter β was applied according to Veloso *et al.* [16] approach, *i.e.*, $\beta = 0.5$.

The post-recommendation filter sorts the active guest hotel rating predictions by descending rating value $\hat{r}_{a,h}$, trustworthiness value $T_{a,h}$ or trustworthiness & similarity value $S_{a,h}$. Equation 2 displays the trustworthiness $T_{a,h}$ of hotel h rating prediction according to the active guest a, which corresponds to the average trustworthiness the active guest gives to the relevant guests who rated hotel h. Equation 3 shows the similarity $S_{a,h}$ between hotel h and the active guest a, which corresponds to the average similarity between the active guest and the relevant guests who rated hotel h. Specifically, $T_{a,g}$ corresponds to the trustworthiness between the active guest a and the relevant guest g, $CPS(\hat{P}_a, \hat{P}_g)$ to the profile similarity between the active guest a and the relevant guest g, $\hat{P}_a$ to the profile of the active guest a, $\hat{P}_g$ to the profile of the relevant guest g, $G_{a,h}$ is the number of relevant guests (from the perspective of active guest a) who rated hotel h and G_h is the total number of guests who rated hotel h.

$$T_{a,h} = \frac{1}{G_{a,h}} \sum_{g=1}^{G_h} \begin{cases} T_{a,g} : T_{a,g} \geqslant 40\% \\ 0 : otherwise \end{cases} \tag{2}$$

$$S_{a,h} = \frac{1}{G_{a,h}} \sum_{g=1}^{G_h} \begin{cases} CPS(\hat{P}_a, \hat{P}_g) : T_{a,g} \geqslant 40\% \\ 0 : otherwise \end{cases} \tag{3}$$

3.1   Evaluation

The evaluation of the proposed method involves predictive accuracy and classification metrics. We calculate the predictive accuracy between the predicted rating and the real guest rating using the Root Mean Square Error (RMSE). In terms of classification metrics, we apply the Recall [6] and the Target Recall (TRecall) [15]. The Recall determines the number of relevant items selected from the total number of relevant items available. The TRecall evaluates recommendation in terms of its closeness to the target rating, *i.e.*, the real rating given by the guest. Specifically, we consider a subset of N predictions. In the case of Recall@N, it counts a hit if the real rating belongs to the list of the top N predictions, whereas, in the case of TRecall@N, counts a hit if the predicted rating lays within the $\pm \frac{N}{2}$ predictions centred on the real rating. These metrics were determined with two different evaluation protocols:

Off-line Protocol – In this case, the full data set is known in advance. The first 80 % of the data set (train partition) is used to build the model and the remaining 20 % (test partition) is for evaluation. This protocol uses the test partition to determine the RMSE and the Recall@10 and TRecall@10.

On-line Protocol – In this case, the model and the evaluation metrics are incrementally updated whenever a new event occurs. The first 20 % of the data was used to create the initial model and the remaining 80 % as stream data. This protocol uses the stream data to determine incrementally the RMSE, as proposed by Takács et al. [14], and the Recall@10 and TRecall@10, as suggested by Cremonesi et al. [2].

4   Experiments and Results

The experiments were conducted on an Openstack cloud instance with 16 GiB RAM, 8 CPU and 160 GiB of hard-disk space. We applied both evaluation protocols to the data sets after ordering the data temporally.

4.1   Data Sets

The selected crowdsourced hotel data sets present the following features:

HotelExpedia was proposed by Leal et al. [10]. It contains 6030 hotels, 3098 reviewers, including anonymous reviewers, and 381 941 reviews from 10 different locations. Since these experiments were performed exclusively with data from the 1089 identified reviewers, *i.e.*, we discarded the anonymous users and their inputs. Each user classified at least 20 hotels and each hotel contains at least 10 reviews. Specifically, we use the user and hotel identification and, as multi-criteria ratings, the *overall*, *cleanliness*, *hotel condition*, *service* and *room comfort*.

TripAdvisor was collected by Wang et al. [17]. It includes 9114 hotels, 7452 users and 127 517 hotel reviews. Each user classified at least 20 hotels and each hotel contains at least 10 reviews. Specifically, we use the user and hotel identification and, as multi-criteria ratings, the *overall, value, rooms, location, cleanliness, service,* and *sleep quality*.

4.2   Recommendation Results

The on and off-line prediction of the unknown hotel ratings is performed by ALS-WR matrix factorisation, a model-based collaborative recommendation filter. The predicted ratings are sorted using the Default (PWRA), Trust and Trust & CPS post-filtering. The threshold value v_t for the inter-guest trust modelling was set to 10 % of the rating scale.

Off-line – The post-processing results are depicted in Table 2. They show the positive impact of the inter-guest trust and similarity post-filtering in the recommendation. In the case of HotelExpedia, the Recall@10 and TRecall@10 improve 81 % and 24 % with Trust post-filtering and 238 % and 77 % with Trust & CPS post-filtering. In the TripAdvisor case, the Trust post-filtering improves the Recall@10 and TRecall@10 by 26 % and 20 %, while the Trust & CPS post-filtering increases Recall@10 and TRecall@10 by 103 % and 53 %, respectively. Since RMSE is determined prior post-filtering, it remains unchanged. The HotelExpedia predictions have an average RMSE of 0.178 and 0.189 with TripAdvisor.

On-line – The incremental updating results are depicted in Table 3. The effectiveness of trust and similarity post-recommendation filtering remains. The HotelExpedia data set has a Recall@10 and TRecall@10 improvement of 23 % and 3 % with the Trust post-filtering and 83 % and 36 % with the Trust & CPS post-filtering. In the TripAdvisor case, the Recall@10 and TRecall@10 increase 25 % and 6 % with Trust post-filtering and 53 % and 45 % with the Trust & CPS post-filtering. The RMSE remains 0.161 with HotelExpedia and 0.170 with TripAdvisor.

Table 2. Results of the evaluation metrics – off-line

Data set	Post-filter	RMSE	Recall@10	TRecall@10
Hotel Expedia	Default: $\hat{r}_{a,h}$	0.178	0.227	0.240
	Trust: $T_{a,h}$	0.178	0.411	0.298
	Trust & CPS: $S_{a,h}$	0.178	0.768	0.425
Trip Advisor	Default: $\hat{r}_{a,h}$	0.189	0.395	0.427
	Trust: $T_{a,h}$	0.189	0.497	0.514
	Trust & CPS: $S_{a,h}$	0.189	0.802	0.654

Table 3. Results of the evaluation metrics – on-line

Data set	Post-filter	RMSE	Recall@10	TRecall@10
Hotel Expedia	Default: $\hat{r}_{a,h}$	0.161	0.282	0.289
	Trust: $T_{a,h}$	0.161	0.348	0.299
	Trust & CPS: $S_{a,h}$	0.161	0.514	0.394
Trip Advisor	Default: $\hat{r}_{a,h}$	0.170	0.399	0.412
	Trust: $T_{a,h}$	0.170	0.497	0.438
	Trust & CPS: $S_{a,h}$	0.170	0.612	0.597

4.3   Discussion

We performed a wide set of experiments involving: (i) Trust and Trust & CPS post-recommendation filtering; (ii) off-line and on-line processing; and (iii) two crowdsourced hotel data sets.

Consistently, the best recommendation performance occurred with Trust & CPS post-filtering and the best predictive accuracy with on-line processing. These results corroborate our claim that data stream mining and profile-based post-recommendation filters contribute to the improvement of personalised hotel recommendations.

Concerning on and off-line processing, on-line presents better predictive accuracy, while off-line shows improved recommendation results. Specifically, the on-line mode displays -10% in RMSE with both data sets, -33% and -7% in Recall@10 and TRecall@10 with HotelExpedia and -24% and -9% in Recall@10 and TRecall@10 with TripAdvisor. The lower recommendation results of the on-line mode can be explained by the fact that the on-line experimental protocol is considerably more demanding – using 20% of the events to build the initial model and the remaining 80% to update and evaluate the model incrementally – than the off-line experimental protocol – using 80% of the events to create an static model and the remaining 20% for model evaluation.

Table 4. Coefficient of variation of the recall-based metrics – on-line.

Data set	Post-filter	Recall@10	TRecall@10
Hotel Expedia	Default: $\hat{r}_{a,h}$	0.009	0.010
	Trust: $T_{a,h}$	0.016	0.002
	Trust & CPS: $S_{a,h}$	0.008	0.012
Trip Advisor	Default: $\hat{r}_{a,h}$	0.017	0.024
	Trust: $T_{a,h}$	0.021	0.037
	Trust & CPS: $S_{a,h}$	0.035	0.031

Finally, all experiments were repeated ten times to determine the Coefficient of Variation (CV) of the different metrics. The RMSE remained invariant and the recall-based metrics displayed low variation. Table 4 displays the CV of the recall-based metrics regarding the on-line experiments.

5   Conclusions

The emergence of tourism crowdsourcing platforms, *e.g.*, Expedia, TripAdvisor, *etc.*, allows tourists to continuously produce and share large volumes of feedback data regarding tourism resources. This crowdsourced information, which classifies prior tourist experiences, influences the behaviour of present and future tourists. However, those tourists are unable to process such large volumes of crowdsourced data or identify trusted raters, much less in real time. To address this problem, this paper presents an on-line hotel recommendation engine, integrating inter-guest trust and similarity post-recommendation filtering, and compares it against the corresponding off-line counterpart.

The proposed method uses ALS-WR collaborative filtering to predict the unknown ratings and post-recommendation filtering to sort the predictions by rating and inter-guest trust and similarity. In terms of performance, it was assessed both off-line (post-processing) and on-line (real time processing). Our results show that incremental updating improves the predictive accuracy, while inter-guest trust and similarity post-filtering increases considerably the quality of recommendations.

Thus, we contribute with a novel method, integrating incremental collaborative filtering and inter-guest trust and similarity post-filtering, for on-line hotel recommendation based on multi-criteria crowdsourced rating streams. As future work, we intend to explore textual reviews and multi-criteria ratings, as well as modulating incoming ratings with the rater reputation.

Acknowledgements. This paper is based upon work from COST Action IC1406 High-Performance Modelling and Simulation for Big Data Applications (cHiPSet), supported by COST (European Cooperation in Science and Technology). This work was partially supported by the European Regional Development Fund (ERDF) through (*i*) the Operational Programme for Competitiveness and Internationalisation - COMPETE Programme - within project «FCOMP-01-0202-FEDER-023151» and project «POCI-01-0145-FEDER-006961», and by National Funds through Fundação para a Ciência e a Tecnologia (FCT) - Portuguese Foundation for Science and Technology - as part of project UID/EEA/50014/2013; and (*ii*) the Galician Regional Government under agreement for funding the Atlantic Research Center for Information and Communication Technologies (atlanTTic).

References

1. Chen, Y.F., Law, R.: A review of research on electronic word-of-mouth in hospitality and tourism management. Int. J. Hosp. Tour. Adm. **17**(4), 347–372 (2016)
2. Cremonesi, P., Koren, Y., Turrin, R.: Performance of recommender algorithms on top-n recommendation tasks. In: Proceedings of the fourth ACM Conference on Recommender Systems, pp. 39–46. ACM (2010)
3. Egger, R., Gula, I., Walcher, D.: Open Tourism: Open Innovation, Crowdsourcing and Co-Creation Challenging the Tourism Industry. Springer, Heidelberg (2016)
4. Farokhi, N., Vahid, M., Nilashi, M., Ibrahim, O.: A multi-criteria recommender system for tourism using fuzzy approach. J. Soft Comput. Decis. Support Syst. **3**(4), 19–29 (2016)
5. Friedman, A., Berkovsky, S., Kaafar, M.A.: A differential privacy framework for matrix factorization recommender systems. User Model. User-Adapt. Interact. **26**(5), 425–458 (2016)
6. Herlocker, J.L., Konstan, J.A., Terveen, L.G., Riedl, J.T.: Evaluating collaborative filtering recommender systems. ACM Trans. Inf. Syst. (TOIS) **22**(1), 5–53 (2004)
7. Jannach, D., Gedikli, F., Karakaya, Z., Juwig, O.: Recommending hotels based on multi-dimensional customer ratings. In Fuchs, M., Ricci, F., Cantoni, L., (eds.) Information and Communication Technologies in Tourism 2012: Proceedings of the International Conference in Helsingborg, Sweden, 25–27 January 2012. Springer, Vienna, pp. 320–331 (2012)
8. Jøsang, A., Ismail, R., Boyd, C.: A survey of trust and reputation systems for online service provision. Decis. Support Syst. **43**(2), 618–644 (2007)
9. Korovaiko, N., Thomo, A.: Trust prediction from user-item ratings. Soc. Netw. Anal. Min. **3**(3), 749–759 (2013)
10. Leal, F., Malheiro, B., Burguillo, J.C.: Prediction and analysis of hotel ratings from crowd-sourced data. In: World Conference on Information Systems and Technologies. Springer, pp. 493–502 (2017)
11. Leal, F., González-Vélez, H., Malheiro, B., Burguillo, J.C.: Profiling and rating prediction from multi-criteria crowd-sourced hotel rating. In: Proceedings of the 31th European Conference on Modelling and Simulation, ECMS 2017, pp. 576–582 (2017)
12. Nilashi, M., bin Ibrahim, O., Ithnin, N., Sarmin, N.H.: A multi-criteria collaborative filtering recommender system for the tourism domain using expectation maximization (EM) and PCAANFIS. Electron. Commer. Res. Appl. **14**(6), 542–562 (2015)
13. Song, W.W., Lin, C., Avdic, A., Forsman, A., Åkerblom, L.: Collaborative filtering with data classification: a combined approach to hotel recommendation systems. In: 25th International Conference on Information Systems Development (ISD 2016), Katowice, Poland, 24–26 August 2016 (2016)
14. Takács, G., Pilászy, I., Németh, B., Tikk, D.: Scalable collaborative filtering approaches for large recommender systems. J. Mach. Learn. Res. **10**, 623–656 (2009)
15. Veloso, B., Malheiro, B., Burguillo, J.C., Foss, J.: Personalised fading for stream data. In: SAC 2017: Symposium on Applied Computing Proceedings, 32nd ACM Symposium on Applied Computing (SAC 2017), Data Streams Track, pp. 1–3. ACM, New York (2017)

16. Veloso, B., Malheiro, B., Burguillo, J.C.: A multi-agent brokerage platform for media content recommendation. Int. J. Appl. Math. Comput. Sci. **25**(3), 513–527 (2015)
17. Wang, H., Lu, Y., Zhai, C.: Latent aspect rating analysis on review text data: a rating regression approach. In: Proceedings of the 16th ACM SIGKDD International Conference on Knowledge Discovery and Data Mining, KDD 2010, pp. 783–792. ACM, New York (2010)
18. Xiu, D., Liu, Z.: A formal definition for trust in distributed systems. In: ISC, pp. 482–489. Springer (2005)

Traffic Flow Forecasting on Data-Scarce Environments Using ARIMA and LSTM Networks

Bruno Fernandes[1(✉)] , Fábio Silva[1,2] , Hector Alaiz-Moretón[3] ,
Paulo Novais[1] , Cesar Analide[1] , and José Neves[1]

[1] Department of Informatics, ALGORITMI Centre,
University of Minho, Braga, Portugal
bruno.fmf.8@gmail.com,
{fabiosilva,pjon,analide,jneves}@di.uminho.pt
[2] CIICESI, ESTG, Polytechnic Institute of Porto, Felgueiras, Portugal
[3] Department of Electrical and Systems Engineering, Universidad de León,
Escuela de Ingenierías, León, Spain
hector.moreton@unileon.es

Abstract. Traffic flow forecasting has been in the mind of researchers for the last decades, remaining a challenge mainly due to its stochastic nonlinear nature. In fact, producing accurate traffic flow predictions would be extremely useful not only for drivers but also for those more vulnerable in the road, such as pedestrians or cyclists. With a citizen-first approach in mind, forecasting models can be used to help advise citizens based on the perception of outdoor risks, dangerous behaviors and time delays, among others. Hence, this work develops and evaluates the accuracy of different ARIMA and LSTM based-models for traffic flow forecasting on data-scarce and non-data-scarce environments. The obtained results show the great potential of LSTM networks while, in contrast, expose the poor performance of ARIMA models on large datasets. Nonetheless, both were able to identify trends and the cyclic nature of traffic.

Keywords: Traffic flow forecasting · Data-scarce environments ·
Long Short-Term Memory · AutoRegressive Integrated Moving Average ·
Road safety

1 Introduction

In our days, road safety has become a major concern of our society. This fact is easily explained with the substantial number of deaths happening on the road every day. Even though many stakeholders are focusing on vehicles and roads, one should not dismiss those more vulnerable at the road, known as Vulnerable Road Users (VRUs). Such road users are typically defined as pedestrians or cyclists, with their vulnerability arising from the lack of external protection, their age or disabilities, among others. Indeed, an approach that aims to increase VRUs' safety focuses on modelling traffic prediction. Works that already produce such outcomes are typically focused on vehicles and drivers. However, such information could be extremely useful for VRUs.

© Springer Nature Switzerland AG 2019
Á. Rocha et al. (Eds.): WorldCIST'19 2019, AISC 930, pp. 273–282, 2019.
https://doi.org/10.1007/978-3-030-16181-1_26

A simple example is the one of a cyclist who pretends to know which hour will have less traffic so that he could go cycling. As in all problems that involve forecast, one is required to have a set of data to work with. However, there may be situations where such data is less substantial. That should not be a motive to dismiss the problem.

Within the Machine Learning field, many different models and algorithms can be used to predict future points. In particular, for time series analysis there are two that stand out: AutoRegressive Integrated Moving Average (ARIMA) and Long Short-Term Memory (LSTM). In this work we develop models based on the referred algorithms, and evaluate and validate its accuracy and performance for traffic prediction in data-scarce and non-data-scarce environments. Such information is then shared with VRUs who can opt to avoid certain roads or certain hours to travel or do sports. Hence, with respect to this work the following research questions were elicited:

- RQ1. Which algorithm, ARIMA or LSTM, has better accuracy on data-scarce environments and which has better accuracy on non-data-scarce ones?
- RQ2. Do these algorithms behave better (in terms of accuracy and performance) on non-data-scarce environments when compared to data-scarce ones?

The remaining of this paper is structured as follows, viz. The next section reviews previous studies on traffic forecasting. The third section introduces the materials and methods, focusing on describing the dataset and data preparation, the assessed metrics and the developed ARIMA and LSTM models. The fourth section focuses on the performed experiments while section five gathers, analyses and discusses the obtained results. Finally, conclusions are gathered and future work is outlined.

2   Literature Review

This work focuses on the problem of traffic modelling on data-scarce and non-data-scarce environments using two distinct models. On the one hand we have ARIMA, developed over the Auto Regressive Moving Average (ARMA) model [1]. ARIMA has been implemented on many domains such as temperature and pollution prediction [2], and short-term traffic flow prediction [3], among others. On the other hand, LSTM consists of a special case of Recurrent Neural Networks (RNN), with recent years showing a significant increase in the use of LSTM in domains such as speech recognition [4], traffic flow prediction [5] and many others.

Researchers have been devoting many efforts to the use of ARIMA, LSTM or other models or algorithms for traffic flow prediction. However, for our knowledge, there is yet to be studied the accuracy and error rates on data-scarce environments against non-data-scarce ones. Nonetheless, an interesting work that studies the use of these models for traffic flow prediction is the one performed by Fu et al. [5], where these authors used LSTM to predict short-term traffic flow, showing that LSTM had a slightly better performance than ARIMA. In fact, short-term traffic forecast focuses in the near future, ranging from some minutes to some dozens of minutes. In terms of data, the authors used PeMS dataset, which has over 15 000 sensors deployed state-wide in California. Another study focused on short-term traffic flow is the one performed by Zhao et al. [6] where these authors propose a model applied to a dataset with 25.11 million records.

To evaluate the performance of the model they used Mean absolute error (MAE), mean square error (MSE) and mean relative error (MRE) as criteria. For their model and datasets, LSTM proved to behave better than ARIMA, specially for long forecast windows. On the other hand, Ma et al. [7] proposed a LSTM model to capture non-linear traffic dynamic in an effective manner. Their dataset consisted of 42387 records. The developed models were used to predict speed in the next 2 min based on speed and volume in the previous period on the same day. Interestingly, LSTM outperformed both RNN and Support Vector Machine models by, at least, 28%. Moreover, considering a multivariate model, LSTM also showed to be superior. The multivariate model also showed to be better than the univariate one. Another approach to the problem being addressed by this work involves the use of ensemble models based on ARIMA for traffic estimation on urban areas [8].

3   Methods and Materials

The next lines describe the materials and methods used in this work with the objective of creating time series forecasting models for traffic in smart cities, enabling the deployment of geographical models to advise VRUs in aspects related to traffic flow.

3.1   Dataset

The dataset was created from scratch and contains real world data. A software was developed to collect data from a set of public APIs. In particular, TOMTOM Traffic Flow API was the one used to create the traffic dataset, which has, as features, the city name; the functional road class describing the road type; the current average speed at the selected point; the free flow speed expected under ideal conditions; current travel time, in seconds; the travel time in seconds which would be expected under ideal free flow conditions; the confidence on the values; the ratio between live and the historical data; the coordinates; and a timestamp. The software went live on 24 July 2018 and has been collecting data uninterruptedly. It works by making an API call every twenty minutes using an HTTP Get request. It parses the received JSON object and saves the records on the database. The software was made so that any other road of any city or country can be easily added to the fetch list. For this work, the dataset included data until 11 October 2018, which consists of 80 days of data. The software is modular, configurable and is able to build more datasets specifically focused on pollution data, the weather and crowd sensing data. An initial approach was to consider a multi-variate model comprising both traffic and the weather. However, since during the 80 days that comprise the traffic dataset the weather was constant in terms of temperature, humidity and precipitation, the final decision was to develop a univariate model.

3.2   Data Preparation and Preprocessing

To answer our research questions, two distinct datasets were built. The first one, Dataset A, was used to evaluate the models on data-scarce environments and consisted of a total of 80 records without missing values. Each record corresponds to an entire day (grouped by day) and has, as main features the city name; the timestamp; the confidence average; the speed difference average, for each day, between the current speed at the selected points and the expected free flow speed; the expected free flow speed; the travel time difference average, for each day, between the current travel time in seconds and the expected travel time under free flow conditions; and the expected free flow travel time. On the other hand, the second dataset, Dataset B, was used to evaluate the models on non-data-scarce environments. The features of this dataset are the same as the ones describe in the previous lines. The difference was on the number of records. While the first dataset contained a record for each day, this second dataset consisted of a total of 1894 records. The same 80 days were used in both datasets, but while the first is grouped by day, the latter is grouped by hour. No missing values were present, but some hours are missing due to situations where the API limits were reached or the API was unavailable.

The criterion for dataset normalization was based on the following equation:

$$\frac{X_i - \min(X)}{\max(X) - \min(X)} \tag{1}$$

The MaxMinScaler data input pre-processing has been implemented with *sklearn. preprocessing.MinMaxScaler* [9]. In the case of ARIMA, data was transformed into a set of differences to make data stationary as requested by the model.

3.3 Assessed Metrics

Aiming to get the best possible combination of parameters for the developed models, we opted to optimize the MSE. The goal was to find the best combination of the parameter grid which would minimize MSE and give the best model. Root Mean Squared Error (RMSE) and MAE were also used.

3.4 Arima

ARIMA is a forecasting algorithm originally developed by Box and Jenkins [10]. It belongs to a class of univariate autoregressive algorithms used in forecasting applications based on timeseries data. The Arima models are generally defined by three parameters (q, d, p), where q is the order of the autoregressive components; d is the number of differencing operators; and p is the highest order of the moving average term. The parameters control the complexity of the model and, consequently, the Auto Regression, Integration and Moving Average components of the algorithm [11], i.e.:

$$\check{y}_t = \Phi_1 \Upsilon_{t-1} + \Phi_2 \Upsilon_{t-2} \ldots + \Phi_p \Upsilon_{t-p} + a_t - \theta_1 a_{t-1} \ldots - \theta_p a_{t-q}. \tag{2}$$

Where:

- y denotes a general time series;
- $\check{y}_t$ is the forecast of the time series y for time t;

- $\Upsilon_{t-1} \ldots \Upsilon_{t-p}$ are the previous p values of the time series y (and form the auto-regression terms);
- $\Phi_1 \ldots \Phi_p$ are coefficients to be determined by fitting the model;
- $a_t \ldots a_{t-q}$ is a zero mean white noise process (and forms the moving average terms);
- $\theta_t \ldots \theta_{t-q}$ are coefficients to be determined by fitting the model.

ARIMA works best when data has stable past correlations with few outliers, where its performance is usually better. On the other hand, the search for optimal parameters is not straightforward and requires analysis and experimentation. Nevertheless, ARIMA models should be kept simple and have no more than the value 3 for Autoregression or Moving Average terms.

3.5   LSTM

LSTM networks are a specific kind of convolutional networks. They are based on the idea that they can connect previous information to the present task, enabling memory capacities. Therefore, LSTM has been designed for long-term dependency problems as they are able to remember information for periods of time. LSTM core is based on contextual state cells that work as long-term or short-term memory cells. Thus, the output of a LSTM is handled by the state of these cells. This feature becomes interesting when it is necessary to develop a forecast model that is time dependent. An LSTM layer consists of a set of recurrently connected blocks, known as memory blocks, memory cells or just cells that are the computational units of the LSTM network [12]. These cells are composed by weights and gates. Each memory block contains one or more recurrently connected memory cells and three multiplicative units: input, output and forget gates. The gates allow the net to interact with the cells. The forget gate and the input gate update the internal state, while the output gate is the final limiter of the cells' output. These gates and the consistent dataflow called CEC, or Constant Error Carrousel, keep each cell stable. For the implementation of this network, we used the Keras framework for python with Tensor flow backend, using GPU processing and an improved performance version that uses CuDNN (CudnnLSTM).

4   Experiments

This section describes the set of experiments that have been implemented in this work, as well as the grid-search associated with each model. In fact, each model and their respective parameter grid were applied to the datasets explained before, i.e.:

- Dataset A: 80 records, grouped by day, describing 80 days;
- Dataset B: 1894 records, grouped by hour, describing the same 80 days:
  - Sub-dataset B0: all the 1894 records;
  - Sub-dataset B1: 1694 records as result of 1894-200;
  - Sub-dataset B2: 1294 records as result of 1894-600.

To obtain the best possible combination of parameters for ARIMA and LSTM, a nested cross-validation approach was implemented. This technique is based in the creation of several subsets. In this case, three was the number of created subsets for Dataset B: the first uses all data available (sub-dataset B0), the second had 200 records removed (sub-dataset B1) and the last had 600 records removed (sub-dataset B2). This strategy was applied to verify the model's forecast, to test how the models handle different amounts of data and if the produced errors are stable. Due to the small size of Dataset A, this sub-division was not implemented there. Dataset A and all sub-datasets B were then split into two slices. For Dataset A, 71 cases were used for training while 8 were used for calculating error metrics. These were the possible numbers due to the few records available in Dataset A. On the other hand, for sub-datasets B, the fist split retains, for training purposes, 75% of data. The second split, which contains the remaining data, is used to test the model and calculate the final error values.

4.1   Arima Experiments

A pre-condition of the ARIMA model is that data must be stationary. So, in order to successfully run the models, the dataset required preparation and its mean and variance needed to be stationary. That was accomplished by turning the original dataset into a dataset of differences values where an input is the difference between two sequential inputs. After successfully preparing the data, the problem is then proposed to different ARIMA models based on different (p, d, q) parameters in order to find the best performing models. This required repeated experiments with the sample dataset in the same conditions to directly compare algorithms errors under different conditions. The experiments were carried out for values in the range [0–10] for all parameters for the autoregressive component and in the range [0–2] for the remaining parameters. The decision to increase the order of the Auto Regression component was made because it yields better results each time, however the time required to run the model was noticeably larger. Safeguards were also implemented since some configurations of the ARIMA model do not converge, resulting in algorithm exceptions.

4.2   LSTM Experiments

Due to the random initialization of LSTM weights, five repetitions of each combination of parameters were performed, taking the mean of RMSE to verify the model's quality. A method to define a seed for eliminating the random feature of LSTM was not chosen mainly due to the small size of the datasets, specifically, Dataset A. The grid of parameters chosen for implementing LSTM in Dataset A is as follows:

- Batch size: for Dataset A = [2, 4, 10, 20]; for sub-datasets B = [12, 24, 48];
- Window size: for Dataset A = [3, 7, 10]; for sub-datasets B = [8, 24, 48].

The batch size parameter defines the number of cases that were included in each epoch. The windows size defines the number of past cases used for learning, considering that a big window may produce worst results before the presence of cyclic behaviours in times series. The remaining parameters are described in the following code:

```
model = Sequential()
model.add(CuDNNLSTM(128, inputshape=(window_size,
n_features=1]),return_sequences=True))
model.add(Dropout(0.2))
model.add(CuDNNLSTM (64, return_sequences=False))
model.add(Dropout(0.2))
model.add(Dense(32,activation='relu',  kernel_initializer
='uniform'))
model.add(Dropout(0.2))
model.add(Dense(1))
model.compile(loss='mse', optimizer='RMSprop')
```

CuDNNLSTM refers to LSTM layers oriented for execution in a GPU-CUNDD environment, while the 'RMSprop' parameter defines the optimizer. This restricts the oscillations in the vertical direction, thus improving the learning rate.

5 Results and Discussion

To validate the results, a set of 4 independent experiments were performed. The experiments made forecasts based on the average speed noticed on roads of a city. For Dataset A, the data-scarce one, the goal was to predict the last 7 days without prior knowledge. Results show that both models performed similarly, with the best ARIMA model being slightly worse than LTSM. However, training time for an ARIMA of order 10 for Auto Regression is more time intensive for each forecast. Table 1 shows the obtained results while Fig. 1 gives a graphic view of predictions.

Table 1. MAE, RMSE and MSE of ARIMA and LSTM models as well as the used parameters.

Dataset	Model	MAE	RMSE	MSE	Parameters
Dataset A	ARIMA	0.66160	0.91824	0.84316	(10, 1, 1)
Dataset A	LSTM	0.70257	0.90454	0.82301	Window size = 7 Batch size = 20
Sub-dataset B0	ARIMA	1.84567	2.79692	7.82278	(3, 1, 1)
Sub-dataset B0	LSTM	1.40967	2.16929	4.70613	Window size = 24 Batch size = 50
Sub-dataset B1	ARIMA	1.85845	2.65893	7.06989	(7, 0, 1)
Sub-dataset B1	LSTM	1.47858	2.30358	5.30649	Window size = 24 Batch size = 50
Sub-dataset B2	ARIMA	1.85311	2.68335	7.20038	(3, 0, 1)
Sub-dataset B2	LSTM	1.40726	1.91171	3.65474	Window size = 48 Batch size = 30

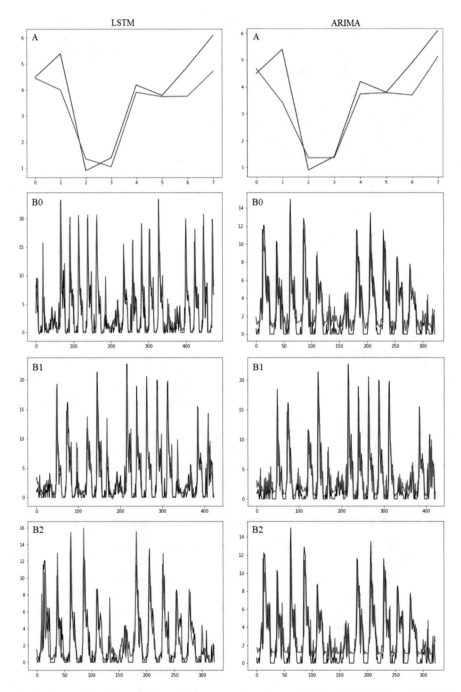

Fig. 1. Comparison of real values (black) vs predicted ones (red) for LSTM (left) and ARIMA (right), and for dataset A and sub-datasets B0, B1 and B2. X axis for time and Y axis for the speed difference average values.

Over the next set of experiments, the dataset was changed to an hourly dataset (Dataset B). Predictions are made using real data and produce one prevision at a time. The real past value is added to the dataset and compared to predicted values to assess MAE, RMSE and MSE. Regarding the obtained results for sub-dataset B0, LTSM showed to have better performance. The ARIMA model suffered from memory overflows, weight calculations and lack of convergence errors, and, so, the best computable model has a lower order of magnitude for the Auto Regressive parameter. Regarding sub-dataset B1, it has fewer instances but LTSM continues to show superiority over ARIMA, which still struggles with memory intensive operations. In the last sub-dataset, B2, even fewer cases were used. ARIMA was still not able to outperform LTSM. This provides evidence that ARIMA models may be better suited when data is scarce as in the case with dataset A. In addition, it becomes evident that the performance of LTSM is more constant over time, with consistently better results. In fact, the larger the dataset the worse ARIMA performs. In both Table 1 and Fig. 1, it is evident that LTSM yields better results than ARIMA, which presents similar performance on sub-datasets B0, B1 and B2. In fact, both Arima and LTSM have a constant error rate throughout the experiments, independently of the size of sub-dataset B.

In terms of computational time, ARIMA models are more time consuming when the (p, d, q) order increases to the point where it is unfeasible to train new models. LTSM optimized networks take significantly less time to train and once trained can yield constant predictions based on the data provided in contrast to ARIMA models, which need to be retrained.

6 Conclusions and Future Work

The use of forecasting techniques applied to specific roads and intersections has interesting applications in the context of smart cites, apart from the usual use case of traffic management. The approach presented in this paper compares two different algorithms used for mean velocity forecasting in real world roads. The results provide evidence that both algorithms can produce meaningful outputs and correct forecasting's using real world data both in data-scarce and non-data-scarce environments. This translates to accurately describing the velocity at which vehicles usually make their movement. To a smart city citizen such information is helpful to choose roads to perform outdoor activities or choose pedestrian routes. It was found that LTSM has better performance than ARIMA, however both identified trends and the cyclic nature of traffic. In the case of ARIMA, it is evident that the time needed to run the model is significantly lower in data-scarce environments when compared to non-data-scarce ones. LSTM has the ability to handle large sets of data and after the training period it runs very fast. Hence, as direct answers to the elicited research questions, we can say that LSTM has better accuracy on data-scarce and non-data-scarce environments. In terms of computational performance, LSTM proved to be better in both datasets, with ARIMA showing a poor performance on non-data-scarce environments.

In the context of traffic forecasting on smart cities, more focus will be given to the use of LTSM models or ensemble algorithms based on LTSM due to the continuous nature of data. Other ensemble methods using multi-variate estimations are also being considered. We will also focus on creating large geographical maps supported by LTSM models to predict traffic flow and velocities. The obtained results are promising and, so, next iterations will include an extension to massive online forecasting systems able to handle the complexity and volume of smart cities data streams.

Acknowledgments. This work has been supported by COMPETE: POCI-01-0145-FEDER-007043 and FCT – *Fundação para a Ciência e Tecnologia* within the Project Scope: UID/CEC/00319/2013, being partially supported by a Portuguese doctoral grant, SFRH/BD/130125/2017, issued by FCT in Portugal.

References

1. Cortez, P., Rocha, M., Neves, J.: Evolving time series forecasting ARMA models. J. Heuristics **10**(4), 415–429 (2004)
2. Babu, C., Reddy, B.: Predictive data mining on Average Global Temperature using variants of ARIMA models. In: IEEE International Conference on Advances in Engineering, Science And Management (ICAESM 2012), pp. 256–260 (2012)
3. Li, K., Zhai, C., Xu, J.: Short-term traffic flow prediction using a methodology based on ARIMA and RBF-ANN. In: 2017 Chinese Automation Congress (CAC), pp. 2804–2807 (2017)
4. Graves, A., Mohamed, A., Hinton, G.: Speech recognition with deep recurrent neural networks. In: IEEE International Conference on Acoustics, Speech and Signal Processing, pp. 6645–6649 (2013)
5. Fu, R., Zhang, Z., Li., L.: Using LSTM and GRU neural network methods for traffic flow prediction. In: 31st Youth Academic Annual Conference of Chinese Association of Automation (YAC), pp. 324–328 (2016)
6. Zhao, Z., Chen, W., Wu, X., Chen, P., Liu, J.: LSTM network: a deep learning approach for short-term traffic forecast. IET Intel. Transp. Syst. **11**(2), 68–75 (2017)
7. Ma, X., Tao, Z., Wang, Y., Yu, H., Wang, Y.: Long short-term memory neural network for traffic speed prediction using remote microwave sensor data. Transp. Res. Part C: Emerg. Technol. **54**, 187–197 (2015)
8. Guan, W., Hua, X.: A combination forecasting model of urban ring road traffic flow. In: Proceedings of the IEEE Intelligent Transportation Systems Conference, pp. 671–676 (2006)
9. Scikit-learn: sklearn.preprocessing.MinMaxScaler (2018). http://scikit-learn.org/stable/modules/generated/sklearn.preprocessing.MinMaxScaler.html. Accessed 10 Oct 2018
10. Box, G., Jenkins, G.: Time Series Analysis: Forecasting and Control (1976)
11. Van Der Voort, M., Dougherty, M., Watson, S.: Combining Kohonen maps with arima time series models to forecast traffic flow. Transp. Res. Part C: Emerg. Technol. **4**(5), 307–318 (1996)
12. Graves, A., Schmidhuber, J.: Framewise phoneme classification with bidirectional LSTM and other neural network architectures. Neural Netw. Off. J. Int. Neural Netw. Soc. **18**(5–6), 602–610 (2005)

Mapping Clinical Practice Guidelines to SWRL Rules

Samia Sbissi[1], Mariem Mahfoudh[2,3](✉), and Said Gattoufi[1]

[1] SMART Laboratory, Tunis University, Tunis, Tunisia
samia.sbissi@gmail.com, algattoufi@yahoo.com
[2] MIRACL Laboratory, University of Sfax, Sfax, Tunisia
mariem.mahfoudh@gmail.com
[3] ISIGK, University of Kairouan, Kairouan, Tunisia

Abstract. Clinical practice guideline is an evolving reference document that contains recommendations and knowledge which aims to assist professionals to master a medical domain. In our work, we are interested in using these documents in order to assist doctors to make decisions about appropriate health care for patients who are at risk of cardiovascular disease. More precisely, our paper proposes an automatic approach that parses and transforms text (clinical practice guideline) into OWL DL (Ontology Language Web Description Logic) axioms and SWRL (Semantic Web Rule Language) rules. To parse the text, we have used an existing ontology of cardiovascular domain and natural language processing tools (NLP). Our work is original in that studies the mapping between text and SWRL rules while the related works are focused only on mapping text on OWL lightweight axioms.

1 Introduction

Given a sentence in natural language, semantic parsing aims to produce a formal representation of its meaning [9]. These meaning representations may be used for example in reasoning and inference or also to determine whether two texts are in an entailment relation [6]. In our work which is in collaboration with the hospital of the "Rabta" (Tunis), we use semantic parsing to parse medical documents. We aim to make an assistance system that helps doctors to keep up with the evolution of diseases, the new medical treatments and to make decisions about patients who are at risk of cardiovascular disease, especially aortic dissection. The aortic dissection is a partial disruption of the wall of the aorta that may at any time evolve towards a complete rupture, with consequent death [1]. It is, therefore, an absolute emergency in its diagnosis and management. It is important to note that in the absence of medical management, the mortality is 95% (2% per hour). The documents that we parse called clinical practice guidelines [3]. It is an evolving reference document that contains a set of recommendations which aim to assist professionals to master a medical domain. Recommendation example: In patients with an abdominal aortic diameter of 25–29 mm, new ultrasound imaging should be considered 4 years later).

© Springer Nature Switzerland AG 2019
Á. Rocha et al. (Eds.): WorldCIST'19 2019, AISC 930, pp. 283–292, 2019.
https://doi.org/10.1007/978-3-030-16181-1_27

Our goal is to transform these recommendations into logical forms, more precisely into Semantic Web Rule Language (SWRL) rules. SWRL [7] is a semantic web language which is integrated directly within OWL (Ontology web language) ontologies. It allows defining rules in the form of logical implications between conditions and conclusions. We think that the use of the semantic web technicals and the ontology inference mechanism could be a good help to well elaborate our medical assistant. Therefore, to parse the text we have used an existing ontology of the cardiovascular domain and natural language processing tools (NLP). The first step focuses on the semantic annotation and consists on annotate the text (the set of the recommendations) with the domain ontology while identifying the similarities between the text and the concepts of the ontology. The second step consists to parse the text not by a simple transformation into logical description but rather by the construction of the SWRL rules. The semantic analysis is led by the semantic annotation and NLP mechanism. We think that our work is original in that studies the mapping between text and SWRL rules while the related works are focused only on mapping text to OWL lightweight axioms.

This paper is structured as follows: Sect. 2 provides an overview on the semantic parsing highlighting some related works. In Sect. 3, we describe our approach for annotation and parsing semantic. Section 4 presents the results of the implementation. Finally, in Sect. 5, we conclude and give some future work.

2   Related Work

The automatic acquisition and extraction knowledge from texts has a long history and a variety of approaches and methods [17]. The reader can find in [4] a comparison of fourteen tools for extracting knowledge on unstructured corpora. However, the extraction association rules field is more recent and complicated due to the complexity of modeling relationships for obtaining rules. Instead of detecting only relationships between pairs of concepts, these approaches must also automatically link a series of these relationships together to build compound requirements and encoded them as rules. Several approaches are focusing then on the extraction of semantic knowledge from more or less structured texts. We present here works closely to the ontology domain as we are interested in the annotation and semantic parsing.

Gangemi et al. [5] proposed a system called FRED that extracts n-ary relations based on discourse representation structures and mapped them to RDF (Resource Description Framework) representation. Park and Lee [14] proposed a semi-automatically method to extract rules from web documents. Their method requires, as inputs, an existing domain ontology and manual selection of relevant web pages. The approach uses only simple NLP techniques which limits it to handle complex texts. Many of the text mining algorithms extensively make use of NLP techniques, such as part of speech tagging (POG), syntactic parsing and other types of linguistic analysis [8]. The recent approaches are trained using data-to-text corpora and evaluated extrinsically, based on how well they help in realizing the target task (e.g., does the meaning representation produced succeed in retrieving the correct answer from the database? Does it provide the correct instruction for the robot?). They are based on higher-order lambda calculus, it includes both symbolic [2] and [12] statistical approaches to determine the mapping between words and lambda terms. Gyawali et al. [6] describe a method that converts natural language definitions to DL axioms, then parse definitions and applying transformation rules for parsing trees. [15] proposed neural semantic analyses that derive DL formulate from natural language definitions.

In our work, we are interested not only in the transformation of text in description logic (OWL axioms) but also in the automatic generation of SWRL rules. We think that parsing a text which is composed by a set of recommendations into a set of SWRL rules is more helpful for elaborate an assistance system.

3 Approach Overview and Cases Study

Our approach is essentially composed of the following steps:

- text preprocessing step: aims to prepare the text to the semantic annotation task. It is based on different technical as tokenization, stop words, tagging, etc.
- semantic annotation step: annotates the text of clinical practice guideline throws a medical domain ontology. We use similarity measures to detect the links between text terms and ontology concepts.
- semantic parsing step is guided by the semantic annotation in order to transform the text into OWL axioms and SWRL rules.

3.1 Text Preprocessing

Our text corpus is composed by a set of medical recommendations defined by the European Society of Cardiology ESC. Some rules of the corpus are presented below (Table 1).

The preprocessing of our text is based on the following methods:

- **Tokenization**: is the task of breaking a character sequence up into pieces (words/phrases) called tokens, and perhaps at the same time throw away certain characters such as punctuation marks [18].

Table 1. Some rules extract from clinical practice guideline.

In complicated Type B AD, TEVAR is recommended
In complicated Type B AD, surgery may be considered
In patients with acute contained rupture of TAA, urgent repair is recommended
If the anatomy is favourable and the expertise available, endovascular repair (TEVAR) should be preferred over open surgery

- **Filtering**: aims to remove some stop words from the text, words which have no significant relevance and can be removed from the documents [16].
- **Lemmatization**: considers the morphological analysis of the words, i.e. grouping together the various inflected forms of a word so they can be analysed as a single item.
- **Stemming**: aims at obtaining stem (root) of derived words. Stemming algorithms are indeed language dependent [11].
- **Part of Speech Tagging**: tags for each word (whether the word is a noun, verb, adjective, etc.), then finds the most likely parse tree for a piece of text. Tagged corpora use many different conventions for tagging words. In Table 2 some ones are represented and used in the following example.

Table 2. Some tagging words

ADJ 1	Adjective
ADV	Adverb
DT	Determiner
NN	Noun

In all patients with AD, medical therapy including pain relief and blood
pressure control is recommended.

⇓

[('In', 'IN'), ('all', 'DT'), ('patients', 'NNS'), ('with', 'IN'), ('AD', 'NNP'), (',',
','), ('medical', 'JJ'), ('therapy', 'NN'), ('including', 'VBG'), ('pain', 'NN'),
('relief', 'NN'), ('and', 'CC'), ('blood', 'NN'), ('pressure', 'NN'), ('control',
'NN'), ('is', 'VBZ'), ('recommended', 'VBN')]

3.2   Semantic annotation

To annotate the text, we propose to use a domain ontology. One of the ontologies we found close to our domain is the CVDO ontology[1]. It is an owl ontology, designed to describe the entities related to cardiovascular diseases. The semantic annotation process is based on the result of the preprocessing task. It takes the pre-processed text of clinical practice guideline and the CVDO ontology and then tries to find the similarities (the links) between them. Recall that an OWL ontology is a set of the concepts (classes, object properties data properties, etc.).

---

[1] http://purl.bioontology.org/ontology/CVDO.

We use the concepts names to produce an expanded list of equivalent or related terms. Each term of the input text may be associated with one or more entities from the ontology. To find the similarities, we have used:

1. exact matching: identifies the identical entities (String) in the text and in the domain ontology;
2. morphological matching: identifies the entities with a morphological correspondence;
3. syntactical similarities: using Levenshtein measure [10];
4. semantic matching: identifies the synonyms relations with WordNet ontology.

3.3 Semantic Parsing

In order to automatise the process of parsing text into SWRL rules, we have defined a set of patterns. These patterns reflect the set of rules of our corpus and give a general classification of them. Thus, a candidate rule will be compared to those patterns in order to determinate which pattern corresponds to the candidate rule.

Construct Pattern. We have identified several patterns for our recommendations rules, we presents here two types of example.

Pattern1: Rules with Class Expressions. SWRL rules with class expressions are used in many cases. For example:

if the patient has a disease related to heart disease.

$$\Downarrow$$

Pattern1: Patient(?p) $\wedge$ (hasRelatedDisease min 1 RelatedDiseases)(?p)
-> RelatedDiseaseHistory (?p, true)

Pattern2: Rules with Data Range Restriction Expressions. Datatypes, such as xsd:string, xsd:integer and literals such as "2":xsd:integer, can be used to express data ranges.

In patients with abdominal aortic diameter of 25 to 29 mm, new ultrasound imaging should be considered 4 years later.

$$\Downarrow$$

Pattern2: $Patient(?p) \wedge hasAbdominalDiametre(?p, ?d) \wedge swrlb :$
$greaterThan(?d, 25) \wedge swrlb : lessThan(?d, 29)->$
$recommendedDiagnosis(?p, \text{``}ultrasounImaging\text{''})$

The data range restriction is satisfied when abdominal diametre is between 25 and 29 variable.

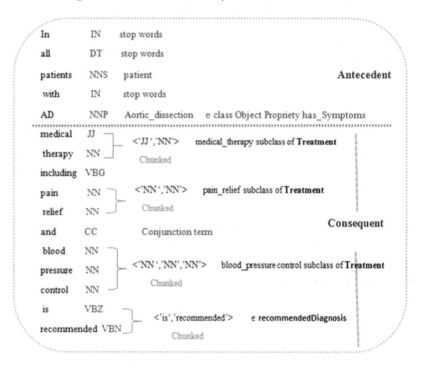

Fig. 1. Chunked tree for <NN/JJ.?>*<NN.?>*<VB>?

Fig. 2. Example 1: component correspondences.

Semantic Parsing. SWRL rules are composed of two parts: the antecedent and consequent. Each rule is an implication between the antecedent (body) and consequent (head). The decision that can be understood as if the body conditions are true then makes the conditions of the head also true.

To parse the recommendations set, we propose in the first step to chunk them (Figs. 1 and 2) then generate the chunked tree of each recommendation that will be converted automatically into an SWRL rule via their correspondence with SWRL patterns. Note that Chunking is the process by which groups various words together with their part of speech tag.

The following steps describe how to parse and chunk our text to produce SWRL rules:

– Divide the sentence (or text) into words (tokenization).
– Assign to each word its part of speech (POS; tagging).

- Eliminating stops words that are frequent and unnecessary.
- Identifying the antecedent and the consequent of the rules.
- Chunk the terms obtained. As it is illustrated in Fig. 1, blood pressure control can be detected as a chunk term.
- Finding for each text component its corresponding entity in the ontology.
- Define the correspondence and types of components in the ontology (class, instance, property, or literal).
- Search the correspondence SWRL pattern for each recommendation in which we are generating its chunked tree and their corresponding matching with the ontology.

After the extraction of the correspondence between chunked trees for each recommendation of medical corpus and ontology as it is illustrated in Fig. 3, we propose to transform each recommendation represented by its chunked tree to SWRL rules. A correspondence search will be conducted by the builder patterns, and the chunked tree (candidate rule). The separation of the condition and the conclusion part of the recommendation is determined in an iterative way by analysing all the recommendations. For example, we have a type of recommendation start with "IN" as it is illustrated in Fig. 4.

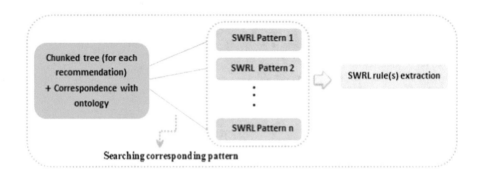

Fig. 3. SWRL rule extraction

We notice that so from in up to the comma we have the condition part. The other is the conclusion containing "is recommended". Once this recommendation is passed through the preprocessing part and matching with the ontology, we start by comparing the correspondence between and the existing SWRL pattern.

Fig. 4. Recommendation example.

In this part, it happens that we do not find an exact match given the difference in instances example or data property or even lack of concepts. This would allow us to enrich the ontology (by adding a new concepts instance...). The task of enrichment is based on the work treated by [13].

4   Implementaion and Result

4.1   Semantic Annotation

In order to establish the correspondence between the ontology and the text, we developed Java-based implementation illustrated in Fig. 5. Stanford CoreNLP is used for identifying morphology matching. NLTK library from Python is used for the text preprocessing. Note that, the CVDO ontology contains 514 concepts and the recommendations text contains 614 words. We have found 30% of similarities between them. It is important to note that CVDO is a cardiovascular domain ontology and contains most of the domain knowledge validated by our doctor. However, it is poor in meaning that it doesn't contain data properties and object properties. Thus the misses concepts will be added during the construction of the SWRL rules.

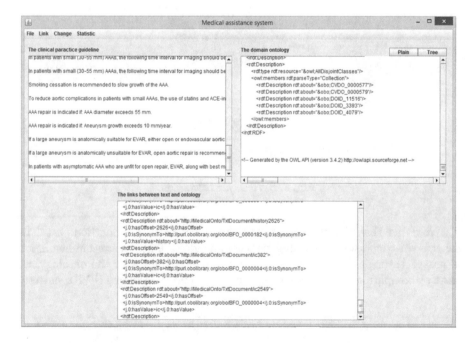

Fig. 5. Our tool for the semantic annotation step.

4.2 Semantic Parsing and Construction SWRL Rules

For the construction of the patterns and the transformation of the rules, we have used the Jena framework and SWRLAPI. 60% of SWRL rules are correctly extracted from the corpus of recommendations. This percentage is explained as, in some cases, one recommendation can correspond to one or two SWRL patterns at the same time which has reduced the accuracy rate. Also, we need to add others patterns to covers all the type of recommendations which still escape our tool.

5   Conclusion

The paper presented a method to annotate and parse medical recommendation texts with a domain ontology. It transforms these recommendations into logical forms, more precisely into Semantic Web Rule Language (SWRL) rules. Our approach addresses an important gap in the existing literature, which knowledges are only transformed in simple logical forms such as first-order logic or OWL lightweight ontologies. The semantic analysis, in this case, is led by a semantic annotation of the recommendation text by the concepts of a domain ontology using similarity measures for the matching. We combined the use of the OWL ontology and the techniques of NLP to capture knowledge and represent its in the form of rules in the SWRL language. Recall that our objective is to elaborate a medical assistance system that helps doctors to take decisions about aortic diseases, we are planning in a future work to incorporate the result of the semantic annotation and parsing (SWRL rules) in the domain ontology in order to enrich and exploit it, in inference tasks.

References

1. Criado, F.J.: Aortic dissection: a 250-year perspective. Tex. Hear. Inst. J. **38**(6), 694 (2011)
2. Curran, J.R., Clark, S., Bos, J.: Linguistically motivated large-scale NLP with C&C and boxer. In: Proceedings of the 45th Annual Meeting of the ACL on Interactive Poster and Demonstration Sessions, pp. 33–36. Association for Computational Linguistics (2007)
3. Erbel, R., Aboyans, V., Boileau, C., Bossone, E., Bartolomeo, R.D., Eggebrecht, H., Evangelista, A., Falk, V., Frank, H., et al.: 2014 ESC guidelines on the diagnosis and treatment of aortic diseases. Eur. Hear. J. **35**(41), 2873–2926 (2014)
4. Gangemi, A.: A comparison of knowledge extraction tools for the semantic web pp. 351–366 (2013)
5. Gangemi, A., Presutti, V., Reforgiato Recupero, D., Nuzzolese, A.G., Draicchio, F., Mongiovì, M.: Semantic web machine reading with FRED. Semant. Web **8**(6), 873–893 (2017)
6. Gyawali, B., Shimorina, A., Gardent, C., Cruz-Lara, S., Mahfoudh, M.: Mapping natural language to description logic. In: European Semantic Web Conference, pp. 273–288. Springer (2017)

7. Horrocks, I., Patel-Schneider, P.F., Boley, H., Tabet, S., Grosof, B., Dean, M., et al.: SWRL: a semantic web rule language combining OWL and RuleML. W3C Memb. Submiss. **21**, 79 (2004)
8. Kao, A., Poteet, S.R.: Natural Language Processing and Text Mining. Springer, Heidelberg (2007)
9. Kiryakov, A., Popov, B., Terziev, I., Manov, D., Ognyanoff, D.: Semantic annotation, indexing, and retrieval. Web Semant.: Sci. Serv. Agents World Wide Web **2**(1), 49–79 (2004)
10. Levenshtein, V.I.: Binary codes capable of correcting deletions, insertions and reversals. Sov. Phys. Dokl. **10**, 707–710 (1966)
11. Lovins, J.B.: Development of a stemming algorithm. Mech. Transl. Comp. Linguist. **11**(1–2), 22–31 (1968)
12. MacCartney, B., Manning, C.D.: Natural logic for textual inference. In: Proceedings of the ACL-PASCAL Workshop on Textual Entailment and Paraphrasing, pp. 193–200. Association for Computational Linguistics (2007)
13. Mahfoudh, M., Forestier, G., Thiry, L., Hassenforder, M.: Algebraic graph transformations for formalizing ontology changes and evolving ontologies. Knowl.-Based Syst. **73**, 212–226 (2015)
14. Park, S., Lee, J.K.: Rule identification using ontology while acquiring rules from web pages. Int. J. Hum.-Comput. Stud. **65**(7), 659–673 (2007)
15. Petrucci, G., Ghidini, C., Rospocher, M.: Ontology learning in the deep. In: European Knowledge Acquisition Workshop, pp. 480–495. Springer (2016)
16. Saif, H., Fernández, M., He, Y., Alani, H.: On stopwords, filtering and data sparsity for sentiment analysis of Twitter (2014)
17. Upadhyay, R., Fujii, A.: Semantic knowledge extraction from research documents. In: 2016 Federated Conference on Computer Science and Information Systems (FedCSIS), pp. 439–445. IEEE (2016)
18. Webster, J.J., Kit, C.: Tokenization as the initial phase in NLP. In: Proceedings of the 14th Conference on Computational Linguistics-Volume 4, pp. 1106–1110. Association for Computational Linguistics (1992)

Towards the Automatic Construction of an Intelligent Tutoring System: Domain Module

Alan Ramírez-Noriega[1(✉)], Yobani Martínez-Ramírez[1],
José Emilio Sánchez García[2], Erasmo Miranda Bojórquez[2],
J. Francisco Figueroa Pérez[1], José Mendivil-Torres[1], and Sergio Miranda[1]

[1] Facultad de Ingeniería Mochis, Universidad Autónoma de Sinaloa,
Los Mochis, Mexico
{alandramireznoriega,yobani,juanfco.figueroa,jose.mendivil,
smirandamondaca}@uas.edu.mx
[2] Universidad Autónoma Intercultural de Sinaloa, El Fuerte, Mexico
{esanchez,emiranda}@uais.edu.mx

Abstract. Intelligent Tutoring Systems (ITS) are software tools that mimic a teacher's teaching methods through artificial intelligence techniques. The generalized model of these systems is divided into four main modules: tutoring, student, domain, and interface. Although it has been shown that ITS is very useful in cases where a teacher cannot be present, the development of these systems is expensive and time-consuming, since it requires experts and available programmers. Therefore, this research proposes a framework to develop an authoring tool to build ITS automatically, with a focus on the domain module. We consider that the domain model represents the most important module of the ITS because it contains the knowledge that should be taught and evaluated. Based on this module, the rest of the modules will make decisions.

Keywords: Intelligent Tutoring System · Authoring tool ·
Domain model · Knowledge representation · Education

1 Introduction

An Intelligent Tutoring System (ITS) is a computer-based teaching-learning system that uses Artificial Intelligence techniques to interact with students to teach them [7]. Currently, ITSs provides effective tracking of the teaching process as it offers personalized tutoring. ITSs offer the following advantages [21]: (1) Constitutes a source of teaching materials. (2) Provides problems for the student to reach a certain level of knowledge. (3) Controls the level of difficulty of the problems so that the students face exercises appropriate to their needs. (4) Helps the student to acquire additional knowledge.

© Springer Nature Switzerland AG 2019
Á. Rocha et al. (Eds.): WorldCIST'19 2019, AISC 930, pp. 293–302, 2019.
https://doi.org/10.1007/978-3-030-16181-1_28

The use of ITS in virtual environments has been shown to improve the teaching-learning process [14]. However, ITS is still not common in classrooms due to time constraints [8], to the intense work [22], to the requirement of experts in both intelligent tutors and in the subject to be developed.

To avoid the above problems, ITS authoring tools have been developed. The aim of this tool is to simplify the development of ITS, to increase the number and diversity of available tutors. It also reduces the complexity of constructing them. Authoring tools also allow rapid ITS prototype design [4,11]. Accomplishing these objectives will assist ITS developers and users who do not have programming skills construct ITSs [11]. Various components of ITS require different techniques in assisting authoring [6]. As a result, the development of ITS authoring tools has progressed slowly [24].

Literature shows different uses for ITS authoring tools [4], however, little research has focused on the domain module [2]. This module is important because it establishes the bases of knowledge that will be taught and evaluated.

The objective of this research is to establish the framework and foundations in developing an authoring tool to build a domain module that serves as an information base for an ITS. The specific objectives of the research are: To design a generalized model to build the ITS domain module automatically, build an internal representation of the ITS knowledge domain, design an algorithm to determine student knowledge based on representation internal, and build a module of queries on the internal representation to answer and increase knowledge.

This paper is organized as follows: Sect. 2 describes related work. Section 3 shows the topic background. Section 4 displays the methodology. The Sect. 5 show conclusions and references.

2   Related Work

The study of the authoring tools for ITS has been investigated by various authors. Dermeval et al. [4] conducted a systematic literature review (from January 2009 to June 2016) finding 33 articles related to the authoring tools for ITS. Most of these articles focused on the pedagogical model and the domain model. Our research focuses on the domain model.

Dermeval et al. identified features offered by the authoring tools, dividing contributions by ITS module. They used the following features related to the domain module: Definition of problem solutions, authoring by demonstration, automatic domain model generation, definition of hints, reuse of learning content, and human computation.

Our research focuses on working the automatic domain model generation, for which the following investigations were detected:

– Brawner [2] developed Tools for Rapid Automated Development of Expert Models (TRADEM). TRADEM uses a domain model built as a summarization of provided content mixed into a set of topics, as a part of the Generalized Intelligent Framework for Tutoring (GIFT) Domain Module [20]. The purpose of the TRADEM project has been to rapidly and mostly-automatically create expert models and sequence domain material from initially provided texts. The information is given by the user so that TRADEM builds the knowledge domain. There are benefits in using the TRADEM tool, including aiding in front end analysis of content, automatically summarizing existing documents, providing the foundation of a course. His project is very good, and has the same basis as our proposal. It automatically builds the knowledge domain of the ITS. However, our research aims to build a representation of knowledge based on nodes of concepts with the aim of inferring and evaluating knowledge.

– Matsuda et al. [6] developed SimStudent. This is a machine-learning agent initially developed to help novice authors to create cognitive tutors without heavy programming. SimStudent helps authors to create an expert model for a cognitive tutor by tutoring SimStudent in solving problems. The expert model represents a how-type of knowledge about the domain. It assumes that the prospective users of the SimStudent authoring system are domain experts who, by definition, know how to solve problems and can identify errors. They consider the expert model as a module that presents problems and provides advice or support in solving them. Matsuda's work in this article, does not focus on Brawner's research and our proposal.

– Mitrovic et al. [8] developed ASPIRE-Author. ASPIRE-Author supports the authoring of the domain model, in which the author is required to provide a high-level description of the domain, as well as examples of problems and their solutions. ASPIRE-Author goal is to reduce the time and effort required for producing ITSs by building an authoring system that can generate the domain model with the assistance of a domain expert and produce a functional system. It describes the processes of the system in seven steps: 1. Specify the domain characteristics 2. Compose the domain ontology 3. Model the problem and solution structures 4. Design the student interface 5. Add problems and solutions 6. Generate syntax constraints 7. Generate semantic constraints 8. Deploy the tutoring system. Step two builds an ontology that represents a hierarchy of concepts, however, it was expected that in later steps this structure would be used to develop the content, this is not the case. In this article, they do not develop concepts of the ontology.

– Suraweera et al. [22] developed Constraint Authoring System (CAS). CAS was developed to generate the knowledge required for constraint-based tutoring systems, which reduced the effort and the amount of engineering knowledge and programming expertise required. ITS authoring using CAS is a semi-automated process requiring the assistance of a domain expert who initiates the process by modelling the domain as an ontology. The experts then defines the form that solutions will take for problems in this domain using concepts from the ontology. CAS's constraint-generation algorithms use both the ontology and the solution structure to produce constraints that verify the syntactic validity of solutions. The domain expert is also required to provide sample problems and their solutions, which are used by CAS generators to produce semantic constraints. These verify that a student's solution is semantically equivalent to a correct solution to the problem. Finally, the author has the opportunity to validate the generated constraints. This work seems to be a continuation of the project of [8], since it is a co-author of Saraweera's work and basically it is the same proposal.

In another paper outside the research of [4], like the research of Romero et al. [19]. They developed Hedea, a tool for the construction of ITS. This article presents preliminary results obtained by an authoring tool that allows a non-expert teacher in the area of intelligent tutors to develop an ITS based on the definition of a course. This research builds the knowledge representation structure that we want to build, however, Romero et al. builds the network based on the hierarchical breakdown of topics, sub-topics, and concepts. The development of the theory of topics is not addressed in the research.

The work of [2] is the closest to our proposal, since the research proposes to develop the domain model of an ITS, which displays the information or knowledge that a student must have. In addition, our proposal includes turning this knowledge into a network of concepts which allows evaluating the student's knowledge. This same structure should allow summarize information, and infer knowledge.

3 Background

3.1 Intelligent Tutoring Systems

ITS are computer-based learning systems, which were designed to impart knowledge guiding the student in the learning process through some form of intelligence. These systems exhibit a similar behavior to a human tutor and assists the student with cognitive help. The system adapts to the student's behavior, identifying the way to solve a problem to offer help when needed [21].

There is a generalized architecture for ITS that considers four basic modules [3]:

- The tutor module generates learning interactions based on the student's learning difficulties.
- The domain module defines the area of knowledge that the ITS teaches.
- The student module defines the student's knowledge in the work session.
- The interface module allows the student to interact with the ITS.

Resources needed to build an ITS come from multiple research fields, including artificial intelligence, the cognitive sciences, education, human-computer interaction, and software engineering. This multidisciplinary foundation makes the process of building an ITS a thoroughly challenging task, given that authors may have very different views of the targeted system [13].

3.2   Authoring Tools for ITS

For several decades developers and researchers have been investigating the possibilities for creating ITS authoring tools with the aim for: (1) reduce the effort and cost of building or customizing ITS, and (2) allow non-programmers, including teachers and domain experts, to participate fully or partly in building or customizing ITS [10].

Authoring tools streamline and accelerate the construction of ITS by providing a framework within which an author can design a learning system. Some authoring systems are general-purpose tools that provide an author with a great deal of leeway. Others embody a set of assumptions about what the authored product will look like and how it will behave. [1]

Murray [9] classifies ITS authoring tools into two main groups:

- Pedagogy-oriented systems focus on instructional planning and teaching strategies, assuming that the instructional content will be fairly simple. Such systems provide support for curriculum sequencing and planning, authoring tutoring strategies, composing multiple knowledge-types and authoring adaptive hypermedia.
- performance-oriented systems focus on providing rich learning environments where students learn by solving problems and receiving dynamic feedback. The systems in this category include authoring systems for domain expert systems, simulation-based learning and some special purpose authoring systems focusing on performance.

Woolf et al. [24] proposed a framework for organizing the necessary building blocks found in authoring systems for building ITS. The author identified four layers, each including specific classes of building blocks:

- Level 1: The knowledge representation level includes tools for easily representing knowledge, the user should adopt the right formalism and select the right language or tool to ease the representation process (Semantic net, constrains, productions rules, frames).
- Level 2: It is about the type of domain and student models (procedural skills, student effect, student misconceptions).
- Level 3: Contains tools for implementing teaching knowledge while (content planning, delivery planning, tutoring decision).
- Level 4: Comprises those for communication knowledge (interface design, pedagogical agent, natural language dialog).

4   Methodology

4.1   Overview of the Project

An ITS is considered a complex system to build, it is necessary to divide the problem for a simpler solution. Thus, the current article focuses on building the domain module. We consider that the domain model represents the most important module of the ITS because it contains the knowledge that should be taught and evaluated. Based on this module, the rest of the modules will take the decision to act as a teacher. Our framework considers the following phases to build a domain module:

- To design a generalized model to build the ITS domain module automatically: Initially, a general view of the architecture of the proposed system should be proposed, with the aim of defining the scope of the project and specifying the modules that will be developed.
- To Build an internal representation of the knowledge domain of the ITS: Each topic of the content must form an internal representation, which not only serves to represent the content but also serves to evaluate knowledge, consult and infer information.
- To design an algorithm that allows evaluating the student's knowledge based on internal representation: An ITS must be able to evaluate the student's knowledge, to propose support or reinforcement issues. There are some techniques in the literature that will serve as a base, however, the proposal must solve the deficiencies of the current techniques.
- To build a query module on internal representation to answer queries and infer knowledge: information on topics must be processed and converted to a structure that allows queries about the information. In addition, the structure must infer knowledge to adequately respond to queries.
- To develop an application for the knowledge domain of the ITS: Develop a web application that allows generating knowledge domains automatically, that is, to develop topics searching the content in information sources.
- To implement the internal representation, the evaluation, and query algorithms: Implement algorithms in the web content generation system to build the internal representation, incorporate the knowledge evaluation algorithm, and the query and inference algorithm.

Figure 1 represents the architecture of an ITS, which considers the 4 basic modules: tutoring, student, domain, and interface. However, the scope of this project focuses on the domain module, for this reason, this module is more developed than the rest of modules. Basically, the domain model will be built with Internet documents or e-books, from these, an internal structure that represents the contents will be created to make queries to the information and knowledge inference. It is contemplated that this representation serves to evaluate the student's knowledge.

4.2   Internal Representation of Knowledge

A computer-based education system teaches what is known as "knowledge domain". When represented in a way that a computer understands, it is called "Knowledge Representation" (KR). It is important to consider the KR processes and automates information management through computers. In addition, makes inferences that allow decision-making in a human manner in order to improve the tutoring task [16].

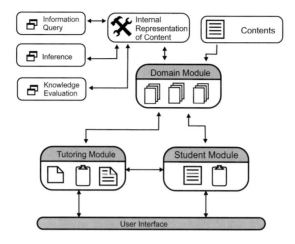

Fig. 1. Architecture of an ITS

Authors such as [17, 18] establish that in the human semantic memory exists a hierarchy of concepts with relations to organize this knowledge. Thus, arises the idea of representing knowledge by means of graphs. There are several techniques to represent this knowledge [16], however, our proposal establishes a combination of Bayesian networks and ontologies. The Bayesian network is a graphical probabilistic model which represents knowledge through nodes and relations [15]. On the other hand, ontology is a representative model of semantic knowledge [5].

We will consider elements which represent knowledge in an educational environment based on [16] as follows:

– Concepts: A concept is an elemental piece of knowledge. According to the domain expert, it cannot be divided into smaller parts. Therefore, a concept is considered the primary unit of knowledge.
– Skills: A skill is a cognitive process that interacts with one or more concepts, usually through an application. It has a particular purpose and produces an internal or external result.
– Relations: Relations are known as links. The relations goal is to know how the concepts are related. Relations can be of three types according to the link direction. These are Unidirectional, bidirectional, and loops.
– Inference: This is also called reasoning. The inference refers to obtaining deductions or conclusions based on knowledge already established. The main types of inference are abductive, deductive, analogy, and inductive.
– Type of Knowledge: There are two types of knowledge: (1) Procedural Knowledge focuses on tasks that must be performed to reach a particular objective or goal. (2) Declarative knowledge refers to the representation of objects and events, and about knowledge facts and its relations. It is the knowing whether "something is true or false". Declarative knowledge is applied in educational institutions. It is easy to represent and structure, so it is the kind of knowledge that is taught by computer-aided systems.

– Hierarchical Structure: Information for these kind of techniques is organized through a hierarchical structure which forms superclasses, subclasses and shared properties. Based on the hierarchical structure, some elements are considered by different authors as important because they establish a structure which maintains the contents. These elements are: Successors, predecessors, classes, inheritance, among others.

The proposed model considers constructing the nodes based on the textual information of the knowledge domain, forming a structure that can be exploited based on statistical and semantic techniques. Some features considered for the domain model are:

– The structure of the topics, can be given by an index in a hierarchical form, as an ontology, or a topic based on the breakdown of topics of encyclopedias as is the case of Wikipedia.
– Important concepts: These concepts are taken based on textual information. Various techniques have been used to consider this concept, such as links, the frequency of the concept, coincidence with title concepts, among others.
– A concept can have relations with other topics, thereby inheriting the properties of the concept with other concepts or related topics.
– Relations: Illustrates between concepts can be of two types: (1) Cause-effect: Determines a weight of relation between two concepts. It is useful in determining the knowledge of the student. (2) Semantics: Regarding the meaning relations that will have a label as is_part_of, is_a, kind_of, among others.

The domain model to be developed serves as the basis for the student and tutoring module with aspects such as:

– Learned topics: A topic learned by the use of examinations and other reinforcement activities.
– Inference of knowledge: The structure generated can make inference of knowledge to give explanations based on a feedback system.
– Probability propagation algorithms: The probability propagation algorithms will update the status of concepts, applying forward and backward propagation.

There are different investigations that have addressed this issue, such as [12, 23, 25, 26]. Therefore, the current research has a solid background, which will allow us to take up ideas from other projects.

5   Conclusions

The literature analysis allowed to determine that there is little research to build the domain module automatically, hence, the proposal in this article establishes a framework to build the domain module of an ITS based on text. The proposal considers forming a map of concepts that represents the knowledge that a student must have, using algorithms to evaluate their knowledge, and make the inference of the information to answer questions or feedback automatically.

The project, in general, considers developing a software tool which builds an ITS automatically, by considering the four basic modules: Student, tutoring, domain, and interface. For the scope of this article only the framework of the domain module is established, and the base literature that will be used.

References

1. Bell, B.: One-size-fits-some: its genres and what they (should) tell us about authoring tools. In: Sottilare, R.A., Graesser, A., Xiangen, H., Brawner, K. (eds.) Design Recommendations for Intelligent Tutoring Systems: Authoring Tools and Expert Modeling Techniques, chap. 3, pp. 31–45 (2015)
2. Brawner, K.: Rapid dialogue and branching tutors. In: CEUR Workshop Proceedings, vol. 1432, pp. 69–76 (2015)
3. Carbonell, J.R.: AI in CAI: an artificial intelligence approach to computer assisted instruction. IEEE Trans. Man Mach. Syst. **11**, 190–202 (1970)
4. Dermeval, D., Paiva, R., Ibert, B., Vassileva, J., Borges, D.: Authoring tools for designing intelligent tutoring systems : a systematic review of the literature. Int. J. Artif. Intell. Educ. **28**, 336–384 (2017)
5. Gruber, T.R.: Toward principles for the design of ontologies used for knowledge sharing. Int. J. Hum.-Comput. Stud. **43**(5–6), 907–928 (1995). citeulike-article-id:230211
6. Matsuda, N., Cohen, W.W., Koedinger, K.R.: Teaching the teacher: tutoring sim-student leads to more effective cognitive tutor authoring. Int. J. Artif. Intell. Educ. **25**(1), 1–34 (2015). https://doi.org/10.1007/s40593-014-0020-1
7. Millán, E., Pérez-De-La-Cruz, J.L.: A Bayesian diagnostic algorithm for student modeling and its evaluation. User Model. User-Adapt. Interact. **12**, 281–330 (2002). https://doi.org/10.1023/A:1015027822614
8. Mitrovic, A., Martin, B., Suraweera, P., Zakharov, K., Milik, N., Holland, J., McGuigan, N.: ASPIRE: an authoring system and deployment environment for constraint-based tutors. Int. J. Artif. Intell. Educ. **19**(2), 155–188 (2009)
9. Murray, T.: An overview of intelligent tutoring system authoring tools: updated analysis of the state of the art. In: Authoring Tools for Advanced Technology Learning Environments, pp. 491–544 (2003). https://doi.org/10.1007/978-94-017-0819-7-17
10. Murray, T.: Theory-based authoring tool design: considering the complexity of tasks and mental models. In: Sottilare, R.A., Graesser, A., Xiangen, H., Brawner, K. (eds.) Design Recommendations for Intelligent Tutoring Systems: Authoring Tools and Expert Modeling Techniques, chap. 2, pp. 9–30 (2015)
11. Naser, S.S.A.: ITSB : an intelligent tutoring system authoring tool. J. Sci. Eng. Res. **3**(5), 63–71 (2016)
12. Niño Zambrano, M., Jimena Pérez, D., Pezo, D., Cobos Lozada, C., Ramírez González, G.: Procedure for building semantic indexes based on domain-specific ontologies. Entramado **9**(1), 262–287 (2013)
13. Nkambou, R., Bourdeau, J., Psyché, V.: Building intelligent tutoring systems: an overview. In: Nkambou, R., Bourdeau, J., Mizoguchi, R. (eds.) Advances in Intelligent Tutoring Systems. Studies in Computational Intelligence, chap. 18, pp. 361–375. Springer, Heidelberg (2010). https://doi.org/10.1007/978-3-642-14363-218

14. Noguez, J., Sucar, L.E.: A probabilistic relational student model for virtual laboratories. In: Sixth Mexican International Conference on Computer Science (ENC 2005), pp. 2–9, September 2005. https://doi.org/10.1109/ENC.2005.7
15. Pearl, J.: Probabilistic Reasoning in Intelligent Systems: Networks of Plausible Inference. Morgan Kaufmann Publishers Inc., San Mateo (1988)
16. Ramírez-Noriega, A., Juárez-Ramírez, R., Jiménez, S., Martínez-Ramírez, Y.: Knowledge representation in intelligent tutoring system. In: Hassanien, A.E., Shaalan, K., Gaber, T., Azar, A.T., Tolba, M.F. (eds.) Proceedings of the International Conference on Advanced Intelligent Systems and Informatics 2016, pp. 12–21. Springer International Publishing, Egypt (2017). https://doi.org/10.1007/978-3-319-48308-5_2
17. Rivas Navarro, M.: Procesos cognitivos y aprendizaje significativo. BOCM, Madrid. Servicio de Documentación y Publicaciones (2008)
18. Rodríguez, R.J.: Herramientas informáticas para la representación del conocimiento software. Subj. Pocesos Cogn. **14**(1712), 217–232 (2010)
19. Romero Inzunza, M.A., Sucar Succar, E., Gómez-gil, P.: Diseño De Hedea: Una Herramienta Para La Construcción De Sistemas Tutores Inteligentes. Memorias del Décimo Encuentro de Investigación. Instituto Nacional de Astrofísica, Óptica y Electrónica, pp. 179–182, Puebla (2009)
20. Sottilare, R.A., Brawner, K.W., Goldberg, B.S., Holden, H.K.: The Generalized Intelligent Framework for Tutoring (GIFT). US Army Research Laboratory–Human Research & Engineering Directorate (ARL-HRED), Orlando, FL, 1–12 October 2012. https://doi.org/10.13140/2.1.1629.6003
21. Suárez Granados, J.J., Arencibia Rodríguez del Rey, Y., Pérez Fernández, A.C.: Methodology for developing an intelligent tutoring system based on the web, for engineering students. Universidad y Sociedad **8**(4), 108–115 (2016)
22. Suraweera, P., Mitrovic, A., Martin, B.: Widening the knowledge acquisition bottleneck for constraint-based tutors. Int. J. Artif. Intell. Educ. **20**(2), 137–173 (2010). https://doi.org/10.3233/JAI-2010-0005
23. Toledo-Alvarado, J.I., Guzmán-Arenas, A., Martínez-Luna, G.L.: Automatic building of an ontology from a corpus of text documents using data mining tools. J. Appl. Res. Technol. **10**(3), 398–404 (2012)
24. Woolf, B.P.: Building Intelligent Interactive Tutors: Student-centered Strategies for Revolutionizing e-Learning. Morgan Kaufmann Publishers Inc., San Francisco (2007)
25. Yarushkina, N., Filippov, A., Moshkin, V., Egorov, Y.: Building a domain ontology in the process of linguistic analysis of text resources. Preprints 2018, pp. 1–10 (2018). https://doi.org/10.20944/preprints201802.0001.v1
26. Zouaq, A., Nkambou, R.: Building domain ontologies from text for educational purposes. IEEE Trans. Learn. Technol. **1**(1), 49–62 (2008). https://doi.org/10.1109/TLT.2008.12

Deep Learning and Sub-Word-Unit Approach in Written Art Generation

Krzysztof Wołk[1(✉)], Emilia Zawadzka-Gosk[1],
and Wojciech Czarnowski[2]

[1] Polish-Japanese Academy of Information Technology, Warsaw, Poland
{kwolk, ezawadzka}@pja.edu.pl
[2] Jatar, Koszalin, Poland
wcz@jatar.com.pl

Abstract. Automatic poetry generation is novel and interesting application of natural language processing research. It became more popular during the last few years due to the rapid development of technology and neural computing power. This line of research can be applied to the study of linguistics and literature, for social science experiments, or simply for entertainment. The most effective known method of artificial poem generation uses recurrent neural networks (RNN). We also used RNNs to generate poems in the style of Adam Mickiewicz. Our network was trained on the 'Sir Thaddeus' poem. For data pre-processing, we used a specialized stemming tool, which is one of the major innovations and contributions of this work. Our experiment was conducted on the source text, divided into sub-word units (at a level of resolution close to syllables). This approach is novel and is not often employed in the published literature. The sub-words units seem to be a natural choice for analysis of the Polish language, as the language is morphologically rich due to cases, gender forms and a large vocabulary. Moreover, 'Sir Thaddeus' contains rhymes, so the analysis of syllables can be meaningful. We verified our model with different settings for the temperature parameter, which controls the randomness of the generated text. We also compared our results with similar models trained on the same text but divided into characters (which is the most common approach alongside the use of full word units). The differences were tremendous. Our solution generated much better poems that were able to follow the metre and vocabulary of the source data text.

Keywords: Deep learning · Machine learning · Style transfer ·
Poetry generation

1 Introduction

Creativity is a magnificent human ability. Domains such as art, music and literature are thought of as requiring great creativity. One of the most demanding is poetry, as it needs to satisfy at least two prerequisites: content and aesthetics. Creating high quality poetry is thus demanding even for humans, and, until recently seemed almost impossible for machines. Until the 90s computer generated poetry was mostly restricted to science fiction such as Stanislaw Lem's 'Fables for Robots'.

© Springer Nature Switzerland AG 2019
Á. Rocha et al. (Eds.): WorldCIST'19 2019, AISC 930, pp. 303–315, 2019.
https://doi.org/10.1007/978-3-030-16181-1_29

Initial attempts of automatic poem creation were made in late 90s, and systems such as Hisar Manurung's application generating natural language strings with a rhythmic pattern were introduced [6].

Automatic poetry generation was practiced as an art [5], hobby or entertainment, but also became a discipline of science, and an area of research in the natural language processing domain.

According to [7], approaches to poetry generation can be classified into four groups. Template based poetry generation, as in the ALAMO group, where the vocabulary of one poem and the structure from another can be used to create a new one. The Generate and Test approach, where WASP is an example. These types of systems produce random word sequences, which meet formal requirements, but are semantically challenged. There is also a group of applications with an evolutionary approach. One such example in this category is Levy's computer poet implemented as a neural network trained on data acquired from human testers. The fourth category includes systems that use a case-based reasoning approach. This strategy retrieves existing poems and adapts them to information provided by the user. One example in this category is the ASPID system.

Nowadays the most broadly used techniques for generating poems are neural-networks-based [3, 4]. Different proposals for the language unit are considered [8]. Words [2, 10] are the most common element of learning and creating process, but there are also many systems based on characters or even phonemes [11]. In our work we introduce a new approach, using sub-word units for our implementation.

2   Experimental Environment

We performed our experiment using the Google Collaboratory environment - a platform provided by Google for machine learning systems where Nvidia K80 graphic cards are available, and developed our models in Python with the PyTorch machine learning library.

3   Data Pre-processing

In the first step, we loaded and pre-processed the corpus data. Uppercase and capitalized words needed to be annotated (during text tokenization) with respectively _up_ and _cap_ tokens. After tokenization the entire training text was lowercased. Our specialized stemmer program was used to divide the corpus into sub-word units [14], the tool was implemented by us especially for such purposes (it supports only Polish language). The program is a proprietary solution performing multiple operations to segment Polish texts into suffix, prefix, core and syllables. Moreover, it annotates texts with grammatical groups and tags it with part of speech (POS) tags. We used the application to split text in Polish into sub-word units. The program was run with the '7683' option, a combination of stemming and division options. According to the application documentation:

7683 = 1+2 + 512 + 1024 + 2048 + 4096 (divide (more) words with dictionary, then algorithmically and unknown words), where: 1 - stem "nie" prefix and 2 - stem extra prefixes are stemming options, 512 - divide words, 1024 - divide with dictionary, 2048 - divide algorithmically, 4096 - divide unknown words that belong to division options category.

In the next step we tokenized our corpus and created a list of all tokens. Additionally, we changed end of line characters into _eol_ token. For the syllables that do not exist in a corpus the _unk_ token was created. For remaining syllables and tokens, we added '++', and '−' as connection symbols (e.g. in '−PO++', the '_' tokens were removed).

Two auxiliary functions were also created, one transforming text into the list of created tokens and the other translating the list of tokens into text.

To prepare input data for the neural network (NN), we divided our data string into chunks of 400 syllables. Each chunk was converted into a LongTensor, by assigning an index from the _all_tokens list to each syllable. This pre-processing allowed us to create a set of input and target tensors.

4   RNN

In our experiment we used a Recurrent Neural Network. The RNN network construction enables the model to use the state from the previous step as its input information for the next step. This characteristic enables the network to learn from its 'experience'. The data from the previous steps affects the next step's results. This attribute of RNNs facilitates the discovery of patterns from sequences provided. Our input data is a poem in Polish, so we can treat it as a string with repeatable patterns (Fig. 1).

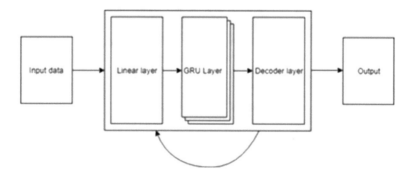

Fig. 1. Overview diagram of RNN used in experiment

Our RNN consists of three layers: a linear layer, one GRU layer and a final decoder layer. The linear layer codes the input signal. The GRU layer has multiple sub-layers, and operates on internal and hidden state events. The decoder layer returns a probability distribution. Our model received step t-1 token as its input with a target to output a token at step t.

5   Experiment

As the corpus for our experiment we used text of 'Pan Tadeusz' (Full title in Polish is: 'Pan Tadeusz, czyli ostatni zajazd na Litwie. Historia szlachecka z roku 1811 i 1812 we dwunastu księgach wierszem', in English: 'Sir Thaddeus, or the Last Lithuanian Foray: A Nobleman's Tale from the Years of 1811 and 1812 in Twelve Books of Verse'). It is an epic Polish poem written by Adam Mickiewicz [1]. The book was published in the 19th century, and consists of over 9000 verses and has Polish alexandrine metrical line, which means that every line has 13 syllables. The poem is regarded as a national epic of Poland and has been translated into many languages. Its first verses are as follows:

KSIĘGA PIÉRWSZA.
GOSPODARSTWO.
TREŚĆ.
Powrot panicza -- Spotkanie się piérwsze w pokoiku, drugie u
stołu -- Ważna Sędziego nauka o grzeczności -- Podkomorzego uwagi
polityczne nad modami -- Początek sporu o Kusego i Sokoła -- Żale
Wojskiego -- Ostatni Woźny Trybunału -- Rzut oka na ówczesny stan
polityczny Litwy i Europy.
Litwo! Ojczyzno moja! ty jesteś jak zdrowie;
Ile cię trzeba cenić, ten tylko się dowie
Kto cię stracił. Dziś piękność twą w całéj ozdobie
Widzę i opisuję, bo tęsknię po tobie.

In our experiment we loaded the corpus (as a big text file), tokenized it and divided it into syllables (sub-word units) with the '7683' option and our stemming procedure. Afterwards, an 'all_tokens' list was created with 5,059 different tokens found in our corpus. Pre-processed text was divided into chunks of 400 syllables. As the result we got 198 chunks (whole text was divided into 79,544 tokens).

To evaluate our network, we input one token at a time, use the outputs as a probability distribution for the next sub-word unit and repeat the action. To start the text generation, we directed the initial series to build the hidden state and generate one token at a time.

We used the following parameters to train our network:

Number of epochs = 15
Hidden layer size = 500
Number of layers = 3.

Additionally, we created a set of functions to monitor the training process. The loss function was used to monitor the training progress, while training time was also monitored. The 15 epochs of training took 96 min using Google Collaboratory platform.

The loss function is an indicator describing a dissimilarity between real and desired output of neural network. Lower values of loss suggest better accuracy for output values. Figure 2 demonstrates values of the loss during the whole training of our network. The system is adapting its parameters into expected outcome with every iteration of the training, therefore it can be observed a decreasing tendency of the loss changes, with the final value below 1.

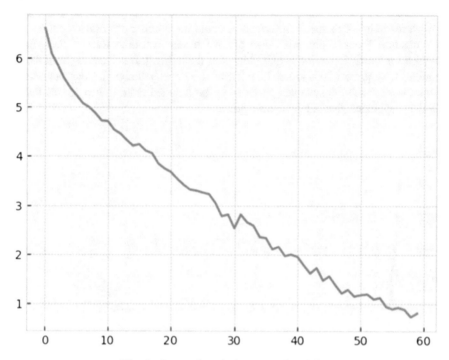

Fig. 2. Loss values during network training

Our trained network generated this poem in the style of Adam Mickiewicz:

> *Litwo! Ojczyzno moja! czaślam w Wilnie,*
> *Lecz niedźwiedź w pole; więc Sędzia z Domcom,*
> *Niech co to nie wiedziacie rzegnąć swą rómą,*
> *Szlachta odwieczna, w któréj otwierzycie?*
> *Czy to Rejent równie podszepnął: już do polu żeca,*
> *Pokojowa zaś znalazł się przeznało*
> *Odbijamy od drzewo Rejent, zawołał Rejent,*
> *Od Bonaparte znowu w kota się przerzuca,*
> *Daléj drzeć pazurami, a Suwarów w kuca.*
> *Obaczcież co się stało w milczeniu głębokiém.*
> *Sywciąż wtenczas widząc, jak od chmielu tyki,*
> *W kurtkach, w budowne jak chartaści i urzęwe dawał.*

This text was written with a metre similar to Polish alexandrine. The generated verses length is mostly thirteen syllables, but does not always meets this criterium. We also can observe proper punctuation and text capitalization. The vocabulary seems to also be similar to Mickiewicz's. It can be seen that although the whole creation does not have an overall coherent theme or clear semantic structure, some of the individual verses do create meaningful sentences.

Another important indicator we decided to monitor during network training was 'bad-words ratio'. This metric informed us about the number of syllables not used for word creation. Figure 3 presents values of this measure over time. Ratio of the syllables not used for words formation is quite low after the training. It decreased (from the first iteration) from almost 30% to less than 2% after 5 epochs. It signifies that our network learned how to use the syllables provided in the inputted data. It also proves that the sub-word approach works as we anticipated.

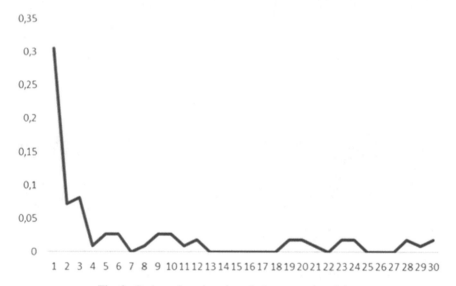

Fig. 3. Bad-words ratio values during network training

6 Results

To control the randomness of the generated poem we used a temperature hyperparameter. Depending on the temperature value, different and interesting results are generated. Lower values make the output data more probable given the training data. Values around 1 or higher make the outputs more creative, whereas values closer to 0 force the network to generate only the most probable results.

To be more precise, the temperature is a hyperparameter of LSTM's (and neural networks generally) used to control the randomness of predictions by scaling the logits before applying softmax. For example, in TensorFlow's Magenta implementation of LSTMs, temperature represents how much to divide the logits by before computing the softmax.

When the temperature is 1, we compute the softmax directly on the logits (the unscaled output of earlier layers) and using a temperature of 0.6 the model computes the softmax on logits/0.6, resulting in a larger value. Performing softmax on larger values makes the LSTM more confident (less input is needed to activate the output layer) but also more conservative in its samples (it is less likely to sample from unlikely candidates). Using a higher temperature produces a softer probability distribution over the classes and makes the RNN more "easily excited" by samples, resulting in more diversity and more mistakes as well.

Neural networks produce class probabilities with logit vector z where $z = z = (z_1, ..., z_n)$ by performing the softmax function to produce probability vector $q = (q_1, ..., q_n)$ by comparing zi with the other logits.

$$q_i = \frac{\exp{(z_i/T)}}{\sum_j \exp{(z_i/T)}}$$

is the temperature parameter, normally set to 1.

The softmax function normalizes the candidates at each iteration of the network based on their exponential values by ensuring the network outputs are all between zero and one at every timestep. Temperature therefore increases the sensitivity to low probability candidates [16].

The effects were surprising and interesting. Setting the temperature to a value of 0.8, the network generated the following poem, the first lines of which are presented below.

Litwo! Ojczyzno moja!
To za Dowejce i stanęli Mateczniku,
Wieczają tylko polski Tadeuszkę ożeką,
I zamiast iść do Panów, hałasy że cieli
Toczyli w krzesłach w powiewawych zadartym, świecie bez rury,
Spowiedzi się; Półkozic jaśniał w las,
A Gerwazy; Skołuba z zapaści przymierzy,
Czyli las był; z ich wszystkich zaoczyny.
Przerwał je stare, albo raz niewiewał uczyć,
Niby krano mumienne po trawie, robacze,
Marciały, odgadniesz, że na charty po chwili.
Czyli też zdarzeczka, czyli widzisz, kieli

The result seems to be diverse. Most of the verses and sentences does not have any sense, although the majority of the words are understandable and exist in Polish vocabulary. Metre is close to Polish alexandrine.

Subsequently, when the parameter was changed to 0.2 the following was obtained.

Litwo! Ojczyzno moja!
pradzi Jeden wielki myślili -- Cham, Skołuba krzyknęła
czówna, ogon nim do mnie się zabawić! -- Ja myśliwiec!
przerwałem (tu do wielkiéj pozwory? Doweyko!
nierwała Tekomości! Podczaszyc na wielkim gniewie, Ale Hrabia,
A Pan Hrabię i Sędzia tuż przy Berdziego Boże. Ja, Sędziecznym młody, pytany o
zdromie
polityczny Litwie monarcha myślani
młowłaśni młodzi rzecz spadał towarzyszy,
przerwała Tekomości! to oni!
przerwałem że czasem wystrzeliła

The Poem generated with temperature = 0.2 consists of more real Polish words and even whole phrases. The metre seems to be more distant from our training data, but still imitates Polish alexandrine.

Selecting higher temperature values results in more random output values. For experimental purposes we decided to apply a temperature as high as 1.4. The resulting poem is presented below.

Litwo! Ojczyzno moja! ty wszystko, usłynił kitdą zostaną, zda ciedzenie przestrzyni
Leste, tém zrarza ku otbią mami rzędawał,
Niby wódz widziałam była po leśka zwierz siedzenie,
Chce je trzewałem ska wielka, będzie o otwiemie!
Bo deusz na może świępać i siebie po dworze
Jéj żeby się pokładał obławą i świamę.
Wstępali prozem o malarstwie w komciadki,
Zosia w lot przywiedzenie się milczą się gniewa,
Tam dęsałam i ze czuły: pas konato,
Pilnami mu po drzeni, owchalby charty,
Szła mamą! cóż sach tysiąc złota już grunt dawem,
Niby Pan krzywał się, lecz

The resultant text consists of valid Polish words, but also a number of invalid expressions. No consistent meaning can be found in the generated text, although it still mostly follows the desired metre.

As our approach to poetry generation is quite novel, we decided to juxtapose our results with text generated character by character by a recurrent neural network. The network was trained with following parameters:

```
n_epochs = 50
hidden_size = 400
n_layers = 2
```

The number of training epochs was increased to 50, but the network consisted of 2 layers. The training time was similar in both experiments and took about 1.5 h.

After pre-processing the necessary data and training the network we generated poems using three different temperatures. Following previous experiments, the temperature indicator was set to 0.8 with the following results presented below.

Pana,
Pajał wim storemymną, tery tejy;
Wychynią szwogole, jać jaka,
Skrzekło udy, wniekoła,
I nudość wisi mała wieryży słajwie wła udawytył oporaźny,
Sona, świ, przykił szeną roż stestwe wydowniewa neny,

The poem generated by the char-based network does not follow the Polish alexandrine metre. Created words are stylized into the poet's vocabulary, but most of them do not qualify as correct words in Polish language. The poem itself cannot be classified as meaningful.

As lower temperatures enable the generation of more probable output, we present below a poem created using temperature 0.2.

Thy, odo się się wie się się szada wi się wie się się wie szała wie się wie wie sza się
wie się wie się szała, wie wyła, wie wie do wie się się wie podała, wie się się się się
wie do wie jak wie szała w

The generated text does not look like a poem at all. Words created by the NN are real Polish expressions, but they do not have any schema or phrasal meaning.

The higher temperatures produce more random results. The last example was created with temperature = 1.4.

Thecegéj utéjégła! nguscy, ażąc. VąrŚni.ż dŻóc nię; wą Dolu chogyśsakiękdy uwy;
o_jicpaże, Ban desie je każymik!
Akyski nie żessył kórzcierbłéj awwi,, hrzeliwachrzeguj.
A rwum mawlakguł łyłrąy, arzymu

The text looks very incoherent. Almost all created strings do not belong to Polish vocabulary. No metre can be noticed.

Additionally, it is worth paying attention to the influence of the sequence length on the recognition of distant relationships. 30–50 tokens is a window in which the network remembers the sequence structure well. Using syllables instead of characters shortens (compresses) the sequence approximately 3 times, which makes training easier. The network can 'analyze' the structure of several rows at once. Potentially using LSTM instead of GRU can give even better results when modeling long-distance dependencies.

7 Discussion and Conclusion

Polish is a complex language with complicated grammar and vocabulary. The language has three tenses. Verbs are in perfect or imperfect form, also they differ for gender and number. There are seven cases of a noun: nominative, genitive, dative, accusative, instrumental, locative, vocative. Nouns also occur in two numbers (singular and plural) and three gender forms (masculine, feminine and neuter). Adjectives also have gender, case and number. All those rules cause the complexity of language. The vocabulary (language corpus) is very large due to the morphological richness of the Polish language. Therefore, any automatic analysis or generation of Polish language is difficult [12].

In this paper we proposed generating author-stylized poems using innovative sub-word units, based on the example of 'Sir Thaddeus'. Our approach focuses on sub-word units. In the solution we created, we tried to generate a poem congenial to Adam Mickiewicz's 'Sir Thaddeus' book. We used an RNN for our work. The neural network was trained on Mickiewicz's poem, based on a sub-word unit (close to syllable) corpus. After preparation of the data with our special stemmer tool and the network training we compared the results of the generated text using three temperatures: 0.8, 0.2 and 1.4. All three cases were able to generate quite promising texts, imitating 'Sir Thaddeus'. The 0.2-temperature generated a poem with most understandable words and phrases, as we expected. In all three cases the metre was at least partially preserved. The general vocabulary was epitomizing Mickiewicz's.

For comparison, we conducted a similar experiment using a classical solution with characters as the unit of text generation. The differences between the results was tremendous. The neural network trained on the poem divided into characters, after 50 epochs of training demonstrated much worse results. The experiment performed with a temperature of 1.4 generated text resembling a random string, with no metre present. Setting temperature to 0.2, the generated text consisted of correct Polish words, but they did not construct any phrases and there was no metre. Only a temperature value equal to 0.8 created an impression of a poem, but its quality was much lower than that generated with sub-words units.

The poems generated by our solution cannot yet be confused with poems generated by humans, but they clearly recognize and follow stylistic and poetic rules applied by the author of the poem used as training data. The attempt to create the resultant text in Polish alexandrine is clearly visible in the shown examples. The word formation is quite good, the number of incorrect phrases is low. Punctuation and capitalization were satisfying. It was also the case that the method we used in our research provides much better results than when the unit of the training data is a character.

We plan to continue our research on automatic poetry generation using various training data. As units different to character or word gave promising results, we will experiment with various language units and the data amount of training data. Modifying the parameters of our network will also be considered in our future work. The temperature dimension might be adjusted more accurately. In our further work, we plan to perform fine-tuning of the hyperparameters of our neural network. We would like to use the Cyclical Learning Rates method [13] to adjust learning rate of our RNN.

Finally, we plan to annotate our training texts with POS tags (and/or grammatical groups) and interpolate the results with the Byte Pair Encoding (BPE) algorithm [16].

Performing our research, we concluded that although generating poetry (or other texts) does not seem to be a pressing concern, it can bring many benefits to several domains. It could be broadly used in entertainment, to generate texts, poems or even whole books or book series. Discovering the pattern of a specific author's style might be used to adapt existing text to the specific language of the selected author, era or type. It can also be used to for author style transfer or style mixing. It allows us to construct any lyric in any style imitating any epoch. Such a solution might also be valuable education. Recognizing the patterns behind a specific author will be useful to discover all the books written by the same author but under different pseudonyms, e.g. Stephen King. This writer published as Richard Bachman, because the editor did not want to agree to publish more than one book a year [15]. Also, Rowling, the author of the 'Harry Potter' series, after a successful saga about wizards, decided to publish as Robert Galbraith. With automatic style recognition it might be much easier to identify a writer. The idea could be especially useful for the authors living in previous ages, as usually the data available for researchers is very limited, so the authorship can be mainly confirmed by the analysis of the text. The main domain where our research can be applied is literature and linguistic. It can be imagined that generating author-stylized text might expand the field of literature analysis.

Finally, we see the potential improvement of quality in the use of a non-direct but general language model of the Polish, projected on the language model of poetry and ultimately projected on a specific author (e.g. "Pan Tadeusz"), completed with fine-tuning. [18] This will allow us to learn more about poetry and language structure (Fig. 4).

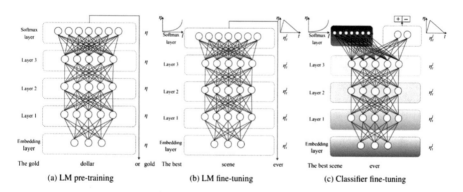

Fig. 4. Language model adaptation.

In such approach we can distinguish three stages: (a) The LM is trained on a general-domain corpus to capture general features of the language in different layers, (b) the full LM is fine-tuned on target task data using discriminative fine-tuning ('Discr') and slanted triangular learning rates (STLR) to learn task-specific features,

(c) the classifier is fine-tuned on the target task using gradual unfreezing, 'Discr', and STLR to preserve low-level representations and adapt high-level ones (shaded: unfreezing stages; black: frozen) [18].

We can also use Transformer architecture to better model distant relationships [17]. Additionally, natural language processing tasks, such as caption generation and machine translation, involve generating sequences of words. Models developed for these problems often operate by generating probability distributions across the vocabulary of output words and it is up to decoding algorithms to sample the probability distributions to generate the most likely sequences of words. For this problem we would like to use in future a popular heuristic, which is the beam search. It expands upon the most common greedy search and returns a list of most likely output sequences. Instead of greedily choosing the most likely next step as the sequence is constructed, the beam search expands all possible next steps and keeps the k most likely, where k is a user-specified parameter and controls the number of beams or parallel searches through the sequence of probabilities. This would most likely lead to better art generation as well [19].

Analysing and recognizing text might be useful in forensics, in verification and evaluation of evidence of crime. Finally, social sciences and art can benefit from using artificially created texts.

References

1. Milosz, C.: The History of Polish Literature. University of California Press (1983)
2. Tikhonov, A., Yamshchikov, I.P.: Guess who? Multilingual approach for the automated generation of author-stylized poetry. arXiv preprint arXiv:1807.07147 (2018)
3. Liu, B., Fu, J., Kato, M.P., Yoshikawa, M.: Beyond narrative description: generating poetry from images by multi-adversarial training. arXiv preprint arXiv:1804.08473 (2018)
4. Cheng, W.F., Wu, C.C., Song, R., Fu, J., Xie, X., Nie, J.Y.: Image inspired poetry generation in XiaoIce. arXiv preprint arXiv:1808.03090 (2018)
5. Lamb, C., Brown, D.G., Clarke, C.L.: A taxonomy of generative poetry techniques. J. Math. Arts 11(3), 159–179 (2017)
6. Oliveira, H.: Automatic Generation of Poetry: An Overview. Universidade de Coimbra (2009)
7. Gervás, P.: Exploring quantitative evaluations of the creativity of automatic poets. In: Workshop on Creative Systems, Approaches to Creativity in Artificial Intelligence and Cognitive Science, 15th European Conference on Artificial Intelligence, Lyon, France. IOS Press (2002)
8. Zhang, J., Feng, Y., Wang, D., Wang, Y., Abel, A., Zhang, S., Zhang, A.: Flexible and creative Chinese poetry generation using neural memory. arXiv preprint arXiv:1705.03773 (2017)
9. Loller-Andersen, M., Gambäck, B.: Deep Learning-Based Poetry Generation Given Visual Input
10. Hopkins, J., Kiela, D.: Automatically generating rhythmic verse with neural networks. In Proceedings of the 55th Annual Meeting of the Association for Computational Linguistics (Volume 1: Long Papers), vol. 1, pp. 168–178 (2017)

11. Wołk, A., Wołk, K., Marasek, K.: Analysis of complexity between spoken and written language for statistical machine translation in West-Slavic group. In Multimedia and Network Information Systems, pp. 251–260. Springer, Cham (2017)
12. Smith, L.N.: Cyclical learning rates for training neural networks. In: 2017 IEEE Winter Conference on Applications of Computer Vision (WACV), pp. 464–472. IEEE, March 2017
13. Wolk, K., Marasek, K.: Survey on neural machine translation into polish. In: International Conference on Multimedia and Network Information System, pp. 260–272. Springer, Cham, online tool version, September 2018 http://ld.clarin-pl.eu:5000
14. Collings, M.R.: Stephen King as Richard Bachman, p. 168. Starmont House (1985)
15. Sennrich, R., Haddow, B., Birch, A.: Neural machine translation of rare words with subword units. In: Proceedings of the 54th Annual Meeting of the Association for Computational Linguistics (ACL 2016), Berlin, Germany (2016)
16. Hinton, G., Vinyals, O., Dean, J.: Distilling the knowledge in a neural network. arXiv preprint arXiv:1503.02531 (2015)
17. Devlin, J., Chang, M.W., Lee, K., Toutanova, K.: Bert: pre-training of deep bidirectional transformers for language understanding. arXiv preprint arXiv:1810.04805 (2018)
18. Howard, J., Ruder, S.: Universal language model fine-tuning for text classification. In: Proceedings of the 56th Annual Meeting of the Association for Computational Linguistics (Volume 1: Long Papers), vol. 1, pp. 328–339 (2018)
19. Tillmann, C., Ney, H.: Word reordering and a dynamic programming beam search algorithm for statistical machine translation. Comput. Linguist. **29**(1), 97–133 (2003)

Toward a Knowledge Sharing-Aimed Virtual Enterprise

Nastaran Hajiheydari[1(✉)], Mojtaba Talafidaryani[2],
and SeyedHossein Khabiri[2]

[1] York Business School, York St John University, York, UK
N.hajiheydari@yorksj.ac.uk
[2] Faculty of Management, University of Tehran, Tehran, Iran
{Mojtabatalafi,Hosseinkhabiri}@ut.ac.ir

Abstract. One of the requirements for a successful collaboration in Collaborative Networked Organizations (CNOs) is the sharing of knowledge resources between members. Knowledge sharing is a core competency for creating competitive advantage in Virtual Enterprises (VEs). Accordingly, the evaluation of the contribution of each VE member organization in sharing knowledge resources, as well as monitoring the distribution of these resources among the people is of great importance. For this purpose, a conceptual framework of a VE with the aim of knowledge sharing (henceforth, a knowledge sharing-aimed VE) is presented in this paper. In this framework, the conceptual structure of a cognitive system is designed to supervise the process of knowledge sharing among organizational staff in each VE organization, and thus to support decision-making by knowledge managers. The system collects log data of various organizational Knowledge Management Systems (KMSs) and then processes it using machine learning algorithms to identify the learning patterns and knowledge application in the organization to ensure employees access to distributed knowledge resources. In addition, by considering each organization member of VE as an individual user of an inter-organizational KMS, the same mechanism can be used at the cross-organizational level. Finally, it should be acknowledged that the implementation of this conceptual framework could play an essential role in improving the knowledge sharing and management process, which is the cornerstone of the organizations co-operation in a VE.

Keywords: Collaborative Networked Organization · Virtual Enterprise · Knowledge sharing · Cognitive system · Log data mining

1 Introduction

Omnipresent knowledge management tools and technologies which enable online and offline communications between members propose more effective learning environments than traditional ones. Today's business complexity urges firms to sustain and manage their knowledge resources through inter-organizational collaboration [9]. Li et al. [20] enumerate enhancing knowledge use rate, reducing knowledge duplication, and managing risk as the benefits of a virtual knowledge alliance but when it comes to a knowledge sharing context (especially a virtual one), an important question arises: How to evaluate each member's contribution (sharing behavior) in the whole network?

© Springer Nature Switzerland AG 2019
Á. Rocha et al. (Eds.): WorldCIST'19 2019, AISC 930, pp. 316–325, 2019.
https://doi.org/10.1007/978-3-030-16181-1_30

Individuals are not interested in sharing their know-how knowledge and expertise spontaneously [13]. Controlling knowledge sharing behavior is a complex challenge for knowledge managers and lots of knowledge sharing platforms face failure due to this issue [28]. At a higher level, the context in which various organizations form a knowledge-sharing platform are interacting, evaluation of each organizations knowledge sharing contribution and behavior becomes a critical inter-organizational concern. On the other hand, within organizations, understating learning experience and behavior is an essential issue for knowledge managers [29].

Narendra et al. [25] address the rise of conflict and the need for conflict management among firms in a VE. This kind of concern is also authentic fact in a knowledge sharing-aimed VE because the contribution of each partner in sharing valuable knowledge is vague. Therefore, applying a framework to evaluate each firm's contribution to sharing valuable knowledge is an essential.

So, in this research, we are looking for an answer to this important question: How can we improve the knowledge sharing process in a VE? So that ultimately, each collaborative network member has an effective partnership in this process and also knowledge sources within each of these organizations will be appropriately distributed.

Designing and proposing a conceptual framework of a cognitive system as a platform to collect and process different KMSs log data in a VE, in the first place, helps us identify learning behaviors within organizations and then monitor each firm's role in sharing valuable knowledge in the whole collaboration network, accordingly paving the way to move toward a knowledge sharing-aimed VE.

2   Literature Review

In today business environment, which is extremely inter-connected, businesses can ally together to form a VE to complement their competencies and/or share their resources collaboratively around the globe by the means of computer networks to seize new opportunities (or react to upcoming threats) in order to remain competitive [5, 14–17]. VEs are considered as the next organizational paradigm [30], and firms prefer to focus on their own core competencies more deeply and be a qualified node of value chain rather than focusing on other steps of a value chain out of their own capabilities [4]. A VE is a sociotechnical system in which geographically dispersed enterprises unite in a business alliance and with a specific aim to collaborate and share their resources in order to confiscate emerging opportunities and compete in the new complex business context [2, 5, 14, 16]. Virtual business alliances form typically with the aim of sharing and complementing either products/services or knowledge [20].

As businesses are becoming more knowledge-intensive, knowledge is considered as an important strategic resource. Therefore, sharing and creating knowledge in a VE is a desired outcome of collaborations [31], and contributes to a VE competence [21]. In an inter-organizational collaboration on knowledge sharing, different participants have access to external knowledge shared in the collaborative network and thus strengthening their knowledge as a strategic resource of the world today [1, 23].

As discussed before, firms are reluctant to share their knowledge in a networked collaboration and evaluation of each firm's efforts in sharing knowledge is an essential. Evaluating firms' knowledge sharing behavior merely by human cognition seems irrational. Cognitive systems are a kind of intellectual systems which can make or support complex decision makings with the help of data analytic techniques like machine learning techniques (known as artificial cognitive systems) [32] to improve human cognition and decision-making effectiveness [18].

This paper considers log data generated by various systems in a VE as an apt source to help humans get insights about each firm's contribution. Systems logs trace and collect all activities within a software system and are in the form of log files, databases or streams of data [24]. Their applications are optimizing and debugging system performance, detecting misdeeds and system intrusions, prediction and the most important to our study: Profiling user behaviors and reporting business analytics [26]. Applying data mining techniques on log data is called log data mining (or log mining) and it outperforms conventional log data analysis methods and even finds hidden ("below the radar") patterns [7].

There are various parts of KMSs contributing to constructing a knowledge sharing collaboration. In addition to Communities of Practice (COP) which are informal communities, Content Management Systems (CMSs), Learning Management Systems (LMSs), and Document Management Systems (DMSs) are among different modules of KMSs available to form an infrastructure for a knowledge sharing-aimed VE. A LMS is an e-learning system, which can support project-based instructions and collaborative learning [27]. Laudon and Laudon [19] define LMS as a consolidation of management, delivery, tracking, and learning effectiveness measurement of employee learning and training. While CMSs make more general structured and semi-structured knowledge available inside the enterprise, COP are cross-organizational informal communities for professionals to share specific best practices [19].

These systems log data are a precious and appropriate source [8, 22]. For instance, LMS log data can be used for analyzing learner performance, satisfaction [12], and engagement because it is on activity level, and scalable [11]. Park and Jo [27] conducted an analysis on LMSs log files and used activity theory [10, 33] as a framework to describe the results and drew some conclusions for improving learning with LMSs in the virtual education context. This claim is also supported in the virtual education context, utilizing data mining techniques on log data generated by Technology-Enhanced Learning (TEL) systems are far more effective [6] than using surveys and interviews for learning context and process analysis [3]. Empirical research related to the aim of this study is limited to the discussed researches to the best knowledge of authors.

The reviewed literature indicates that although the formation of a VE with the aim of sharing knowledge has attracted researchers' attention, no effort has done to effectively evaluate how each participant behaves in sharing knowledge. In addition, earlier studies have focused on analyzing the log data of LMSs, while all parts of KMSs (LMS, DMS, and CMS) produce their own specific log data and analyzing their log data can be as beneficial. Therefore, the contribution of this research is, first, to suggest utilizing and analyzing all relevant systems to knowledge management as an infrastructure for forming a knowledge-aimed VE and secondly, providing a preventative approach for conflict management in assessing the contribution of each partner in sharing knowledge. According to the authors, filling these two gaps can accelerate moving toward a knowledge sharing-aimed VE, which is the ultimate goal of this research.

3   Knowledge Sharing-Aimed Virtual Enterprise

In this section of the research, the conceptual framework of a knowledge sharing-aimed VE is proposed. This framework is shown in Fig. 4. As represented in this figure, geographically dispersed organizations collaborate with each other in the form of a VE to share knowledge through an inter-organizational information system to achieve a common goal. Within each of these organizations, there are also different information systems related to knowledge management that provide and share knowledge flow from outside the organization among employees. In other words, at the level of the collaborative network (Fig. 1), the knowledge resources of the partners are provided to each organization through an inter-organizational information system and then at the organizational level (Fig. 2), these resources are provided through organizational information systems to knowledge workers so that they can use it in the context of organizational operations.

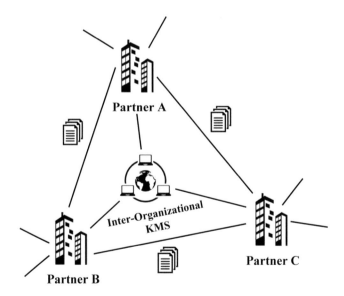

Fig. 1. Knowledge sharing in a collaborative network of business partners

The interaction of the organizations users with different parts of organizational KMSs, such as CMS, DMS, LMS, etc., produces many log files that their data analysis yields valuable results for tracking and obtaining the organizational desired knowledge goals. Log files related to KMSs consist of valuable information on how users are using the distributed knowledge. This information provides details about which knowledge worker has done which activity at what time, these activities include searching, viewing, labeling, recording, storing, retrieving, browsing, distributing, publishing, and even producing content in different formats. Therefore, the analysis of these files leads to the identification of learning patterns and the use of knowledge by individuals, which can ensure the proper sharing of knowledge among employees and their access to distributed knowledge resources.

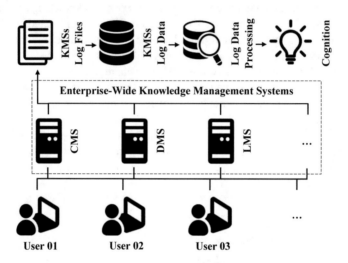

Fig. 2. The conceptual structure of a cognitive system that recognizes knowledge sharing behaviors

In this regard, in order to monitor the knowledge management and sharing process in each organization member of the VE, a cognitive system can be used so that its outputs help the organization's knowledge managers to make their decisions. As it can be seen in Fig. 3, this system connects to log files repository related to KMSs to intelligently fetch and store various items from these files. Then, in the preprocessing and cleanup stage, these raw and initial data are converted into aggregated and appropriate format for analyzing data. In the next stage, the system will use different data mining techniques to explore the log data provided so that rules and patterns can be extracted that will lead to profound knowledge and insight into the way knowledge workers work. Also, these systems will be able, over time and after interacting with data and individuals, to learn the desired knowledge patterns during the data mining process using the methods of machine learning and to offer practical suggestions. Definitely, machine learning techniques to be used depend on the aims of log files analysis which knowledge managers follow. For instance, machine learning algorithms which are used for anomaly detection (to detect abnormal behaviors) or clustering (to cluster users based on desirable and undesirable behaviors) can be applied to log files. With this regard, an example is provided in the following.

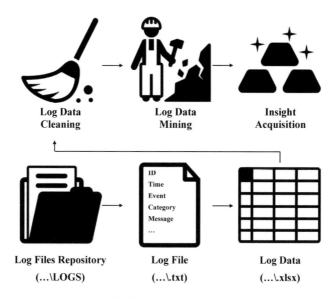

| Log Data Cleaning | Log Data Mining | Insight Acquisition |

| Log Files Repository (...\LOGS) | Log File (...\.txt) | Log Data (...\.xlsx) |

Fig. 3. Log files analysis

For example, a cognitive system, after analyzing log data, finds that a group of people in an organization still do not see new documentation published about a particular part of the project or that they have not spent enough time watching the uploaded video. To this end, the system can alert the organization's knowledge managers to follow up the issue. After an investigation, knowledge managers find that the content is not available for one of the organization's units or that it has not been created in the process of sharing. Sometimes, during the process of collaborating between organizations, new information is provided to the top managers of each organization and managers are required to convey this information to their employees so that they will be informed of the new directions of the project. Otherwise, employees of the organization, based on previous information, will perform their duties, which are not consistent with emerging plans. In such circumstances, the organization's knowledge managers provide new information to their subordinates so that they share content with other people. However, this information seems to be not available to the knowledge workers and while knowledge managers imagine that information was distributed in the organization, on the other hand, knowledge workers are completely unaware of the existence of such information. As a result, due to the heterogeneous distribution of information at the organization level, there are different orientations that ultimately don't lead to a common goal. In this situation, the importance of using the affronted cognitive system becomes increasingly essential.

Whether the entered knowledge into the organization at what time and through which channel is available to each user and how each user used or changed the knowledge is among the achievements of these cognitive systems. Now if the same mechanism is applied on the inter-organizational KMS, each organization in the VE is considered as a user of this system, as a result, it is possible to evaluate each partner in the process of sharing knowledge and measuring the extent of its participation in this collaboration. Consequently, it becomes clear that how log data analysis in the context of a cognitive system, leads to insights for learning patterns inside the organization and also, knowledge sharing behaviors between organizations, hence, the process of managing and sharing knowledge, which is the foundation of the organizations collaboration in a VE can be improved.

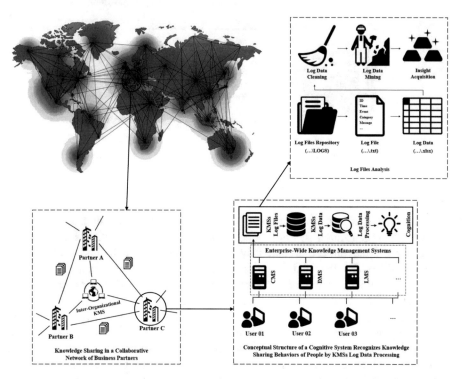

Fig. 4. The conceptual framework of a knowledge sharing-aimed virtual enterprise

4   Conclusion

Sharing knowledge resources is a key stand of collaboration in VEs in the knowledge era. Therefore, any research that contributes to improving the knowledge sharing process in such a VE will be of great importance. In this research, a conceptual framework of a VE with the aim of knowledge sharing is presented, which its implementation has two aspects of importance: Firstly, monitoring the knowledge distribution process within each of the member organizations of the network is possible. Secondly, it evaluates the participation and contribution of each VE member in the creating and sharing of knowledge resources.

The main limitation of the researchers in this study is the lack of access to the data needed to implement and test the conceptual framework provided. For future research and in order to validate this framework, researchers might collect log data from organization's KMSs. Afterward, using data mining techniques and analyzing log data would adapt and compare the results to the goals represented in this paper. In the case of matching findings with expectations, researchers can use the machine learning algorithms to implement and develop the cognitive system described in the suggested conceptual framework. Also, in the case of having access to an inter-organizational network and possessing sufficient facilities, they can implement this mechanism in one of the member organizations of the network. In addition, this step should be further extended and by using the log data analysis of the inter-organizational KMSs, conditions described in this study should be tested and evaluated by simulation.

Eventually, future research can report the results of the implementation of this framework, so that it gets out of just being a concept and proceeds into action. Also interested in this field of study can identify and report on new applications for the framework through a more detailed review of the contents of the log files of KMSs. In addition, researchers in a more comprehensive view can map out and implement this framework beyond the knowledge management domain for other types of information systems, such as Business Process Management Systems (BPMSs), in which business processes at the level of inter-organizational collaboration (VE business processes) could also be amended.

References

1. Andriessen, D.: Designing and testing an OD intervention: reporting intellectual capital to develop organizations. J. Appl. Behav. Sci. **43**(1), 89–107 (2007). https://doi.org/10.1177/0021886306297010
2. Baxter, G., Sommerville, I.: Socio-technical systems: from design methods to systems engineering. Interact. Comput. **23**(1), 4–17 (2011). https://doi.org/10.1016/j.intcom.2010.07.003
3. Brill, J., Park, Y.: Evaluating online tutorials for university faculty, staff, and students: the contribution of just-in-time online resources to learning and performance. Int. J. E-learn. **10**(1), 5–26 (2011)
4. Camarinha-Matos, L.M., Afsarmanesh, H., Garita, C., Lima, C.: Towards an architecture for virtual enterprises. J. Intell. Manuf. **9**(2), 189–199 (1998). https://doi.org/10.1023/A:1008880215595
5. Camarinha-Matos, L.M., Afsarmanesh, H.: A comprehensive modeling framework for collaborative networked organizations. J. Intell. Manuf. **18**(5), 529–542 (2007). https://doi.org/10.1007/s10845-007-0063-3
6. Chatti, M.A., Dyckhoff, A.L., Schroeder, U., Thüs, H.: A reference model for learning analytics. Int. J. Technol. Enhanc. Learn. **4**(5–6), 318–331 (2012). https://doi.org/10.1504/IJTEL.2012.051815
7. Chuvakin, A., Schmidt, K., Phillips, C.: Logging and log Management: The Authoritative Guide to Understanding the Concepts Surrounding Logging and Log Management. Syngress, Waltham (2013)
8. Cocea, M., Weibelzahl, S.: Disengagement detection in online learning: validation studies and perspectives. IEEE Trans. Learn. Technol. **4**(2), 114–124 (2011). https://doi.org/10.1109/TLT.2010.14

9. Cricelli, L., Grimaldi, M.: Knowledge-based inter-organizational collaborations. J. Knowl. Manag. **14**(3), 348–358 (2010). https://doi.org/10.1108/13673271011050094

10. Engeström, Y., Miettinen, R., Punamäki, R.L.: Perspectives on Activity Theory. Cambridge University Press, Cambridge (1999)

11. Henrie, C.R., Bodily, R., Larsen, R., Graham, C.R.: Exploring the potential of LMS log data as a proxy measure of student engagement. J. Comput. High. Educ. **30**(2), 344–362 (2017). https://doi.org/10.1007/s12528-017-9161-1

12. Henrie, C.R., Bodily, R., Manwaring, K.C., Graham, C.R.: Exploring intensive longitudinal measures of student engagement in blended learning. Int. Rev. Res. Open Distrib. Learn. **16**(3), 131–155 (2015). https://doi.org/10.19173/irrodl.v16i3.2015

13. Huysman, M., Wulf, V.: IT to support knowledge sharing in communities, towards a social capital analysis. J. Inf. Technol. **21**(1), 40–51 (2006). https://doi.org/10.1057/palgrave.jit.2000053

14. Januska, M., Chodur, M.: Virtual enterprise network. In: 32nd International Spring Seminar on Electronics Technology, pp. 1–5. IEEE Press, Brno (2009). https://doi.org/10.1109/isse.2009.5206949

15. Januska, M., Kurkin, O., Miller, A.: Communication environment for small and medium enterprises. IBIMA Bus. Rev. **2010**, 1–8 (2010). https://doi.org/10.5171/2010.270762

16. Januska, M., Palka, P., Sulova, D., Chodur, M.: Value chain of virtual enterprise: possible modern concepts and value drivers identification. Ann. DAAAM Proc. **20**(1), 469–471 (2009)

17. Januska, M.: Communication as a key factor in virtual enterprise paradigm support. Innov. Knowl. Manag. A Glob. Compet. Advant. **3**, 1–9 (2011)

18. Kamesh, D.B.K., Sumadhuri, D.S.K., Sahithi, M.S.V., Sastry, J.K.R.: An efficient architectural model for building cognitive expert system related to traffic management in smart cities. J. Eng. Appl. Sci. **12**(9), 2437–2445 (2017). https://doi.org/10.3923/jeasci.2017.2437.2445

19. Laudon, K.C., Laudon, J.P.: Management Information Systems: Managing the Digital Firm. Pearson, London (2016)

20. Li, Z., Yang, F., Zhang, D.: The virtual alliance knowledge sharing model and selection strategy. Procedia Comput. Sci. **91**, 276–283 (2016). https://doi.org/10.1016/j.procs.2016.07.075

21. Liu, P., Raahemi, B., Benyoucef, M.: Knowledge sharing in dynamic virtual enterprises: a socio-technological perspective. Knowl.-Based Syst. **24**(3), 427–443 (2011). https://doi.org/10.1016/j.knosys.2010.12.004

22. Macfadyen, L.P., Dawson, S.: Mining LMS data to develop an "early warning system" for educators: a proof of concept. Comput. Educ. **54**(2), 588–599 (2010). https://doi.org/10.1016/j.compedu.2009.09.008

23. Mentzas, G., Apostolou, D., Kafentzis, K., Georgolios, P.: Inter-organizational networks for knowledge sharing and trading. Inf. Technol. Manag. **7**(4), 259–276 (2006). https://doi.org/10.1007/s10799-006-0276-8

24. Nagappan, M.: Analysis of execution log files. In: 32nd ACM/IEEE International Conference on Software Engineering, pp. 409–412. IEEE Press, Cape Town (2010). https://doi.org/10.1145/1810295.1810405

25. Narendra, N.C., Norta, A., Mahunnah, M., Ma, L., Maggi, F.M.: Sound conflict management and resolution for virtual-enterprise collaborations. Serv. Oriented Comput. Appl. **10**(3), 233–251 (2016). https://doi.org/10.1007/s11761-015-0183-0

26. Oliner, A., Ganapathi, A., Xu, W.: Advances and challenges in log analysis. Commun. ACM **55**(2), 55–61 (2012). https://doi.org/10.1145/2076450.2076466
27. Park, Y., Jo, I.H.: Using log variables in a learning management system to evaluate learning activity using the lens of activity theory. Assess. Eval. High. Educ. **42**(4), 531–547 (2017). https://doi.org/10.1080/02602938.2016.1158236
28. Probst, G., Borzillo, S.: Why communities of practice succeed and why they fail. Eur. Manag. J. **26**(5), 335–347 (2008). https://doi.org/10.1016/j.emj.2008.05.003
29. Psaromiligkos, Y., Orfanidou, M., Kytagias, C., Zafiri, E.: Mining log data for the analysis of learners' behaviour in web-based learning management systems. Oper. Res. **11**(2), 187–200 (2011). https://doi.org/10.1007/s12351-008-0032-4
30. Putnik, G.D., Cruz-Cunha, M.M.: A contribution to a virtual enterprise taxonomy. Procedia Technol. **9**, 22–32 (2013). https://doi.org/10.1016/j.protcy.2013.12.003
31. Rasmussen, L.B., Wangel, A.: Work in the virtual enterprise - creating identities, building trust, and sharing knowledge. AI Soc. **21**(1), 184–199 (2007). https://doi.org/10.1007/s00146-005-0029-y
32. Sica, F.C., Guimarães, F.G., De Oliveira Duarte, R., Reis, A.J.R.: A cognitive system for fault prognosis in power transformers. Electric Power Syst. Res **127**, 109–117 (2007). https://doi.org/10.1016/j.epsr.2015.05.014
33. Vygotsky, L.S.: Mind in Society: The Development of Higher Psychological Processes. Harvard University Press, Cambridge (1978)

Towards the Digital Transformation: Are Portuguese Organizations in This Way?

Carla Santos Pereira[1,3,5], Fernando Moreira[1,2,3,4(✉)],
Natércia Durão[1,3], and Maria João Ferreira[1,3,6]

[1] DCT, Universidade Portucalense,
Rua Dr. António Bernardino de Almeida 541, 4200-070 Porto, Portugal
{carlasantos,fmoreira,natercia,mjoao}@upt.pt
[2] IJP, Universidade Portucalense, Porto, Portugal
[3] REMIT, Universidade Portucalense, Porto, Portugal
[4] IEETA, Universidade de Aveiro, Aveiro, Portugal
[5] CEMAT, Instituto Superior Técnico, Universidade Técnica de Lisboa,
Lisbon, Portugal
[6] Centro Algoritmi, Universidade do Minho, Braga, Portugal

Abstract. Digital transformation is increasingly an element to be considered in the new paradigm of organizations. This transformation increases business opportunities and open a window for new approaches to the same business, both internally and externally, including changes in business processes and relationships with the organization's stakeholders. Due of this, several concepts like digital transformation and enablers among others have emerged, and occurred in this new era. Being this in mind, it is fundamental to see if organizations in Portugal are already living in the aforementioned digital transformation or if they are aware of the need to adapt to this new reality. Thus, the objective of this research is to perceive the state of the art of Portuguese organizations in light of digital transformation.

Keywords: Digital transformation · Agility · Four pillars ·
Innovation accelerators · Technology · Business processes

1   Introduction

In a time of great technological development and globalization, Digital Transformation (DT) is already a global reality that encompasses a large number of organizations. This transformation becomes disruptive in many areas, namely business, government, and society [1].

The DT is based on four pillars – Mobility, Cloud, Big data & Analytics, and Social media – and it is driving organizations to the next level of digital customer engagement and Information Technology (IT)-enabled business processes, products, and services. In the majority of the organizations, digital technologies are opening unprecedented transformations and changing the ways of working, the learning approaches and the people life in ways, that have never been anticipated [2]. In Rowe [3] a case is presented of the impact of DT's adoption on two well-known companies in the

© Springer Nature Switzerland AG 2019
Á. Rocha et al. (Eds.): WorldCIST'19 2019, AISC 930, pp. 326–336, 2019.
https://doi.org/10.1007/978-3-030-16181-1_31

audiovisual sector, Netflix and Blockbuster. According to the author "*Netflix embraced digital transformation and succeeded while Blockbuster fought it and failed*". The same author asserts that people by nature are reluctant to change, but DT implies profound changes, not only in technology adoption, but mainly in organizational structure, because "*if you don't transform, your competitors surely will, and you could end up like Polaroid, Blockbuster, or Kodak—a diminished or nonexistent brand.*" In conclusion, the same author quotes Daniel Newman, to point out that the main challenge organizations are facing today, regarding digital transformation, "*is building a culture that can change.*"

From the stated, it can be claim that DT, nowadays, is an unavoidable reality, it is not a passing trend. DT is the new reality, and the major reason for the implementation of DT in organizations is the rapid advancement of technology, which allows new business models to be introduced at an ever-increasing rate with rapidly declining costs. The challenges are high with expectations from customers growing. In this context, the only sustainable advantage that an organization can have over their competitors is agility.

According Kane et al. [4], the 21st century is about agility, adjustment, adaptation and creating new opportunities under the digital transformation. Agility in organizations is an integrated ecosystem that can provide a significant boost in customer engagement and value growth in organizations. As mentioned earlier, the technologies allow the emergence of innovative products and services with added value and, on the other hand, that the targeted data analyzes offer more space to leverage new product and service offerings for existing customers, or directed in order to "conquer" new audiences. Since traditional revenue models and business strongholds may face redundancy, digital tools and technologies can be used to explore new revenue sources and plan future-ready portfolio expansion. Technologies in the context of DT take, not only into account the four pillars, but also the broad spectrum of technologies that are at the service of organizations, namely artificial intelligence, virtual and augmented reality, blockchain, among others.

In this regard, the aim of this paper is to investigate the perception of digital technology adoption as a support for business transformation in Portugal and for that was constructed a questionnaire with the title "*Digital Transformation in Portugal*".

2   Background

Digital transformation has become a topic of discussion within the strategic initiative of organizations. According to the Economist Intelligence Unit (EIU), 77% of companies say DT is their first strategic priority. Both native and non-digital companies are working to reinvent today's technologies to the fullest – all to stay competitive and profitable in an increasingly dynamic environment. Customer experience, business agility, and operational efficiency are the primary goals that drive organizations through DT.

In this section, we intend to introduce the following topics, Digital transformation and Organizational agility, the basic concepts of this research work.

2.1 Digital Transformation

Lucas et al. [5], define DT as transformation *"precipitated by a transformational information technology"*. This transformation involves fundamental changes in business processes, operational routines and organizational capacity. However, DT is based on the technological pillars as well as the innovation accelerators, which impose an alignment between IT and business.

In the study presented, in [6], the authors conclude that *"executives are digitally transforming three key areas of their enterprises: customer experience, operational processes and business models"*. However, Gruman [7] defines DT as *"the application of digital technologies to fundamentally impact all aspects of business and society"*.

According to [8], the main challenges of DT are: (1) Priorities; (2) Aggregate data or customize; (3) Providing more resources to IT staff vs. more self-service analytics; (4) Storing all data vs. selecting data to store that serves a specific purpose; (5) Work performed by people vs. computing machines; (6) Security vs. accessibility; (7) Privacy of individuals vs. understanding of an individual.

As mentioned in the previous section, DT is supported not only on four technological pillars, but also other technologies, called innovation accelerators that act as DT drivers. As an example of innovation accelerators, we can refer IoT, Robotics, 3D Printing, Artificial Intelligence, Augmented and Virtual Reality, Cognitive Systems and Next Generation (NextGen) Security and Blokchain. However, such technologies cannot be used without careful consideration of the organization's needs and strategy.

2.2 Organizational Agility and Digital Transformation

Considering the current context, organizations face daily challenges that require them to have a constant capacity for change, often unpredictable in several areas, namely technology, social, legislative, competitiveness and globalization. Thus, to ensure their place in the constantly evolving environment, organizations must be agile and ensure their sustainability through continuous improvement. Organizational agility must therefore be one of the main objectives of any organization [9]. Therefore and according to Sambamurthy et al. [10] *"Firms are increasingly relying on information technologies, including process, knowledge, and communication technologies, to enhance their agility."*

Organizational agility is defined, in the Business Dictionary, as the capability of an organization to rapidly change or adapt in response to changes in the market. As referred a high degree of organizational agility can help an organization to react successfully to the emergence of new competitors, the development of new industry-changing technologies or sudden shifts in the overall market conditions. Further, agility encompasses organizations' capabilities related to interactions with customers, orchestration of internal operations, and utilization of its ecosystem of external business partners. Specifically, agility comprises three interrelated capabilities: customer agility, partnering agility, and operational agility [10].

In this respect, and taking into account what has been presented in Sect. 2, it can be said that the commitment to DT should be done in order to simplify the business and make it more agile. The introduction of new technologies, as they are popular in the industry, will hardly result in a successful digital transition. Companies should evaluate how each system can improve its agility, before making it the cornerstone of a DT plan [11].

In the era of the native digital client and the ever-changing landscape, digital transformation has become one of the most viable strategies to accelerate business activities, processes, business growth, and fully leverage available opportunities. Most companies have gone through some degree of digital transformation. However, simply adding improved software is not, by itself, a significant change.

3 State of the Art

The literature in the area of digital transformation is vast, since 2016. For the construction of the state of the art, the B-on portal (www.b-on.pt) was used, which is an Online Library of Knowledge that provides unlimited and permanent access to thousands of international scientific journals and e-books. The research was carried out for the period of 2016–2018 with the following queries search: (1) *"(Digital AND Transformation AND SMEs AND Portugal)"*; (ii) *"(DIGITAL AND TRANSFORMATION AND SMEs AND Portugal)"*; (iii) *"(Digital AND Transformation AND SMEs AND PORTUGAL)"*. The results obtained indicate a near absence of studies presenting which direction digital transformation is having in SMEs in Portugal. There are only two scientific papers [12, 13], the first one surveys the relationships between the enablers of digital transformation, while the second presents a benchmark of digital transformation best practices in the Tourism industry. None of the papers include a survey of the Digital Transformation in SMEs, independently of the activity area.

In order to ensure that there are already studies performed when the search query is *"(Digital AND Transformation AND SMEs)"*, a search was conducted which proved that there were already 4,250 entries, even though most of them are not directly related to the entire search query. Considering the above mentioned research, it was possible to conclude that there are no studies on Digital Transformation in Portuguese SMEs, thus making it relevant and justified.

4 Research Methodology

The main feature of the scientific method is an organized research, strict control of the use of observations and theoretical knowledge. For the present study, it was used the quantitative research methodology, since it is more appropriate to determine the opinions of the respondents, based on structured questionnaires.

The aim of this study is to investigate the perception of adoption of digital technology as a support for business transformation in Portugal, and for that, a questionnaire was built with the title *"Digital Transformation in Portugal"*. Before being online available, the questionnaire was subjected to an evaluation of four experts in the field.

The questionnaire consists of 3 Sections which include: *"Organization characterization"* (Sect. 1, with four questions), *"Current organization characterization regarding Digital Transformation"* (Sect. 2, with nine questions), and *"Organization's future in relation to Digital Transformation"* (Sect. 3, with three questions). To achieve this paper's main goal, it was only analyze the data from the two first sections to perceive the current state.

For Sect. 1, question A2 ("*What role do you play in the organization?*"), and A4 ("*What is the general feeling of your organization when it comes to technological disruption?*"), the respondents could only choose one of eight available options, and one of the four available options, respectively. Due to General Data Protection Regulation (GDPR), questions A1 and A3 could not be treated. Section 2, questions B1 ("*The organization has explored how Digital Transformation impacts suppliers, distributors and other partners*"), B2 ("*The organization's leadership has considered the costs, savings and return on investment associated with Digital Transformation*") and B3 ("*The organization has, a plan, or strategy, to implement Digital Transformation*") use a five-point Likert scale ranging: "*Strongly disagrees*" (1), "*Disagree*" (2), "*Neutral*" (3), "*Agree*" (4) and "*Strongly Agree*" (5). For questions B4 ("*What is the most important goal of the Digital Transformation strategy in your organization?*"), B5 ("*Who leads the Digital Transformation initiative in your organization?*"), B6 ("*What are the main factors that currently help your organization implement Digital Transformation?*") and B7 ("*What are the biggest obstacles that prevent your organization from implementing Digital Transformation?*"), the respondents could choose more than one option. For the remaining questions, B8 ("*Evaluate the state of the organization's current digital adoption for the following technology categories*") and B9 ("*Classify the various departments of the organization based on their ability to adapt to technological change*") the respondents must classify seven technologies and nine departments within a specific scale. For question B8, the respondents must classify eleven technologies within the following scale ranging: (1) "*Nothing prepared*", (2) "*Unprepared*", (3) "*Prepared*", (4) "*Fully prepared*", and (N/A) "*Not applicable*". Regarding the last question (B9), the respondents must classify nine departments within the following scale ranging: (1) "*Not agile*", (2) "*Not very agile*", (3) "*Agile*", (4) "*Extremely agile*", and (N/A) "*Not applicable*".

The questionnaire was online for 60 days and 77 valid responses were received. Data collected were pooled and treated by using the IBM SPSS Statistics 24.0 software. The statistical analyses [14] used for the data analysis were Descriptive Analysis (frequency analysis, descriptive statistics and graphical representations), Inferential Analysis (Spearman's ordinal correlation) and Reliability Analysis (Cronbach's alpha).

5   Analysis and Discussion of Results

Relatively to question (A2) of Sect. 1, what is highlighted is the Senior executive (27.3%), Senior manager (27.3%), CIO (15.6%) followed by Manager, CEO and General manager.

The general feeling about organization technological disruption (A4) reveals that 61% of the respondents think, that it "*Provides new opportunities to improve business*", being of little relevance the other options ("*Helps in the conquest of new markets*" – 18.2%; "*Eventually the organization will adapt*" – 14.3% and "*Represents a threat to the survival of the organization*" – 6.5%). The high percentage (14.3%) of respondents who say that, eventually, their organization will adapt is worrying.

As the percentage of those who stated that the technological disruption "*Provides new opportunities to improve business*" is so high, we find it pertinent to assess the percentage of individuals who indicated this option according to the role they play in the organization (Fig. 1).

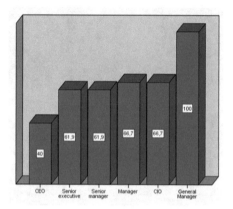

Fig. 1. Percentage of individuals who think that technological disruption "*Provides new opportunities to improve business*" by function in the organization.

As shown at Fig. 1, all General Manager consider that this is the general feeling of their organizations, as well as most of the individuals who perform other functions.

Regarding to questions (B1) and (B2) from Sect. 2, we found that: 68.9% of respondents agree/strongly agree that their organization has explored how DT impacts suppliers, distributors and other partners; 62.4% of respondents agree/strongly agree that their organization's leadership considers the costs, savings, and return on investment associated with implementing DT. It is important to observe that, a considerable percentage of organizations (20.8%) have a neutral opinion on this issue.

Question (B3) allows us to conclude that 63.7% of respondent organizations state that they agree/strongly agree that the organization has a plan, or strategy, to implement DT. It is also important to mention the high percentage of organizations (19.5%) that showed no opinion.

The most important goal of the DT strategy in organization (B4) was "*Reach and engage with customers more effectively*" (39%), followed by "*Modernize legacy IT systems and processes and reduce costs*" (29.9%), and "*Improve business visibility and increase income*" (19.5%).

Figure 2 shows, for each role the respondent plays in the organization, which objectives are pointed out as the most important of the DT strategy applied to your organization.

According to Fig. 2, it is clear that for "*CEOs*" and "*Senior managers*", it is a primary objective to modernize legacy IT systems/processes and reduce costs. For "*CIOs*" and "*Senior executives*", the most important goal pointed out is to reach and engage with customers more effectively. "*General Managers*" consider that achieving

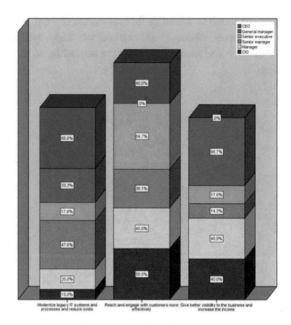

Fig. 2. Percentages of objectives by each of the roles played in the organization

better visibility to the business and increasing the income, it is the most important objective. Finally, for the managers is equally important, in the strategy of DT in their organization to reach and engage with customers more effectively, and achieve better visibility to the business and increase the income.

Concerning who leads the DT initiative in the organization (B5), there is massive leadership from the "*Senior executive*" team and the "*CEOs*".

When questioned regarding the main factors that help implementation of DT (question B6), the "*Leadership Vision*" factor pointed out by 64.9% of the organizations, stands out significantly. It can also highlight the "*Culture of the organization*" (48.1%), the "*Support of the organization's managers*" (42.9%) and the "*Technological partners*" (40.3%) as relevant factors. It should also be mentioned the low result (27.3%), which was surprising for us, regarding the factor "*collaborators with knowledge*".

The most pointed obstacles that prevent organization from implement DT (B7), with approximately equal percentages are the "*Culture of the organization*", and "*Inadequate budgets with the values*" 42.9% and 40.3%, respectively.

The organizations in analysis it was found that the leadership and culture of the organization are the most important factors for the implementation of DT, therefore, it makes perfect sense that the inadequate budget is a relevant obstacle.

The percentage of organizations that indicate that one of the biggest obstacles is "*Managers resistance*", "*Employees do not have the necessary skills*" and "*Confused leadership on what to do*" is similar (between 20% and 30%), percentages that, in our opinion, are worrying. Among the various options to choose the obstacle "*Few technological partners*" was the least chosen (6.5%).

The current state of the organization's digital adoption (B8), on a 4-point Likert scale and N/A (Not Applicable), is evaluated by a set of ten technologies (Table 1). The test of validity and reliability was performed for this items and a high internal reliability was obtained (Cronbach´s alpha = 0,852) [15]. Table 1 shows the mean (*m*) and standard deviation (*sd*) for each item in question (B8).

Table 1. Descriptive measures for items of question (B8)

Items	*m*	*sd*
Mobility	3.21	.778
Cloud solutions	3.18	.770
Big data & analytics	3.16	.861
IoT technology/sensors	2.57	1.003
3D printing	1.68	.872
Virtual reality/augmented reality	1.90	.911
Robotics/automation	2.69	.988
Agile collaboration tools	3.36	.804
AI	2.45	.968
Blockchain	1.80	.943

The technological categories that the respondents most pointed out as a response for (N/A) (Not Applicable), to the state of the organization's current digital adoption, are "*3D Printing*", "*Blockchain*", "*Virtual Reality Technology/Augmented Reality*", "*IoT Technology/Sensors*" and "*Robotics/Automation*". It is interesting to see that it is in these categories that organizations are less prepared, even though these technologies are the main innovation enablers.

In relation to categories which the organizations have a greater degree of preparation, it is verified that the one for which the organizations are more prepared is to "*Agile Collaboration Tools*" followed by "*Mobility*", "*Cloud Solutions*" and "*Big Data & Analytics*", the 4 pillars of DT. In addition to these results, it was also found that there is homogeneity in the responses given by the organizations, since the variation coefficients are low (maximum 27%). The analysis that was made also allows us to identify a respondent outlier, a respondent whose role in the organization (A2) is Specialist, and for all the pillars of DT, attributes the minimum value to the degree of agility of his organization.

We also found that, among the respondents who most agree that their organization has a plan or strategy to implement DT (B3), there are also those who consider that the organization is better prepared with respect to Mobility and "*Big Data & Analytics*" (r_s = 0.424; p-value = 0.000 and r_s = 0.472, p-value = 0.000, respectively).

The results of organizations' assessment of their adaptability to technological change in the various departments of their organization (question B9), are presented in Table 2.

Table 2. Descriptive measures for items of question (B9)

Item	m	sd
IT	3.37	.670
Marketing	3.07	.804
Sales/business development	2.76	.755
Manufacturing/logistics	2.55	.891
Customer service	2.85	.765
Product management	2.68	.850
Human resources	2.54	.958
Legal department	2.32	.918
Retail	2.53	.983

The test of validity and reliability was performed on 9 departments and a very high internal reliability was obtained (Cronbach's alpha = 0,903).

It should be considered that, regarding the adaptability to technological changes, the department of the organization which most points out opinion (N/A) is the retail department (58%). This result is, perhaps, due to the fact that adopting DT does not make much sense in such departments. The departments that have a greater degree of agility in the adaptation to technological change are IT, followed by Marketing. In addition to these results, it was also found that there is homogeneity in the responses given by the organizations, since the variation coefficients are low (maximum 26%).

With regard to the technological categories (B8) that constitute the four pillars of DT, we evaluated the association of each of them with (B9). We conclude that, after the correlation analysis was performed, there is only a moderate positive correlation between Big Data & Analytics technology and the ability to adapt to change in the Marketing department and in the Sales/Business development department. For the technological category "*Agile tools for collaboration*", there is also a moderate positive correlation between this and the ability to adapt to change in the Manufacturing/Logistics, Human Resources, Legal and Retail departments. All these correlations are statistically significant with p-value < 0.1.

6   Conclusions

Digital Transformation is becoming, more and more an expression of the everyday live, due to its relevance for the life of organizations. As a consequence of not observing and integrating their implications, it has led large companies, with a consolidated market, to disappear. This reluctance to change is an intrinsic factor of the human being, and it is not recognized at the deepest level of the change that organizations have to make for DT, because technology adoption is not enough, it has to be deeper, it has to be organizational. In order to understand the perception of Portuguese organizations regarding the adoption of DT, a questionnaire was created.

The results presented and discussed in Sect. 5, have showed that the awareness of the importance of DT in organizations begins to be noticeable, both the importance of organizational awareness and by the adoption of technology. However, we are aware that the sample is small relatively to the high amount of survey questions. These results show that DT in Portugal is still at a relatively mature stage. This conclusion is based, on the one hand, on the existence of a significantly high percentage of responses to organizational awareness that is necessary for this adoption 14.3% of respondents who say that eventually their organization will adapt (A4), 20.8% of organizations that have neutral opinion on this issue (B2), and 19.5% of organizations without opinion (B3) and, on the other hand, on the point of view of the technological adoption by technological inductors/accelerators which is presented at very low adoption values (B8).

As future work, an analysis of the data obtained from Sect. 3 *"Organization's future in relation to Digital Transformation"* will be carried out, in order to understand how Portuguese companies are preparing to respond to this incoming challenge. In addition, this study will be extended broader audience, by evaluating the Iberian Peninsula organizations.

References

1. Schuelke-Leech, B.: A model for understanding the orders of magnitude of disruptive technologies. Technol. Forecast. Soc. Change **129**, 261–274 (2018)
2. Uhl, A., Gollenia, L.: Digital Enterprise Transformation: A Business-Driven Approach to Leveraging Innovative IT. Routledge, Taylor & Francis Group (2016)
3. Rowe, S.: Digital transformation needs to happen: the clock is ticking for companies that have been unwilling to embrace change, Customer Relationship Management, pp. 30–33 (2017)
4. Kane, G., Palmer, D., Phillips, A., Kiron, D., Buckley, N.: Strategy, not technology, drives digital transormation: becoming a digitally mature enterprise, MIT Sloan Management Review and Deloitte University Press (2015)
5. Lucas, H., Agarwal, R., Clemons, E., El Sawy, O., Weber, B.: Impactful research on transformational information technology: an opportunity to inform new audiences. MIS Q. **37**(2), 371–382 (2013)
6. Westerman, G., Bonnet, D., McAfee, A.: The nine elements of digital transformation. MIT Sloan Manag. Rev. **55**(3), 1–6 (2014)
7. Gruman, G.: What digital transformation really means. InfoWorld. (2016). http://www.infoworld.com/article/3080644/it-management/what-digital-transformationreally-means.html. Accessed: 10 Oct 2018
8. Heavin, C., Power, D.: Challenges for digital transformation –towards a conceptual decision support guide for managers. J. Decis. Syst. **27**(1), 38–45 (2018)
9. Imache, R., Izza, S., Ahmed-Nacer, M.: Clustering-based urbanisation to improve enterprise information systems agility. Enterp. Inf. Syst. **8**(9), 861–877 (2015)
10. Sambamurthy, V., Bharadwaj, A., Grover, V.: Shaping agility through digital options: reconceptualizing the role of information technology in contemporary firms. MIS Q. 237–263 (2003)
11. Kane, G.C., Palmer, D., Phillips, A.N., Kiron, D., Buckley, N.: Strategy, not technology, drives digital transformation - becoming a digitally mature enterprise. MIT Sloan Manag. Rev. Deloitte Univ. Press **14**, 1–25 (2015)

12. de Mendonca, C., de Andrade, A.: Microfoundations of dynamic capabilities and their relations with elements of digital transformation in Portugal. In: Proceedings of the 13th CISTI, vol. 1, pp. 1–6 (2018)
13. Pereira, N.: Digital transformation at turismo de Portugal (Doctoral dissertation) (2017)
14. Maroco, J.: Análise Estatística com o SPSS, 7a edição. ed. ReportNumber, Lda (2018)
15. Pestana, M., Gageiro, J.: Análise de dados para Ciências Sociais. a complementaridade do SPSS, 6a Edição. ed. Edições Sílabo, Lisboa (2014)

Validation and Evaluation of the Mapping Process for Generating Ontologies from Relational Databases

Bilal Benmahria, Ilham Chaker$^{(\boxtimes)}$, and Azeddine Zahi

Faculty of Science and Technology, 2202 Fez, Morocco
{bilal.benmahria,ilham.chaker,
azeddine.zahi}@usmba.ac.ma

Abstract. In this paper, we propose a generalized approach for mapping a relational database into ontology. This proposed approach is composed of two main phases. The first one consists of generating the ontology from the relational database, based on a direct mapping process by using R2RML language. The second one is devoted to evaluating the correctness of the generated ontology, with regards to two validation criteria, Information preservation, and Query preservation. In addition, we evaluate the computational performance of the proposed mapping process in terms of storage space and execution time.

Keywords: Ontology · Direct mapping · R2RML

1 Introduction

During the last decade, the growing interest in the semantic web has attracted the attention of the database research community, since a huge amount of data information on the web is stored in Relational Database (RDB). This is also associated with the fact that most of these data information is heterogeneous [1]. Therefore, the scientific community infers that the success of the semantic web is strongly related to the issues of preserving adaptability and consistency with Relational Database Management Systems (RDBMS) [2]. For this reason, the mapping of a relational database to ontology has become one of the most challenging tasks in the semantic web, where the goal is to provide the user with a consolidated view of information residing in different sources, through which these data sources can be represented as a global target schema by using ontologies [3].

Formally, generating ontologies from a relational database, which is also named a mapping process, refers to the process of extracting metadata from database models and generating their corresponding rules according to the ontology language [4]. In this respect, the W3C RDB2RDF (Relational database to Resource Description Framework) Working Group defined two recommendations: Direct Mapping (DM) [5] and RDB to RDF Mapping Language (R2RML) [6]. The former provides a set of fixed transformation rules and defines an ontology representation for the data in the relational database. The latter is a language for defining a customized direct mapping. In fact, such mapping technique does not only provides the capacity to define the traditional

© Springer Nature Switzerland AG 2019
Á. Rocha et al. (Eds.): WorldCIST'19 2019, AISC 930, pp. 337–350, 2019.
https://doi.org/10.1007/978-3-030-16181-1_32

direct mapping but also, add a set of knowledge encoded in another source of information and target vocabularies, such as Wordnet [7], CIDOC CRM [8] and Music-Brainz [9].

The DM approach has been adopted by many research studies [10–16], where the SQL Definition Description Language (DDL) script is mechanically transformed into an ontology and the relational data is exposed as instances of the generated ontology. While, the R2RML has recently become a standard language for building ontology from RDB, and has been extensively studied in the literature [6, 17–19]. The reason for using R2RML is justified by the powerful support of manipulating the M:N relationships, the project attributes and the select conditions, etc., more detailed features can be found in [20].

Despite the significant progress made during the last few years and the wide number of proposed approaches, there are still many issues that has not been sufficiently addressed. First, no correctness criteria for validating the resulted ontology has been investigated (i.e. information preservation and query preservation). Second, no effective study with respect to the computational aspects and storage requirement has been reported. To the best of our knowledge, there are only very few papers have appeared with this concern [21, 22]. Tarasowa et al. [21] proposed a novel method for measuring the quality of mapping between RDB to Ontology based on 14 requirements and described ways for measuring them. Sequeda et al. [22] introduced an effective approach for validating the mapping process with regard to four properties. Nevertheless, those studies mainly focused on the validation and evaluation of the mapping process, while they have largely ignored the performance measurement as the storage requirement and the generation time of the resulted ontology. In addition, they did not indicate which type of mapping is used, and also the mapping process has not been clearly stated.

Motivated by the issues encountered in mapping relational databases to ontology. In this paper, we propose a generalized approach for mapping the relational database to ontology. This proposed approach is composed of two main components: The first one consists of generating the ontology from the relational database, based on a direct mapping process by using R2RML language. The second one is devoted to evaluating the correctness of the generated ontology, with regards to two validation criteria, Information preservation and Query preservation. Consequently, several numerical experiments are conducted in this study, to validate the effectiveness of the proposed approach. Initially, the correctness of the generated ontology is investigated. Then, we evaluate the performance of the mapping process in terms of storage space and execution time. Furthermore, these experiments are performed on four of the most commonly used databases.

The remainder of this paper is organized as follows. In Sect. 2, we introduce two recommendations for mapping the relational database into ontology. In Sect. 3, we present correctness criteria for measuring the quality of the resulted ontology. In Sect. 4, we detail our proposed approach and algorithms. Section 5 contains the experimental results and discussions. Finally, Sect. 6 contains a short background of the important points emphasized in this paper and suggests some future works.

2 From a Relational Database to Ontology: Recommendations

The W3C foundation defined two recommendations for mapping the relational database into ontology: A Direct Mapping [5] and R2RML [23].

2.1 A Direct Mapping (DM) Description

The purpose of DM is to specify a mapping of a relational database to RDF and OWL in a straightforward manner with no user interaction in the mapping process. As depicted in Fig. 1, the direct mapping takes as input a relational database, including schema and instances and generates an ontology that is called the direct labeled graph. Accordingly, to the W3C, a direct mapping defines a set of fixed rules to be applied for generating automatically an RDF labeled graph that reflects the structure and content of the relational database [5]. A direct mapping maps relational tables to classes in an RDF Vocabulary, relational attributes to properties in that vocabulary accompanied by some heuristics to figure out the other rules. In the literature, Direct Mapping is often used as a synonym for Automatic Mapping, Local Ontology mapping, or Ad-hoc ontology mapping.

2.2 An R2RML Mapping Description

R2RML is the mapping language proposed by the W3C RDB2RDF Working Group in 2009, to standardize the RDB-To-RDF mappings. It offers more flexibility characteristics than the traditional direct mapping. The latter does not support the vocabulary reuse, a logical table to class, a project attributes as well as a select condition features, whilst the former covers all of these features in addition to others features that are described in details in [20]. Consequently, we can use the R2RML mapping language to define a direct mapping. In this case, the structure of the resulting ontology reflects the structure of the relational database. On the other hand, with R2RML we can define a customized mapping from the relational database to ontology. As can be seen in Fig. 2, after generating the R2RML file, the process engine such as Morph-RDB [24], Nknos [25], Ontop [26], can be used to convert R2RML content to RDF triples.

In brief, the Direct Mapping is the usual choice when the purpose is to create data sources in a machine-readable format but when the aim is to support efficient integration between information systems, the direct mapping must be customized by reusing vocabularies, and other sources of information such as WordNet.

3 Correcteness Criteria for the Direct Mapping

To measure the quality of the generated from the direct mapping, two fundamental properties are adopted: Information preservation, and Query Preservation.

3.1 Information Preservation

Information preservation addresses the capacity to reconstruct the original relational database from the resulted ontology. It is a fundamental property for verifying if the mapping does not lose information. Concisely, Information preservation consists of finding a way to retrieve the relational database individuals from the resulted ontology [27].

3.2 Query Preservation

A direct mapping is a query preserving implies that each query over a relational database can be converted into an appropriate query over the resulted ontology. Therefore, to specify a query preservation, the emphasis is on relational database queries that can be defined in SQL language and RDF queries that can be defined in SPARQL language. Thus if the result of SQL query is equal to the result of SPARQL, then the direct Mapping is query preserving [27].

4 Our Proposed Approach and Algorithms

In this section, we present an approach that deals with the automatic mapping of the RDB to ontology. As depicted in Fig. 1, our proposed approach is not only for generating the ontology from the RDB but also validating the mapping process. Therefore, our process is composed mainly of Database Metadata Extraction Engine (DMEE), Mapping Generator Engine (MGE) and R2RML Engine.

Firstly, the system receives as an input SQL file that contains statements represented by SQL DDL (Data Definition Language). Secondly, the DMEE analyzes this file and extracts automatically the metadata from it. The extracted metadata includes tables, columns, primary keys (PKs), and Foreign Keys (FKs). Secondly, MGE exploits the extracted metadata and build a mapping file (R2RML file). Lastly, R2RML engine takes as input, the database model(Schema + Instances) and the generated mapping document that contains a set of rules representing the database schema, then provides an output represents the RDF dataset (triples). Several tools for generating ontology from the R2RML mapping file and database model are proposed in the literature such as morph-RDB [24] and Nknos [25], etc. The R2RML Engine used in our approach is r2rml-kit-master.

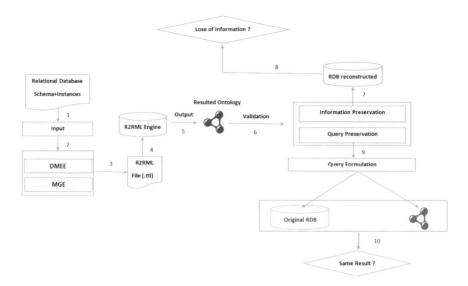

Fig. 1. Our proposed approach for generating ontology from relational database.

4.1 Database Metadata Extraction Engine(DMEE)

Our process for generating metadata from a relational database consists of four steps:

Step1: Identification of tables info. We introduce getTableInfo method that consists of extracting the name of tables residing in the relational database.

Step2: Identification of Column info. The method getColumnInfo is programmed for extracting the column information such as column name, type, and verify if the column is not null for each column.

Step3: Identification of primary keys. The method getPrimaryKeys returns for each table, a list of its primary keys.

Step4: Identification of foreign keys. The method getForeignKey returns for each table, a list of its foreign keys.

4.2 Mapping Generator Engine (MGE)

In this phase, the R2RML triples maps that correspond to the metadata identified in the previous phase are produced. Concisely, the main steps for implementing the MGE is depicted in the following algorithm.

```
Algorithm: Mapping Generator Engine (MGE)
Input  : metadata;
output: R2RML Mapping file;
var:
       triplesMap: TriplesMap;
       collectionMap: Collection<TriplesMap>;
       metadata: DButils;
       table_names: ArrayList<String>;
       colInfo: Map<String, List<TableComponent>>
   Begin
 1:    table_names←metadata.getTablesInfo();
 2:    for each i<table_names.size();
 3:       logicalTable←mf.createTableOrView()
 4:       colInfo←data.getColumnInfo(tbl_names.get(i))
 5:       for each key in colInfo.KeySet() do
 6:          for each k < colInfo.get(key).size()
 7:             if(metadata.isPrimarykey()) then
 8:                templat←mf.createTemplate()
 9:                subjetcMap←mf.createSubjectMap(template)
10:             else
11:                pm←mf.createPredicateMap("ex:"+colInfo.getColName);
12:                objectMap←mf.createObjectMap(colInfo.getColumnName());
13:                prdicateObjectMap←mf.createPredicObjectMap(pm,oMap);
14:                triplesMap←mf.createTriplesMap(ltable,sMap,pOMap);
15:             if(metadata.isForeignKey())
16:                tmap←getRefObjetMap(getPtb, getPtC,getCht(),getChC());
17:       triplesMap.setResource(createResource(url));
18:       collectionMap.add(triplesMap);
19:       R2RMLFileMapping←write(collectionMap)
20:    End
```

- **LogicalTable**: is the SQL table that the RDF triples will be generated from. The logical table can either be an SQL table, an SQL view, or an R2RML view.
- **Template**: a template consists of a series of string segments and column names. A template is useful to create URIs based on the values of one or more columns.
- **SubjectMap**: A subject map that generates the subject of all RDF triples that will be generated from a logical table row. The subjects often are IRIs that are generated from the primary key column(s) of the table.
- **PredicateMap**: A predicate map maps a column of the table to a predicate in the RDF graph. There may be several predicate maps in order to specify more predicates.
- **ObjectMap**: An object map maps the values of the database to objects in the RDF graph. A referencing object map is useful to map foreign key relations.
- **PredicateObjectMap**: A predicate object map specifies predicates and objects of the RDF triples that will be generated. A predicate object map has at least one predicate map and one object map.
- **TriplesMap**: The triples map is the main part of an R2RML mapping. A triples map specifies RDF triples corresponding to a logical table.
- **DBUtils**: Is the class that contains the DMME.

5   Experimental Results and Discussion

In this section, several experiments are provided to validate the proposed approach. Therefore, this section is divided into two sub-sections. In the first one, we describe some experimental studies to validate the quality of the resulted ontology. In the second one, a set of appropriate experiments have been arranged in order to study the performance of the mapping process, developed in the previous sections. We selected four real-world databases which covering various of important factors used in the mapping process such as tables, views, columns, foreign keys (FKs) and primary keys (PKs). As depicted in Fig. 2, these factors represent the important metadata extracted automatically from RDB by our algorithms for building R2RML mapping file. The information detailed of these data sets are summarized in Table 1. The two fields namely size and instances show respectively the size of the RDB (Schema + Rows) file and the number of rows for each dataset. An experimental prototype is implemented on eclipse photon 4.8.0 (J2SE, JDK 1.8), MYSQL Workbench 8.0, and Apache Jena 2.9.4 api. Finally, it is important to note that all the experimental simulations were conducted on a personal computer under Windows 10, with Intel core i7 2.70 GHz processor and 16 GB RAM.

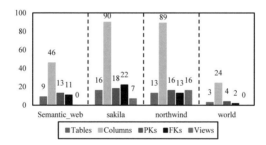

Fig. 2.   Important metadata extracted from RDBs to build R2RML mapping file.

Table 1.   A list of metadata extracted from RDB

RDB	Size (kb)	Tables	Columns	PKs	FKs	Views	Instances
Sakila	40375	16	90	18	22	7	47273
Northwind	506	13	89	16	13	16	3310
world	240	3	24	4	2	0	5302
Iswc (sematic_web)	23,5	9	46	13	11	0	79

5.1   Validation of the Resulted Ontology

To demonstrate the quality of our resulted ontology, two correctness criteria are adopted: Information preservation and Query preservation. The information preservation ensures that the mapping does not lose information, where query preservation guarantees that every piece of information that can be extracted from the relational database by SQL query, can also be extracted from the resulting ontology by a SPARQL query.

Table 2. Different queries over the used four datasets in natural language, SQL language and SPARQL language.

	Query	SQL	SPARQL
ISWC	**Q1**: What is the title of the document 8	select papers.title from papers where paperId=8;	select ?title where { :8 dc:title ?title. }
	Q2: what is the name of the conference that accepts the papers with the Semantic Web as concept	Select conferences.name from conferences, papers,topics, rel_paper_topic where papers.conference=conferences.confId And rel_paper_topic.PaperID=papers.PaperID And rel_paper_topic.TopicID=topics.TopicIDand topics.TopicName="Semantic Web";	select ?conference where { ?document conf:conference ?conference. ?document skos:subject ?concept. ?concept skos:prefLabel "Semantic Web".}
	Q3: Find out all persons that belong to the Department of Computer Science and their research of interest is a semantic web.	select persons.FirstName, persons.LastName from persons, organizations,topics, rel_person_topic, rel_person_organization where persons.PerID=rel_person_organization.PersonID And rel_person_organization.OrganizationID = organizations.OrgID And organizations.Name= "Department of Computer Science" And persons.PerID=rel_person_topic.PersonID And rel_person_topic.TopicID = topics.TopicID And topics.TopicName="Semantic Web";	select ?names where{?persons foaf:name ?names. ?persons conf:has_affiliation ?organization. ?organization rdfs:label Department of Computer Science. ?persons conf:research_interests ?concept. ?concept skos:prefLabel "Semantic Web".}
	Q4: Find out all persons that belong to the Department of Computer Science and their research of interest is a semantic web.	select persons.FirstName,persons.LastName, organizations.Name, topic.topicName from persons, organizations,topics, rel_person_topic, rel_person_organization where persons.PerID=rel_person_organization.PersonID And rel_person_organization.OrganizationID=organizations.OrgID And organizations.Name="Department of Computer Science" And persons.PerID=rel_person_topic.PersonID And rel_person_topic.TopicID = topics.TopicID And topics.TopicName="Semantic Web";	select ?n ?orga ?con where{?doc dc:creator ?persons. ?persons foaf:name ?n. ?persons conf:has_affiliation ?orga. ?organization rdfs:label "Department of Computer Science". ?persons conf:research_interests ?con." ?con skos:prefLabel "Semantic Web". }
	Q5: Find out all the titles and subjects of the papers.	select title, subject from papers.	select ?title ?subject where { ?document dc:title ?title. ?document skos:subject ?concept. ?concept skos:prefLabel ?subject.}
Sakila	**Q1**: which actors have the first name = 'Scarlett'.	select last_name from actor where first_name = 'Scarlett';	select ?last_names where { ?person rdf:type actor. ?person j.0:actor_first_name "SCARLETT". }
	Q2: which actors have appeared in the most films	select actor.actor_id, actor.first_name, actor.last_name, count(actor_id) as film_count from actor join film_actor using (actor_id) group by actor_id order by film_count desc limit 1;	Select ?actor_id ?firstname ?lastname (COUNT(?actor_id) AS ?filmCount) Where{ ?person rdf:type j.0:actor. ?person j.0:actor_actor_id ?actor_id. ?person j.0: actor_first_name ?firstname. ?person j.0:actor_last_name ?lastname. ?actor_id j.0:film_actor_actor_id ?film_actor_id.} Group By ?film_actor_id ?firstname ?lastname ORDER By DESC(?filmCount) LIMIT 1;
	Q3: which copies at store 1	select film.film_id, film.title, store.store_id, inventory.inventory_id from inventory join store using (store_id) join film using (film_id) where film.title = 'Academy Dinosaur' and store.store_id = 1;	select ?film_id ?film_title ?store_id ?inventory_id where { ?film rdf:type j.0:film. ?store rdf:type j.0: store. ?film j.0:film_film_id ?film_id. ?film j.0:film_title "Academy Dinosaur". ?film_id j.0:inventory_film_id ?iv_id. ?store_id j.0:inventory_store_id "1 ".}
	Q4: List all English movies	select film.title,language.name from film,language where film.language_id=language.language_id and language.name="English" ;	select ?title ?langauge where { ?film rdf:type j.0:film. ?film j.0:film_title ?title. ?film j.0:film_language_id ?lan. ?lan j.0:language_name "English".}
	Q5: what is the rental duration of the film 1	select rental_duration from film where film_id = 1;	select ?duration where { ?film rdf:type j.0:film. ?film j.0:film_id "1". ?film j.0:film_rental_duration ?duration.}
Northwind	**Q1**: get current Product list (Product ID and name).	select ProductID, ProductName FROM Products WHERE Discontinued = "False" ORDER BY ProductName;	select ?id ?n where { ?product rdf:type j.0:products. ?product j.0: products_ProductName ?n. ?product j.0:products_ProductID ?id.}
	Q2: get Product name and quantity/unit.	select ProductName, QuantityPerUnit FROM Products;	Select ?Prn ?Qpunit where { ?product rdf:type j.0:products. ?product j.0:products_ProductName ?Prn. ?product j.0:products_QPerUnit ?Qpunit.}
	Q3: get most expense and least expensive Product list (name and unit price).	select ProductName, UnitPrice FROM Products ORDER BY UnitPrice DESC;	Select ?Prn ?Up where{ ?product rdf:type j.0:products. ?product j.0:products_ProductName ?Prn. ?product j.0:products_UnitPrice ?Up.} ORDER BY DESC(?UnitPrice)
	Q4: Write a query to get Product list (id, name, unit price) where current products cost less than $20.	SELECT ProductID, ProductName, UnitPrice FROM Products WHERE (((UnitPrice)<20) AND ((Discontinued)=False)) ORDER BY UnitPrice DESC;	Select ?p_id ?pn ?pup where { ?product rdf:type j.0:products. ?product J.0:products_ProductId ?p_id. ?product j.0:products_ProductName ?pn. ?product j.0:products_UnitPrice ?pup. ?product j.0:products_Discontinued "False". Filter(?p_unitPrice<20).} ORDER BY DESC(?p_unitPrice)
	Q5: count current and discontinued products.	SELECT Count(ProductName) FROM Products GROUP BY Discontinued	Select ?dis ?p_id (COUNT(?p_id)As ?productCount) Where{ ?product rdf:type j.0:products. ?product j.0:products_ProductId ?p_id. ?product j.0:products_Discontinued ?dis.} GROUP BY ?dis ?p_id

(continued)

Table 2. (*continued*)

Q1: get the number of cities in China.	SELECT COUNT(*) FROM world.city WHERE CountryCode ="CHN"	Select ?china Count((?china) As ChinaCities) where{?china rdf:type j.0:city. ?china j.0:city_CountryCode "CHN".}
Q2: List all countries and their language	select country.name,countrylanguage.Language from country, countrylanguage where country.code=countrylanguage.countrycode	Select ?country ?l Where {?c df:typ j.0:country. ?c j.0:country_Name ?country. ?country j.0:country_Code ?co. ?co j.0:countrylanguage_Language ?l.}
Q3: what country has the highest population	select Population, Name FROM world.country WHERE Population IS NOT NULL ORDER BY Population DESC Limit 1;	Select ?pop ?Name where { ?country rdf:type j.0:country. ?country j.0:country_Population ?pop. ?country j.0:country_Name ?Name.} ORDER BY DESC(?population)
Q4: List all cities that have Zambia as country.	select city.name from city, country where country.code=city.countrycode and country.name="Zambia"	Select ?name where { ?city rdf:type j.0:city. ?city j.0:city_Name ?name. ?city j.0:city_CountryCode ?code. ?code j.0:country_Code ?c. ?c j.0:country_Name "Zambia".}
Q5: List All cities that have countryCode=AFG.	select city.name from city where CountryCode="AFG";	Select ?name where { ?city rdf:type j.0:city. ?city j.0:city_Name ?name. ?city j.0:city_CountryCode "AFG".}

Concisely, the resulted ontology is said to be information preserving if all instances including in the relational database are also included in the resulted ontology. Therefore, to prove the information preservation criteria, a reconstruction of the relational database from the resulted ontology is needed. It is important to note that the process of the reconstruction is automatic. Accordingly, Fig. 3 shows the number of instances of the original database and of the reconstructed database. It is evident from the results of Fig. 3 to conclude that the instances of the original database are included in the reconstructed database, and therefore the newly introduced mapping approach is information preserving.

In order to prove that our resulted ontology is query preserving, for each dataset we formulate five complicated queries. Specifically, the queries over the dataset are summarized in Table 2. The purpose of this table is to show the ability to query relational database by using as a context the resulted ontology. As depicted in this table, every query over a Relational Database can be translated into an equivalent query over the result of the direct mapping. In addition, to bring a significant reason about the query preservation of resulted graph, we must compare the result generated by the SQL query and the other result generated by the SPARQL query. For this, Table 3 contains the result provided by the two query languages presented in Table 2. Consequently, we can note that the two query languages provide the same result.

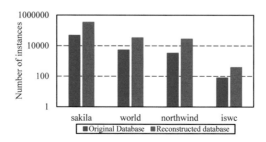

Fig. 3. The number of instances generated from the original database and reconstructed database.

Table 3. The results of the SQL and SPARQL queries presented in Table 2.

	Query	Original Database Result	Resulted ontology Result
ISWC	Q1	D2R Server - Publishing Relational Databases on the Web as SPARQL Endpoints	D2R Server - Publishing Relational Databases on the Web as SPARQL Endpoints
	Q2	International Semantic Web Conference 2002, Sardinia.	International Semantic Web Conference 2002, Sardinia.
	Q3	Borys Omelayenko.	Borys Omelayenko.
	Q4	"Borys Omelayenko" <http://data.example.com/organization/5> <http://data.example.com/concept/5>	"Borys Omelayenko" | <http://data.example.com/organization/5> | <http://data.example.com/concept/5>
	Q5	Integrating Vocabularies: Discovering and Representing Vocabulary Maps Three Implementations of SquishQL, a Simple RDF Query Language Three Implementations of SquishQL, a Simple RDF Query Language ...	Integrating Vocabularies: Discovering and Representing Vocabulary Maps Three Implementations of SquishQL, a Simple RDF Query Language Three Implementations of SquishQL, a Simple RDF Query Language ...
Sakila	Q1	DAMON BENING	DAMON BENING
	Q2	107 GINA DEGENERES	107 GINA DEGENERES
	Q3	1 ACADEMY DINOSAUR 1 1 1 ACADEMY DINOSAUR 1 2 1 ACADEMY DINOSAUR 1 3 1 ACADEMY DINOSAUR 1 4	1 ACADEMY DINOSAUR 1 1 1 ACADEMY DINOSAUR 1 2 1 ACADEMY DINOSAUR 1 3 1 ACADEMY DINOSAUR 1 4
	Q4	'ACADEMY DINOSAUR', 'English' 'ACE GOLDFINGER', 'English' 'ADAPTATION HOLES', 'English' 'AFFAIR PREJUDICE', 'English' ...	'ACADEMY DINOSAUR', 'English' 'ACE GOLDFINGER', 'English' 'ADAPTATION HOLES', 'English' 'AFFAIR PREJUDICE', 'English' ...
	Q5	6	6
Northwind	Q1	3 Aniseed Syrup 40 Boston Crab Meat 33 Geitost 15 Genen Shouyu ...	3 Aniseed Syrup 40 Boston Crab Meat 33 Geitost 15 Genen Shouyu ...
	Q2	Chai 10 boxes x 20 bags Chang 24 - 12 oz bottles Mishi Kobe Niku 18 - 500 g pkgs. Ikura 12 - 200 ml jars ...	Chai 10 boxes x 20 bags Chang 24 - 12 oz bottles Mishi Kobe Niku 18 - 500 g pkgs. Ikura 12 - 200 ml jars ...
	Q3	Sir Rodney's Marmalade 81.0000 Carnarvon Tigers 62.5000 ...	Thringer Rostbratwurst 123.7900 Sir Rodney's Marmalade 81.0000 Carnarvon Tigers 62.5000 ...
	Q4	57 Ravioli Angelo 19.5000 44 Gula Malacca 19.4500 2 Chang 19.0000 36 Inlagd Sill 19.0000 40 Boston Crab Meat 18.4000 39 Chartreuse verte 18.0000 1 Chai 18.0000 ...	57 Ravioli Angelo 19.5000 44 Gula Malacca 19.4500 2 Chang 19.0000 36 Inlagd Sill 19.0000 40 Boston Crab Meat 18.4000 39 Chartreuse verte 18.0000 1 Chai 18.0000 ...
	Q5	'69' '8'	'69' '8'
World	Q1	363	363
	Q2	Aruba Dutch Aruba English Aruba Papiamento Aruba Spanish Afghanistan Balochi	Aruba Dutch Aruba English Aruba Papiamento Aruba Spanish Afghanistan Balochi ...
	Q3	'1277558000', 'China'	'1277558000', 'China'
	Q4	Lusaka Ndola Kitwe ...	Lusaka Ndola Kitwe ...
	Q5	'Kabul' 'Qandahar' 'Herat' ...	'Kabul' 'Qandahar' 'Herat' ...

As a final comment, the information preservation and query preservation are comparable properties. If a direct mapping is information preserving, it guarantees that every SQL query translated into SPARQL query provides the same result. Nevertheless, finding the equivalent SPARQL query based on the SQL still the main challenge for demonstrating that the direct mapping is query preserving.

5.2 Computational Performance of the Mapping Process

To show the efficiency and the robustness of the proposed approach, it is important to discuss its computational performance with respect to execution time and storage requirement. Figure 4 depicts the execution time in second (s) taken by the proposed method on each database, where we present the time for generating the ontology and the number of instances of corresponding databases. As shown in this Fig. 4, The execution time on each database increases with its number of instances.

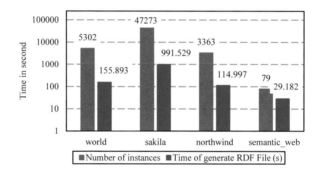

Fig. 4. The execution time of the generated RDF file and number of instances for each database

To further illustrate the performance of the proposed method on different databases, Fig. 5 (4 figures) shows the execution time in second for an increasing number of instances. From the results of Fig. 5, we can notice a strong relationship between the execution time of the generated ontology and the number of instances in each data set, as the number of instances increases the execution time increases.

It is commonly known that an efficient size of the resulted ontology should offer a best performance for processing and retrieving the data. Therefore, the performance of storage requirement depends on various factors such as the size of the used dataset, the efficiency of the query engine, and the performance of the inference engine [28]. In fact, we are interested in this sub-section to the first criteria, which is the size of the data set. Figure 6 displays a comparison in terms of memory space kilobyte (Kb) required for storing the generated RDF file and the original relational database.

From this figure, one can observe that the size of the generated RDF file is relatively greater than the size of the original RDB. However, it is important to note that this augmentation in size is not higher than 6% over for larger databases, like Sakila. Showing that for larger databases the sizes of the RDB and the RDF file become very comparable. Finally, we can conclude that our proposed approach preserves the information, but in terms of performance, we need to search the best solution for optimizing the execution time and the storage requirement.

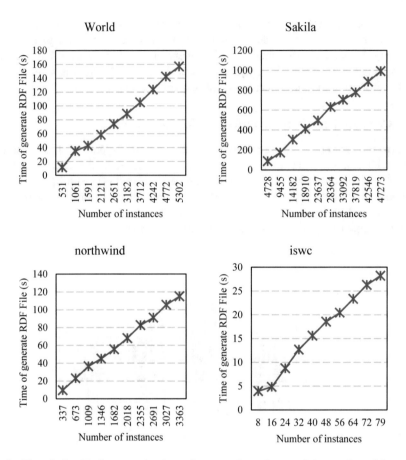

Fig. 5. The relationship between the time of generated ontology and the number of instances

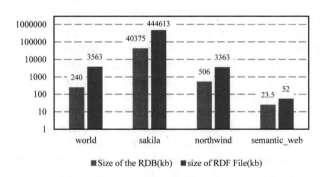

Fig. 6. Illustration of the size of resulted ontology and its corresponding database

6 Conclusion and Future Work

In this paper, we proposed a generalized approach for generating ontology from a relational database. This approach is composed of two main components. The first one consist of mapping the relational database into ontology by using R2RML language. The second one is devoted to evaluating the quality of the generated ontology based on two fundamental criteria: Information Preservation and Query Preservation. As far as we know, our proposed approach is the first one that combine these two components, in order to indicate that the mapping process alone is not sufficient, it must be followed by the validation process. To show the efficiency and the robustness of the proposed approach, the experimental studies have been divided into two sections. The first section consisted of proving that our proposed approach is information preserving and Query preserving. The second section has been devoted to discussing computational performance with respect to the storage requirement and the execution time. In future works, we aim to add an optimization component to enhance the RDF data storage of the proposed approach. Also, we plan to focus on the query optimization techniques to find the best method that allows extracting information in an efficient and optimal way. Finally, we plan to move toward the unstructured dataset for constructing ontologies.

References

1. De Giacomo, G., Lembo, D., Lenzerini, M., Poggi, A., Rosati, R.: Using ontologies for semantic data integration. In: A Comprehensive Guide Through the Italian Database Research Over the Last 25 Years, pp. 187–202. Springer (2018)
2. Spanos, D.-E., Stavrou, P., Mitrou, N.: Bringing relational databases into the semantic web: a survey. Semant. Web **3**, 169–209 (2012)
3. Touma, R., Romero, O., Jovanovic, P.: Supporting data integration tasks with semi-automatic ontology construction. In: Proceedings of the ACM Eighteenth International Workshop on Data Warehousing and OLAP, pp. 89–98. ACM (2015)
4. Myroshnichenko, I., Murphy, M.C.: Mapping ER schemas to OWL ontologies. In: 2009 IEEE International Conference on Semantic Computing, ICSC 2009, pp. 324–329. IEEE (2009)
5. Arenas, M., Bertails, A., Prud'hommeaux, E., Sequeda, J.: A direct mapping of relational data to RDF. W3C Recomm. **27** (2012)
6. de Medeiros, L.F., Priyatna, F., Corcho, O.: MIRROR: automatic R2RML mapping generation from relational databases. In: International Conference on Web Engineering, pp. 326–343. Springer (2015)
7. Suchanek, F.M., Kasneci, G., Weikum, G.: Yago: a large ontology from wikipedia and wordnet. Web Semant.: Sci. Serv. Agents World Wide Web **6**, 203–217 (2008)
8. Bountouri, L., Gergatsoulis, M.: The semantic mapping of archival metadata to the CIDOC CRM ontology. J. Arch. Organ. **9**, 174–207 (2011)
9. Swartz, A.: Musicbrainz: a semantic web service. IEEE Intell. Syst. **17**, 76–77 (2002)
10. Dadjoo, M., Kheirkhah, E.: An approach for transforming of relational databases to OWL ontology. arXiv preprint arXiv:1502.05844 (2015)
11. Sedighi, S.M., Javidan, R.: A novel method for improving the efficiency of automatic construction of ontology from a relational database. Int. J. Phys. Sci. **7**, 2085–2092 (2012)

12. Cullot, N., Ghawi, R., Yétongnon, K.: DB2OWL: a tool for automatic database-to-ontology mapping. In: SEBD, pp. 491–494 (2007)

13. Sane, S.S., Shirke, A.: Generating OWL ontologies from a relational databases for the semantic web. In: Proceedings of the International Conference on Advances in Computing, Communication and Control, pp. 157–162. ACM (2009)

14. Buccella, A., Penabad, M.R., Rodriguez, F.J., Farina, A., Cechich, A.: From relational databases to OWL ontologies. In: Proceedings of the 6th National Russian Research Conference (2004)

15. Zhang, H., Diao, X., Yuan, Z., Chun, J., Huang, Y.: EVis: a system for extracting and visualizing ontologies from databases with web interfaces. In: 2012 International Symposium on Information Science and Engineering (ISISE), pp. 408–411. IEEE (2012)

16. Tirmizi, S.H., Sequeda, J., Miranker, D.: Translating SQL applications to the semantic web. In: International Conference on Database and Expert Systems Applications, pp. 450–464. Springer (2008)

17. Kyzirakos, K., Vlachopoulos, I., Savva, D., Manegold, S., Koubarakis, M.: GeoTriples: a tool for publishing geospatial data as RDF graphs using R2RML mappings. In: TC/SSN@ISWC, pp. 33–44 (2014)

18. Vidal, V.M.P., Casanova, M.A., Neto, L.E.T., Monteiro, J.M.: A semi-automatic approach for generating customized R2RML mappings. In: Proceedings of the 29th Annual ACM Symposium on Applied Computing, pp. 316–322. ACM (2014)

19. Hazber, M.A., Li, R., Xu, G., Alalayah, K.M.: An approach for automatically generating R2RML-based direct mapping from relational databases. In: International Conference of Young Computer Scientists, Engineers and Educators, pp. 151–169. Springer (2016)

20. Hert, M., Reif, G., Gall, H.C.: A comparison of RDB-to-RDF mapping languages. In: Proceedings of the 7th International Conference on Semantic Systems, pp. 25–32. ACM (2011)

21. Tarasowa, D., Lange, C., Auer, S.: Measuring the quality of relational-to-RDF mappings. In: International Conference on Knowledge Engineering and the Semantic Web, pp. 210–224. Springer (2015)

22. Sequeda, J.F., Arenas, M., Miranker, D.P.: On directly mapping relational databases to RDF and OWL. In: Proceedings of the 21st International Conference on World Wide Web, pp. 649–658. ACM (2012)

23. Richard Cyganiak, D.: R2RML: RDB to RDF mapping language (2012)

24. Priyatna, F., Corcho, O., Sequeda, J.: Formalisation and experiences of R2RML-based SPARQL to SQL query translation using morph. In: Proceedings of the 23rd International Conference on World wide web, pp. 479–490. ACM (2014)

25. Konstantinou, N., Kouis, D., Mitrou, N.: Incremental export of relational database contents into RDF graphs. In: Proceedings of the 4th International Conference on Web Intelligence, Mining and Semantics (WIMS14), p. 33. ACM (2014)

26. Rodriguez-Muro, M., Kontchakov, R., Zakharyaschev, M.: Ontology-based data access: ontop of databases. In: International Semantic Web Conference, pp. 558–573. Springer (2013)

27. Sequeda, J.F.: Integrating Relational Databases with the Semantic Web. IOS Press (2016)

28. Harris, S., Gibbins, N.: 3store: efficient bulk RDF storage (2003)

ERP Conceptual Ecology

Fernando Bento[1], Carlos J. Costa[2], and Manuela Aparicio[1(✉)]

[1] Instituto Universitario de Lisboa (ISCTE-IUL) ISTAR-IUL, Lisbon, Portugal
`fbentodsca@gmail.com`, `manuela.aparicio@iscte-iul.pt`
[2] Advance/CSG, ISEG, Universidade de Lisboa, Lisbon, Portugal
`cjcosta@iseg.ulisboa.pt`

Abstract. The technological evolution of recent years has made that information systems frequently adapt to the market realities to fulfill the improvements of the company's organizational processes. In this context, new paradigms, approaches, and concepts were disseminated through the new realities of information systems. This study aims to verify how ERP (Enterprise Resource Planning) has been related to other information systems within its ecosystem. For this purpose, we have reviewed the literature based on 650 publications whose central theme was the ERP. The data were treated through a graphical analysis, inspired by SNA (Social Network Analysis), represented by related ERP concepts. The study results, determine the connection degree between the concepts that emerged with the technological evolution and the ERP, thus representing the ERP interoperability tendencies, over the last years. The study concludes that ERPs have been improving and substantially increasing the conditions of interoperability with other information systems and with new organizational concepts that have emerged through the technological availability. This fact led to a better organizational process's adoption and more organizational performance.

Keywords: ERP · Enterprise Resource Planning · Information systems · Systems integration · SNA · ERP evolution

1 Introduction

Many studies produced by several authors have pointed to ERPs as an essential information system to the service of organizations. Also, new business strategies became mechanisms that generated more challenges for companies and consequently new information systems requirements. These data inputs (internal and external) have been increased business intelligence levels [1], which contributed to a better optimization of the relationship between management processes and processed output, where the ERP was the critical element of an information systems infrastructure that supported the organizational processes [2–4]. This study has as main objective shows that nowadays the ERP ecosystem moves a new paradigm based on intermobility principles that it strongest the relation between the ERP, new organizational needs concepts and other information systems. Thus, with this work, we also want to demonstrate how the ERP accompanied the challenges posed by technological evolution, in the sense of the organization's needs that suggested new system updates or

© Springer Nature Switzerland AG 2019
Á. Rocha et al. (Eds.): WorldCIST'19 2019, AISC 930, pp. 351–360, 2019.
https://doi.org/10.1007/978-3-030-16181-1_33

even new information systems implementation to guarantee that organizations had aligned with market needs and updated with the last technological trends. To reach the objectives of this study, a review of the literature based on scientific articles on ERP was carried out. The research was done through ACM's digital library (ACM, 2018) between the years 1991 and 2015. We believe that the results presented in this study contribute to a better understanding of the importance of ERP to its ecosystem, namely the degree of relationship of ERP with other systems and other concepts, in IT scope. The structure of this article focuses on five sections, which reflect the development of the work involved in achieving the proposed goal of this research. Section 1 describes the purpose of this work and used methodology. In Sect. 2, the theoretical foundations and concepts related to the ERP ecosystem are systematized through a literature review that filtered the relevant theoretical aspects, needs to the study. Section 4 describes the empirical research. Data was collected by a social network analysis tool, that was used to process the information in the direction of results and study conclusions.

2   Background

2.1   Information System

An organization can be defined as an intentional combination of people and technologies [5] with the purpose of achieving specific objectives. The full organization knowledge reality is fundamental to information systems design that can contribute to the materialization of its mission [6]. The information system can be defined as a sociotechnical entity that, through a set of equipment and logical supports, can perform tasks such as acquisition, transmission, storage, retrieval and data exposure [7]. We understand that this definition is the one that best fits within the scope of computer science and specifically in the interests of this work. The information systems are firmly integrated with the organizations [8], making that practically all organization activities depend on the information systems to ensure their proper functioning [9, 10].

2.2   Enterprise Resource Planning (ERP)

Traditionally, ERP is a solution that integrates business functions into a single system and can be shared across all of the organization [10–12]. An ERP can be defined as an integrated commercial software package that flows through the entire enterprise [12–16] and which are used to gain operational and strategic competitive advantage. A strong ERP market growth is expected in the next 7 years through new paradigms, as the cloud [17] and IOT (internet of the things) ERP facilitate the information and communication flow between different organizational units [18], but only show their real potential, if the ability to integrate with other subsystems is feasible [12, 16, 19–22]. In recent years ERP has incorporated other extensions of business, such as logistics, customer relationship management, information mobility among others, thus becoming increasingly competitive software, in the sense of its completeness needs and market requirements [23].

3   Methodological Approach

This study intended to understand the ERP evolution on the point of view of the integration with other information systems and subsystems. As mentioned above, ERP has emerged as a response to the intense competitiveness of firms in the market, at a time than stock control in industrial management was carried out in a very traditional way, without any control as regards the production needs. It was then possible through these programs to calculate the active material needs, necessary to produce the finished products in a controlled manner, in addition to achieving a global and integrated view of most of the organizations' business processes through ERP [24]. We collected 560 publications from all the authors who published their research work ACM, between 1991 and 2015. The data collection focused only on publications that in keyword list has the word ERP or Enterprise Resource Planning. Next, we collect all the abstracts of those studies, and we made filter mechanisms that were used by development tools (see Fig. 1). The objective was to understand what the most common concepts involved with the word ERP are. Thus, we decide it excludes the ERP word and its synonyms from our selection data. We found 44 keywords that which were shown to be more common in our data collected universe. The information was further separated by years groups [25], 1991–1995, 1996–2000, 2001–2005, 2006–2010, 2011–2015, but searchable by years, concepts or authors. The last process phase was determined on the concepts (TOP 44) represented by a matrix that evidence a combinatorial frequency between pairs of concepts. It should be noted that the concepts were transcribed from the publications to the repository of collected data. We use Social Network Analysis (SNA) techniques to obtain the empirical results.

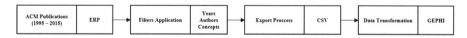

Fig. 1. Study methodology.

After filtered data, were exported to CSV (Comma-separated values) and made available for import through SNA tool. The software that we select was Gephi. This software is one of the most used open source tools intended for data network analysis graphs. The metrics provide results, such as the number of times the concepts (Table 1) appear in the network, the level of centrality, that is, the level by which the concept relates to other concepts and the proximity degree of each concept to others. The social networks analysis, associated with graph theory has been prevalent among researchers, especially in the engineering field [26]. Through the measures used by social networks analysis and in relation to this study, specifically the measures of centrality, we were able to describe the network structural properties (concepts and their relations) that determine indicators in a sense to perceiving for example, the cohesion degree between the different nodes (in the case of this study, the concepts). The edges number incident on a given concept is called by the degree of the concept (vertex). A graph whose number of edges is zero is called a depleted or empty graph [27]. The more connections there are in a concept, the more central is the concept of the network itself. This metric determines an index that reflects the quality by which the network is interconnected between the various concepts (nodes) [28]. The minimum length path is defined as the smallest (geodesic) path between two vertices of a given network. If there is no path between two concepts, then there are connected subgraphs, also known as, related components [27, 29]. Closeness is another centrality measure that measures the proximity of a concept to other concepts. The closer a specific concept is to others, the higher the degree of closeness, and therefore, the higher the relationship between concepts [30]. The betweenness measure allows measuring the capacity that a particular concept must influence other concepts of the network. The larger this capability is the more central and essential the concept in the network [30]. The measures of degree, proximity, and betweenness are considered the primary measures of centrality [30] and for this study, were considered necessary, for this investigation.

Table 1. Study concepts

Concepts				
Adoption	Collaboration	Decision support	Integration	SAP
AHP	Cloud computing	E-business	Knowledge management	Simulation
Balanced scorecard	Control	E-commerce	Knowledge transfer	SME
Business information systems	Critical success factors	Enterprise systems	MES	SMEs
Business intelligence	CRM	ERP implementation	Open source	SOA
Business process	CSFs	Evaluation	Organizational change	Supply chain management
Case studies	Customization	Information systems	Organizational culture	Survey
Case study	Data envelopment analysis	Information technology	Project management	Web service
Change management	Data mining	Implementation	Risk management	

4 Results

The Gephi software was the tool selected to analyse the data on SNA. As already mentioned, the study in question only concentrates its interest in centrality measures. It is important to mention that for a better understanding of this empirical study, that the objective was to identify concepts directly related to the "ERP" word or acronym synonym. Thus, it is true that only data nodes (concepts) with a degree higher than zero should be considered for data analysis. The matrix developed focuses on the intersection of each one of the concepts (top 44) with all others (n−1), in this way we can see that dimension has the relationship between all the concepts considered. Between 1996 and 2015, the following concepts (nodes) and connections (edges) were considered in Table 2. From the observation of this table (Table 2), it can be deduced which concepts were more clearly related to each other, between 2006 and 2010. It is also noticeable that between 1996 and 2000 the relations between the concepts were practically non-existent, suggesting that the concept of ERP began to create its ecosystem from the year 2001. After processing the data with Gephi, we obtained analyses individualized by period and by different metrics. The next table (Table 3) presents the results centrality measures of concepts that were shown more evident. Periods A, B, and C correspond respectively to the years 2001–2005, 2006–2010 and 2011–2015. The period 1996–2000, although analysed, is not presented in the table, because it presents a weak expression in relation to the concept's relationship. However, during this period authors began to relate the ERP with other concepts, such as "integration," "customization," "SME," contributing to the growth of the ERP ecosystem of next years. According to Table 3 and specifically with regard to the intermediation measure, we discover which concepts have marked its presence in all periods, thus marking a constant degree influence on ERP ecosystem. Examples of this are, "CRM," "Information Systems," "Integration," "Supply Change Management," "ERP Implementation" and "Implementation." It is also visible through the results presented in the degree centrality measure (Table 2) that, there are evident relationships between different types of other information systems (Table 3) and ERP systems, examples of this are, "CRM", "Supply Chain Management", "Business Intelligence", "Data Mining", "E-commerce", "Decision Support" and "MES". However, "Information Systems," "Enterprise Systems," "ERP implementation," "Integration," "MES" and "Change Management" are the most popular concepts of these networks since they have a greater centrality degree (Fig. 2). According to the Closeness centrality measure, although the values are very close to each other, the non-rounded values of "Implementation" and "Integration" are the highest values with 0.707 and 0.683 respectively, which were verified in the period 2006–2010 (Fig. 2). Thus, being the concepts with the shortest (geodesic) distance to all members of the network, which makes them closer to all. Analysing the values of the intermediation centrality measure, it is also the "Implementation" and "Integration" concepts that registered the highest values, with 163, 933 and 139, 686 respectively (Fig. 2). Also, through graphical analysis (Fig. 2), we can perceive what the computational results describe, that is, a visible growth related to the relationship of the concept throughout all periods. We note too that network has an increasing dimensional evolution from period to period and with stronger links between concepts. Figure 3 represents the type of relationship that exists between the concepts most used by the various studies selected for this research in respect to centrality degree between 1996 and 2015. The degree of centrality shows the number of connections of each concept (node) [31]. Thus, it is possible to understand the involvement that each concept has between itself and in a global way, the importance that each concept assumes by all ERP ecosystem. It is clear from the graphic evolution that over the years that the ERP ecosystem, evolved by diversified with other concepts and other subsystems, in organizational management scope [31].

Table 2. Evolution of the concepts (Nodes) and relations (Edges) in the period of analysis (1996–2015).

Period	Nodes	Edges	Average degree	Average weighted degree
1996–2000	10	8	1.60	1.6
2001–2005	27	45	3.33	5.11
2006–2010	42	194	9.24	21
2011–2015	41	128	6.24	12

Table 3. Centrality measure results

Concepts	Centrality measures								
	Degree			Proximity			Betweenness		
	2001 a 2005	2006 a 2010	2011 a 2015	2001 a 2005	2006 a 2010	2011 a 2015	2001 a 2005	2006 a 2010	2011 a 2015
Business intelligence	–	5.0	8.0	–	0.5	0.5	–	1.1	23.5
Business information systems	–	13.0	4.0	–	0.6	0.4	–	18.1	–
Business process	–	7.0	5.0	–	0.5	0.4	–	5.1	39.0
Change Management	1.0	21.0	–	0.3	0.7	–	–	53.4	–
CRM	4.0	7.0	5.0	0.4	0.5	0.4	19.4	5.1	18.7
Data envelopment analysis	1.0	5.0	2.0	0.3	0.4	0.4	–	40.7	–
Data mining	–	7.0	1.0	–	0.5	0.3	–	6.0	–
Decision support	1.0	4.0	10.0	0.3	0.4	0.5	–	–	37.0
E-commerce	2.0	3.0	3.0	0.4	0.4	0.4	–	0.5	41
Enterprise systems	5.0	18.0	12.0	0.5	0.6	0.5	44.9	35.5	–
ERP implementation	6.0	20.0	11.0	0.5	0.7	0.5	30,0	63.4	24.4
Evaluation	–	13.0	10.0	–	0.6	0.5	–	21.3	101.9
Information systems	4.0	22.0	18.0	0.4	0.7	0.6	25.4	96.7	168.3
Implementation	12.0	25.0	18.0	0.6	0.7	0.6	165.0	139.0	101.3
Integration	7.0	23.0	9.0	0.5	0.7	0.5	86.6	139.7	35.8
MES	–	17.0	8.0	–	0.7	0.5	–	93.1	7.5
SME	–	15.0	16.0	–	0.6	0.6	–	56.7	162.2
Supply change management	2.0	8.0	4.0	0.3	0.5	0.4	25.0	8.6	7.9

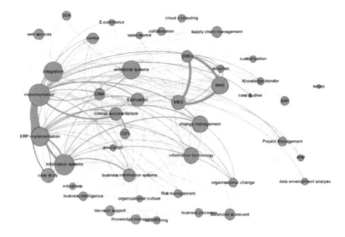

Fig. 2. Graph of the concepts by higher weight of the centrality measures between 1996 and 2015

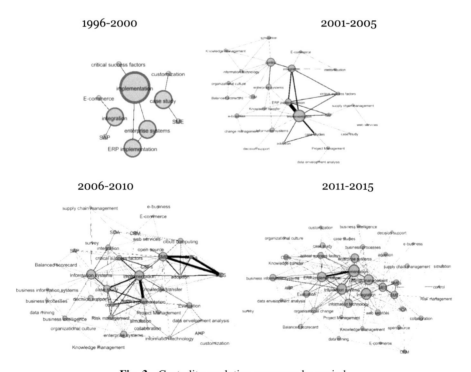

Fig. 3. Centrality evolution measures by period,

4.1 Discussion

Many studies have been developed about the ERP context in recent years [32–34]. There is also a growing interest by the information technology researchers regarding the socio-technical information systems analysis, namely applying methodologies in their studies through SNA [27, 35–39]. However, we did not find any study that used SNA that related ERP with associated concepts, in the sense of interpreting the interrelation between them. The work developed is in this study and in line with the literature review. The SNA (Social Networks Analyzes) has proved to be an effective method with respect to the comprehension of the complexities of a social network [40], namely in the case of this work, the perception of the degree of connectivity between the concepts associated to the studies of the most diverse authors about the ERP. The social bonds are established under concrete prisms and with common interests of each concept [27]. From this point of view, it seems to be clear that the concepts closest to the ERP have had a more continuous relationship than the more distant concepts (over the years) for this reason they were many times natively integrated with the ERP itself. This is the case of the concepts as "integration," "SME," "CRM," "SCM." However, the study presents concepts such as "implementation," "information systems," "enterprise systems," "project management" which are not precisely subsystems or information systems, are abstract and generic concepts that provide an important proof of the existence of a strong multidisciplinary environment, around the ERP ecosystem. As we have already mentioned, although there are many studies on the evolution of ERP and other information systems, we did not find any study to specifically use SNA to demonstrate the relationship between all the actors across ERP (subsystems and concepts). On the other hand, the SNA has been presenting a growing demand by the researchers, as a research method, in the field of studies in the area of sciences and information technologies thus legitimizing, the methodology chosen for this study [41].

5 Conclusions and Future Work

In this study, it was tried to show that the ERP developed a complex ecosystem over the years, based on the own technological evolution and consequently of many and new organizational challenges. The analysed data confirm that the ERP has been following the new technological paradigms and organizational challenges. The integration capability that ERP has been developing over the years with other information systems or subsystems has made ERP one of the most versatile and popular solutions on the market. Also, over the years, the implementation of ERP has been a subject increasingly discussed by researchers. The study showed the centrality measures importance, in the case of this study, regarding the relationship that exists in the ERP with the organizational processes and the technological evolution. We conclude that today, ERPs continue to be very current solutions and with expectations of growth vis-à-vis their ecosystem developed over more than 25 years, given its capacity for interoperability between technical evolution and organizational evolution. As future work, we understand that it would be important in the ERP context, the evolution of the ERP itself and the ability to integrate it with the impact of the new technological paradigms, namely the issue of IOT (Internet of the things) and artificial intelligence. Also, will be useful as scientifically contribution that within the organizational management scope, a study that investigates the impact that top management has on the technology capture for new business processes, and understands the effect that this fact has on the ERP success.

Acknowledgement. We gratefully acknowledge financial support from FCT- Fundação para a Ciencia e Tecnologia (Portugal), national funding through research grants UID/Multi/04466/2019, and UID/SOC/04521/2019

References

1. Lee, J., Siau, K., Hong, S.: Enterprise integration with ERP and EAI. Commun. ACM **46**, 54–60 (2003)
2. Abd Elmonem, M.A., Nasr, E.S., Geith, M.H.: Benefits and challenges of cloud ERP systems – a systematic literature review. Future Comput. Inf. J. **1**, 1–9 (2016)
3. Costa, C.J., Aparicio, M.: ERP and assistance systems. In: Proceedings of the 11th WSEAS International Conference on Applied Informatics and Communications, pp. 216–221. WSEAS (2011)
4. Costa, C., Aparicio, M.: Organizational tools in the web: ERP Open Source. In: Proceedings of the IADIS International Conference on WWW/Internet, pp. 401–408 (2006)
5. Hampton, D.R.: Management. Mcgraw-Hill College, New York (1985)
6. Amaral, L., Varajão, J.: Planeamento de Sistemas de Informação. FCA (2007)
7. Alter, S.: Information Systems: A Management Perspective. Addison Wesley, Boston (1999)
8. Tricker, R.I.: The management of organizational knowledge (1992)
9. Costa, C.J., Ferreira, E., Bento, F., Aparicio, M.: Enterprise resource planning adoption and satisfaction determinants. Comput. Hum. Behav. **63**, 659–671 (2016)
10. Bento, F., Costa, C.J., Aparicio, M.: SI success models, 25 years of evolution. In: 2017 12th Iberian Conference on Presented at the Information Systems and Technologies (CISTI) (2017)
11. Boersma, K., Kingma, S.: From means to ends: the transformation of ERP in a manufacturing company. J. Strateg. Inf. Syst. **14**, 197–219 (2005)
12. Bento, F., Costa, C.J.: ERP measure success model; a new perspective. In: Presented at the Proceedings of the 2013 International Conference on Information Systems and Design of Communication (2013)
13. Davenport, T.H., Prusak, L.: Working Knowledge: How Organizations Manage What They Know. Harvard Business Review Press, Boston, Mass (2000)
14. Batista, M., Costa, C.J., Aparicio, M.: ERP OS localization framework. In: Proceedings of the Workshop on Open Source and Design of Communication, pp. 1–8. ACM, New York (2013)
15. Lopes, N.G., Costa, C.J.: ERP localization: exploratory study in translation: European and Brazilian Portuguese. In: Proceedings of the 26th Annual ACM International Conference on Design of Communication, pp. 93–98. ACM, New York (2008)
16. Bento, F.: Open Source ERP's I18n. New York (2010)
17. Pinheiro, P., Aparicio, M., Costa, C.: Adoption of cloud computing systems. In: Proceedings of the International Conference on Information Systems and Design of Communication, pp. 127–131. ACM, New York (2014)

18. Amoako-Gyampah, K.: Perceived usefulness, user involvement and behavioral intention: an empirical study of ERP implementation. Comput. Hum. Behav. **23**, 1232–1248 (2007)
19. Leon, A.: ERP Demystified. McGraw Hill Education, New Delhi (2014). 3/e
20. Hau, E., Aparício, M.: Software internationalization and localization in web based ERP. In: Proceedings of the 26th Annual ACM International Conference on Design of Communication, pp. 175–180. ACM, New York (2008)
21. Costa, C.J.: Testing Usability of ERP Open Source Systems. New York (2010)
22. Costa, C.J.: ERP Open Source. In: Costa, C.J. (ed.) Information Technology, Organizations and Teams. ITML Press (2007)
23. Costa, C.J.: Testing usability of ERP Open Source systems. In: Proceedings of the Workshop on Open Source and Design of Communication, pp. 25–30. ACM (2010)
24. Orlicky, J., Plossl, G.: Orlicky's Material Requirements Planning. McGraw Hill Professional, New York City (1994)
25. Aparicio, M., Bacao, F., Oliveira, T.: Trends in the e-learning ecosystem: a bibliometric study. In: AMCIS 2014 Proceedings, Savannah (2014)
26. Kim, Y., Choi, T.Y., Yan, T., Dooley, K.: Structural investigation of supply networks: A social network analysis approach. J. Oper. Manag. **29**, 194–211 (2011)
27. Cerqueira, C.H.Z., de Souza Costa, J.M., de Araujo Carvalho, D.M.: Aplicação de análise de redes sociais em uma cadeia de suprimentos de uma epresa do setor elétrico brasileiro. Sist. Gest. **9**, 418–429 (2014)
28. Nieminen, J.: On the centrality in a graph. Scand. J. Psychol. **15**, 332–336 (1974)
29. Bondy, J.A., Murty, U.S.R.: Graph Theory With Applications. Elsevier Science Ltd/North-Holland, New York (1976)
30. Freeman, L.C.: Centrality in social networks conceptual clarification. Soc. Netw. **1**, 215–239 (1978)
31. Yustiawan, Y., Maharani, W., Gozali, A.A.: Degree centrality for social network with opsahl method. Proc. Comput. Sci. **59**, 419–426 (2015)
32. Addo-Tenkorang, R., Helo, P.: Enterprise resource planning (ERP): a review literature report. In: Presented at the Proceedings of the World Congress on Engineering and Computer Science (2011)
33. Bosch, J.: From software product lines to software ecosystems. In: Presented at the Proceedings of the 13th International Software Product Line Conference (2009)
34. Shaul, L., Tauber, D.: Critical success factors in enterprise resource planning systems: review of the last decade. ACM Comput. Surv. CSUR **45**, 55 (2013)
35. van der Aalst, W.M.P., Reijers, H.A., Song, M.: Discovering social networks from event logs. Comput. Support. Coop. Work CSCW **14**, 549–593 (2005)
36. Dado, M., Bodemer, D.: A review of methodological applications of social network analysis in computer-supported collaborative learning. Educ. Res. Rev. **22**, 159–180 (2017)
37. Han, E.J., Sohn, S.Y.: Technological convergence in standards for information and communication technologies. Technol. Forecast. Soc. Change **106**, 1–10 (2016)
38. Kazienko, P., Michalski, R., Palus, S.: Social network analysis as a tool for improving enterprise architecture. In: Agent and Multi-Agent Systems: Technologies and Applications, pp. 651–660. Springer, Heidelberg (2011)
39. Latorre, R., Suárez, J.: Measuring social networks when forming information system project teams. J. Syst. Softw. **134**, 304–323 (2017)
40. Scott, J.: Social Network Analysis: A Handbook. Sage Publications, Thousand Oaks (2000)
41. Rainie, L., Wellman, B.: Networked: The New Social Operating System. The MIT Press, Cambridge, Mass (2014)

Data Quality Mining

Alexandra Oliveira[1,2(✉)], Rita Gaio[3,4], Pilar Baylina[1], Carlos Rebelo[5],
and Luís Paulo Reis[2,6]

[1] School of Health, Polytechnic Institute of Porto,
Porto, Portugal
aao@ess.ipp.pt
[2] LIACC - Artificial Intelligence and Computer Science Laboratory,
Porto, Portugal
[3] DM - FCUP - Department of Mathematics,
Faculty of Sciences of the University of Porto, Porto, Portugal
argaio@fc.up.pt
[4] CMUP - Center of Mathematics of the University of Porto,
Porto, Portugal
[5] 3Decide, Porto, Portugal
[6] DEI - FEUP - Department of Informatics Engineering,
Faculty of Engineering of the University of Porto, Porto, Portugal
lpreis@fe.up.pt

Abstract. We are living in a world of information abundance, surplus, and access. We have technologies to acquire any type of information but we still face the challenge of extracting the underlying valuable knowledge. Data analyses and mining processes may be severely impaired whenever data are corrupted by noise, ambiguity and distortions.

This paper aims to provide a systematic procedure for data cleaning in single files data sources without schema that may be corrupted by the most common data problems. The methodology is guided by the dimensions of data quality standards and focuses on the goal of performing reasonable posterior statistical analyses.

Keywords: Data quality · Data mining · Validation · Repair · Improve quality

1 Introduction

We are living in a world of information abundance, surplus, and access. We have technologies to acquire any type of information but we still face the challenge of extracting the underlying valuable knowledge. Data analyses and mining processes may be severely impaired whenever data are corrupted by noise, ambiguity and distortions [1]. Data depends largely on its quality which is build upon the way the collection process is implemented and managed [2,3]. Most of data analyses and mining processes focus on extracting knowledge from data; whenever the latter is of poor quality, the objective may be severely impaired and can even be beyond the scope of statistical analysis [4–6].

© Springer Nature Switzerland AG 2019
Á. Rocha et al. (Eds.): WorldCIST'19 2019, AISC 930, pp. 361–372, 2019.
https://doi.org/10.1007/978-3-030-16181-1_34

Data quality is an ever gradually developing concept, with roots in measurement error and survey uncertainty, that encompasses multiple disciplinary fields such as commerce, engineering, medicine, public health and policy making. This complex concept spotlights a rich set of scientific, technological and process control challenges [4,7,8].

There are several definitions of data quality (DQ), almost as much as the study fields. It can be defined as:

the capability of data to be used effectively, economically and rapidly to inform and evaluate decisions (with roots in decision theory field) [4]

or as

the degree to which a set of characteristics of data fulfils requirements (ISO/IEC 25012 standard, with roots in engineering and computer science) [8]

or even as

accurate, reliable, valid, timely and trusted data (with roots in integrated public health informatics network) [2].

Despite the diversity, many definitions became aligned with the engineering, computer science and decision theory definitions. This was the situation in health information and survey research [7].

Statistics provides valuable contributions such as outliers detection, statistical data editing, probabilistic record linkage and the measurement error and survey methodology. Total quality measurement uses concepts such as multidimensional DQ, DQ metrics, evaluation of the user's assessment of DQ and data production maps [4]. Computer science is exposed to DQ issues since organizations started collecting and storing their data electronically. So, it has well-developed technologies to address issues such as data standardization and duplicate detection and elimination, and data parsing, among others.

It is recognized that DQ is a concept with multi-dimensions (characteristics or attributes) and depends on a sustainable work-flow model integrated with minimal disruption into the day-to-day life of all the relevant stakeholders [2,4,8]. These characteristics are fundamental drivers that can be corrupted by specific problems with genesis on each of the steps of the work-flow. As such, the best practice process for improving and ensuring high DQ follows the so-called DQ cycle. The cycle is made up of an iterative process of analysing, cleansing and monitoring DQ [9–11].

It is very rare to find the raw data in the correct format, without errors, complete and with all the correct labels and codes that are needed for analysis [14]. In single data collections, such as files and databases, several problems commonly arise due to misspellings during data entry, missing information, lack standardization and/or validation of self-reported information, the existence of duplicate or redundant information, or the existence of anomalous events [2,7,12].

Data quality mining deals with detecting and removing errors and inconsistencies from data [12,13]. Moreover, developing a data cleaning procedure guided by the dimensions that characterize the quality of data is not only essential but also a quite difficult task [8].

Given the spread of DQ definitions and the time that take to prepare a data set to proper conduct a statistical research, it is important to stablish a systematic, effective and sustainable approach to identify potential problems, evaluate The DQ and the impact on the research goals and develop mechanisms to resolve them. So, this paper proposes an approach to unify the concept of DQ integrating computer science and statistical concepts. It extends the map between common problems and DQ dimensions analysing which dimension can be affected. Finally, systematize some state of the art repair and improve methods. The procedure applies to single files data sources without schema that may be corrupted by the most common data problems. The methodology is guided by the dimensions of DQ standards and focuses on the goal of performing reasonable posterior statistical analyses.

In Sect. 2, we present a DQ definition and corresponding dimensions; also the map of common problems and DQ dimensions ; in Sect. 3 we provide a summary of some advanced statistical methods for data anomalies detection and data editing and the last Sect. 4 contains the concluding remarks.

2   Data Quality Dimensions and Issues

At a database level, we align the DQ definition by the ISO/IEC 25012 but reflected by the accessibility, accuracy, completeness, consistency, timeliness, and relevance dimensions [8,15,16]. A precise definition of each dimension can be found in ISO/IEC 25012 catalogues or in Laranjeiro *et al.* or in The Quality Assurance Framework of the European Statistical System (ESS QAF) [4,8]. Briefly, we have the following:

Accessibility: The degree to which data can be accessed in a specific context of use, which includes suitability of representation.

Accuracy: The degree to which data attributes or variables correctly represent the true value of the intended object or population.

Completeness: The degree to which subject data associated with an entity has values for all expected attributes and related entity instances in a specific context of use.

Consistency: The degree to which data has attributes that are free from contradiction and are coherent with other data in a specific context of use. It can be either or both among data regarding one entity and across similar data for comparable entities.

Timeliness: Time-related dimension, reflecting if data is up to date.

At a statistical level, we add the relevance dimension defined by the Eurostat as:

Relevance: The degree to which statistics meet the needs of current and potential research objective [16].

Data validation confronts a data set with a group of desired properties. It provides awareness for the major issues in the dataset, commonly classified as being at a structure (schema) level or at an instance-level. The latter type reflect sampling and non-sampling errors such as data processing errors (includes collection, coding and entry) or inconsistencies while the former are associated with lack of integrity constraints or poor structure design. For data sources without schema, such as single files, there are few restrictions on how data can be entered and stored, turning errors and inconsistencies more probable [12]. The most common data problems and issues as well as the corresponding affected data quality dimensions are described in Table 1 [8,12,17]. We notice that if data are inconsistent then they are also inaccurate; indeed, if values are represented in different formats and/or structures, then it is difficult to determine their true representation. So, it was extended the correspondences surveyed by Laranjeiro. Moreover, we added the dimension 'relevance' and the most commonly encountered file and statistical problems: proprietary file format, wrong variables type, tabular design, missing variables, lack of information diversity, outliers and coverage errors.

A proprietary format may contain data that is ordered and stored according to a particular encoding-scheme and so the decoding and interpretation of the stored information may only be accomplished with the use of a particular software or hardware. If the specification of the data encoding format is not released, the accessibility of the information is compromised. Also, if a variable is assigned to a wrong type, the information may be difficult to access.

Tabular design problems occurs when variables form both rows and columns and/or column' headers are values, not variables names. This inconsistencies violate the dimensions accessibility (i.e. data needs to be manually inspected and reasoning about the problem) and accuracy (i.e., the variables do not correctly represent the true population value). The issue of missing variables occurs when important variables were not included in the dataset. This compromises the dimensions completeness and accuracy, once the true object is not represented. Lack of information diversity occurs when a variable has few and under-represented unique values. This affects accessibility since the variable is not suitably represented and accuracy because the true value of the intended object may be missing. Finally, we point out the existence of outliers, corresponding to extreme values of, at least, one variable that are *far apart* from the remaining values. It may or may not be an incorrect value (although very suspicious) and by definition affects the accuracy.

Relevance may be affected by issues such as incorrect values, coverage, missing variables and lack of information diversity since the objective of the statistical methods may not be accomplished due to the lack of the necessary information. Also, illegal values, violation of logical dependencies, wrong data type or syntax violation, missing data, incorrect data, misspellings, ambiguous data, extraneous data, misfield values may impair statistical inferences and therefore affect relevance. This dimension may be affected by the presence of duplicates because they may impair representativeness of the data.

Illegal values, violation of logical dependencies, wrong data type or syntax violation, missing data, incorrect data, misspellings, ambiguous data, extraneous data, misfield values may impair statistical inferences and therefore affect relevance may also affect consistency.

Table 1. Extended version of the mapping between data problems and Data Quality (DQ) dimensions [5, 8]

Problem	Description	Example	Accessibility	Accuracy	Completeness	Consistency	Timeliness	Relevance
a. Proprietary File Format	Occurs when a file is designed for a specific application / system.		X					
b. Wrong Variables Type	Occurs when a variable is assumed to be of a wrong type.			X				
c. Tabular data design	Occurs when variables form both rows and columns and/or column headers are values, not variable names			X	X			
d. Illegal values	Occurs when variables have a value that is out of range	Diagnosis date = 30/02/2006		X		x		X
e. Violation of values logical dependencies	Occurs when some logical dependency between values is broken	Patient(age = 35, sex = female, HIV risk group = 'MSM')		X		X		X
f. Wrong data type or syntax violation	Occurs when a value does not respect the data type constraints	Hospital = '42'		X		X		X
g. Missing data	Occurs when data is not present or coded with a dummy value	Probable year of infection = 9999, 'Not mentioned'		X	X	X	X	
h. Incorrect data	Occurs when data contain valid values that do not correspond to real values	HIV risk group = 'Heteressexual' instead of 'IDUs'		X		X	X	X
i. Misspelled	Occurs when data are misspelled	Nationality = 'Portgal' instead of 'Portugal'		X				X
j. Ambiguous data	Occurs when data values can have different interpretations. It can be due to abbreviations or an incomplete context	Hospital = 'Maria', it can be 'Hospital Santa Maria' or 'Hospital Santa Maria Maior' or even 'Hospital Santa Maria Pia' (Hospitals in Portugal)	X	X		X		X
k. Extraneous data	Occurs when additional data is represented	Name = 'Mrs. Maria'	X	X		X		X
l. Outdated temporal data	May be valid for a time point or interval	Hospital = 'Desterro' (this hospital was closed in 2007)		X			X	
m. Misfielded values	Occurs when the data values are stored in the wrong column	CD4 count in the variable of T4/T8 ratio	X	X	X	X		X
n. Duplicates	Occurs when the same data appear more than once but with a different identifier and may have contradicting information	Patient1(ID = 'AERER', age = 31, sex = 'male', risk group = 'MSM') and Patient2(ID= 'AERER', age = 31, sex = 'male', risk group = 'IDU')	X	X		X		
o. Outliers	Occurs when a variable has an outlier. It may or may not be an incorrect or dummy value (although they are very suspicious)	Age=88,age=99		X				
p. Missing Variables	Occurs when an important variable is missing.	Individual records of HIV patients without clinical variables.		X	X			X
q. Lack of information diversity	Occurs when a variable has few and under-represented unique values.	All observations of variable HIV type equal to HIV1.		X	X			X
r. Coverage errors	Occurs when a unit in the sample is incorrectly excluded or included, or is duplicated in the sample.				X	X*		X

* Just in case of excluding a sample.

It is important to point that data validation adds value to the dataset by rising awareness about the presence of issues that may impair its usefulness for the intended purposes [18].

3 Procedure for Improving Data Quality

Data Quality Mining (DQM) can be defined as the deliberate application of data mining techniques for the purpose of data quality improvement [13]. The main goals in any data quality improvement process are: the detection, the explanation of the source and correction of deficiencies such that the 'improved' dataset is close as possible to the original collected data. Other important goal is to confidently allow for posterior statistical analysis.

In this paper we follow Wickham and Jogen definitions of dataset [5,14]: it is a collection of values (qualitative or quantitative), each belonging simultaneously to a variable and an observation. A variable contains all values that measure the same underlying attribute across objects. An observation contains all values measured on the same object across attributes [5].

The central problem for data improvement is how to make data consistent, accurate, accessible, timeliness, complete and relevant while keeping the edits to as few as possible, respecting the constraints and context [14,19,20].

A common approach to address the problem of dirty data is to apply a set of data quality rules or constraints over a target database, to 'detect' and to eventually "repair" data issues [19], but first one must be able to read and understand the data. After that, four stages can be identified: data structuring, data validation, error localization and repair.

3.1 Data Validation and Structuring

The credibility of the data may be assessed by data validation activities whose negative outcome guarantees the data have quality issues, mainly accuracy and consistency dimensions. Data validation can be defined as *an activity verifying whether or not a combination of values is a member of a set of acceptable combinations* [18,21].

Data validation is an iterative procedure based on the tuning of validation or edit rules that will converge to a minimal set that must be necessarily satisfied [21]. Moreover this set must be complete, concise and consistent [22].

This property is hard to quantify but may be assessed by knowledgeable peers review or by checking whether at least one of the explicitly defined rules is valid (assuming that the set of rules is on a standard and minimal form) [22].

Statistical data editing is the automated process of stepping through the data observations and correcting them whenever they violate the validation rules [4]. The validation activities may be grouped, from a research perspective, in six ordered levels starting with simple single value checks and moves to more complex checking involving more observations, variables and data sets [23,24]:

Level 0: Validation of the Information Technology structural requirements. At this level the data set is checked format and file structure and variables types (issues **a, b** and **c** from Table 1).

Level 1: Validation for consistency of single data points. For example comparing a single data point with constants. At this level the data set is confronted with rules for checking the presence of issues as described in **d, f, g, i, j, k** and **m** on Table 1.

Level 2: Validation for consistency of multiple variables of the same statistical object. At this level the data set is confronted with rules for checking the presence of issues as in **e** above.

Level 3: Validation for consistency of multiple statistical objects of the same variable. At this level the data set is confronted with rules for checking the presence of issues such as **o** and **q**.

Level 4: Validation for consistency of repeated measures of the same variable and same statistical object. At this level the data set is confronted with rules for checking the presence of issues such as **l**.

Level 5: Validation for consistency of multiple variables and multiple statistical objects. At this level the data set is confronted with rules for checking the presence of issues such as **n**.

Level 6: Validation for consistency based on the comparison of the file content with the content of other files. By other files we mean other versions of exactly the same file, same data set but referring to a different time period, different data set but collected from the same domains or even from correlated domains. At this level the data set may be confronted with rules for checking the presence of issues such as **h, p** and **r**.

3.2   Error Localization

Data editing is a complex process that often involves cross-variable and interrelated contingency rules. Fellegi and Holt formulated a theoretical model for this process with the following goals [25]:

1. in each record or observation, data should satisfy all validation rules by changing the fewest possible items (variables or fields);
2. imputation rules should be derived automatically from validation rules;
3. when imputation is necessary, it is desirable to maintain the marginal and joint frequency distributions of variables.

There are several algorithms for solving this problem, the two most popular being the branch-and-bound and mixed-integer programming, described in [26].

3.3   Repair and Improve

In this subsection, methods for repairing the main issues presented on Table 1, guided by the validation steps given on page 7, will be over-viewed. Note that the correction steps are not in the exactly same order as the checking steps; more precisely, all the imputation methods are done after all the errors have been addressed.

It is recommended that data is in a text-based format because this has several favourable properties over other formats, including being human readable, the possibility to represent any type of value, and avoiding underlying stored data issues.

More than 80% of all data is unstructured [3]. For statistical analysis it is desirable that the values are organized in a rectangular table with rows and columns such that: each column corresponds to a single variable, with a proper label; each row has just one observed case, and each type of observational unit forms a table.

Since data improvement is a resource consuming process, it is important to chose wisely which variables need to be cleaned taking into consideration the context relevance. For the selected variables we should proceed with a uniformization of the values. A simple scan for unique values and its corresponding distribution in data can detect several problems such as: misspellings, extraneous data, misfield values, wrong data type or syntax violation, ambiguous data and illegal values.

An observation (or set of observations) which appears to be inconsistent with that set of data is called an outlier [14]. For detection of outliers, several approaches have been developed, such as error bounds, tolerance limits and control charts; model-based regression depth and residual analysis, distributional representations and time-series analysis.

Outliers can appear from several different mechanisms or causes: they can be simply an error or arise from the inherent variability of the data. When an observation is clearly an error, simply delete the value but, if it is legitimate or of unclear source, then the problem becomes difficult. One way for keeping an outlier is to transform the variable either by applying a function or by truncation to the closest extreme. If none of these transformations turn out to be helpful, then one may need to use "robust" parameters estimate methods for data mining [27].

Violations of logical dependencies between variables are easily detected with data visualization tools and with cross tabulations. Each dependence between variables, also known as editing rule, require the definition of a proper pre-specification of specify domain-knowledge-based constraints.

As a violation may be resolved in more than one way, an immediate question is which one to choose? Solutions may include repairing values that require the least number of operations, or repairing values according to a pre-specified cost model.

After the cleaning process, one should turn his/her attention to missing values. Since attempts to recover missing values may impair inferences (the main goal of a statistical procedure), a missing value treatment cannot be properly evaluated apart from the modelling, estimation, or testing procedure in which it is embedded [28]. Several methods for data imputation have been proposed and they can be divided into two categories: single variable or considering relationships among variables [6]. For single variable methods it is assumed that the distribution of the missing values is the same as the non-missing values and so the missing values are replaced by the non-missing values mean, median or other point estimates. The methods that consider relationships among variables use regression (parametric) or propensity scores (nonparametric). These techniques assume that the explanatory variables are non-missing. After any imputation procedure it is necessary to perform a sensitivity analysis, comparing the pre-imputation with the post-imputation scenarios.

Duplicated records may be introduced in a data set through several different mechanisms. Typically, by typographical and data entry errors. An individual's information may change over time with some life events like moving or deliberately reporting false informations [29]. Due to these idiosyncrasies, the previous data cleaning and standardization steps are critical. It is also important to point that the definition of duplicate change according to the research hypothesis.

The duplicate detection is typically performed by applying similarity functions to pairs of observations. If the values of two records are sufficiently similar, they are assumed to be duplicates' [30].

This task is highly data-dependent and therefore choosing a detection technique is similar to model selection and performance prediction for data mining tasks [31,32].

For correcting ambiguous data and missing variables it is useful to check different versions of the same data set, or to match the data with correlated data sets from the same or different sources or even different domains.

Incorrect data rising from random intentional errors may be treated as outliers but if these errors are systematic, it is very hard to detect and correct them.

Coverage errors can be minimized by capture - re-captures techniques, matching the data set with a correlated data set from a different domain.

A relevant audit is if any important variable is missing in the dataset. If so, the main goal for which the data was collected may be irreparably compromised.

In this section, it is proposed a macro sequence across the levels of validation, which follows from level 0 until 6 but the contained errors checking may be of any order. The repair process, however, must follow a more strict order, the last addressed errors being the ones for which the correction involves imputation and outliers handling due to the obvious influence over the results of a subsequent statistical analyses.

Once the repair and improve process is ended, data is then ready for statistical analysis and model selection.

4   Conclusions

Data cleaning refers to the correction or amelioration of data problems, including missing values, incorrect or out-of-range values, responses that are logically inconsistent with other responses in the database, and duplicate observations. Though we dream with perfection, in reality, 'clean data' is a relative term. Metadata documents are of crucial importance and documents explaining which data elements should be cleaned, with the "description of data validation rules or logical checks for out-of-range values, how missing values and values that are logically inconsistent can be handled, and discussing how duplicate patient observations can be identified and managed" are of great importance. [2].

Data pre-processing and knowledge retrieving, is a multidisciplinary discipline involving topics from statistics, computer science and domain knowledge experts, it is always on data pre-processing and analysis [4].

This paper describes methods for identifying the most common problems in a dataset that influence the data quality dimensions and provides some correction methods. It proposes two processes: one process for detection of data issues and the another for repair and improve of the information in the dataset.

Acknowledgements. Luís Paulo Reis and Alexandra Oliveira were partially founded by the European Regional Development Fund through the programme COMPETE by FCT (Portugal) in the scope of the project PEst-UID/CEC/ 00027/2015 and QVida+: Estimação Contínua de Qualidade de Vida para Auxílio Eficaz à Decisão Clínica, NORTE010247FEDER003446, supported by Norte Portugal Regional Operational Programme (NORTE 2020), under the PORTUGAL 2020 Partnership Agreement.

Rita Gaio was partially supported by CMUP (UID/MAT/00144/2019), which is funded by FCT with national (MCTES) and European structural funds through the programs FEDER, under the partnership agreement PT2020.

References

1. Balasingam, B., Mannaru, P., Sidoti, D., Pattipati, K., Willett, P., Pedrycz, W., Chen, S.-M. (eds.): Online anomaly detection in big data. In: The First Line of Defense Against Intruders Data Science and Big Data: An Environment of Computational Intelligence, pp. 83–107. Springer International Publishing (2017)
2. Gliklich, R.E., Dreyer, N.A., Leavy, M.B. (eds.) Registries for Evaluating Patient Outcomes: A User's Guide, 3rd edn., 11 April 2014. Data Collection and Quality Assurance 2014
3. Cai, L., Zhu, Y.: The challenges of data quality and data quality assessment in the big data era. Data Sci. J. **14** (2015)
4. Karr, A.F., Sanil, A.P., Banks, D.L.: Data quality: a statistical perspective statistical methodology. Elsevier **3**, 137–173 (2006)
5. Wickham, H.: Tidy data. J. Stat. Softw. **59**, 1–23 (2014). Foundation for Open Access Statistics
6. Dasu, T., Johnson, T.: Exploratory Data Mining and Data Cleaning. Wiley, Hoboken (2003)

7. Keller, S., Korkmaz, G., Orr, M., Schroeder, A., Shipp, S.: The evolution of data quality: understanding the transdisciplinary origins of data quality concepts and approaches. Ann. Rev. Stat. Appl. **4**, 85–108 (2017)
8. Laranjeiro, N.; Soydemir, S.N., Bernardino, J.: A survey on data quality: classifying poor data. In: 2015 IEEE 21st Pacific Rim International Symposium on Dependable Computing (PRDC), pp. 179–188 (2015)
9. BARC (Business Application research Center) - a CXP Group Company, Data Quality and Master Data Management: How to Improve your data quality (2017)
10. EUROSTAT, Handbook on Data Validation in Eurostat -Practical Guide to Data Validation in EuroSttat (2010)
11. Azimaee, M., Smith, M., Lix, L., Burchill, C., Orr, J.: MCHP data quality framework. Manitoba Centre for Health Policy, University of Manitoba, Winnipeg (Manitoba) (2015)
12. Rahm, E., Do, H.H.: Data cleaning: problems and current approaches. IEEE Data Eng. Bull. **23**, 3–13 (2000)
13. Hipp, J., Guntzer, U., Grimmer, U.: Data quality mining-making a virute of necessity. In: DMKD (2001)
14. De Jonge, E., van der Loo, M.: An introduction to data cleaning with R Heerlen, Statistics Netherlands (2013)
15. Taleb, I., Dssouli, R., Serhani, M.A.: Big data pre-processing: a quality framework Big Data (BigData Congress). In: IEEE International Congress on 2015, pp. 191–198 (2015)
16. ESS Task Force Peer Review, Quality Assurance Framework of the European Statistical System- Version 1.2, European Statistical System (2015)
17. Barateiro, J., Galhardas, H.: A survey of data quality tools. Datenbank-Spektrum **14**, 48 (2005)
18. van der Loo, M.: A formal typology of data validation functions (2015)
19. Chalamalla, A., Ilyas, I.F., Ouzzani, M., Papotti, P.: Descriptive and prescriptive data cleaning. In: Proceedings of the 2014 ACM SIGMOD International Conference on Management of Data, pp. 445–456 (2014)
20. Cong, G., Fan, W., Geerts, F., Jia, X., Ma, S.: Improving data quality: consistency and accuracy. In: Proceedings of the 33rd International Conference on Very Large Data Bases, pp. 315–32 (2007)
21. Zio, M., Fursova, N., Gelsema, T., Giebing, S., Guarnera, U., Petrauskiene, J., Kalben, Q., Scanu, M., Bosch, K., van der Loo, M., Walsdorfer, K.: Methodology for data validation 1.0. (2016)
22. van der loo, M.: Properties of validation rules. In: Methodology for Data Validation 1.0 (2016)
23. van der Loo, M.: Validation levels based on decomposition of metadata - Essnet Validat Foudation. In: Methodology for data validation 1.0 (2016)
24. Giessing, S., Walsdorfer, K.: Validation levels from a business prespective - Essnet Validat Foudation. In: Methodology for data validation 1.0 (2016)
25. Winkler, W.E.: Inf. Syst. Methods for evaluating and creating data quality **29**, 531–550 (2004)
26. de Waal, T., Pannekoek, J., Scholtus, S.: Handbook of Statistical Data Editing and Imputation. Wiley, Hoboken (2011)
27. Osborne, J.W., Overbay, A.: The power of outliers (and why researchers should always check for them) Practical assessment, research and evaluation, vol. 9, pp. 1–12 (2004)
28. Schafer, J., Graham, J.: Missing data: our view of the state of the art. Psychol. Methods **7**, 147 (2002)

29. Dusetzina, S., Tyree, S., Meyer, A., Green, L., Carpenter, W.: Linking data for health services research: a framework and instructional guide. Agency for Healthcare Research and Quality (US), Rockville (MD) (2014)
30. Forchhammerl, B., Papenbrockl, T., Steningl, T., Viehmeierl, S.: Duplicate detection on GPUs. HPI Future SOC Lab: Proc. 2011 **70**, 59 (2013)
31. Elmagarmid, A.K., Ipeirotis, P.G., Verykios, V.S.: Duplicate record detection: a survey. IEEE Trans. Knowl. Data Eng. **19**, 1–16 (2007)
32. Christen, P. Data matching: concepts and techniques for record linkage, entity resolution, and duplicate detection, Springer Science and Business Media (2012)

Contextualising the National Cyber Security Capacity in an Unstable Environment: A Spring Land Case Study

Mohamed Altaher Ben Naseir[1(✉)], Huseyin Dogan[1], Edward Apeh[1],
Christopher Richardson[2], and Raian Ali[1]

[1] Bournemouth University, Poole, UK
{mnaseir, hdogan, eapeh, rali}@bournemouth.ac.uk
[2] Digital Smart Solutions Limited, London, UK
christopher@digitalsmart.solutions

Abstract. Threats to global cyber security, including physical, personnel, and information, continue to evolve and spread across a hyper-connected world, irrespective of international borders, in both their elaboration and the scale of their impact. This cyber domain represents a constant challenge to national security, as its socio-technical components are both real and cognisant. The exacerbation of cyber-attacks undermines countries' stability, its escalation produces a landscape of genuine global threat, and the magnitude of its expanding attack mechanisms creates a '*tsunami effect*' on national cyber defenses. This paper reviews the current politically unstable state of Spring Land's cyber security capacity, utilising Interactive Management (IM) approach. It reports the findings of an IM session conducted during a workshop involving a total of 26 participants from the Spring Land National Cyber Security Authority (NCSA), other government agencies. The workshop utilised different IM techniques, such as Idea Writing (IW), Nominal Group Technique (NGT), and Interpretive Structural Modelling (ISM). Using trigger questions, based on the dimensions of the Cybersecurity Capacity Maturity Model for Nations (CCMM), a set of objectives was derived to contextualise and support identified the key initiatives for the development of national cyber security capacity in the country.

Keywords: Cyber security · Cyber security maturity models
Cyber security in Spring Land · Interactive Management

1 Introduction

Over the last decades the global security environment has been characterised by several security insufficiencies, which are defined as a government's inability to meet its national security onuses [1]. The security insufficiencies lead to the state instability. Unstable states are clear and often dramatic examples of unsuccessful governance and public supervision failure [2]. Generally, an unstable state is characterised by: civil war; political and economic upheaval-al; absence of law; lack of a reliable body that rep-resents the state beyond its borders at the inter-national level [2, 3]. Global Security

© Springer Nature Switzerland AG 2019
Á. Rocha et al. (Eds.): WorldCIST'19 2019, AISC 930, pp. 373–382, 2019.
https://doi.org/10.1007/978-3-030-16181-1_35

(Physical, Personnel and Information) threats are continuing to evolve and spread across our hyperconnected world, irrespective of any international borders, in both their elaboration and scale of impact. The threats to any nation's infrastructure of networked information systems fluctuate from degrees of disablement to complete debility [4, 5]. Annual Global Risk reports published by the World Economic Forum (WEF) demonstrate an increased annual technological risk, such as data fraud, cyber terrorism, cyberattacks, and Critical Information Infrastructure (CII) breakdown [6]. Therefore, it is crucial identify potential cyber threats with the potential to have a detrimental rippling effect on various aspects of society and global security.

The aim of this paper is to contextualise the state of the national cyber security capacity maturity levels within unstable environments and provides guidance on moving forward to the higher levels, employing Spring Land as an exemplar case study. Spring Land is a fictional name given to the country from which this real case study is conducted. The paper utilizes the Cybersecurity Capacity Maturity Model (CCMM) for Nations, originally proposed by the Cyber Security Capacity Centre at the University of Oxford [7], as a baseline. The ultimate aim of the paper is to provide benchmark for measuring and planning cyber security for unstable environments.

The paper is structured as follows: Sect. 2 discusses related research, Sect. 3 presents the approach of the present study, and the results of the Interactive Management (IM) sessions are presented in Sect. 4. Finally, Sect. 5 provides a discussion and conclusions.

2 Related Research

2.1 Cybersecurity at a Nation-State Level

Cyber strategic stability has become a central issue for many countries, and it is increasingly imperative that it is strategically correlated, in leading economies, such as the United Kingdom (UK), NATO and the United States (US) [4]. Cyber security was defined by the International Telecommunication Union (ITU) as: the collection of tools, policies, security concepts, security safeguards, guidelines, risk management approaches, actions, training, best practices, assurance and technologies that can be used to protect the cyber environment and organisations and user's assets [8]. US President Obama's (2013) Executive Order 13636 "Improving Critical Infrastructure Cybersecurity" addressed the threats the US faces in cyberspace [9]. President Obama stressed that nations must remain vigilant, and ensure the resilience of their complex critical infrastructure systems, whether physical or cyber, by mitigating the threats and fissures that can weaken them. This Executive Order altered the approach of many security practices in terms of how and where cyber issues are addressed, improving the resilience of the national critical infrastructure.

Meanwhile, in 2013, allowing for the significance of Critical Information Infrastructures (CIIs), the European Union (EU) created a practical guide concerning national cyber security strategies (NCSS), later updating it in 2016 to solidify the CIIs' status in neutralising terrorist cells [10]. This CIIs guide helped EU member states to develop their own robust national cyber resilience capability, thereby acknowledging the existence of cyber threats and their risk to national security.

Cyberspace is the 5th domain, alongside land, sea, air, and space, in modern warfare at the operational level, as soldiers are increasingly reliant on digital capacity, and also at the strategic level, since a state's weaknesses and strengths in cyberspace can be employed to deter and affect the strategic balance of power. Nation-states employ cyber weapons directly to disrupt other nation-states' critical infrastructure and computer systems [11].

Researches demonstrated that the motivation behind most cyber-attacks in 2017 was driven by cybercrime, hacktivism, cyber espionage, cyber terrorism, and cyber warfare [12]. The upsurge in cyber espionage has become a significant factor informing Diplomatic, Information, Military, and Economic in the regard to the art of war, due to the development of cyber technology, and the transformation of traditional means of intelligence into cyber espionage. Terrorists and spies can now employ Open Source Intelligence (OSINT) in the cyber domain, as a means to gather information that is not disclosed publicly, as world leaders are acutely aware of, and complain publically about, the potential damage to intellectual property posed by cyber espionage [13]. With regards to cyber-attacks we differentiate: 'cyber terrorism' which uses computers as weapons, or targets, by politically motivated international, or sub-national groups who cause violence and fear in order to influence an audience, or cause a government to change its policies [14]; and cyber warfare, which references attacks conducted by nation-state actors [15].

2.2 Cyber Security in Spring Land

To date, Spring Land authorities have been unable to reinforce their security position, or to enhance their cyber security to meet this demand and its associated criminal, malicious, or state-inspired risks of increased online activity. In 2013, the Spring Land government officially established the National Cyber Security Authority (NCSA), the primary mission of which was to encourage and sustain the secure use of digital services, together with preventing, detecting, and responding effectively to the associated cyber risks [16]. In the same year, with the support of (ITU), Spring Land Computer Emergency Response Team (CERT) was established with national-level responsibilities, and is in charge of prevention, detection, and mitigation of cyber threats. Due to the current political conflict and austerity measures, NCSA faces a lack of funding, which hinders its attempts to advance cyber security [17]. Thus, its ability to address cyber security concerns at any level does not inspire sufficient public confidence. The onus is now on the Spring Land Homeland Security apparatus to prevent any possible terrorist threats, and to preserve and protect the country's critical infrastructure through applying coherent strategy shared by all the relevant departments.

2.3 Cybersecurity Capacity Maturity Model for Nations (CCMM)

Assessing the risk of national critical infrastructure has gained increasing attention. The assessment and detection of cyber threats is conducted though CCMMs [18]. Various types of CCMM exist, such as the International Organisation for Standardisation's Systems Security Engineering Capability Maturity Model (SSE-CMM), the National Institute of Standards and Technology (NIST) Cybersecurity framework, and the US Department of Energy's Cybersecurity Capability Maturity Model (C2M2) [18]. The majority of these frameworks are employed at an organisational level; whereas the CCMM proposed by the Cyber Security Capacity Centre at the University of Oxford is employed at a national level, and has been deployed to review cyber security capacity in over 40 countries [7]. Developed through collaboration with international stakeholders, this academic model is politically neutral, offering a comprehensive analysis of cybersecurity capacity through five different dimensions: (i) Cyber Security Policy and Strategy; (ii) Cyber Culture and Society; (iii) Cyber Security Education, Training, and Skills; (iv) Legal and Regulatory Frameworks; and (v) Standards, Organisations, and Technologies. Each dimension includes multiple factors and attributes, each making a significant contribution to capacity building. Meanwhile, each factor, involves five stages of maturity, with the lowest indicator implying a non-existent, or inadequate, level of capacity, and the highest indicating both a strategic approach, and ability to dynamically enhance against environmental considerations, including operational, socio-technical, and political threats [7]. These dimensions were employed when establishing the trigger questions in the present study's IM workshop, in order to capture feedback from the participants, and to contextualise the problem space, centred on the Spring Land case study.

3   Research Method

The aim of this case study was to review the current state of Spring Land's cyber security capacity utilizing an approach called Interactive Management (IM). This case study is an example of a Socio-Technical System (STS) of unstable environment. According to Baxter and Sommerville [19], the STS considers human, social and structural factors, as well as technical factors in the design of organisational systems. This is supported by Appelbaum and Trist [20] by claiming that STS design functions on the presumption that an organisation is a combination of social and technical parts open to its environments. IM was chosen for this reason for the present study, and is discussed further in the following section.

3.1   Interactive Management (IM)

The IM technique concerns complex situations requiring a group of people, who are knowledgeable in terms of the situation, to collaborate in tackling the main aspects of an issue, to develop a deep understanding of the situation under analysis, and to detail the basis for effective action [21]. The concept was developed by Warfield and Christakis [22] in 1980. IM involves three phases; the planning phase: in this phase, the situation is defined, and the scope of the issue clarified [21].The workshop phase: this phase involves uniting a group of participants in an understanding of the issue, or situation. According to Ward et al. [23], the IM workshop involves three procedures: Idea Writing (IW), Nominal Group Technique (NGT), and Interpretive Structural Modelling (IS) [21, 24]. In IW, a trigger question is provided to the participants, about which they are invited to silently compose their ideas in a written form. This is followed by the NGT, in which the participants generate further ideas, based on the more holistic view of the problem gained from the IW. The final part of the workshop involves transforming the idea statements into objectives, and then building an

Interpretive Structural Model (ISM) to identify the relationships between the various items surrounding the problem. The follow-up phase: in this phase, the outcome and the objectives derived from the previous phase are initiated, commencing the implementation plan towards a solution.

3.2 Participant's Profile

In this present study, a one-day workshop hosted by NCSA was conducted with a total of 26 participants (25 male, 1 female) from different stakeholders. The age of participants was between 25–55 years old. NCSA issued an invitation letter to all of the stakeholders to help the researcher to contextualise the problem space, which featured the current state of Spring Land's homeland security. The participants were selected due to their contributions in their decision making roles, and included government officials, managers, and general employees participating in security development from areas such as Defence, e-services, Private Sector, Banking, Digital Crime Unit, Oil and Gas sector and Intelligence agency.

4 Results

4.1 Idea Writing (IW) Results

IW was employed to reveal the issues relating to a given a trigger question, providing the participants with a forum to brainstorm and exchange ideas. The participants were divided randomly into three groups to discuss the question, and to provide their views concerning the issues relating to cyber security in Spring Land. The trigger question employed was: What are the current issues of cyber security in Spring Land? After the session, the statements produced were numbered, merged and organised, then sorted into categories, according to each of the CCMM dimensions. Table 1, below, presents the list of shortcomings to face in unstable environments taking Spring Land as a case study.

Table 1. Unstable environments Vs CCMM dimensions

D1 - Cyber Security Policy and Strategy
D1.1. Lack of a national cyber security strategy;
D1.2. Unavailability of a national risk management plan, and threat of cyberspace, has not been identified on the national or sector-specific level;
D1.3. Deficiency of a national roadmap for a cyber defence strategy;
D1.4. Difficulty in implementing the cyber security strategy, due to political issues, and scarcity of resources;
D1.5. Absence of a public and private partnerships for sharing information;
D1.6. Miss of a national crisis management protocol and incident response plan for national critical infrastructure assets, and this has not been prioritised;
D1.7. Lack of a national cyber security framework for monitoring the adoption of international cybersecurity standards in the government sectors

(continued)

Table 1. (*continued*)

D2 - Cyber Culture and Society
D2.1. Lack of a cyber security culture, and the absence of an understanding of cyber-risk and its consequences in public and private sectors, and decision makers;
D2.2. Lack of awareness-raising programmes on the governmental level;
D2.3. Citizens' confidence in the use of e-government services is weak
D3 - Cyber Security Education, Training, and Skills
D3.1. Dearth of experienced people to train and teach cyber security programmes, and migration of experiences, due to the security situation in the country;
D3.2. Lack of a national plan or curriculum in the education system that meets the needs of the cyber security environment;
D3.3. Education outputs in the cyber security domain are weak, and focus only on technical issues;
D3.4. Absence of training collaboration between the public and private sector;
D3.5. Lack of a strategic view of cybersecurity capacity building
D4 - Legal and Regulatory Frameworks
D4.1. Non-existence of cybersecurity legislation or regulations to protect personal, commercial, and governmental data. In addition, the initiatives to issue laws related to cyber security face difficulties, resulting from the political situation;
D4.2. Lack of legislation or regulations for reporting breaches and abuses of cyberspace;
D4.3. Absence of a legislative system, due to unrest in the political situation;
D4.4. Poor cooperation between the authorities in the Ministries of Justice and Interior, especially in the field of digital criminal investigation;
D4.5. Absence of human rights law concerning cyberspace;
D4.6. Lack of an official national framework for the reporting or sharing of technical vulnerabilities;
D4.7. Insufficiency of specific legislation concerning cybercrime, and lack of courts to handle cybercrime cases;
D4.8. Shortage of resources and expertise for digital crime investigation
D5 - Standards, Organisations, and Technologies
D5.1. Lack of use of the information security management systems (ISMS), in all governmental sectors, except for a telecommunications provider
D5.2. Most government sectors use technologies and applications provided by third parties and international companies, without heeding the need to review the security vulnerabilities in the systems;
D5.3. Lack of a national agency for digital certification;
D5.4. Absence of national benchmarking, auditing, and risk assessment policy;
D5.5. Lack of a national infrastructure resilience plan. Military and political conflicts have severely affected the resilience of the infrastructure, and exposed the telecommunications, electricity, and water sectors to destruction or theft

4.2 Nominal Group Technique (NGT) Results

The NGT technique was employed to generate and obtain an initial rating of a set of objectives. Following the organising and numbering of the IW, the participants were required to transform their ideas into a set of objectives, which were used to create an interpretive structural model (ISM), and to summaries the interactions between them.

The final part of the workshop employed the NGT, requiring the participants to select their top three objectives from the list for each dimension, with one being the least important, and three the most important. A total of 19 of the participants then voted on the objectives, although seven of the participants failed to vote, owing to external commitments or issues. Table 2 shows the three most important objectives from each CCMM dimension.

Table 2. CCMM Vs three top priority objectives

Dimension	Objective	Total
D1	**D1.1.** Adopt a national cyber security framework	41
	D1.3. Establish a central committee to design a national roadmap for a cyber defence strategy	37
	D1.6. Create a national list of CNI assets, and identify the risk priorities	13
D2	**D2.2.** Develop a national awareness program that is compatible with the current situation, targeting all of society	40
	D2.3. Encourage all stakeholders to run regular awareness-raising campaigns	39
	D2.4. Improve e-services, in order to promote the required level of trust, and improve the application of security measures	20
D3	**D3.1**. Develop national cyber security education and cyber security modules	40
	D3.2. Provide a sufficient budget for capacity building	39
	D3.4. Classify training needs, and develop cyber exercises and drills	12
D4	**D4.1**. Draft national laws and regulations relating to digital crime	40
	D4.2. Create a strong national legal framework for the sharing of information incidents, vulnerability disclosure, and reporting	31
	D4.3. Build and strengthen national capacity in law enforcement	27
D5	**D5.1**. All stakeholders to adapt and adopt international standards, such as ISO27000	39
	D5.2. Create a national risk assessment, crisis management, and auditing framework	31
	D5.5.Enhance physical security	26

4.3 Interpretive Structural Modelling (ISM)

The ISM technique helped the participants to examine the inter-relationships between the elements gained through the NGT process, and provided a structure for tackling its complexity [25]. The ISM is an acknowledged methodology for classifying relationships among a set of interconnected criteria, which define a problem or an issue [26]. In order to create a clear ISM, the objectives were grouped by similarity, to facilitate the identification of the three most important objectives from each dimension, which are presented in Table 2. The ISM, derived from the objective statements and their interactions based on the dimensions of the CCMM, is represented in Fig. 1.

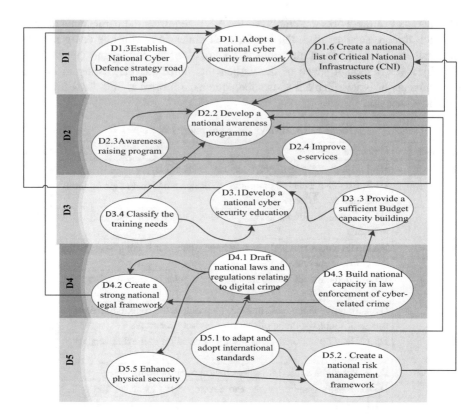

Fig. 1. Interpretive structural modelling for unstable environment

As can be observed from Fig. 1, the development of a national blueprint is deemed to be important because current state interactions in cyberspace manifest the lack of national cyber frameworks. The results in Fig. 1 also show that the group considered that the provision of a robust national awareness programme, which targets the whole society, would be a significant factor in improving national cyber security. The group believed that the creation of a national strategy framework would drive the creation of an effective national legal framework, which would assist in the improvement of information sharing, incident vulnerability disclosure, and reporting between governmental sectors. Furthermore, the group decided that enhancing physical security would also help to increase the national and organisational capability to resist and react to internal and external threats.

5   Conclusion and Future Work

In this paper we explored the main characteristics of Spring Land cyber security as an example of unstable STS. We used IM-based approach. IM provided a rational grounding in the current cyber security challenges in Spring Land, and how they should be addressed. The set of problem statements and objectives derived from the IM approach can be employed to support the management of a national cyber security capacity in an unstable environment, similar to the case study exemplar presented herein. However, these results require further validation and generalisable data, which will be addressed in a future study. The relationships between objectives from ISM will be analysing using the adjacency matrix and create a reachability matrix. In addition, development of a meta-model based on the Interpretive Structural Model (ISM) developed for an unstable environment. The modelling approaches such as IDEF0, UML, SysML, Data Flow Diagrams and Flow Charts can be used to decompose the ISM into further applied functional models.

References

1. McCrabb, M.: Rough waters. Nav. War Coll. Rev. **70**, 141–145 (2017)
2. DeRouen Jr., K., Goldfinch, S.: What makes a state stable and peaceful? Good governance, legitimacy and legal-rationality matter even more for low-income countries. Civil Wars **14**, 499–520 (2012)
3. Brinkerhoff, D.W.: Rebuilding governance in failed states and post-conflict societies: core concepts and cross-cutting themes. Public Adm. Dev.: Int. J. Manag. Res. Pract. **25**, 3–14 (2005)
4. Nissenbaum, H.: Where computer security meets national security. Ethics Inf. Technol. **7**, 61–73 (2005)
5. Bush, G.W.: President George W. Bush: The National Strategy to Secure Cyberspace. Morgan James Publishing (2003)
6. W.E.F. Weforum: The Global Risks Report 2018 13th (edn.) (2018). http://www3.weforum.org/docs/WEF_GRR18_Report.pdf
7. GCSCC: Cybersecurity Capacity Maturity Model for Nations (CMM) (2017). https://www.sbs.ox.ac.uk/cybersecuritycapacity/system/files/CMM%20Version%201_2_0.pdf
8. ITU: Definition of Cybersecurity (2016). http://www.itu.int/en/ITU-T/studygroups/2013-2016/17/Pages/cybersecurity.aspx
9. Orded, E.: Executive order–improving critical infrastructure cybersecurity (2013)
10. Argomaniz, J.: The European union policies on the protection of infrastructure from terrorist attacks: a critical assessment. Intell. Natl. Secur. **30**, 259–280 (2015)
11. Hjortdal, M.: China's use of cyber warfare: Espionage meets strategic deterrence. J. Strateg. Secur. **4**, 1 (2011)
12. Passeri, P.: Cyber Attacks Statistics. http://www.hackmageddon.com/category/security/cyber-attacks-statistics/. Accessed 28 Aug 2017
13. Geers, K.: The cyber threat to national critical infrastructures: Beyond theory. Inf. Secur. J.: Glob. Perspect. **18**, 1–7 (2009)
14. Wilson, C.: Computer Attack and Cyberterrorism: Vulnerabilities and Policy Issues for Congress. CRS Report for Congress. Congressional Research Service-The Library of Congress (2005)
15. Prichard, J.J., MacDonald, L.E.: Cyber terrorism: a study of the extent of coverage in computer science textbooks. J. Inf. Technol. Educ.: Res. **3**, 279–289 (2004)
16. NCSA: The National Cyber Security Authority (NCSA) (2013)
17. Symantec: Cyber crime and cyber security trends in Africa Report, Symantec (2016)

18. Miron, W., Muita, K.: Cybersecurity capability maturity models for providers of critical infrastructure. Technol. Innov. Manag. Rev. **4**, 33 (2014)
19. Baxter, G., Sommerville, I.: Socio-technical systems: From design methods to systems engineering. Interact. Comput. **23**, 4–17 (2011)
20. Appelbaum, S.H.: Socio-technical systems theory: an intervention strategy for organizational development. Manag. Decis. **35**, 452–463 (1997)
21. Warfield, J.N, Cárdenas, A.R.: A Handbook of Interactive Management. Iowa State Press (2002)
22. Banathy, B.A.: Information-based design of social systems. Syst. Res. Behav. Sci. **41**, 104–123 (1996)
23. Ward, J., Dogan, H., Apeh, E., Mylonas, A., Katos, V.: Using human factor approaches to an organisation's bring your own device scheme. In: International Conference on Human Aspects of Information Security, Privacy, and Trust, pp. 396–413 (2017)
24. Janes, F.R.: Interactive management: framework, practice, and complexity, pp. 51–60 (1995)
25. Dogan, H., Pilfold, S.A., Henshaw, M.: The role of human factors in addressing systems of systems complexity. In: 2011 IEEE International Conference on Systems, Man, and Cybernetics, pp. 1244–1249, (2011)
26. Shahabadkar, P.: Deployment of interpretive structural modelling methodology in supply chain management–an overview. Int. J. Ind. Eng. Prod. Res. **23**, 195–205 (2012)

EU General Data Protection Regulation Implementation: An Institutional Theory View

Isabel Maria Lopes[1,2,3(✉)], Teresa Guarda[1,4], and Pedro Oliveira[3]

[1] Centro ALGORITMI, Guimarães, Portugal
tguarda@gmail.com
[2] UNIAG (Applied Management Research Unit), Bragança, Portugal
[3] School of Technology and Management, Polytechnic Institute of Bragança,
Bragança, Portugal
{isalopes,pedrooli}@ipb.pt
[4] Universidad Estatal Peninsula de Santa Elena – UPSE, La Libertad, Ecuador

Abstract. The General Data Protection Regulation entered into force on 25 May 2018, but was approved on 27 April 2016. The General Data Protection Regulation (GDPR) aims to ensure the coherence of natural persons' protection within the European Union (EU), comprising very important innovative rules that will be applied across the EU and will directly affect every Member State. The organizations/Institutions had two years to implement it. Despite this, it has been observed that, in several sectors of activity, the number of organizations having adopted that control is low. This study aimed to identify the factors which condition the implementation the GDPR by organizations. Methodologically, the study involved interviewing the officials in charge of information systems in 18 health clinics in Portugal. The factors facilitating and inhibiting the implementation of GDPR are presented and discussed. Based on these factors, a set of recommendations to enhance the implementation of the measures proposed by the regulation is made. The study used Institutional Theory as a theoretical framework. The results are discussed in light of the data collected in the survey and possible future works are identified.

Keywords: Regulation (EU) 2016/679 · General Data Protection Regulation · Institutional Theory · Health clinics

1 Introduction

Existing European data protection rules, mainly expressed via the EU Directive 95/46/EC, laid out a respectable foundation for the development of EU member states' national legislations.

However, new legislation (GDPR) was needed since the original directive of 1995 was formulated in what now appears to be a different technological era. Back then, just 1% of the world's population was using the Internet, but today it is almost ubiquitous across the EU. Cloud computing and social media were not known then, nor were smartphones or tablets. Today, the vast majority of information is produced and consumed electronically, making it harder to protect [1].

© Springer Nature Switzerland AG 2019
Á. Rocha et al. (Eds.): WorldCIST'19 2019, AISC 930, pp. 383–393, 2019.
https://doi.org/10.1007/978-3-030-16181-1_36

The GDPR aims to improve the level of personal data protection and harmonization across the EU as Directive 95/46/EC no longer meets the privacy requirements of the present-day digital environment.

The GDPR aims to meet the current challenges related to personal data protection, strengthen online privacy rights and boost Europe's digital economy. It specifically aims to provide individuals with better capabilities for controlling and managing their personal data [2], hence striving to reinforce the data subjects' trust in personal data collecting companies. Within the new data protection framework, individual service users may also benefit from the free movement of data if it results in growing businesses with improved and personalized services [3].

The definition of personal data has been expanded. It states that personal data includes information from which a person could be identified, either directly or indirectly. Under the new definition, identifiers such as IP addresses and cookies are included as personal information [1]. In GDPR (Article 4 (1)) "Personal data" means any information relating to an identified or identifiable natural person ("data subject"); an identifiable natural person is one who can be identified, directly or indirectly, in particular by reference to an identifier such as a name, an identification number, location data, an online identifier or to one or more factors specific to the physical, physiological, genetic, mental, economic, cultural or social identity of that natural person [4]. In other words, it is any data that can lead to the identification of a specific (living) person. It can be as obviously identifiable data as a name, but it can also be a combination of "innocent" data such as age, height/weight, wealth, job position, company, city, etc. which when combined can allow the identification of a person.

GDPR defines special categories of personal data (sensitive data) that should be protected with additional means, and should not be collected without explicit consent, good reason, or a few other exceptions. Those categories are: racial or ethnic origin, political opinions, religious or philosophical beliefs, trade union membership, genetic data, biometric data, health data, sex life and sexual orientation.

Businesses that understand their data protection obligations and seek to meet them in an intelligent way will be best placed to unlock the benefits of the personal data that they hold. Getting data protection right is not just a matter of legal compliance. It also makes sound business sense. So, whether evolutionary or revolutionary, the GDPR requires a step change for businesses in their management and delivery of personal data and privacy. Planning is required, priorities need to be set and resources allocated, but no responsible business can afford to turn a blind eye to the GDPR's many requirements [5].

As a European regulation, the GDPR becomes the law in all EU member states. The GDPR reinforces the rights of Personal Data Holders, increases the accountability of those who manipulate such data, all aiming to protect citizens' Privacy. Besides increasing organizations' accountability, it applies heavy penalties to all misuses and wrong uses of personal data. By approving this law, the EU intended to increase its citizens' trust in the growing digitalization we live in.

In a study [6] conducted in 57 health clinics (the survey was sent to 190 clinics, but only 57 gave an effective answer) about whether they had started or completed the process of implementation of the measures laid down in the GDPR, 39 (69%) gave a negative answer and 18 (31%) said to have started or completed the adoption of such measures.

Among the 18 clinics which are in the implementation stage, 7 (39%) consider that the process is completed, while the remaining 11 (61%) still have some measures to implement. These figures seem insufficient when bearing in mind that the GDPR had already come into force 6 months before the conduction of this study.

Thus, the following two research questions were formulated in order to guide the research work:

1. Which factors condition the implementation the GDPR?
2. Which recommendations might be put forward as to enhance the implementation of the GDPR by health clinics?

The answer to the first question aims to know the positive and negative conditioning factors influencing the implementation of the GDPR by health clinics. In the possession of these elements, it will be relevant to produce a set of recommendations which enable the implementation of the GDPR measures by clinics.

The structure of the present work consists of an introduction, followed by a desk review on the general data protection regulation and its implementation. The following section focuses on the research methodology, identifying the target population and the structure of the survey. The conditioning factors are discussed in Sect. 4, followed by the guidelines for the GDPR institutionalization in Sect. 5 and the conclusions drawn from the study. Finally, the limitations of this research work are identified and possible future studies are proposed.

2 Institutional Theory

Changes in legislation, technology and in the economy generate modifications in the organizational environment. In the face of this, the search for innovation represents one way for the survival of organizations. The success of the organization is then measured by the capacity to survive, change, and anticipate the market needs [7]. Therefore, organizations gradually institutionalize organizational practices in order to face new realities, which cannot be faced using the previously existing organizational practices.

The Institutional Theory considers the processes through which structures (e.g., frameworks, rules, norms, and routines) are established as trustworthy guidelines for social behavior. Also, it accounts for the way these elements are created, spread, adopted, and adapted throughout time and space, as well as the way they fall into decline and disuse [8]. Institutions may be conceived as high resilient social structures that enable and constrain the behavior of social actors and that provide stability and meaning to social life [9, 10], and [11].

The authors [12] outlined the processes inherent to institutionalization as consisting of four stages, namely innovation, habitualization, objectification, and sedimentation. The institutionalization process starts at "Innovation", which occurs due to external forces such as technological change, legislation, or market forces. In this sense, the word innovation means structural rearrangements or new organizational practices aimed at solving organizations' problems. Following this comes a sequential process of three stages which enables the evaluation of the institutionalization degree of a certain social reality.

In an organizational context, the process of "Habitualization" involves the creation of new structural arrangements in response to specific organizational problems or sets of problems, shaped through policies and procedures of a specific organization or set of organizations with similar problems. Hence, this is the pre-institutionalization stage.

After the solution for the problem has been generated, it is possible to move on to the "Objectification" process, which accompanies the spreading of the new structure, expanding its use. Objectification implies the development of a certain degree of social consensus regarding the structure and its growing adoption, based on that consensus, by the organization. This process configures the semi-institutional stage.

The stage in which institutionalization is complete is called "Sedimentation" and it is characterized by the adoption of the structure or organizational practice by the whole organization for a long period of time.

The author [11] discusses the distinction between studies focusing on the creation of institutions and studies focusing on the change of institutions. The first ones focus on the process and the conditions which give place to new rules, understandings, and practices. The second ones examine the way a set of beliefs, norms, and practices is attacked, becomes "non-legitimate" or falls into disuse, being then replaced by new rules, ways, and scripts. Deep down, these two processes are related, as the institutional creation implies the change of the existing institutions and the institutional change implies the creation of new institutions.

The Institutional Theory classifies into three pillars the way structures or mechanisms of diverse nature, which are essential for the creation of new institutions or for the change of existing institutions, can be created, maintained, altered, or destroyed. Those three pillars of institutions are the regulative, normative, and cultural-cognitive pillars, and their main features are indicated in Table 1.

Table 1. Pillars of institutions. source [11]

	Regulative	Normative	Cultural-cognitive
Basis of compliance	Expedience	Social obligation	Taken-for-grantedness shared understanding
Basis of order	Regulative rules	Binding expectations	Constitutive schema
Mechanisms	Coercive	Normative	Mimetic
Logic	Instrumentality	Appropriateness	Orthodoxy
Indicators	Rules Laws Sanctions	Certification Accreditation	Common beliefs shared logics of action isomorphism
Affect	Fear guilt/innocence	Shame/honor	Certainty/confusion
Basis of legitimacy	Legally sanctioned	Morally governed	Comprehensible recognizable culturally supported

The regulative pillar constrains and regulates behavior through formal rules, sanctions and punishments. Therefore, the legitimacy of actors' actions is based on the compliance with the legally sanctioned instruments. In the normative pillar, emphasis is given to a deeper moral legitimating basis, in which values and norms are highlighted as elements capable of pressing organizational action, thus turning into a social obligation through daily use. The third pillar, the cultural-cognitive structures, sustains meanings which are shared among the actors about the regulative and normative structures, that is to say, about the reality which surrounds the actors while they continuously build and negotiate that social reality, within a context that includes symbolic, objective and external structures which offer guidance for understanding and action.

Just as it is possible to analyze the evolution of a certain institution within an organization, it is also possible to interpret evolutions in other levels of analysis, such as in industrial sectors and societies. Since its inception, Institutional Theory has been used to analyze and make sense of institutionalization processes in organizations, industries and societies. Viewed as projections of organizations in what concerns their information manipulating activities [13], information systems have also constituted fertile ground for the application of Institutional Theory. Illustrative studies of this application are the works by [14–21].

3   Research Methodology

The survey (first study) was sent to 190 clinics, but only 57 gave an effective reply, which corresponds to a response rate of 30%. The sample subjects were selected randomly based on the kind of clinic and their location distributed throughout the 18 inland Portuguese districts as well as Madeira and the Azores [6].

Considering the research questions previously formulated, and in articulation with the selected research method (Institutional Theory), a primary data collection tool emerged: the interview.

For a better understanding of the results, we can group the clinics into three clusters (Table 2):

Table 2. Clusters of GDPR implementation

Cluster	Stage	Number of Clinics n (%)	Number of Clinics interviewed
1	In compliance with the regulation	7 (12,2)	6
2	Which are implementing the measures imposed by the regulation	11 (19,3)	6
3	Which are not in compliance with the regulation	39 (68,5)	6

- Cluster 1 – Clinics in compliance with the regulation;
- Cluster 2 – Clinics which are implementing the measures imposed by the regulation;
- Cluster 3 – Clinics which are not in compliance with the regulation.

Altogether 18 health clinics were interviewed, distributed equitably among the three clusters (each cluster contributed with 6 interviews).

The Table 3 shows the characterization of the clinics involved in the study according to their type.

Table 3. Clinics Surveyed and interviewed

Type of clinic	Clinics surveyed	Clinics interviewed
Nursing	12	4
Dental	16	6
Ophthalmology	4	1
Medical and diagnostics	15	4
Orthopedics	3	1
Physiotherapy	7	2

The structure of the survey resulted from a desk review on personal data protection and the study of the legal framework Regulation (EU) 2016/679 of the European Parliament and of the Council of April 27 – General Regulation on Data Protection [4].

As far as the process is concerned, the field study was developed through the following steps:

1. Elaborating the interviews guides – three guides were drawn, one for each cluster.
2. Elaborating the codebook – in order to guide the interview codification process, a codebook containing 26 codes was designed according to the previously defined interviews guides.
3. Elaborating coding instructions – along with the codebook, a set of coding instructions was defined describing the procedures that operationalized the codification work.
4. Doing the interviews – all interviews were audio recorded, after obtaining the interviewees' authorization.
5. Transcribing the interviews – all interviews were fully transcribed.
6. Codifying the interviews – the codification of all interviews was done with the support of a data analysis application.
7. Analyzing results – after the interviews codification, the results were analyzed in light of Institutional Theory.

4 Conditioning Factors

The analysis of the interviews led to the identification of various conditioning factors to the implementation of GDPR by the health clinics. Part of these factors is positive, facilitating the implementation of such measures. Another part is negative, inhibiting the implementation of measures. According to the nature of the identified factors, it was possible to categorize them according to the three pillars of institutions, as shown in Table 4.

Table 4. Conditioning factors in the adoption of ISS policies

Institutional pillar	Factors	
	Facilitators	Inhibitors
Regulative	Defining policies, processes and procedures Proactive and knowledgeable staff and manager with regard to the GDPR Risk assessment Monitoring the compliance with the GDPR. Maping of all personal data	Absence of initiative from the management Disobedience from staff Shortage of technicians Articulation of the policy with the law Absence of outsourcing
Normative	Budget and organizational strengthening Reinforcement of data treatment security. Training sessions and staff awareness raising Adoption of solutions capable of handling the detected non-conformities Measuring the level of adequacy and efficiency of the implementation process	Low budget for investing in IT equipment Absence of a clear definition of a course of action Changes in the technological and procedural structure
Cultural-cognitive	Ensuring that the personal data protection policy is transversal to the whole organization Explaining the advantages of personal data protection to staff Involving staff in the implementation of the GDPR Understanding the implementation as na opportunity for organizations to review their processes	Complexity of creating and scheduling a course of action Staff's resistance to change Unawareness of the risks of not protecting personal data Primacy of technology over the GDPR measures Lack of time. Organizations' small size and low number of staff

With regard to the regulative pillar, the definition of targets and refinement of processes stand out among the facilitating factors to enable organizations to reach a good level of compliance with the measures to be implemented. The definition of such targets and policies takes place after a rigorous analysis of all the personal data handled daily by the organization.

The inhibiting factors highlighted in the same pillar are the difficulty to implement the GDPR whenever the managers and headships are not involved in the whole process. Another inhibiting factor is the non-compliance of the policy drafted by the organization with the regulation.

As far as the normative pillar is concerned, the facilitating factors which stand out are the redefinition of the internal mechanisms and procedures and the adoption of solutions capable of handling the non-conformities detected. Such solutions may be computing, documental or procedural-related. Note that this is possibly the most sensitive key point regarding the GDPR implementation, since each organization has its specificities and it is paramount to guarantee that the solutions implemented are feasible within the organizational routine.

Concurrently, the need to provide staff with training and awareness raising sessions must be pondered.

In this pillar, the inhibiting factors we highlight are the clear absence of courses of action as well as the definition of the work team and their respective functions. Such work team should be as multidisciplinary as possible.

With regard to the cultural-cognitive pillar, the noteworthy facilitating actions towards the GDPR implementation are the staff's involvement in the implementation as well as the acknowledgement of such action, so as to guarantee that the personal data protection policy adopted is transversal to the whole organization.

The inhibiting factor highlighted in this pillar is the unawareness of the risks of not protecting personal data, which is oftentimes a consequence of the fact that the company is viewed as small, handling little personal data, and with few staff, which leads the headships to mistakenly consider data protection as a considerably easy process. The right to privacy is one of the cornerstones of democratic societies, thus playing a major role in all types of organization, regardless of their size.

5   Guidelines for the GDPR Institutionalization

Considering the identified factors influencing the implementation of the GDPR, we argue that the institutionalization of GDPR in health clinics will be a process of several stages, shaped by pressures of regulative, normative, and cultural-cognitive nature.

As far as the regulative pillar is concerned, we consider that the implementation of the GDPR is in most cases a lengthy, complex and costly process, but an indispensable one too in order to decrease the risk of non-compliance with the new rules, and a fundamental one to secure the rights and guaranties of each one of us. For that, the definition of a set of measures to be followed by organizations might be an asset for the successful implementation of the GDPR. Obviously, each organization has its own specificities and such model will not totally adapt to all organizations. However, the guidelines it comprises may be of great help to those who are in the implementation stage, therefore contributing for the adoption of the GDPR by the largest possible number of organizations, be it in the health sector or not, since all organizations must ensure the personal data protection and secure the rights of the holders of such data.

With respect to the normative pillar, we must bear in mind that the law regulates data protection but does not solve the issue of personal data protection if people are not aware or sensitive to this matter, to the rights they have and the precautions they must take. Therefore, we suggest the definition of an awareness raising program addressed to all the workers who handle personal data.

As far as the cultural-cognitive pillar is concerned, the most immediate measure which could be adopted is programming training sessions in the scope of the GDPR, in which users are trained to have behaviors which protect personal data. Therefore, we suggest the creation of sections aiming to provide workers with clarification, guidance and training regarding the norms imposed by the GDPR. However, it is not enough to train and enlighten workers towards the GDPR and personal data treatment. It is also necessary to learn how to deal with the cultural change that a project of this kind implies.

The conjunction of these actions to enhance the implementation of the GDPR in Health clinics can be summarized in six essential points:

1. Review periodically - Data protection is not a destination, it is a journey! So, review on a regular basis and correct your course accordingly
2. Raise Awareness - Conduct staff training and awareness sessions. Most breaches occur due to staff ignorance so make them aware and mitigate the risk.
3. Review policies & related documents - Review existing policies and create new ones where needed. These policies allow you to formulate processes and procedures for staff to follow
4. Make a plan - Based on the audit and risk assessment, determine a roadmap to achieve compliance. While it is important to address high risk areas, do not ignore the low hanging fruit. Small, easy wins can get the project off to a positive start!
5. Identify & Assess privacy related risks - Assess the risks associated with the ways of processing the data. Compile a risk register.
6. Conduct an audit - Know your Data - how you got it; who can access it, where it is stored; how long you keep it etc. Create a list of recommendations to address areas of concern.

According to the institutionalization stages proposed by [12], the institutionalization process starts at "Innovation". In the case of the GDPR implementation, innovation was triggered by external forces, namely the approval of the new General Data Protection Regulation on April 27 2016. The GDPR 2016/679 aims to ensure the coherence of natural persons' protection within the EU, comprising very important innovative rules that will be applied across the EU and will directly affect every Member State. For the subsequent stages – Habitualization, Objectification and Sedimentation – we suggest that the mechanisms brought forward by the regulative, normative and cultural-cognitive pillars can support the organizations' evolution throughout those stages.

In the field of action, the institutionalization process can occur essentially according to two formats: in a naturalist way or based on agents' action [11]. The first format matches a situation in which the phenomenon is gradually institutionalized in a natural way, which normally represents a slow and long process. The second format, based on agents' action, introduces a catalyzing element – the agent – which enables the acceleration of the institutionalization process. Contrarily to what happens in the naturalist way, in the institutionalization based on agents' action, the normative frameworks are designed, created and modified rationally, through conscientious and deliberate processes, the same happening with cultural-cognitive elements which, in this case, also tend to be conscientiously conceived and spread by certain agents.

The strategy based on agents is a way to enhance the institutionalization of the GDPR. The main agents who may play an active part in this process are the organizations' headships, the workers in charge of implementing the GDPR and the Government which proposes and passes the draft law ensuring the national execution of the GDPR.

6 Conclusion

Most of the information currently shared digitally was previously shared in paper. This change brings along new digital challenges and threats as far as security and privacy are concerned, namely with regard to the protection of personal data in a society increasingly more digital [8, 22, 23].

Although the GDPR came into force two years ago, most organizations only awoke to the complexity of this issue on last May 25, when the GDPR effectively started to have total legal effect.

This study identified a set of factors which condition the implèmentation of the GDPR in the health clinics. Besides this contribution, this paper brought forward guidelines which are believed to enhance the institutionalization of measures imposed by the GDPR in organizations. We also argue that the use of Institutional Theory as a support to the interpretation of the adoption stage of implementation by organizations and as a support to the projection of guidelines which enhance the institutionalization of these GDPR measures in organizations represents a promising means for research.

The implementation of the regulation implies the definition of procedures, records and policies. Both people and technologies represent critical success factors to its implementation. Therefore, it might be relevant to carry out further research to determine to what extent this GDPR, although targeted at data protection, might not be as well a booster for the digital transformation of health clinics.

Acknowledgments. UNIAG, R&D unit funded by the FCT – Portuguese Foundation for the Development of Science and Technology, Ministry of Science, Technology and Higher Education.. Project n. ° UID/GES/4752/2019.

This work has been supported by FCT – Fundação para a Ciência e Tecnologia within the Project Scope: UID/CEC/00319/2019.

References

1. Tankard, C.: What the GDPR means for businesses. Netw. Secur. **2016**(6), 5–8 (2016)
2. Mantelero, A.: The EU proposal for a general data protection regulation and the roots of the 'right to be forgotten'. Comput. Law Secur. Rev. **29**(3), 229–235 (2013)
3. Tikkinen-Piri, C., Rohunen, A., Markkula, J.: EU General data protection regulation: changes and implications for personal data collecting companies. Comput. Law Secur. Rev. **34**, 134–153 (2018)
4. European Parliament and Council, Regulation (EU) 2016/679 of the European Parliament and of the Council of 27 April 2016, Official Journal of the European Union (2016)
5. Allen & Overy: Preparing for the General Data Protection Regulation (2018)
6. Lopes, I.M., Oliveira, P.: Implementation of the general data protection regulation: a survey in health clinics. In: 13th Iberian Conference on Information Systems and Technologies, June 2018, pp. 1–6 (2018)
7. Brown, S.L., Eisenhardt, K.M.: Competing on the Edge: Strategy as Structured Chaos. Harvard Business School Press, Boston (1998)
8. Scott, W.: Institutional Theory, pp. 408–414. Encyclopedia of Social Theory, Sage, Thousand Oaks (2004)

9. DiMaggio, P., Powell, W.: Introduction. In: Powell, W.W., DiMaggio, P.J. (eds.) The New Institutionalism in Organizational Analysis, pp. 1–38. University of Chicago Press, Chicago (1991)
10. North, D.: Institutions Institutional Change and Performance. Cambridge University Press, Cambridge (1990)
11. Scott, W.R.: Institutions and Organizations: Ideas and Interests, 3rd edn. Sage, Thousand Oaks (2008)
12. Tolbert, P.S., Zucker, L.G.: The institutionalization of institutional theory. In: Handbook of Organization Studies. Sage, London (1996)
13. Carvalho, J.A.: Strategies to deal with complexity in information systems development. In: Proceedings of the ISAS-CSI 2002 – 6th World Multiconference on Systems, Cybernetics and Informatics, Orlando, pp. 42–47 (2002)
14. Orlikowski, W.J.: The duality of technology: rethinking the concept of technology in organizations. Organ. Sci. **3**(3), 398–426 (1992)
15. King, J., et al.: Institutional factors in information technology innovation. Inf. Syst. Res. **5** (2), 139–169 (1994)
16. Premkumar, G., et al.: Determinants of EDI adoption in the transportation industry. Eur. J. Inf. Syst. **6**(2), 107–121 (1997)
17. Chatterjee, D., et al.: Shaping up for E-commerce: institutional enablers of the organizational assimilation of web technologies. MIS Q. **26**(2), 65–89 (2002)
18. Teo, H., et al.: Predicting intention to adopt interorganizational linkages: an institutional perspective. MIS Q. **27**(1), 19–49 (2003)
19. Baptista, J.: Institutionalisation as a process of interplay between technology and its organisational context of use. J. Inf. Technol. **24**(4), 305–320 (2009)
20. Bharati, P., Chaudhury, A.: Technology assimilation across the value chain: an empirical study of small and medium-sized enterprises. Inf. Resour. Manag. J. **25**(1), 38–60 (2012)
21. Alvarenga, T., Rodriguez, C.: Institutional theory and its applicability in research related to operations management, VII Congresso Brasileiro de Engenharia de Produção, Brasil (2017)
22. Martins, J., Gonçalves, R., Oliveira, T., Cota, M., Branco, F.: Understanding the determinants of social network sites adoption at firm level: a mixed methodology approach. Electron. Commer. Res. Appl. **18**, 10–26 (2016)
23. Martins, J., Gonçalves, R., Branco, F., Peixoto, C.: Social networks sites adoption for education: a global perspective on the phenomenon through a literature review. In: 2015 10th Iberian Conference on Information Systems and Technologies (CISTI), pp. 1–7. IEEE (2015)

Design of a Situation-Awareness Solution to Support Infrastructure Inspections

Carme Vidal Quintáns and Gabriel Pestana[✉]

UNIDCOM – Unidade de Investigação em Design e Comunicação do IADE,
Universidade Europeia, Lisbon, Portugal
carme.vidal.quintans@gmail.com,
gabriel.pestana@universidadeeuropeia.pt

Abstract. This paper demonstrates how an interactive design approach, complemented with information visualization techniques, might contribute to reducing maintenance costs associated to inspections performed in industrial infrastructure, equipped with wind turbines, in particular by triggering events outlining for those situations requiring the intervention of the decision maker (context-awareness). The goal is to mitigate operational costs associated to blade inspections, lessening the time to conclude each inspection without compromising the quality and level of accuracy in identifying possible safety breaches that may compromise the operability of the blades, assuring in this way the sustainability of such critical infrastructures.

The proposed solution addresses a digital cloud-based platform providing information to different stakeholders, enabling those who contracted the inspection services to follow undergoing or programmed actions and to receive notifications whenever their intervention is required. The solution also makes use of autonomous drones to collect on-site data and in this way support the information needs related to blade inspections. From a service-design perspective, a user-centered approach was followed to improve the User Experience (UX) in analyzing events related to blade inspections, complemented with information visualization techniques to streamline their interactions whenever required. The Agile Design Science Research Model (ADSR) was applied to the blade inspection process for a demonstration of the primary outcomes.

Keywords: Interaction Design · Service Design Thinking ·
Information visualization (interactive dashboard) · User interface ·
Data-as-a-Service

1 Introduction

Interaction Design (IxD) is an essential component within the umbrella of user experience (UX) design. It focuses on creating user experiences to support and enhance the way users work, communicate and interact, for instance when dealing with software products. In simple terms, IxD is defined as the design of the interaction between the user and the system of interest, with the goal to improve the user perception and to understand how the designed product addresses their information needs [1].

© Springer Nature Switzerland AG 2019
Á. Rocha et al. (Eds.): WorldCIST'19 2019, AISC 930, pp. 394–404, 2019.
https://doi.org/10.1007/978-3-030-16181-1_37

The paper presents a case study related to inspections performed to blades, describing how IxD was applied to keep decision-makers well informed about events detected during blade inspections – adoption of a context-awareness approach to alert whenever the user intervention is required. In this domain, the operational context is the one provided by ProDrone - a small and medium size enterprise (SME), created in 2015 and based in Lisbon (Portugal) that operates as a service provider in the wind power industry.

ProDrone solution did not support an interactive dashboard to enable their clients (i.e., turbine owner) to get a clear perception about the status of the blades. Besides having a high operational cost, the lack of such situational awareness prevented the client from accessing critical information regarding the risk of degradation of the blades, with a significant impact on the longevity of the equipment. The solution also did not provide a set of services designed to simplify searching and follow-up of specific information about the blades status and the corresponding inspection actions/reports. Such lack of features was compromising the data quality, causing unnecessary redundancies and favoring critical data to become outdated when the client accessed the data. The paper explains the theoretical context that was considered to approach the identified challenges, with a focus on the following main topics:

- Empower the user to receive and monitor information about events related to the status of the blades.
- Trigger situation-awareness to decision makers, streamlining their perception about operations requiring their intervention.

In order to construct the informational artifacts associated with the design of the proposed solution, concepts from Information Visualization domain were used to help in defining the visual representations of the specified informational artifacts. The research also addresses aspects from the Service Design Thinking (SDT) domain, to model the behavior of the system based on ProDrone business goals and user's information needs. A User Model was specified to characterize the actors interacting with the system together with the corresponding requirements to specify the system structure and behavior. Figure 1 presents a graphical representation of the predominant research areas that were considered, outlining how they interact with each other. The concept of UI is defined as the space where interaction takes place, where the user performs tasks in order to achieve goals; a close attention to human factors and ergonomics concerning usability was considered. To quantify the awareness, we took advantage of the use of interactive dashboards on the definition of the UI. The system behavior was modeled using a SDT approach, complemented with the description of the information flow using BPMN diagrams to characterize each data object. All data objects have a visual representation on the dashboard, enhancing situation awareness. Figure 1 also outlines the interaction between the SDT and Information Visualization areas; such interaction streamlines the way the user access to data, following the principles of Data-as-a-Service (DaaS). Overlapping all these areas is the User Model (UM), meaning that each of them will finally depend on the characterization of the different user roles and profiles to be considered and the specifications of the requirements for each of them.

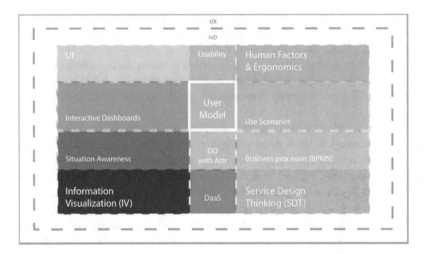

Fig. 1. Diagram outlining the interactions between the research areas

As the approach was to adopt an applied research addressing ProDrone specific business model and since ProDrone followed an agile development methodologies within their internal software development processes, this paper presents the Agile Design Science Research Methodology (ADSRM), proposed by Donnellan [2], to integrate the research process within the company processes, providing a ground-breaking approach to the proposed solution.

2   State of the Art

This section presents a literature review related to the three predominant research areas under the umbrella of UX design. The project proposal addresses aspects related to Interaction Design (IxD), with a more profound contribution from the Information Visualization (IV), Service Design Thinking (SDT) and User Model (UM), which can be considered as sub-domains within the IxD theoretical framework. The paper also addresses aspects related to situation awareness while monitoring activities and events on the system's user interface (UI).

Information Visualization, concept relevant to stimulate the user interest and awareness for those situations requiring intervention or to keep the user well informed about events that require specific attention – for simplicity, such features will be designated just as Situation Awareness (SA). Considering the perception and comprehension of the elements in the environment within a volume of time and space, and the projection of their future status and its potential to influence on effective human decision-making [3]. The objective is to optimize the degree to which information is processed, and interactions are integrated into the problem-solving domain. While data visualization is purely quantitative and numerical, it can also be useful to improve perceptions. Narratives and styles reflected on the content [4], can be used as mental models and schemata, tailoring the way information is presented to users concerning their immediate goals, while providing SA in real time, with predictive insight regarding future events.

The BladeInsight's visual-communication strategy follows the IV capabilities to deal with the complexity of what is represented, reframing the user understanding [5], not only when trying to discover a problem but also to seek for a potential solution. The primary goal of this strategy is to represent the KPIs and highlight specific aspects of the data requiring user attention, allowing the user to get the proper feedback within a reasonable time. The strategy is to present a Style Guide in order to promote visual consistency within the system not only in its current state but also in possible future iterations.

Service Design Thinking, In the era of digital transformation, according to [6], IT services are considered to be the primary driving force within the digital economy. The proposed solution has analyzed and aggregates the processes and actors involved in the inspection process and has created a set of IT services, mapped into a service catalog. The baseline was to provide dynamic interactions to enhance the value proposition of each service. In this context, the SDT approach supported the design of business processes, mapping the user journey, improving the user experience by anticipating and meeting both the user needs and ProDrone's business goals [7]. For the proposed solution, it was possible to align key tech trends related to digital transformation, namely:

- Consider the user as a critical source for innovation, shifting the focus from technical innovations to a service orientation [8] (*the internet of me*);
- Focusing on the value in use (the ability to supply the user with the tools and functionalities needed to accomplish his goals while providing a meaningful experience), rather than product or service characteristics [9] (*outcome economy*);
- Taking advantage of cloud technology to deliver the service, next-generation services (*the platform revolution*);
- Content delivery as a service-oriented feature in order to prove valuable, and therefore, monetizable [10] (*the intelligent enterprise*);
- Automation of processes (and sub-processes) forcing machines and humans to learn to work together effectively (*workforce reimagined*); and
- Process-based vision instead of traditional optimization approach [11], using BPMN modeling tools.

User Model, following a user-centered methodology, it was possible to tackle the definition of a User Model; considering the existence of different individual users (or groups of users) with different roles and therefore, different tasks and goals requiring different semantic representations of the user interface (spatial dashboard) [12]. For the specifications of the requirements for each of them, we took advantage of context diagrams, personas, scenarios and CRUD[1] matrixes. These requirements are valuable inputs to the information visualization and the service design process; insights based on goals rather than actions with emphasis on the user's activities rather than on technical aspects of the system [13]. Apart from user requirements, we also considered business requirements/goals to express the intentions or motivations of the stakeholders regarding the operations supporting business offerings (products or services).

---

[1] CRUD – Create, Read, Update and Delete.

3   Research Challenges

This research followed the Agile Design Science Research Methodology (ADSRM) defined in [14]. The ADSRM aims to find a balance between procedural rigor and a more empirically-driven problem/solution coevolution approach, allowing researchers to engage problems closer to industrial practice, that we believed it is helpful to streamline the integration of the proposed solution within ProDrone's business processes.

The approach followed a methodology structured into two main phases, presented in Fig. 2. During the AS-IS phase, the ProDrone business context and the existing solution were analyzed (as it was before the beginning of this project). The primary goal was to clarify the identification of existing gaps and constraints both from a technical point of view and most importantly from a business perspective. This analysis was supported by a heuristic evaluation by UX experts and usability testing with the end-users; namely, card sorting and tree-map tests; along with, ProDrone's business goals regarding the solution. After that, the attention was on characterizing the customer segment, including a better specification of the user profiles and their informational needs; in parallel, a broader review to the business processes was accomplished and formalized using BPMN[2]. The second phase (TO-BE), can be generically classified as the knowledge acquisition and management of the user expectation in order to achieve a better alignment between the user information needs and the data workflow to be in place to adequately address the identified informational needs. Figure 2 presents a diagram of how the research methodology was applied, outlining the steps followed on each phase. From a SDT perspective, the research challenges were defined as:

- Design an interface for a web-based application to support the business processes and cope with the information needs for each actor involved in the blade inspection process;

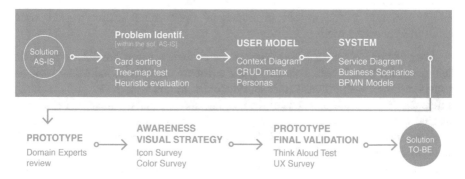

Fig. 2. Diagram with the development timeline of the proposed solution.

---

[2] BPMN – Business Process Model and Notation (http://www.bpmn.org).

- Specify a set of interfaces to provide an interactive context-awareness Dashboards, following an innovative *information visualization* approach to empower the user experience in analyzing events; and
- Streamline the blade inspection process by adopting a user-centered approach to comply with usability standards.

By applying a heuristic evaluation to the AS-IS solution, it was possible to identify a set of gaps that had to be filled in order to overcome the lack of informational artifacts to characterize inspection events accurately. The info-structure of metadata to characterize the existing entities was designed in close cooperation with the development team, integrating feedback from key-users and information acquired from the literature review. The design approach also considered the visual communication strategy to represent the key performance indicators (KPI) of each process/service, highlighting specific aspects of the data requiring the user attention, outlining for the changes in the system status (color codes, iconography, alerts, and notifications). Thereby, mitigating the risk derived from the lack of a situational awareness strategy within the AS-IS solution. Following a user-centered methodology, different user profiles (or groups of users) with different roles were also considered, therefore different tasks and goals were identified requiring different semantic representations of the user interface. Heuristic evaluations from domain experts and stakeholders and iterative user testing with target users were applied to evaluate the results of each phase; online surveys to validate the main elements of the visual communication strategy (color codes and icons); and think aloud task-based tests for final validation of the results.

4   Results Addressing the Research Challenges

This section presents the main achievements of the proposed solution (TO-BE) and how the research challenges were attained. The design of the solution was presented in the form of a clickable (demonstrable) prototype, using the online tool *Invision app*; and a style guide for the layout, navigation patterns, typeface styles, palettes, iconography and interaction components.

The proposed solution presents an interactive and context-awareness dashboard to monitor events related to the status of assets, triggering events for those situations requiring intervention from the decision makers at the office. Issues related to inspections performed at the field or inspection requirements regarding a specific wind blade equipment are mapped as alerts; enhance for the rigor and objectivity of maintenance plans, while preventing from any safety hazard and complex inspection logistics. The non-existence of such interactive interface represents a significant operational issue from both a final perspective and form the lack of information awareness about equipment degradation.

A comparative analysis of current solutions stood out that; drawing a parallelism between emergency management tasks and the ones integrated into infrastructure inspection procedures; mainly, the focus is on the response and recovery phases. Thus, the industry requires data that facilitates the creation of predictive maintenance plans, regarding the principles of economy and prudence. The proposed solution follows a preventive approach with a set of functionalities aiming to provide the user with the tools needed to carry out a predictive maintenance plan, rather than support maintenance activities of potential issues; addressing services such as:

- Alerts (or reminders) based on recommended actions from inspectors (e.g., a particular issue should be monitored, and a new inspection should be scheduled in the short-term)
- Inspection history (e.g., a particular asset has not been inspected in a long time).
- Provide the user with a set of services designed to promote efficient searching, consulting and monitoring of information about the status of specific assets or enable the selection of assets marked (by the system) with a red flag and the corresponding inspection actions.

In this domain, data is not only available but also presented on an interactive way and under the control of the user, enabling to switch between information perspectives to explore the information or to get a better understanding of specific aspects within the data - to support analytical reasoning. To facilitate this analytical approach, we took advantage of mechanisms like sortable and searchable columns, plain filtering tools and view-modes switching.

Aligned with the principles of DaaS (data as a service), users do not passively receive the information, but participate in the production of data. The proposed solution was not designed based on the available data but taking into account the flow of data through the system to address the informational need of the user, giving the user an active role in the creation of value. In order to facilitate collaboration and knowledge transfer, between the different roles and stakeholders, the proposed solution integrates services for social interaction (e.g., comment and tagging other users on particular issues to call their attention or share opinions); and customizable (e.g., expiration date) shareable-links for non-registered users.

In order to ensure the solution scalability and its potential adaptation to other business models or operational domains; the info-structure of the metadata was designed using BPMN, in order to standardize the modeling of the information flows; identifying and characterizing all the informational artifacts needed to provide the services.

Due to the complexity of the system, the number of different profiles identified in the user model (Super Admin, Licensee, Client, Survey Team, Inspector and Third Parties) and the heterogeneity in the requirements for each one of them, for simplicity all upcoming examples relate to the Licensee profile. This profile (Licensees) and the Client profile are the two profiles that have more interaction with the system. It is also important to mention that the Licensee profile integrates all the client's functionalities with their own, more advanced ones (e.g., create a client user account).

A Service Diagram was used to define all the informational artifacts needed to support services, meaning what data is necessary for an activity to execute its function correctly; and, to understand and communicate how services should be organized (structured) within the system. Figure 3 presents an overview of the service diagram developed for the Licensee's user interface. Each data object was characterized by a unique name and a set of attributes, including a status that can also change. All data objects have a visual representation on the dashboard, enhancing situation awareness. The design of the UI took into consideration the structure and behavior defined in the Service Diagram.

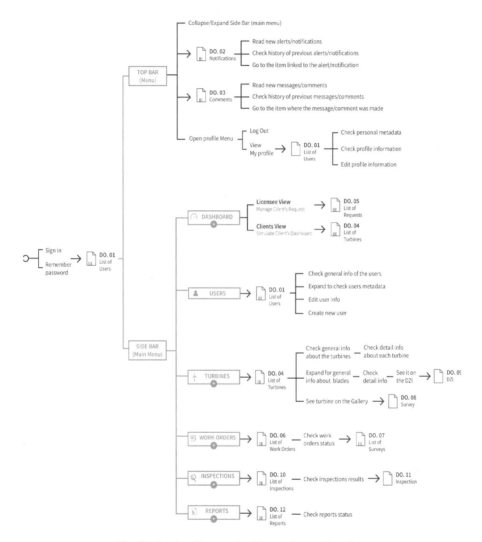

Fig. 3. Service diagram for Licensee's user interface.

Figure 4 shows an example of how the Service Diagram for the interface "*Users*" was implemented on the UI. The data object *"DO.01 – List of Users"* indicates the list of attributes that are represented on the UI (e.g., name, email, etc.). The color code shows the current status (situation awareness) of the DO.01 and the four different services define the UI functionalities (e.g., top right button to create a new user or expand button at the beginning of each row to check the user profile metadata).

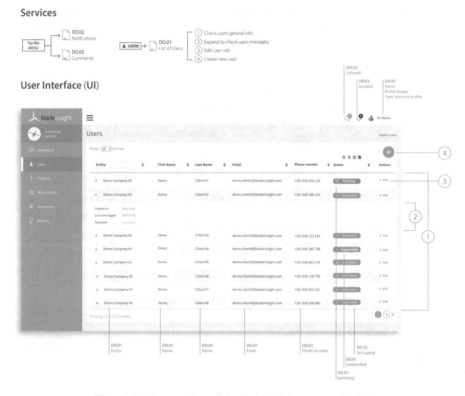

Fig. 4. Implementation of the Service Diagram on the UI

To summarize, the solution does not only contribute to reduce costs by streamlining the processes involved in inspection procedures, but it also optimizes the way users access information (in a maximum of 3 clicks), decreasing the total time between performing an on-site blade inspection and the time a new report is generated with a diagnose from an expert (Inspector).

Besides improving the process with new services addressing the reliability of the outcomes, the proposed solution also improves the global user experience. By defining a visual-communication strategy (color codes and iconography), to highlight specific aspects of the data requiring attention, the user is able to get the proper feedback in a meaningful and straightforward manner, within a reasonable time (e.g., color codes to represent status of processes or severity of damages), and following a preventive approach. The solution provides a set of functionalities aiming to provide the user with the tools needed to carry out a predictive maintenance plan, rather than support mending activities of potential issues (e.g., alerts or reminders, inspection history, etc.).

5 Conclusions and Future Work

In the era of digital transformation, designers assume a new role as system designers; defining the behavior and info-structure of the system and turning essential the exploration of new communication tools that will support their integration in development teams. In particular, interaction designers must try to understand the problems and needs of real users. In this domain, understanding how technology might help to improve the user experience while interacting with the environment can be considered as one of the core contribution of designers in the digital economy; preventing this process to turn out to be a pure technology ostentation, and ensuring instead that the designed solution provides a valuable contribution in addressing real problems.

There is still a lack of understanding of the relevance of the designer's role in the digital transformation process and their involvement in the ideation and conceptualization stages. Designers should work as an interface between users and developers, bringing important conceptual insights that should lead to early developing stages.

More work is still to be done to convince companies about the impact of a good investment in resources and time in order to characterize the project scope adequately and in complying with the role of the users, instead of presenting a standard solution. There is also a lack in bridging the knowledge transfer between the IT industry and Universities in particular in the design domain.

During the development of this research, some features identified as potential improvements were left out of the project scope, in order to achieve its conclusion within the available time and with the available resources. A short overview summarizes the main pending issues for future work:

- User model should be reviewed and supported by in-depth user research, against the current marketing approach. The solution involves a complex value-constellation, with multiple actors and stakeholders.
- From the six different profiles identified in the user model, the focus was only on Licensee. In the future, the analysis and characterization of the other profiles (e.g., Inspectors) should be done.
- None of the current solutions cover all the phases within inspection procedures cycles, leaving some improvement space for the solution to stand above its competitors, namely to consider features for predictive modeling and Artificial Intelligence to improve the proactivity behavior of the system. Key issues to reduce (even more) operational costs and extend the longevity of the equipment.
- There is also a big spectrum of improvement for analytical process modeling in order to automate functionalities with predictive and learning algorithms that would improve the user experience, for instance, to manage inspection plans based on gamification parameters, stimulating competition between inspectors and the supply of equipment components from multiple suppliers.

To conclude, in the future, the relevance of design should be considered to increase the value proposition of the final solution, something that was not a priority in the last years. ProDrone places its main market advantage on the delivery of services, defining themselves as IT service provider focused on technology developments, neglecting the product design component. Future technological improvements should be led by user experience goals; giving the service an actual priority position.

References

1. Interaction Design Foundation: Encyclopedia of Human-Computer Interaction (2014)
2. Stickdorn, M., Schneider, J., Rittel, H.W.J., Webber, M.M.: Dilemmas in a general theory of planning **4**(2) (2011)
3. Endsley, M.R.: Toward a theory of situation awareness in dynamic systems. Hum. Factors J. Hum. Factors Ergon. Soc. **37**(1), 32–64 (1995)
4. Boehnert, J.: Ecological perception: seeing systems abstract the emergence of visual intelligence. In: Proceedings of DRS 2014 Design's Big Debates, pp. 425–438 (2014)
5. Boehnert, J.: Data visualisation does political things. In: Proceedings of DRS 2016 Design + Research + Society - Focused Thinking (2016)
6. World Economic Forum: World Economic Forum White Paper: Digital Transformation of Industries - Healthcare Industry, no. January, p. 15 (2016)
7. Maffei, S., Milano, P., Mager, B., Sangiorgi, D.: Innovation through service design. from research and theory to a network of practice. A users' driven perspective. In: Joining Forces. University of Art and Design Helsinki (2005)
8. Rytilahti, P., Rontti, S., Jylkäs, T., Alhonsuo, M., Vuontisjärvi, H.: Making service design in a digital business. In: DRS 2016, pp. 1–15 (2016)
9. Sakao, T.: What is PSS design? – explained with two industrial cases. Procedia - Soc. Behav. Sci. **25**(2011), 403–407 (2011)
10. Newman, D.: Digital intelligence: the heart of successful digital transformation. Futur. Res. LLC, no. October, pp. 1–11 (2017)
11. Emili, S., Ceschin, F., Harrison, D.: Supporting SMEs in designing sustainable business models for energy access for the BoP: a strategic design tool. In: DRS 2016 (2016)
12. Pereira, J.: Handbook of Research on Personal Autonomy Technologies and Disability Informatics (2010)
13. Jang, S., Nam, K.Y.: Storytelling technique for building use-case scenarios for design development. In: IASDR 2017 (2017)
14. Donnellan, B., Helfert, M., Kenneally, J., Vandermeer, D., Rothenberger, M., Winter, R.: Agile design science research. In: Lecture Notes in Computer Science (including Subseries Lecture Notes in Artificial Intelligence and Lecture Notes in Bioinformatics), vol. 9073, pp. 168–180 (2015)

A Systematic Literature Review in Blockchain: Benefits and Implications of the Technology for Business

João Pedro Marques Ferreira[(✉)], Maria José Angélico Gonçalves, and Amélia Ferreira da Silva

CEOS.P, Porto Accounting and Business School,
Polytechnic of Porto, Porto, Portugal
joao.f.12@outlook.com, {mjose,acfs}@iscap.ipp.pt

Abstract. Blockchain is a core, underlying technology with promising applications to disrupt the world of business. Using a systematic review, in this paper we provide an overview of the concept of blockchain technology, identify some advantages and barriers for business use and find that while some industries seem to benefit more from the technology (healthcare and those that depend on supply-chains like physical retail stores), others may suffer a blow (banks, energy and other service providers).

Keywords: Business innovation · Blockchain architecture · Public ledger · Computational trust

1 Introduction

A blockchain is a digital ledger that records information in what's called "blocks" and protects it using cryptography by linking new information to the one previously stored. Due to rapid developments in blockchain we seek to identify advantages and disadvantages surrounding the technology as well as identify were value is being created, by using a systematic review methodology [1] around business, management and accounting journals.

Blockchain has also seen some recent development in regards to corporate incorporation [2] but there are still many questions left to be answered. In this paper we try to look at benefits and problems as well as different types of applications that blockchain can have.

In this work, we begin by describing the methodology applied. Next, we present the results defining blockchain and pointing advantages and disadvantages provided by the technology, then we look at current research interests and some possible impacts its application can have. We conclude by summarizing some problems blockchain faces coupled with research ideas.

© Springer Nature Switzerland AG 2019
Á. Rocha et al. (Eds.): WorldCIST'19 2019, AISC 930, pp. 405–414, 2019.
https://doi.org/10.1007/978-3-030-16181-1_38

2   Methodology

A systematic review was the method chosen for this study, because it allows us to obtain a wide view of the research available [3]. This methodology also enables the structuring of a solid scientific ground to stand in the analysis of the current state of art surrounding blockchain technology. Using Budgen and Brereton [1] method, in the first phase of the study we define a number of questions that we want answered with the literature review. After that, on the second phase a protocol is defined to help evaluate what studies are relevant to the research. The last step involves the process of answering the initial questions based on the papers collected. Figure 1 presents the steps followed in the application of the methodology.

Fig. 1.   Steps followed in the systematic review

The initial group of questions were:

1. How is blockchain defined?
2. What are advantages and disadvantages of the technology?
3. What are current interests in research surrounding this technology?
4. Will existing business models benefit or be hurt by blockchain?

From that, 146 journals where selected from the "SCImago Journal & Country Rank" portal, from which papers would be extracted. The selection was made with the follow criteria:

- All subject categories: Business, Management and Accounting;
- Rank indicator: Q1–Q2.

The reason for this choice, was to circumvent the search tools presented by b-on, where the paper search was conducted. SCImago rankings offer a way to filter for quality not presented in the final database, from where the papers where extracted.

The search for articles was conducted in the Online Knowledge Library "b-on", with the following string: "TX All Text: "blockchain" and ISSN: "the number corresponding to each journal"". The literature search was carried out between 29/10/2018 and 06/11/2018.

During the search, papers that did not meet the following parameters were immediately excluded:

- The paper is not in English.
- The paper is not available for viewing.

3   Results

After the search was completed 55 papers were collected, with 28 journals remaining from among the initial 146. The next step was the application of the first filter, which consisted of an analysis of the title and abstract of the paper, in order to exclude irrelevant studies. From then a second filter was applied, that involved the reading of the introduction and conclusion of each paper and the final number achieved was 22.

3.1   How Is Blockchain Defined?

Millard [4] says that due to the unorthodox way in which blockchain was developed, it's hard to pin a decisive definition of a what a blockchain is. He considers the technology to be a system, that uses cryptography to record data with shared copies of the ledger maintained via an agreed upon consensus mechanism. Savelyev [5] offers a concise definition based on definite characteristics: blockchains are digital ledgers that offer transparency, redundancy, immutability and disintermediation. Blockchains can record the information about the ownership of a specific asset, who bought a particular product or service, who sold it and even about who has the right to make a decision, working as an evolution of the previous role ledgers fulfilled [6, 7].

Because blockchains are distributed, the system where a blockchain operates runs on several computers controlled by individuals or organizations. Kamel Boulos, Wilson, and Clauson [8] explain that to validate the information there are consensus protocols built on the system to find agreement among all the participants. The two most common type of consensus protocol are Proof-of-Work (PoW) and Proof-of-Stake (PoS).

According to Savelyev [5] in PoW, a miner (a node on the network) competes with others on the network to validate transactions by solving a complex mathematical formula and if successful in return receives a compensation in the form of new created coins and a fee paid by the users who sent the transaction. This is also the process by which new coins are created. It is necessary to understand that the transactions validated by the miners are grouped in what's called blocks, and what users validate is that aggregate of transactions. In PoS nodes mine a percentage of the transaction according to the ownership stake of the coins of each node [9].

3.2   What Are the Advantages and Disadvantages of the Technology?

Soto [7] considers property ledgers to be one of the most important factors in the development of civilizations and the consensus mechanisms found in blockchains, talked about in the previous chapter, provide a step forward for new developments. The difference between countries and civilizations who have harnessed ledgers in the past, versus does who didn't is highlighted and it opens speculation to what blockchains can do. Places like Africa can bypass the lack of infrastructure necessary for old business models and instead take advantage of new digital mediums, that require less infrastructure due to their decentralized nature, with blockchain being one of them but not the only [10, 11].

The biggest advantages mentioned in the papers are related to the elimination of intermediaries, data reliability, trust and transparency. The permanent keeping of transactions without the need of third-party entities, allows for faster transactions at a reduced cost [2]. The most prominent example presented are banks, but since blockchain technology can be used in the transfer of all sorts of information it doesn't seem to stop there, with applications in the health sector and supply chain management [12]. Figure 2 presents the main advantages of the blockchain architecture.

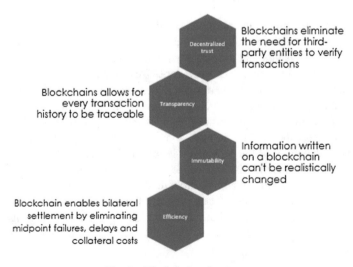

Fig. 2. Blockchain advantages

Dierksmeier and Seele [13] make the case that privacy and anonymity can be both a positive and a negative when it comes to decentralized data, depending on the practices applied. This is illustrated through a list of pros and cons associated with cryptocurrencies, which are currently the biggest application of blockchain technology. This are as such:

Pros:

- Low cost of transfer;
- Low barrier of entry – in many cases only requiring internet access and phone;
- Censorship resistance – no one can be denied access based on their personal entity;
- No infrastructure or documentation needed – could be very advantageous for developing nations.

Cons:

- Allows the ease of transactions in the dark web;
- No deposit insurance – because there is no entity keeping track of your money, if you lose access to the account for some reason, you lose the money;
- High volatility – this can be dangerous from both an investment and a currency point of view.

With that said an interesting idea modeled by Hendrickson [14] is that a bitcoin ban is not sufficient to discourage its use. The model presented shows that users who prefer bitcoin simply exclude those that favor other type of currency, making it a close system difficult to track.

One of the biggest disadvantages pointed out in the papers, has to do with the network distribution. To truly be decentralized and trustworthy a blockchain network needs to have many participants, not only so that it can keep functioning in case of attacks, but to also create trust between the participants and avoid a 51% attack [14].

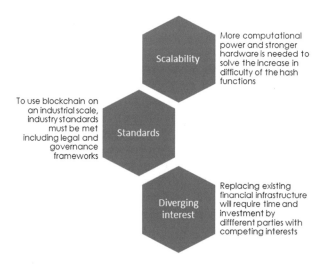

Fig. 3. - Blockchain barriers

There are other issues that bring to question the capacity for the technology as a long-term solution for maintaining trust over digital records [4, 10, 15]. This is due to the possibility of having unreliable or inauthentic outside information anchored to the blockchain, which makes it hard if not impossible to verify. Radziwill [16] also questions if the technology is ready to be implemented, with issues like network distribution, scalability and energy consumption present in its current state. Figure 3 presents the main barriers of the blockchain architecture.

3.3 What are Current Interests in Research Surrounding this Technology?

Fraud prevention like in the case of insurance claims is an example of the advantages in trust and auditability provided by the technology [8, 17]. Despite the fact that the first blockchain applications were in cryptocurrencies, which are financial in nature, the technology can be applied to the transfer of any digital asset [8]. Wang and Kogan [9] take this further with the tokenization of real-world assets and resources (e.g. finished goods, raw materials, labor, etc.) in its blockchain model for accounting.

Kamel Boulos et al. [8] present a case for blockchain application in the health industry, acting on clinical data sharing, as a way to secure patient and provider identities as a way to optimize management, to run clinical trials, for public health surveillance and medical records management. The concept of monetization is also discussed, because due to patients now having control over their data it allows users to directly sell this information. Seidel [2] presents the same concept when applied to other industries. An example of that is social media. From the moment users are in control of their data, they can now sell it directly to advertisers if they which to do so.

O'Leary [18] looks at the differences between different types of blockchain architecture when applied to accounting, auditing and supply chains. Blockchains can be either private or public depending on the needs of the business. Public blockchains like bitcoin allow for anyone to participate, while a private blockchains could be used in a consortium of companies like banks in order to mediate transactions between themselves. The level of information available can also be adjusted depending on what's deemed necessary [19]. There are a number of reasons as to why enterprises would not be willing to participate in open transactions with a blockchains:

Wash trading - which involves the act of an entity generating transactions with itself in order to create the illusion of market movement.

Spoofing – is the creation of false information to obtain some sort of advantage.

Off-chain transactions – blockchain only guarantees the integrity of registered information, transactions that occur outside the blockchain cannot be validated.

There are also blockchain models for accounting which would allow for continuous monitoring and fraud prevention [9]. This would be done through the tokenization of real-world assets and resources (e.g. finished goods, raw materials, labor, etc.). The real-world transaction would be accompanied by a token in a blockchain and he would receive the token representing the cash in return. Due to the ledger nature of the blockchain it stays registered forever. The data is always available and is distributed through all the participants. Fraud becomes more difficult in financial statements due to permanence of records.

3.4   Will Existing Business Models Benefit or be Hurt by Blockchain?

Blockchain technology can impact companies like UBER and Facebook. Users can own their own data and sell it directly to advertisers instead of having companies do it for them. Or in the case of uber, the company now acts as an intermediary between the driver and the customer, but a decentralized ledger with smart contracts can do the same thing. The same is true for stock brokers and some roles financial banks play, like custody [20]. Another example is the energy industry, were some users produce solar energy and instead of selling it directly to the energy company, they can sell it directly to other people on an open market [2].

There is also the potential for decentralized autonomous organizations. "The DAO", was one of these types of organization that failed due to a hack that stole $50 million from the company. And although no new attempts at such an endeavor have been made since, it brings attention to the fact that technology limitations that caused the situation can be solved in the future and so this new type of organization needs to be studied [2].

Blockchain may also provide new ways of fund raising [21]. Initial Coin Offerings (ICOs), which are a way of raising capital similar to Initial Public Offerings (IPOs), involve the transfer of a token in exchange for money. This form of capital raising allows the avoidance of third-party entities such as crowdfunding platforms and payment agents like banks.

Current methods of fundraising are very inefficient and localized when compared to ICOs as the project has to be presented to many different individuals until one decides to invest while ICOs are almost global. There are some downsides, as ICOs can be tax inefficient if the amount raised is treated as revenue and it also has high risk, due to many projects not surviving the hardship of the early stages. They also face regulatory uncertainty. There are different approaches being made. For example, the USA is creating stricter regulations while others like Switzerland and Singapore are friendlier towards this type of crowdfunding [22].

Real estate crowdfunding has been gaining traction as a way for real estate developers to fund their projects in ways similar to those of the ICOs, by pooling small amounts of capital from a big group of investors [23]. This directly competes with traditional financing institutions, and the inclusion of a token in a blockchain could be a logical step in this process, due to all the benefits it provides.

Intellectual property will also be able to be tracked through blockchain [6]. With blockchain it is possible to track and account for the contribution of each member in a project. For example, in a team of coders developing a new software each time a new member uploads a new piece of content to the network it's tracked and with the help of smart contracts he can receive a token that has some monetary value in exchange.

For this to happen a metric of value needs to be created but it may not be possible to apply in every scenario, because it could be hard to measure the value of any given contribution. The other issue has to do with the size of the data, which would be enormous and only technological innovation beyond what is currently available would solve this [24].

In short, some industries seem to benefit more from the technology (healthcare and those that depend on supply-chains like physical retail stores), while others may suffer a blow (banks, energy and other service providers).

4 Conclusions and Future Work

The objective of this paper was to answer a number of questions selected beforehand by applying a systematic review. Through the review of the papers we were able to do that, and we identified new fields of research and opportunities to develop surrounding blockchain technology.

Public blockchains could bring many benefits such as allowing those who have no access to any type of digital information either by lack of infrastructure or due to censorship. At the same time, it has some downsides mostly related to regulation and illegal activity, although it seems like blockchains is more a gateway then a catalyst for this, further research on this topic is necessary.

It currently faces some technological difficulties like scalability that need to be addressed in order for them to be useful and have mass adoption. In the same note different blockchain architectures need to be studied so it becomes easier to identify which models fit best for each economic and social necessity.

Companies also need to be able to adapt and be able to incorporate decentralized business models and the process of tokenization if appropriate. Currently there is a big lack in research of the impact this has on economic business models, but it may simply be due to lack of applications to study.

Our biggest limitation in doing this paper was the access to research tools. The one used for this (b-on), doesn't offer the robustness that something like "Scopus" does in regards to quality control, but it's the only one the researchers currently have access too. Because of this and to circumvent the quality issue surrounding the database, we looked selected the top journals from "SCImago Journal & Country Rank" and conducted the search from articles within those journals.

As future work we intend to implement a prototype that allows the tracking of purchase and sale of products. This permits the buyer to know the way products were grown, either biologically or in other types of production and gives sellers and producers access to various metrics that can aid with management, accounting and marketing.

References

1. Budgen, D., Brereton, P.: Performing systematic literature reviews in software engineering. In: Proceeding of the 28th International Conference on Software Engineering - ICSE 2006, p. 1051. ACM Press, Shanghai (2006)
2. Seidel, M.-D.L.: Questioning centralized organizations in a time of distributed trust. J. Manag. Inquiry **27**, 40–44 (2018). https://doi.org/10.1177/1056492617734942

3. Greenhalgh, T.: How to read a paper: papers that summarise other papers (systematic reviews and meta-analyses). BMJ **315**, 672–675 (1997). https://doi.org/10.1136/bmj.315.7109.672

4. Millard, C.: Blockchain and law: incompatible codes? Comput. Law Secur. Rev. **34**, 843–846 (2018). https://doi.org/10.1016/j.clsr.2018.06.006

5. Savelyev, A.: Copyright in the blockchain era: promises and challenges. Comput. Law Secur. Rev. **34**, 550–561 (2018). https://doi.org/10.1016/j.clsr.2017.11.008

6. Lakhani, K., Felin, T.: What problems will you solve with blockchain?. https://sloanreview.mit.edu/article/what-problems-will-you-solve-with-blockchain/ (2018)

7. de Soto, H.: A tale of two civilizations in the era of Facebook and blockchain. Small Bus. Econ. **49**, 729–739 (2017). https://doi.org/10.1007/s11187-017-9949-4

8. Kamel Boulos, M.N., Wilson, J.T., Clauson, K.A.: Geospatial blockchain: promises, challenges, and scenarios in health and healthcare. Int. J. Health Geogr. **17** (2018). https://doi.org/10.1186/s12942-018-0144-x

9. Wang, Y., Kogan, A.: Designing confidentiality-preserving blockchain-based transaction processing systems. Int. J. Account. Inf. Syst. **30**, 1–18 (2018). https://doi.org/10.1016/j.accinf.2018.06.001

10. Bounfour, A.: Africa: the next frontier for intellectual capital? J. Intellect. Cap. **19**, 474–479 (2018). https://doi.org/10.1108/JIC-12-2017-0167

11. Hain, D.S., Jurowetzki, R.: Local competence building and international venture capital in low-income countries: exploring foreign high-tech investments in Kenya's Silicon Savanna. J. Small Bus. Enterp. Dev. **25**, 447–482 (2018). https://doi.org/10.1108/JSBED-03-2017-0092

12. Kearney, A., Harrington, D., Kelliher, F.: Executive capability for innovation: the Irish seaports sector. Eur. J. Train. Dev. **42**, 342–361 (2018). https://doi.org/10.1108/EJTD-10-2017-0081

13. Dierksmeier, C., Seele, P.: Cryptocurrencies and business ethics. J. Bus. Ethics **152**, 1–14 (2018). https://doi.org/10.1007/s10551-016-3298-0

14. Hendrickson, J.R., Hogan, T.L., Luther, W.J., College, K.: The political economy of bitcoin, p. 30 (2015)

15. Lemieux, V.L.: Trusting records: is blockchain technology the answer? Rec. Manag. J. **26**, 110–139 (2016). https://doi.org/10.1108/RMJ-12-2015-0042

16. Radziwill, N.: Mapping Innovation: A Playbook for Navigating a Disruptive age, 2017. Greg Satell. New York: McGraw-Hill Education, 240 pages. Qual. Manag. J. **25**, 64 (2018). https://doi.org/10.1080/10686967.2018.1404372

17. Eling, M., Lehmann, M.: The impact of digitalization on the insurance value chain and the insurability of risks. Geneva Pap. Risk Insur.-Issues Pract. **43**, 359–396 (2018). https://doi.org/10.1057/s41288-017-0073-0

18. O'Leary, D.: Configuring blockchain architectures for transaction information in blockchain consortiums: The case of accounting and supply chain systems. Intell. Syst. Account. Financ. Manag. **24**, 138–147 (2017). https://doi.org/10.1002/isaf.1417

19. O'Leary, D.: Open information enterprise transactions: business intelligence and wash and spoof transactions in blockchain and social commerce. Intell. Syst. Account. Financ. Manag. **25**, 148–158 (2018). https://doi.org/10.1002/isaf.1438

20. Akdere, Ç., Benli, P.: The nature of financial innovation: a post-schumpeterian analysis. J. Econ. Issues **52**, 717–748 (2018). https://doi.org/10.1080/00213624.2018.1498717

21. Adhami, S., Giudici, G., Martinazzi, S.: Why do businesses go crypto? An empirical analysis of initial coin offerings. J. Econ. Bus. (2018). https://doi.org/10.1016/j.jeconbus.2018.04.001
22. Chen, Y.: Blockchain tokens and the potential democratization of entrepreneurship and innovation. Bus. Horiz. **61**, 567–575 (2018). https://doi.org/10.1016/j.bushor.2018.03.006
23. Montgomery, N., Squires, G., Syed, I.: Disruptive potential of real estate crowdfunding in the real estate project finance industry: a literature review. Prop. Manag. **36**, 597–619 (2018). https://doi.org/10.1108/PM-04-2018-0032
24. Gürkaynak, G., Yılmaz, İ., Yeşilaltay, B., Bengi, B.: Intellectual property law and practice in the blockchain realm. Comput. Law Secur. Rev. **34**, 847–862 (2018). https://doi.org/10.1016/j.clsr.2018.05.027

SPAINChain: Security, Privacy, and Ambient Intelligence in Negotiation Between IOT and Blockchain

Mohamed A. El-dosuky[1(✉)] and Gamal H. Eladl[2]

[1] Computer Sciences Department, Faculty of Computer and Information,
Mansoura University, P.O 35516, Mansoura, Egypt
mouh_sal_010@mans.edu.eg
[2] Information Systems Department, Faculty of Computer and Information,
Mansoura University, P.O 35516, Mansoura, Egypt
gamalhelmy@mans.edu.eg

Abstract. Vulnerability in Internet-of-Things (IoT) necessitates the existence of a safeguard for both security and privacy without sacrificing "smartness" of the environment, or its Ambient Intelligence in scientific terms. Blockchains, underpinning crypto-currency, could be the remedy for this vulnerability. This paper quickly overviews security, privacy, Ambient Intelligence, and the use of Blockchain as a safeguard for IOT, before proposing and evaluating SPAINChain:Security, Privacy, and Ambient Intelligence in Negotiation between IOT and Blockchain. IOT-Blockchain mapping is proven to be feasible. The gamut of blockchain is based on a new concept called solidus. Solidus is basically the sum of blockchain favorability dimensions, adjusted by preference of blockchain type.

Keywords: Security · Privacy · Ambient Intelligence · IOT · Blockchain

1 Introduction

The home sweet home is a smart one indeed. Environment "smartness", or Ambient Intelligence [1] in scientific terms, comes at the cost of vulnerability due to:

– security attacks [2], and
– manipulation of privacy-sensitive data [3].

This vulnerability in Internet-of-Things (IoT) necessitates the existence of a safeguard for both security and privacy without sacrificing "smartness" of the environment [4].

Blockchains, underpinning crypto-currency [5], could be the remedy for this vulnerability.

This paper quickly overviews security, privacy, Ambient Intelligence, and the use of Blockchain as a safeguard for IOT (in Sect. 2), before proposing, evaluating and concluding SPAINChain:Security, Privacy, and Ambient Intelligence in Negotiation between IOT and Blockchain in subsequent sections.

© Springer Nature Switzerland AG 2019
Á. Rocha et al. (Eds.): WorldCIST'19 2019, AISC 930, pp. 415–425, 2019.
https://doi.org/10.1007/978-3-030-16181-1_39

2 Previous Work

This section quickly overviews security, privacy, Ambient Intelligence, and the use of Blockchain as a safeguard for IOT.

2.1 Security

Security is modeled based on CIA model which is an amalgamation of Confidentiality, Integrity, and Availability [6]. The former ensures that solely authorized users have access. Integrity ensures that data is received as sent, without modification, The latter ensures data availability when needed.

Non-bespoke IOT devices usually lack a safeguard for security, that they can be easily hacked [7]. Smart home vulnerability is inevitable, even with the existence of gateways controlling packets exchange to and fro [8] Object security, as implemented in OSCAR architecture [9], is powerful that IoT researchers commence utilizing it as in [10].

Cryptography seems to be a key ingredient in many confidentiality solutions [11]. Recently, a clever Variable Key-length cryptography (VKLC) solution is proposed based on the motto, "A valuable thing needs higher security" [12]. It constructs a dynamic model for the estimated economic value of data, and enforces the appropriate security accordingly. The more "valuable" the data, the bigger the cryptography key length.

Bitcoin, or blockchain v.1, introduces Proof-of-Work (POW) cryptographic puzzles [5]. Not only a blockchain ensures transparency, but also it can be an integrity-assurance ingredient [13].

2.2 Privacy

Access control, usually implemented as an access matrix of privileges [14], is not enough to reaching the holy-grail of privacy [15]. Preserving IOT privacy raises many challenges [16]. Blockchains are at the rescue to provide a decentralized access [17], and credibility verification [18]. A middleware is proposed between sender and receiver to control access [3]. IGOR is recently proposed [19] and augmented to provide a unified IOT access control [20].

2.3 Ambient Intelligence

There is a plethora of smart home projects, leveraging Ambient Intelligence [21]. Implementation usually takes the form of a context-aware middleware as shown in Fig. 1.

The cycle is self-describing. Usually a context is related to energy consumption, and this paper shall stick to that.

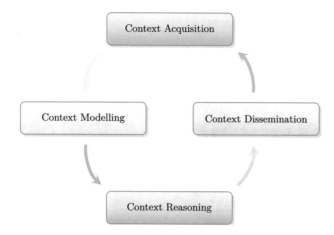

Fig. 1. Middleware context life-cycle [21]

2.4 Blockchain-IOT Integration

For a detailed survey of possible blockchain-IOT integration scenarios, kindly refer to [22]. Integration involves inevitable challenges [23] such as scalability and latency. Blockchain Platforms are meticulously analyzed for IoT [24]. Based on that analysis favorable platforms (Table 1, showing ranking of Security, Scalability, Suitability for smart contract on a scale of 1 to 5) and selection criteria (Fig. 2) are determined.

3 Proposed System

Let us first set the mathematical foundation of IOT-Blockchain Mapping.

Theorem 1 (IOT-Blockchain Mapping)

$$\lambda_B \leftarrow \frac{\delta_B(\varepsilon_T - \lambda_T)}{\varepsilon_T} + \xi_B \tag{1}$$

where λ_B is blockchain feature, usually platform rank, in interval $[\xi_B, \xi_B + \delta_B]$. ξ_B is the minimum and is not necessarily a zero, allowing negative values. $\xi_B + \delta_B$ is the maximum. Hence, δ_B is the range. λ_T is IOT feature, usually energy consumption, in interval $[0, \varepsilon_T]$. Clearly, ε_T is the maximum. $\square$

Proof

$$0 \leq \lambda_T \leq \varepsilon_T \tag{2}$$

$$0 \geq -\lambda_T \geq -\varepsilon_T \tag{3}$$

$$\varepsilon_T \geq \varepsilon_T - \lambda_T \geq 0 \tag{4}$$

$$1 \geq \frac{\varepsilon_T - \lambda_T}{\varepsilon_T} \geq 0 \tag{5}$$

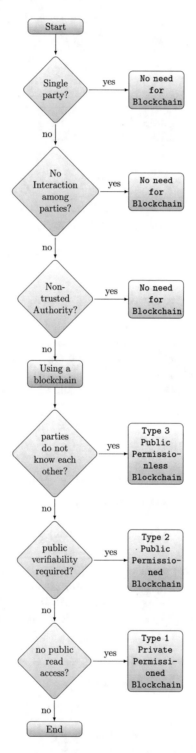

Fig. 2. Blockchain platform selection [24]

Table 1. Blockchain platform favorability [24]

Blockchain	(Security, scalability, smart contract)
Bitcoin (Type 3)	(3, 1, 1)
Ethereum (Type 3)	(3, 3, 4)
Hyperledger (Type 2)	(4, 4, 4)
Ripple (Type 2)	(4, 4, 4)
Multichain (Type 1)	(4, 4, 1)
Eris (Type 1)	(4, 3, 4)

$$0 \leq \frac{\varepsilon_T - \lambda_T}{\varepsilon_T} \leq 1 \tag{6}$$

$$0 \leq \frac{\delta_B(\varepsilon_T - \lambda_T)}{\varepsilon_T} \leq \delta_B \tag{7}$$

$$\xi_B \leq \frac{\delta_B(\varepsilon_T - \lambda_T)}{\varepsilon_T} + \xi_B \leq \xi_B + \delta_B \tag{8}$$

Note that the middle expression:

$$\frac{\delta_B(\varepsilon_T - \lambda_T)}{\varepsilon_T} + \xi_B \tag{9}$$

is a blockchain feature in interval $[\xi_B, \xi_B + \delta_B]$. Denoting it λ_B,

$$\lambda_B \leftarrow \frac{\delta_B(\varepsilon_T - \lambda_T)}{\varepsilon_T} + \xi_B \tag{10}$$

□

Based on this mapping theorem, it is possible to seek the optimal blockchain-IOT pairing as in Fig. 3, in which device T (seeking context C) is paired with blockchain B (with configuration suitable for the same context C). Hence, shifting from access matrix to a dichotomy of B matrix and T matrix. The former determines the context for each blockchain (Table 1 and Fig. 2 help in constructing such a matrix), and the latter determines the context for each IOT device (It could be dynamically constructed, utilizing cycle shown in Fig. 1). Context elicitation and blockchain matching are depicted in Fig. 4. Now the time is ripe

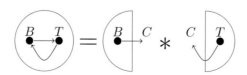

Fig. 3. Blockchain-IOT pairing

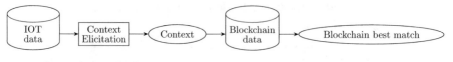

Step 1: Ambient Intelligence   Step 2: Blockchain Matching

Fig. 4. Context elicitation and blockchain matching

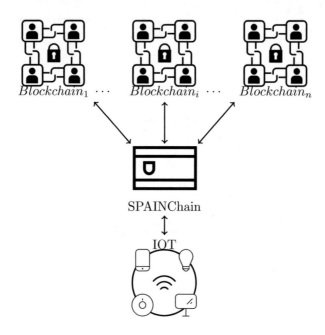

Fig. 5. Conceptual model of SPAINChain

to propose SPAINChain conceptual model (Fig. 5), implemented as a set of layers (Fig. 6). SPAINChain elects the best blockchain to safeguard a specific IOT device in a certain context.

SPAINChain is decomposed of $4+1$ layers. The lowest layer, L_0, is the basic blockchain scanner. The top layer, L_4, is for interaction with entities outside SPAINChain. It integrates IGOR ([19] for authentication and access control, striving for privacy), an updated ontology of IOT devices, an updated ontology of available Blockchains, Data Importer, and Data Exporter. L_3 is for user convenience. It integrates an updated Users' list, Context Aquisitioner, Preprocessor, and Visualizer. The layer beneath, L_2, is for maintaining ambient intelligence. It integrates an updated ontology of Contexts, Context Modeler, and Context Disseminator. L_1 integrates the core Negotiator, Context Reasoner, OSCAR ([9] for security), and VKLC ([12], determining sufficient degree of security based on context).

4   Case Study and Evaluation

Assume an arbitrary smart home shown in Fig. 7. The gamuts of blockchains and devices are shown in Figs. 8, 9 respectively.

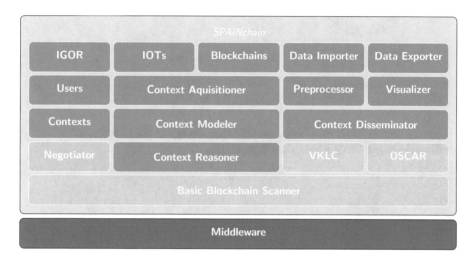

Fig. 6. SPAINChain layers

Fig. 7. Arbitrary smart home

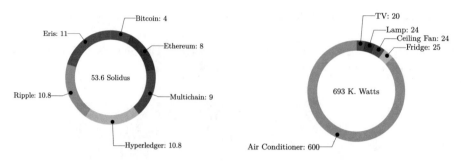

Fig. 8. Blockchain gamut   **Fig. 9.** Device gamut

The gamut of devices is based on monthly energy consumption, as estimated by Electric Regulatory Agency in Egypt [25].

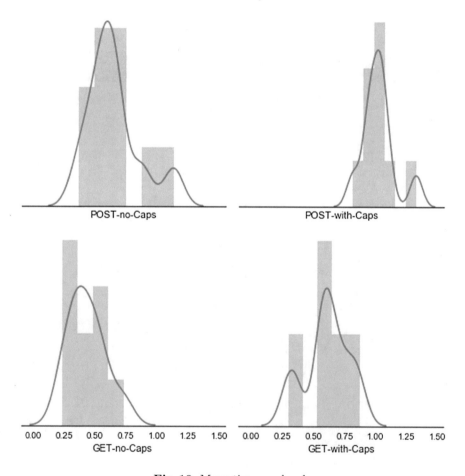

Fig. 10. Mean time overheads

The gamut of blockchain is based on a new concept called solidus. Solidus is basically the sum of favorability dimensions shown in Table 1, adjusted by preference of blockchain type, as follows.

$$Solidus(B) = (1 - \gamma * (B_{Type} - 1)) * \sum_{dim \in FavDims} B^{dim} \qquad (11)$$

where γ is a degradation factor, set to 0.1. For Eris with $B_{Type} = 1$ and favorability $(4, 3, 4)$, Solidus(Eris) $= 11$.

Distribution of mean time overheads either with or without capabilities are shown in Fig. 10. Overheads are bearable. The key merit here is the dynamic determination of sufficient degree of IOT privacy and security based on context. This allows for a dynamic overhead compared to fixed blockchain-based overhead solutions such as [26].

5 Conclusion and Future Work

IOT-Blockchain mapping is proven to be feasible. Based on this mapping theorem, it is possible to seek the optimal blockchain-IOT pairing, in which device T (seeking context C) is paired with blockchain B (with configuration suitable for the same context C). Hence, shifting from access matrix to a dichotomy of B matrix and T matrix. The former determines the context for each blockchain, and the latter determines the context for each IOT device. The gamut of devices is based on monthly energy consumption, while the gamut of blockchain is based on a new concept called solidus. Solidus is basically the sum of favorability dimensions, adjusted by preference of blockchain type.

Then came SPAINChain, decomposed of $4 + 1$ layers. The lowest layer,L_0, is the basic blockchain scanner. The top layer, L_4, is for interaction with entities outside SPAINChain. It integrates IGOR [19], devices, Blockchains, Data Importer, and Data Exporter. L_3 is for user convenience. The layer beneath,L_2, is for maintaining ambient intelligence. L_1 integrates the core Negotiator, Context Reasoner, OSCAR [9], and VKLC [12].

SPAINChain attempts to determine sufficient degree of IOT privacy and security based on context, by allowing an automatic negotiation between IOT and blockchains.

Future direction shall consider producing SPAINChain based on the proof-of-concept provided in this paper.

References

1. Ramos, C., Augusto, J.C., Shapiro, D.: Ambient intelligence-the next step for artificial intelligence. IEEE Intell. Syst. **23**(2), 15–18 (2008)
2. Sicari, S., Rizzardi, A., Grieco, L.A., Coen-Porisini, A.: Security, privacy and trust in internet of things. Comput. Netw. **76**, 146–164 (2015)

3. Chakravorty, A., Wlodarczyk, T., Rong, C.: Privacy preserving data analytics for smart homes. In: IEEE 2013 Security and Privacy Workshops (SPW), pp. 23–27. IEEE (2013)
4. Roman, R., Zhou, J., Lopez, J.: On the features and challenges of security and privacy in distributed internet of things. Comput. Netw. **57**(10), 2266–2279 (2013)
5. King, S.: Primecoin: cryptocurrency with prime number proof-of-work, 7 July 2013
6. Komninos, N., Philippou, E., Pitsillides, A.: Survey in smart grid and smart home security: issues, challenges and countermeasures. IEEE Commun. Surv. Tutor. **16**(4), 1933–1954 (2014)
7. Notra, S., Siddiqi, M., Gharakheili, H.H., Sivaraman, V., Boreli, R.: An experimental study of security and privacy risks with emerging household appliances. In: 2014 IEEE Conference on Communications and Network Security (CNS), pp. 79–84. IEEE (2014)
8. Sivaraman, V., Chan, D., Earl, D., Boreli, R.: Smart-phones attacking smart-homes. In: Proceedings of the 9th ACM Conference on Security and Privacy in Wireless and Mobile Networks, pp. 195–200. ACM (2016)
9. Vučini ć, M., et al.: OSCAR: object security architecture for the Internet of Things. Ad Hoc Netw. **32**, 3–16 (2015)
10. Alphand, O., et al.: IoTChain: a blockchain security architecture for the Internet of Things. In: 2018 IEEE Wireless Communications and Networking Conference (WCNC), pp. 1–6. IEEE (2018)
11. Delfs, H., Knebl, H.: Introduction to Cryptography, vol. 2. Springer, Heidelberg (2002)
12. Khorsheed, N.K., et al.: Management of data security based on data cost evaluation. J. Comput. Theor. Nanosci. **14**(10), 4964–4969 (2017)
13. Weizhi, M.E.N.G., et al.: When intrusion detection meets blockchain technology: a review. IEEE Access **6**, 10179–10188 (2018)
14. Sandhu, R.S., Samarati, P.: Access control, principle and practice. IEEE Commun. Mag. **32**(9), 40–48 (1994)
15. Colombo, P., Ferrari, E.: Privacy aware access control for Big Data, a research roadmap. Big Data Res. **2**(4), 145–154 (2015)
16. Ziegeldorf, J.H., Morchon, O.G., Wehrle, K.: Privacy in the Internet of Things, threats and challenges. Secur. Commun. Netw. **7**(12), 2728–2742 (2014)
17. Ali, M.S., Dolui, K., Antonelli, F.: IoT data privacy via blockchains and IPFS. In: Proceedings of the Seventh International Conference on the Internet of Things, p. 14. ACM (2017)
18. Qu, C., et al.: Blockchain based credibility verification method for IoT entities. Secur. Commun. Netw. (2018)
19. Pemberton, S.: An architecture for unified access to the internet of things. In: XML London 2017 Conference Proceedings, vol. 1, pp. 38–42 (2017)
20. Shieng, P.S.W., Jansen, J., Pemberton, S.: Fine-grained access control framework for igor, a unified access solution to the internet of things. Procedia Comput. Sci. **134**, 385–392 (2018)
21. Che-Bin, F., Hang-See, O.: Research article. A review on smart home based on ambient intelligence, contextual awareness and Internet of Things (IoT). In: Constructing Thermally Comfortable and Energy Aware House (2016)
22. Panarello, A., et al.: Blockchain and IoT integration, a systematic survey. Sensors **18**(8), 2575 (2018)
23. Dorri, A., Kanhere, S. S., Jurdak, R.: Blockchain in internet of things: challenges and solutions. arXiv preprint arXiv:1608.05187 (2016)

24. Pahl, C., El Ioini, N., Helmer, S.: A decision framework for blockchain platforms for IoT and edge computing. In: International Conference on Internet of Things, Big Data and Security (2018)

25. Egyptian Electric Utility and Consumer Protection Regulatory Agency (EgyptERA). http://egyptera.org/ar/. Accessed 1 Oct 2018

26. Dorri, A., et al: Blockchain for IoT security and privacy: the case study of a smart home. In: 2017 IEEE International Conference on Pervasive Computing and Communications Workshops (PerCom Workshops), pp. 618–623. IEEE (2017)

Ontology Supporting Model-Based Systems Engineering Based on a GOPPRR Approach

Hongwei Wang[1], Guoxin Wang[1], Jinzhi Lu[2(✉)], and Changfeng Ma[3]

[1] School of Mechanical Engineering,
Beijing Institute of Technology, Beijing, China
[2] KTH Royal Institute of Technology, Brinellvgen 83,
100 44 Stockholm, Sweden
jinzhl@kth.se
[3] ZhongKe Fengchao Technology Ltd., Beijing, China

Abstract. Model-based systems engineering (MBSE) has become a new trend in systems engineering. The core of MBSE is the use of models to formalize product development and the product itself (requirement, architecture, etc.) using different languages throughout the entire lifecycle. However, different tools based on different modeling languages in each development phase lead to considerable deviations during information sharing between the models. In this paper, a GOPPRR approach is proposed to describe system characteristics and system development based on meta-models for different architecture descriptions using meta-meta models including *Graph*, *Object*, *Point*, *Property*, *Role*, and *Relationship*. Moreover, ontology based on W3C Web Ontology Language (OWL) is designed to support the GOPPRR approach for unified model representations and further to implement model integration in the entire lifecycle. Finally, the feasibility of the GOPPRR approach and the ontology formalism supporting MBSE is verified through an auto-braking case of an autonomous driving system.

Keywords: MBSE · GOPPRR · OWL · Model transformation

1 Introduction

The continuous evolution of modern technology has led to increased complexity and R&D cost of system development. Model-based systems engineering (MBSE) is proposed to address the challenges brought by the increasing complexity which formalize system characteristics and system development across lifecycle using models. However, stakeholders across the lifecycle have different viewpoints, which are represented by architecture models based on different modeling languages. Different syntax and data structure of these languages lead to model integration difficult across the lifecycle.

Currently, it is common that stakeholders formalize their own specific problems by using their own domain-specific modeling languages. Moreover, such languages are often supported by different tools which the syntax and semantics are also different. Consequently, different stakeholders related to system development are unable to

© Springer Nature Switzerland AG 2019
Á. Rocha et al. (Eds.): WorldCIST'19 2019, AISC 930, pp. 426–436, 2019.
https://doi.org/10.1007/978-3-030-16181-1_40

understand all the information in each other's models. Moreover, it is difficult to standardize the information representations of these models from different tools or to create a complete model flow between the tools.

This paper focuses on a modeling approach based on *Graph*, *Object*, *Point*, *Property*, *Role*, and *Relationship* (GOPPRR) to formalize MBSE with a unified model representations and ontology design based on OWL for model integration between different tools. The GOPPRR approach is defined based on a M3–M0 modeling framework [1] including meta-meta models, meta-models, models and system views. Moreover, by using the OWL-based ontology, meta-models in GOPPRR approach are represented as class; the semantic relationships between the meta-models are represented by the object property. For instance, one *Graph* includes meta-models, such as *Objects* and *Relationships*. An individual of class is used to represent model compositions mirrored with one meta-model. For example, a requirement *Object* created in a requirement *Graph* refers to an instance of requirement *Object*.

The contributions of this paper are as follows:

- **Adopt GOPPRR to support a unified representation of MBSE in the entire lifecycle.** Using the GOPPRR approach, meta-meta models including *Graph*, *Object*, *Point*, *Property*, *Relationship* and *Role* are used to develop meta-models based on different MBSE languages. Then the models built on these meta-models are used to represent viewpoints of different stakeholders. Thus, a unified representation of models across the entire lifecycle is defined based on the meta-meta models.

- **OWL-based ontology for the GOPPRR approach to support model integration throughout the lifecycle**. During the lifecycle, models built based on the GOPPRR approach are transformed to the OWL-based ontology. Such ontology provides the same syntax and data structure in order to realize model integration across tools.

The rest of this paper is as follows. Firstly, the related work is described in the Sect. 2. The GOPPRR approach, ontology design, and transformation between model and ontology are explained in the Sect. 3. In Sect. 4, an auto-braking system is used as a case study to verify the feasibility of GOPPRR approach and ontology design. Finally, the conclusions are offered in the Sect. 5.

2   Related Work

MBSE has become a new trend in systems engineering [2], which modeling approaches have also become the focus of attention. Currently, there are many general modeling languages supporting system development based on system engineering. For example, the Unified Modeling Language (UML) [3], are used in software and system development conducting the inspection and analysis of models based on consistency rules. Systems Modeling Language (SysML) which is an extension of UML, has defined nine types of graphs to assist the structural, behavior and requirement modeling, and to specify models for system design [4]. Business Process Modeling Notation (BPMN) [5] offers unified graphical symbols to provide process modeling for the complex system development. Though these languages formalize different views of systems engineering, there are also some challenges when such languages are adopted in the industry: (1) a general modeling approach is difficult to support the formalisms of specific domains; (2) models of different languages are difficult to integrate across different modeling tools; (3) it is strongly needed to formalize different viewpoints of systems engineering [6].

Except for modeling languages, researchers proposed a M0–M3 modeling framework to support domain-specific modeling [7]. Based on this framework, *Kelly et al.* proposed a domain-specific modeling approach based on the six key meta-meta models [8]. A VM GME refers to a meta-modeling language based on UML notation and OCL constraints [9]. In this paper, the GOPPRR approach is adopted to support meta-model development in order to realize MBSE formalisms across lifecycle.

When the GOPPRR approach is used to support MBSE formalisms, model integration is needed across modeling tools. Currently, several techniques are used to integrate models from different tools. For example, XMI is used to support data exchange between SysML models across tools [10]. In addition, an ontology is also adopted to formalize the domain-specific concepts and their relationship between different languages [11]. *Van R. et al.* proposed an ontology-based method according to international standards [12]. *Walter et al.* designed a framework allowing for ontology to describe domain-specific languages [13]. *Panov et al.* designed a general ontology for representing data types [14]. A domain-specific modeling tool based on GOPPRR approach, MetaEdit+ [15], adopts database to manage its model data. Using these methods, model integrations across tools are difficult to realize because of gaps between different syntax and data structures. In this paper, OWL is used to design the ontology for GOPPRR approach to support model integration across tools.

From literature reviews, we designed a GOPPRR approach in a M3–M0 modeling framework to formalize unified representations of MBSE in the entire lifecycle and designed ontology based on OWL for model integrations. Compared with the existing research, the GOPPRR approach supports domain-specific views, which the meta-meta models are used to develop different languages. Moreover, the meta-meta models are the basic for model integration allowing OWL for representing meta-models and models in a unified way so that model integrations between different tools are implemented in the entire lifecycle.

3 Ontology Design Supporting a GOPPRR Approach for Model-Based Systems Engineering

3.1 Overview

In this chapter, a GOPPRR approach and ontology design based on OWL are introduced. We first introduce the relevant concepts of meta-meta model, meta-model, and model and explain how to implement MBSE formalisms using the GOPPRR approach. Then OWL-based ontology is introduced aiming to support GOPPRR approach. Finally, we introduce the transformation mechanism between models and ontology in a developed multi-architecture modeling tool, *MetaGraph*[1].

---

[1] A tool developed by http://www.zkhoneycomb.com/.

3.2   GOPPRR Modeling Approach

The GOPPRR approach is implemented based on a M3–M0 modeling framework as shown in Fig. 1. M3 represents a meta-meta model layer, including key meta-meta models and complementary meta-meta models; M2 represents a meta-model layer where meta-models based on the languages such as SysML, are developed according to the meta-meta models; M1 represents a model layer where the models are developed based on meta-models aiming to represent real views of systems engineering in the M0 layer.

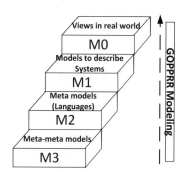

Fig. 1. M3–M0 modeling framework

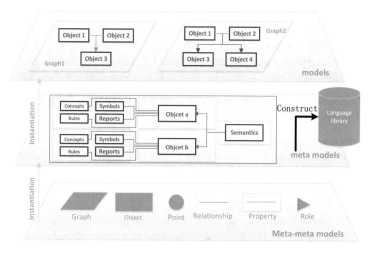

Fig. 2. GOPPRR approach

Figure 2 shows a GOPPRR approach supporting MBSE formalisms. The meta-models are instantiated based on six key meta-meta models and some complementary meta-meta models. The six key meta-meta models of GOPPRR are introduced based on the related definitions proposed by *Kelly et al.* [8].

- *Graph* is a collection of *Objects*, *Relationships* and their connections.
- *Object* refers to a basic entity in the *Graph* representing one thing independent of the relationship and the *Role*, such as a button, etc.
- *Property* refers to one attribute of the other five meta-meta models describing their characteristics.
- *Relationship* is used to indicate how *Objects* are connected. Each *Relationship* is connected through *Roles* and *Objects*. The connections between *Objects* are defined by bindings shown in Fig. 3.

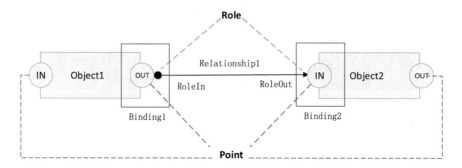

Fig. 3. Connections between objects

- *Role* is used to indicate one connection of *Objects* and *Relationships*. As shown in the above figure, each connection is determined *Objects*, *Points*, *Relationships* and *Roles*. Icon • specifies *OUT* port of *Object1* as the starting point of the *Relationship1*; Icon → specify the *IN* port of *Object2* as the end of the *Relationship1*.
- *Point* describes one port in the *Object* that is connected to the *Relationship*.

   In addition, some complementary meta-meta models support the six key meta-meta models to develop meta-models.

- *Inclusion* refers to one "include" relationship between elements and its compositions. For example, each *Graph* includes different *Objects* and *Relationships*.
- *Binding* defines connections between *Object*, *Point*, *Role* and *Relationship*. For example, as shown in Fig. 3. The *Point: OUT* of *Object1* is defined to connect to the *Role: RoleIn* of *Relationship1*.
- *Explosion*: One *Object*, *Role*, or *Relationship* is explored into one or more *Graphs*.
- *Decomposition*: An *Object* is decomposed into one new *Graph*.

   During meta-model definitions and development, domain knowledge is used to define semantics domain referring to the meanings of meta-models represented by the rules and symbols. Then, based on semantics domain, syntax of meta-models is designed including concrete syntax (descriptions and display of one meta-model) and abstract syntax (how meta-models construct models). Such meta-models construct different modeling languages, which are used to build models.

3.3 Ontology Design Based on OWL

In order to support model integrations, we use OWL to design the ontology for model integrations across the entire lifecycle. As shown in Fig. 4, based on GOPPRR approach, a transformation rule is defined to design the ontology. (1) key meta-meta models are transformed to *Class* concepts in OWL; (2) complementary meta-meta models are transformed to *Object-property* concepts in OWL; (3) Meta-models are defined as *sub-class* in OWL; (4) Models for describing systems are defined as *individuals*.

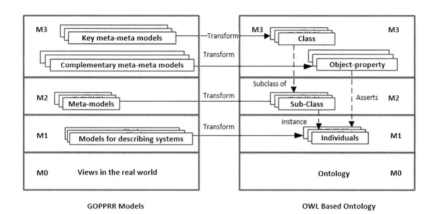

Fig. 4. Transformations between GOPPRR models and OWL-based ontology

Fig. 5. OWL model for GOPPRR approach

Based on the transformation rules defined in Fig. 4, an OWL model are developed in Protégé [16], as shown in Fig. 5. In the ontology model, there are six classes defined to represent the key meta-meta models in Fig. 5-1. Moreover, in order to describe engineering perspectives, additional *Project* class including *Language* and *Model* subclasses is defined.

- *Project* refers to one project for developing meta-models and models.
- *Language* refers to one modeling language constructed by developed meta-models. Each *Project* includes one or more languages.
- *Model* refers to one model constructed by defined meta-models.

The relationship between meta-models is linked by object property which representative complementary meta-meta models as shown in Fig. 5-2. In Fig. 5-3, restrictions between key meta-meta models are described by the object properties in OWL, such as one *Point* in an *Object* is connected with one *Role* of a *Relationship*. The restrictions in each meta-meta models are described by object restrictions, such as one Graph includes one *Object*. Figure 5-4 shows individuals referring to compositions to construct one model. Figure 5-5 shows the property assertion and data assertion of each composition in order to describe topologies in the models and the *Properties* of other meta-models.

3.4 Conversion Process Between GOPPRR Models and Ontology in *MetaGraph*

In order to support transformations between models based on GOPPRR approach and ontology based on OWL, a workflow is designed in a multi-architecture modeling tool, *MetaGraph*, as shown in Fig. 6.

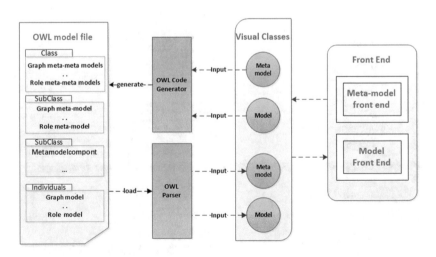

Fig. 6. GOPPRR modeling and ontology conversion process

- Front-end supports development of meta-models and models. During writing the OWL models, developed meta-models and models are first conversed into visual classes in Java from front-end. Then the visual classes are transformed to OWL model files through an OWL code-generator developed based on Protege-OWL API [16].
- During reading the OWL models, OWL models are first loaded by an OWL parser and then transformed to visual classes, which are accessed in the front-end.

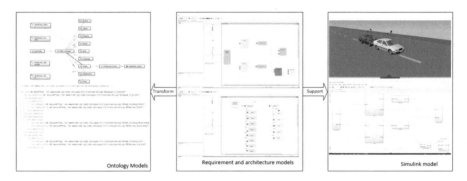

Fig. 7. Auto-braking case to formalize the simulation process

4   Case Study and Evaluation

In this section, an auto-braking case of an autonomous driving system [17] is taken as an example to illustrate the flexibility of the GOPPRR approach supporting MBSE formalism and the conversion workflow for generating the ontology. In this case, the simulation models represent two cars where a controller of braking system in the second car is verified in Simulink if the second car can brake automatically when the distance between them is too short. As shown in Fig. 7, models based on GOPPRR are developed to formalize one simulation based on Simulink models of the auto-braking system. Then one ontology file is generated from the GOPPRR models based on the developed conversion workflow.

As shown in Fig. 8, we construct a graph of overview (Fig. 8-A), requirement description graph (Fig. 8-B), architecture description graph (Fig. 8-C) and verification & validation graph (Fig. 8-D). The graph of overview describes the relationships of other three graphs represented by the related *Objects*. These *Objects* are decomposed into their own Graphs: In the requirement graph, requirements for compositions and their relationships in the structure referring to the Simulink model are defined. As shown in Fig. 8-B, three meta-models were created to describe the requirements: the blue meta-model represent a generic requirement used to represent the requirements for building the models in this case. The red one refers to one requirement for adding blocks during modeling. The green meta-model is used to represent the requirement for adding relationships between blocks during modeling. In the architecture description graph, meta-models of different colors referring to different blocks in Simulink are

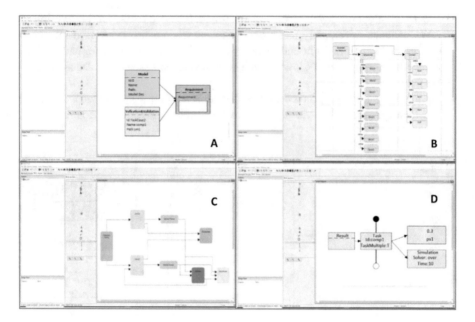

Fig. 8. Four graphs to describe one simulation for auto-braking case

created representing model structure of the Simulink model. In the verification & validation graph, one task referring to one simulation is defined associated with simulation and parameter configurations.

Fig. 9. OWL files formalizing the GOPPRR models of auto-braking case

As shown in Fig. 9, an ontology file is transformed from the models shown in Fig. 8 where meta-models (Class), relationships between the meta-models (Object Property) and models (Individuals) are described. In Fig. 9, each individual with properties is described. Moreover, the data properties are defined to describe the *Properties* in models.

According to the case study, a GOPPRR approach formalizes meta-models and models for the auto-braking case, which we can infer it supports MBSE formalisms for the entire lifecycle. Moreover, the generated ontology shows the conversion from the models to ontology successful for model integration with other tools.

5 Conclusion

In this paper, we introduce a GOPPRR approach supporting MBSE formalisms and ontology design based on OWL to support model integrations for the entire lifecycle. First, we demonstrate GOPPRR concepts in a M3–M0 modeling framework. Then, we introduced a designed transformation rule between the GOPPRR concepts and OWL concepts. Moreover, based on a transformation rule, one OWL model is developed in Protégé and a conversion workflow is implemented in a multi-architecture modeling tool, *MetaGraph*. Finally, the feasibility of the GOPPRR approach supporting MBSE formalisms and ontology supporting model integration is verified by an auto-braking case.

References

1. Aßmann, U., Zschaler, S., Wagner, G.: Ontologies, meta-models, and the model-driven paradigm. In: Ontologies for Software Engineering and Software Technology, Springer, pp. 249–273 (2006)
2. INCOSE: A world in motion: systems engineering vision 2025. International Council on Systems Engineering, San Diego, CA, USA (2014)
3. Torre, D., Labiche, Y., Genero, M., Elaasar, M.: A systematic identification of consistency rules for UML diagrams. J. Syst. Softw. **144**(June), 121–142 (2018)
4. Silva, R.F., Gimenes, I.M.S., Oquendo, F., Fragal, V.: SyMPLES - a SysML-based approach for developing embedded systems software product lines, pp. 257–264 (2013)
5. Chinosi, M., Trombetta, A.: BPMN: an introduction to the standard. Comput. Stand. Interfaces **34**(1), 124–134 (2012)
6. Lee, E.A., Tripakis, S., Törngren, M.: Viewpoints, formalisms, languages, and tools for cyber-physical systems, vol. 931843, no. 212 (2011)
7. Bapty, T., Neema, S., Sztipanovits, J.: Meta II: Multi-model language suite for cyber physical systems (2013)
8. Kelly, S., Tolvanen, J.P.: Domain-Specific Modeling: Enabling Full Code Generation. IEEE Xplore (2008)
9. Walker, J.D., et al.: Survivability modeling in DARPA's adaptive vehicle make (AVM) program. In: Proceedings of the 27th International Symposium on Ballistics, Freiburg, Germany (2013)
10. Huang, E., Ramamurthy, R., Mcginnis, L.F.: System and simulation modelling using SYSML. In: Henderson, S.G., Biller, B., Hsieh, M.-H., Shortle, J., Tew, J.D., Barton, R.R. (eds.) Proceedings of the 2007 Winter Simulation Conference, pp. 796–803 (2007)
11. Dinar, M., Rosen, D.W.: A design for additive manufacturing ontology. In: Volume 1B: 36th Computers and Information in Engineering Conference, vol. 17, no. June 2017, p. V01BT02A032 (2016)

12. van Ruijven, L.C.: Ontology for systems engineering. Procedia Comput. Sci. **16**, 383–392 (2013)
13. Walter, T., Parreiras, F.S., Staab, S.: An ontology-based framework for domain-specific modeling. Softw. Syst. Model. **13**(1), 83–108 (2014)
14. Panov, P., Soldatova, L.N., Džeroski, S.: Generic ontology of datatypes. Inf. Sci. (Ny) **329** (October), 900–920 (2016)
15. Kelly, S., Lyytinen, K., Rossi, M.: Metaedit+ a fully configurable multi-user and multi-tool case and came environment, vol. 1080. Springer, Heidelberg (1996)
16. Musen, M.A., Stevens, R.D.: The protege OWL experience. In: Proceedings of the OWLED, Workshop on OWL: Experiences and Directions, no. January (2005)
17. Jinzhi, L.: A model-driven and tool-integration framework for whole vehicle co-simulation environments. In: 8th European Congress on Embedded Real Time Software and Systems (ERTS 2016) (2016)

A Multi-agent System Framework for Dialogue Games in the Group Decision-Making Context

João Carneiro[1]([✉])(iD), Patrícia Alves[1](iD),
Goreti Marreiros[1](iD), and Paulo Novais[2](iD)

[1] GECAD – Research Group on Intelligent Engineering and Computing
for Advanced Innovation and Development, Institute of Engineering,
Polytechnic of Porto, 4200-072 Porto, Portugal
{jrc, prjaa, mgt}@isep.ipp.pt
[2] ALGORITMI Centre, University of Minho, 4800-058 Guimarães, Portugal
pjon@di.uminho.pt

Abstract. Dialogue games have been applied to various contexts in computer science and artificial intelligence, particularly to define interactions between autonomous software agents. However, in order to implement dialogue games, the developers need to deal with other important details besides what is presented in the model's definition. This is a complex work, mostly when it is expected that the agents' interactions correctly represent a human group behavior. In this work, we present a multi-agent system framework specifically designed to facilitate the implementation of dialogue games under the context of group decision-making in which agents interact as the humans do in face-to-face meetings. The proposed framework, named MAS4GDM, encapsulates the JADE framework and provides a layer that allows developers to easily implement their dialogue models without being concerned with some complex implementation details, such as: the communication model, the agents' life cycle, among others. We ran an experimental evaluation and verified that the proposed framework allows to implement dialogue models in an easier way and abstract the developers from important implementation details that can compromise the application's success.

Keywords: Multi-agent systems · Dialogue games · Group decision-making · JADE

1 Introduction

The ultimate goal of a group decision-making process is to select one or more alternatives as the solution for a certain problem [1]. However, due to the problems self-nature, and considering it involves several decision-makers with different preferences and beliefs, it is a complex process to deal with. In the last decades, different technological mechanisms have been proposed to help decision-makers in the group decision-making processes [2]. One well-known strategy consists in using autonomous software agents to represent decision-makers in terms of their preferences, beliefs and interactions [3, 4]. The use of a multi-agent system in the group decision-making context facilitates the representation of

© Springer Nature Switzerland AG 2019
Á. Rocha et al. (Eds.): WorldCIST'19 2019, AISC 930, pp. 437–447, 2019.
https://doi.org/10.1007/978-3-030-16181-1_41

the several entities involved in the process, as well as benefits from the agents' characteristics (autonomy, pro-activity, etc.) in important tasks, such as reasoning and communication [5]. In order to make agents capable of communicating and interacting with each other, the researchers have been developing formal models of several dialogue types [6, 7]. The deliberation dialogue type is one of the six primary dialogue types identified by Walton and Krabbe [8] and represents the scenario where participants collaborate to reach a decision [9]. Some of these models tend to represent the communication performed by the decision-makers in face-to-face scenarios where the decision evolves over time. However, in real face-to-face scenarios when a decision-maker speaks, all other decision-makers receive the message at the same time, which is not the case of the communication performed by the agents. When an agent sends a message to all other agents, the message is received in different time instants by each of them. This time difference, even extremely small (milliseconds), is enough to make agents act in the possession of different amount of knowledge, which can compromise the quality of the decision made. In addition, this technological limitation can deceive researchers in the dialogue models' evaluation phase.

In this work, we present the Multi-Agent System for Group Decision-Making (MAS4GDM) framework, which aims to facilitate the implementation of dialogue game models that somehow intend to simulate or virtualize the interaction of groups of decision-makers in face-to-face contexts. The MAS4GDM framework allows programmers to focus only on the implementation of dialogue models and to abstract themselves from other complex implementation aspects, such as: managing the agents lifecycle and their ids, and system failures due to incomplete or incorrect messages. The framework characteristics help programmers in the definition of the agent's behavior, as well as to access information. In addition, the communication flow that guarantees security, integrity and that agents always act in the possession of the same knowledge, is of the framework's responsibility.

The rest of the paper is organized in the following order: in the next section, the MAS4GDM framework is described and in Sect. 3 we can find the experimental evaluation. In Sect. 4 the discussion is presented. Finally, in Sect. 5 some conclusions are taken along with the work to be done hereafter.

2   MAS4GDM

MAS4GDM is a framework that allows the implementation of dialogue models that intend to simulate/represent/virtualize the type of interaction practiced by decision-makers in face-to-face group decision-making processes. The MAS4GDM framework encapsulates the JADE framework, implements a communication model that virtualizes a face-to-face context and allows the developers to focus only on the implementation of the agents' behavior. In addition to the communication model, the MAS4GDM framework creates the agents automatically, manages their ids, prevents the system from failing due to incomplete or incorrect messages, and assists the developers in defining the agents' behavior as well as to access information.

2.1   Architecture

In Fig. 1, the "*external Application*" represents an application using the MAS4GDM framework. The layer provided by the framework for the developers consists on the following classes: *AbstractMeeting*, *AbstractDecisionMaker* and *Message*.

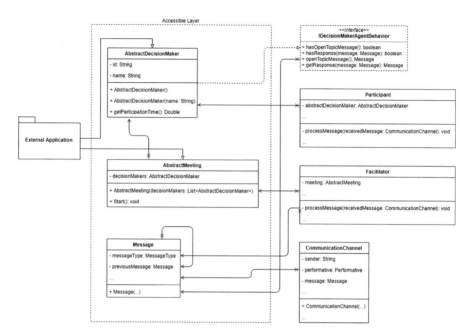

Fig. 1.   Short version of the MAS4GDM framework's class diagram.

These are the 3 classes that the developers can extend and that allow them to implement their business:

- *AbstractMeeting*: From the programmer's point of view, this class is the heart of the framework, since it implements the "*start()*" method that allows to start the whole process. It would not be possible to start a group decision-making process without a "Meeting". For a *Meeting* object to be created, it is necessary to send a list of *AbstractDecisionMaker* as an argument, since there are no group decision processes without participants;

- *AbstractDecisionMaker*: This class represents a decision-maker. The developers can create as many extensions of *AbstractDecisionMaker* as the types of decision-makers they wish to represent. The fact that it is possible to create several types of decision-makers, allows the agents representing each of those decision-makers to make use not only of different levels of knowledge but also to implement different behaviors. This class implements the *IDecisionMakerAgentBehavior* interface, which means that any *AbstractDecisionMaker* that the programmer defines will always know how to execute the 4 core behaviors of the framework:

– *boolean hasOpenTopicMessage()*: The agent checks if it is interested in starting a new topic;
– *Message getOpenTopicMessage()*: The agent returns a message to start a new topic;
– *boolean hasResponse()*: The agent checks if it is interested in responding to any message received in the current topic;
– *Message getResponse()*: The agent returns a message to respond to a previous message from the topic that is open.

• *Message*: This class represents the messages that are exchanged by the agents from the programmer's point of view.

As agents exchange messages of type *Message* and all messages are encapsulated in the *CommunicationChannel* of the framework, there are never any errors that affect the correct functioning of the framework derived from flaws in the construction of objects of type *Message*, such as: incorrect definition of the sender and receivers, or at the limit, even if the object is null.

2.2 Agents

The management of the agents' lifecycle is of the MAS4GDM framework complete responsibility. Thus, with the use of the MAS4GDM framework, developers have no need to manage/create agents.

The MAS4GDM framework considers the existence of two types of agents:

• Participant Agent: Each participant agent represents a decision-maker. It means that the number of agents involved will be equal to the number of decision-makers who are part of a decision-making process;
• Facilitator Agent: There is a facilitator agent for each decision process being executed. The facilitator agent is responsible for managing the communication process and creating the participant agents in the same *AgentContainer* in which it was created.

Whenever a decision process is initiated (*AbstractMeeting.start()*) a Facilitator Agent is created, receiving the *AbstractMeeting* and the *AgentContainer* in which it was created as arguments. In turn, the Facilitator Agent uses the *AbstractDecisionMaker* list that is contained in the *AbstractMeeting* to create the Participant Agents (in the *AgentContainer* that it received as argument). For each *AbstractDecisionMaker* in the list, a Participant Agent is created. Each Participant Agent receives as an argument the *AbstractDecisionMaker* that it will represent and where the behaviors, defined in the *IDecisionMakerAgentBehavior*, are implemented. When the process ends, the Facilitator Agent notifies the Participant Agents and after they respond with an acknowledgment, they terminate. When all the Participant Agents have responded, the Facilitator Agent also terminates.

2.3   Communication

One of the most important components of this framework is the virtualization by the communication model used by agents of the type of interaction between real decision-makers in face-to-face contexts. The type of communication implemented in this framework guarantees that all agents always act in possession of the same knowledge (when it is public). In addition, the framework is prepared to deal with agents that may have different levels of activity (different probability of starting dialogues or responding to a message), which means that it is possible to define agents that represent more or less "talking" styles. The framework does not oblige to implement any hard communication format, as is the case of "question-answer" models. On the contrary, it allows complete freedom in the implementation of the type of dialogue to be carried out by the agents. It is possible to implement dialogues that represent complex dialogues (via audio or writing) such as those performed on a social network like Facebook, where anyone can open a new topic and each message can successively originate several other messages, allowing at the same time to come up with successive answers on a certain subject and go back to resume a "reasoning".

The MAS4GDM framework organizes dialogues by topics. Each topic is related to a specific subject; however, the subject's detail level can be defined according to the model that is being implemented. Which means that in the limit, there can only be 1 topic. Each topic consists of an indefinite number of messages. The first message of a topic is the root message and has no previous messages. All the other messages in the topic are a response to any previous message, and as such, point to a previous message. The messages that constitute a topic are then represented in the form of a k-ary tree.

All messages exchanged by the agents are of the *CommunicationChannel* type. These messages may or may not carry messages of the type *Message*. Since all messages of type *Message* are encapsulated in the messages exchanged by the agents, the framework is immune to possible failures in the incorrect implementation of a dialogue model. All the messages exchanged by the agents have a defined performative:

- ENTER_DIALOGUE: ask about the interest in responding to some message of the topic being debated;
- ENTER_DIALOGUE_RESPONSE: Respond to ENTER_DIALOGUE with a participation time or null if not interested;
- OK: acknowledgment;
- OPEN_DIALOGUE: Ask about the interest in starting a new topic;
- OPEN_DIALOGUE_RESPONSE: reply to OPEN_DIALOGUE with a participation time or null if not interested;
- PROCESS_ENTER_DIALOGUE: notify the selected agent of the right to communicate a new message about the topic being debated;
- PROCESS_ENTER_DIALOGUE_RESPONSE: reply to PROCESS_ENTER_DIALOGUE with a Message or null;
- PROCESS_OPEN_DIALOGUE: notify the selected agent of the right to start a new topic;

- PROCESS_OPEN_DIALOGUE_RESPONSE: reply to PROCESS_OPEN_DIA-LOGUE with a Message or null;
- PROCESS_OVER: notify that the process has finished;
- READY: notify READY state.

Figure 2 represents (in a non-formal format) the flow of communication between the Facilitator Agent and the Participant Agents. As noted above, the Facilitator Agent is responsible for creating the Participant Agents. Whenever a Participant Agent is created, it sends a message to the Facilitator Agent with the READY performative. When the Facilitator Agent receives READY from all Participant Agents, the process is ready to begin. To initiate the process, the Facilitator Agent sends a message to the Participant Agents with the OPEN_DIALOGUE performative. All Participant Agents that are interested in starting a new topic respond with a participation time (by default, they return a value between [0; 1], otherwise they will reply null). The message with the participation time is then sent to the Facilitator Agent and has the OPEN_DIALOGUE_RESPONSE performative.

When the Facilitator Agent receives the message with the participation time of all Participant Agents, it selects the Participant Agent that sent the highest participation time to start a new topic (if it received null from all Participant Agents, it means that the decision process has ended, since the agents had anything else to add). After selecting the Participant Agent that has earned the right to start a new topic, the Facilitator Agent sends a message with the PROCESS_OPEN_DIALOGUE performative, notifying all Participant Agents of the Participant Agent who has gained the right to start a new topic (including the selected Participant Agent). Then, the selected Participant Agent sends a message to the Facilitator Agent with the content with which it wants to start a new topic, with the performative PROCESS_OPEN_DIALOGUE_RESPONSE. If at the moment that the Participant Agent is informed that he has gained the right to speak, he no longer has interest in starting a new topic, it may respond with a null Message object. In this case, when the Facilitator Agent verifies that the Message is null, it re-sends a Message with the OPEN_DIALOGUE performative to all Participant Agents. After receiving the message from the Participant Agent with the performative PRO-CESS_OPEN_DIALOGUE_RESPONSE, the Facilitator Agent broadcasts the Message to all Participant Agents using the performative ENTER_DIALOGUE.

This Message, in addition to disseminate knowledge by all Participant Agents, also serves to question them about their interest in responding to that or any other message of the topic being debated. When the Participant Agents receive the message with the ENTER_DIALOGUE performative, they check to see if they are interested in responding to the topic being debated and, once again, respond with a message containing the participation time (or null) and the ENTER_DIALOGUE_RESPONSE performative. When the Facilitator Agent receives the message with the performative ENTER_DIALOGUE_RESPONSE from all Participant Agents, it again selects the Participant Agent with the highest participation time (if all Participant Agents have sent null, it means that there is nothing else to add about the topic being debated and the Facilitator Agent re-sends a message with the OPEN_DIALOGUE performative to the Participant Agents). After selecting the Participant Agent that has earned the right to respond, the Facilitator Agent sends a message to all Participant Agents with the performative PROCESS_ENTER_DIALOGUE informing them of the selected agent.

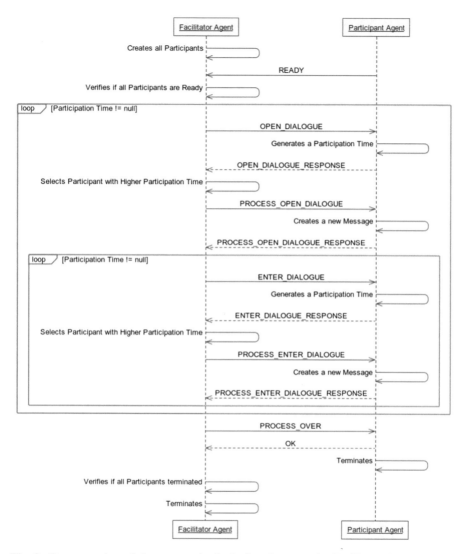

Fig. 2. Representation of the communication's flow between the Facilitator Agent and the Participant Agents.

Then, the selected Participant Agent sends a message to the Facilitator Agent with the PROCESS_ENTER_DIALOGUE_RESPONSE performative with the content with which it wants to respond (if at the moment the Participant Agent is informed that it has gained the right to speak but it is no longer interested in responding, it may respond with a null Message and the Facilitator Agent will resend a message to all agents with the performative ENTER_DIALOGUE containing the last valid message it received). After replying to a message with the PROCESS_ENTER_DIALOGUE_RESPONSE performative, the Facilitator Agent re-sends a message with the performative

ENTER_DIALOGUE to all Participant agents. Finally, when no Participant Agent wants to respond to the current topic or start a new topic, the process ends. The Facilitator Agent then sends a message with the PROCESS_OVER performative to all Participant Agents and they respond (before they finish) a message with the OK Performative. When the Facilitator Agent receives a message from all the Participant Agents with the OK performative, it also terminates.

3 Experimental Evaluation

In order to empirically evaluate the proposed framework, we implemented an argumentation-based dialogue model [10, 11] designed to the group decision-making context. In order to make the scenario more complex, the agents were defined with different social aspects: behavior styles, levels of expertise and credibility [12]. A decision scenario that consisted of the acquisition of 100 cars to renew the fleet of an organization was also defined (the considered multi-criterion problem can be found in Table 1). The existence of 5 decision-makers was considered, and preferences regarding the different criteria and alternatives were defined for each of them.

Table 1. Multi-criteria problem.

	Price	HP	Trunk	Consumption	Style	GPS	Color
Ford Fiesta	9900	70	292	3.5	2	True	Green
SEAT Ibiza	9900	75	355	3.8	2	False	Blue
Opel Corsa	8800	75	285	3.2	1	False	Red

The dialogue carried out by the agents was quite long and as such, only 4 of the more than 20 generated topics are presented below.

1. ag1: I agree with being in favor of Opel Corsa 1.4 90 Selective 5p S/S MTA because very good recent experience with their assistance service.
   a. ag2: I agree.
   b. ag3: I disagree.
   c. ag4: I agree.
   d. ag5: I disagree because, I think that argument can be applied to every brand.
      (1) ag3: I agree.
      (2) ag4: I agree.
      (3) ag1: I agree.
      (4) ag2: I disagree because, that is not true.
        i. ag1: I do not have that information.
        ii. ag3: I do not have that information.
        iii. ag5: I do not have that information.
        iv. ag4: I disagree because, if you check car magazines you will see that the other brands also have good marks in terms of assistance service.

    (a) ag3: I agree.

    (b) ag1: I disagree.

    (c) ag5: I disagree because, that was verified on previous models. However, if you check the tests regarding the newest models, you will see that the difference is considerable.

        A. ag2: I agree.

        B. ag1: I agree.

        C. ag4: I agree.

        D. ag3: I agree.

    (d) ag2: I disagree because, this particular brand has always the highest scores and more importantly we have a positive past experience with this brand.

        A. ag5: I am not sure about that.

        B. ag3: I disagree.

        C. ag1: I do not have that information.

        D. ag4: I agree.

2. ag5: Which criterion/a do you consider most important?

    a. ag3: For me the most important criterion/a is/are: consumption.

    b. ag4: For me the most important criterion/a is/are: color.

    c. ag2: For me the most important criterion/a is/are: price.

    d. ag5: For me the most important criterion/a is/are: price.

    e. ag1: For me the most important criterion/a is/are: consumption.

3. ag5: Hey ag1, do you accept this alternative as the solution? Opel Corsa 1.4 90 Selective 5p S/S MTA (Appeal to Self Interest)

    a. ag1: I accept.

4. ag5: Hey ag4, do you accept this alternative as the solution? Opel Corsa 1.4 90 Selective 5p S/S MTA (Appeal to Common Sense)

    a. ag4: I accept.

In this small excerpt, as in the full version of the dialogue, some agents were more active, which was due to the use of different styles of behavior (defined with different levels of activity). As we have seen previously, the MAS4GDM framework allows to define different probabilities of activity in the agents. In the first topic created by the agents it was verified that the agents "resumed a previous reasoning" when after having responded to the message iv - (c), an agent sent a message (d) to respond to the message iv. In each topic a different number of messages were exchanged according to the interest of the agents in sharing or not new knowledge, as well as in their interest in defending a certain idea.

4   Discussion

In the light of the current state of the art, the MAS4GDM framework provides the discussion of several important aspects. Next, we will see those we considered most relevant to the context of this publication. It seems clear to us that although the context of the framework is that of group decision-making, it is possible to use it in other contexts. Any type of dialogue defined by Walton and Krabbe [8] can be implemented using the MAS4GDM framework. However, since this framework implements an acknowledgment system that guarantees that agents always behave in equality of knowledge, they tend to exchange more messages, which consequently affects performance, which is unnecessary (but possibly not relevant) in certain contexts.

Another relevant aspect is that, the MAS4GDM framework helps, even intuitively, in the definition of the dialogue game protocol as proposed by McBurney and Parsons [13]. The structure of communication flow that composes it, based on topics and responses not necessarily sequential (as verified in the experimental evaluation), facilitates the implementation of elements such as commencement rules, commitments termination rules.

5   Conclusions and Future Work

At this moment we can observe the MAS4GDM framework in terms of the perception the experimental evaluation transmitted to us. We concluded that the MAS4GDM framework greatly facilitates the implementation of dialogue models, since it abstracts the developers from a series of implementation details. In addition, it makes the code much more legible because it completely separates the intelligent layer (dialogue model's definition) from the communication layer and the agents' lifecycle management. The fact that the framework is responsible for the agents' lifecycle and communication flow facilitates the detection of implementation errors. As the MAS4GDM framework was designed to operate in the group decision-making's context, it allows to obtain richer outputs when compared to systems that aim to virtualize face-to-face scenarios and that do not include mechanisms that guarantee that all agents always act in equality of knowledge, and as a consequence, behaviors, interactions and outputs do not represent properly the conceptual definition of those models.

As future work, we want to transform the architecture of the MAS4GDM framework into a micro services architecture in order to make the framework completely independent. We also intend to conduct performance tests to study the behavior of the framework in complex scenarios that involve many agents exchanging a large volume of messages simultaneously.

Acknowledgments. This work was supported by the GrouPlanner Project (POCI-01-0145-FEDER-29178) and by National Funds through the FCT – Fundação para a Ciência e a Tecnologia (Portuguese Foundation for Science and Technology) within the Projects UID/CEC/00319/2013 and UID/EEA/00760/2013.

References

1. Pérez, I.J., Cabrerizo, F.J., Alonso, S., Herrera-Viedma, E.: A new consensus model for group decision making problems with non-homogeneous experts. IEEE Trans. Syst. Man Cybern.: Syst. **44**, 494–498 (2014)

2. Alonso, S., Herrera-Viedma, E., Chiclana, F., Herrera, F.: A web based consensus support system for group decision making problems and incomplete preferences. Inf. Sci. **180**, 4477–4495 (2010)

3. Groeneveld, J., Müller, B., Buchmann, C.M., Dressler, G., Guo, C., Hase, N., Hoffmann, F., John, F., Klassert, C., Lauf, T.: Theoretical foundations of human decision-making in agent-based land use models–a review. Environ. Model Softw. **87**, 39–48 (2017)

4. An, L.: Modeling human decisions in coupled human and natural systems: review of agent-based models. Ecol. Model. **229**, 25–36 (2012)

5. Russell, S.J., Norvig, P.: Artificial Intelligence: A Modern Approach. Pearson Education Limited, Malaysia (2016)

6. Walton, D., Toniolo, A., Norman, T.J.: Towards a richer model of deliberation dialogue: closure problem and change of circumstances. Argum. Comput. **7**, 155–173 (2016)

7. Thimm, M.: Strategic argumentation in multi-agent systems. KI-Künstliche Intell. **28**, 159–168 (2014)

8. Walton, D., Krabbe, E.C.: Commitment in Dialogue: Basic Concepts of Interpersonal Reasoning. SUNY press, Albany (1995)

9. McBurney, P., Hitchcock, D., Parsons, S.: The eightfold way of deliberation dialogue. Int. J. Intell. Syst. **22**, 95–132 (2007)

10. Carneiro, J., Martinho, D., Marreiros, G., Jimenez, A., Novais, P.: Dynamic argumentation in UbiGDSS. Knowl. Inf. Syst. 1–37 (2017)

11. Carneiro, J., Martinho, D., Marreiros, G., Novais, P.: Arguing with behavior influence: a model for web-based group decision support systems. Int. J. Inf. Technol. Decis. Mak. (2018)

12. Carneiro, J., Saraiva, P., Martinho, D., Marreiros, G., Novais, P.: Representing decision-makers using styles of behavior: an approach designed for group decision support systems. Cogn. Syst. Res. **47**, 109–132 (2018)

13. McBurney, P., Parsons, S.: Dialogue games for agent argumentation. In: Argumentation in Artificial Intelligence, pp. 261–280. Springer (2009)

Employee Performance Evaluation Within the Economic Management System of the Spanish Air Force: Development of a Methodology and an Optimization Model

Manuel A. Fernández-Villacañas Marín[1,2]([⊠])

[1] Spanish Air Force, Madrid, Spain
mfermarl@ea.mde.es
[2] Department of Organization Engineering,
Business Administration and Statistics,
Technical University of Madrid, Madrid, Spain

Abstract. The economic administration and procurement system of the Spanish Air Force is composed of a series of management units, whose staff needs to be optimized. In order to accomplish this, the study analyses the concept of performance evaluation, and a methodology has been designed to evaluate performance which is improved through systemic analysis and organizational behavior. The traditional models of performance evaluation are based on the psycho-sociological research of the workers. They are static and require a lot of management effort. Faced with these limitations, the new model that has been implemented to evaluate performance and measure workload, is interdisciplinary, dynamic, it can be used with high periodicity, and demands little management effort. Moreover, it serves to guide decision-making regarding the reassignment of jobs and personnel, and allows an improvement in effectiveness, efficiency, economicity and the service levels of the system. Above all, it has been developed and implemented as a "pilot project".

Keywords: Organizational performance evaluation · Workload measurement · Human resources optimization

1 Introduction

The Spanish Air Force (SAF) is designated to defend Spanish airspace and contributes to the maintenance of international security through its participation in peace operations and humanitarian aid. At present, the human resources available amount to some 23,000 employees. In recent years, there has been a reduction of about 4,000, with a growing tendency to continue reducing this figure by another 3,000 during the next 3 years. This significant reduction, which contrasts with the increase in demand for operational activity and training, logistics and administration, has led to personnel becoming an increasingly scarce resource and its optimization has become an essential initiative. About 500 employees are directly assigned to the organic elements that make up the economic management and procurement system of the SAF, a collective which

© Springer Nature Switzerland AG 2019
Á. Rocha et al. (Eds.): WorldCIST'19 2019, AISC 930, pp. 448–455, 2019.
https://doi.org/10.1007/978-3-030-16181-1_42

includes another 500 who work part-time in the operational, training and logistic support Units, Centres and Organizations (UCOs). Both groups should be significantly reduced over the next few years. The system is currently made up of the Directorate of Economic Affairs and the Procurement Directorate, 14 Economic-Administrative Sections (SEAs), 7 Economic Support Organs (OAEs) in national deployment and 4 in international deployment, and 218 Units that correspond to the UCOs of the SAF that configure their deployment.

2 Background and Literature Review

2.1 Performance Evaluation: Alternative Reference Approaches and Its Application in the Spanish Public Administration

Despite the growing importance that most public and private organizations have been giving to both the management of human resources and the evaluation of their performance, the more frequent ones are those that continue to maintain traditional approaches and whose evaluative activities mainly focus on the simple unilateral judgment of the boss regarding the work of his subordinate (Harris 1986).

However, regardless of this non-evolved way of operating, the last decades have been very prolix as to the conceptual development of increasingly sophisticated and interdisciplinary evaluation models. For example, Ployhart, Schneider and Schmitt (2006) consider that the evaluation represents the conceptualization, measurement and analysis of the level at which employees do their work and the satisfaction they obtain regarding their work situation. Sánchez and Bustamante (2008) point out that those organizations need to know how employees are performing their tasks, in order to identify who effectively adds value and who does not. Consequently, there is a shared approach among the relevant authors, both in relation to the purposes that the concept of performance evaluation raises, as well as in the need to measure the contribution of each worker as to their achievements regarding the objectives of the organization. However, there is no homogeneity in terms of how this is evaluated, or on the effects that can or could derive from the selected evaluation system selected.

Perhaps for this reason, new disruptive tendencies in the evaluation of performance based on the global analysis of management data and continuous evaluation are being imposed in the business world. These new approaches reduce the use of complicated traditional methods and make a final qualification prevail, without relying on voluminous audit reports (Barceló 2018). These evaluations are developed around a continuous dialogue between bosses and subordinates, where the centre of the permanent evaluation is the person, not with the idea of measuring but rather promoting their continuous improvement.

In relation to the Public Sector in Spain, Law 7/2007, 12 of April, of the Basic Law of Public Employees, the concept of performance evaluation was introduced. However, until now, this Law has not been applied or deployed, and only some training initiatives and "pilot projects" between 2008 and 2012 have been proposed in the General State Administration, to which the SAF belong (See Gorroti 2007 and Grupo de Trabajo ad hoc 2015).

2.2 Systemic Analysis and the Organizational Behaviour

The planning of social and organizational change, with the intention of solving a problem or developing a view, should predict and keep in mind the changes in the circumstances that will predictably occur, and when they will do so. It is necessary to bear in mind that every change that is generated in an organization will affect various variables of its internal social system. The detailed knowledge of the nonlinear relationships between the variables that explain it offers those responsible for the planning process, the possibility of achieving great changes in a desired variable through relatively small changes in another one (Meyer 1983).

In this sense, interdisciplinary field studies of organizational behaviour are very useful, since it lets us know the impact that individuals, groups and structures have on human behaviour within the organizational scope, creating more effective and efficient organizations, thus optimizing personal and group work performance (Mullins 1996). These studies allow the approaches to be enhanced, provide references on the dimensions which is necessary to take into account, and make the guidelines to be followed more visible. To begin with, the specific corporate culture of each management unit plays an important role in the way in which individuals interact with each other, they also set up formal and informal groups and develop their work activities. In addition, success through achieving a high level of professional satisfaction requires the existence of an intrinsic and extrinsic rewards system, which should be felt as being fair by the majority. Likewise, the development of adequate labour standards, the existence of an appropriate style of management and supervision, satisfactory working conditions, as well as a leadership that is totally aligned with the approach of the implemented management system, should all promote a participative, effective, efficient and service-oriented style. Finally, power and authority, which should operate in an interdependent manner with ethical and legal guidelines proper to the functioning of the Public Administration, through which these elements are exhibited and used, represent key components to manage a cohesive and sound workplace.

3 The Problem Case: The Strategic Concept and Design of the Specific Model Used

The aim of this work has been to analyse the problem as well as develop and implement a dynamic model that allows group performance evaluation and workload measurement of each of the management units of the economic administration system and acquisitions of the SAF. The management of this model only requires a reduced consumption of resources, essentially making use of the data of the computerized management system, thus making it possible to guide decision-making regarding staff and personnel reallocation, training and changes in organizational culture and leadership, and moving towards the improvement of the effectiveness, efficiency, economic viability and service to the UCOs.

The Computer System of Economic Management and Administration of the Ministry of Defence of Spain (SIDAE) manages all financial resources and management activities thereof, including the SAF: budget, contracting and procurement, accounting, financial management and treasury processing service commissions, as well as B2B electronic management with various suppliers.

3.1 Reference Performance Evaluation Models in the Spanish Public Administration

Three generic performance evaluation models have been preferentially used in Spanish Public Administrations:

- The contextual performance model by Borman and Motowidlo (1993), is based on both individual and group analysis, the persistence of extra effort and enthusiasm to successfully complete the activities required by the tasks, the voluntarism in the individual tasks, support and cooperation are other ones, alignment of the rules, procedures and organizational policies, and ongoing support to fulfill the objectives of the organization.
- The performance techniques and core tasks model by Campbell, Gasser and Oswald (1996), generic tasks of similar posts as well as specific ones of particular posts, tasks that need verbal and written communication to variable sized audiences, maintaining personal discipline, facilitating team tasks, monitoring activities regarding influence, management and monitoring of workers as well as management tasks related to the organization, implementation and resourcing.
- The task productivity model of Viswesvaran and Ones (2005), which is based on their analysis of the behaviours that are considered constituent of the central tasks of the positions, interpersonal competence, leadership, effort, knowledge of the position and the counterproductive conducts.

The specificity of the problem under study recommended the creation of an ad hoc dynamic model that would integrate the main aspects of the previous three, and oriented towards the generation of strategic intelligence produced by a monitor on objective quality regarding management and productivity of the assigned resources, as well as on subjective quality and satisfaction of the users.

3.2 The Specific Model Used: Empirical Analysis

The analyses has been developed with a multidisciplinary approach that integrates the creation of a matrix of the four functional areas of activity, global management, contracting and procurement, accounting and finance, and treasury management, with the three types of continuous improvement established for performance evaluation (Fig. 1): EFFICIENCY, quality of the management and productivity of the resources allocated; ECONOMICITY, internal control of the management activities; and EFFECTIVENESS, subjective quality and user satisfaction of the services provided.

The specific model proposed incorporates an approach that evaluates continuous improvement, with a vocation to export "best practices" and avoid "inappropriate practices", concentrating the effort on the most important aspects of the four functional areas, depending on their deviation from the standard or their variability and risk. For this purpose, an integrated monitor of quantitative and qualitative indicators was developed, for which a classification of their values was proposed in conditions of "normality", of "insignificant abnormality" and of "very significant abnormality", as well as their objective values. It was decided that the studies should begin by evaluating the performance of the 14 SEAs like a "pilot project".

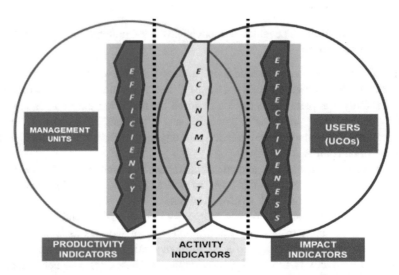

Fig. 1. Types of indicators for the performance evaluation of the economic management and procurement system of the SAF.

In relation to the ACTIVITY indicators and part of the PRODUCTIVITY indicators, 21 KPIs were defined within the possibilities offered by the SIDAE for the integration and monitoring of quantitative information related to the four established functional areas. It was necessary to integrate them, by deliberating the ones that were most adjusted to the average reality of the set of SEAs, even though there was proof that there were differences in the actual procedures that apply to each of them.

A Bottom-Up study was carried out estimating from an "IT-technical" approach, a first determination of such weight only taking into account the average efforts made in the management of the SIDAE information processes. However, it was complemented with a Top-Down study, from a "user manager" approach, to integrate in each indicator the comprehensive effort of the managers, that is, in addition to the computer processes, other processes, documentaries and cognitive activities, coordination, archiving processes, etc. of both management personnel and support staff. An average estimation of the weights of the 14 SEAs was taken, and the variability of the weights was studied, measuring their variances to determine the congruence of the results obtained, both in comparative terms of what is proposed by the rest of the SEAs, as in relation to the weights obtained from the previous Bottom-Up approximation.

The information on PRODUCTIVITY was complemented by qualitative research, through a survey carried out on all of the staff of the SEAs, by means of questions whose answers were tabulated with Likert scales, in relation to the main attributes of leadership, motivation, training and corporate culture.

Finally, in relation to the IMPACT indicators, the main aspects of the quality perceived by the users of the services provided by the SEAs to the UCOs were studied. A survey evaluated statistically representative samples, and likewise, a Likert scale evaluated the importance, assessment and diversity of the satisfaction attributes. These included: advice and use of clear and transparent information; knowledge of the needs of the UCO; professionalism, efficiency and reliability in meeting the needs of the UCO; availability, speed, efficiency and flexibility in management responses; loyalty and treatment in the personal interaction regarding services provision; and an overall level of satisfaction.

The model developed is synthesized in Fig. 2, which is summarized below:

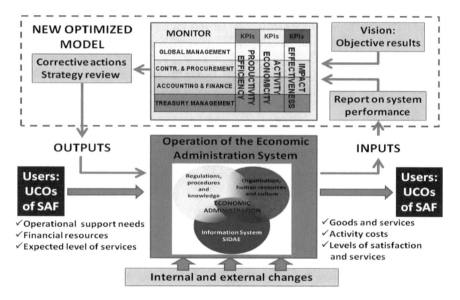

Fig. 2. Synthesis of the dynamic model developed for the performance evaluation of the economic management and procurement system of the SAF.

A variety of tools have been used to extract and process the SIDAE data. Most of the extraction has been done through Query String, directly made against the Database in SQL language (Structured Query Language) through the Oracle SQL Developer tool. These queries have been downloaded to Excel files for processing in the Microsoft Excel Tool through macros and programming in Visual Basic for Applications (VBA). My thanks to the engineers of ALHAMBRA SYSTEMS S.A. in charge of the maintenance of the SIDAE, for their support to this study.

Regarding qualitative aspects of overall satisfaction, the study resulted in average values higher than 4 (1 min./5 max.), both in importance and in the assessment of all its attributes, with very low diversity of opinions. Likewise, the levels of personnel motivation and the exercising of leadership among the intermediate commanders, gave average result levels regarding valuation superior to 3.5 but, on the other hand, identifying a lack of training in more than 80% of the employees.

Regarding the quantitative aspects of workload, there is a great divergence in the workload levels supported in absolute terms by each of the SEAs. These were estimated according to the result obtained from the integration of the KPIs used, which is especially significant in terms of the unit levels of workload of jobs in the existing workplaces, for each job covered and each employee allocated.

4 Conclusions and Generalization of the Developed Model

This dynamic model has turned out to be sufficiently effective and explanatory. After verifying the significantly high satisfaction levels of the UCOs, it has been possible to confirm the existence of a distribution of jobs whose load is very unbalanced. Therefore, it is necessary to adjust the demand and supply of labour, promoting a reallocation of personnel, or reassigning the UCOs that are managed by each SEA.

Following the generalization of the analysis to complete SAF, it is necessary to develop standardized actions and intensify the deployment of the SIDAE, thus promoting a more homogeneous corporate culture based on participative leadership and reviewing the functions and tasks of each organic element of the system as well as providing a higher level of integration to it.

References

Barceló, J.M.: Nuevas tendencias en evaluación del desempeño, IMF Business School (2018). https://blogs.imf-formacion.com/blog/recursos-humanos/evaluacion-desempeno/nuevas-tendencias-en-evaluacion-del-desempeno/. Accessed 03 Nov 2018

Borman, W.C., Motowidlo, S.J.: Expanding the criterion domain to include elements of contextual performance. In: Schmitt, N., Borman, W.C. (eds.) Personnel Selection in Organizations. Jossey-Bass, San Francisco (1993)

Campbell, J.P., Gasser, M.B., Oswald, F.L.: The substantive nature of job performance variability. In: Murphy, K.R. (ed.) Individual Differences and Behavior in Organizations. Jossey-Bass, San Francisco (1996)

Gorroti, M.: La Evaluación del Desempeño en las Administraciones Públicas Españolas. Revista de Psicología del Trabajo y de las Organizaciones, vol. 23 núm. 3, pp. 367–387. Colegio Oficial de Psicólogos de Madrid, Madrid (2007)

Grupo de Trabajo ad hoc: Modelo de Evaluación del Desempeño en el Marco de la Planificación de Objetivos y Estratégica en las Administraciones Públicas, Instituto de Estudios Fiscales, Ministerio de Hacienda y Administraciones Públicas, Madrid (2015)

Harris, O.: Administración de Recursos Humanos: Conceptos de Conducta Interpersonal y casos, 1st edn. Limusa, México DF (1986)

Meyer, J.W., Scott, W.R.: Organizational Environment, Ritual and Rationality. Sage Publications, Beverly Hills (1983)

Mullins, L.J.: Management and Organizational Behaviour. Pitman Publishing, London (1996)

Ployhart, R.E., Schneider, B., Schmitt, N.: Staffing Organizations: Contemporary Practice and Theory, 3rd edn. Lawrence Erlbaum Associates Publishers, Mahwah (2006)

Sánchez, J., Bustamante, K.: Auditoría al proceso de evaluación del desempeño. Contabilidad y Auditoría **198**, 105–133. Facultad de Económicas de la Universidad de Buenos Aires, Buenos Aires (2008)

Viswesvaran, C., Ones, D.S.: Job performance: assessment issues in personnel selection. In: Evers, A., Anderson, N., Voskuijl, O. (eds.) The Blackwell Handbook of Personnel Selection. Blackwell, London (2005)

Improving Control Effectiveness in IS Development Projects Through Participatory Implementation

Roman Walser$^{(\boxtimes)}$ (iD)

Vienna University of Economics and Business, 1020 Vienna, Austria
roman.walser@wu.ac.at

Abstract. Efficient Information Systems (IS) adequately serving business needs can be seen as an important prerequisite for the survival of most organizations. Newly developing or adopting those systems is a challenging task. To increase probability of success in often complex development projects, managers usually implement controls. However, most controls require support and active commitment of the controllees in order to be effective. To better understand the concept of control legitimacy and resulting control responses, researchers have recently shifted their focus on the controllee perspective. However, the process of choosing and enacting control is still considered top-down. In this poster paper, we propose to move away from this assumption. Instead, project managers should make the process more participatory by involving the controllees already at the beginning of IS development projects. This would help to choose and implement a set of controls which is perceived as legitimate and thus supported by the controllees.

Keywords: Control legitimacy · Control emotional appraisal · Resistance · Top-down · Detrimental behavior

1 Introduction and Relevancy

Information Systems (IS) have become the backbone of our modern and globalized economy. Continuously changing business needs make the adoption or development of IS an indispensable responsibility of today's information officers. However, successfully managing IS development (ISD) projects is a challenging task because they are complex in many respects. Simply put, ISD projects involve stakeholders of various organizational departments with diverging knowledge and needs, producing highly abstract outcome, which is intangible and hard to measure [1]. To meet those challenging characteristics of ISD projects and make software development undertakings a success, project managers typically implement controls.

Formally, controls are any attempts "to align individual behaviors with organizational objectives" [2, p. 742]. This traditional view includes a dyadic relationship between controller (a manager implementing control) and controllees (a subordinate being controlled). A control in software development could be pair programming as a means to improve software quality. Over the last decades, control in the context of ISD

© Springer Nature Switzerland AG 2019
Á. Rocha et al. (Eds.): WorldCIST'19 2019, AISC 930, pp. 456–459, 2019.
https://doi.org/10.1007/978-3-030-16181-1_43

projects has gained increasing attention among researchers. However, the resulting common body of knowledge is strongly controller-centric, neglecting that there are also controllees with their individual perception of control. This is problematic, since control can only be effective when being endorsed and supported by both the controller and the controllees. Only recently, researchers have started addressing controllees' perceived level of legitimacy (also: emotional appraisal) and its potential effects on resulting behavior, such as resistance [3, 4]. This trend towards including also the controllee perspective is important. However, choosing and enacting control is still clearly seen as a top-down process. We argue that controllees' appraisal is not generalizable and that controllers should not only try to empathize with the controllees. Rather, the controllees should be actively involved in the process of control choice and enactment already at the beginning of an ISD project.

In this poster paper, we therefore want to highlight the need for a participatory control definition process, where controller and controllees together find appropriate ways of ensuring ISD project success. The deduced research question looks as follows: *"How does a participatory control choice and enactment process change control effectiveness compared to a traditional top-down approach?"* By maximizing congruence between controller and controllee through transparency (exchanging knowledge about individual needs) and active involvement, we aim to avoid misjudgments of the controller and, in turn, prevent undesired controllee behavior.

For this purpose, we will briefly summarize the results of two relevant studies, which have been published recently. In a subsequent chapter, we will then point out the potential advantages of a more participatory control definition process.

2 Perceived Legitimacy, Emotional Appraisal and the Potential Consequences for Effectiveness of ISD Project Control

Only recently, first researchers started elaborating on factors influencing the extent to which controllees perceive controls implemented by their controllers to be legitimate. Drawing on earlier findings from Bijlsma-Frankema et al. [5], Cram and Wiener [3] argue that four factors shape control legitimacy: (1) the perceived justice in terms of fairness and reasonability, allowance for (2) autonomy, (3) group identification and (4) competence development. In a multi-case study, they find that not only the mode of control (i.e. behavior, outcome, clan, self-control [6]), but also the degree (tight vs. relaxed) and style (unilateral vs. bilateral) of control enactment influence the controllees' level of perceived legitimacy. For instance, their results suggest that formal controls enacted in a bilateral style correspond with higher perceptions of justice and autonomy, when compared to formal controls enacted in a unilateral style. Even though a bilateral (also: enabling) control style would be characterized by appreciation of controllee feedback and collaboration, the controllees would be involved only at a later stage of the project.

In a similar vein, Murungi et al. [4] find a relationship between control styles and controllee emotional appraisal. They see dynamics between controller and controllee activities, appraisal, emotional response and resulting behavior and propose that controllees' emotions are relevant for better understanding control processes. In their case study, the authors could observe that "in the case of a shift in control styles over time, these negative emotions were associated with resistance behaviors" [4, p. 1]. This is problematic, since the effectiveness of most controls is highly dependent on the controllees' willingness to actively commit: attendance and active participation in meetings, timely and correct submission of time sheets or compliance with prescribed processes (e.g. pair programming) - just to name a few.

To sum up, it can be argued that project managers who are choosing and enacting control would perceive their approach as being appropriate. However, this does not mean that also the subordinates (controllees) perceive the same level of legitimacy. A low level of perceived legitimacy might lead to bad emotions and, in turn, result in detrimental behavior negatively affecting control effectiveness and project success. Thus, the gap between the controller's and the controllees' level of perceived legitimacy should be minimized.

3   Implementing Control in a Participatory Way

Although research has started to recognize the importance of the controllees' view on control, we can still not find any claims to involve the controllee in the control design phase at the beginning of ISD projects. Instead, the controllers are advised to ensure that controls address the four previously mentioned dimension of legitimacy when choosing and enacting controls. This is surprising, since involving the controllee at the earliest possible stage would probably be the easiest way to maximize the level of congruence between controller and controllee. Moreover, it is arguable that the currently prevailing model of control legitimacy is not able to fit every ISD project, as the perceived legitimacy might be highly dependent on the people involved - all of them having their own needs. By establishing a common understanding of (usually fixed) project goals and possible ways of getting there, controller and controllees could agree on a shared set of suitable controls during the ISD project definition phase. This approach could not only work with traditional waterfall models, but also with agile software development because both methods include some forms of control.

Involving the controllees when implementing control would also take findings from Murungi et al.'s emotion-centered model of IS project control dynamics into account. Assuming that the controllees show some response to the project manager's controls, there will be a higher tension in case of strongly diverging legitimacy perceptions. In other words, the deviation between expected and observed controllee behavior will be higher whenever controls are being perceived as illegitimate. One could think of it as an "activity-response pendulum" with a given amplitude. By moving controller and controllees closer together, the emotional responses and activities of both parties could be lowered, making the course of the project more stable and foreseeable.

There might be many ways of involving the controllee in the process of control choice and control enactment. Presenting them or only a selection would already go beyond the scope of this paper. We understand that the illumination of the controllee perspective is still in its beginnings. Nonetheless, we see the necessity to not generalize the dimensions of controllees' perceived legitimacy as they are, more likely than not, individually different. It is also arguable that those individual perceptions depend on the project context (e.g. project size, complexity, team), so that the only solution can be to involve the controllees already at the beginning of each ISD project. In a participatory process, controller and controllees together should find a shared set of controls, which both parties perceive as legitimate and which is capable to promote projects success. In the end, this would also bring more satisfied employees, who are willing to actively commit to implemented controls.

4 Conclusion

The aim of this poster paper was not to come up with a concrete approach for involving controllees when creating and enacting a control portfolio. Rather, we wanted to point out the necessity to reassess the currently prevailing top-down approach for implementing control in ISD projects. We acknowledge that over the last years, researchers have shown efforts to better understand factors shaping control legitimacy and controllees' emotional appraisal of control. So far, however, research is neglecting the possibility to involve the controllee already before an ISD project is executed. Closing this gap could help researchers to better understand what makes control legitimate from the perspective of a controllee in a given situation. In practice, findings in this regard might help project managers to significantly improve control effectiveness and, in turn, increase overall ISD project success rates.

References

1. Remus, U., Wiener, M., Mähring, M., Saunders, C., Cram, A.: Why do you control? The concept of control purpose and its implications for is project control research. In: 2015 International Conference on Information Systems: Exploring the Information Frontier, ICIS 2015, pp. 1–19 (2015)
2. Wiener, M., Mährich, M., Remus, U., Saunders, C.: Control configuration and control enactment in information systems projects: review and expanded theoretical framework. MIS Q. 40(3), 741–774 (2016)
3. Cram, W.A., Wiener, M.: Perceptions of control legitimacy in information systems development. Inf. Technol. People 31(3), 712–740 (2018)
4. Murungi, D., Wiener, M., Marabell, M.: Project control and emotions: understanding the dynamics of controllee resistance behaviors. In: Academy of Management - Best Paper Proceedings (2018)
5. Bijlsma-Frankema, K.M., Costa, A.C., Sitkin, S.B., Cardinal, L.B.: Consequences and antecedents of managerial and employee legitimacy interpretations of control: a natural, open system approach. In: Organizational Control, pp. 396–434 (2010)
6. Kirsch, L.S.: Portfolios of control modes and IS project management. Inf. Syst. Res. 8(3), 215–239 (1997)

DSL Based Automatic Generation of Q&A Systems

Renato Preigschadt de Azevedo[1]([✉]), Maria João Varanda Pereira[2], and Pedro Rangel Henriques[1]

[1] Centro Algoritmi (CAlg-CTC), Dep. Informática, Universidade do Minho, Braga, Portugal
renato@redes.ufsm.br, prh@di.uminho.pt
[2] Research Centre in Digitalization and Intelligent Robotics (CeDRI), Instituto Politécnico de Bragança, Bragança, Portugal
mjoao@ipb.pt

Abstract. In order to help the user to search for relevant information, Question and Answering (Q&A) Systems provide the possibility to formulate the question freely in a natural language, retrieving the most appropriate and concise answers. These systems interpret the user's question to understand his information needs and return him the more adequate replies in a semantic sense; they do not perform a statistical word search, thus differing from the existing search engines. There are several approaches to develop and deploy Q&A Systems, making hard to choose the best way to build the system. To turn easier this process, we are proposing a way to create automatically Q&A Systems based on DSLs (Domain-specific Languages), thus allowing the setup and the validation of the Q&A System to be independent of the implementation techniques. With our proposal, we want the developers to focus on the data and contents, instead of implementation details.

Keywords: Q&A Systems · NLP · DSL

1 Introduction

The increase in the processing power and the availability of information gave rise to Q&A Systems and consequently augmented the need for such systems. Q&A Systems dialog with the end-user in a more natural way accepting questions formulated in natural language and providing more accurate answers when compared with the traditional search engines.

With the advent of smartphones with personal assistants, which allow the user to ask questions and get answers from various subjects, these systems are being used by a large number of people. Approximately forty or fifty years ago the study about Q&A Systems began [16], but because of computational limitations, these systems had limited scope. Some Q&A Systems were more or less successful, some of them were discontinued, demonstrating the difficulty of

© Springer Nature Switzerland AG 2019
Á. Rocha et al. (Eds.): WorldCIST'19 2019, AISC 930, pp. 460–471, 2019.
https://doi.org/10.1007/978-3-030-16181-1_44

building and maintaining a system capable of understanding natural language queries as a human can do, and provide the appropriate answers. Recently more efficient systems have appeared featuring real applicability. However, to improve these tools more research is necessary.

Questions are asked and answered several times per day by a human. Q&A Systems try to do the same level of interaction between computers and humans. This approach differs from standard search engines (Google[1], Bing[2], etc.) because it makes an effort to understand the intention that the question expresses and try to give concise answers instead of using only keywords from the question asked and provide documents as results.

Unlike standard search engines that retrieve documents based on the keywords expression provided in the input, Q&A Systems aim to recognize the input sentence written in a high-level natural language enabling the construction of concise answers instead of a set of possibly related documents.

A simple Q&A System is composed of several processes: question interpretation, query processing, and answer building [9]. Question interpretation is done by analyzing the user's input text to extract its meaning; it can be implemented with Natural Language Processing (NLP) techniques. The query processing phase aims at recovering the information necessary to answer the question from relevant documents or Knowledge Base (KB); information retrieval techniques or knowledge base querying can be applied. In the third phase, using the collected information, a list of answer candidates is built and the elements are ranked according to the probability to better satisfy the user needs.

To be able to create a successful question and answering system, all these processes have to be carefully specified by the domain specialist and implemented by the programmer. The programmer has to be an expert in the chosen programming language, as well as he needs to master the various libraries required to implement the system (Natural Language Processing, Knowledge Processing and Inference Mechanism, Database or Triple Storage Access, among others). The complexity of such systems components makes the implementation process hard and error-prone; it is indeed a time consuming and costly task.

The use of an approach based on formal Specifications written in Domain-specific Languages (DSLs) can simplify and accelerate the development of applications. The design of a specific language to support the development of a Q&A System allows the user to specify its components in a more abstract and concise way, avoiding implementation details. This approach makes the process of implementing the system more straightforward, and less error-prone. To the best of our knowledge, this is the first work that uses a DSL to create Q&A Systems. We did not identify any similar work to compare and discuss.

The structure of the paper is as follows: Implementation of Q&A Systems are discussed in Sect. 2; in Sect. 3 some relevant aspects on the use of DSLs for automatic construction of language-based tools are presented; The proposed

---

[1] https://www.google.com.
[2] https://www.bing.com.

system is presented in Sect. 4 and in Sect. 5 a Q&A System Specification in AcQA DSL is presented; Sect. 6 presents conclusions and future works.

2   Question & Answering Systems

The wideness of information available associated with the demand for direct answers from the users requires a different approach from standard search engines. Q&A Systems provide a way to process natural language inputs from a user extracting their meaning and providing concrete and concise answers. These systems allow a user to make questions more naturally and get concise and straightforward answers, thus decreasing the effort necessary to find the right answer.

As examples of Q&A Systems through natural language processing we have WolframAlpha, a mathematical Q&A System [21], which offers knowledge by analyzing the collection of information it possesses in its local database; IBM Watson: A system that was initially used to answer generic questions from the American TV show Jeopardy! [15, 32], but today it is used in several domains.

A basic Q&A System have to process the user's input questions and be able to respond with an answer or a rank-ordered list of answers candidates. Q&A System is composed of at least three main methods to process the user's question: question analysis, extract of potential answers and answer formulation.

To implement Q&A Systems is required distinct techniques or approaches from three broad areas: question interpretation, query processing, and answer building. The techniques used in question interpretation usually employ natural language processing (NLP), which seeks to recover meaning from the input text. Natural language processing is an area of computation that involves machine learning techniques and parsing [22]. Pattern matching and the use of tags can also be used to process the input text. In the area of question interpretation, it is important to discover the user's intention to understand what is being asked. This intention must be represented internally to be used later by the other modules, thus generating a more relevant answer. Query processing approaches are responsible for handling the input text in order to create the queries necessary to extract relevant information from the knowledge repository.

The techniques necessary to build answers rely on the methods used in the query processing and question interpretation. They usually use fragments of documents and sentences to define the most appropriated answers and present them to the user. Also, a succinct answer approach can be used, where the technique tries to present a concise answer.

There are several approaches in the literature explaining the construction of Q&A Systems [15, 24, 35, 37], demonstrating the various stages required for construction. Several technical approaches should be carefully studied to allow the understanding and processing of natural language. In the past small KBs were used, such as simple schemas, a small number of entities and relations, limited set of sensitive issues and utilized ad-hoc approaches (manually constructed rules) among other simpler ones, thus not allowing the construction of scalable systems and making complex the development of open domain systems. Currently, there are several approaches to the question interpretation such as Query Graphs, Topic Entity Linking, Relation Matching using Deep Convolutional Neural Network.

2.1 Q&A Systems Classification

Q&A System are classified according to the kind of domain they are able to deal with, being of two types: closed domain, or open domain. Closed domain (or General domain) Question Answering Systems aims to answer anything that the user asks. The questions are domain independent. This type of Q&A System works with a large repository of information to be able to answer questions of all kind of subject. This type of system is supported by large sets of information and generic ontologies.

Examples of open domain Q&A Systems are Intelligent Q&A System based on Artificial Neural Network [2], Automatic Question-Answering Based on Wikipedia Data Extraction [20], and SEMCORE [19]. Restricted domain Question Answering Systems works only with a specific domain, not being able to answer questions outside the proposed field. The information repository is made of data only related to the area, being able to achieve better accuracy than general domain Q&A System. The restricted domain is also known as a closed-domain system, usually based on well-defined, structured databases, and ontologies. These systems are limited to the particular domain implemented. Some examples of Closed domain Q&A Systems are Question Answer System on Education Acts [26], Python Question Answer System (PythonQA) [34], and K-Extractor [4].

2.2 Related Works

In this subsection, the most relevant Q&A Systems so further developed and described in the literature will be briefly introduced. The PythonQA [34] system was developed using the Python programming language, together with some libraries such as Natural Language ToolKit (NLTK) [7], Django, among others. To process the input from the user, a module called Phrase Analysis divides a phrase into several components and tries to identify three elements: action, keywords, and question type. These three elements are then compared to the knowledge base to retrieve and show answers to the users of the Q&A System.

In MEANS [5] the authors propose a semantic approach to a medical Q&A System. They apply NLP to process the corpora and the user questions. The sources documents are annotated with RDF, based on an ontology. The authors propose ten question types to classify the questions.

In work proposed by [26], a Q&A System to handle education acts is presented. The knowledge base is created from the data publicly available from the United Kingdom parliament using NLP techniques. Only keywords are extracted from the user question, ignoring the question type and possible actions present in the text input from the user.

The authors in [8] created AskHERMES, a Q&A System for complex clinical questions that uses five types of resources as a knowledge base (MEDLINE,

PubMed, eMedicine, Wikipedia, and clinical guidelines). The user question is classified by twelve general topics, made by a support vector machine (SVM). To process the possible answers, the authors developed a question summarization and answers presentation based on a clustering technique. In work proposed by Weissenborn et al. [38] the authors propose a fast neural network Q&A System. The system uses a simple heuristic, and their results show that the proposed system can achieve the same performance compared to more complex systems.

In work proposed by [13] is introduced an R-NET, and neural network model for answering questions. The neural network tries to answer questions from a given text. The work proposed in [2] creates a deep neural network from documents provided by the user. They use deep cases along with artificial neural networks models to understand the contents of the information provided by the user.

WolframAlpha [21] is a well established open domain Q&A System that initially was a closed domain system for mathematics. It allows the user to use the version available online with the pre-existing knowledge base or to upload data through a paid subscription.

IBM Watson [15] is an open domain Q&A System that was initially created to compete in the Jeopardy TV quiz program. Watson is currently an Artificial Intelligence framework provided by IBM for a variety of areas, one of which is the Q&A Systems and natural language processing. Watson is made available through paid subscriptions.

In the work of Jayalakshmi and Sheshasaayee [23], they use a similarity measure based on the user-written question and discover the appropriate meaning between the words. The authors propose the WAD Q&A System. It uses an ontology and hierarchical web documents to perform entity linking to predict the answers, based on the user question.

Rajendran and Sharon [33] propose a Q&A System that uses ontology assistance, template assistance, and user modeling techniques to achieve 85% of accuracy in their experiments. The authors of [29] also use ontologies to assist the Q&A System. In this work, they propose an algorithm to automatically update the ontology used by the system and use a semantic analyzer that operates on an ontology to extract answers.

In work [6] the authors improve question interpretation and the representation of question structure using typed attributed graphs and a question ontology. They also state that using domain ontologies and lexico-syntactic patterns improves the results.

It is proposed in [25] a graph matching algorithm for query matching with an ontology using a spread activation algorithm. The spread activation algorithm uses the WordNet [28] to calculate semantic similarity.

An approach to automatically generate image descriptions is proposed in [14]. Firstly words describing the image is detected. Secondly, sentences relating to the objects in the picture are produced. The final step is to rank the phrases according to the MERT [31] model and present the best-ranked sentence to the user.

The authors of [36] introduces an implicit reasoning neural networks (IRNs) to infer implicitly missing relations among the data present in the knowledge base, allowing the Q&A System to outperform other approaches in the FB15k benchmark.

In work proposed by [30], they introduce a new dataset to assess machine reading comprehension. The questions are a sample from a real user dataset, and the answers were generated by humans. Some questions have multiple answers to access Q&A Systems.

3 Domain Specific Languages (DSL) and Code Generation

Domain Specific Languages (DSLs) can simplify and accelerate the development of applications [1]. This advantage comes with the disadvantage of learning a new language [27]. According to Fowler [17], Domain-Specific Language is a computer programming language of limited expressiveness focused on a particular domain. DSLs are relevant for two main reasons: improve programmer productivity and allow programmers and non-programmers to read and understand the source code. The improved programmer productivity is achieved because DSLs try to resolve a minor problem than general-purpose programming languages (GPL) [18] making more straightforward to write and modify programs/specifications.

Since DSLs are smaller than GPLs, they allow domain specialists to see the source code and get a more abstract view of their business. DSLs offer the capacity to domain specialists to create a functional system, with no prior knowledge of GPLs.

What distinguishes DSLs from GPLs is the expressiveness of the language: instead of providing all the features that a GPL must contain, such as supporting diverse data types, control, and abstraction structures the DSL has to support only elements that are necessary to a real domain. Examples of commonly used DSLs, according to [18], are SQL, Ant, Rake, Make, CSS, YACC, Bison, ANTLR, RSpec, Cucumber, HTML.

Generative programming concerns the construction of specialized and highly optimized systems through combination and design of modules. According to [11] the goals of generative programming are to decrease the conceptual gap between coding and domain concepts, achieve high reusability and adaptability, simplify the management of several components, and to increase efficiency (space and execution time).

To be able to achieve the goals proposed by [11], generative programming recommends to apply separation of concerns, that is, deal with one important issue at a time, and combine these issues to generate a component; Parameterization of the components to be able to deal with families of components, allowing the use of the developed component in different scenarios; Separation from the problem space to solution space; Dependencies and interactions management to allow combination of components that have parameters that differ and imply in another component; Perform domain-specific optimizations through generation of some components statically, or making transformations to allow distributed processing.

Generative programming uses DSL at a modeling level [10] to allow users to operate directly with the domain concepts, instead of dealing with implementation details of GPLs.

According to [12] generative programming is a system-family approach, which allows the automatic generation of a system-family member, that is, a system that can be automatically generated from a textual or graphical DSL specification. In this work, the concepts related to generative programming through the use of DSL is used to allow the creation of Q&A Systems. An engine is used to process the DSL formal specification (grammar) and generate a Q&A System according to the written description of the DSL proposed. Also through the written description of the DSL is generated a complete Q&A System, without the need to build the system line-by-line by the developer. Section 4 present the proposed approach.

4   AcQA - The Proposed System

As stated in the Introduction, the use of DSLs and generative programming allows domain specialists to build entire systems without the need for GPLs knowledge. We propose to design the AcQA (Automatic creation of Q&A Systems) domain-specific language that allows a specialist to develop a Q&A System focusing on the knowledge associated with the closed domain, on the sources to be used to access that knowledge, on the process of answer formulation and on the formula to rank answer candidates, rather than how to implement them. When a developer implements a Q&A System, several issues should be addressed, such as backend technologies to be able to process the user's questions and to retrieve the appropriate answers, or frontend technologies to get input from the user and to display the built answers. As an example of the complexity underlying the development of a Q&A System, in [3] Python was used as a GPL for the engine in conjunction with Django and the Python Natural Language Toolkit [7].

Our proposal apply the concepts discussed in the Sects. 2 and 3 to achieve the objective of generating a Q&A System. The proposed architecture is depicted in Fig. 1.

The diagram in Fig. 1 shows the three main modules: Core Module, Data Module, and Presentation Module. The Core Module is comprised of the AcQA specification and a collection or library of techniques available to use in the construction of the Q&A System. The Q&A System specification is written by the domain expert, according to the AcQA grammar, and define which techniques will be used to produce the Q&A System. These techniques are initially based on our previous works described in [3,34], where we implemented a closed-domain Q&A System to answer Python related questions. This initial Q&A System that will support our framework is presented in the paper [3,34].

The Data Module provides a way to build a connection between the Q&A System that will be generated and the data repository (the knowledge base) provided by the user. This module resort to general purpose parsers available to import XML, or database engines to process queries in SQL/SPARQL, or even to customized parsers to deal with proprietary data formats.

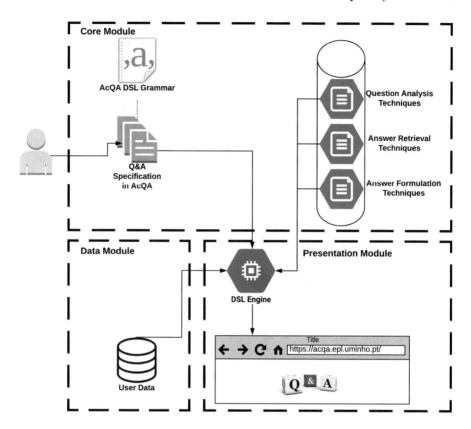

Fig. 1. Proposed architecture to create Q&A Systems based on a DSL.

The Presentation Module includes the AcQA compiler and all the processors around it needed to generate the desired Q&A System. It is responsible for creating the frontend that will be made available to the users of the Q&A System. It defines the UI that will be deployed. The UI can be Web or a REST API, in the initial version of the proposed architecture.

The AcQA language is an external DSL, containing a custom syntax to make easier the formal specification task necessary to build the aimed Q&A System automatically. To give a feeling about the AcQA DSL, Sect. 5, presents an example of a Q&A System implemented in AcQA.

The system that we have previously developed, PythonQA [3,34], support the implementation of this third module.

5   Q&A System Specification in **AcQA** DSL

As already stated in Sect. 3 this work use concepts of generative programming and domain-specific languages. This section introduces the under development AcQA DSL.

AcQA DSL has four code blocks which specify: data import, question analysis, answer retrieval, and presentation. Figure 2 presents a code fragment of AcQA DSL to configure a Q&A System similar to PythonQA. The code in line 1 configures which parser to use to process the data that will be used to create the knowledge base and receive as a parameter the file with the full path. Lines 2 and 3 configure the processors to deal, respectively, with question analysis and answer retrieval. Note that in this specification the processors' are not personalized. After line 5 we have the configuration of the presentation module where is instantiated a Django user interface and customization of the parameters of UI.

```
DataImporter = PyFaqParser('/path/to/input/file')
QuestionAnalysis = TripletsProcessor()
AnswerRetrieval = TripletsRetrieval()

Presentation = DjangoUI()
Presentation.header = "PythonQA"
Presentation.footer = "Copyright gEPL - UMinho - 2018"
Presentation.about = "About text"
Presentation.admin = "name of admin user"
Presentation.password = md5("encoded password")
```

Fig. 2. PythonQA specification in AcQA

Despite the simple example presented in Fig. 2, the AcQA DSL allows the user to change the behavior of the pre-defined processors used. For example, the user can change the language of the system to Portuguese instead of English, and the changes into the tokenizer, POS tagger, lemmatizer, and Wordnet language are applied automatically to the generated system.

6   Conclusion

Based on the expertise of our research team in grammar design, language processing, and generative approaches to software development, in this position paper, the architecture of a DSL-based generator to create Q&A Systems is presented. This DSL allow non-programmer end-users to create and test Q&A Systems, speeding up the development and validation phases. We should also carry out an in-depth analysis of distinct techniques needed to process natural languages and build/explore (consult) knowledge bases. New artificial intelligence techniques, like deep learning, to model and represent knowledge and to process it to create powerful inference engines will be investigated and tested. For the execution of experiments, different domains will be set up to allow the comparison among several approaches to the implementation of Q&A Systems.

Acknowledgement. This work has been supported by FCT - Fundação para a Ciência e Tecnologia within the Project Scope UID/CEC/00319/2013.

References

1. Adam, S., Schultz, U.P.: Towards tool support for spreadsheet-based domain-specific languages. In: ACM SIGPLAN Notices, vol. 51, pp. 95–98. ACM (2015)
2. Ansari, A., Maknojia, M., Shaikh, A.: Intelligent question answering system based on Artificial Neural Network. In: 2016 IEEE International Conference on Engineering and Technology (ICETECH), pp. 758–763. IEEE, March 2016. http://ieeexplore.ieee.org/document/7569350/
3. Azevedo, R., Henriques, P.R., Pereira, M.J.V.: Extending PythonQA with knowledge from StackOverflow. In: Rocha, Á., Adeli, H., Reis, L.P., Costanzo, S. (eds.) Trends and Advances in Information Systems and Technologies, WorldCist 2018, Advances in Intelligent Systems and Computing, 1st edn., vol. 745, pp. 568–575. Springer, Heidelberg (2018)
4. Balakrishna, M., Werner, S., Tatu, M., Erekhinskaya, T., Moldovan, D.: K-extractor: automatic knowledge extraction for hybrid question answering. In: Proceedings - 2016 IEEE 10th International Conference on Semantic Computing, ICSC 2016 (2016)
5. Ben Abacha, A., Zweigenbaum, P.: MEANS: a medical question-answering system combining NLP techniques and semantic Web technologies. Inf. Process. Manag. **51**(5), 570–594 (2015). https://doi.org/10.1016/j.ipm.2015.04.006
6. Besbes, G., Baazaoui-Zghal, H., Ghezela, H.B.: An ontology-driven visual question-answering framework. In: Proceedings of the International Conference on Information Visualisation, September 2015, pp. 127–132 (2015)
7. Bird, S., Klein, E., Loper, E.: Natural Language Processing with Python, 1st edn. O'Reilly Media, Inc., Newton (2009)
8. Cao, Y.G., et al.: AskHERMES: an online question answering system for complex clinical questions. J. Biomed. Inf. **44**(2), 277–288 (2011). https://doi.org/10.1016/j.jbi.2011.01.004
9. Clark, A., Fox, C., Lappin, S.: The Handbook of Computational Linguistics and Natural Language Processing. Wiley-Blackwell, Hoboken (2010)
10. Cointe, P.: Towards generative programming. In: Unconventional Programming Paradigms, pp. 315–325. Springer (2005)
11. Czarnecki, K.: Generative programming: principles and techniques of software engineering based on automated configuration and fragment-based component models. Ph.D. thesis, Technical University of Ilmenau (1999)
12. Czarnecki, K.: Overview of generative software development. In: Unconventional Programming Paradigms, pp. 326–341. Springer (2005)
13. Etworks, S.E.L.F.A.N.: R-Net: Machine Reading Comprehension With Self-Matching Networks*, pp. 1–11 (2017). https://www.microsoft.com/en-us/research/wp-content/uploads/2017/05/r-net.pdf
14. Fang, H., Gupta, S., Iandola, F., Srivastava, R.K., Deng, L., Dollár, P., Gao, J., He, X., Mitchell, M., Platt, J.C., Zitnick, C.L., Zweig, G.: From captions to visual concepts and back. In: Proceedings of the IEEE Computer Society Conference on Computer Vision and Pattern Recognition, 07–12 June, pp. 1473–1482 (2015)
15. Ferrucci, D.: Build Watson: an overview of DeepQA for the Jeopardy! Challenge. In: 2010 19th International Conference on Parallel Architectures and Compilation Techniques (PACT), p. 1 (2010)
16. Fortnow, L., Homer, S.: A short history of computational complexity. Technical report, Boston University Computer Science Department (2003)
17. Fowler, M.: Domain-Specific Languages. Pearson Education, London (2010)

18. Ghosh, D.: DSLs in Action. Manning Publications Co., Shelter Island (2010)
19. Hoque, M.M., Quaresma, P.: A content-aware hybrid architecture for answering questions from open-domain texts. In: 19th International Conference on Computer and Information Technology (2016)
20. Huang, X., Wei, B., Zhang, Y.: Automatic question-answering based on Wikipedia data extraction. In: 10th International Conference on Intelligent Systems and Knowledge Engineering, ISKE 2015, Taipei, Taiwan, 24–27 November 2015, pp. 314–317 (2015). https://doi.org/10.1109/ISKE.2015.78
21. Wolfram Research Inc.: Wolfram Alpha (2018)
22. Jain, A., Kulkarni, G., Shah, V.: Natural language processing. Int. J. Comput. Sci. Eng. (2018)
23. Jayalakshmi, S., Sheshasaayee, A.: Automated question answering system using ontology and semantic role. In: International Conference on Innovative Mechanisms for Industry Applications (ICIMIA 2017), pp. 528–532. No. Icimia (2017)
24. Kaisser, M., Becker, T.: Question answering by searching large corpora with linguistic methods. In: TREC (2004)
25. Kalaivani, S., Duraiswamy, K.: Comparison of question answering systems based on ontology and semantic web in different environment. J. Comput. Sci. 8(8), 1407–1413 (2012)
26. Lende, S.P., Raghuwanshi, M.M.: Question answering system on education acts using NLP techniques. In: IEEE WCTFTR 2016 - Proceedings of 2016 World Conference on Futuristic Trends in Research and Innovation for Social Welfare (2016)
27. Mernik, M., Heering, J., Sloane, A.M.: When and how to develop domain-specific languages. ACM Comput. Surv. (CSUR) 37(4), 316–344 (2005)
28. Miller, G.A.: WordNet: a lexical database for English. Commun. ACM 38(11), 39–41 (1995). http://portal.acm.org/citation.cfm?doid=219717.219748
29. Mochalova, V.A., Kuznetsov, V.A.: Ontological-semantic text analysis and the question answering system using data from ontology. ICACT Trans. Adv. Commun. Technol. (TACT) 4(4), 651–658 (2015)
30. Nguyen, T., Rosenberg, M., Song, X., Gao, J., Tiwary, S., Majumder, R., Deng, L.: MS MARCO: a human generated MAchine reading COmprehension dataset. In: CEUR Workshop Proceedings 1773 (Nips), pp. 1–10 (2016)
31. Och, F.: Minimum error rate training in statistical machine translation. In: Proceedings of the 41st Annual Meeting on Association for Computational Linguistics, vol. 1, pp. 160–167 (2003). http://dl.acm.org/citation.cfm?id=1075117
32. Packowski, S., Lakhana, A.: Using IBM watson cloud services to build natural language processing solutions to leverage chat tools. In: Proceedings of the 27th Annual International Conference on Computer Science and Software Engineering, No. November, IBM Corp., Markham, Ontario, Canada, pp. 211–218 (2017). http://dl.acm.org/citation.cfm?id=3172795.3172819
33. Rajendran, P.S., Sharon, R.: Dynamic question answering system based on ontology. In: 2017 International Conference on Soft Computing and its Engineering Applications (icSoftComp), pp. 1–6. IEEE, December 2017. http://ieeexplore.ieee.org/document/8280094/
34. Ramos, M., Pereira, M.J.V., Henriques, P.R.: A QA system for learning Python. In: Communication Papers of the 2017 Federated Conference on Computer Science and Information Systems, FedCSIS 2017, Prague, Czech Republic, 3–6 September 2017, pp. 157–164 (2017). https://doi.org/10.15439/2017F157
35. Sasikumar, U., Sindhu, L.: A survey of natural language question answering system. Int. J. Comput. Appl. 108(15) (2014). ISSN 0975-8887

36. Shen, Y., Huang, P.S., Chang, M.W., Gao, J.: Modeling large-scale structured relationships with shared memory for knowledge base completion. In: Proceedings of the 2nd Workshop on Representation Learning for NLP (2017). http://arxiv.org/abs/1611.04642

37. Vargas-Vera, M., Lytras, M.D.: AQUA: a closed-domain question answering system. Inf. Syst. Manag. **27**(3), 217–225 (2010)

38. Weissenborn, D., Wiese, G., Seiffe, L.: FastQA: a simple and efficient neural architecture for question answering. arXiv preprint arXiv:1703.04816 (2017)

On Semantic Search Algorithm Optimization

Alexander Gusenkov[(✉)] and Naille Bukharaev[(✉)]

Kazan Federal University, Kazan, Russia
gusenkov.a.m@gmail.com, boukharay@gmail.com

Abstract. In the article we consider, on the example of development of a relational database (RDB) information system for Tatneft oil and gas company, an approach to organization of effective search in large arrays of heterogeneous data, satisfying the following essential requirements.

On the one hand, the data is integrated at the semantic level, i.e. the system supports the presentation of data, describing its semantic properties within an unified subject domain ontology. Accordingly, end user's request are formulated exclusively in the subject domain terminology.

On the other hand, the system generates unregulated SQL-queries, i.e. the full text of possible SQL-queries, not just values of particular parameters, predefined by the system developers.

Considered approach includes both the possibilities of increasing the reactivity of the universal SQL queries generation scheme as well as more specific optimization possibilities, arising from the particular system usage context.

Keywords: Semantic search · Intellectual search · Ontology ·
Algorithm optimization

1 Introduction

With all diversity of its aspects, the core nature of the search problem in heterogeneous resources is determined by the type of data integration. Kogalovsky in [1] proposed to distinguish physical, logical and semantic levels of integration.

Integration at the physical level presumes converting data from various sources into a unified physical presentation format. The logical level supports the ability to access data from various sources in common terms of some global logical scheme. The semantic level supports access to data exclusively in terms of its semantic properties, described in a subject domain ontology.

The main advantage of the semantic approach lies in its obvious proximity to the task of intellectual search in multiple data sources of various logical structure and physical organization, related to one subject domain [2, 3]. The description of the domain is considered in this case as a configurable parameter of the corporate information system.

The inevitable price of the advantages of the semantic approach is the greater complexity of its implementation, which means

© Springer Nature Switzerland AG 2019
Á. Rocha et al. (Eds.): WorldCIST'19 2019, AISC 930, pp. 472–481, 2019.
https://doi.org/10.1007/978-3-030-16181-1_45

- theoretical novelty of the methods used;
- structural complexity of the system architecture, in which all three levels of data integration are present;
- computational complexity of the main semantic search algorithm.

Many of the arising problems can be solved at the general theoretical level. Currently, research in the field of computer linguistics is actively developing. Significant progress has been made in the development of electronic dictionaries, thesauruses, ontologies and algorithms for automatic extraction of information from the natural language text. Within the framework of this direction, a large number of specialized search systems for various subject domains have been developed. In particular, in [4–6] the semantic approach was applied to the integration of relational databases (RDB).

In the approach described by the authors in [4–6], physical database models, logical subject domain model and thesaurus of user terminology form the basis of RDBs integration. All of these information resources are presented uniformly in the ontology formalism. To build the ontologies, natural language text extracted from the names and descriptions of the RDBs tables and their attributes was used as a source of information. The proposed approach was successfully implemented to develop Tatneft oil and gas company intellectual search system, which showed high relevance of results for overwhelming majority (over 90%) of standard user queries.

Above we've mentioned the problem of computational efficiency as inevitably arising in development of the genuine intelligent search systems, which do not restrict the end user's language to a small set of predefined parametric queries. Indeed, since the user's queries are formulated exclusively in semantic terms, then the main algorithm, which has to locate the relevant information in the RDBs, in that case is of exponential complexity.

Further we explore the ways to increase the search algorithm efficiency without losing the expressive power of the end user query language.

The content of the article is structured as follows. In Sect. 2 we analyze the main existing approaches to RDB intelligent search systems development. Section 3 provides a general description of the authors' approach to organization of the search procedure and related data structures. Section 4 contains an overview of the search system architecture. In Sect. 5 the concept of intelligent search is specified through description of the end user query language. Section 6 describes preprocessing stage as a process of pre-tuning and initial optimizing further queries execution. Section 7 proposes two approaches to the search algorithm optimization; the first of them is based on the storage of the query history and the second one utilizes specifics of the subject area.

2 Intelligent Search RDB Systems

Since the mid-1990s several techniques to make full unregulated interaction with databases affordable for the users without knowledge of the SQL language and its many DBMS specific dialects have been developed. These techniques initially were mainly centered around the idea of visual programming and they generated a large number of products for constructing queries and generating reports. Now there is a lot of products of this type; among classic examples are Crystal Reports [7] and Oracle Discoverer, which is currently is a part of Oracle Fusion Middleware [8].

One of the most famous systems of this type is Microsoft Semantic Search [9], which is the successor of Microsoft English Query. The system is based on syntactically-oriented templates, associated with the subject domain model and, through it, with the database schema. To configure the system, one needs to specify at first the models of the database and the subject domain and then select from the list of English grammar templates an appropriate one for each database relation.

Although the report generators really allow to build any database query, attempts to make them serve as an end user tool have not been successful. These analysis and reporting instruments are aimed mainly on advanced users who should know well the database structure and have sufficiently good understanding of how SQL-queries are built.

An end user usually knows the subject domain much better than the database developers and that's why he/she may need to formulate complex queries which are not supported by the standard interface. Obviously, for such a professional, the main obstacle in such cases will be necessity to know the database definition and to be able to build complicated SQL queries. In connection with this, attempts to develop a natural language (NL) based interface for database access have repeatedly been made.

From remarkable examples of the kind for the Russian language, it is possible to single out the InBase system, developed by the school of Narinyani [10]. A distinctive feature of this approach is the usage of outrunning semantic analysis during parsing and understanding of user queries. The parsing here is based on the object model of the subject domain, linked by the designer to the database model.

Let's note that the development of such systems is often negatively affected by too straightforward understanding of what is actually a "natural language interface". In our opinion, it's at first is an user friendly interface. On the other hand, technical ability to generate arbitrarily complex grammatically correct, but poorly structured and hard to understand sentences is not advantage at all - whether they are expressed in natural or formal language.

For example, in practice it is hardly possible to be sure in correctness of a query in natural language, equivalent in direct or reverse translation to a SQL query with 5–6 attributes, referring to 2–3 database tables. Such considerations lead us to prefer the types of interface based on clearly structured forms of communication, already established in practice. In our case, these are tables containing NL expressions denoting the terms of the subject area.

3   Semantic Search Algorithm

Let's consider now the questions of representing in ontology formalism semantics of RDBs data, critical for the main search algorithm definition. In general, the variety of options here can be reduced to two basic approaches.

The first approach [11] presumes a straightforward conversion of a set of RDB tables into a set of the ontology concepts with slots, corresponding to the table columns, and projecting relations between tables (in the form of migrating keys) into relations between the corresponding table concepts. An ontology in this case includes also concepts to describe data types. In other words, this approach implies creating an unique ontology for each RDB, which makes process of ontology creation laborious and less universal.

The second approach, which is adopted here, presumes a higher degree of abstraction, in which all the basic theoretical concepts of RDB are described as ontology concepts. Objects (tables, columns, keys and domains) of a particular database in this case are represented as instances of universal concepts of the corresponding type.

Namely, the ontology contains the universal concepts TABLE, COLUMNS, KEY and DOMAIN, corresponding to the main database objects, and the following universal relations:

- TABLE contains the COLUMN;
- TABLE has the primary KEY;
- TABLE has the foreign KEY;
- KEY contains the COLUMN;
- COLUMN values belong to the DOMAIN.

The universal ontology definition includes also two following interpretation functions, playing important role in the semantic search algorithm:

$\varphi 1$: If TABLE1 has primary KEY1 and TABLE2 has foreign KEY1, then there exists TABLE3, containing all columns of TABLE1 and TABLE2.

$\varphi 2$: If TABLE1 contains COLUMN1, then there exists TABLE2, containing all columns from TABLE1 except COLUMN1.

The first interpretation function describes the table join operation and the second one describes the relation projection operation, necessary to reduce the set of columns obtained by joining tables to the desired one.

Thus, the task of extracting information from integrated data sources for a given user query can be reduced to finding all the ways to extract specified attributes from RDB tables. In other words, in this case we need to find such sequences of application of $\varphi 1$ and $\varphi 2$ functions, which result in the desired set of columns {C} from the ontology O.

Let's especially note here a new point, which significantly distinguishes our approach from the existing works in the field of semantic search [11]. Apart from the existing subject domain ontology and strictly formalized data properties information (such as the RDB schemes), for creating subject domain ontology content we also use information, extracted from informal natural language texts. Namely, those are comments on the names of tables and their attributes. The same information is also used to build the subject domain language thesaurus, which serves as the natural language interface basis. The general approach adopted to solve arising problems of analysis of weakly structured data, using methods of mathematical linguistics, is described in [12].

4   Main System Components

To understand more clear the nature of arising problem of computational complexity, let's have a quick look on the system architecture. The main components of the system are:

- *the RDB ontology*, describing the basic concepts of relational databases in the ontology formalism. For describing ontologies, the OWL language [13], developed by the Semantic Web Activity working group and recommended by the W3C consortium was used [14];
- *the subject domain ontology*. As an initial ontology prototype, the Epicentre data model of Petrotechnical Open Software Corporation (POSC) [15, 16] was used. A general scheme for converting this type of models into the OWL ontology description language using methods of the formal grammars theory has been developed [4, 6];
- *the linguistic thesaurus* of the subject domain language, defining formally the language of the user-system communication;
- *the algorithm of unregulated access to the set of RDBs*, accepting user queries on that natural language dialect.

More detailed description of the general problematic, functionality and architecture of the system can be found in [4, 6].

5   Semantic Search Algorithm Data

The core task of the algorithm is the generation of the text of the SQL-query for a given end user request. The latter has a structure of a table, containing phrases from the linguistic thesaurus in its left column and condition on the values of the corresponding notion in its right column (see Table 1).

Table 1. End user query example.

Professional term	Condition
Oilfield	Novo-Elkhovskoe
Well No.	*
Period	From 01.2010 to 12.2015, monthly
Volume of production, in tons	Sum
Water cut percentage	Average

Thus, the user query language can be considered as a kind of professional dialect of natural language, presented in a form familiar to the end users. Note that such a request, formulated as a sentence in natural language, would be very cumbersome not only for machine analysis, but also for human understanding.

Let's note also, that the user requests refer just to the semantic properties, defined by the subject domain ontology, and contains no references to the logical structure and location of data stored in RDBs.

The most significant and also costly, in terms of computational complexity, part of the algorithm is related to the search of data location, using information on the logical structure of data, described in terms of the RDB ontology.

6 Preprocessing

At this stage, the information on the table joins is extracted from the RDB schemes. As the result, the following graph is constructed from the RDB ontology instance. The vertices of the graph present the "table" RDB ontology concepts, and its arcs are determined by the presence of the common key in the tables; the arcs are oriented in accordance with the "has primary key - has foreign key" relation (see Fig. 1).

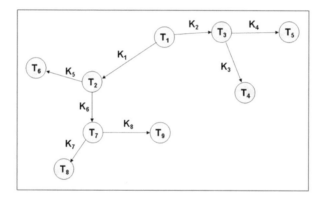

Fig. 1. Graph representation of potential table joins; here T_i are table names and K_i are table joins.

The constructed graph is supplemented by information on possible table joins (up to key migration). This is an algorithmically simple, but resource-intensive procedure, which can be called RDB markup. If the RDB scheme does not change, then this markup is executed once. Otherwise, the above graph must re-build from the new RDB ontology instance.

7 Semantic Analyses of the End User Query

Recall that the subject domain is represented in the system as a semantic network (ontology). During user request analyses we identify in that network all subgraphs, connecting the ontology concepts, corresponding to the phrases of the linguistic thesaurus, used in the user query.

Thus, semantic analyses of the user query is reduced to enumeration of all simple paths, corresponding to some subgraph of the domain ontology in the graph, constructed from the RDB ontology instance (see Fig. 2).

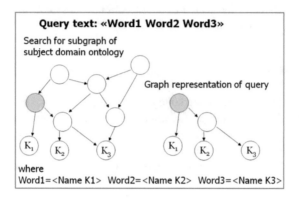

Fig. 2. Process of "understanding" the end user query.

As a result of the semantic analysis, for each column of the user's query several the most relevant locations are found. The corresponding columns can be contained in tables of various RDBs (see Fig. 3). If there are several combinations of the relevant columns, then the number of tables in the constructed join is also taken into account.

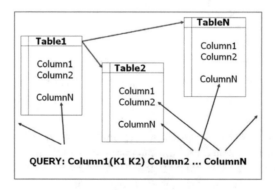

Fig. 3. Search of table joints.

8   On "Fit-for-Purpose" Optimization Approach

The search algorithm described above can be characterized as a "semantically restricted" full search procedure. From a formal point of view, it belongs to the class of graph wandering algorithms, having exponential complexity [17]. Evidently the speed of generation and subsequent execution of the SQL-query essentially depends on the number of columns, contained in it. The main time of the algorithm execution is spent on finding out the set of RDB tables joins when searching for data location. So the proposed technology in its current implementation works most effectively with the user queries, which refer to a small number of columns; let's note though, that it is quite typical case for corporate RDB queries.

As it follows from said above, the main problem of increasing the system reactivity is connected with the task of reducing the time spent on enumeration of the table joins options.

Let's describe two pragmatic ways to optimization of the alike algorithms, following the "fit-for-purpose" principle, i.e. those aimed on specific situation and purpose of usage. The first, more traditional and universal way presumes improvement of some characteristics of the algorithm (in our case, the speed) for arbitrary input data; it this case we can also take into account expected statistics of usage. The second is based on more specific knowledge of the input data content. Which are in our case the queries made by an oil and gas production specialist.

Optimization of the first type can be implemented as follows. We already store in the system user context, including the history of each user's activity. Practice of the system exploitation shows, that each user usually uses the small number of typical queries related to his/her job duties.

Let's store for each user the sequences of table joins, generated by his/her queries. The analysis of the system functioning shows, that in the overwhelming majority of cases the stored sequence of table joins is sufficient to obtain the desired result in the future. As a rule, the user changes only the query parameters; so the cases, when the search of table joins is required at the last stage of the algorithm execution, are quite rare.

Optimization of the second type takes into account the following specifics of the subject domain. From the beginning of an oilfield development, oil producing companies collect on the regular basis geological survey data to model wells functioning and predict oil reservoir release. As a result RDBs contain a large number of databases and/or tables (up to key migration), that have the same structure and contain the same type of information about the wells exploitation on various calendar periods.

When constructing the graph of table joins during preprocessing, let's store information on presence of date-related (i.e. of year or year-month type) key fields in the considered tables. If later on an user query refers to the temporal characteristics (such as year, year-month, or range of such values), then we will not include into enumeration of table joins the ones which do not match those key values.

9   Conclusions

In the article the rationale for the main semantic search algorithm of data integration systems is given on the example of the actual development of an information system of a large oil mining company. It is noted that there exist theoretical constraints, in the form of the exponential complexity of the algorithm, following from the very statement of the problem. Nevertheless, that leaves room for the successful application of more pragmatic approaches to increase the reactivity of such systems, It can be done either by implementing the well-known effects of "re-pumping complexity" (in our case, storing the actual query history), or by taking into account the type of information specific to a given subject domain. As the results of experiments show, if without the use of the methods described above, queries containing 6–8 columns of various tables were executed in real time, then with their help it is possible to increase the corresponding number up to 12–15 columns. That is more than enough for operational queries reference.

Acknowledgments. This work was funded by the subsidy allocated to Kazan Federal University for the state assignment in the sphere of scientific activities, grant agreement 1.2368.2018 and subsidy of the Russian fund of fundamental research, grant agreement 18-07-00964.2018.

References

1. Kogalovsky, M.R.: Methods of data integration in information systems. Institut problem rynka RAN, Moscow (2010). http://www.ipr-ras.ru//articles/kogalov10-05.pdf. Accessed 30 Nov 2018
2. Kogalovsky, M.R.: Ontology-based data access systems. Program. Comput. Softw. **38**(4), 167–182 (2012). https://doi.org/10.1134/s0361768812040032. https://link.springer.com/article/10.1134/S0361768812040032. Accessed 30 Nov 2018
3. Kogalovsky, M.R.: Data access systems based on ontologies. Programming, MAIK. "Nauka. Interperiodika" **38**(4), 55–77 (2012). http://www.ipr-ras.ru/articles/kogalov12-03.pdf. Accessed 30 Nov 2018
4. Birialtsev, E., Bukharaev, N., Gusenkov, A.: Intelligent search in big data. J. Phys.: Conf. Ser. **913**, Conf. 1 (2017). http://iopscience.iop.org/article/10.1088/1742-6596/913/1/012010/pdf. Accessed 30 Nov 2018
5. Gusenkov, A.M.: Intelligent search for complex objects in big data arrays. Electron. Lib. **19**(1), 3–39 (2016)
6. Gusenkov, A., Birialtsev, E., Zhibrik, O.: Intelligent search in structured data arrays. LAP LAMBERT Academic Publishing, Deutschland: OmniScriptum Marketing DEU GmbH (2015). ISBN 978-3-659-76919-1
7. SAP Crystal Reports. http://www.crystalreports.com/emea/. Accessed 30 Nov 2018
8. Oracle Fusion Middleware. https://docs.oracle.com/cd/E28280_01/index.htm. Accessed 30 Nov 2018
9. Semantic Search. https://docs.microsoft.com/en-us/previous-versions/sql/sql-server-2012/gg492075(v=sql.110). Accessed 30 Nov 2018
10. Zhigalov, V.A., Sokolova, E.G.: InBASE: technology of building NL-interfaces to databases. Moscow, ROSNII Artificial Intelligence (2001). http://www.dialog-21.ru/digest/2001/articles/zhigalov/. Accessed 30 Nov 2018
11. Zhuchkov, A.V.: New technologies for conceptual networks created in the framework of the ICST "New generation vaccines and diagnostic systems of the future". Electron. Lib. **6** (2003). https://elbib.ru/ru/article/244. Accessed 30 Nov 2018
12. Birialtsev, E.V., Gusenkov, A.M., Mironov, S.V.: One approach to implementing unregulated access to relational databases. In: Trudy Kazanskoj shkoly po komp'yuternoj i kognitivnoj lingvistike TEL-2008, pp. 10–23. Kazanskij gosudarstvennyj universitet, Kazan (2009)
13. OWL Web Ontology Language. https://www.w3.org/TR/2004/REC-owl-features-20040210/. Accessed 30 Nov 2018

14. World Wide Web Consortium (W3C). https://www.w3.org/. Accessed 30 Nov 2018
15. Epicentre v3.0. http://www.energistics.org/energistics-standards-directory/epicentre-archive. Accessed 30 Nov 2018
16. Petrotechnical open standards consortium (Energistics). http://www.energistics.org. Accessed 30 Nov 2018
17. Anderson, J.A.: Discrete Mathematics with Combinatorics, 2nd edn., p. 784. Prentice Hall (2003). ISBN 0130457914

Predict the Personality of Facebook Profiles Using Automatic Learning Techniques and BFI Test

Graciela Guerrero$^{(\boxtimes)}$, Elvis Sarchi, and Freddy Tapia

Universidad de las Fuerzas Armadas ESPE, Av. General Rumiñahui s/n,
171-5231B Sangolquí, Ecuador
{rgguerrero, ersarchi, fmtapia}@espe.edu.ec

Abstract. The present research work aims to predict the personality of a user's Facebook profile. To do this, we have identified the attributes that are extracted from Facebook, with which the prediction of personality was possible. The data was extracted using the Graph API of Facebook, which was implemented in a web page. To achieve the knowledge base of machine learning, the BFI personality test is implemented for 118 users. In order to perform the training and classification of the automatic mode of learning by using the Weka tool, the degree of accuracy of the algorithms used in the prediction of the user's personality was verified. The evaluation was carried out with two scenarios: using supervised learning and not using unsupervised learning. The work done yields results that indicate that it is necessary to increase the dictionary of data of the Spanish language, another result obtained is that in supervised learning, they gave data in which women have tendencies to be of neurotic personality compared to men. These data also determined that women are more difficult to predict their personality.

Keywords: Personality prediction · Machine learning · Facebook · Supervised learning · Unsupervised learning

1 Introduction

Social networks are changing the way people communicate and interact. Social networks have become the letter of presentation for many [1], and it has also become a way to find out about someone who has just met, an example of this is the social network Facebook, thanks to the "likes" that the user performs such as: books, movies, music, interests, politics, etc. All these interactions that the user performs on his Facebook profile and share it with the public, let him know what his tastes and tendencies are.

The purpose of this research is to predict the personality of a person's Facebook profile, this is intended to be done through automatic learning as a support tool for the selection process of candidates in the area of Human Resources that has been taken as a case of application.

The structure of this research paper is presented as follows: (i) In Sect. 2, two essential topics are presented: (a) the theoretical analysis of the relationship between

© Springer Nature Switzerland AG 2019
Á. Rocha et al. (Eds.): WorldCIST'19 2019, AISC 930, pp. 482–493, 2019.
https://doi.org/10.1007/978-3-030-16181-1_46

personality groups that have been social networks, and (b) the analysis of works related to the purpose of identifying the attributes that are necessary to predict the personality of the user of Facebook. (ii) Sect. 3 presents the proposed architecture, which seeks to obtain information on the publications made on the user's wall, through the development of a Facebook application. (iii) In Sect. 4, two types of evaluation are carried out: (a) the training is done by applying a BFI personality test to 118 users, (b) the training and classification are applied through machine learning models applied to the users on Facebook. (iv) Finally, Sect. 5 presents the conclusions and the lines of future work.

2   Related Works

2.1   Five Groups of Personalities in Convergence with Facebook

According to Adali et al. [2], the personality of people can be divided into five main axes, which are: openness, responsible, extraversion, kindness and neuroticism. These 5 large groups of exposed personalities have been accepted by professionals in the field of study of psychology, becoming the definitive model of personality. The five personality traits are characterized by the following:

Openness to experience represents curiosity, open-mindedness and the willingness of individuals to explore new ideas. People with an openness to the experience and use of Facebook, according to Moore et al. [3] are involved in greater sociability online through Facebook. Correa et al. [4] indicates that people with high openness scores are more open to revealing personal information about themselves on their Facebook profile.

Kind people, according to Wehrli [5] are nice people with tendencies to be understanding, polite, flexible, trusting and clement. Moore et al. [3] talks about **kindness** and its use in Facebook so that the friendliest users use Facebook less frequently, make fewer posts on their wall about others and express more anguish for their Facebook activity.

Extraversion refers to the extent to which people are sociable, cheerful, optimistic, active and communicative [5]. Introverts earn more in the use of social networks because this communication is indirect to compensate for their lack of interpersonal skills, in addition to using the platform to communicate with their friends and contacts.

Those **responsible** tend to be reliable, responsible, organized and self-disciplined [5]. Responsible people are negatively related to the use of Facebook [3], so people with greater responsibility spend less time on Facebook, use it less often, have fewer friends, publish fewer photos, publish fewer ads on its walls and they will express more repentance in their publications.

Neurotics are individuals who show negative attributes such as distrust, sadness, anxiety, shame and difficulty handling stress. Neurotic people like chat rooms and instant messages, they publish private information but fewer photos on their Facebook profile [5]. This has a related explanation where neurotic people have more time to contemplate what they are going to say rather than communicate face to face, since neurotic people are anxious and nervous by nature, they are more likely to get angry and regret publishing something like that.

2.2   Words Most Used in the Classification of Personalities

People differ when expressing their thoughts, feelings and actions, according to studies of Hirst et al. [6] have identified systematic associations between personality and language use, in a variety of different contexts. In the study of Yarkoni [7] the relationship between personality and language used in participants whose writing sample is large, of diverse and easily accessible topics. This study he made it with 700 blogs that contained an average of 115,423 words each, this allowed him to model reliably between the personality and the use of words not only at the level of semantic categories but also at the level of individual words. The data obtained from this study resulted in a top of words according to each personality as shown in Table 1. Therefore, the analysis performed with the words obtained is supported by the BFI evaluation test to improve the predictability of the user's personality type.

Table 1. Top 20 of the most used words of the 5 personality groups [7].

5 personality groups	Top 20 most used words
Neuroticism	Terrible, though, lazy, worse, depressing, irony, road, terrible, south, stressful, horrible, order, visited, annoyed, embarrassed, ground, prohibition, old, guest, completed
Extraversion	Bar, other, drinks, restaurant, dance, restaurants, cats, grandpa, Miami, innumerable, drink, shot, computer, girls, glorious, minor, pool, crowd, sang, roasted
Openness to experience	Folk, human, of, poet, art, by, universe, poetry, narrative, culture, draw, century, sexual, movies, novel. decades, ink, passage, literature, blues
Kindness	Wonderful, together, visit, morning, spring, porn, walked, beautiful, stay, felt, cost, share, gray, joy, afternoon, day, moments, hug, happy, fuck
Responsible	Completed, adventure, stupid, bored, adventures, desperate, enjoy, say, Hawaii, pronounce, those, extreme, deck

There is an application called Apply Magic Sauce [8], developed at the University of Cambridge, currently there are no other applications similar to this one. The tool is capable of making a psychological report based on the user's tastes on Facebook, but not in their comments. The application accesses the profile of the user of Facebook under his permission to analyze the activity of the "likes" and this compares with the information of more than six million people. In this way, the application predicts age, gender, marital status, religion, political affinity, intelligence, kindness, responsibility, shyness or extroversion, and how happy a person is the data extracted from Facebook.

What is intended in this proposal is to improve the way to show the data intuitively and quickly with a few clicks, you can already predict the personality, this study adds personality analysis from the likes, post and/or user comments and a text learning that describes each of the personality groups to which a user belongs, is also added mathematical calculations and machine learning algorithms that are used to predict the personality without the need to compare millions of profiles.

3   Proposed Architecture

The present research work developed the architecture shown in Fig. 1 and its functionalities are the following activities: (i) Implementation of a web application with access to Facebook that allows access and collection of user data through Graph API. (ii) Evaluation of the personality using the BFI test to fill the knowledge base necessary for automatic learning. (iii) Realization of the training and classification by means of models of automatic learning that allow to carry out the analysis of the prediction of the personality according to the five great personalities.

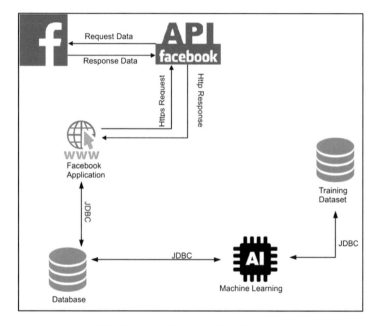

Fig. 1. General proposed architecture.

In Fig. 2 the classification scheme is observed, in which automatic learning is implemented for the construction of models, these models will be trained according to the data of the Facebook user account, taking the comments and the tastes of users, with this information from a classification system is generated according to the five great personalities. For the validation of the model, test examples are presented that will be tested to determine the percentage to which it belongs according to the classification of the system.

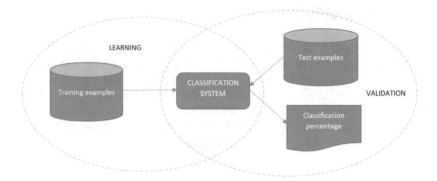

Fig. 2. Classification scheme [9].

3.1 Development of the Application on Facebook

The Facebook application allows you to obtain the Facebook user data, specifically the publications made on the wall of each user are necessary for a more detailed analysis. The first step is to create a Facebook application, since it is necessary to have an account for certain Facebook requirements that are: (i) Have a domain, (ii) the website must have a certificate and (iii) the site must have security policies. These requirements are fundamental for Facebook to allow access to user data and for the user to feel secure when accessing their data.

By complying with the aforementioned requirements and after obtaining approval from Facebook, it is possible to access the following user data such as: email, friend list, name, surname, profile photo, gender and age.

In the case of this project it is necessary to obtain access to the information of the publications in the user's wall, to have this permission it is necessary to access the application's revision tab and to manage a new element for approval, Facebook reviews the proposal that is sent, this proposal can be a video where it is explained where the data will be used and in which part the user benefits from the use of this data, if the user cannot visualize the use of their data, Facebook will not give permission to the request.

3.2 Web Interface

For the development of the website, several platforms, Heroku service platform [10] was chosen, which supports several languages, such as Java, Node.js, Ruby, Python, PHP, etc. The basis of its operating system is Debian and it is based on a managed container system, with integrated services. Heroku also facilitates that a domain or subdomain in the case of this project is https://personalityprediction2018.herokuapp.com/, also provides a certificate so that all connections are secure and have databases that are useful for this project, such as Postgres, with a limit of 20 simultaneous connections, this database will contain the information extracted from the Facebook user.

To make the connection of the website with the Facebook application it is necessary to use the Facebook SDK that is written in JavaScript. The first step when loading the web page is to determine if the user has a session started in the application, to determine this process a call is made to the function 'FB.getLoginStatus', which makes a call to Facebook to obtain the status of login (Fig. 3a), returning a response object that has several fields that are: (i) state: specifies the login status of the user that uses the application. The state is subdivided into (a) connected: the user has logged in to Facebook and your application. (b) not_authorized: the user has started the edition on Facebook, but not the application. (c) Unknown: the user has not logged in to Facebook and it is not known if he has done so in the application. (ii) authResponse: this object is included if the state is connected and contains the following elements: (a) accessToken: contains an access identifier for the user that uses the application. (b) expiresIn: indicates the time in UNIX when the identifier expires. (c) signedRequest: is a signed parameter that contains information about the user that uses the application. (d) User_ID: the unique identifier of the user that uses the application.

Fig. 3. (a) Login to the application. (b) Facebook dialog window.

In the creation code of the Facebook is necessary permissions are specified (Fig. 3b), in the "scope" section enter the necessary permissions, for this project you need two permissions that are "public_profile" that return the data of the user as are the names and the photo of the profile, the other permission is "user_posts" that allows the entrance to the states in the wall of the user, which can be their publications, profile updates, photos loaded on Facebook or shared states.

The publications were published in the last two years, which allows obtaining updated data, since people over the years change their habits and their way of presenting themselves in social networks, publishing their wall in a different way to previous years.

The publications are delivered by Facebook in Json format, which contain the following data: (i) id: is the identifier of the publication. (ii) message: is the text of the publication on the Facebook wall. (iii) story: is the text of the user's story, such as when two friends become friends, a photo is uploaded, a publication is shared or the user's profile is updated. (iv) created_time: time at which the publication was initially published, this will be the date and time of the event.

To define which publications have been shared on the Facebook wall, take the text of "history" and look for the reserved word "shared", this word means that the publication has been shared elsewhere. To identify the number of times that one or several photos have been uploaded, the text of "history" is taken and the reserved word "new photo" is searched. Finding this text string means that the publication is related to the publication of photos.

To define the number of times the user updates their profile image, the "history" text is taken and the text string "cover photo" is searched. Finding this string means that the publication is done through a profile photo update.

3.3 Creation of the Database

The next step is to extract the user's information at the moment of logging into the web page that has been stored in a Postgres database, for which new data is created in the Heroku resource panel and the base is assigned of data to the created application. Then the tables with the necessary attributes for this project are created, to administer the database the "pgAdmin" tool was used. To establish the connection, you must obtain the credentials granted by Heroku, it must be mentioned that the connections to the Postgres database are made through the SSL protocol, which allows the data to travel safely through the network of encrypted form. With Postgres pgAdmin the necessary tables for this project are created.

3.4 Insertion of Words According to Personality

With the extracted data, the analysis of the text string of all the advertisements made by the user that are now stored in the user table in the column 'user_posts' is performed, first the table of the words corresponding to the personality with their respective score. A total of 288 words was obtained, these words were extracted from the work done by Yarkoni [7] where each word that was obtained was added its synonyms, plus each word corresponds to a weight, which can be positive or negative. If the weight is negative, it means that the word corresponds to the opposite personality to the one that is defined, it was also considered the same word with and without a tilde and in its singular form. The words have been translated into Spanish because the area of the research evaluation environment is of Spanish-speaking origin.

3.5 Analysis of Texts and Learning of Publications

With the words in the database the analysis of the text of the Facebook user's publications is done, following the process: (i) counting all the words in the text string, omitting the non-alphabetic characters and inserting them in a list. (ii) Load the list of words that correspond to the personality with their respective weights. (iii) Using the method of stemming, which is to find the word that contains its root, for example, the word family is contained in relatives, with this method it is compared if the word that belongs to the personality is contained in the list of words that were determined from the text of the publications. (iv) If the word is contained or is the same, it is verified to what type of personality it corresponds and the summation is made depending on the weight of the word. (v) At the end of the process of adding all the words, the weights of each personality are stored in the database.

4   Evaluation

The evaluation consists of two sub-sections: the first sub-section is the evaluation of the personality test that consists of the performance of the test that yields a score that is stored in a matrix, adding the contribution of the weights of each word obtained from the comments of Facebook in relation to each personality. The second sub-section deals with the evaluation of the algorithms used and indicates which algorithm yields the best results and then recommends it in the next section.

4.1   Personality Test

In order to make the prediction of the user's personality, a knowledge base was needed consisting of the words extracted of Facebook comments and also with the help of the BFI test [11]. BFI is a personality test consisting of 41 questions (Fig. 4), these results are stored in a table of the database and a percentage of each personality is obtained. Those percentages that reflect being older will indicate the orientation of the user's personality.

Fig. 4. (a) Results of the BFI personality test. (b) Users performing the personality test.

Result of the Personality Test
The result of the test is composed of the five factors of each personality, is determined by the highest score obtained from the sum of the five factors that corresponds to each personality. Figure 4 shows the five personalities in percentages according to the test responses. To define the percentage first you get the maximum possible score that each personality can obtain depending on the evaluation of the BFI test.

For data collection people were needed who are in search or future job search, since it is the target population for this case study, those involved were 118 users. The scenario was based on entering the web page of the personality test application, logging into the application and this redirects them to solve the personality test.

After analyzing the results of the 118 users, in Fig. 5 showed that the highest score is 41% of the users have a personality open to experience, then users with extraversion personality with 37% and third with responsible with 12%, the last two personalities have 5% kindness and neuroticism.

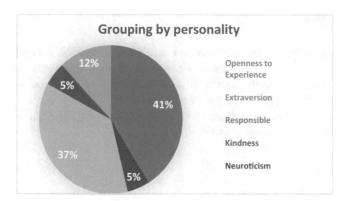

Fig. 5. Grouping by personality.

Regarding the classification by gender, where the 118 users who performed the personality test, 74 of them were male and the remaining 44 were female. The classification by gender and personality is observed where the male gender presents a higher result. In the personality open to experience 33 were obtained by male and 15 female.

For the personality of kindness, 3 male users and 3 female users were registered. Extraversion personality was obtained 27 male users and 17 female users. In the personality of neuroticism is the only case in which the female gender is greater than the male, with 4 users for women and 2 users for men. Finally, for the responsibility personality, it was obtained with 9 male users and 5 female users.

The result was the knowledge base, which is essential for machine learning, which is explained in the next chapter. In this way, we described how the data was obtained and the entire process to obtain the knowledge base. It was explained how the process was carried out and the personality test was carried out and it was finalized showing the results of the tests in the graphs.

4.2   Evaluation of the Best Algorithm

Pre-processing of Data
The pre-processing of data is done by obtaining a flat file, since this flat file must have the .csv or .arff extensions. For processing there are 118 instances and 11 attributes, from which we obtain the number for each attribute in relation to the personality of the users.

Classification
The objective is to build a model that allows to predict the category of the instances based on a series of input attributes. Sometimes, the classification problem is formulated as a refinement in the analysis, once unsupervised association and grouping algorithms have been applied to describe the relationships of interest in the data.

To perform the construction of the data classifier, we proceed to evaluate different classification algorithms, this will be done in two scenarios. The two scenarios evaluate 12 algorithms that are located under the different classifiers (Table 2) that the Weka application possesses.

Table 2. Classification of algorithms.

Bayes	Functions	Lazy	Rules	Trees
NaiveBayes	Logistic	IBK	ZeroR	J48
	Multiplayer perceptron	Kstar	Decision table	Ramdom forest
				DecisiónStump
				Ramdom tree
				REPT tree

Scenario 1: All 118 instances were used to perform the training to build the classifier. After carrying out the execution of the algorithms proposed in Table 2, the algorithms with the highest precision in terms of correct classification are: Ramdom Tree, Ramdom Forest, IBK and Kstar, however, in the case of considering the errors, the algorithms Ramdom Forest and Kstar are discarded, because what is intended in this research is to obtain the least possible error. In Table 3 it is observed that the algorithm of the Ramdom Tree shows 0% error after learning 118 instances. It is necessary to mention that the unknown instance is the row of the data that represents each field.

Table 3. Classification of algorithms results of the algorithms executed in scenarios 1 and 2.

Scenario 1	Correctly classified instances	Incorrectly classified instances	Scenario 2	Correctly classified instances	Incorrectly classified instances
ZeroR	41.03%	58.97%	ZeroR	38.24%	61.76%
Decision table	42.74%	57.27%	Decision table	44.12%	55.88%
J48	80.34%	19.66%	J48	50.00%	50.00%
Ramdom forest	100.00%	0.00%	Ramdom forest	70.59%	29.41%
DecisiónStump	41.03%	58.97%	DecisiónStump	41.18%	58.82%
Ramdom tree	100.00%	0.00%	Ramdom tree	67.65%	32.35%
REPT tree	41.03%	58.97%	REPT tree	44.12%	55.88%
NaiveBayes	29.91%	70.09%	NaiveBayes	47.06%	52.94%
Logistic	51.28%	48.72%	Logistic	44.12%	55.88%
Multiplayer perception	62.39%	37.61%	Multiplayer perception	44.12%	55.88%
IBK	100.00%	0.00%	IBK	67.65%	32.35%
Kstar	100.00%	0.00%	Kstar	76.47%	23.53%

Scenario 2: To execute scenario 2, the training set comprising the 118 instances will be divided and divided into two parts: the first 70% of the data will be used to build the classifier and the final 30% to perform the test.

In the execution of scenario 2 (Table 3), the results obtained after learning 34 instances corresponding to 30% of the total data, are the algorithms: Ramdom Forest and Kstar that are in 70% success after the learning obtained from 30%.

4.3   Discussion

With the analysis of the data obtained from the 118 participants, it was shown that the words obtained are not reliable in predicting personality, since the results of the words for the personalities of responsibility and open to the experience were more negative, but corresponded to the extraversion of the personality.

The favorable result was that for the words that correspond to the personality of extraversion, the greater its weight, the latter corresponds to its personality. This is due to the fact that the words entered were obtained from a work in which people of the English language participated, while the tests were for users of the Spanish language, however prior to the selection an analysis of neutral words was carried out. they are included in the different Hispanic dialects with the aim of having a better precision in the results.

With the data obtained from the personality test and the algorithms used (Table 3), the personality can be predicted, but not in all cases were favorable, only with the algorithm "Random tree", an error of 0% was obtained, this algorithm works in a way that a tree is built and it is considered a random number of given characteristics in each node, in this way the prediction is adjusted to the needs of the algorithm to give the correct result of the personality.

It was also shown that the female gender has characteristics that make it difficult for algorithms to predict their personality.

5   Conclusions and Future Work Lines

Data extraction using the Facebook API requires the developer to take security measures such as SSH protocols and demonstrate that the data extracted is to improve the user experience. This gives the user a security in which the data given to the application will be for their benefit. It is necessary to consider that at the moment there have been made changes in the API of Facebook and apply new rules to accede to the permissions of visualization of the commentaries of the users.

The words that correspond to each personality according to the algorithm is not confirmed when predicting the personality, this is due to the translation of the words English to Spanish and that in the country of evaluation (Ecuador) different jargons are used in the case of study where the words were obtained. It is suggested to carry out a study of the most used words according to the personality in the region where the test is applied, since different slang is used to those of other regions, in this way a prediction of the personality can be made with greater reliability.

The results of supervised learning yielded data in which women have tendencies to be neurotic personality compared to men. These data also determined that women are more difficult to predict than men with a greater margin of error.

For a future study it is proposed to increase different age ranges of the participants, in order to get closer to more accurate predictions and obtain a greater number of personalities in the different traits. Since this study is based on an age range of 20 to 25 years and most of these have extroverted personalities and open to experience.

References

1. Bachrach, Y.: Human judgments in hiring decisions based on online social network profiles. In: IEEE International Conference on Data Science and Advanced Analytics (DSAA), Campus des Cordeliers, Paris, France (2015)
2. Adali, S., Golbeck, J.: Predicting personality with social behavior. In: International Conference on Advances in Social Networks Analysis and Mining (ASONAM), Istanbul (2012)
3. Moore, K., McElroy, J.C.: The influence of personality on Facebook usage, wall postings, and regret. Comput. Hum. Behav. **28**(1), 267–274 (2012)
4. Correa, T., Hinsley, A., De Zúñiga, H.: Who interacts on the Web?: The intersection of users' personality and social media use. Comput. Hum. Behav. **26**(2), 247–253 (2010)
5. Wehrli, S.: Personality on Social Network Sites: An Application of the Five Factor Model. ETHZ, Switzerland (2008)
6. Hirsh, J.B., Peterson, J.B.: Personality and language use in self-narratives. J. Res. Pers. **43**(3), 524–527 (2009)
7. Yarkoni, T.: Personality in 100,000 words: a large-scale analysis of personality and word use among bloggers. J. Res. Pers. **44**, 363–373 (2010)
8. Stakhova, B., Kielczewski, A.: Apply Magic Sauce-Prediction API-Account Information Reminder, Cambridge (2018). https://applymagicsauce.com. Accessed 28 Nov 2018
9. García, S., Luengo, J., Herrera, F.: Data Preprocessing in Data Mining. Springer, New York (2015)
10. Heroku: The Heroku platform as data services and services, Heroku. https://www.heroku.com/platform. Accessed 21 June 2018
11. Oliver, R.W., John, P.: Handbook of Personality, Third Edition: Theory and Research, 3rd edn. The Guilford Press, New York (2008)

A Review on Relations Extraction in Police Reports

Gonçalo Carnaz[1,2(✉)], Paulo Quaresma[1,2], Vitor Beires Nogueira[1,2],
Mário Antunes[3,6], and Nuno N. M. Fonseca Ferreira[4,5,7]

[1] Informatics Department, University of Évora, Évora, Portugal
d34707@alunos.uevora.pt, pq@uevora.pt, vbn@di.uevora.pt
[2] LISP, Évora, Portugal
[3] School of Technology and Management,
Polytechnic Institute of Leiria, Leiria, Portugal
mario.antunes@ipleiria.pt
[4] Institute of Engineering of Coimbra, Polytechnic Institute of Coimbra,
Coimbra, Portugal
nunomig@isec.pt
[5] GECAD, Institute of Engineering, Polytechnic Institute of Porto, Porto, Portugal
[6] INESC-TEC, CRACS, University of Porto, Porto, Portugal
[7] LIACC, Porto, Portugal

Abstract. Relation Extraction (RE) is part of Information Extraction
(IE) and aims to obtain instances of semantic relations in textual documents. The countless possibilities of relations, the myriad of subjects,
the difficulty in identifying emotions and the amount of unstructured and
heterogeneous data, have challenged the researchers to define innovative
and even more accurate methodologies. This paper presents the evaluation results obtained with a set of RE systems on identifying semantic
relations in criminal police reports. We have evaluated different applications with documents in English and Portuguese. The results obtained
give us useful insights to continue the research work, and to design the
relation extraction system applied to related domain.

Keywords: Natural language processing · Information extraction ·
Relation Extraction · Final police reports

1 Introduction

The deluge of data produced by the "digital society" has grown exponentially
over the years, and several studies began to grow in some specific domains, such
as the criminal investigation. An essential piece of these data is compound by
text, published every day across the heterogeneous sources and different languages. Over the years, systems have been proposed by several researchers, and
have demonstrated that is possible to process corpus efficiently and at the same
time extract millions of facts or relations with accuracy. Nevertheless, the size of

© Springer Nature Switzerland AG 2019
Á. Rocha et al. (Eds.): WorldCIST'19 2019, AISC 930, pp. 494–503, 2019.
https://doi.org/10.1007/978-3-030-16181-1_47

these data makes impossible their processing by reading. An example that could benefit from IE is the criminal police departments, which deal with forensic information. Aimed at simplifying this process, this paper has the primary goal to study the related works, tools for the automatic extraction of semantic information as well as their evaluation applied to texts in English and Portuguese, and to propose a system designed regarding RE.

The remainder of this paper is organized as follows. First, in Sect. 2 we discuss RE concepts and methods. Then, Sect. 3 details relevant research works. Next, in Sect. 4 we proposed a conceptual system to RE from crime police reports. In Sect. 5 we describe the systems selected and their results, and set up and analyze the results. Finally, in Sect. 6 we present the conclusions and future work.

This work was partially funded by the Agatha Project SI&IDT n° 18022.

2   Relation Extraction Background

In every text retrieved from different sources, we can extract relations between named-entities, such as Organizations or Locations, that could help us to represent relevant knowledge regarding multiple domains, like the criminal domain [1]. IE can be defined as *"automatic extraction of structured information such as entities, relationships between entities, and attributes describing entities from unstructured sources"* [2]. Martinez et al. [3] refer to IE are related to three elements: *Entities*, such as Michael Jordan or 1975; *Concepts*, named set of individuals, such as the President of Republic, or Topics, that refer to categories to which individuals or documents relate, such as Portugal Politics; *Relations*, the relation between *Entities* or *Concepts*, such as *plays(Michael Jordan, NBA)* and generating a triple relation *(Subject, Predicate, Object)*.

In the context related to the RE, we need to define the *Relations* concept, which is defined by Gruber [4] as a set of tuples that represent relations between entities. Mainly, semantic relations are relations between concepts and meanings that involve concepts related to linguistic units and components. For example, *"captured by"* relation relates to *Organization* entity (Police) with *Person* entity (Suspect). Throughout the literature, several types of relationships have been enumerated for the most distinct languages. However, during an extraction process, a question must be done: *"how can identify a relevant relation?"*, it is the type of information that will be analyzed, the study domain and its conditioning, or the objective of the extraction task.

According to Culotta et al. [5], we can define relation extraction as *"the task of discovering semantic connections between entities. In the text, this usually amounts to examining pairs of entities in a document and determining (from local language cues) whether a relation exists between them"*. RE could be applied to question answering (QA) systems, information retrieval, summarization, semantic web annotation, construction of lexical resources and ontologies. To perform RE in sentences retrieved from texts, we need to realize the analysis of several features regarding the syntactic and semantic structures of the sentence. There are several methods for the development of RE systems: from

Traditional Relation Extraction and Open Information Extraction (OIE), using rule-base, supervised, semi-supervised and distantly supervised methods. In an evaluation study, it is essential to provide measures in a way that supports the comparison of used methods.

We need to extract metrics from RE system for their evaluation, to do that, measures were defined such as: *Precision (P)*, *Recall (R)*, and the F_1-*Measure (F1)*. That is described below:

$$P = \frac{CR}{IR} \quad (1) \qquad\qquad R = \frac{CR}{TR} \quad (2) \qquad F_1 = \frac{2 * P * R}{P + R} \quad (3)$$

Where, CR (Correct Relations), IR (Identified Relations) and Total Relations (TR).

3   Related Work

In this section we focus on relations or events extraction approaches from unstructured data.

In 2008, the SEI-Geo System [6] recognized *part-of* relationships between geographic entities, using hand-crafted patterns based on linguistic features to detect geographic entities in text documents. Bruckschen et al. [7] proposed a system named by SeRELeP to recognize three different types of relationships: *occurred*, *part-of*, and *identity*, supported by an heuristic rules applied over linguistic and syntactic features. Cardoso [8] proposed a system called *REM-BRANDT*, that identifies 24 different relations types using hand-crafted rules based on linguistic features and supported by two knowledge bases: DBpedia[1] and Wikipedia. Garcia and Gamallo [9] proposed in 2011, a system to extract occupation relationship instances over Portuguese texts. Each training sentence, using Support Vector Machines (SVM) classifier, are evaluated by extracting each word, the lemma and the PoS-tag, computing the syntactic dependencies between words, using a syntactic parser. Souza and Claro [10] in 2014, aim to develop a supervised OIE approach for extracting relational triples from Portuguese texts. Using Corpus CETENFolha[2]. Collovini et al. [11] in 2016, aims to evaluate the Conditional Random Fields (CRF) classifier to extract relations between named-entities, such as Organizations, Locations, or Persons from Portuguese texts. In 2017, Sena et al. [12] aim to extract facts in Portuguese without pre-determining the types of facts. Furthermore, the authors used an inference approach (identification of transitive and symmetric issues) to increase the quantity of extracted facts using open IE methods.

Fader et al. [13] proposed an OIE system, called *Reverb*, that extracts relationships based on following constraints, where every phrase that has a sequence of words connecting two entities, must be either: (1) a verb, e.g., created; (2) a verb followed immediately by a preposition, e.g., located in; (3) a verb followed by nouns, adjectives, or adverbs ending in a preposition, e.g., has atomic

---

[1]   https://wiki.dbpedia.org/.
[2]   https://www.linguateca.pt/cetenfolha/.

weight of. With many probable matches for a single verb, the highest possible match is chosen. Akbik and Broß [14] proposed the *Wanderlust* algorithm that automatically extracts semantic relations from plain text into tiples. Using linguistic patterns that are defined over the dependency grammar of sentences, this method performs an unsupervised method without any specific type of semantic relation, using English Wikipedia corpus. Chambers and Jurafsky [15] aims to propose a template-based IE that removes the requirement of a predefined template structure in advance. Their algorithm learns the template structure from raw text, producing template schemes assets of linked relations, e.g., bombings include detonate, set off, and destroy events associated by semantic roles.

Akbik and Löser [16] proposed in 2012 the *KRAKEN* system, an OIE system, that aims to detect N-ary facts, performing an experimental study was performed on extracting facts from Web text, comparing their system with ReVerb. Concluding, that system could be advantageous if: (1) noise free and grammatically correct, to avoid incorrect parsing; (2) each sentence must be fact-rich. Schmitz et al. [17] describe a system called *OLLIE*, that expands the syntactic scope of OIE systems by identifying relationships mediated by nouns and adjectives, improving results obtained by other, such as *Reverb*.

In 2013, Corro and Gemulla [18] proposed an approach to OIE, called *ClausIE* which extracts relations and their arguments from text documents. Separating the detection of relevant information pieces, by exploring semantic and syntax knowledge by identifying the type of each sentence according to the grammatical function of its constituents.

In 2014, Ta and Thi [19] proposed to identify semantic relations, using algorithms that combine machine learning and Natural Language Processing (NLP), such as OpenNLP, Stanford Lexical Dependency Parser to explore sentences, using the following algorithms: Identifying Semantic Relations, Identifying Synonyms, Hyponyms and Hypernyms Relations and Identifying Semantic Relations based on Syntax Patterns and Linguistic.

In 2016, Augenstein et al. [20] presented an approach to reduce data sparsity impact, improving the robustness of named-entity recognition tools across domains, and extracting relations over sentence boundaries using unsupervised co-reference resolution methods. Gamallo and González [21] proposed a suite of tools, called *Linguakit* for extraction, analysis or annotation with integration into a Big Data infrastructure. Using several tasks, such as POS-tagging, syntactic parsing, co-reference resolution, and including applications for relation extraction, sentiment analysis, summarization, extraction of multiword expressions, or entity linking to DBpedia. Linguakit could be applied to four languages: Portuguese, Spanish, English, and Galician.

The framework presented in the Sect. 4 is based on traditional extraction techniques using semantic parsers, that are also used in approaches [9,14,19]. Other NLP techniques are common, such as tokenization, sentence detection, named-entity recognition, POS Tagging and Lemmatization.

4   Relation Extraction Framework

Figure 1 depicts the proposed approach to extract relations from criminal police reports, defining each module and outcomes that will be used on RE between named-entities.

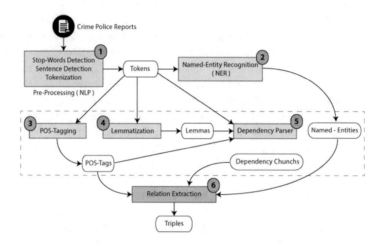

Fig. 1. System proposal for RE applied to criminal police reports.

Our approach is divided into the following modules: (1) Preprocessing, using stop-words detection, sentence detection and tokenization that generates defined tokens to be used in following modules; (2) Named-Entity Recognition (NER) module to extract the named-entities (NEs), such as Persons, Organizations or Locations. Using the tokens retrieved from Preprocessing module and a trained model; (3) Part-of-Speech (POS) Tagging module is used for syntax identification, specifying parts of speech to each word, such as noun, verb, adjective. Using tokens as input and a tagging system to generate tags for each token; in (4) the Lemmatization module removes the inflectional ends and to return the base or dictionary form of a word, known as the lemma; (5) Dependency Parser module, analyzes the grammatical structure of a sentence; and finally (6) Relation Extraction module that will generated the triples relations extracted from criminal police reports. The output results, the triple relations, will be used to populate an ontology, representing the relevant knowledge extracted from the criminal police reports.

5   Experiments

Two evaluation texts, in English and Portuguese language, were used to evaluate the RE task, retrieved from Wikipedia. A sentence detection task was performed, using the OpenNLP Toolkit[3], to detect the sentences contained in texts, resulting

---
[3]   https://opennlp.apache.org/.

in 50 sentences for each document. Notice that the obtained results are merely illustrative, because our dataset is composed of 50 sentences. In further work, we will add 500 sentences to our dataset, certainly obtaining different RE measures.

5.1 Selected Systems

Severel RE systems are described in Sect. 3 for English and Portuguese languages. Our RE systems selection was based on the following parameters:

- Open-source tools, and available for download;
  In case of using trial versions, we need to know which IE methods have been applied to extract relations;
- RE tools must support English or Portuguese languages or both;

The selected RE systems met the following requirements:

- Ollie[4]: a tool that automatically identifies and extracts binary relationships from English sentences;
- Reverb[5] [13]: this tool extracts relationships based on following constraints, where every phrase that has a sequence of words connecting two entities;
- Linguakit[6]: a multilingual toolkit for natural language processing with a NER system incorporated, created by the ProLNat@GE Group[7] (CITIUS, University of Santiago de Compostela);
- Rapport - A Portuguese Question-Answering System [22]: that uses a NLP pipeline with a RE task;

We have used Ollie and Reverb for the in English language, and Linguakit and Rapport to sentences in Portuguese language.

5.2 Results Obtained

In this section we discusse the results obtained after evaluation on selected systems for each language. For the experiments we have selected two types of sentences: long sentence and short sentences. The reason for selecting these sentences is to discover the system reaction for long-range relations, e.g., multiple verbs and named-entities, and relations based on one relation, like one verb and two named-entities. In this phase, we only consider for evaluation purposes the patterns where the relation verb, noun, and its named-entities are explicitly present in the text.

We have applied Ollie and Reverb to evaluate RE for English sentences, the Table 1 details the outcome of each tool. Results obtained by Ollie, for the longer sentence (Sentence 1), generated more relationships in opposition to Reverb system, mainly because Reverb analysis sequences of tokens for RE and

---

[4] http://knowitall.github.io/ollie/.
[5] http://reverb.cs.washington.edu/.
[6] https://github.com/citiususc/Linguakit.
[7] https://gramatica.usc.es/pln/.

Ollie works with the tree-like (graph with only small cycles) representation using Stanford's compression of the dependencies. The Ollie tool captures relations that ReVerb misses, such as long-range relations. For short sentence (Sentence 2) the results obtained by Ollie tool generates similar relationships in opposition to Reverb system because, with short sentences and without long-range relations, the differences are blurred.

Table 1. Tools (Ollie and Reverb) results with English sentences.

English Sentences
Sentence 1
"Alphonse Capone was an Italian-American gangster who led a criminal group dedicated to the smuggling and sale of alcoholic beverages among other illegal activities, during the dry law in the United States in the decades of 20 and 30."
Ollie output
a criminal group ; dedicated to ; the smuggling and sale of alcoholic beverages Alphonse Capone ; was ; and Italian-American gangster who led a criminal group dedicated to the ... a criminal group ; dedicated during ; the dry law a criminal group ; dedicated in ; the decades of 20
Reverb output
an Italian-American gangster ; led ; a criminal group Alphonse Capone ; be ; an Italian-American gangster
Sentence 2
"Al was known by Scarface's nickname."
Ollie output
Al ; was known by ; Scarface's nickname
Reverb output
Al ; known by ; Scarface's nickname

For the Portuguese language, we used Linguakit and Rapport to evaluate our sentences and to evaluate the selected tools, Table 2 details the outcome of each tool. Therefore, the Linguakit tool generates more relationships than the other tool because are supported by an open IE (similar to Ollie), which amplifies RE with multiple proposes for relations. Regarding short sentence, the results are similar because Rapport uses a dependency parser (Maltparser) connected to named-entity recognition model, and its limited by entities and relations (verb) detection.

Table 3 shows the results obtained by evaluating the measures, such as the *Correct Relations (CR)*, *Relations Identified (RI)* and *Total Relations (TR)*. The Ollie and Linguakit tools have reached the highest F-measure result for RE, having the best trade-off regarding both measures (*Precision* and *Recall*). Ollie obtained a lower *Precision*, sometimes due to the incorrect identification of the

Table 2. Tools (Linguakit and Rapport) results with Portuguese sentences.

Portuguese Sentences
Sentence 1
'Alphonse Capone foi um gângster italo-americano que liderou um grupo criminoso dedicado ao contrabando e venda de bebidas entre outras atividades ilegais, durante a Lei Seca que vigorou nos Estados Unidos nas décadas de 20 e 30.'
Linguakit output
Alphonse@Capone ; foi ; um gangster. a Lei@Seca ; vigorou em ; os Estados@Unidos
Rapport output
Alphonse_Capone ; ser; um gangster italo-americano que liderou um criminoso...
Sentence 2
'Al era conhecido pelo apelido Scarface.'
Linguakit output
Al ; era ; conhecido por o apelido Scarface
Rapport output
Al ; ser ; conhecido por o apelido Scarface

Table 3. Results of the evaluation for all tools.

Systems		CR	RI	TR	Measures P	R	F1
EN	Ollie	41	113	75	0.36	0.55	0.44
	Reverb	21	48	75	0.44	0.28	0.34
PT	Linguakit	26	46	75	0.57	0.35	0.43
	Rapport	12	43	75	0.28	0.16	0.20

arguments. Rapport obtained a lower *Precision*, sometimes due to the incorrect identification of the arguments.

Concluding, performing RE applied to the Portuguese language, some difficulties will be found related to language syntax and semantics. In general, and applied to all domains, such as news or medical, or particularly to our domain, the criminal domain. Therefore, the difficulties that may affect all domains: (1) the difficulty of taking advantage of the vast amount of data that is being produced, regarding RE; (2) lack of tools and resources for RE in Portuguese; (3) decrease errors concerning text reference relations, such as co-reference resolution or lexical relations; (4) handling the semantic ambiguity. Regarding the criminal domain, such as the crime police reports, we can add other difficulties: (1) the use of slang (figures of speech) text change the text patterns and conduct to a domain derivation; (2) the use of nicknames, changing the "regular" named-entity to a nickname, such as"Paul aka Alien"; (3) in a single report, we have three types of discourse: narrative, expository-descriptive and argumentative which increases difficulties related to semantic meaning.

6   Conclusion and Future Work

The work developed in this paper tries to achieve an analysis of RE systems to support a future system for relevant information that exists in the criminal police reports retrieved from heterogeneous data sources. Analyzing the state of the art and the obtained results, even a reduced dataset, allows us to prove the need for improvement of existing or development of RE systems. Therefore, achieving relevant results about IE regarding named-entities, e.g., Persons, Locations, Organizations, and Dates, for criminal analysis and to support a forensic decision support system for police investigations that will be usefully for our proposal RE system. For future work, we define the following goal: to develop the RE system proposed in Sect. 4 for extracting relations from criminal police reports in Portuguese language.

References

1. Carnaz, G., Nogueira, V., Antunes, M., Ferreira, N.: An automated system for criminal police reports analysis. In: 14th International Conference on Information Assurance and Security, IAS 2018 (2018)
2. Sarawagi, S., et al.: Information extraction. Found. Trends® Databases 1(3), 261–377 (2008)
3. Martinez-Rodriguez, J.L., Hogan, A., Lopez-Arevalo, I.: Information extraction meets the semantic web: a survey. Semantic Web (Preprint), 1–81 (2018)
4. Gruber, T.R.: Ontolingua: a mechanism to support portable ontologies (1992)
5. Culotta, A., McCallum, A., Betz, J.: Integrating probabilistic extraction models and data mining to discover relations and patterns in text. In: Proceedings of the Main Conference on Human Language Technology Conference of the North American Chapter of the Association of Computational Linguistics, pp. 296–303. Association for Computational Linguistics (2006)
6. Mota, C., Santos, D.: Desafios na avaliação conjunta do reconhecimento de entidades mencionadas: O Segundo HAREM. In: Desafios na avaliação conjunta do reconhecimento de entidades mencionadas: O Segundo HAREM, chapter: Geo-ontologias e padrões para reconhecimento de locais e de suas relações em textos: o SEI-Geo no Segundo HAREM, p. 436 (2008)
7. Bruckschen, J.G.M., Souza, R.V., Rigo, S.: Desafios na avaliação conjunta do reconhecimento de entidades mencionadas: O Segundo HAREM. In: Desafios na avaliação conjunta do reconhecimento de entidades mencionadas: O Segundo HAREM, Chapter 14, p. 436 (2008)
8. Cardoso, N.: Rembrandt - reconhecimento de entidades mencionadas baseado em relações e análise detalhada do texto (2008)
9. Garcia, M., Gamallo, P.: Evaluating various linguistic features on semantic relation extraction. In: Proceedings of the International Conference Recent Advances in Natural Language Processing 2011, pp. 721–726 (2011)
10. Souza, E.N.P., Claro, D.B.: Extração de relações utilizando features diferenciadas para Português. Linguamática 6(2), 57–65 (2014)
11. Collovini, S., Machado, G., Vieira, R.: A sequence model approach to relation extraction in Portuguese. In: LREC (2016)

12. Sena, C.F.L., Glauber, R., Claro, D.B.: Inference approach to enhance a Portuguese open information extraction. In: Proceedings of the 19th International Conference on Enterprise Information Systems, vol. 1, pp. 442–451 (2017)

13. Fader, A., Soderland, S., Etzioni, O.: Identifying relations for open information extraction. In: Proceedings of the Conference on Empirical Methods in Natural Language Processing, pp. 1535–1545. Association for Computational Linguistics (2011)

14. Akbik, A., Broß, J.: Wanderlust: extracting semantic relations from natural language text using dependency grammar patterns. In: WWW Workshop, vol. 48 (2009)

15. Chambers, N., Jurafsky, D.: Template-based information extraction without the templates. In: Proceedings of the 49th Annual Meeting of the Association for Computational Linguistics: Human Language Technologies-Volume 1, pp. 976–986. Association for Computational Linguistics (2011)

16. Akbik, A., Löser, A.: Kraken: N-ary facts in open information extraction. In: Proceedings of the Joint Workshop on Automatic Knowledge Base Construction and Web-scale Knowledge Extraction, pp. 52–56. Association for Computational Linguistics (2012)

17. Schmitz, M., Bart, R., Soderland, S., Etzioni, O., et al.: Open language learning for information extraction. In: Proceedings of the 2012 Joint Conference on Empirical Methods in Natural Language Processing and Computational Natural Language Learning, pp. 523–534. Association for Computational Linguistics (2012)

18. Corro, L.D., Gemulla, R.: Clausie: clause-based open information extraction. In: Proceedings of the 22nd International Conference on World Wide Web, pp. 355–366. ACM (2013)

19. Ta, C.D.C., Thi, T.P.: Identifying the semantic relations on unstructrured data. Int. J. Inf. **4**(3), 1–10 (2014)

20. Augenstein, I., Maynard, D., Ciravegna, F.: Distantly supervised web relation extraction for knowledge base population. Semant. Web **7**(4), 335–349 (2016)

21. Gamallo, P., González, M.G.: Linguakit: uma ferramenta multilingue para a análise linguística ea extração de informação (2017)

22. da Conceição Rodrigues, R.M.: RAPPORT: a fact-based question answering system for Portuguese. Ph.D. thesis, Universidade de Coimbra (2017)

Towards a Personalised Recommender Platform for Sportswomen

Juan M. Santos-Gago[1](✉) , Luis Álvarez-Sabucedo[1] ,
Roberto González-Maciel[1] , Víctor M. Alonso-Rorís[2] ,
José L. García-Soidán[3] , Carmina Wanden-Berghe[4] ,
and Javier Sanz-Valero[5]

[1] School of Telecommunication Engineering,
Campus Lagoas-Marcosende, 36310 Vigo, Spain
{jsgago,lsabucedo,rmaciel}@gist.uvigo.es
[2] DataSpartan, Wood Street, London, UK
victor.roris@dataspartan.com
[3] Faculty of Education and Sports Sciences,
Campus A Xunqueira, 36005 Pontevedra, Spain
jlsoidan@uvigo.es
[4] Health and Biomedical Research Institute of Alicante,
University General Hospital of Alicante,
Pintor Baeza, 03010 Alicante, Spain
carminaw@telefonica.net
[5] Faculty of Medicine, Miguel Hernández University,
Sant Joan D'Alacant, 03550 Elche, Spain
jsanz@umh.es

Abstract. Currently, there are many software applications to support
sports practice and fitness. Although a good number of them provide personalised services to their users, such as training plans adapted to the
athlete's condition, very few of these applications take into account the
particular casuistry of women. Moreover, as far as the authors have been
able to find, there are no sports applications that take into account the
menstrual cycle of women and how this cycle affects them individually.
This paper presents a proposal for a telematics platform, SportsWoman,
which allows daily recording of information about the menstrual cycle
and how it affects the athlete and, based on it, offers personalised recommendations. SportsWoman has been designed as an Expert System
based on semantic technologies. In the proposed platform, the knowledge
of specialists (physicians and researchers of sports science) is expressed
using rules that, in turn, determine the daily recommendations for each
user. SportsWoman has been tested and evaluated by 34 athletes through
the well-known System Usability Scale, obtaining a value of 86, which
corresponds to an acceptable level of usability with a grade B.

Keywords: Sports · Women · Expert System · Menstrual cycle

© Springer Nature Switzerland AG 2019
Á. Rocha et al. (Eds.): WorldCIST'19 2019, AISC 930, pp. 504–514, 2019.
https://doi.org/10.1007/978-3-030-16181-1_48

1   Introduction

In recent years, tools for monitoring sports activities have become popular. Both professional and amateur athletes, and even people who perform physical activities sporadically, often make use of mobile devices and applications that allow recording training sessions. This information is beneficial to know the sports performance of the user and serves as a support to set or adapt the training routines and, in this way, to achieve the pursuit sport or fitness goals.

Many of these tools are sophisticated. They allow collecting information of different nature that can be used to customise training plans according to the particular conditions of the athletes at each moment. In most cases, these tools consist of applications that run on mobile devices (smartphones and smartbands) and collect data from sensors within these devices, such as location, altitude, cadence, but also heart rate when using pectoral bands or appropriate smartbands. Also, the athletes themselves introduce data manually (such as weight, height, perceived effort after performing a physical activity, mood, injuries produced, among others). All this information characterise the actual state of the athlete and determines what the training should be like in each moment. It is clear that the physical condition and the mental status (such as stress or depression) drives sports performance. So these conditions are taken into account, especially those that refer to the physical state, by the current sports monitoring tools. However, to this day, there are no tools in this category that make use of information about the menstrual cycle to make personalised recommendations specific to women.

Different studies have highlighted the interrelation between the menstrual cycle and sport. On the one hand, high-intensity training can cause disorders in the menstrual cycle [1,2]. On the other hand, the cycle can influence the performance of athletes [3,4]. The hormonal change linked to the menstrual cycle involves significative changes in the body and the mood of women. Although the effects of these changes manifest themselves very differently from one woman to another, none is free from them. And although in most cases the influence of these changes does not dramatically affect the performance in sports, it is advisable to adjust the type and intensity of training depending on the menstrual cycle and how this cycle particularly affects each particular athlete. Generically, during the follicular phase, i.e., the first half of the cycle, there is a progressive increase in estrogen, increases pain tolerance, resistance, and the ability to exert force. In turn, the luteal phase, i.e., the second half of the cycle, is characterised by a progressive increase of progesterone and a decrease in estrogen and serotonin. Also, it can be noted an increase in body temperature and fluid retention. In the luteal phase, hormonal changes can lead to fatigue and less ability to exert force. Therefore, it is typically advisable to introduce higher training loads towards the second and third week of the menstrual cycle, and reduce them during the week before the period. Although this will depend to a large extent on how the cycle affects each woman in particular.

This article presents SportsWoman, a telematics platform whose objective is to facilitate the monitoring and collection of information related to the menstrual cycle and, based on it, generate daily personalised recommendations for women who practice sports. The article is organised into 6 sections. Section 2 includes a brief description of the study of the state of the art that was carried out in order to identify solutions related to the ideas above mentioned. Sections 3 and 4 present the conceived proposal and its architecture. Section 5 explains the validation process that has been carried out. Finally, Sect. 6 briefly discusses the results obtained and future lines of the work.

2   State of the Art

The achievement of the objective raised in this research departs from an initial study of the state of the art related to technological solutions aimed at training support for female athletes that take into account their menstrual cycle. In this sense, a systematic review was carried out, on the one hand, of the commercial tools available in mobile app stores such as Google Play or App Store and, on the other hand, of works published in the scientific literature.

Concerning commercial software solutions, it is worth mentioning the existence in the market of a multitude of applications aimed at monitoring and analysing sports activities. Remarkable companies of sports equipment and manufacturers of smartphones, smartbands or smartwatches have made available to amateur and professional athletes products specifically aimed at the management of sports activities and fitness. These products range from simple applications that allow registering sports activity, to sophisticated solutions to analyse these activities in detail and even provide recommendations focused on improving performance. However, the number of tools that take into account the particularities of women is minimal and, among them, those considering the menstrual cycle are practically anecdotal.

Among the applications intended to be used by women, taking into account its popularity and the assessment provided by its users, the following three are worth noting: *Female Fitness - Women Workout* [5], *Workout for Women - Weight Loss Fitness App* [6] and *Women Workout: Home Fitness, Exercise & Burn Fat* [7]. These apps are downloaded by millions of women and are valued very positively in stores. In essence, they are tools that suggest exercise routines aimed at reducing weight and working on certain parts of the body that may be of greater interest to women. In some cases this kind of apps are useful tools to record training and motivate users. However, they are simple and have little capacity for personalisation and adaptation to the peculiarities of a particular female athlete. And more importantly, none of them takes into account the menstrual cycle in the proposal of exercise routines.

A more remarkable tool with greater potential in this sense is the platform for sports activities management from Fitbit [8]. In mid-2018, Fitbit announced the launching of a new service for its app aimed at supporting women health. This service facilitates the registration of generic parameters about the menstrual cycle and its symptoms. Fitbit specifies that it is oriented "to help you learn more about your menstrual cycle–and your body–so you can better understand how it affects other aspects of your health and fitness" [9]. However, to this day, the information collected about the menstrual cycle is not used to adapt the functions offered by its platform.

The new service offered by Fitbit was studied in the scope of our research project in order to identify useful elements of characterisation of the menstrual cycle. Other popular menstrual cycle management applications in mobile stores, such as *Period Tracker, Ovulation Calendar & Fertility* [10], *Period Tracker Flo, Pregnancy & Ovulation Calendar* [11] o *Period Tracker: Monthly Cycles* [12], were also analysed with that same purpose. Although the latter are not specifically intended for the sports domain, they help women to trace their menstrual cycles and facilitate a successful conception (or, where appropriate, pregnancy avoidance). The apps allow recording daily parameters such as colour or flow of the menstruation, temperature of the woman, related symptoms or mood.

With regard to the analysis of works related to our proposal in the scientific literature, the search in the most relevant databases was unsuccessful. This search was carried out in the following repositories: Google Scholar, Springer Link, MEDLINE through PubMed and JMIR Publications. The query launched on the different search engines was: (menstruation OR menstrual) AND (sport OR fitness) AND (software OR application OR app). As a result, no publication describing any proposal similar or related to ours or that studied any analogous system was obtained. Hardly, a very limited number of papers dealing with the subject tangentially could be identified. Among the latter is the work of Sohda et al. [13], which presents a study aimed at clarifying how the data obtained from a self-tracking health app for female users (an app similar to the aforementioned for the management of the menstrual cycle) can be used to improve the accuracy of prediction of the date of next ovulation.

From the study of the state of the art conducted, we can conclude the nonexistence of a platform like the one postulated in the previous section. While there are systems that recommend exercise routines specifically for women who practice sports (mainly casual sportswomen), these are very simple tools that do not take into account the current phase of the menstrual cycle or how this cycle specifically affects them.

3   SportsWoman Platform

The SportsWoman telematics platform was conceived within the scope of a multi-disciplinary research project, involving researchers in the fields of sports science, medicine and telematics. The final aim is to support woman's specific sports practice, mainly professionals, but also amateurs or sporadic practitioners. The platform has a double purpose: (i) it is aimed to facilitate the collection of data that allow characterising the menstrual cycle of a woman and determine how this cycle influences the physical and emotional state of the woman, as well as her sports performance; and (ii) it is also designed to provide personalised recommendations to women regarding their sports training, taking into account, among other aspects, the particularities of their menstrual cycle.

For the UI of the platform, it was opted for a sober and functional design, in which the necessary options and functionalities are shown with the least possible number of distractions. To make use of the platform, a woman must register using a web form intended to collect data of a static nature, such as demographic data, physiological parameters, history of the menstrual cycle and habitual symptomatology, parameters related to rest and general characteristics of the sports activity practised. It should be noted that the data is stored anonymously on the platform. Once registered, the user accesses the platform through a Progressive Web App, typically through a mobile device, but could also do so through the web. On the initial screen of the app (cf. Fig. 1a), it is presented the actual point in her menstrual cycle (either the real one, if the necessary data is registered, or the estimated point according to the previous history), as well as an estimation about the starting of the following cycle. Also, a list with the personalised recommendations deduced by the platform for that day is shown.

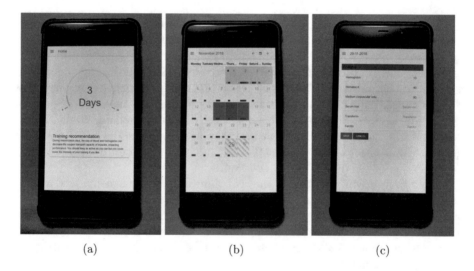

(a) (b) (c)

Fig. 1. SportsWoman user interfaces.

To add new data, see previously recorded data, or modify it, the user accesses a calendar (cf. Fig. 1b). The calendar allows visualising with colour codes the evolution of previous menstrual cycles, as well as an estimate of future cycles. In addition to this information, it is also shown if there is data recorded for each day. The user can select the day for which she wants to enter data or modify already existing one (cf. Fig. 1c). Through these services, users can register five categories of data: (i) data related to menstruation, (ii) biometric and physiological data, (iii) information regarding the training exercises carried out, (iv) data of medical tests, both blood and urine and, (v) physiological and psychological symptomatology.

All this information is used to accurately characterise each woman's menstrual cycle and estimate how this cycle affects her physical and psychological condition. The app also offers functionality for the visual representation of the evolution of different parameters, such as weight, body mass index or hematocrit, among several others. This functionality is very appreciated as it presents simply and graphically the trends of these parameters.

4   SportsWoman Architecture

The SportsWoman platform has been designed as a Knowledge-Based System (KBS) [14] and, more specifically, as an Expert System. Accordingly, information is mainly stored using semantic technologies, and part of the knowledge is stored as logic rules defined by specialists in the domain. These rules determine the recommendations presented to the user.

The authors have successfully applied this design model in several systems for different fields [15–17]. Based on these previous experiences, it was considered an appropriated approach as it allows to include new rules and new conceptual elements in the platform without making significative changes in the code. Following the construction paradigm of KBSs, the main architectural components of the platform are the following (cf. Fig. 2):

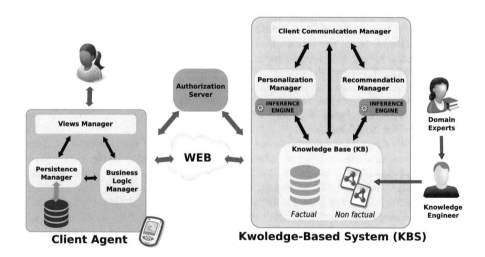

Fig. 2. SportsWoman architecture.

– **Knowledge Base.** This component records all the information required for the proper functioning of the platform. This includes both factual information, i.e. the data registered daily by the users, as well as nonfactual information, that is registered using a Semantic Model. Nonfactual information forms a Semantic Model composed of two elements:

- *Ontology.* It represents the conceptual knowledge of the domain in the form of RDF/OWL triples. This element manages the terms used to describe and represent the domain knowledge, including computer-usable definitions of basic concepts and the relationship among them.
- *Rule sets.* This element represents the heuristic information of the domain. It is made up of two sets of rules. The main group includes the rules about the recommendations daily presented to the users. This set of rules has been defined and agreed upon by a group of domain specialists, in particular, by experts in Medicine and Sports Science. It is a dynamic set of rules, which grows as new information is discovered. A second set of rules determines the users' monitorable parameters.

– **Recommendations Manager.** It is the component responsible for generating personalised recommendations for users. A logic rule execution engine guides its behaviour. This engine applies to the set of factual data in the knowledge base the first rule set above mentioned. This process generates a potential list of recommendations that are appropriate at that moment. The recommendations are prioritised; and only the first one, or a reduced number of them, is presented to the user. As an example, we show the following rules, expressed in logical notation (in the KBS are stored following the Node.JS JSON-rules-engine model):

```
hemoglobineLevel(?u) < hemoglobineMin => RiskOfAnemia(?u)
lastFlowLevel(?u) = HIGH => RiskOfAnemia(?u)
cyclePhase(?u,Lut3) AND RiskOfAnemia(?u) => suggestion(?u,R12)
```

These rules would trigger the generation of the R12 recommendation when the user is in the last third of the luteal phase, and she is at risk of anemia, either due to having a low hemoglobin level or having had high bleeding in the last menstruation. Subsets of similar rules, defined for different points of the menstrual cycle, lead to the generation of other recommendations.

– **Client Communication Interface.** REST API to access the features within the KBS from a Client Agent. This API follows the scheme of verbs and HTTP addressing described in Table 1.

– **Personalisation Manager.** One of the design premises of the proposed platform is to ensure that the interaction interfaces in the client are dynamic. This means that the parameters included in the data forms (cf. Fig. 1c) can be changed and adapted as the needs of the system evolve. A basic example of personalisation occurs according to the phase of the menstrual cycle in which the user is located: when the user declares to start the menstruation, the manager includes among the parameters to be monitored those related to bleeding (e.g., viscosity, colour, end of menstruation, among others). The mechanism of operation of this component is similar to that of the Recommendation Manager.

– **Authorisation Server.** An OAuth server authenticates the communication between the Client Agent and the KBS. This server is usually a trusted third party, such as Google. In this way, the KBS can anonymise the recorded data, just by linking it with a non-personal code provided by the OAuth server. Meanwhile, the Client Agent can customise the user session with the user profile data in the authorisation service, such as the name or the photo, among others.

– **Client Agent.** This component was developed as a Progressive Web Application. In this way it is easy to provide support for a variety of popular smartphones both in iOS or Android systems. The main functional elements of the Client Agent are:

- *Views Manager.* Functional element responsible for generating the user interfaces at runtime. The parameters required for the monitoring forms are retrieved from the KBS (see the Personalisation Manager). To generate a form it receives a JSON object in which each monitorable parameter is described (name, datatype, description, etc.) according to the current conditions. This element processes this description, so that if there is any parameter that can be auto-completed, it does it automatically (e.g., current location). Using the remaining parameters, a graphical interface for the final user is generated.
- *Persistence Manager.* Element responsible for improving the performance, avoiding, if possible, invoking services from the KBS and guaranteeing the operativeness of the client agent in offline mode.

Table 1. KBS Client access REST API

HTTP	URI	Description
GET	/params	Get the list of monitorizable params
GET	/param/{key}	Get information about a param
GET	/users/{id}	Get user info
GET	/users/{id}/logged_value	Get user param values (by dates)
POST	/users/{id}/logged_value	Send params for a specific user-date
DELETE	/users/id/logged_value	Remove param values by user-dates
GET	/users/{id}/logged_value/{key}	Get the user historic information about one param
DELETE	/users/{id}/logged_value/{key}	Remove all the param values
GET	/users/{id}/recommendations	Get the list of recommendations

5 Validation

To validate the usability of the proposed platform, a group of athletes affiliated with the Univ. of Vigo FCCED centre were invited to participate in a pilot test of the platform. These women, who regularly practice sports, were able to use the tool developed for three weeks. After this period, the athletes completed the popular and well-known System Usability Scale (SUS) questionnaire [18]. SUS is a 10-item questionnaire that uses a 5-level Likert scale (from $1 =$ "strongly disagree" to $5 =$ "strongly agree"). The questionnaire combines positive items (in particular odd items) and negative items (even items). For positive items, the best possible response, from the point of view of usability, is reached when responding with a 5; and, conversely, for negative items, the best possible answer is 1. 34 athletes answered the questionnaire. In Table 2 the different statistics obtained after the analysis of these answers are presented. At first glance, the results allow inferring that the perception of athletes on the usability of the platform is clearly high. As the reader can see in the table, the odd items concentrate answers between 5 and 4, while the even items concentrate the responses between 1 and 2. In particular, the resultant SUS score obtained was 86, which corresponds to a range of acceptable usability. That value is associated with a grade B, according to the scale proposed in [18].

To estimate the reliability of the SUS Score, the Cronbach's alpha was calculated from the answers obtained. Its value was 0.96, surpassing the barrier of 0.9, which corresponds to a excellent internal consistency of the answers [19].

Table 2. Responses to the SUS questionnaire

	Item	Mean	Mode	SD
Q1	I think that I would like to use this app frequently	4.588	5	0.492
Q2	I found the app unnecessarily complex	2.029	2	0.664
Q3	I thought the app was easy to use	4.618	5	0.486
Q4	I think that I would need the support of a technical person to be able to use this app	1.206	1	0.404
Q5	I found the various functions in this app were well integrated	4.353	4	0.476
Q6	I thought there was too much inconsistency in this app	2.235	2	0.644
Q7	I would imagine that most people would learn to use this app very quickly	4.676	5	0.468
Q8	I found the app very cumbersome to use	1.971	2	0.514
Q9	I felt very confident using the app	4.765	5	0.424
Q10	I needed to learn a lot of things before I could get going with this app	1.118	1	0.322

6   Conclusions

In this article it has been presented SportsWoman, a telematics platform conceived as an Expert System that facilitates the collection of data related to the menstrual cycle and emulates the duty of a training assistant providing recommendations that take into account the menstrual cycle for the sports domain.

Today, this type of coaching service is only available to elite athletes, who can take advantage of personal trainers. SportsWoman aims to democratise this situation by making available to non-elite sports professionals and amateurs women the possibility of having personalised suggestions. At the same time, SportsWoman can be a useful resource for coaches to improve their work.

To carry out this task, SportsWoman codes the knowledge of specialists in the form of inference rules that generate these recommendations. Currently, SportsWoman's knowledge base contains approximately 50 rules of this nature. Despite not being a large number, the platform has shown to be solvent, as proven by the test group, a group of women regular practitioners of sports.

The platform has been designed so that the incorporation of new rules and new monitoring parameters is simple. This will facilitate the growth of the knowledge base as new recommendations and rules of interest to users are identified.

In the future, it is expected to be able to use Machine Learning techniques in the platform. In particular, the possibility of including inductive inference techniques is planned to generate automatically new rules from the data collected. These new rules must be validated, in any case, by the specialists before forming part of the production rule set.

Acknowledgments. This work has been partially funded by the Spanish EAI and ISCIII and the ERDF "A way of making Europe" under projects TIN2016-80515-R (AEI/EFRD, EU) and PI16/00788 (CWB, MABM, LAS, JSV).

References

1. Loucks, A.B.: Effects of exercise training on the menstrual cycle: existence and mechanisms. Med. Sci. Sports Exerc. **22**(3), 275–280 (1990)
2. Dusek, T.: Influence of high intensity training on menstrual cycle disorders in athletes. Croat. Med. J. **42**(1), 79–82 (2001)
3. Kishali, N.F., Imamoglu, O., Katkat, D., Atan, T., Akyol, P.: Effects of menstrual cycle on sports performance. Int. J. Neurosci. **116**(12), 1549–1563 (2006). https://doi.org/10.1080/00207450600675217
4. Ozbar, N., Kayapinar, F.C., Karacabey, K., Ozmerdivenli, R.: The effect of menstruation on sports women's performance. Stud. Ethno-Med. **10**(2), 216–220 (2016). https://doi.org/10.1080/09735070.2016.11905490
5. Female Fitness - Women Workout. Leap Fitness Group. Mobile app. https://play.google.com/store/apps/details?id=women.workout.female.fitness
6. Workout for Women: Fitness App. Fast Builder Ltd. Mobile app. https://itunes.apple.com/us/app/workout-for-women-fitness-app/id83928568
7. Women Workout: Home Fitness, Exercise & Burn Fat. Fast Builder Ltd. Mobile app. https://itunes.apple.com/us/app/women-workout-exercise-by/id909610529
8. Fitbit Homepage. https://www.fitbit.com/
9. Kosecki, D.: One of Your Most Requested Features is Here! Introducing Female Health Tracking. Fitbit blog. https://blog.fitbit.com/female-health-tracking/
10. Period Tracker, Ovulation Calendar & Fertility. Leap Fitness Group. Mobile app. https://play.google.com/store/apps/details?id=periodtracker.pregnancy.ovulation-tracker
11. Period Tracker Flo, Pregnancy & Ovulation Calendar. Flo Health Inc. Mobile app. https://play.google.com/store/apps/details?id=org.iggymedia.periodtracker
12. Period Tracker: Monthly Cycles. Deltaworks. Mobile app. https://itunes.apple.com/us/app/period-tracker-monthly-cycles/id368868193
13. Sohda, S., Suzuki, K., Igari, I.: Relationship between the menstrual cycle and timing of ovulation revealed by new protocols: analysis of data from a self-tracking health. App. J. Med. Internet. Res. **19**(11), e391 (2017). https://doi.org/10.2196/jmir.7468
14. Akerkar, R., Sajja, P.: Knowledge-Based Systems. Jones & Bartlett Publishers, Burlington (2010)

15. Cañas, A., Santos, J.M., Anido, L., Pérez, R.: A recommender system for non-traditional educational resources: a semantic approach. J. Univ. Comput. Sci. **21**(2), 306–325 (2015). https://doi.org/10.3217/jucs-021-02-0306
16. Rorís, V.M., Álvarez, L.M., Santos, J.M., Ramos, M.: Towards a cost-effective and reusable traceability system. A semantic approach. Comput. Ind. **83**, 1–11 (2016). https://doi.org/10.1016/j.compind.2016.08.003
17. Cervera, M., Alonso, V.M., Santos, J.M., Álvarez, L.M., Wanden-Berghe, C., Sanz-Valero, J.: Management of the general process of parenteral nutrition using mHealth technologies: evaluation and validation study. JMIR mHealth uHealth **6**(4), e79 (2018). https://doi.org/10.2196/mhealth.9896
18. Bangor, A., Kortum, P., Miller, J.: Determining what individual SUS scores mean: adding an adjective rating scale. J. Usab. Stud. **4**(3), 114–123 (2009)
19. Gliem, J.A., Gliem, R.R.: Calculating, interpreting, and reporting Cronbach's alpha reliability coefficient for Likerttype scales. In: Midwest Research-to-Practice Conference in Adult, Continuing, and Community Education (2003)

Trusted Data's Marketplace

António Brandão[1]([⊠]), Henrique São Mamede[2],
and Ramiro Gonçalves[3]

[1] UAb e UTAD, Vila Real, Portugal
ajmbrandao@gmail.com
[2] UAb - Universidade Aberta, INESC-TEC, Lisbon, Portugal
hsmamede@gmail.com
[3] UTAD - Univ. Trás-os-montes e Alto Douro, INESC-TEC, Vila Real, Portugal
ramiro@utad.pt

Abstract. This article presents a literature review and the discussion about the key concepts associated with data markets. Data markets have multiple centralized and decentralized approaches. The main problem is the trust and reliability of supplies, inflows, and suppliers. The proposed study object is the decentralized marketplace data supported by Blockchain technology to ensure confidence in the supply chain of data, in the actors involved in the market and the data sources. The application scenarios are proposed in a model with four levels, data provision, data delivery, rights management, and producer internal sources. That will be done with Blockchain technology, through contracting using smart contracts, the controlled delivery of data by the data producers, the management of flows of data, and access control to data.

Keywords: Data Marketplace · Blockchain · Open data · IoT data

1 Introduction

This work analyzes how data markets can increase the data value assigned to the supply chain by increasing the confidence of the sources and provided data.

The problem arises from a lack of recognition of the data providers or owners of the data and so on the data that can be exploited. There is a need for data owners wish to maintain some degree of control over their rights and restrictions on access and use.

The different approaches to create data markets may be to set the architecture (centralized or decentralized), the data dynamically (streams or sets of static data), the access to data (external or internal API, download or Blockchain), reliability (by user feedback or data integrity and data origin) and the opening (open or closed).

The most important features for participants in data markets pass by a data catalog with search functions, access control, tools to create appropriate arrangements, appropriate monetization for transactions, monitoring of compliance with the SLA and dispute resolution, and tools to assess the quality and reliability of data and its suppliers. Reliability is central to assessing the quality and trust of data and its suppliers. The security management, authentication and authorization identities, is integrated into market solutions.

© Springer Nature Switzerland AG 2019
Á. Rocha et al. (Eds.): WorldCIST'19 2019, AISC 930, pp. 515–527, 2019.
https://doi.org/10.1007/978-3-030-16181-1_49

The data are essential and critical making it difficult to involve enterprises of different sizes, with different collection needs with their data and need to analyze them together with other sources static, dynamic and IoT [1].

The value that can be attributed to the data can come from different services and forms, which can be through filtered access to raw data, processing, and analysis of raw data, deriving business ideas through data mining and predictive modeling, and contexts that can trigger new business.

The quality and availability of data used to develop the commitment to better administration and governance. Technological developments open up new sources of information and analytical innovations [2].

Data delivery models and the underlying architectures rely mainly on the availability of resources, data security and privacy, and pricing.

The capabilities of a data market go through several characteristics, such as the description of the Data, categorization, definitions, tags, and forms of research, and the creation of metadata that facilitate the creation of data catalogs and business glossary.

From a different perspective, we have data security and privacy concerns that involve interfaces of various natures, file, service, SQL access, analytics, data insertion by the owner and the clients.

In the following sections we begin by reviewing the key concepts about the structure of the data markets, present the review of the literature that claims to give state of the art, the problem to be solved, the proposed solution, methodology and conceptual model support, and conclusions.

2   Key Concepts

In this section, we discuss the key concepts that we consider essential to understanding the data markets, seeking to present the various perspectives we consider appropriate for the analysis of this area.

2.1   The Data Delivery Chain

The data supply chain can be presented in a simplified manner (Fig. 1) by suppliers who deliver raw data by producers (Data Generation) that convert the data into products for data warehouses (Data Transformation) that store data for distribution centers that provide retailers (Data Processing) Data to deliver the user (User Analytics).

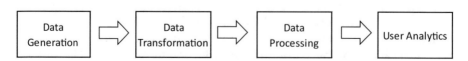

Fig. 1. Data supply chain

The data production process must be monitored, controlled and optimized through control dashboards, data flow control, and security mechanism, to improve the quality of the data supply chain [3].

2.2   Actors in Data Marketplaces

Figure 2 presents a model of IoTs data market system, with various actors in this market, the Data Owner (DO), the Data Broker (DB), the Value-Added Services Provider (VASP) and the User. The model is for each actor a set of activities, paper business, and interests.

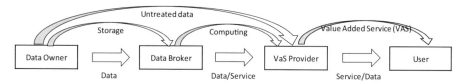

Fig. 2. Data market (Adapted from [1])

The DO of the activities deploy and operate the sensors, and update data. DB pass to store data, perform access control policies and provide computing power to the VASP. The VASP have as activities access to data, the implementation of the changes and provide value-added services. The user uses the value and services.

The business role for the DO refers to investment in infrastructure, the purchase storage services and intermediation, and data sale. The DB also invests in infrastructure, in selling storage services, computing and intermediation. The VASP purchase raw data, computing resources and sells value-added services. The user buys the value and services.

The DO specify access control policies according to its interests and to hide data from other parties. DB controls access rights to data and controls the scope of computable functions. The VASP seeks to maintain its secrecy computing logic, hide the result of your computer and receive significant results.

Stakeholders in data provisioning are referred to in [4] as Data Provider (People - individual/crowds/organization), software and stuff, Service Provider (Software and people), Data Consumer (people, software and Things), data aggregator/integrator (software people more software) and data evaluation (software and people).

The actors in the supply chain in the data market can have others functions, more granular, like the data producer, the data provider, the data distributor, the reseller data, the data amplifier, data buyer and data consumer.

2.3   Data Service

The conceptual data service unit presented in [4] states that as the service data model (consumption, property, provisioning, price) plus the concept of unit (essential component, primary function, modeling, and description) make a data service unit, which can be used privately or publicly, and can be elastic or not.

The data service units in the cloud provide data resources instead of providing computing resources or software in business environments and e-science with a data combined with the analysis of data, combined with AI (Artificial Intelligence) imply the provision of data assets.

Service models in the context of Data-as-a-Service, go through middleware publication/data subscription as a service, the sensor-as-a-Service, Database-as-a-Service (consultation systems structured and unstructured) and storage-as-a-Service (essential functions of storage), deployed in IoT, Edge & Cloud Systems. [4]

Data as a service, according to the definition of the NIST[1], has the self-service demand characteristics in which resources request data to be provisioned at different granularities. The resource pool is with the various types of data, large, static or near real-time, raw data and high information one level, extensive access network, as it can be accessed from anywhere, rapid elasticity as it is easy to add or remove data sources, and the service measured as the assessment, monitoring and publication focused on data use.

The dichotomy of the roles of the different actors in the data market goes through a data market in which the company and the individual can be the seller or buyer.

The data market is a showcase where users can purchase data products, private data, and public data, whether internal and external users, provided they find the data of products that reflect the needs of users and users can develop their Data products to share with others.

The challenges inevitably pass through research and development, prioritizing the development of new data, data products, with the marketing and sale of data products with newly discovered data products, looking for those who pay or go, which implies proper sizing architecture and organization.

2.4   The Value of Data

The amount of data is the critical issue for the success of a data market and trust is central to the assigned value. In this section, we present a set of four points that determine the creation of the data value.

- **Point 1.** The role and potential of creating data of the market value of the company or participating data brokers are different depending on whether the market is in technical startup phase or reached full commercialization.
- **Point 2.** The market relates to the various licensing models to deal with data owners and suppliers to maintain specific information without access and do not share in the market.
- **Point 3.** The terms, the general market conditions, and the data categories are the technical configurations of the market phase, in which the data providers define the licensing conditions independently, and the platform provides references to the licensing conditions.

---

[1] https://www.nist.gov/.

- **Point 4.** Customers and Data market suppliers pass initially set by the market can be commercially viable. Customers include all types of data providers, and the market system should actively seek new types of data to be more attractive to customers.

2.5   The Application Scenarios of Blockchain Technology

As the application of Blockchain technology is taking an increasing importance in new areas of research, in addition to the crypto-currency applications, comes applied in IoT, with 28% in the financial sector, 14% in e-governance, 12% in intelligent contracts, with 10% in smart Cities and businesses with 9% and 5% health [5].

The Blockchain technology may eventually replace systems that rely on third parties by eliminating intermediaries, to confirm third-party transactions, to enhance the process of disintermediation, and the consensus mechanisms distributed and transmit the clear and safe value.

The data market can use the features and mechanisms of Blockchain like the elimination of intermediation, providing a direct link between consumers and data providers, with more confidence between the parties and the available data.

The application scenarios of Blockchain technology proposed in this work involve the use of Smart Contracts in Blockchain in hiring. Blockchain will support the market operations, and the external data flows between data providers for the various data markets and the internal flows within each one of the data providers to insure the origin and quality of data and its access by the participants.

3   Literature Review

The literature review shows a context of exponential increase in data volume and the increasing commercialization of data that thickens the interest to the data markets. The players in this market need to have access to vast amounts of data processed using Big Data concepts, with algorithms that are more sophisticated and use Artificial Intelligence (AI), which can turn the centralized approach to become unworkable [6].

The Blockchain [7] technology can act as a catalyst for growth and can provide a platform where innovative practices may create a truly global and collaborative economy, with common goals for the benefit of the community at large. Trust is at the heart of the EU economy and seeks to incorporate the transition to a single digital market. The Blockchain can simplify interactions between the various actors and provide opportunities for consumers and businesses.

Other fields of application in data markets implies the development of algorithmic solutions incorporating AI and Machine Learning, as work [8], which sought to apply algorithmic solutions to the data market, with a real-time mechanism to efficiently buy and sell data.

3.1   Analysis Applied to Data Markets

Article [9] continues to market research area and the study of data vendors and data markets area, revealing the following trends: the raw and unstructured data became less often available, with data enrichment services and the offer of processed data most sought after, with the data available in more languages, and more amount of updated data offerings increased. Target customers spent focus on more or less relevant consumers and business customers. The fields of application have not changed much, and the sources of data are increasingly growing. Pay per use has increased, and at technical level, the technologies supported on the Web exceeded more traditional exchange formats.

3.2 Blockchain Applications

Blockchain technology is a decentralized technology that allows to perform transactions through shared network participants, supported by new forms of distributed software architectures. This technology improves the transparency of products, the management of the supply chain and more efficient data chain, better loyalty management system, improving customer profiles and preventing counterfeiting [10].

The Blockchain technology has the potential to be applied to supply chain management [11]. This text identifies four categories of questions to explore: How can physical products be tied to the digital diary? How can networks activate by Blockchain be linked to other foreign markets? How can the Blockchain be improved to meet the complicated structures of the supply chain? How can reserve enough space to store the amount of information required by the supply chain?

The management of the supply chain should be able to eliminate the need for third parties in its trading and adapt to the specific needs of supply chains, both concerning data requirements and terms of supply chains of complex structures [12].

In [13] a new Common Information Model Vehicle model is presented (CISG) for open and harmonized data markets, which allows aggregation and sets of generic data. The Automat European project has developed an open market by providing a single point of access for independent Data of the vehicle brand and model. The CISG prototype enables service providers and applications can access data for the vehicle through a single point of access to the weather forecast and monitoring of road quality.

3.3 Trust, Security and Privacy

The commodification of supported confidence in Blockchain technology is evaluated in [14] and suggests an opportunity for regulators and policymakers to shape the development and commercialization of disruptive innovation. Privacy and security are two of the questions that can be applied object of the more widespread use Blockchain technology.

The text [15] proposes a data protection framework based on distributed Blockchain technology to improve the data security of the cyber-attacks against the power system and to increase the robustness of the power systems. In this framework, it is evaluated from cyber-attacks efficiency point-of-view and shows that Blockchain can be considered as a solution to the data security of the power system. However, still wait developments in Blockchain based technology that improve the connection speed of the blocks, the acceleration of reliability and security, reducing the investment risk, and improving the consensus algorithm.

The Enigma [16] platform is a decentralized computing platform that guarantees privately, based on a peer-to-peer network allows different parts store together and that the calculations performed on data remain private. The computer model used is based on an optimized version of secure multiparty computation, secured by a verifiable secret through full verification and public nature of Blockchain.

Article [17] seek to apply the certification mechanisms following Art. 42 of the Data Protection Personal General Regulations. Although Blockchain not the main component of the proposed architecture considers that enables decentralized transactions, reliable among a crowd of pseudonymous participants, to adopt and implement the paradigm of "Privacy by Design".

3.4 Data Marketplaces

Article [18] analyzes the motivations and techniques for the exchange of scientific data and proposes the use of Data Marketplace service model to facilitate the exchange of data between producers and consumers. The service of a business model for the data appears in the commercial sector and can be adopted as a data-sharing platform for the scientific community.

The DataBright [19] prototype it is a system that transforms the creation of training examples and sharing of trusted computing, an investment mechanism that pays dividends whenever the hardware or contributions data is used in training a learning model machine. This system allows the data creators to retain ownership of their contribution and assigns measurable value. The amount of data is given by its utility in distributed computing later performed in the computing market.

The work [20] has a Marketplace decentralized data based on Blockchain (Datapace), where the data is your base and grows with new devices, sensors, and new technologies. Data may be perishable or durable, but with a common problem, how to draw an additional profit from these data. The TAS token will be used as a utility token of this system to allow a fair and functioning insurance system and allow the system. Also uses AI and Machine Learning algorithms to ensure that data streams can be unified in a format and prepared for easy consumption.

The value of an open Data Marketplaces is analyzed in the article [21] through a value proposition for open data users, in which five amounts received are identified, the less complicated the task, the greater access to knowledge, the most significant possibilities to influence, the lower risk and greater visibility. It was concluded that open data markets, through a central portal, could provide better access to associated data and support services and increase the transfer of knowledge within the ecosystem.

3.5 Decentralized Repositories

Article [6] has a market data based on decentralized data repositories data sharing in various scientific fields. The central portal provides access to multiple distributed repositories, with a reliable and consistent basis with standardized metadata and semantically enhanced.

The work [22] presents the demonstration of Sterling, a decentralized market of private data, which allows the distribution of privacy preservation and use of data by using Smart Contracts based on Blockchain. The focus goes through a mechanism of data providers control the use of their data.

A Blockchain-based Database is presented in [7] to ensure the integrity of data in cloud computing environments. In cloud computing environments, the data owners do not control the fundamental aspects of the data, such as the physical storage of data and management of access.

3.6   IoT Data Markets

This paper [23] summarizes the features that are available through a cloud-based platform to support the IoT and describes the federation of eleven Internet deployments of things in various application fields, including smart cities, the intelligent building in crowdsensing and the smart grid.

The work [24] shows the IoT market for smart communities (I3), with dynamic data and motion data, which allows data flows of different entities to be mixed and analyzed, processed, and to support a diverse range of applications. The Internet promotes scalability and interoperability in IP and I3 is a data exchange middleware layer above the transport layer.

Article [25] has a data IoT market supported Blockchain (IDMoB). The solutions of artificial intelligence (AI) and machine learning (ML) are being integrated into the various types of services, products with more "intelligent" in which the centerpiece is the data. A decentralized platform is applied to a data market as proof of concept, implemented in Ethereum platform for Smart Contracts and the Swarm as a storage system. It involves multiple parties, not data in real-time applications and non-critical IoT.

The work [26] shows the dynamic data market (MARSA) based on a cloud of human sensing data (IoT) in near real time for different stakeholders sell and buy these data. The Data Marketplace prototype presents techniques to select the types of data and management of data contracts based on different types of costs, quality of data, and the rights to the data. The text [27] intends to ensure data exchange in a market in the cloud, and provide an overall and practical encryption solution for Secure Data Exchange (SDE).

Article [28] proposed a three-level hierarchy consisting of multiple data sources, a service provider and customer. The Blockchain technology can be applied to create a decentralized structure with producers or consumers.

The [29] work presents the Wibson, a Blockchain based data market for individuals selling safe and anonymous way information in a trusted environment. Combines Smart Contracts and Blockchain to allow sellers and buyers data of personal information and transact directly maintaining anonymity as required.

4   Problem

The main problem is to get a solution that can ensure the confidence and reliability of the sources, flows, and suppliers. This issue is critical because it is to establish a reliable network a data supply chain.

Like any problem more complex, this problem has many challenges that require a structured and coherent solution to establish the various levels of trust dimensions systematically.

5   Solution

The proposed solution involves four levels of performance that go through contracting through smart contracts, the controlled delivery of data by the data producers, rights and access to data, and the management of data flows. The model proposed and represented in Fig. 3 aims to present the architecture of a decentralized data market supported by Blockchain technology to ensure confidence in the data supply chain, the actors involved in the market and the data sources.

5.1   Methodology

The literature review allowed characterize a wide range of applications of the data market and application possibilities of Blockchain technology in this field. This review revealed the growing importance of data, Data markets, and marketplace, some limitations of Blockchain technology, consensus mechanism, and scalability.

The approach is based on research Design Science Research (DSR) [30] with the use of research methodologies supported on conceptual models [31].

The implementation of this model will be the project TDMP - Trust Data Marketplace. This project will be a prototype based on a generic data market portal for ecosystems that make up a Smart Place, in the specific case, a Smart City.

5.2   The Supply and Delivery Model in the Data Marketplace

In this section, we present the model of supply and delivery in the data market, which summarizes the four Blockchain application levels needed to solve the various components of the supply chain, delivery of proprietary rights, user access permissions and audit and trust the source of the data.

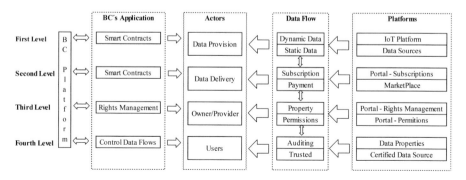

Fig. 3. Data Marketplace model

- **First Level - Data Provision.** The model considers this level by consensus acceptance of actors who can intervene in the data supply chain and contracting by Smart Contracts based on Blockchain.
  The actors' contractual each other via Smart Contracts based Blockchain according to the possible relationships supply.
- **Second Level - Data Delivery.** The model considers this level by consensus acceptance of actors who can intervene in data delivery chain and contracting by Smart Contracts based on Blockchain. The certificates Data deployments spend such suppliers and subscribing entities already accepted and contracted.
  This field up perspective using Blockchain to certify the origin and reliability of the sources.
- **Third Level - Rights Management and Access Permissions.** The model considers this level the management of supplier entities rights and access to data for users of data.
  In this perspective, the classification domain (e.g., free, payment for use or other restrictions to consider) the supplier/producer holder of rights of data sources with the use of Blockchain monitor the implementation of these rights. For users or VAS, apply permissions arising from its profile and contractors access.
- **Fourth Level - Producer guarantees the Internal Sources.** The model considers this level the producer or supplier of data ensures the audit of the origin of data and internal sources.
  This level features to implement mechanisms within the data producer to allow the traceability of the origin and processing of data to the supply, using the Blockchain as Ledger Distributed.

Based on reliable data and using the supply and delivery model, in the data marketplace, Organizations can introduce new key indicators, with better decision support for identification of investment priorities [32]. The implementation of a Decision Support System (DSS), strengthened by high quality data, allows the increase of the levels of performance and precision for the decision-making [33]. Organizations, in regional spaces, should focus on concepts such as Web 2.0 features coupled with a coherent online organizational identity [34], which reveals the need to obtain data that increases e-commerce transactions supported by Blockchain.

6 Conclusions

The information and knowledge economy make trusted data critical to empowering decision-making, rule automation with interaction machine-to-machine, machine learning, and the application of AI algorithms, to create new business opportunities. Data become a new product of value. The pricing of data will be inflated by possible cyber violations, which enhance financial losses, involving data from insurance policies for business protection.

The support infrastructure will require dynamic and secure cloud processing capabilities, resilient network, application code's quality with the paradigms "secure by design" and "privacy by design", to create value with exploring and synthesis of trusts data, that involve the concepts of Big Data and the use of Artificial Intelligence algorithms.

Data marketplaces will foster cross-sourcing with various data producers, involving different formats of data, IoT data, and crowdsourcing. The control of all supply chain begins in the trusted source, continue in the process of treating and create value, and finish with the delivery's satisfaction. To ensure security and reliability, we proposed four levels of application, goes by adding specific features, supported in the blockchain technology.

To solve the problem of confidence in the origin of the data we propose a working model supported Blockchain to address the four levels of this problem, Data Provision, Data Delivery, Rights Management, and Access Permissions, and Producer guarantees the Internal Sources.

Future work will be focused in the development of the prototype that allows operationalizing the proposed conceptual model and revises many of the aspects that may affect the scalability of this solution and may go through adopting different confidence and security mechanisms and treatment of data, involving machine learning and artificial intelligence.

Other lines of research that we can foresee are the review of ways to secure multiparty computation and cloud and look for ways to extend the value of data, especially the data is not shareable.

References

1. Horváth, M., Buttyán, L.: Problem domain analysis of IoT-driven secure data markets. In: Gelenbe, E., Campegiani, P., Czachórski, T., Katsikas, S.K., Komnios, I., Romano, L., Tzovaras, D. (eds.) Security in Computer and Information Sciences, pp. 57–67. Springer, Cham (2018)
2. OECD: The value of data for development. In: Development Co-operation Report 2017. OECD Publishing (2017)
3. Hazen, B.T., Boone, C.A., Ezell, J.D., Jones-Farmer, L.A.: Data quality for data science, predictive analytics, and big data in supply chain management: an introduction to the problem and suggestions for research and applications. Int. J. Prod. Econ. **154**, 72–80 (2014)
4. Truong, H.-L.: Data as a Service, Data Marketplace and Data Lake–Models, Data Concerns and Engineering, 94 p. (2018)
5. Brandão, A., Mamede, H.S., Gonçalves, R.: Systematic review of the literature, research on blockchain technology as support to the trust model proposed applied to smart places. In: Rocha, Á., Adeli, H., Reis, L.P., Costanzo, S. (eds.) Trends and Advances in Information Systems and Technologies, pp. 1163–1174. Springer, Cham (2018)
6. Ivanschitz, B.-P., Lampoltshammer, T.J., Mireles, V., Revenko, A., Schlarb, S.: A Data Market with Decentralized Repositories, 6 p. (2018)
7. Gaetani, E., Aniello, L., Baldoni, R., Lombardi, F., Margheri, A., Sassone, V.: Blockchain-based Database to Ensure Data Integrity in Cloud Computing Environments, 10 p. (2017)

8. Agarwal, A., Dahleh, M., Sarkar, T.: A Marketplace for Data: An Algorithmic Solution. arXiv:1805.08125 [csa] (2018)
9. Stahl, F., Schomm, F., Vossen, G.: The Data Marketplace Survey Revisited, 25 p. (2014)
10. Chakrabarti, A., Chaudhuri, A.K.: Blockchain and its Scope in Retail, vol. 04, 4 p. (2017)
11. Chakrabarti e Chaudhuri - Blockchain and its Scope in Retail.pdf
12. Jabbari, A., Kaminsky, P.: Blockchain and Supply Chain Management, 13 p. (2018)
13. Pillmann, J., Wietfeld, C., Zarcula, A., Raugust, T., Alonso, D.C.: Novel common vehicle information model (CVIM) for future automotive vehicle big Data marketplaces. In: 2017 IEEE Intelligent Vehicles Symposium (IV), pp. 1910–1915. IEEE, Los Angeles (2017)
14. McKnight, L.W., Etwaru, R., Yu, Y.: Commodifying Trust, 23 p. (2017)
15. Liang, G., Weller, S.R., Luo, F., Zhao, J., Dong, Z.Y.: Distributed blockchain-based data protection framework for modern power systems against cyber attacks. IEEE Trans. Smart Grid 1–1 (2018)
16. Zyskind, G., Nathan, O., Pentland, A.: Enigma: Decentralized Computation Platform with Guaranteed Privacy. arXiv:1506.03471 [csa] (2015)
17. Wirth, C., Kolain, M.: Privacy by Blockchain Design: A Blockchain-enabled GDPR-compliant Approach for Handling Personal Data (2018)
18. Ghosh, H.: Data marketplace as a platform for sharing scientific data. In: Munshi, U.M., Verma, N. (eds.) Data Science Landscape, pp. 99–105. Springer, Singapore (2018)
19. Dao, D., Alistarh, D., Musat, C., Zhang, C.: DataBright: Towards a Global Exchange for Decentralized Data Ownership and Trusted Computation. arXiv:1802.04780 [csa] (2018)
20. Draskovic, D., Saleh, G.: Decentralized Data marketplace based on Blockchain, 16 p. (2017)
21. Smith, G., Ofe, H.A., Sandberg, J.: Digital service innovation from open data: exploring the value proposition of an open data marketplace. In: 2016 49th Hawaii International Conference on System Sciences (HICSS), pp. 1277–1286. IEEE, Koloa (2016)
22. Hynes, N., Dao, D., Yan, D., Cheng, R., Song, D.: A demonstration of sterling: a privacy-preserving data marketplace. Proc. VLDB Endowment 11, 2086–2089 (2018)
23. Sánchez, L., Lanza, J., Santana, J., Agarwal, R., Raverdy, P., Elsaleh, T., Fathy, Y., Jeong, S., Dadoukis, A., Korakis, T., Keranidis, S., O'Brien, P., Horgan, J., Sacchetti, A., Mastandrea, G., Fragkiadakis, A., Charalampidis, P., Seydoux, N., Ecrepont, C., Zhao, M.: Federation of Internet of Things testbeds for the realization of a semantically-enabled multi-domain data marketplace. Sensors 18, 3375 (2018)
24. Krishnamachari, B., Power, J., Kim, S.H., Shahabi, C.: I3: an IoT marketplace for smart communities. In: Proceedings of the 16th Annual International Conference on Mobile Systems, Applications, and Services-MobiSys 2018, pp. 498–499. ACM Press, Munich (2018)
25. Özyılmaz, K.R., Doğan, M., Yurdakul, A.: IDMoB: IoT Data Marketplace on Blockchain. arXiv:1810.00349 [csa] (2018)
26. Cao, T.-D., Pham, T.-V., Vu, Q.-H., Truong, H.-L., Le, D.-H., Dustdar, S.: MARSA: a marketplace for realtime human sensing data. ACM Trans. Internet Technol. 16, 1–21 (2016)
27. Gilad-Bachrach, R., Laine, K., Lauter, K., Rindal, P., Rosulek, M.: Secure Data Exchange: A Marketplace in the Cloud, 30 p. (2017)
28. Jang, B., Park, S., Lee, J., Hahn, S.-G.: Three hierarchical levels of big-data market model over multiple data sources for Internet of Things. IEEE Access 6, 31269–31280 (2018)
29. Travizano, M., Minnoni, M., Ajzenman, G., Sarraute, C., Penna, N.D.: Wibson: A Decentralized Marketplace Empowering Individuals to Safely Monetize their Personal Data, 18 p. (2018)

30. Gregor, S., Hevner, A.R.: Positioning and presenting design science research for maximum impact. MIS Q. **37**, 337–355 (2013). https://doi.org/10.25300/MISQ/2013/37.2.01

31. Johnson, J., Henderson, A.: Conceptual models: core to good design. Synth. Lect. Hum. Centered Inf. **4**, 1–110 (2011). https://doi.org/10.2200/S00391ED1V01Y201111HCI012

32. Pereira, J., Martins, J., Santos, V., Gonçalves, R.: CRUDi framework proposal: financial industry application. Behav. Inf. Technol. **33**, 1093–1110 (2014)

33. Branco, F., Gonçalves, R., Martins, J., Cota, M.P.: Decision support system for the agri-food sector – the sousacamp group case. In: Rocha, A., Correia, A.M., Costanzo, S., Reis, L. P. (eds.) New Contributions in Information Systems and Technologies, pp. 553–563. Springer, Cham (2015)

34. Gonçalves, R., Martins, J., Pereira, J., Cota, M., Branco, F.: Promoting e-Commerce software platforms adoption as a means to overcome domestic crises: the cases of Portugal and Spain approached from a focus-group perspective. In: Mejia, J., Munoz, M., Rocha, Á., Calvo-Manzano, J. (eds.) Trends and Applications in Software Engineering, pp. 259–269. Springer, Cham (2016)

A NoSQL Solution for Bioinformatics Data Provenance Storage

Ingrid Santana[1]([✉]), Waldeyr Mendes C. da Silva[2], and Maristela Holanda[1]

[1] UnB, University of Brasília, Brasília, Brazil
ingrid95sl@gmail.com, mholanda@unb.br
[2] IFG, Federal Institute of Goiás, Goiânia, Brazil
waldeyr.mendes@ifg.edu.br

Abstract. Provenance data can support the reproducibility of experiments providing the history of the data in a scientific workflow. Bioinformatics generates an increasing amount of data, which are often analyzed employing workflows. This paper proposes a way to manage automatic executions of Bioinformatics workflows, storing their provenance and raw data in the MongoDB NoSQL database system. It uses a program that manages three different data models, a referenced, an embedded, and a hybrid data model for purposes of comparison. The results showed general advantages and disadvantages for each data model and some particularities of Bioinformatics.

Keywords: NoSQL · MongoDB · Big Data · Bioinformatics · Data provenance

1 Introduction

With the accelerated rate of advances in technology, the need has arisen for databases that are capable of both efficiently storing and processing massive amounts of data, as well as writing and reading them with high performance [1]. This scenario has given rise to the concept of Big Data, which is a term used to describe massive volumes of structured or non-structured data that can be managed mostly with NoSQL Databases [2].

In addition to managing the data storage of scientific *in silico* experiments, it is important to handle their production to support reproducibility. Data provenance is the history or the profile of an execution, capable of answering questions related to the origin of data [3].

This article presents a computational solution for managing the data provenance of Bioinformatics workflows using a NoSQL document-based database. The solution can be implemented using different data models, which are shown and discussed.

© Springer Nature Switzerland AG 2019
Á. Rocha et al. (Eds.): WorldCIST'19 2019, AISC 930, pp. 528–537, 2019.
https://doi.org/10.1007/978-3-030-16181-1_50

This article is organized as follows: Sect. 2 reviews some important concepts in Big Data, NoSQL, MongoDB, data provenance and Bioinformatics workflows. Section 3 discusses some related research works. The method used is presented in Sect. 4, and the results are shown in Sect. 5. The main conclusions and future work are listed under Sect. 6.

2 Background

2.1 Big Data and NoSQL

Nowadays, data generation is growing in velocity, volume, and variety. It has created a demand for data processing and storage with high performance written and read, creating the phenomenon known as Big Data [1]. NoSQL was born in this scenario and is an eclectic and unstructured database model capable of providing storage with flexibility [4,5]. It has been shown to be able to collect a vast and variable volume of data in distributed environments and provide answers to scalability problems [6].

MongoDB is a document-based NoSQL database capable of providing relatively powerful query capabilities, document-level atomicity, and support for complex data types [1,7]. It has been developed for the storage of information with high performance and scalability [5]. In addition, it is currently the most popular application in the category of document-oriented databases [5]. MongoDB organizes the documents in collections of documents, which are stored in the database under two primary storage formats [5,8]:

- Embedded: a denormalized model that stores the structure by grouping data according to its familiarity in a large document.
- Referenced: a standardized model that works with standardized formats that use connections to create relationships with other documents.

There is also the possibility of creating a hybrid model that has both characteristics [5]. One factor to note is that the size of MongoDB documents is limited to 16 MB [8]. As an alternative to creating larger documents, the convention uses the GridFS concept as an alternative [8]. With GridFS, it is possible to insert larger documents by dividing the file into parts called chunks [8].

2.2 Data Provenance and Bioinformatics Workflows

Data provenance is characterized as the history or profile of workflow executions, capable of answering questions related to the origin of data, providing the lineage of data generation, use, and processing [3,9]. Essential for tracking data among the different stages of a workflow [10], its usefulness lies in the fact that, once the data provenance is recorded, it is easier to create, recreate, evaluate or modify computational models or scientific experiments based on the accumulation of knowledge of what was done [11]. Databases to store data provenance has scientific importance because they enable the use of the data's origin as a raw material for reproducing the experiments [12].

In the field of Bioinformatics, most of the data generated come from the high-throughput of Next Generation Sequencing (NGS) [6,13]. Genomics is a Big Data science and will expand and achieve 1 Zbp in the next years according to the historical growth rate [14].

A workflow is a set of tasks with input and output information to be followed in order to achieve a final goal [15]. In Bioinformatics, the input files are the sequences generated by the NGS sequencers. Bioinformatics workflows are often organized in well-defined phases that make use of both these files and several programs or tools and libraries [12,16]. The purpose of creating a workflow for biological data analysis and interpretation is to acquire specific, unambiguous and consistent knowledge that is repeatable under the same conditions [16].

3   Related Works

The usefulness of data provenance and its application in NoSQL databases based on documents is a point of discussion in Mattoso *et al.* [11]. However, there are also several other works with different concepts and ideas worth highlighting. Li *et al.* [3] proposes a structure called ProvenanceLens, which provides provenance management in cloud environments and compares its performance by using MySQL, MongoDB, and Neo4J. Tao Li *et al.* also enumerates characteristics that are desirable for its systems and categorize the provenance into types.

Costa *et al.* [17] were able to capture the data provenance of a Workflow using the PROV-Wf model. However, the provenance was captured and stored in an RDBM, and the workflow was run in both local and cloud environments. Hondo *et al.* [18] present a comparative study between the NoSQL Cassandra, MongoDB and OrientDB databases through the execution of a bioinformatics workflow and the storage of its data provenance. Continuing this work, Hondo *et al.* [19] also conduct a study of bioinformatics workflows by storing data provenance in the previously mentioned databases and then conducting queries. Reis *et al.* [5] carried out a comparative analysis between embedded, referenced and hybrid data models in MongoDB. The models, although not having bioinformatics data, were compared to each other with queries and were in the Big Data domain.

This work differs from previous ones by obtaining the data provenance of Bioinformatics workflows when executing it automatically with a program created to, not only perform the steps of the workflow, but also to acquire its data provenance. The data provenance information collected is then inserted into MongoDB by the same program in three different models (embedded, referenced and hybrid). These three data approaches are then compared to each other, taking into account the time the program took to create each database and also the capability and facility to make queries.

4   Method

Reis *et al.* [5] carried out a comparison and performance analysis of three different variations of a data model. The first variation used fully embedded documents. The second used fully referenced documents, and the third variation considered a hybrid form between the two previous ones. Hondo *et al.* [19] proposed a data provenance storage model for Bioinformatics workflows, which can be seen in Fig. 1.

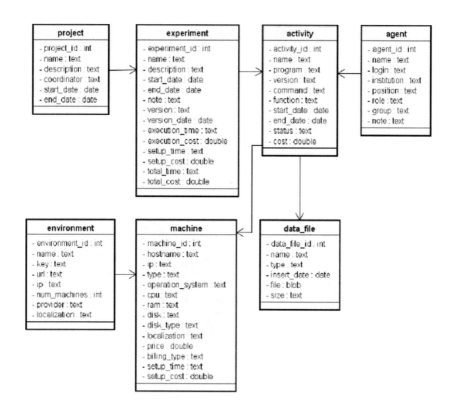

Fig. 1. Data model previously proposed by Hondo *et al.* [19].

In this work, inspired by the approach of Reis *et al.* [5], we adapted the model proposed by Hondo *et al.* [19] to create different document-based data models to store data provenance in MongoDB. We created three data models: a referenced model, an embedded model, and a hybrid model. The proposed models kept some features from [19] while others were adjusted. With the initial focus on the local storage of data provenance for general Bioinformatics workflows, the data models were updated to hold environment data, which are not present in [19]. Figures 2, 3 and 4 show the three proposed data models.

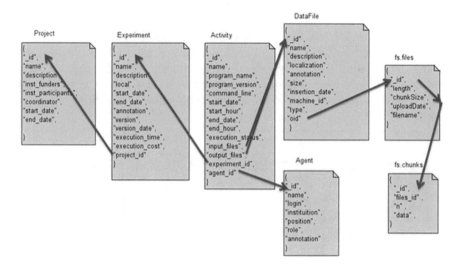

Fig. 2. Referenced data model for data provenance storing in MongoDB. Red arrows represent the references.

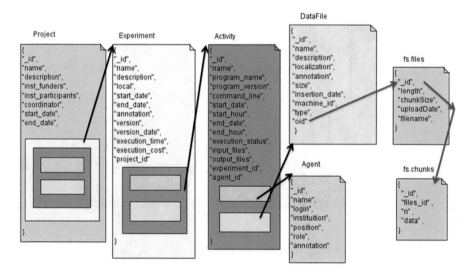

Fig. 3. Embedded data model for data provenance storing in MongoDB. Red arrows represent the references, and the black arrows represent embedded documents.

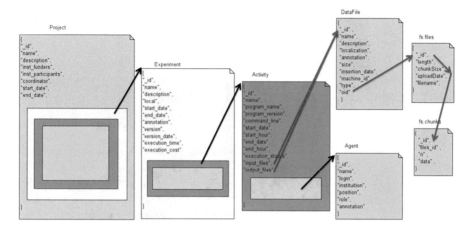

Fig. 4. Hybrid data model for data provenance storing in MongoDB. Red arrows represent the references, and the black arrows represent embedded documents.

We have created a program able to manage the workflow execution, retrieving and storing the data provenance. This program gets the required data for the workflow execution, with the ability to separate the input and output files. Then, it executes the commands for the workflow activity through system calls. For each workflow step, the program retrieves the data provenance and store it in several structures and variables. After executing all the workflow commands and their completion, the provenance data and the file's raw data are inserted into the MongoDB according to the three different proposed data models. The program inserts the data provenance in using each proposed data model sequentially so that each data model can have its exclusive database.

As the biological raw data are often bigger than the MongoDB size limit (16 MB) for documents, the program uses a conversion strategy (chunk) forcing the creation of GridFS collections. In the three models it is also possible to see the *fs.chunks* and *fs.files* collections referring to the collections automatically created by GridFS to support these biological raw data files.

After the creation of all three databases, the program finishes its execution. In addition, the program closes the connection with MongoDB after each database accomplishment.

Figure 5 summarizes a Bioinformatics workflow, which was used as a case study. In this workflow, using the Hisat tool [20], reads (sequences of DNA obtained from NGS sequencers) were mapped in a reference genome, which in this case was the 22nd chromosome of the human genome. After conversion of the mapping file format by the Samtools toolkit [21], a count of the number of mapped reads in gene regions was made using the HTSeq tool [22]. The experiment was repeated four times in order to minimize variances in the analyses. Before each run, the database was completely cleaned to avoid data overlapping and the data provenance was inserted into the MongoDB.

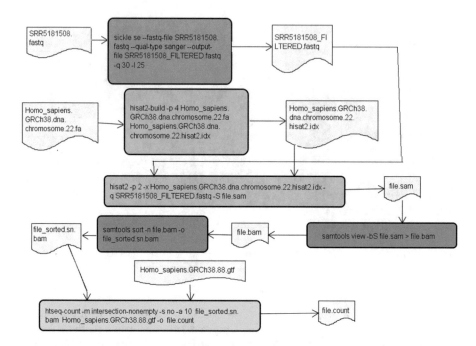

Fig. 5. Executed Bioinformatics workflow. Colors represent a specific phase: pink for filtering, blue for mapping, purple for file conversion and orange for counting.

5   Results

Results showed that it was possible to adequately capture and store the workflow data provenance in all three proposed data models through the built program.

The data provenance, both directly related to the experiment and for the computational environment, was suitably stored into MongoDB applying the three formats: referenced, embedded, and hybrid. The insertion times for each model - shown in Fig. 6 - were noted with the help of the log file generated by the built program, and they do not include the time it took to execute the workflow. When comparing the referenced, embedded and hybrid models, the difference in the amount of time it took to create each model was minimal. Nevertheless, the hybrid model took the shortest time to be created.

We designed and performed a set of queries to collect examples of relevant data provenance information, which allowed us to ensure the consistency of the data among the proposed data models. Also, it was possible to compare the complexity of query construction among the proposed models. For the sake of comparison, the queries answered the same questions asked in [19], and they quickly retrieved consistent data provenance information as can be seen in Table 1. Upon comparing the queries, the fully referenced model queries can be at least a little harder to create, and the information more, if the required information includes two or more documents and their relationships.

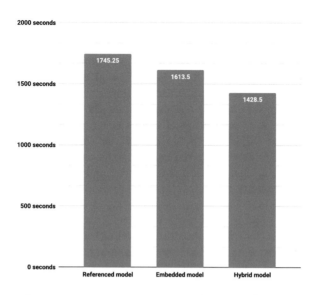

Fig. 6. Average data provenance insertion times (in seconds) into MongoDB for the three proposed data models.

Table 1. Queries for each model.

What are the names and versions of the used programs?	
Referenced	`db.activity3.aggregate([{$match:{"experiment_id":"1"}}, {$unwind:"$_id"},{$group:{_id:{program: "$program_name",version:'$program_version'}}}])`
Embedded	`db.default.aggregate([{$match:{$and:[{id:"1"}, {"experiment.id":"1"},]}},{$unwind: "$experiment.activity"},{$group:{_id: program: "$experiment.activity.program_name", version:'$experiment.activity.program_version'}}}])`
Hybrid	`db.project1.aggregate([{$match:{$and:[{id:"1"}, {"experiment.id":"1"},]}},{$unwind: "$experiment.activity"},{$group:{_id: program: "$experiment.activity.program_name", version:'$experiment.activity.program_version'}}}])`
How many activities were performed in the first experiment?	
Referenced	`db.activity3.count({experiment_id: "1"})`
Embedded	`db.default.aggregate([{$match:{"experiment_id": "1"}}, {$project:{numberOfActivities: {$size:"$experiment.activity"}}}])`
Hybrid	`db.project1.aggregate([{$match:{"experiment_id": "1"}}, {$project:{numberOfActivities: {$size:"$experiment.activity"}}}])`

6 Conclusion

The proposed data models were able to store data provenance from a Bioinformatics workflow in MongoDB, through a program that managed the whole process. The storage embraces both raw data and data provenance, including metadata of the environment.

Regarding the database size, the provenance data played a less important role representing a lower percentage of the data compared to the workflow raw data. Therefore, the difference in the amount of time it took to create each model was minimal. Nevertheless, the hybrid model showed better data insertion performance due to the combination of referenced and embedded documents that created a set of documents and collections which could be inserted into MongoDB more quickly.

Upon comparing the queries, we considered it to be at least a little harder to create queries for the referenced data model. Also, the information could be limited if the required data involved two or more documents and their relationships.

Future work includes the execution of distinct workflows in order to test and improve the power of the proposed program. It is also proposed to add the data and documents to cloud information, execute the project in a cloud environment, retrieve the cloud provenance data and, once again, compare the performance for each model. In addition, these studies may include an analysis of other data provenance information that is relevant to researchers and their work and comparison between different NoSQL databases.

References

1. Han, J., Haihong, E., Le, G., Du, J.: Survey on NoSQL database. In: 6th International Conference on Pervasive Computing and Applications (ICPCA), pp. 363–366. IEEE (2011)
2. Erturk, E., Jyoti, K.: Perspectives on a big data application: What database engineers and it students need to know. Eng. Technol. Appl. Sci. Res. 5(5), 850–853 (2015)
3. Li, T., Liu, L., Zhang, X., Xu, K., Yang, C.: Provenancelens: service provenance management in cloud. In: 10th IEEE International Conference on Collaborative Computing: Networking, Applications and Worksharing (2014)
4. Moniruzzaman, A., Hossain, S.A.: NoSQL database: new era of databases for big data analytics-classification, characteristics and comparison, arXiv preprint arXiv:1307.0191 (2013)
5. Reis, D.G., Gasparoni, F.S., Holanda, M., Victorino, M., Ladeira, M., Ribeiro, E.O.: An evaluation of data model for NoSQL document-based databases. In: World Conference on Information Systems and Technologies, pp. 616–625. Springer (2018)
6. Bellazzi, R.: Big data and biomedical informatics: a challenging opportunity. Yearb. Med. Inf. 9(1), 8 (2014)
7. Gessert, F., Ritter, N.: Scalable Data Management: NoSQL Datastores in Research and Practice (2016)

8. The MongoDB 4.0 Manual. https://docs.mongodb.com/manual. Accessed 23 June 2018

9. Buneman, P., Khanna, S., Wang-Chiew, T.: Why and where: a characterization of data provenance. In: International Conference on Database Theory, pp. 316–330. Springer (2001)

10. Guimaraes, V., Hondo, F., Almeida, R., Vera, H., Holanda, M., Araujo, A., Walter, M.E., Lifschitz, S.: A study of genomic data provenance in NoSQL document-oriented database systems. In: IEEE International Conference on Bioinformatics and Biomedicine (BIBM) 2015, pp. 1525–1531. IEEE (2015)

11. Mattoso, M., Werner, C., Travassos, G.H., Braganholo, V., Murta, L.: Gerenciando experimentos científicos em larga escala, SEMISH – Seminário Integrado de Software e Hardware (2008)

12. De Paula, R., Holanda, M., Gomes, L.S., Lifschitz, S., Walter, M.E.M.: Provenance in bioinformatics workflows. BMC Bioinf. **14**(11), S6 (2013)

13. Abdrabo, M., Elmogy, M., Eltaweel, G., Barakat, S.: Enhancing big data value using knowledge discovery techniques. IJ Inf. Technol. Comput. Sci. **8**, 1–12 (2016)

14. Stephens, Z.D., Lee, S.Y., Faghri, F., Campbell, R.H., Zhai, C., Efron, M.J., Iyer, R., Schatz, M.C., Sinha, S., Robinson, G.E.: Big data: astronomical or genomical? PLoS Biol. **13**(7), e1002195 (2015)

15. Mattoso, M., Dias, J., Costa, F., de Oliveira, D., Ogasawara, E.: Experiences in using provenance to optimize the parallel execution of scientific workflows steered by users. In: Workshop of Provenance Analytics, vol. 1 (2014)

16. Kanwal, S., Khan, F.Z., Lonie, A., Sinnott, R.O.: Investigating reproducibility and tracking provenance-a genomic workflow case study. BMC Bioinf. **18**(1), 337 (2017)

17. Costa, F., Silva, V., De Oliveira, D., Ocaña, K., Ogasawara, E., Dias, J., Mattoso, M.: Capturing and querying workflow runtime provenance with PROV: a practical approach. In: Proceedings of the Joint EDBT/ICDT 2013 Workshops, pp. 282–289. ACM (2013)

18. Hondo, F., Wercelens, P., da Silva, W., Lima, I., Santana, I., de Araujo, G., Araujo, A., Walter, M.E., Holanda, M., Lifschitz, S.: Uso de bancos de dados nosql para gerenciamento de dados em workflow de bioinformática. In: Proceedings of 32nd Brazilian Symposium on Databases, pp. 310–317 (2017)

19. Hondo, F., Wercelens, P., da Silva, W., Castro, K., Santana, I., Walter, M.E., Araujo, A., Holanda, M., Lifschitz, S.: Data provenance management for bioinformatics workflows using NoSQL database systems in a cloud computing environment. In: 2017 IEEE International Conference on Bioinformatics and Biomedicine (BIBM), pp. 1929–1934. IEEE (2017)

20. Kim, D., Langmead, B., Salzberg, S.L.: HISAT: a fast spliced aligner with low memory requirements. Nat. Methods **12**(4), 357 (2015)

21. Li, H., Handsaker, B., Wysoker, A., Fennell, T., Ruan, J., Homer, N., Marth, G., Abecasis, G., Durbin, R.: The sequence alignment/map format and samtools. Bioinformatics **25**(16), 2078–2079 (2009)

22. Anders, S., Pyl, P.T., Huber, W.: HTSeq-a python framework to work with high-throughput sequencing data. Bioinformatics **31**(2), 166–169 (2015)

DOORchain: Deep Ontology-Based Operation Research to Detect Malicious Smart Contracts

Mohamed A. El-Dosuky[1](✉) ⓘ and Gamal H. Eladl[2]

[1] Computer Science Department, Mansoura University,
Mansoura P.O. 35516, Egypt
Mouh_sal_010@mans.edu.eg
[2] Information Systems Department, Mansoura University,
Mansoura P.O. 35516, Egypt
gamalhelmy@mans.edu.eg

Abstract. Blockchains have become of great vogue in different fields after the introduction of Bitcoin. There are some inherent problems that need to be solved. One of these problems is to ensure that secured transactions in blockchain are checked if they are malicious or not. This paper proposes DOORchain that combines three powerful approaches of detecting intrusions and maliciousness. They are Deep learning, Ontology, and Operation Research. This uses the advantage of constraints from operation research to formalize and detect network maliciousness, and ontology to detect behavioral maliciousness in particular. Then it feeds this formalization to deep learning in order to check if the transactions in blockchain are malicious or not. After applying the proposed DOORChain, the final results affirm that accuracy and recall are enhanced with a slight inescapable trade-off in precision.

Keywords: Smart contracts · Deep learning · Ontology · Operation Research · Blockchain

1 Introduction

Blockchain, a distributed ledger, could be contemplated as a data structure that allows for creating signing, sharing, and storing digital transactions in a manner that makes it tremendously difficult to alter or delete blocks once recorded on ledger [1]. A smart contract is capable of both defining agreement obligations, and enforcing those [2].

The *problem* is that blockchain users may make malicious transactions and we don't know if it trustworthy or not. This paper proposes a solution to this problem by developing a methodology that combines three powerful approaches of detecting intrusions and maliciousness. This uses the advantage of constraints from Operation Research (OR) to formalize and detect network maliciousness, ontology to detect behavioral maliciousness, and deep learning, as overviewed in Sect. 2, before presenting, evaluating, and concluding proposed methodology in subsequent sections.

© Springer Nature Switzerland AG 2019
Á. Rocha et al. (Eds.): WorldCIST'19 2019, AISC 930, pp. 538–545, 2019.
https://doi.org/10.1007/978-3-030-16181-1_51

2   Previous Work

Vulnerabilities of smart contracts are investigated [3]. A recent list of attacks and their counter-measures is devised [4]. An audacious attempt to model contracts is proposed [5]. Decentralization scanning of anti-malware is possible [6].

Among the powerful approaches of detecting intrusions and maliciousness there are constraints from OR, ontology-based approaches to detect behavioral maliciousness, and deep learning, as overviewed in the following subsections.

2.1   Operation Research

Constraints, from OR: a Swiss-knife of solving problems [7], can be used in detecting network intrusions [8] as shown in both NeMODe [9] for DSL [10], and PRIDE [11] for wireless networks. This is comparable and yet superior to packet classification [12] built on Snort [13].

2.2   Ontology

Ontology provides semantics-awareness to applications. Recently, it is proposed for malware behavioral analysis and attack detection [14]. For a recent detailed survey on its application in security, refer to [15].

2.3   Deep Learning

Deep learning, the state-of-the-art in multimedia processing, comprehend data with multiple levels of abstraction [16]. Regarding paper scope, there are DeepChain [17], and a clever inspection algorithm that is inspired by colors [18].

3   Proposed Methodology

The proposed methodology can make a transaction and test if it is trustworthy or malicious based on Deep learning, Ontology, and OR, as shown in Fig. 1.

If the transaction is benign (trustworthy), the network will pass it and it will be completed. If the transaction is malicious, the network will stop it because of its maliciousness. DOORchain acts as a 3-layer filter. Decision is based on OR first, then Ontology-based, and Deep learning at last.

The order of Deep [Ontology[OR[chain]]] is logical since the transaction is generated first (as shown in Subsect. 3.1), then the network packets are filtrated using OR (as in Subsect. 3.2), then behavioral maliciousness is detected by Ontology-based approaches (as shown in Subsect. 3.3), and finally the bytecode of smart contract per se is transformed into an encoded image and classified into either benign or malicious based on Deep learning (as shown in Subsect. 3.4).

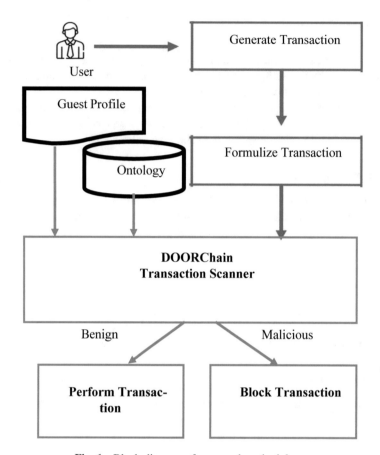

Fig. 1. Block diagram of proposed methodology.

3.1 Generating Transaction

Beside the standard Javascript and Solidity [19], the authors have a tendency to use Python wherever possible. There are a plethora of Python packages to operate on Ethereum, such as web3.py [20] and Pyethereum [21]. Basic log analysis is implemented based on OYENTE [2]. The goal is to detect abuse attempts based on outlier patterns in gas consumption and transaction longevity.

3.2 Operation Research

The mathematical formalization is based on nomenclature shown in Table 1.

Equation (1) formalizes transaction flow to be maximized.

$$\Phi = \mu B + (1 - \mu)M \tag{1}$$

μ is a model parameter taking a value in the range [0..1]. Let us contemplate on its value at the two extremes of the business-as-usual scenario (at $\mu = 1$, the case of full

Table 1. Nomenclature

Symbol	Quantity
Φ	Transaction flow
B	Benign (Trust) mode
M	Maliciousness mode
μ	Model parameter
T_i	Transaction with id i
N	Total number of transactions

trust, with no maliciousness detected) and the paranoia scenario (at μ = 0, the case of no trust, with high rate of maliciousness detection, or susceptibility of suspicion).

To filter out intrusion signatures, network packets are scanned in the network traffic logs, either generated from previous step (Sect. 3.1) or those downloaded from tcp-dump [22]. Each Transaction is generated with benign default mode (B). Equation (2) formalizes maliciousness mode.

$$\forall i, 1 \leq i \leq N \ (Ti \cap MODEL = ? \ \varphi) \rightarrow MODE[Ti] := M \qquad (2)$$

Where φ is empty set, and MODEL is an OR model built in NeMODe front-end to recognize the attack of SYN flood [23] and two Man-in-The-Middle attacks [24], namely DHCP and DNS spoofing. Figure 2 depicts filtrating network packets.

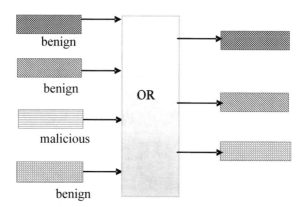

Fig. 2. Modeling the filtration of network packets.

3.3 Ontology

The ontology is built with Protégé [25] after [14], taking into account common backdoor, Trojans, and worms, that infect files, registries, or networks as shown in Fig. 3. A key incorporated concept in the ontology is the elusive Ransomware [26].

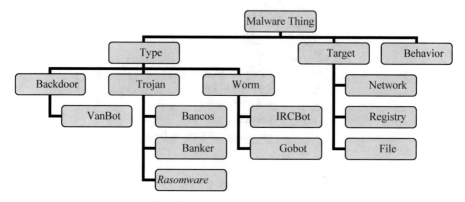

Fig. 3. Malware ontology

3.4 Deep Learning

This component follows the steps from [18]. However, the architecture of the built and deployed convolutional neural network (CNN) to the backend is adopted from [27]. The flow chart is depicted in Fig. 4.

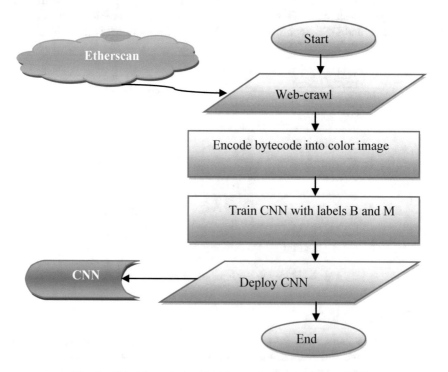

Fig. 4. Flow chart of deep learning component; based on [18]

4   Experiments and Results

Unless otherwise stated, development of proposed methodology took place on either of relatively similar machines whose specifications are shown in Table 2.

Table 2. Hardware specs

Feature	MachineA	MachineB
Manufacturer	Dell	Acer
Model	Inspiron N5040	Extensa 5630
Processor	Intel Pentium 2.1 GHz	Intel Pentium 2.2 GHz
Memory	2048 Mb	3027 Mb
Operating system	Windows 7 32bits	Windows 10 32bits

Smart contract web-crawling is achievable using pyspider [28], then the bytecode can be encoded as in Fig. 5.

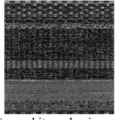   

(**a**) an arbitrary benign case   (**b**) an arbitrary malicious case

Fig. 5. Arbitrary encoded images of smart contract bytecode. Dataset from [29].

To compare DOORchain, CNN is built on the same configuration in [18] on the same dataset [29]. The comparison is depicted in Fig. 6.

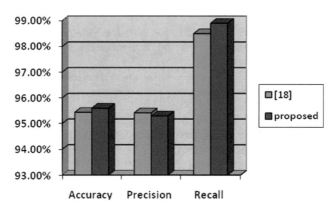

Fig. 6. Comparison between DOORChain and a previous work

After applying the proposed DOORChain, the final results affirm that accuracy and recall are enhanced with a slight inescapable trade-off in precision.

5 Conclusion and Future Directions

The proposed methodology tries to scan smart contracts and filtrate out malicious transactions. It acts as a 3-layer filter based on Deep learning, Ontology, and OR.

One possible future direction could be to combine decentralized firewall for malware detection [30], or considering a way for healing and backwashing alarmed transactions such as [31]. For the OR component, there is a room for speedup [32], and recognizing other attacks such as brute-force.

A possible direction may be to overload the existing OR component in optimizing the benefits of the blockchain financial aspects, based on integer programming for instance [33]. The ontology component may be extended [34]. For the Deep learning component, CapsNet [35] may be considered.

References

1. Swan, M.: Blockchain: Blueprint for a New Economy. O'Reilly Media Inc, USA (2015)
2. Luu, L., et al.: Making smart contracts smarter. In: Proceedings of the 2016 ACM SIGSAC Conference on Computer and Communications Security. ACM (2016)
3. Atzei, N., Massimo, B., Cimoli, T.: A survey of attacks on ethereum smart contracts (SoK). In: Principles of Security and Trust, pp. 164–186. Springer, Heidelberg (2017)
4. Xu, J.J.: Are blockchains immune to all malicious attacks? Financ. Innov. 2(1), 25 (2016)
5. Frantz, C.K., Nowostawski, M.: From Institutions to Code: Towards Automated Generation of Smart Contracts (2016)
6. Noyes, C.: BitaV: Fast anti-malware by distributed blockchain consensus and feed-forward scanning. arXiv preprint arXiv:1601.01405 (2016)
7. Hillier, F.S.: Introduction to Operations Research. McGraw-Hill Edu (2012)
8. Salgueiro, P., Abreu, S.: On using Constraints for Network Intrusion Detection. INForum (2010)
9. Salgueiro, P., et al.: Using constraints for intrusion detection: the NeMODe system. In: International Symposium on Practical Aspects of Declarative Languages. Springer, Heidelberg (2011)
10. Salgueiro, P.D., Abreu, S.P.: A DSL for intrusion detection based on constraint programming. In: Proceedings of the 3rd International Conference on Security of Information and Networks. ACM (2010)
11. Hassanzadeh, A., et al.: PRIDE: practical intrusion detection in resource constrained wireless mesh networks. In: International Conference on Information and Communications Security. Springer, Cham (2013)
12. Song, H., Lockwood, J.W.: Efficient packet classification for network intrusion detection using FPGA. In: Proceedings of the 2005 ACM/SIGDA 13th International Symposium on Field-Programmable Gate Arrays. ACM (2005)
13. Roesch, M.: Snort: lightweight intrusion detection for networks. In: Lisa, vol. 99, no. 1 (1999)
14. Huang, H.-D., et al.: Ontology-based intelligent system for malware behavioral analysis. In: IEEE International Conference on Fuzzy Systems (FUZZ) 2010. IEEE (2010)

15. Luh, R., et al.: Semantics-aware detection of targeted attacks: a survey. J. Comput. Virol. Hacking Tech. **13**(1), 47–85 (2017)
16. LeCun, Y., Bengio, Y., Hinton, G.: Deep learning. Nature **521**(7553), 436 (2015)
17. Weng, J.-S., et al.: DeepChain: Auditable and Privacy-Preserving Deep Learning with Blockchain-based Incentive. Cryptology ePrint Archive, Report 2018/679 (2018). https://eprint.iacr.org/2018/679
18. Huang, T.H.-D.: Hunting the Ethereum Smart Contract: Color-inspired Inspection of Potential Attacks. arXiv preprint arXiv:1807.01868 (2018)
19. Dannen, Chris: Introducing Ethereum and Solidity. Apress, Berkeley (2017)
20. Metwally, O.: On the economics of knowledge creation and sharing. arXiv preprint arXiv: 1709.07390 (2017)
21. Delmolino, K., et al.: A programmer's guide to ethereum and serpent (2015). https://mc2-umd.github.io/ethereumlab/docs/serpent_tutorial.pdf. Accessed 23 Oct 2018
22. Jacobson, V., Leres, C., McCanne, S.: TCPDUMP public repository (2003). http://www.tcpdump.org. Accessed 23 Oct 2018
23. Eddy, W.M.: Syn flood attack. In: Encyclopedia of Cryptography and Security, pp. 1273–1274. Springer, Boston (2011)
24. Ornaghi, A., Valleri, M.: Man in the middle attacks. In: Blackhat Conference Europe (2003)
25. Noy, N.F., et al.: Creating semantic web contents with protege-2000. IEEE Intell. Syst. **16** (2), 60–71 (2001)
26. O'Gorman, G., McDonald, G.: Ransomware: A Growing Menace. Symantec Corporation (2012)
27. Zeiler, M.D., Fergus, R.: Visualizing and understanding convolutional networks. In: European Conference on Computer Vision. Springer, Cham (2014)
28. pyspider.org. Accessed 23 Oct 2018
29. http://mit.twman.org/TonTon-Hsien-De-Huang/research/deeplearning/R2D2. Accessed 23 Oct 2018
30. Raje, S., et al.: Decentralized firewall for malware detection. In: 2017 International Conference on Advances in Computing, Communication and Control (ICAC3). IEEE (2017)
31. Continella, A., et al.: ShieldFS: a self-healing, ransomware-aware filesystem. In: Proceedings of the 32nd Annual Conference on Computer Security Applications. ACM (2016)
32. Schulte, C., Stuckey, P.J.: Speeding up constraint propagation. In: International Conference on Principles and Practice of Constraint Programming. Springer, Heidelberg (2004)
33. Mitchell, S., O'Sullivan, M., Dunning, I.: PuLP: A Linear Programming Toolkit for Python (2011)
34. Mundie, D.A., Mcintire, D.M.: An ontology for malware analysis. In: Eighth International Conference on Availability, Reliability and Security (ARES), 2013. IEEE (2013)
35. Sabour, S., Frosst, N., Hinton, G.E.: Dynamic routing between capsules. In: Advances in Neural Information Processing Systems (2017)

A Smart Cache Strategy for Tag-Based Browsing of Digital Collections

Joaquín Gayoso-Cabada, Mercedes Gómez-Albarrán,
and José-Luis Sierra$^{(\boxtimes)}$

Universidad Complutense de Madrid, 28040 Madrid, Spain
{jgayoso,mgomeza,jlsierra}@ucm.es

Abstract. Tag-based browsing is a common interaction technique in business, the culture industry and many other domains. According to this technique, digital resources have a set of descriptive *tags* associated, which can be used to perform an exploratory search, letting users focus on interesting resources. For this purpose, a set of tags is collected sequentially, and, at each stage, the set of resources described by all the selected tags is filtered. This browsing style can be implemented using *inverted indexes*. However, this implementation requires a considerable amount of set operations during navigation, which can have a negative impact on user experience. In this paper we propose addressing this shortcoming by using a cache that makes it possible to identify *equivalent* browsing states (i.e., states yielding the same set of filtered resources), which in turn will avoid redundant computations. The technique proposed will be compared with more basic implementations using a real-world web-based collection in the field of digital humanities.

Keywords: Browsing · Indexing · Inverted indexes · Cache strategies ·
Web-based digital collections · Tag-based systems · Digital humanities

1 Introduction

Tag-based systems, which consist of associating descriptive *tags* with resources, have a long tradition in many different computing fields (e.g., library science [12], information retrieval [13], file systems [8], and social and collaborative tagging systems [15]). In these settings *tag-based browsing* [5, 16] emerges as a natural navigation style.

During tag-based browsing, users can add and/or remove tags to filter the available resources. For this purpose, each time a tag is added/removed, in addition to updating the set of filtered resources, the system also provides the user with all the additional *selectable* tags that he/she can use to further refine such a set. Therefore, a convenient way of supporting tag-based browsing is through an *inverted index* (a standard artifact used for information retrieval [17]). Indeed, the inverted index for a tag-based digital collection makes it possible to retrieve the resources filtered by each tag.

Inverted indexes can be used in a straightforward way to support tag-based browsing. Indeed, the resources filtered and the selectable tags can be computed by carrying out several set operations involving the information provided by the current browsing state, the inverted index, and, eventually, the overall collection. However, in

© Springer Nature Switzerland AG 2019
Á. Rocha et al. (Eds.): WorldCIST'19 2019, AISC 930, pp. 546–555, 2019.
https://doi.org/10.1007/978-3-030-16181-1_52

many situations the number of operations required may be considerable. While there has been extensive research in performing these operations efficiently [4], the cost may not be negligible. Therefore, in this paper we propose a technique oriented to decreasing the number of required operations. The technique uses a *browsing cache* that not only avoids re-computing the information associated with already-considered browsing states, but also makes it possible to identify when two states filter the same resources (i.e., when two states are *equivalent*), saving further computations.

The rest of the paper is organized as follows. Section 2 makes the concepts behind tag-based browsing more precise. Section 3 describes the basic implementation based on inverted indexes and analyzes its shortcomings. Section 4 describes how the basic technique can be improved by caching browsing results and identifying equivalent browsing states. Section 5 provides some evaluation results. Section 6 summarizes some work related to ours. Finally, Sect. 7 provides some conclusions and lines of future work.

2 Tag-Based Browsing

In this section we describe the aspects concerning tag-based browsing in more detail. We will begin by clarifying what we understand by a tag-based digital collection[1]. From the perspective of this paper, a *tag-based digital collection* will be modelled by a trio $(\mathfrak{R}, \Sigma, \tau)$, where: (i) $\mathfrak{R}$ is a finite set of *digital resources*; (ii) Σ is a finite set of *tags*; and (iii) $\tau \in \mathfrak{R} \to 2^{\Sigma}$ is a *tagging function*, which assigns a set of tags to each resource. From now on, we will assume that Σ does not contain unused tags, in the sense that, for any tag t in Σ, there is at least one resource r in $\mathfrak{R}$ such as t is in $\tau(r)$. In addition, it will be useful to introduce some notations concerning filtering resources by tags, as well as concerning selectable tags associated with a set of resources. Given a set of resources $\mathbf{R} \subseteq \mathfrak{R}$: (i) by $\mathbf{R}_t$, with t a tag, we will denote the resources in $\mathbf{R}$ tagged by t, i.e., $\mathbf{R}_t = \{r \in \mathbf{R} \mid t \in \tau(r)\}$; (ii) by $\mathbf{R}_T$, with T a set of tags, we will denote the resources in $\mathbf{R}$ filtered by T, i.e., $\mathbf{R}_T = \{r \in \mathbf{R} \mid T \subseteq \tau(r)\}$; and (iii) by $\sigma\mathbf{R}$ we will denote the set of tags that can shrink (but not vanish) $\mathbf{R}$, i.e., those tags tagging some (but not *all*) of the resources in $\mathbf{R}$ (formally, $\sigma\mathbf{R} = \cup_{r \in \mathbf{R}} \tau(r) - \cap_{r \in \mathbf{R}} \tau(r)$).

Now, let $(\mathfrak{R}, \Sigma, \tau)$ be a tag-based digital collection. Tag-based browsing of this collection will involve two different types of *user actions*: (i) adding a tag t to the set of *active tags* (this action will be denoted by $+t$); and (ii) removing the tag t from the set of active tags (this action will be denoted by $\times t$). The browsing process itself will be

---

[1] Along the paper we adopt standard set-based notation as a basic descriptive language. In particular, $x \in \Theta$ denotes that the element x belongs to the set Θ. $\Theta_0 \subseteq \Theta_1$ (respectively, $\Theta_0 \subset \Theta_1$) means that Θ_0 is included in, or it is equal to (respectively, is included in, but it is *not* equal to) Θ_1. $\cup$, $\cap$ and - denotes respectively set union, intersection and difference. $\{x \in \Theta \mid \Phi(x)\}$ denotes the set of all the elements x in Θ satisfying the condition $\Phi(x)$. 2^{Θ} denotes the *powerset* (i.e., the set of all the sets) of Θ. $\varnothing$ denotes the empty set. $|\Theta|$ denotes the *cardinality* (i.e., the number of elements) of Θ. $\cup_{x \in \Theta} \Psi(x)$ (respectively, $\cap_{x \in \Theta} \Psi(x)$) denotes: (i) $\varnothing$ if $\Theta = \varnothing$; (ii) $\Psi(x)$ if $\Theta = \{x\}$; and (iii) the union (respectively, intersection) of the sets $\Psi(x_i)$ for each element x_i in Θ if $|\Theta| \geq 2$. Finally, $\Theta_0 \to \Theta_1$ denotes the set of functions from Θ_0 to Θ_1 (if ϕ is one such function and x is an element of Θ_0, then $\phi(x)$ denotes the value given to x by ϕ).

modelled by a *labelled transition system* [1] made of *browsing states* and transitions labelled with user actions:

- Browsing states will be sets of active tags $\mathbf{F} \subseteq \Sigma$. In this way, with each browsing state $\mathbf{F}$ it will be possible to associate: (i) the set $\mathbf{R^F}$ of resources filtered by $\mathbf{F}$, i.e., $\mathbf{R^F} = \mathfrak{R}_F$; and (ii) the set of *selectable* tags $\mathbf{S^F}$: those tags that the user can use to further refine $\mathbf{R^F}$, i.e., $\mathbf{S^F} = \sigma \mathbf{R^F}$.
- There will be an *initial browsing state* in which there are no active tags, i.e., the $\varnothing$ state. Notice that in this state the system will expose the overall set of resources, i.e., $\mathbf{R^\varnothing} = \mathfrak{R}$.
- Transitions, as well as the resulting legal states, will be conditioned by the *allowable user actions* for each state $\mathbf{F}$: (i) $+t$ for any t in $\mathbf{S^F}$ (i.e., the user can add any selectable tag); and (ii) $\times t$ for any t in $\mathbf{F}$ (i.e., the user is allowed to remove any active tag). In the first case there will be an *add* transition $\mathbf{F} \mapsto^{+t} \mathbf{F} \cup \{t\}$; in the second case there will be a *remove* transition $\mathbf{F} \mapsto^{\times t} \mathbf{F} - \{t\}$. Notice that, given a transition $\mathbf{F} \mapsto^{+t} \mathbf{F'}$, it is possible to determine $\mathbf{R^{F'}}$ directly from $\mathbf{R^F}$ by excluding all those resources that are not tagged by t (in other words, $\mathbf{R^{F'}} = (\mathbf{R^F})_t$). This shortcut is not possible in the case of $\mathbf{F} \mapsto^{\times t} \mathbf{F'}$ transitions, however.

3   Tag-Based Browsing with Inverted Indexes: The Naive Approach

Since the set of tags in a collection is finite, the tag-based browsing process as characterized in the previous section will be a finite-state transition system. However, even for small collections the number of states may be prohibitively large. Therefore, it can be necessary to compute the browsing states dynamically during each browsing session. Of course, the computation of these states can be substantially speeded up by precomputing appropriate indices. As aforementioned, inverted indexes provide a convenient solution. An *inverted index* for a tag-based digital collection $(\mathfrak{R}, \Sigma, \tau)$ can be modelled as a function $\varphi \in \Sigma \to 2^\mathfrak{R}$ which, for each tag t, provides the set of resources in $\mathfrak{R}$ tagged by t, i.e., $\varphi(t) = \{r \in \mathfrak{R} \mid t \in \tau(r)\}$.

The inverted index φ for a collection $(\mathfrak{R}, \Sigma, \tau)$ can be used to recreate the relevant information associated to browsing states. For this purpose, it is possible to take advantage of the following results, whose proofs are straightforward (we omit these for the sake of brevity):

- $\mathbf{S^\varnothing} = \{t \in \Sigma \mid |\varphi(t)| < |\mathfrak{R}|\}$.
- Let $\mathbf{F} \mapsto^{+t} \mathbf{F'}$ be an *add* transition. Then $\mathbf{R^{F'}} = \mathbf{R^F} \cap \varphi(t)$. In addition, $\mathbf{S^{F'}} = \{t' \in \mathbf{S^F} - \{t\} \mid 0 < |\mathbf{R^{F'}} \cap \varphi(t')| < |\mathbf{R^{F'}}|\}$.
- Let $\mathbf{F} \mapsto^{\times t} \mathbf{F'}$ be a *remove* transition. Then $\mathbf{R^{F'}} = \bigcap_{t \in \mathbf{F'}} \varphi(t)$ if $\mathbf{F'} \neq \varnothing$, and $\mathbf{R^{F'}} = \mathfrak{R}$ otherwise. In addition, $\mathbf{S^{F'}} = \{t \in \Sigma - \mathbf{F'} \mid 0 < |\mathbf{R^{F'}} \cap \varphi(t)| < |\mathbf{R^{F'}}|\}$.

Algorithm 1. Tag-based browsing strategy based on an inverted index

---

Inputs: (i) a tag-based digital collection $(\mathfrak{R}, \Sigma, \tau)$; and (ii) an inverted index φ for this collection
Method:
$\mathbf{F} \leftarrow \varnothing; \mathbf{R^F} \leftarrow \mathfrak{R}; \mathbf{S^F} \leftarrow \{t \in \Sigma \mid |\varphi(t)| < |\mathfrak{R}|\}$
do
    Wait for the next user action a
    if *the system is not shut down*
        if $a = +t$ **then**
            $\mathbf{F} \leftarrow \mathbf{F} \cup \{t\}; \mathbf{R^F} \leftarrow \mathbf{R^F} \cap \varphi(t); \mathbf{S^F} \leftarrow \{t' \in \mathbf{S^F} - \{t\} \mid 0 < |\mathbf{R^F} \cap \varphi(t')| < |\mathbf{R^F}|\}$
        else let $a = \times t$ **in**
            $\mathbf{F} \leftarrow \mathbf{F} - \{t\}$
            if $\mathbf{F} = \varnothing$ **then**
                $\mathbf{R^F} \leftarrow \mathfrak{R}$
            else
                $\mathbf{R^F} \leftarrow \bigcap_{t \in \mathbf{F}} \varphi(t)$
            endif
            $\mathbf{S^F} \leftarrow \{t \in \Sigma - \mathbf{F} \mid 0 < |\mathbf{R^F} \cap \varphi(t)| < |\mathbf{R^F}|\}$
        endif
    endif
until the system is shut down

The resulting browsing strategy is described in Algorithm 1. Notice that this algorithm makes the shortcomings anticipated in Sect. 1 apparent. Indeed, to determine the resources filtered for a browsing state $\mathbf{F}$, 1 or $|\mathbf{F}|$-1 intersections are needed, depending on the action type (adding or removing a tag). In addition, to update the set of selectable tags an additional intersection will be needed for each *candidate* tag. In the case of *add* actions, candidate tags are the currently selectable ones (excluding the last selected one), while in the case of *remove* actions this set can be considerably larger (all the tags in the collection, excluding the active ones). In any case, the cost of computing this set is not negligible.

In addition, the naive implementation outlined by Algorithm 1 exhibits two clear sources of inefficiency that accentuate these shortcomings:

- On the one hand, it does not take into account that a browsing state $\mathbf{F}$ can be visited multiple times during browsing. It can lead to $\mathbf{R^F}$ and $\mathbf{S^F}$ being computed again and again.
- On the other hand, the strategy does not take into account the equivalency among browsing states, in the sense of filtering the same set of resources.

These shortcomings can be alleviated by an adequate caching strategy. The details are examined in the next section.

4 The Enhanced Approach

In order to avoid the shortcomings of the naive browsing approach described in the previous section we propose using a suitable cache not only to recover the information from already-visited browsing states but also to store information on browsing states to detect *equivalent* states. Let $(\mathfrak{R}, \Sigma, \tau)$ be a tag-based digital collection, and $\mathbf{F}$ and $\mathbf{F}'$ two browsing states for a tag-based browsing process running in this collection. We will say that $\mathbf{F}$ and $\mathbf{F}'$ are *equivalent* (and we will denote it by $\mathbf{F} \sim \mathbf{F}'$) when $\mathbf{R}^{\mathbf{F}} = \mathbf{R}^{\mathbf{F}'}$. Notice that $\sim$ is an *equivalence relation* between browsing states. Therefore, for each browsing state $\mathbf{F}$ it makes sense to refer to the *equivalence class* $[\mathbf{F}]$ of $\mathbf{F}$ (i.e., the set of all those states that are equivalent to $\mathbf{F}$). During the browsing process, the first time that a state $\mathbf{F}$ filters a set $\mathbf{R}^{\mathbf{F}}$, the *representative* of $[\mathbf{F}]$ will be set to $\mathbf{F}$.

In this way, the *browsing cache* proposed for supporting the enhanced technique will be modelled, for a collection $(\mathfrak{R}, \Sigma, \tau)$, by a trio of functions (ρ, ϕ, δ) in the following way:

- $\rho: 2^{\Sigma} \rightarrow 2^{\mathfrak{R}} \cup \{\bot\}$ will model the *resource set store*, a table which maps browsing states $\mathbf{F}$ to the filtered resources $\mathbf{R}^{\mathbf{F}}$. Thus, given a browsing state $\mathbf{F}$, $\rho(\mathbf{F}) = \mathbf{R}^{\mathbf{F}}$ will mean that this state has already been visited, and that its associated filtered resources are $\mathbf{R}^{\mathbf{F}}$. On the contrary, $\rho(\mathbf{F}) = \bot$ will mean that there is no information for $\mathbf{F}$ in the store.
- $\phi: 2^{\Sigma} \rightarrow 2^{\Sigma} \cup \{\bot\}$ will model the *selectable tag store*, a table which maps $\mathbf{F}$ states to the selectable tags $\mathbf{S}^{\mathbf{F}}$. Its purpose is analogous to ρ, but mapping states to a set of selectable tags instead of sets of filtered resources.
- $\delta: 2^{\mathfrak{R}} \rightarrow 2^{\Sigma} \cup \{\bot\}$ will model the *representative store*, a table which maps sets of resources $\mathbf{R}^{\mathbf{F}}$ to representatives for $[\mathbf{F}]$. $\delta(\mathbf{R}) = \mathbf{F}$ will mean that a state $\mathbf{F}$ such as $\mathbf{R} = \mathbf{R}^{\mathbf{F}}$ has been cached (and, therefore, this state will be a representative of the equivalence class $[\mathbf{F}]$). In turn, $\delta(\mathbf{R}) = \bot$ will mean $\mathbf{R}$ has not yet been cached.

Therefore, ρ and ϕ will be used to recover the information from already-visited browsing states, while δ will be used to identify equivalent states. Additionally, with $\rho[\mathbf{F}/v]$ (respectively, $\phi[\mathbf{F}/v]$ or $\delta[\mathbf{R}/v]$) we will denote the new *resource set* (respectively, *selectable tag* or *representative*) *store* that results from replacing the entry for $\mathbf{F}$ (respectively for $\mathbf{F}$ or $\mathbf{R}$) with v in ρ (respectively ϕ or δ). Finally, empty stores will be denoted by $\varnothing$.

Equipped with a browsing cache like the one described, the enhanced strategy will avoid redundant computations by caching the information for the already-visited browsing states. More specifically:

- At the beginning of the process the stores will be initialized to $\varnothing$. The browsing state $\mathbf{F}$, and the associated information $\mathbf{R}^{\mathbf{F}}$ and $\mathbf{S}^{\mathbf{F}}$ will be initialized as in the naive approach. Finally, this information will be cached in the stores.[2]

---

[2] Thus, since information for $\varnothing$ is always cached, it is not necessary to distinguish an special case for $\mathbf{F} = \varnothing$ when computing $\mathbf{R}^{\mathbf{F}}$ after a *remove* action.

- The processing of an action begins by adding or erasing the corresponding tag t to/from $\mathbf{F}$, depending on whether the action is $+t$ or $\times t$. Then the updated $\mathbf{F}$ will be used to query the *resource set store* in order to detect whether the new state has already been visited. If this is the case, $\mathbf{R}^{\mathbf{F}}$ and $\mathbf{S}^{\mathbf{F}}$ will be updated respectively to $\rho(\mathbf{F})$ and $\phi(\mathbf{F})$. Otherwise, the set of resources $\mathbf{R}^{\mathbf{F}}$ will be computed and used to query the *representative store* to see whether there is a cached representative for $\mathbf{F}$. If this is the case, $\mathbf{S}^{\mathbf{F}}$ will be updated to $\phi(\delta(\mathbf{R}^{\mathbf{F}}))$. Otherwise, $\mathbf{S}^{\mathbf{F}}$ will be computed as in the naive strategy. Finally, the stores will be updated as needed.

Algorithm 2. Enhanced tag-based browsing strategy

Inputs: (i) a tag-based digital collection $(\mathfrak{R}, \Sigma, \tau)$; and (ii) an inverted index φ for this collection
Method:
$\rho \leftarrow \varnothing; \phi \leftarrow \varnothing; \delta \leftarrow \varnothing; \mathbf{F} \leftarrow \varnothing; \mathbf{R}^{\mathbf{F}} \leftarrow \mathfrak{R};$
$\mathbf{S}^{\mathbf{F}} \leftarrow \{t \in \Sigma \mid |\varphi(t)| < |\mathfrak{R}|\}; \rho[\varnothing/\mathbf{R}^{\mathbf{F}}]; \phi[\varnothing/\mathbf{S}^{\mathbf{F}}]$
do
    Wait for the next user action a
    if *the system is not shut down*
      if $a = +t$ **then**
        $\mathbf{F} \leftarrow \mathbf{F} \cup \{t\}$
      else let $a = \times t$ **in**
        $\mathbf{F} \leftarrow \mathbf{F} - \{t\}$
      endif

      if $\rho(\mathbf{F}) \neq \perp$ **then**
        $\mathbf{R}^{\mathbf{F}} \leftarrow \rho(\mathbf{F}); \mathbf{S}^{\mathbf{F}} \leftarrow \phi(\mathbf{F})$
      else
        if $a = +t$ **then**
          $\mathbf{R}^{\mathbf{F}} \leftarrow \mathbf{R}^{\mathbf{F}} \cap \varphi(t);$
        else let $a = \times t$ **in**
          $\mathbf{R}^{\mathbf{F}} \leftarrow \cap_{t \in \mathbf{F}} \varphi(t);$
        endif
        if $\delta(\mathbf{R}^{\mathbf{F}}) \neq \perp$ **then**
          $\mathbf{S}^{\mathbf{F}} \leftarrow \phi(\delta(\mathbf{R}^{\mathbf{F}}))$
        else
          if $a = +t$ **then**
            $\mathbf{S}^{\mathbf{F}} \leftarrow \{t' \in \mathbf{S}^{\mathbf{F}} - \{t\} \mid 0 < |\mathbf{R}^{\mathbf{F}} \cap \varphi(t')| < |\mathbf{R}^{\mathbf{F}}|\}$
          else let $a = \times t$ **in**
            $\mathbf{S}^{\mathbf{F}} \leftarrow \{t \in \Sigma\text{-}\mathbf{F} \mid 0 < |\mathbf{R}^{\mathbf{F}} \cap \varphi(t)| < |\mathbf{R}^{\mathbf{F}}|\}$
          endif
          $\delta[\mathbf{R}^{\mathbf{F}} / \mathbf{F}]$
        endif
        $\rho[\mathbf{F} / \mathbf{R}^{\mathbf{F}}]; \phi[\mathbf{F} / \mathbf{S}^{\mathbf{F}}];$
      endif
    endif
until the system is shut down

The resulting enhanced strategy is described in Algorithm 2. This strategy solves the shortcomings of the naive strategy. The price to pay for these enhancements is:

- On the one hand, the time overhead associated with cache management. However, as the browsing process evolves and the number of cache hits increases, the savings provided by the enhancements will pay off.
- On the other hand, the memory footprint caused by the cache. Nevertheless, if required, the strategy could be adapted to limit the cache size and to apply a suitable eviction strategy (e.g., an LRU-based policy).

5   Evaluation Results

In order to evaluate the browsing approach described in this paper we have implemented it in *Clavy*, an experimental system for managing digital collections with *reconfigurable* metadata schemata [6, 7]. Digital resources in *Clavy* are described in terms of *attribute-value* pairs. In turn, the browsing style in *Clavy* is tag-based, where the *tags* are identified with the *attribute-value* pairs.

In this context, we set an experiment concerning *Chasqui* [14],[3] a digital collection on Pre-Columbian American archeology. The current *Chasqui* version in *Clavy* consists of 2060 resources. Using this collection, we randomly generated a browsing session consisting of 10000 user actions. For this purpose, we expanded a random browsing process. In each state **F** in which both *add* and *remove* actions were allowed, a first, random decision on whether to choose one or another type of action was made (equal probability of 0.5 for each choice). The generation of *add* actions was performed by performing a second random decision, oriented to deciding whether to pick a tag between the 20 first selectable ones (probability of 0.8), or where to pick a less significant tag (those placed from position 21 onwards). Finally, once the segment was determined, a tag in this segment was picked with equal probability. Concerning *remove* actions, random selection prioritized the most recent added tags: the last k-th added tag was picked with probability $0.8^{k-1} \times 0.2$ (i.e., following a geometric distribution with p = 0.2).

The experiment was run on a machine with an Intel® Core™ i5-5200U 2.20 GHz processor, RAM 8 GB and Windows 10 OS. Browsing software was programmed in Java. The browsing cache was maintained in memory using Java's HashMaps. Sets were managed using *roaring bitmaps* [3]. In addition to the basic and enhanced strategy described in this paper, we also included a simpler caching strategy without detection of equivalent states, in order to assess the effect of equivalency detection in the final performance.

Figure 1 shows the results obtained for each of the three strategies. As it makes apparent, introduction of a caching strategy, even without detection of equivalent states, dramatically improves browsing performance (a total of 1.9 s for the simple caching strategy vs. 14.3 s for the basic, un-cached one, approximately a 7.5-fold decrease). Moreover, the addition of equivalency detection makes it possible to reduce

---

[3] http://oda-fec.org/ucm-chasqui.

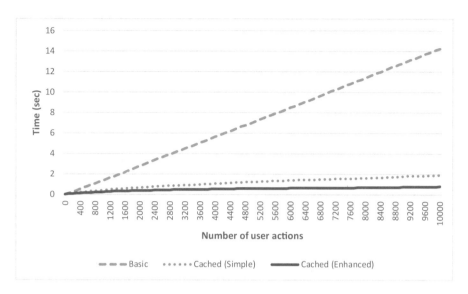

Fig. 1. Cumulative browsing times for: (i) the basic strategy; (ii) a simplified caching strategy without detection of equivalent states; (iii) the full, enhanced, cached strategy

the total time to 0.78 s (approximately a 2.4-fold decrease with respect to the simple cached strategy, an 18.3-fold decrease with respect to the basic one).

6   Related Work

Tag-based browsing is a typical interaction technique in semantic file systems, in which resources are described using tags instead of being placed in particular folders or directories. In this way, tag-based browsing is used to resemble conventional directory-based navigation. Some works adopting this organization are [2, 5, 8]. Compared to our proposal, none of these works analyze particular caching strategies to improve browsing performance.

Social tagging systems also exhibit different kinds of tag-based browsing models. Some examples of works following this navigation model are [9–11]. All these systems are supported by inverted indexes. Again, none of these works place special emphasis on cache strategies to browse intermediate results.

The work described in this paper is closely related to our previous work on *navigation automata* [6, 7] for speeding up tag-based browsing. While the more sophisticated indexes used in [6, 7] can speed up browsing, they can generate a considerably larger memory footprint than the dynamic, cache-based approach described in this paper. In addition, they also end up being considerably more difficult to construct and maintain than inverted indexes. Nevertheless, the marriage between the two approaches (i.e., to explicitly compute a navigation automaton or a non-deterministic equivalent and to use it to provide all the information required by each browsing state, or to adopt a fully dynamic approach based on inverted indexes and browsing caches) appears to be a highly promising solution.

A preliminary version of the work presented in this paper is described in [18]. The current paper provides a more detailed and formal description of the approach and empirically evaluates it. In [19] we discuss two other types of cache strategies to speeding-up tag-based browsing systems. The approach described in this paper can be conceived as a suitable combination of those two ones. In addition, as demonstrated in [20], it can be readily extended to manage modifications in the underlying collection without the need of discarding all the content of the cache after each modification operation.

7   Conclusions and Future Work

In this paper we have shown how the performance of tag-based browsing can be enhanced with a suitable caching strategy. This strategy not only consists of saving the information associated with each browsing state (i.e., with each set of active tags) and trying to recover this information before explicitly computing it, but also of identifying equivalent states (i.e., sets of active tags filtering the same resources). The experiment carried out in *Chasqui*, a real-world collection in the field of archeology, has shown how the technique can substantially speed up navigation with respect to the basic, uncached strategy. The price to pay is the overhead caused by cache management, as well as the higher memory footprint caused by the technique. However, the experiment with *Chasqui* has made it apparent how: (i) on the one hand the cache management overhead is compensated by eliminating the explicit computation of resources filtered and/or selectable tags on many states, and (ii) on the other hand the cache size is maintained within reasonable ranges, even when it is not upper-bounded.

Currently we are working on improving the cache strategy by combining it with our previous work in navigation automata. The main idea is to conceive the cache as a navigation automaton that is lazily built and maintained. We are also working on generalizing the browsing strategy to support, for instance, navigation through links between resources. In addition, we plan to combine browsing and Boolean search, allowing users to browse search results according to the tag-based browsing model. Finally, we also plan to carry out a more exhaustive evaluation comprising additional collections and the simulation of more realistic user browsing behavior models, as well as real-users browsing traces.

Acknowledgements. This research is supported by the research projects TIN2014-52010–R and TIN2017–88092–R.

References

1. Arnold, A.: Finite Transition Systems. Prentice-Hall, Englewood Cliffs (1994)
2. Bloehdorn, S., Görlitz, O., Schenk, S., Völkel, M.: TagFS-tag semantics for hierarchical file systems. In: Proceedings of the 6th International Conference on Knowledge Management (I-KNOW 2006) (2006)
3. Chambi, S., Lemire, D., Kaser, O., Godin, R.: Better bitmap performance with roaring bitmaps. Software Pract. Exper. **46**(5), 709–719 (2016)

4. Culpepper, J.-S., Moffat, A.: Efficient set intersection for inverted indexing. ACM Trans. Inf. Syst. **29**(1), (2010). Article no. 1

5. Eck, O., Schaefer, D.: A semantic file system for integrated product data management. Adv. Eng. Inf. **25**(2), 177–184 (2011)

6. Gayoso-Cabada, J., Rodríguez-Cerezo, D., Sierra, J.-L.: Multilevel browsing of folksonomy-based digital collections. In: Proceedings of the 17th Conference on Web Information Systems Engineering (WISE 2016) (2016)

7. Gayoso-Cabada, J., Rodríguez-Cerezo, D., Sierra, J-L.: Browsing digital collections with reconfigurable faceted thesauri. In: Proceedings of the 25th International Conference on Information Systems Development (ISD 2016) (2016)

8. Gifford, D.K., Jouvelot, P., Sheldon, M.A., O'Toole, J.W.: Semantic file systems. SIGOPS Operating Syst. Rev. **25**(5), 16–25 (1991)

9. Helic, D., Trattner, C., Strohmaier, M., Andrews, K.: On the navigability of social tagging systems. In: Proceedings of the 2010 IEEE Second International Conference on Social Computing (SocialCom 2010) (2010)

10. Hernandez, M.-E., Falconer, S.-M., Storey, M.-A., Carini, S., Sim, I.: Synchronized tag clouds for exploring semi-structured clinical trial data. In: Proceedings of the 2008 Conference of the Center for Advanced Studies on Collaborative Research: Meeting of Minds (CASCON 2008) (2008)

11. Kleinberg, J.: Navigation in a small world. Nature **406**, 845 (2000)

12. Redden, C.S.: Social bookmarking in academic libraries: trends and applications. J. Acad. Librarianship **36**(3), 219–227 (2010)

13. Salton, G., McGill, M.J.: Introduction to Modern Information Retrieval. McGraw-Hill, New York (1986)

14. Sierra, J.-L., Fernández-Valmayor, A., Guinea, M., Hernanz, H.: From research resources to learning objects: process model and virtualization experiences. Educ. Technol. Soc. **9**(3), 56–68 (2006)

15. Sinclair, J., Cardew-Hall, M.: The folksonomy tag cloud: when is it useful? J. Inf. Sci. **34**(1), 15–29 (2008)

16. Trattner, C., Lin, Y., Parra, D., Yue, Z., Real, W., Brusilovsky, P.: Evaluating tag-based information access in image collections. In: Proceedings of the 23rd ACM Conference on Hypertext and Social Media (HT 2012) (2012)

17. Zobel, J., Moffat, A.: Inverted files for text search engines. ACM Comput. Surv. **33**(2) (2006). Article 6

18. Gayoso-Cabada, J., Gómez-Albarrán, M., Sierra, J.-L.: Tag-based browsing of digital collections with inverted indexes and browsing cache. In: Proceedings of the 5th International Conference on Technological Ecosystems for Enhancing Multiculturality (TEEM2018) (2018)

19. Gayoso-Cabada, J., Gómez-Albarrán, M., Sierra, J.-L.: Query-based versus resource-based cache strategies in tag-based browsing systems. In: Proceedings of the 20th International Conference on Asia-Pacific Digital Libraries (ICADL 2018) (2018)

20. Gayoso-Cabada, J., Gómez-Albarrán, M., Sierra, J.-L.: Enhancing the browsing cache management in the clavy platform. In: Proceedings of the XX International Symposium on Computers in Education (2018)

Identifying Most Probable Negotiation Scenario in Bilateral Contracts with Reinforcement Learning

Francisco Silva[1], Tiago Pinto[1,2(✉)], Isabel Praça[1], and Zita Vale[3]

[1] GECAD - Research Group on Intelligent Engineering
and Computing for Advanced Innovation and Development,
Polytechnic of Porto (ISEP/IPP), Porto, Portugal
{fspsa,tmcfp,icp}@isep.ipp.pt
[2] BISITE – Research Centre, University of Salamanca, Salamanca, Spain
tpinto@usal.es
[3] Polytechnic of Porto (ISEP/IPP), Porto, Portugal
zav@isep.ipp.pt

Abstract. This paper proposes an adaptation of the Q-Learning reinforcement learning algorithm, for the identification of the most probable scenario that a player may face, under different contexts, when negotiating bilateral contracts. For that purpose, the proposed methodology is integrated in a Decision Support System that is capable to generate several different scenarios for each negotiation context. With this complement, the tool can also identify the most probable scenario for the identified negotiation context. A realistic case study is conducted, based on real contracts data, which confirms the learning capabilities of the proposed methodology. It is possible to identify the most probable scenario for each context over the learned period. Nonetheless, the identified scenario might not always be the real negotiation scenario, given the variable nature of such negotiations. However, this work greatly reduces the frequency of such unexpected scenarios, contributing to a greater success of the supported player over time.

Keywords: Automated negotiation · Bilateral contracts ·
Decision Support System · Electricity Markets ·
Reinforcement learning algorithm

1 Introduction

The world is constantly changing. Nowadays, these changes happens a lot faster than before and the tendency is for this rate to keep increasing. This is mainly due to the easier access to information and increased interaction and change

---

This work has received funding from National Funds through FCT (Fundaçao da Ciencia e Tecnologia) under the project SPET – 29165, call SAICT 2017.

© Springer Nature Switzerland AG 2019
Á. Rocha et al. (Eds.): WorldCIST'19 2019, AISC 930, pp. 556–571, 2019.
https://doi.org/10.1007/978-3-030-16181-1_53

of knowledge between people. As the world always seeks equilibrium, when something changes, all other things will change too in order to re-establish the equilibrium.

The Electricity Markets (EMs) are not any different. They have been constantly changing to keep up with the society needs. However, sometimes the change takes too long to take place and when it happens, it requires more profound changes. That is what happened to the EMs back in 2000 [1]. Those changes allowed the sector liberalization, introducing free competition in its various segments such as production, transportation and energy distribution. Nowadays there is another key change in the EMs paradigm that needs to be addressed, which is the increased use of energy from renewable sources. The use of this source is being highly encouraged with the aim of adopting a more sustainable growth (by reducing CO_2 emissions), as well as archiving energy independence [2]. The "20-20-20" program [3], which has been introduced by the European Union, is a good example of such encouragements. The mentioned program sets ambitious energy goals to be met by 2020, contributing to the large scale implementation of distributed generation.

While the evolution of EMs contributed to keep their stability and address society needs, it also brought new challenges for the participating entities. The restructuring of the sector introduced new entities with complex interactions that increased its unpredictability. That unpredictability only got bigger with the introduction of renewable energy sources, due to their intermittent nature. Consequently, the participating entities face higher risks and a lot more variables, increasing the importance and impact of decision-making. This way, the participating entities in EMs need proper tools to keep up with the changes, increasing their knowledge and improving their participation. The literature presents several simulators with focus in modelling EMs [4]. However, they are mostly focused on auction-based market models such as Day-ahead spot and Intra-day, overlooking Bilateral Contracts model (negotiation between players). The process of negotiation itself is a subject, common to several different domains, that has been widely explored. A relevant review in automated negotiation identifies the main phases of negotiation and exposes the features that are partially or completely missing in current models [5]. One of the identified gaps is the poor opponents analysis in the pre-negotiation phase, where the negotiator has to define its objectives and the opponent(s) to trade with. In EMs domain, players can establish bilateral contracts with several different players and the selection of the right opponent(s), according to its objectives, can have a great impact in the negotiation outcome. For this purpose, the negotiator needs to be able to identify the most probable negotiation scenario that it can face, among all the different scenarios that can occur in its current negotiation context. Each scenario is composed by the expected prices that each opponent may offer for each power amount.

To address the identified gap, this work presents a methodology, based on a reinforcement learning algorithm (Q-Learning [6]), to determine the most probable scenario that the supported player can face, in a future negotiation. The methodology is integrated in a Decision Support System (DSS) for the pre-negotiation of bilateral contracts, which is able to analyse the opponents and generate multiple negotiation scenarios.

2 Bilateral Contracts in Electricity Markets

The EMs are usually composed of several market types [7,8], based on several different models such as: day-ahead spot; intra-day, both usually auction based; and bilateral contracts.

In the scope of EMs, bilateral contracts are long-term contracts established between two entities, buyer and seller, for energy transaction, without the involvement of a third entity. The transaction is usually carried out several weeks or months after the contract is made [9] and usually has the following specifications: start and end dates and times; Price per hour (€/MWh) and amount of energy (MW), variable throughout the contract and, finally, a range of hours relative to the delivery of the contract. Players can use customized long-term contracts, trading "over the counter" and electronic trading to conduct bilateral transactions [10]. In MIBEL, there are four types of bilateral contracts: the first type are Forward Contracts, that consist in energy exchange between a buyer and a seller for a future date, for the price negotiated at that moment; the second type are Future Contracts, which are similar to Forward Contracts except that they are managed by a third party responsible for ensuring compliance with the agreement; the third type are Option Contracts, that are similar to the Forward and Future contracts with the difference that the two entities only guarantee a buy/sell option; the last one are Contracts for Difference, that allows concerned entities to protect themselves from the energy price change between the agreement establishment date and the agreed exchange date.

With the exception of Contracts for Difference, this type of negotiation allows players to control the price at which they will transact energy, in contrast to what happens in spot markets, due to the proposals' instability. In establishing a Forward or Future contract, players are committing themselves to transact energy for a given price at a future time, with the risk of making a transaction at a lower price than the expected and lose competitive power. Option Contracts or Contracts for Difference can avoid this risk. The first allows the player to choose not to go through with the exchange while the second ensures that the transaction is carried out at the market price. However, the first option also has the risk of not guaranteeing whether or not the other party will exercise their option to exchange and the second option does not allow better prices than the market. This way, it is possible to understand the risk associated with the negotiation of bilateral contracts and the need that players have of tools that help them reduce this risk and even optimize their profits.

Automated Negotiation

The process of negotiation itself has been widely explored in the literature of several different domains such as social psychology [11], economics and management science [12], international relations [13] and artificial intelligence [14,15]. The last one, artificial intelligence, is the most related area with the present work. As this area itself is also very rich in research about the process of negotiation, it has motivated the conduction of a very thorough review [5].

In the review, the authors present the state-of-art of the existing negotiation models and, as result of their study, they are able to present the most common phases of automated negotiation for computational agents: (I) Preliminaries, (II) Pre-Negotiation, (III) Actual Negotiation and (IV) Renegotiation. However, it is important to note that the Preliminaries and Pre-Negotiation phases are often joined together as the Preliminaries can be considered as part of the Pre-Negotiation phase. Despite being a common phase, the Pre-Negotiation is often very simple, not exploring its full potential. The last phase, Renegotiation, is also not present in all models, as some do not allow the final agreement modification. This way, it is possible to verify that the main focus of the existing models is the Actual Negotiation phase. However, the other phases are also important and can have a great impact in the negotiation process.

Decision Support Systems for the Negotiation of Bilateral Contracts
Some DSS for the negotiation of bilateral contracts can be found in the literature, such as Electric Market Complex Adaptive Systems (EMCAS) [16], General Environment for Negotiation with Intelligent Multi-purpose Usage Simulation (GENIUS) [17], and Multi-Agent Negotiation and Risk Management in Electricity Markets (MAN-REM) [18].

EMCAS [16] is a multi-agent simulator that aims to simulate various EMs market models, including Bilateral Contracts. The tool considers the objectives of all the participating players. The players can be either demand or generation company agents. The demand agents formulate their proposals and then each generation agents decide the price for the amount of power they want to sell. At last, the demand agent decides to accept or reject the generation agent conditions.

GENIUS [17] is a multi-agent simulator with the main focus of facilitating and evaluating automated negotiators strategies. The tool main features are: bilateral and multilateral negotiations; agent-to-agent and human-to-agent negotiations; domain independent; negotiators performance analysis, including comparison between results and optimal solution. The negotiation process follows three phases: Preparation, when the agents, protocol and domain of the negotiation are defined; Negotiation, when the actual negotiation occurs; and Post-negotiation, when the negotiation is analysed in detail.

MAN-REM [18] is a framework that combines small multi-agent EMs simulators for the simulation of bilateral contracts. In the simulations, the framework models two agents, besides the expected Seller and Buyer agents, which are: the Trader agent, which distributes the energy; and the Market Operator agent, which validates the contracts. The negotiation process follows three phases: Pre-Negotiation phase, when the proposer agent defines its contract preferences and its response to counter-offers; Actual Negotiation phase, when the Buyer and Seller agents trade offers; and Post-Negotiation phase, when the two entities reach an agreement.

The presented tools are mainly focused in the actual negotiation, being in accordance with the analysis of the previous subsection. Regarding Pre-Negotiation phase, EMCAS has a basic approach while the others explore it further, specially GENIUS. However, these tools are not capable to address the identified gap of poor opponents analysis in the Pre-Negotiation phase which does not allow the identification of the most probable negotiation scenario.

3   Proposed Methodology

Reinforcement Learning Approach

This paper proposes the adaptation of Q-Learning [6], a reinforcement learning algorithm, for the identification of the most probable negotiation scenario that the supported player can face, under a certain context. A negotiation scenario is composed by the expected prices that each opponent may offer for each power amount.

The Q-Learning is a very popular reinforcement learning algorithm. The concept of this algorithm is that an agent can take an action, from a set of possible actions, in each state that it can be. The aim of the algorithm is to help the agent identify the best action to take in each state. For that purpose, every state-action pair have an Q value that represents the utility of taking that action in that state. The agent will always choose the action with the highest Q value. Being a reinforcement learning method, Q-Learning is able to update the Q value of each state-action pair. Every time the agent repeats an action in a given state, the Q value of that state-action pair gets a reward r. The algorithm contains two variables related to the future learning: Learning Rate (α), which defines the contribution of the reward to the previous Q value; and Discount Factor (γ), which defines if the algorithm should only consider the current reward or look forward for highest rewards in the future.

The proposed methodology is an adaptation of Q-Learning that, instead of evaluating the best action for each state, evaluates the closest negotiation scenario to the one that the player will face in reality, under a certain context. In this adaptation, the agent is the supported player, the states are each different context c, in which the player may trade, and the actions are each scenario s that the player may face. The Eq. 1 presents the mapping performed by the Q function.

$$Q : c \times s \to U \tag{1}$$

The U is the expected utility value, which represents how close the scenario s is to the real negotiation scenario, that the supporter player may face under context c. The utility value (Q value) of each context-scenario gets rewarded once the information about the real scenario is available. This way, it is possible to evaluate how close each scenario was to the real negotiation scenario. The reward r is defined in Eq. 2.

$$r_{s,c,t} = 1 - norm|RP_{c,t,a,p} - EP_{s,c,t,a,p}| \tag{2}$$

The $RP_{c,t,a,p}$ is the real price negotiated by opponent p, in contract c, at the time t, for the amount of power a. The $EP_{s,c,t,a,p}$ is the price expected in scenario s under context c for the same opponent, amount of power and time.

To simplify the analysis of each Q value, the r values are normalized, in a scale between 0 and 1, to keep the Q value inside that interval. This way, the Q value resulting from the $Q(c, s)$ function, can be interpreted as the probability of occurrence of the scenario s under context c. After the calculation of the reward of a context-scenario pair, its current Q value gets updated by following Eq. 3.

$$Q_{t+1}(c_t, s_t) = Q_t(c_t, s_t) + \alpha(c_t, s_t)$$
$$\times [r_{s,c,t} + \gamma U_t(c_{t+1}) - Q_t(c_t, s_t)] \tag{3}$$

The α is the learning rate, γ is the discount factor and $U_t(c_{t+1})$ is the maximum utility expected in all of the scenarios of c_{t+1}, as represented by Eq. 4.

$$U_t(c_{t+1}) = \max_s Q(c_{t+1}, s) \tag{4}$$

Every time the Q value of a context-scenario pair is updated, its value is normalized, in a scale between 0 and 1, to represent the proximity of the given scenario to the actual negotiation scenario, under the given context. This normalization is represented in Eq. 5.

$$Q'(c, s) = \frac{Q(c, s)}{\max[Q(c, s)]} \tag{5}$$

This way, the scenario of each context with the highest Q value will have the value 1, after normalization, as it is the closest scenario to the actual negotiation scenario under the same context.

The adapted Q Learning execution is presented in Algorithm 1.

Algorithm 1. Adapted Q-Learning Execution

---

initialize $Q(c, s) \leftarrow 0$
repeat
    wait for new event $\triangleright$ (new established contract)
    for all scenarios of current contract **do**
        calculate reward $\triangleright$ (equation 2)
        update $Q(c, s)$ $\triangleright$ (equation 3)
    end for
    normalize $Q(c, s)$ $\triangleright$ (equation 5)
until stopping criteria

Decision Support System

The proposed methodology presents itself as a good solution for the determination of the most probable scenario that the supported player may face in a future negotiation. However, this methodology requires a tool that is capable to detect different negotiation contexts and generate alternative negotiation scenarios per context. The proposed methodology has been included in a DSS, whose architecture is presented in Fig. 1, which meets the identified prerequisites.

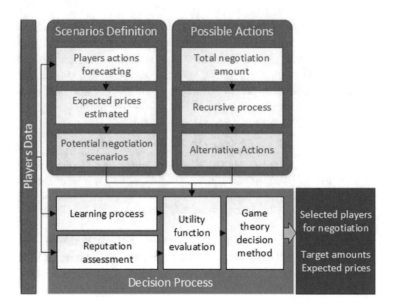

Fig. 1. Architecture of the DSS

As it can be seen in Fig. 1, the DSS is composed by three main components: Scenarios Definition, Possible Actions and Decision Process.

The Scenarios Definition is responsible for the generation of several alternative scenarios for each different context. A scenario is composed by the expected price that each one of the possible opponents find acceptable for the negotiation of different amounts of power (from the minimum negotiable amount to the amount that the supported player wants to trade). The expected price is obtained through forecasting, considering the historical data of the possible opponents. However, it is not always possible to forecast all the power amounts as the historical data might not have enough information about those quantities. In this cases, an estimation is performed instead, based on the prices that were possible to forecast. Each scenario uses a different forecast algorithm to make it possible to generate different scenarios.

After the Scenarios Definition, the generation of all Possible Actions, that the supported player can take, is performed. For this purpose, it is generated all the possible distributions, of the desired power amount to trade, among all the possible opponents. Each possible distribution is one action that the supported player can take. The actions can range from trading all the desired power amount, with only one of the possible opponents, to trade with all of the possible opponents, an equally split amount. Then, to evaluate each action individually and make it possible to select the best action of each scenario, the utility of each action is calculated. The utility of each action takes into account how profitable the action can be (taking into account the expected prices) and the reputation of the involved opponents. The impact of each component depends on the risk that the supported players is willing to face. The lower the impact of the reputation component, the higher the risk. Each action has an utility value per each possible scenario that the supported player can face under the current context.

The generation of all possible actions per each possible scenario creates the need of a decision method that enables the selection of the best action that supported player can take. The third component, Decision Process, is responsible for that final decision. For that matter, the system contains three different decision methods: Optimistic, Pessimistic and Most Probable (made possible by the proposed methodology). The optimistic method chooses the action with the highest utility value among all the scenarios. On the other hand, the pessimistic method, follows the mini-max game theory approach, where the selected action is the one with the highest utility value of the scenario with the lowest global utility (sum of the utility of all the actions of a scenario). At last, the contribution of the proposed methodology to this DSS: the Most Probable decision method, which selects the action with the highest utility, of the scenario that is most probable to occur in reality.

As result of the simulation, the supported player obtains the opponent(s) to trade with and how much power to trade with each one, considering the amount of power that it wants to trade, the list of possible opponents, the risk that it is willing to take and the preferred decision method.

4   Experimental Findings

This section presents a case study that has been conducted to test the proposed methodology as well as the impact of its integration in the presented DSS. For that purpose, the DSS will be run to aid the supported player in the scenario presented in Table 1.

Table 1. Case study scenario

Power amount	40
Transaction type	Purchase
Context	Weekday
Possible opponents	5
Reputation calculation	50% Personal Opinion and 50% Social Opinion
Risk	50% *(The economical and reputation components have the same weight)*
Decision method	Most Probable *(Proposed Methodology)*
Possible actions	135 751

The DSS provides, in this scenario, two different contexts: Weekday and Weekend. For each context, the DSS will generate five different scenarios, provided by three different methods: Artificial Neural Network (ANN),

Support Vector Machine (SVM) and Average. The last two consider the last 1000 contracts in their training while ANN has three different methods: ANN1 (500 contracts), ANN2 (1000 contracts) and ANN3 (1500 contracts).

This is a simple amount of scenarios that are capable to test the proposed methodology without being too complex to analyse. However, the DSS is capable to execute a much higher number of scenarios, which makes it possible to use other forecasting techniques and test different configurations (as exemplified with the ANN). Three different versions of the ANN algorithm will be used, where the only difference is the amount of contracts used in its training.

Besides the presented DSS internal definitions for this case study (contexts and scenario generation methods), there is also another very important definition: the historical data of the possible opponents. The data of the five possible opponents could be generated and still be able to test which scenario is capable to better represent the generated negotiation scenarios. However, the optimal test can only be performed with real data. Therefore, a real dataset is used instead, which contains executed physical bilateral contracts declared in the Spanish System Operator (SO) [19]. The dataset data ranges from 1 July 2007 to 31 October 2008 (16 months/488 days) and each day contains 24 negotiation periods (one per hour), in a total of 11 712 periods. The negotiations were performed by 132 different players (88 Buyers and 44 Sellers) which established 1 797 996 contracts. The Table 2 presents a detailed overview of the dataset.

Table 2. Dataset overview

	MIN	AVG	STDEV	MAX
Contracts/Period	128	157	17,78	180
Contracts/Day	147	3 753	485,78	4 287
Contracts/Player	2	27 244	58 653,22	288 160
Contracts/Player/Period	1	5	6,83	29
Power/Period/Contract	1	69,04	6,25	3 575
Power/Player/Contract	1	89,05	223,17	3 575
Power/Period	7 718	10 813	1 346,38	14 128
Power/Day	8 210	258 405,89	34 317,46	316 801
Power/Player	30	1 875 400,33	4 503 101,94	26 081 833

However, as it can be seen in Table 2, there is not information about contract prices, because the dataset only contains the traded power amount. The established price of a contract is a key information and the involved entities avoid sharing it. The share of such information can negatively affect their future negotiations. This way, there is the need to generate a price for each contract present in the dataset.

To guarantee an increased realism, the contracts price are generated taking into account the market price of the same negotiation period of the Spanish Day-ahead Market [20]. Nevertheless, the contracts established in the same negotiation period, between different players, can not have the same price. It would not make sense. Therefore, each player can have one of five different negotiation profiles: Profile 1, in which the player defines a minimum price, based on the market price, and keeps increasing it according to the power amount increase, until reaching a maximum price; Profile 2, which is similar to Profile 1 but with an higher minimum and lower maximum prices; Profile 3, which follows market price; and Profiles 4 and 5 which are the reverse of Profiles 2 and 1, respectively, in which the price keeps decreasing according to the decrease in the traded power amount.

Besides the contracts price definition, the dataset also needs another complement, to make it possible to use the full capabilities of the DSS: the reputation assessment. The DSS is capable of calculating the reputation of each opponent, based on various components, but it requires the personal opinion of each player about the other players. Therefore, the personal opinions of each player have been defined taking into account three components: personal experience with the evaluated player; number of opponents that traded with that player; and number of contracts established by the player. Then, there is also the need to define groups of players, as the DSS uses that information in the social component of its reputation assessment process. The players are divided into four groups according to their average traded power amount.

After the analysis of the dataset, there is the need to select: (I) the period of time that the proposed methodology will start learning; (II) the negotiation period when the supported player will attempt to trade; (III) the five possible opponents, which the supported player may trade with; and (IV) the supported player. First, as the EM will consider the contexts Weekday and Weekend, a good starting point is the most recent Sunday in the dataset. This way, the proposed methodology can learn both contexts. It also allows the supported player to attempt to negotiate on the next Tuesday, in the Weekday context, as specified in this case study. The last day of the dataset (31 October 2008) is a Friday, therefore the chosen date is 26 October 2008, the previous Sunday. This way, the proposed methodology will be able to learn during 48 negotiation periods, evaluating more than 2000 contracts per context. Second, the negotiation period of the supported player is the first period of 28 October 2008, the following Tuesday. Third, the possible opponents selection is made by identifying the players with more contracts in the learning period and with power amounts ranging from 1 to 40 (amount of power that supported player wants to trade). Therefore, the following players are selected: Player 1 (Profile 2, Group 3); Player 2 (Profile 3, Group 4); Player 3 (Profile 1, Group 2); Player 4 (Profile 4, Group 4); and Player 5 (Profile 3, Group 3).

At last, the selected player to support is a buyer that established contracts with each one of the possible opponents in the learning period.

After the preparation of the case study, the DSS used the proposed methodology to learn what is the most probable scenario, during the selected period (26 and 27 October 2008). For that purpose, the proposed methodology is run with a Learning Rate of 0.3, allowing a slow learning, and Discount Factor of 0.8, favouring future rewards, considering the available amount of data. The final Q values for each scenario under each context are presented in Table 3.

Table 3. Final results (Q) of the learning process

Context	ANN1	ANN2	ANN3	SVM	Average	Contracts
Weekday	4,862	4,858	4,845	**4,883**	4,762	2 714
Weekend	**4,902**	4,876	4,713	4,801	4,585	2 671

As it can be observed in Table 3, the proposed methodology learned over 2 500 contracts for each context. At the end of the learning process, the most probable scenario in a Weekday is SVM and, in a Weekend, is ANN1. The results shows that, in both contexts, the ANN results improve with the reduction of the number of contracts considered in its training. The ANN1, which has the lower number of contracts (500), presents better results than ANN2 (1000) and even better than ANN3 (1500). However, ANN1 is slightly surpassed by SVM (1000) in the Weekday. On the other hand, the ANN1 and ANN2 are better than SVM in the Week, with an higher distance. The ANN proves to be a better algorithm overall, when considering the same amount of contracts, and even better with a lower number. The Average method presents the worst results in both contexts, representing the great uncertainty present in the real negotiation scenarios, where the players keep changing their behaviours. The Q values for each scenario under each context are very close, ranging from 4.585 to 4.902, which is caused by the converging nature of Q-Learning, over its iterations.

The Figs. 2 and 3 presents the normalized Q value of each scenario under Weekday and Weekend contexts respectively, over all the analysed contracts.

The Fig. 2 proves that SVM scenario is really dominant, being the scenario with the maximum Q value during more contracts. However, it is not always the most probable scenario. In fact, the SVM scenario were very far from reality in the first 708 contracts, the period in which ANN1 dominated [102, 707], after the initial success of the Average scenario [1, 101]. Then, the SVM scenario is only surpassed by ANN1 [1314, 1516], and ANN2 [2122, 2425]. As seen in Table 3, the success of each ANN method is measured by the amount of contracts considered. The fewer the number of contracts, the better the results. The Average scenario only had success in the beginning of the learning process, as it is a simple average, which does not requires much learning to know its potential, contrary to the other scenarios. Having seen the learning process of the Weekday context, it may be interesting to see how it compares to a different context, which in this case is the Weekend (Fig. 3).

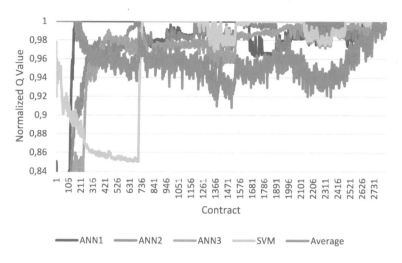

Fig. 2. The learning process for the Weekday context

It is visible in Fig. 3 that SVM scenario does not have as much success as the one presented in the Weekday context. Nonetheless it still presents good results, being the second most successful. The Table 3 shows that ANN1 finished the learning process as the most probable scenario. However, it was not just a momentary achievement since it was the most dominant scenario. The Average scenario only had success in the first 200 contracts, like has been verified in the Weekday context. The ANN2 only had a few successful periods but it is the most regular scenario, after the ANN1.

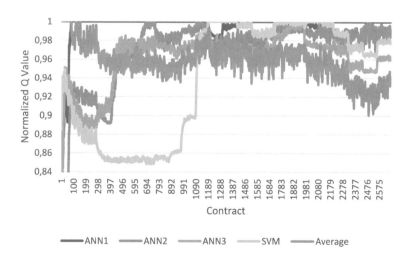

Fig. 3. The learning process for the Weekend context

After the learning process, the DSS is ready to provide the supported player with the best action to take under the most probable scenario under Weekday context (SVM). The Table 4 presents the best action of SVM scenario as well as a comparison with the best action of the other scenarios.

Table 4. Best action per scenario

Scenario	Player 1	Player 2	Player 3	Player 4	Player 5	Total price
ANN1	**1**	**37**	**1**		**1**	1809,08
	36,87	45,88	32,61		42,07	
ANN2	**38**		**1**		**1**	2556,80
	65,08		39,45		44,21	
ANN3	**1**		**39**			2492,46
	40,56		62,87			
SVM	**1**	**37**	**1**		**1**	**1935,06**
	40,60	49,03	35,34		44,88	
Average	**40**					1199,01
	29,98					

It is possible to observe in Table 4 that the most probable scenario (SVM) and the second most successful scenario (ANN1) identified the same best action, with a slight price variance. The ANN1 presented a less expensive action (about 9% less). However, the less expensive action was identified by the Average scenario while ANN2 and ANN3 present the most expensive ones. If the supported player selected the Optimistic decision method, instead of the Most Probable, the selected action would be the Average scenario one as it present the highest utility (which in this case is the highest profit, as all the possible opponents have similar reputations - about 0.62). On the other hand, if the Pessimistic decision method were selected, the best Action would be the one of ANN3 scenario, as it is the scenario with the lowest global utility (ANN3: 39 905.94; ANN2: 41 521.69; ANN1: 52 674.94; SVM: 60 689.93; Average: 73 904.44).

The supported player could take three different actions depending on the decision method but which one will be the closest to reality? Will SVM confirm itself as the most probable? After learning the real results of the negotiation period, for which supported player required decision support, an unexpected result is observed. The Average scenario presented the lowest average error per contract (Average: 9.31%; ANN1: 12.38%; ANN2: 16.55%; SVM: 18.30%; ANN3: 18.57%). Consequently, after learning from the real contracts, the ANN1 and

Average scenarios surpassed the Q value of SVM scenario (ANN1: 4.920; Average: 4.784; SVM: 4.780; ANN2: 4.777; ANN3: 4.767). The improvement of the Average scenario shows that the participating players have been more regular, being closer to their average behaviour. The ANN1 is once again the leading scenario resultant of a slow, but constant, growing verified in the most recent negotiation periods.

This case study scenario could not detect the most probable scenario for the negotiation period in matter. However it could do it for most of the preceding negotiation periods. The reason is that the chosen negotiation period coincided with a turning point in the learning process. The ANN1 and Average scenarios were slowly improving while SVM was slowly decaying. As seen in this case study, there is always room for the unexpected to happen but that is exactly why the proposed methodology arose: to reduce the frequency of such situations as much as possible.

5   Conclusion

Nowadays, the EMs are constantly facing new challenges which result in constant changes and, consequently, an increased complexity for the involved entities. There is a growing need of tools that are capable to ease the experience of those entities, providing them with a better insight of what is going on as well as supporting them in their decisions. Various tools have emerged in the literature but they do not cover all of the current needs.

This paper identifies the lack of decision support for the pre-negotiation phase of bilateral contracts negotiation in the EMs. Although there are some tools, as analysed in this work, they do not address one of the key aspects of the pre-negotiation phase: the possible opponents analysis. For that purpose, a DSS has been developed which allows a good analysis of the possible opponents that the supported player may trade with. However, with the use of such tool, another problem arises: it is capable to generate various alternative negotiation scenarios under different contexts but it does not know which one is the most probable to occur in reality, under each context.

An adaptation of the Q-Learning algorithm is proposed to address that problem. Through its use, it is possible to learn through time, what is the most probable scenario to occur under a given context. The methodology confronts the generated scenarios with the real negotiation scenarios, being able to update their probability of occurrence.

By executing the proposed methodology in the presented case study, it can be concluded that it fulfils its purpose. By analysing real contracts under two different contexts (Weekday and Weekend), it was capable to determine which of the five scenarios had the highest probability of occurrence for each context, over the simulated period. The most probable scenario for each context kept changing over time and the scenarios had different results according to the context in which they were inserted. This way, it is possible to verify the importance of the presented methodology. There is not a scenario that will be always the

F. Silva et al.

most probable and it will also always depend on the context. The only way to guarantee a good selection is to keep learning through time, taking into account previous information without underestimating the new information.

References

1. Shahidehpour, M., Yamin, H., Li, Z.: Market operations in electric power systems: forecasting, scheduling, and risk management. Institute of Electrical and Electronics Engineers. Wiley, New York (2002)
2. Wüstenhagen, R., Menichetti, E.: Strategic choices for renewable energy investment: conceptual framework and opportunities for further research. Energy Policy **40**, 1–10 (2012)
3. European Commission: "The 2020 climate and energy package" (2009)
4. Ringler, P., Keles, D., Fichtner, W.: Agent-based modelling and simulation of smart electricity grids and markets - a literature review. Renew. Sustain. Energy Rev. **57**, 205–215 (2016)
5. Lopes, F., Wooldridge, M., Novais, A.Q.: Negotiation among autonomous computational agents: principles, analysis and challenges. Artif. Intell. Rev. **29**(1), 1–44 (2008)
6. Rahimi-Kian, A., Sadeghi, B., Thomas, R.: Q-learning based supplier-agents for electricity markets. In: IEEE Power Engineering Society General Meeting, pp. 2116–2123. IEEE (2005)
7. Silva, F., Teixeira, B., Pinto, T., Santos, G., Vale, Z., Praça, I.: Generation of realistic scenarios for multi-agent simulation of electricity markets. Energy **116**, 128–139 (2016)
8. Sheblé, G.B.: Computational Auction Mechanisms for Restructured Power Industry Operation. Springer, Boston (1999)
9. Algarvio, H., Lopes, F.: Risk Management and Bilateral Contracts in Multi-agent Electricity Markets, pp. 297–308. Springer, Cham (2014)
10. Kirschen, D.S., Strbac, G.: Fundamentals of Power System Economics. Wiley (2004)
11. Thompson, L.: Mind and Heart of the Negotiator, 2nd edn. Prentice Hall Press, Upper Saddle River (2000)
12. Snyder, G.H., Diesing, P.: Conflict Among Nations: Bargaining, Decision Making, and System Structure in International Crises. Princeton University Press (1977)
13. Jennings, N.R., Faratin, P., Lomuscio, A.R., Parsons, S., Wooldridge, M., Sierra, C.: Automated negotiation : prospects, methods and challenges. Group Decis. Negot. **10**(2), 199–215 (2001)
14. Rahwan, I., Ramchurn, S.D., Jennings, N.R., Mcburney, P., Parsons, S., Sonenberg, L.: Argumentation-based negotiation. Knowl. Eng. Rev. **18**(4), 343–375 (2003)
15. Veselka, T., Boyd, G., Conzelmann, G., Koritarov, V., Macal, C., North, M., Schoepfle, B., Thimmapuram, P.: Simulating the Behavior of Electricity Markets with an Agent-based Methodology: the Electric Market Complex Adaptive Systems (EMCAS) Model (2002)
16. Koritarov, V.: Real-world market representation with agents. IEEE Power Energ. Mag. **2**, 39–46 (2004)
17. Lin, R., Kraus, S., Baarslag, T., Tykhonov, D., Hindriks, K., Jonker, C.M.: GENIUS: an integrated environment for supporting the design of generic automated negotiators. Comput. Intell. **30**, 48–70 (2014)

18. Lopes, F., Rodrigues, T., Sousa, J.: Negotiating bilateral contracts in a multi-agent electricity market: a case study. In: 2012 23rd International Workshop on Database and Expert Systems Applications, pp. 326–330 (2012)
19. OMIE, "ejecucioncbfom" (2017). http://www.omie.es/aplicaciones/datosftp/datosftp.jsp?path=/ejecucioncbfom/. Accessed 18 Oct 2017
20. OMIE, "marginalpdbc" (2017). http://www.omie.es/aplicaciones/datosftp/datosftp.jsp?path=/marginalpdbc/. Accessed 18 Oct 2017

Organizational Models and Information Systems

Using the Technology Acceptance Model (TAM) in SAP Fiori

Daniela Beselga[1(✉)] and Bráulio Alturas[2]

[1] Instituto Universitário de Lisboa (ISCTE-IUL),
University Institute of Lisbon, Av. Forças Armadas, 1649-026 Lisbon, Portugal
dfpalminha@gmail.com
[2] Instituto Universitário de Lisboa (ISCTE-IUL), ISTAR-IUL,
University Institute of Lisbon, Av. Forças Armadas, 1649-026 Lisbon, Portugal
braulio.alturas@iscte-iul.pt

Abstract. This article presents a case study that was carried out in two companies that have implemented SAP Fiori. The As-Is and To-Be description of the process in which SAP Fiori was implemented was performed. The advantages and disadvantages of using SAP Fiori were also identified. The Technology Acceptance Model (TAM) has been used in order to understand the aspects that most influence users to consider SAP Fiori as an added value, and how it optimizes the tasks of users. TAM has two variables that will influence the acceptance of a technology, which are: perceived ease of use and perceived utility.

Keywords: ERP · SAP Fiori · TAM · Mobile · Technology

1 Introduction

Nowadays Technologies and Information Systems have a key role, improving the competitiveness of a company. The efficient management of a company is essential and since there are several software packages for this purpose, where Enterprise Resource Planning (ERP) is one of the most recognized because of its potential to promote more efficiency in decision-making. One of its characteristics is the ability to automate and integrate the business processes of organizations [1].

An ERP system is a technology infrastructure that can assist a company in integrating information from all internal departments with suppliers and customers. It links all areas of a company's internal functions and processes with the external ones in order to create a close relationship between customers and suppliers. ERP also allows information to be shared between different partners, supports the effectiveness of the supply chain management, and improves the flow of information. These should enable managers to make better decisions based on more accurate and up-to-date information [2].

To improve the efficiency and effectiveness of ERP system use, organizations need to research the factors that impact user satisfaction. In this area, the Technology Acceptance Model (TAM) is one of the most widely used models for explaining the behavioral intention and actual usage and can improve our understanding of how

© Springer Nature Switzerland AG 2019
Á. Rocha et al. (Eds.): WorldCIST'19 2019, AISC 930, pp. 575–584, 2019.
https://doi.org/10.1007/978-3-030-16181-1_54

influence on actual usage could help increase efficiency and effectiveness of ERP system use.

Using ERP reduces significantly the time needed to complete a business process and helps sharing the information in the organization. The work environment for the employees is better when an ERP is implemented and more efficiency [3].

Martins & Alturas (2016) concluded that although an ERP is a representative investment in costs and time for companies, it has benefits in several perspectives, in the short and long term and in external and internal relations to the company [4].Thus, the ERP is a useful tool that companies use to improve Performance, make the best decisions and achieve competitive advantage.

This study intends to answer the research question: "What are the main aspects in which SAP Fiori is accepted as an added value by users?"

To answer this question, five objectives have been defined:

1. To understand and analyze the functionality of the application;
2. Assessing the advantages and disadvantages;
3. To understand if some of the daily tasks of users have become more effective with the SAP Fiori tool;
4. To understand if certain tasks of a user are carried out through the application and the limitations of the same;
5. Verify the acceptance of SAP Fiori application by applying the TAM model.

2   Literature Review

2.1   ERP

The ERP system is defined as a large applicational software package which helps to solve the fragmentation information issues in the organization. The ERP enables an automatic integration of the business key-process, providing a real time information of these process therefore, is possible to link the information between departments, sharing data and organization best practices [5, 6].

The ERP system was introduced by ERP providers, such as SAP (Systeme, Anwendungen, Produkt in der Datenverarbeitung), Oracle, PeopleSoft and others to replace old systems, providing integrated resources and a technology platform, that help companies to gain competitive advantage and allows to compete globally. To implement an ERP system is required to change the organizational behavior, takes a long time to be implemented and consumes a considerable amount of financial resources. Therefore, companies need to know clearly what type of ERP is necessary and how the system can affect the company before thinking about implementing the system [7].

2.2   SAP ERP

SAP is an abbreviation for the German company name "Systeme, Anwendungen, und Produkte in Datenverarbeitung". Nowadays, SAP leading ERP, Supply Chain

Management (SCM) e Customer Relationship Management (CRM). The main direct competitors are Oracle, Lawson, Infor, Sage, Microsoft Dynamics e NetSuite.

According to Sharma (2010), the SAP ERP can contribute for sales raising, increased productivity, reduction of purchase costs, reduction of inventories, reduction of assets, reduction of quality costs, elimination of physical inventory, improving cash flows and improving the productivity of indirect labor flows [8].

SAP R/3 consists of several modules and contains core features of a business. ERP is a system that integrates all the departments and functions of an organization into a single, central database. This allows to meet all the specific needs of each of the different sections of an organization (sales, billing, purchasing, warehouse management, production, maintenance, costs, among others).

2.3 Technology Acceptance Model (TAM)

The technology acceptance model was developed by Davis (1989), one of the main researchers in accepting the use of technologies. TAM has been widely developed and there has been an existence of different studies examining the acceptance behavior of the technology in different information systems.

TAM consists in two variants: perceived utility (PU) and perceived ease of use (PEOU) are the key to understand the computer acceptance behavior (Davis, 1989) PU is defined as "the degree to which a person believes that using a particular system would increase your work performance" [9] and PEOU is "the degree to which a person believes that using a particular system would be effortless" [9].

The two central hypotheses in TAM are PU and PEOU that influence positively attitude towards of using a new technology, which in influences their intention to use it. Finally, the intention is positively related to the actual use (see Fig. 1). TAM also believes that PEOU influence PU; as Davis explain in 1989: "the effort saved by improving perceived ease of use can be redistributed, allowing a person doing more work with the same effort" [9].

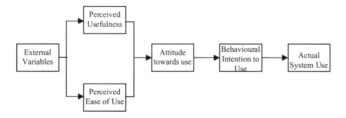

Fig. 1. Technology Acceptance Model (TAM) Davis (1989).

These two factors are influenced by variables. The main external factors that are usually manifested are social factors, cultural factors and political factors. Social factors include language, skills, and facilitating conditions. Political actors are mainly the impact of the use of technology on politics and political crises.

At the beginning of the 21st century and based on TAM, Venkatesh et al. (2003), studied previous models and theories and formed the Unified Theory of Acceptance and Use of Technology (UTAUT). The UTAUT has four predictors of the behavioral intention of the users, being: performance expectancy; effort expectancy; social influence and facilitating conditions [10].

2.4   SAP Fiori

The world is constantly changing, and technology is at the center of everything we do. Younger generations have been born in a generation of rapid evolution of technology, and Millennium generation are integrated from the workforce with the mindset that simplicity is the key. We are a society owner of smartphones that instantly allows us to access social networks. Consequently, end users are accustomed to a new and improved experienced user (EX). The business must quickly follow this new requirement and realize that enterprise software can't be an exception. The business was forced to change over time to keep on leading, and SAP emerged as a leader in this evolution.

SAP quickly realized that it was necessary to create a task-based application that would easily complete the job, be consistent with the offer, and intuitive to work on any device. In May 2013, SAP was present with the launch of SAP Fiori as the new user experience in SAP.

Fiori was based on a framework known as UI5 that was developed on top of HTLM5, which emphasizes the user interface UI designed with user-centric applications. SAP has created a responsive and responsive UI for all screen sizes and runs on any device with an HTML5 compatible browser.

SAP heard the feedback that from almost the world and developed a set of applications, which includes the most frequently used transactions such as: Approval of purchase orders; creation of sales orders; information research, among others. HTML5 is easy to access seamlessly across desktops; tablets and smartphones. This collection of apps is called Fiori. The term Fiori is derived from the Italian word "flowers". SAP wants a beautiful UI, and it also intends to create user experience that is simple and elegant.

The new UX for SAP simplifies the old GUI and makes it compatible with any device and any screen size that supports HTML5, including mobile devices. This change is characterized by a big step for SAP, which has adopted a new technology and entire strategy for the UI.

With SAP Fiori, the applications have been developed from a user perspective, applications are user-centric, and characterized by being simple and relevant to a user role and designed to perform certain activities and tasks. Basically, the Fiori is an application that replaces SAP standard transactions and turns them into multiple, easy-to-use small applications becoming more fitted to each individual user use (Fig. 2).

Fig. 2. SAP Fiori APP. SAP Product Road Map SAP User Interface Technologies (2015).

3 Methodology

After the theoretical framework of ERP systems, the Technology Acceptance Model (TAM) and SAP Fiori technology, it will be studied how the implementation of SAP Fiori is accepted or not by its users in accordance with TAM.

As previously mentioned, the research question of this study is: What are the mains aspects in which SAP Fiori is accepted as an added value by users?

This is an exploratory study, which intends to make a first approach to the problem. The research method adopted was the case study.

The research methodology adopted was qualitative. Based on the study of "Critical factors for successful implementation of enterprise systems" [11] and "Determinants of acceptance of ERP software training in business schools: Empirical investigation using UTAUT model" [12].

It was created an interview script and was applied to the two case studies "BeHealthy" and "ItSolutions". In the interview, questions were analyzed to characterize the sample (Gender, age, department, functions and academic background), and questions to verify the acceptance of SAP Fiori by users.

The answers to these questions were obtained through interviews. To perform the analysis of results, we used a qualitative analysis software: "QDA Miner". It was made the analysis of interviews with key users about the implementation of SAP Fiori in ItSolutions and BeHealthy.

For the interviews, we used structured interviews with the purpose of collecting the perceptions of the users about SAP Fiori and the acceptance of this technology.

Table 1. Departments description of the respondents.

Interviewed	Department
PC_1	Compras
PC_2	Compras
PC_3	Administração
ANALYST_1	IT
ANALYST_2	Comercial
ANALYST_3	Comercial

We conducted a total of six individual interviews (Table 1), three of them about the implementation of SAP Fiori in ItSolutions and remaining three to the BeHealthy company. The interviews were made during April and May 2018. The interviews conducted at Itsolutions were face-to-face and the ones made at BeHealthy were conducted by mobile phone. The average duration of each interview was about 15 min and all interviews were recorded in audio format.

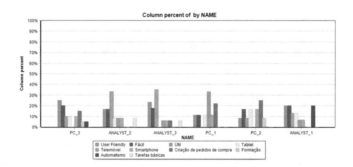

Fig. 3. Percentage of keywords referred by interviewed - Vertical bar chart Source: QDA Miner.

According to what is presented in the vertical bar graph (Fig. 3), the interviews were carried about the implementation of SAP Fiori (creation and approval of purchase orders) are: PC_1; PC_2 and PC_3. About 10% words that fell into the category of "user friendly", the word "useful" was mentioned in the 1st and 2nd interview about 10%, the word "easy" was also identified as being one of the words mentioned, about 16% in average of words mentioned. In addition to these, other words with expression, such as "creation of purchase orders", were also mentioned, about 20%, on average of the three interviews; finally, "mobile phone" and "tablet", with about 13%.

Regarding the interviews carried out of "HANALYST" project, there are: "ANALYST_1; ANALYST_2 and ANALYST_3".

The first interview (ANALYST_1) the words with the highest percentage of reference are: the word " automatism" with 20%, followed by the word "user friendly" with 18%, followed by the word "easy" either with 18%.

Therefore, the words that are included in the category of "user friendly", "useful", "easy", have great expression in the 2nd and 3rd interviews.

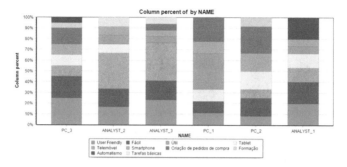

Fig. 4. Percentage of keywords referred by interviewee - Stacked bar chart Source: QDA Miner

According to the stacked bar graph (Fig. 4), the interviews conducted in the context of the creation and approval of purchase orders, in the 3rd interview (PC_3), the word user friendly was referred to about 25%; easy, 20%; useful, 10%; tablet, 10%; smartphone, 10%; creation of purchase orders, 15%; training, 5%; and finally, automatism was referred in 5%.

In the second interview under the "HANALYST" project (ANALYST_2), about 18% of the words mentioned were user friendly; 14%, easy; 31% useful; 7%, tablet; 7%, mobile phone; 8% smartphone; and lastly, 9% basic tasks.

In the 3rd interview of the project "HANALYST" (ANALYST_3) about 22% of the words were user friendly; 17%, easy; 27%, useful; 5%, mobile phone; 5%, smartphone; 5%, creation of purchase orders; and lastly, 5%, basic tasks.

In the first interview of the project to create and approve purchase orders (PC_1), about 11% of the words were user friendly; 10%, easy; 10%, tablet; 33%, mobile phone; 10%, smartphone; finally, 22%, creation of purchase orders.

In the second interview of the project to create and approve purchase orders (PC_2), about 8% of the words were: user friendly; 16%, easy; 10%, useful; 18%, tablet; 18%, smartphone; 23%, creation of purchase orders; finally, 9%, training.

In the third interview of HANALYST (ANALYST_3) about 20% of the words referred to was automatism; 8%, smartphone; 10%, mobile phone; 15%, tablet; 10%, mobile phone; 8%, smartphone; and finally, 20%, automatism.

4   Conclusions

This study intends to answer the initial question: "What are the main aspects in which SAP Fiori is accepted as an added value by users?"

The proposed objectives were all achieved, including:

1. To understand and analyze the functionality of the application;
2. Assessing the advantages and disadvantages;

3. To understand if some of the daily tasks of users have become more effective with the SAP Fiori tool;
4. To understand if certain tasks of a user are carried out through the application and the limitations of the same;
5. Verify the acceptance of SAP Fiori application by applying the TAM model.

All the objectives have been met, namely:

1. The main functionalities of the SAP Fiori APP

In the HANALYST implementation are:

(a) The identification of the necessary corrections to make possible migrate to a new database version;
(b) The identification of the effort required for this migration;
(c) Layout adaptable to any size of screen and mobile or fixed device;
(d) Detailed information on the problems to be solved in each object;
(e) Menu with options of easy access to different views, tables and graphs.

In the implementation of the creation and approval of purchase orders are:

(a) Creation and/or approval of purchase orders in a more intuitive way;
(b) Possibility to create and/or approve purchase orders through a mobile device;
(c) Reduction of training time for the creation and/or approval of purchase orders;
(d) Increase of the automation in the process of creation and/or approval of purchase orders.

2. Assess the advantages and disadvantages
   The main advantages have already been mentioned previously for both implementations of SAP Fiori. As main disadvantages, we haven't the possibility to do all tasks in Fiori, but only the most basic tasks are performed in SAP Fiori.
3. To understand if some of the daily tasks of users become more effective with the SAP Fiori tool
   As mentioned earlier, in the analysis of interviews, it is easier nowadays to perform tasks such as creating purchase orders and approving than before. Nowadays, with a simpler and intuitive layout, this task became easier.
   Also, in the case of HANALYST, nowadays it is easier from a commercial point of view, to present a client a report with all the necessary changes to the migration to a HANA database. With an easy graphical presentation and the possibility of this presentation being demonstrated by a mobile device.
4. To understand if certain tasks of a user are carried out through the application and the limitations of the same
   The process of creating and approving purchase orders started being made by SAP Fiori.
   However, in the case of migration to a HANA database, one of the limitations of SAP Fiori is the impossibility of changing the code, it must be modified in a classic version (SAP GUI).
5. Verify acceptance of the SAP Fiori application by applying the TAM model.

According to TAM, there are several factors that will influence the acceptance of a technology. According this model, the perceived utility and perceived ease of use will influence the attitude towards the acceptance of a technology and, later, the behavioral intention to use that same technology.

In the case of perceived ease of use, what contributed most was in both implementations of SAP Fiori, is to be: intuitive, simple and easy to use this application. After the implementation of SPA Fiori, the tasks that were previously performed with the classic version (SAP GUI), are now carried out in a simplest and quickest way.

In case of SAP Fiori the implementation for create and/or approve purchase orders, the most important factor was the fact that these two processes are simpler, faster and practical. These two tasks can be carried via mobile in any place.

In the case of the SAP Fiori HANALYST implementation, the usefulness of this new technology is that in nowadays is much easier and faster to obtain a list of all the modifications required for code correction when it is intended to migrate to a HANA database. The previously process was carried out much more slowly. With Fiori it was possible to realize how much time is spent on each activity needed to proceed with the correction of the code.

Having said this, there is a clear identification of utility and ease of use of the new technology, we can verify that the attitude towards use, and the behavioral intention of use, is effectively reflected in the use.

The main limitations to this work were: the fact that it was carried out with a reduced number of cases; The impossibility of a quantitative data processing due to the reduced number of cases is also identified as another of the limitations; The three interviews were made by phone, since the geographical distance made it impossible for them to be carried out in person.

The results of this work offer important implications as they can help managers better choose the ERP to implement in their company.

As a proposal of future work, it would be interesting to carry out the same study with more companies from different sectors; Realization of questionnaires instead of interviews, enabling a quantitative analysis; Validation of the TAM model in a more objective and not so generic way.

References

1. Nah, F., Lau, J., Kuang. J.: Critical factors for successful implementation of enterprise systems. Bus. Process Manage. J. **7**(3), 285–296 (2001)
2. Al-Mashari, M., Zairi, M.: Supply-chain re-engineering using enterprise resource planning (ERP) systems: an analysis of a SAP R/3 implementation case. Int. J. Phy. Distrib. Logistics Manage. **30**(3), 296–313 (2000)
3. Lee, H., Lee, M., Olson, L., Chung, H.: The effect of organizational support on ERP implementation, industrial. Manage. Data Syst. **110**(2), 269–283 (2010)
4. Martins, A.R., Alturas, B.: Organizational impact of implementing an ERP module in Portuguese SME. In: CISTI 2016 - 11ª Conferência Ibérica de Sistemas e Tecnologias de Informação, Gran Canaria, Espanha (2016)

5. Gefen, D., Pavlou, P.: The moderating role of perceived regulatory effectiveness of online marketplaces on the role of trust and risk on transaction intentions. In: Proceedings of the ICIS, Milwaukee, Wisconsin, USA (2006)
6. Khaparde, V.: Barriers of ERP while implementing ERP: a literature review. J. Mech. Civil Eng. **3**(6), 49–91 (2012)
7. Loonam, J., McDonagh, J.: Principles, Foundations, & Issues in Enterprise Systems, Dublin. Ideal Group Inc., Ireland (2005)
8. Sharma, K.: Selection, implementation & support of SAP ERP system approach in manufacturing industry. Glob. Digital Bus. Rev. **4**(1), 1931–8146 (2010)
9. Davis, F.D.: Perceived usefulness, perceived ease of use, and user acceptance of information technology. MIS Q. **13**(3), 319–340 (1989)
10. Venkatesh, V., Morris, G., Davis, B., Davis, D.: User acceptance of information technology: toward a unified view. MIS Q. **27**(3), 425–478 (2003)
11. Sternad, S., Bobek, S.: Impacts of TAM-based external factors on ERP acceptance. In: CENTERIS 2013 - Conference on ENTERprise Information Systems, Lisbon, Portugal (2013)
12. Chauhan, S., Mahadeo, J.: Determinants of acceptance of ERP software training in business schools: empirical investigation using UTAUT model. Int. J. Manage. Educ. **14**(3), 248–262 (2016)

Evaluation of Local E-government Maturity in the Lima Metropolitan Area

Gonçalo Paiva Dias[1(⊠)], Manuel Tupia[2],
and José Manuel Magallanes Reyes[2]

[1] ESTGA/GOVCOPP, Universidade de Aveiro,
Rua Comandante Pinho e Freitas, 28, 3750–127 Águeda, Portugal
gpd@ua.pt
[2] Pontifícia Universidad Católica del Peru,
Av. Universitaria 1801, San Miguel 15088, Peru
{tupia.mf,jmagallanes}@pucp.edu.pe

Abstract. This article presents a preliminary evaluation of e-government maturity for 37 district municipalities of the Lima Metropolitan Area, in Peru. A three-dimensional maturity model (with dimensions e-information, e-services; and e-participation) previously used in other studies was applied. Empirical data was obtained using Website content analysis. Descriptive statistics and clustering technics were used to analyze results. The results support the conclusions that the Lima municipalities tend to privilege simpler developments in detriment of more complex ones; that, consequently, the e-information dimension is more developed than the e-service dimension which, in turn, is more developed that the e-participation dimension; and that different profiles of municipalities coexist in what concerns the balance between the three different dimensions of the maturity model. These findings are consistent with other studies that used the same method in a different geography. The results also indicate the need to update the maturity model used and adapt it to the specific Peruvian case.

Keywords: E-government · Local government · Digital government · Peru

1 Introduction

According to the 2018 UN e-government survey [1], Peru is still in a medium level of development in what concerns e-government maturity. It ranks 77 among the 193 World countries and 13 in Latin America. This happens despite, as we will address later, the country having a fairly developed governance and regulatory framework for e-government. As we will also address in the text, most of the e-government services available to citizens and companies in the country correspond to municipal procedures. Thus, it is pertinent to study local e-government maturity in Peru, which is precisely the objective of the preliminary study presented in this article. To the best of our knowledge, this is the first study to address this subject.

© Springer Nature Switzerland AG 2019
Á. Rocha et al. (Eds.): WorldCIST'19 2019, AISC 930, pp. 585–594, 2019.
https://doi.org/10.1007/978-3-030-16181-1_55

Internationally, studies of local e-government maturity are fairly common (see, for example, [2–5]). The great majority of those studies use Website content analysis to collect empirical data and maturity models to subsequently categorize the observations. The same approach was used in this study to evaluate the district municipalities of the Lima Metropolitan Area.

The remaining of this article is organized as follows: in Sect. 2 we address the Peruvian e-government context; in Sect. 3 we address local government in Peru; in Sect. 4 we describe the methods used; in Sect. 5 we present and discuss the results; and in Sect. 6 we addresses the main conclusions and the future work.

2   The E-government Context in Peru

For the Peruvian State, e-government involves "the use of information and communications technology (ICT) to streamline the processes and services of public administration in order to promote the country's competitiveness and bring the State closer to citizens, gradually reducing the so-called digital divide" [6].

For approximately two decades, the central government of Peru has been establishing structures and systems for e-government that aim at implementing this vision. The responsibility to lead these efforts is committed to the National Office of Electronic Government and Information Technology (ONGEI), an agency that reports directly to the Presidency of the Council of Ministers (PCM). Formally, this agency is responsible for regulating, coordinating, integrating, and promoting the development of information technology in the Public Administration[1].

Peruvian e-government regulation is primarily based on the Law for the Modernization of State Management[2]. Further relevant regulations include the following:

- The resolution on the mandatory use of the Peruvian Technical Standard 'NTP-ISO/IEC 12207:2004 – Information Technology, Software Life Cycle Processes, 1st Edition' in the entities of the National Computing System[3];
- The Development Plan of the Information Society in Peru - The Peruvian Digital Agenda 2.0[4];
- The Law on Digital Signatures and Certificates[5];
- The Guidelines that establish the minimum content of the Strategic Plans of Electronic Government[6];
- The creation of the State Interoperability Platform (PIDE)[7];
- The National E-Government Policy 2013–2017[8];

---

[1] Supreme Decree No. 066-2003-PCM, Supreme Decree No. 067- 2003-PCM.
[2] Law No. 27658.
[3] Ministerial Resolution No. 179-2004-PCM,
[4] Approved by Supreme Decree No. 066-2011-PCM.
[5] Law No. 27269.
[6] Ministerial Resolution No. 61-2011-PCM.
[7] Supreme Decree No. 083-2011-PCM.
[8] Supreme Decree No. 081-2013-PCM.

- The National Electronic Government Plan [6];
- The memorandum recommending the application and use of the Peruvian Technical Standard 'NTP-ISO/IEC 27001:2014 – Information Technology, Security techniques, Information security management systems, Requirements, 2nd Edition' in all the entities of the National Computing System[9];
- The Peruvian Technical Standard NTP-ISO 37001: 2017 on Anti-bribery management systems[10];
- The Digital Government Law[11].

Particularly, The Digital Government Law defines e-government as "the strategic use of digital technologies and data in Public Administration for the creation of public value", stating that it is based "on an ecosystem composed of public sector actors, citizens and other stakeholders, who support the implementation of initiatives and design actions, creation of digital services and content, ensuring full respect for the rights of citizens and people in general in the digital environment". However, in Peru, as in Latin America in general, the attempts to establish e-government structures have mainly focused on creating digital services through the web pages of public institutions and making information available in public portals, in a mixed scheme of static information and interactive transactional services [7, 8].

For countries such as Peru to reach the same levels of efficiency and flexibility in e-government that the developed countries have it is necessary to equip the State with the technological and, above all, methodological resources for its implementation [9]. This call for the adoption of IT governance, understood as the set of responsibilities and practices exercised by the top management of the organizations in order to direct and generate value in the business through the administration of ICT [10].

Because of the inherent size and complexity, this difficulty is probably less felt in municipalities than at central government. This might explain the fact that most of the services available through the Peruvian Portal of Services to Citizens and Companies[12] correspond to municipal procedures, although many of them are purely based in information queries [11].

3   The Local Government Context

Peru is a unitary but decentralized country. It has three different levels of government: national government, regional government and local government. At the local level, two types of municipalities can be distinguished: provincial municipalities and district municipalities. These two types of entities have political, economic and administrative autonomy on matters within their competences. However, district municipalities take relevance because they are the lowest level of government, being the closest link between the State and its citizens.

---

[9] Memorandum No. 152-2015-PCM/ONGEI, Ministerial Resolution No. 004-2016-PCM.
[10] Directorial Resolution No. 012-2017-INACAL/DN.
[11] Legislative Decree No. 1412-2018.
[12] http://www.serviciosalciudadano.gob.pe/.

The most important metropolis in Peru is Lima Metropolitana. The importance of this city relies in the fact that it is the capital of the Republic. Moreover, historically, this city has concentrated the largest levels of resources and population. In total, Lima Metropolitana has 42 district municipalities.

District municipalities are composed by the municipal council (mayor and its council members), the mayor's office (executing agency), coordination bodies and an administrative structure[13]. However, depending on their needs and resources, each entity can create new management departments. Thus, in practice, each municipality may have a differentiated structure.

For example, the number of management departments in the districts of Lima Metropolitana varies considerably. While San Miguel and Independencia show the largest number of management departments (17 and 15, respectively) in their organizational charts, Villa El Salvador and Punta Negra show the smallest number of management departments (6 and 5, respectively).

The mayor and local authorities are elected democratically for a period of four years and access power after obtaining the highest number of votes. Any citizen with the right to suffrage can apply to the posts in question[14]. The immediate reelection of mayors is currently prohibited[15].

District municipalities play a fundamental role in the provision of services and infrastructure. Their exclusive competences are, for example, granting licenses for business operations, garbage collection, provision of local security, and maintenance of roads, schools, hospitals, among others.

As noted by [12], the budget allocation process involves matching a source of income with a cost item, until the resources are exhausted. Once the budget is approved, it represents a limit on how much the municipality can spend. In practice, entities may not spend their entire budget. Thus, the percentage of budget execution has been taken as a measure of the management capacity of municipalities [13].

In 2018, the aggregate budget of all the district municipalities in Lima Metropolitana amounted to more than 3 billion Soles[16]. At the end of October 2018 these municipalities had executed an average of 64.92% of their budget, with the municipalities of Punta Hermosa and Jesús María showing the highest level of budget execution (83.3% and 81.5%, respectively).

---

[13] Divided into municipal management, internal audit, municipal public prosecutor's office, legal advice and planning and budget, however, it is up to each local government to establish its administrative structure according to its economic resources.

[14] The only requirement is to have residence in the jurisdiction to which you apply for at least two years prior to the election.

[15] According to Law No. 30305.

[16] The exact figure of the aggregate budget for the district municipalities was S/. 3,945,207,116. This amount is equivalent to 1,175,585,715 dollars.

District municipalities have a limited capacity to collect taxes[17]. Of the taxes that can be collected by these entities, the one that represents the greater proportion of municipal budget is the Property Tax[18] [12]. To give an example, for the year 2018[19], the resources coming from locally collected taxes represent, on average, 31.6% of the total budget of the district municipalities of Lima Metropolitana.

Most of district municipalities budget come from National Government transfers [12]. The main transfers are FONCOMUN[20] and resources allocated by different types of Canon[21]. Thus, for example, in 2018, 55.7% of the total budget of district municipalities in Lima Metropolitana was transferred from the central government.

Finally, it must be noted that there are mechanisms for citizenship participation in the allocation of budget, through Participatory Budget[22]. This management mechanism incorporates the organized population in the prioritization of investments on each municipality, through dialogue and consultation with local authorities[23]. The amounts that will be submitted to the Participatory Budget are determined by the Communal Council at the beginning of each year. For example, one of the biggest district municipalities in Lima Metropolitana, San Juan de Miraflores assigned two million Soles in 2018 to 17 investment projects, to be executed within the framework of the participatory budget[24].

4   Methods

In this study we applied the method proposed by Dias and Costa [5] for their study of local e-government maturity in Portugal. This method was selected because it has previously been tested in the Latin American context [14]. The first step of the process included a content analysis of the Websites of all the district municipalities of Lima in order to identify the presence or absence of 12 key characteristics. This information was then combined using a multidimensional maturity model in order to access e-government maturity according to three different dimensions (key characteristics for each dimension listed in parentheses): e-information (general information; documents

---

[17] The types of taxes charged by Municipalities are specified between Articles 8 and 59 of the Municipal Taxation Law.

[18] Land, buildings and fixed and permanent facilities that constitute integrated parts thereof, which can not be separated without altering, deteriorating or destroying the building are considered.

[19] Considering the closing of the month of October.

[20] The FONCOMUN is a fund established in the Constitution, which aims to promote investment in different municipalities of the country, with a redistributive approach in favor of the most remote areas, prioritizing the allocation to the rural and urban-marginal localities of the country.

[21] There are five types of Canon: the Mining Canon, the Hydroenergetic Canon, the Gas Canon, the Fishing Canon and the Forest Canon. To these, the Canon and Oil Sobrecanon are added.

[22] According to Law No. 28056, Law No. 29298 and Art.19 of the Law on Decentralization.

[23] Representatives of the population, previously registered, will participate.

[24] Minutes of agreements and commitments of the 2018 participatory budget http://www.munisjm. gob.pe/archivos/2017/actacompromiso2018.pdf.

for public access; text based search tools; semantic search tools); e-services (information on services; query status of service provision; form submission; complete online transaction) and e-participation (complains/suggestions; opinion pull/free discussion; procedures for public discussion; participatory budget).

Each characteristic was valued 1 if observed in the Website and with 0 if not. The scores for each municipality in each dimension were then computed by adding these values (thus ranging theoretically between 0 and 4). Those were then added to compute the final e-government index (ranging theoretically between 0 and 12) and rank the municipalities.

The empirical data was collected in October, 2018. The first phase included the observation of the Websites by students of one Peruvian university. One to two municipalities were assigned to each group of students. The tool described in [5] was translated to Spanish and used for this purpose. It allows the registration of the URLs of the Web pages where evidences for the presence of the key characteristics are observed. In the second phase, the data was aggregated into a single database and validated by an independent team. The previously registered URLs were used for this purpose and corrections where made whenever errors or misconceptions were identified. A total of 37 valid observations were obtained[25].

The final data was then analyzed using descriptive statistics in order to access the validity of the instruments used and the general degree of local e-government sophistication. A clustering analysis was also performed in order to identify the most significant profiles of municipalities to what concerns the distribution of their scores in the three dimensions of analysis. The hierarchical clustering procedure with the average linkage approach was selected and the square Euclidian distance was used as a measure of disparity, as in [5]. Five clusters were used based on dendrogram analysis.

5   Results and Discussion

Table 1 presents the results obtained for all district municipalities of Lima Metropolitana having functional Websites. The clusters the municipalities belong to and the scores they obtained for the three dimensions of analysis and for the final e-government index are presented. The municipalities are order by the score they obtained in the latest and, subsequently, by alphabetical order.

In Table 2 the mean, median, standard deviation and range for each index are shown. By comparing Tables 1 and 2 it can be observed that there are 21 municipalities (57%) having scores for the e-government index that are over the mean for that index and 16 (43%) that are bellow that value. Four municipalities (11%) have scores of 9 or higher (Miraflores, San Borja, San Isidro, and Los Olivos) while six (16%) have scores of 5 or lower (Punta Hermosa, Santa Maria del Mar, Santa Rosa, Chorrilhos, Lurín, and

---

[25] The Websites of the municipalities of Ancón, Carabayllo, Chosica, San Miguel and San Juan de Miraflores were not accessible in at least one of the two phases of the process and were therefore excluded from the analysis.

Pachacámac). Those constitute the groups of more sophisticated and less sophisticated municipalities, respectively, to what concerns global e-government maturity. The remaining 27 municipalities (73%) score between 6 and 8 in this index and thus constitute the great majority of cases.

Table 1. Results for district municipalities of the Lima Metropolitan Area.

Municipality	Cluster	E-inf.	E-serv.	E-part.	E-gov.
Miraflores	1	4	4	2	10
San Borja	1	4	4	2	10
San Isidro	1	4	4	2	10
Los Olivos	1	4	3	2	9
Ate	2	4	2	2	8
Independencia	1	3	3	2	8
Jesús María	3	2	4	2	8
La Molina	1	3	3	2	8
La Victoria	2	4	2	2	8
Lince	2	4	2	2	8
Puente Piedra	4	2	2	4	8
San Martin de Porres	1	3	3	2	8
Santiago de Surco	3	2	4	2	8
Villa El Salvador	3	2	4	2	8
El Agustino	3	2	4	1	7
Pucusana	2	4	1	2	7
Pueblo Libre	2	3	2	2	7
Rímac	2	3	2	2	7
San Bartolo	2	3	2	2	7
Surquillo	2	3	2	2	7
Villa María del Triunfo	2	4	2	1	7
Barranco	5	2	2	2	6
Breña	5	2	2	2	6
Chaclacayo	5	2	2	2	6
Cieneguilla	2	3	1	2	6
Comas	2	3	1	2	6
Magdalena del Mar	2	4	1	1	6
Punta Negra	2	4	1	1	6
San Juan de Lurigancho	2	3	1	2	6
San Luis	5	2	2	2	6
Santa Anita	5	2	2	2	6
Punta Hermosa	5	3	1	1	5
Santa María del Mar	5	3	1	1	5
Santa Rosa	5	2	2	1	5
Chorrillos	5	3	1	0	4
Lurín	5	2	1	1	4
Pachacámac	5	2	1	1	4

Table 2. Mean, median, standard deviations and range for each index.

Index	Mean	Median	Std. dev.	Range
E-information	2.95	3	0.81	2-4
E-services	2.19	2	1.08	1-4
E-participation	1.76	2	0.64	0-4
E-government	6.89	7	1.58	4-10

Table 3. Statistical characterization of clusters (medians are presented for e-information, e-service, e-participation and e-government).

Cluster	N	Percentage	E-information	E-services	E-participation	E-government
1	7	19%	4	3	2	9
2	14	38%	3.5	2	2	7
3	4	11%	2	4	2	8
4	1	3%	2	2	4	8
5	11	30%	2	2	1	5

Still referring to Table 2, it can be observed that the mean and range vary significantly depending on the dimension of analysis. Indeed, e-information is the most developed dimension, with the higher mean and a range between 2 and 4, which means that all municipalities in the dataset exhibit at least two characteristics in this dimension. The e-services dimension follows, with a mean of 2.19 and a range of 1 to 4. Finally, the e-participation is the least developed dimension, having the lowest mean and exhibiting a range of 0 to 4. It is interesting to notice that this hierarchy of e-government dimensions is compatible with the results obtained for Portugal by Dias [15], Dias e Costa [5] and Dias and Gomes [16], despite the different contexts and the temporal gap between studies.

Table 3 resumes the results of the clustering analysis. This analysis allows the identification of different profiles of municipalities concerning the different contributions of each dimension for the overall e-government score. Four main clusters are visible: Cluster 1 groups municipalities with good global results that follow the previously identified hierarchy between the three dimensions of analysis; Clusters 2 groups municipalities that exhibit a more balanced contribution of the different dimensions to their global result, although with a slightly greater relevance of the e-information dimension; Cluster 3 groups municipalities that obtained their best score for the e-services dimension; and Cluster 5 groups municipalities with global results below the average and a greater weight of the e-information in relation to the other two dimensions. Cluster 4 includes the sole municipality for which the participation dimension is more significant than the others (Puente Piedra).

Table 4. Frequency of the characteristics and percentage of the municipalities exhibiting them.

Dimension	Characteristic	Frequency	Percentage
E-information	General information access	37	100%
	Documents for public access	37	100%
	Text based search tools	24	65%
	Semantic search tools	11	30%
E-services	Information on services	37	100%
	Query status of service provision	22	59%
	Form submission	8	22%
	Complete online transaction	14	38%
E-participation	Complains/suggestions	25	68%
	Opinion pull/free discussion	1	3%
	Procedures for public discussion	1	3%
	Participatory budget	34	92%

Table 4 includes information about the frequency of the characteristics used to access the three dimensions of analysis. As can be observed, items that relate to the publication of information and documents are the most frequently observed in the Websites. This tendency for municipalities "to anticipate the development of simpler facilities in detriment of more complex ones or that may have more serious political or administrative implications" was also observed by Dias and Costa [5]. Another relevant aspect is the high frequency of observation of the 'participatory budget' characteristic. This has not been observed for the Portuguese case and is probably related to the specific Peruvian mechanisms for citizenship participation in the allocation of budget, as discussed in Sect. 3. This and the fact that three other characteristics were observed by all municipalities indicate that there might be room for updating the observation method in order to improve its discriminative power, while adapting it to the specific Peruvian context.

6   Conclusions and Future Work

The study presented in the article focus specifically in the Lima Metropolitan Area and it is the first one to address local e-government maturity in Peru. Despite its preliminary nature, the results allow some conclusions that are compatible with previous studies performed for Portugal: (i) municipalities tend to privilege simpler developments in detriment of others that are technically more complex or have more relevant political or administrative implications; consequently, (ii) the e-information dimension is typically more developed than the e-service dimension which, in turn, is more developed that the e-participation dimension; but, besides that general trend, (iii) several different profiles of municipalities coexist in what concerns the balance between the three different dimensions of analysis.

Future direction of the present study will include updating and adapting to the Peruvian context the multidimensional maturity model used in the study; extend the analysis to other Peruvian regions; find the main determinants of local e-government maturity in Peru and compare it to other countries or regions of the globe; and extend the study to other countries with similar characteristics, namely in Latin America.

References

1. UN: United Nations e-Government Survey 2018. UN Department of Economic and Social Affairs (2018)
2. Pina, V., Torres, L., Royo, S: e-Government evolution in EU local government: a comparative perspective. Online Inf. Rev. **33**(6), 1137–1168 (2009)
3. Armstrong, C.L.: Providing a clearer view: an examination of transparency on local government websites. Gov. Inf. Quart. **28**(1), 11–16 (2011)
4. Sandoval-Almazan, R., Gil-Garcia, J.R.: Are government internet portals evolving towards more interaction, participation, and collaboration? revisiting the rhetoric of e-government among municipalities. Gov. Inf. Quart. **29**(SUPPL. 1), S72–S81 (2012)
5. Dias, G.P., Costa, M.: Significant socio-economic factors for local e-government development in Portugal. Electron. Gov. **10**(3–4), 284–309 (2013)
6. ONGEI: Una mirada al Gobierno Electrónico en el Perú: La oportunidad de acercar el Estado a los ciudadanos a través de las TIC. ONGEI, Lima, Peru (2013)
7. Grönlund, Å., Lindblad-Gidlund, K.: Electronic identity management in Sweden: governance of a market approach. Identity Inf. Soc. **3**(1), 195–211 (2010)
8. Tupia, M.F., Villena, M.A., Bruzza, M.A.: Implementation of an E-Government plan in the peruvian public sector: a case study. Publ. Policy Adm. Rev. **2**(3), 23–39 (2014)
9. Bruzza, M.A., Tupia, M.F.: Implementation model of the electronic government. phase II - processes and implementation guide. AISS: Adv. Inf. Sci. Serv. Sci. **10**(2), 1–9 (2018)
10. ISACA International: COBIT 5 for information security. ISACA Publishing, Rolling Meadows, Ilinois (2012)
11. Bruzza, M.A., Tupia, M.F.: A systematic review based on Kitchengam's criteria about use of specific models to implement e-government solutions. In: 3th International Conference on eDemocracy & eGovernment, pp. 81–86. IEEE, New Jersey (2016)
12. Aragón, F., Casas, C.: Local governments' capacity and performance: evidence from Peruvian municipalities. CAF Working paper No. 2008-06. Caracas (2008)
13. Sánchez, J., Wong, S.: De la ejecución al desempeño municipal: una propuesta alternativa a la tipología del MEF. Universidad del Pacífico, Facultad de Economía y Finanzas, Lima, Peru (2015)
14. Maciel, G., Gomes, H., Dias, G.P.: Assessing and explaining local e-government maturity in the Iberoamerican community. J. Inf. Syst. Eng. Manage. **1**(2), 91–109 (2016)
15. Dias, G.P.: Local e-government information and service delivery. In: Proceedings of the 6th Iberian Conference on Information Systems and Technologies (CISTI), Art. No. 5974287 (2011)
16. Dias, G.P., Gomes, H.: Evolution of local e-government maturity in Portugal. In: Proceedings of the 9th Iberian Conference on Information Systems and Technologies (CISTI), Art. No. 6877041. IEEE (2014)

An Approach to GDPR Based on Object Role Modeling

António Gonçalves[1(✉)], Anacleto Correia[1], and Luís Cavique[2]

[1] CINAV, Alfeite, 2810-001 Almada, Portugal
agoncalves@tecnico.ulisboa.pt,
cortez.correia@marinha.pt
[2] BioISI-MAS, Universidade Aberta, 1269-01 Lisbon, Portugal
luis.cavique@uab.pt

Abstract. The General Data Protection Regulation 2016/679 (GDPR) is a set of legal rules to attain the privacy of people in the handling of their personal data and the movement of such data across countries. When those rules are considered in the operation of information systems, the one becomes attainable for legal approval within that scope. This paper presents a model we are developing to help enterprises do align their information system with the GDPR requirements. The model shall serve the purpose of analyzing the enterprises in what concerns the use of the subject's personal data, allowing to capture and improve data protection capabilities placed in the GDPR. The main issue of our approach is to set a baseline to define the requirements for establishing, implementing, maintaining and continually improving data protection management system on organizations.

Keywords: Personal data protection · Regulation (EU) 2016/679 · GDPR

1 Introduction and Motivation

The General Data Protection Regulation (GDPR) was stated by the European Union. It is a legal document consisting of a set of rules to achieve a high level of protection of natural persons in what personal data processing and the free movement of such data is concerned. The assurance of compliance with the GDPR, at the level of enterprise operation, demands an effort from people when analyzing and understanding such regulation since it is a mix of legal rules, organization rules, and technical rules.

We output that, an organization, to be compliance with GDPR, requires a dynamic approach where protection of personal data is achieved in a continuous way and personal data should be considered a valued organization information assets [1]. Personal data protection is part of a complex organization privacy process that encompasses the preservation of personal data from unauthorized access, use, modification, recording or destruction. Since this kind of process is offered in a continuous way, it is important to measure the effectiveness of this process, i.e., the information security quality of process [2].

© Springer Nature Switzerland AG 2019
Á. Rocha et al. (Eds.): WorldCIST'19 2019, AISC 930, pp. 595–602, 2019.
https://doi.org/10.1007/978-3-030-16181-1_56

The goal of this paper is to propose a model that describes the concepts captured from the analyses of GDPR documentations. It is our goal to promote an improved understanding of the legal, organization and technical concepts present in the GDPR document.

From the breakdown of the GDPR document, we aim to elaborate a model of personal data protection, based on the capture of main concepts from the regulation.

We aim to use the model to help the operation of enterprise information systems, to show it possible for legal approval inside the scope of personal data protection, specifically within the requirements of the GDPR.

This paper also details some guidelines to assess the concept of personal data security risk assessment through the identification of threats and vulnerabilities that are carried out by the risk team. The goal is to operate the organization information system supported by a service based on risk management, in order to maximize the organization's output, while at the same time decreasing unexpected negative outputs generated by potential threads.

The analysis of the GDPR is a complex mission. It involved several tasks, namely reading, manual knowledge extraction, and characterization of many concepts and sentences expressed in that legal document.

The paper is structured as follows. Section 2 outline GDPR principles, that we used to present a methodology to analysis it. Section 3 related the GDPR data protection principles and information security principles. Section 4 explains our ontological model of analyzing the GDPR. Section 5 examines some challenge and future direction for our work.

2   GDPR Analysis

The GDPR is aimed at the protection of natural persons regarding the processing of their personal data. Across GDPR documentation is specified a set of definitions, such as personal data, sensitive data, and data processing. It also defines the actors involved in data processing such as: controller, processor, operator and subject, their roles, and responsibilities. Finally, GDPR portrays the obligations identifying with information controllers and processors with explicit reference to the legitimate motivation behind information handling and the reception of wellbeing safety efforts to control the risk unauthorized use of data.

According to the GDPR, a broad concept of the information system should be considered. For example, GDPR applies to a filing system if it is a set of structured personal data manageable via certain criteria. GDPR also mentions the possibility of data subjects getting direct access to their own personal data in order, for example to rectify personal data.

To implement GDPR on an organization a set of very important governance process may also be considered, such as process where controllers or processors could record and manage the way personal data is processed. This kind of process is very important since data subjects could request for the rectification of their own personal data, where controllers or processors could record processing activities and where data protection officers could monitor that registration.

GDPR, in the context of this paper, is a source to drive business process and also system requirements to support the analysis of information systems requirements, in order to capture the regulatory data protection capabilities disposed in the GDPR on the organization systems. The analysis of the GDPR involved a set of tasks as showed in Fig. 1:

1. Task 1: Ontological Data Protection concepts. This task comprised of the familiarization of the organization with the domain of personal data protection. The output of this task is a set of GDPR domain terms definition. This task will allow the organization to understand regulatory systems, from a business perspective.
2. Task 2: Stakcholders Capture Model. This task comprised of identifying and typifying the parties interested in the development and operation of regulatory systems.
3. Task 3: Structure Model. This task comprised of an assembly composed of Data Protection concepts and stakeholders. This task consisted of modeling the relationship between the data protection concepts identified in the task1 and Stakeholders Capture identified in the task2, along with their attributes.
4. Task 4: Context Model. This task comprised of identifying each organization artifacts involved in the regulatory protection systems. The output of this task will be a list of in-scope artifacts and the characterization of its type.
5. Task 5: Regulatory protection business goals Model. This task comprised of identifying and characterizing the business processes, based on an ordering of tasks concerned with regulating the processing of personal data and identifying and characterizing the business goals to be achieved with the application of the provisions in the GDPR to regulatory systems.

The execution of these tasks is directed by a workflow that suggests the adopted GDPR analysis methodology.

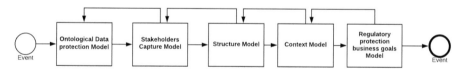

Fig. 1. BPMN model of GDPR methodology.

3 The Relation Between Data Protection and Information Security Models

Normally organization uses the term data protection, derived from the information security, to designate any artifact that has personal information with value for the organization, and for that reason, it needs to have adequate data protection [3].

While information security and data protection have a significant similarity, these two terms are not analogous. The general definition of information security comprises the CIA model of Availability, Integrity, and Confidentiality. Data protection includes other aspects that extend beyond the limits of information security, including the concept of data processing and genuine reason for information handling and the acceptance of security measures to limit information.

According to GDPR the aim of personal data protection is to guarantee business continuity and minimize business damage by controlling the impact of data protection incidents preserving the confidentiality, integrity, availability of personal data.

Additionally, it is also necessary to keep analyzing the legitimate purpose of personal data processing [4]. Data protection is not only technology but always encompass a process. For example, if we start information security as a strictly technical issue, we also must take care of the process of securing these technical issues. Hence, it is necessary to evolve in order to extend beyond only the technical [5].

Information security introduces an important definition in terms of the properties that data processing should have. These include the availability, confidentiality, and integrity of information. Availability refers to the fact that the personal data used by an organization remain accessible when they are needed. Thereafter, system failure is an organizational security issue [6].

Confidentiality refers mostly to restricting personal data access to those who are authorized. Organizations strive to control personal data access since technology offers developments aimed at making information accessible to the many [6].

Integrity refers to maintaining the values of the personal data stored and manipulated, such as maintaining the correct meaning. Personal data integrity is commonly assured by using cryptographic or/and information replication strategies. The cryptographic tools are indeed used to sign single pieces of data so that any faking information can be detected through signature validations [6].

According to ISO 31000, we can define personal data protection risk as the effect of uncertainty due to a deficiency of information that hinders achieving organizational objectives. Personal data protection risk management is a permanent challenging process which allows an understanding of the potential risks to the organization's valuable personal data assets and the tools to address them [7].

4   Proposed Ontological Data Protection Model

The integrated modeling of different aspects of an organization is only in the context of the emerging discipline of Organizational Engineering that concerns itself essentially with the modeling in three aspects:

- The core, essential model of an organization from the point of view of business (e.g., ontological model);
- The Business integration, information, and documents (e.g., the achievement of the Organization) and
- The model by which an organization is operated by all the people and information technologies (e.g., the implementation of the Organization).

An ontological data protection model is related to the construction and operation of an information system in the context of privacy. An ontological model has many advantages, such as it can help improve organizational consciousness; it enables sharing of knowledge between individuals through the representation of different organizational aspects, such as business processes, resources (technological artifacts, people, materials, etc.) and organizational structure.

From the lookout of the proposed ontological model of data protection, we support the proposal made by Dietz [8] where the ontological model abstracts from all realization and implementation issues.

The ontological model describes the essence of the business aspects of an organization. Realization model describes the detailed integration of all aspects of business, through the layered nesting, of information and document necessaries to the operation of the organization [8].

According to our suggestion, we can analyze the ontological of an organization with the use of fact model and use facts to find useful measures, viability norms, and dysfunction.

A Fact Model structures essential business knowledge about business concepts and business operations. It is occasionally named a business entity model. The main purpose of a fact model is to create a standard vocabulary by which all stakeholders can use to communicate clearly. Therefore, we can use facts that reproduces the real world.

The fact model focuses on the core business concepts, and the logical connections between them, which are named facts.

The facts are usually verbs which designate how one concept link to another. For example, the two concept Person and Personal Data may have a fact connecting them called have (a Person have Personal Data).

We agreed that the organization implementation is a result of an engineering process that can be analyzed from a fact model. This model can be used to understand technology (i.e., people, rules, a division of work and tools) that is part of organization operation. For that, we propose an ontological model to capture the essential structure of activities from ontological organization model. The model is present in Fig. 2.

The model is composed of the following entity: E01 to E07. All entities are artifacts under this model. We identified the following entities: E0: data, E01: processing, E02: transfer, E03: data quality, E04: Risk, E05: Person, E06: lawfulness, fairness, and transparency, E06: purpose limitation, E07: data minimization, and E08: accuracy.

We identified the following facts that result in the relationship between two or more Entities: All relevant data we consider is data that is related to a person (i.e., that we can uniquely identify a person from that). Associated with personal data is the risk entity.

Risk entity is associated with uncertainty from an expected organization processing and can be quantified as a positive or negative deviation. There are several ways to output a risk. One of the possible ways is by linking it to the events that may happen, their consequences and the likelihood of the occurrence. The lack of information regarding the event occurrence, its consequence, or likelihood, is what drives to the state of uncertainty that underlies risk [9, 10].

The definition of Data Quality is a multi-dimensional concept that varies depending on context, stakeholder interests. Data Quality depends on both subjective perceptions of the individuals involved with the data, and the objective measurements based on the data set in question. General Data Quality privacy is related with the following facts: comprise the lawfulness, fairness, and transparency of data processing, the purpose limitation that allows future processing if the new purpose if either is compatible with your original purpose. Data minimization creates the fact that data is acceptable,

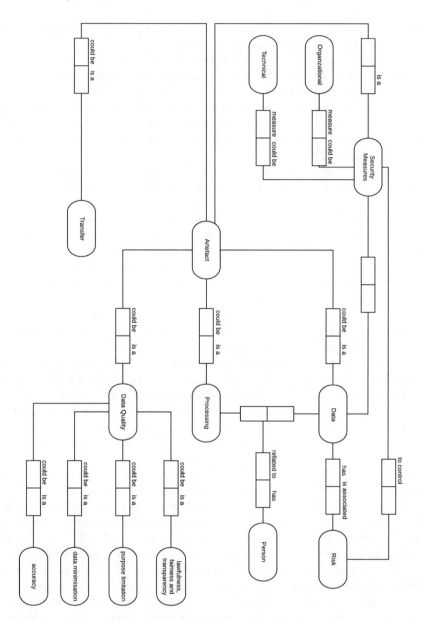

Fig. 2. Ontological model of data protection.

important and restricted for the end for which they are processed. Accuracy allows people to gain control about their personal data and choose when, how and to what degree the personal data is conveyed to other people.

A threat denotes a likely violation of the security of data privacy with some negative impact [11]. The vulnerability is a real security flaw which is an open door to an attack. So, an attack is a use of a vulnerability to realize a threat.

The Data protection facts model can be useful to measure the importance of risk. We use a taxonomy of privacy proposed by Pfitzmann [12], as follows: We categorize the entity unlikability, as a means to measure the relation between set of personal data outside of a specific domain; (ii) Transparency, as a means that we can measure the likelihood that data, can be understood and reconstructed at any time. The information should be available before, during, and after the processing takes place; This allows that the data subject could have access to information requested from an organization; (iii) Intervenability ensures intervention is possible concerning all ongoing or planned privacy-relevant data processing, by those persons whose data are processed. It allows the possibility of a data subject to a request to rectification and erasure of data; (iv) Anonymity refers the measure of the set of subjects with potentially the same attributes and (v) confidentiality refers to hiding the data content or controlled release of data content.

5   Conclusion and Future Work

This paper describes an ontological model on data protection and free movement according to European regulation 2016/679.

According to Dietz [8], it will only be possible to manage the complexity of an organization and reduce and manage its entropy through its ontological model. The ontological model, being coherent, comprehensive, consistent and concise, relies only on the essence of an organization and enables it to deal with the current and future problems of the business challenge. The assumption made in his proposal is that communication between the people of an organization provides the necessary and sufficient support to develop an organization theory.

In this context, an application of the holistic regulation is certainly useful and beneficial, in that its practice cannot be done in a disorganized and disenchanted way from the organization reality. However, to describe an organization is a complex task, since its representation must be done in an integrated way, which, if not supported in an organized approach, will translate into distinct and uncoordinated perceptions that manifest themselves internally and externally, so that the capture of relevant activities in an environment faces a high number of challenges, highlighting: the creation of methods that allow harmonious and cooperative apprehension by the people, the relevant activities and their respective articulation.

A number of approaches seek to capture activities for complex and dynamic domains such as teaching, information integration, the development of the Man-Machine interface, the description of system requirements. Unfortunately, however, uncertainty in these dynamic and complex domains prevents coherent understanding, in particular, because of the fact that there are sectarian and often inconsistent conspectations of their environment.

A starting point is an ontological approach, based on the description of the essence of the operation in an organization.

In this article, a fact-based model is proposed that describes and relates the main artifacts. To this end, interpretations of the main concepts of regulation and information security were suggested.

In the future it will be possible, from the proposed model, to advance to proposals for operation of treatment protection in organizations taking into account other socio-technical models, namely activity theory.

The theory of activity will allow the explicit description of the articulation of activities, based on the casual connection of the fundamental elements of the onto-logical model, enables to obtain a congruent model of the organization under the aspects of identification of the operations involved and the structuring of the activities in its nuclear elements

Acknowledgements. The work was funded by the Portuguese Ministry of Defense and by the Portuguese Navy/CINAV.

References

1. Peltier, T.R.: Information Security Policies, Procedures, and Standards: Guidelines for Effective Information Security Management. CRC Press (2016)
2. Von Solms, R., Van Niekerk, J.: From information security to cyber security. Comput. Secur. **38**, 97–102 (2013)
3. Cherdantseva, Y., Hilton, J.: A reference model of information assurance & security. In: 2013 International Conference Availability, Reliable Security, pp. 546–555 (2013). https://doi.org/10.1109/ares.2013.72
4. Whitman, M., Mattord, H.: Principles of Information Security. Cengage Learning (2011)
5. Laudon, K.C., Laudon, J.P.: Management Information Systems, 13e (2013)
6. Andress, J.: The basics of information security: understanding the fundamentals of InfoSec in theory and practice. Syngress (2014)
7. Purdy, G.: ISO 31000:2009 - setting a new standard for risk management: perspective. Risk Anal. **30**, 881–886 (2010)
8. Dietz, J.: Enterprise Ontology: Theory and Methodology. Springer (2006)
9. ISO I 31000: 2009 Risk management–Principles and guidelines. Int Organ Stand Geneva, Switz (2009)
10. Guide ISO 73: 2009. Risk Manag (2009)
11. Oladimeji, E.A., Supakkul, S., Chung, L.: Security threat modeling and analysis: a goal-oriented approach. In: Proceedings of 10th IASTED International Conference on Software Engineering Application SEA 2006, pp. 13–15 (2006)
12. Pfitzmann, A., Kiel, U.L.D.: Pseudonymity, and identity management – a consolidated proposal for terminology. Management, 1–83 (2008)

A Study over NoSQL Performance

Pedro Martins[1], Maryam Abbasi[2], and Filipe Sá[1,3]([⊠])

[1] Department of Computer Sciences, Polytechnic Institute of Viseu, Viseu, Portugal
{pedromom,filipe.sa}@estgv.ipv.pt
[2] Department of Computer Sciences, University of Coimbra, Coimbra, Portugal
maryam@dei.uc.pt
[3] Câmara Municipal de Penacova, Penacova, Portugal

Abstract. Large amounts of data (BigData) are nowadays stored using
NoSQL databases and typically stored and accessed using a key-value
format. However, depending on the NoSQL database type, different performance is offered. Thus, in this paper, NoSQL database performance
is evaluated and compared in aspects relating with, query performance,
based on reads and updates.

In this paper, the five most popular NoSQL databases are tested and
evaluated: MongoDB; Cassandra; HBase; OrientDB; Voldemort; Memcached and Redis. To assess the mentioned databases, are used, workloads represented by Yahoo! Cloud Serving Benchmark.

Results allow users to choose the NoSQL database that better fits
application needs.

Keywords: NoSQL · Performance · YCSB · Database · Benchmark

1 Introduction

In the last few years, databases became present in all systems, small or big. Up to
a certain point on data dimension and complexity, relational databases, alongside
with the standard SQL language, were the main logical choice for enterprises. The
essential operations were to allow storage, extraction, and manipulation of data.
Although, as data volume grows, relational databases have several limitations
which bring up limitations of storage and management, for instance: scalability;
limited storage size; query efficiency related with data volume. To solve the
problems mentioned above, new databases models emerged, known as NoSQL
databases [8].

NoSQL databases appeared as a solution to complement relational databases
and not to replace them. This complementing includes, performance increase by
adding a set of new key characteristics, such as schema flexibility; horizontal
scalability [15]; transparent clusters of nodes, that do not require manual data
distribution or high maintenance; recover from faults; and automatic repair in
case of faults [7]. This characteristics and others easy the task of database administration when faced with large data-sets.

© Springer Nature Switzerland AG 2019
Á. Rocha et al. (Eds.): WorldCIST'19 2019, AISC 930, pp. 603–611, 2019.
https://doi.org/10.1007/978-3-030-16181-1_57

Depending on what each system needs, different NoSQL engines provide different Consistency, Availability, and Partition Tolerance (CAP) feature combination [6]. For instance, SimpleDB, DynamoDB and MongoDB, have support for eventual consistency. With the objective of speeding up query execution NoSQL engines started to make use of in-memory storage, mapping parts of the database into volatile memory.

Despite non-relational databases evolution, for specific purposes/business, it is crucial to understand its main features and characteristics, as in relational systems. NoSQL databases have mechanisms to insert and retrieve data, oriented to performance optimization, resulting in different data load times, and execution times for both updates and read operations.

In this paper are tested seven NoSQL database engines which were considered to be the most popular during the related work review: Redis; Memcached; Voldemort; OrientDB; MongoDB; HBase; and Cassandra. All seven non-relational engines are tested regarding their execution speed upon different benchmark workloads, which provide execution parameters to get and put operations [4], that allow to understand and compare the performance of the tested databases (i.e., is the non-relational database faster for reads or writes/inserts).

In this paper, different features that impact different non-relation databases performance are studied and compared, allowing to understand which NoSQL database performs better facing specific workload scenarios.

This paper is organized as follows. Section 2, makes resumed review of the related work. Section 3, describes the applied experimental setup. Section 4, shows the experimental evaluation and results. Section 5, concludes the work and presents future research guidelines.

2   Related Work

NoSQL refers to an open database model which does not use SQL interface [12]. The origin can be related to Google's BigTable, used to store projects developed by Google, such as Google Earth [2]. Amazon, faced with the evolution, developed by Dynamo [5]. Over the last decades, these model of databases have been perfected, and their performance evaluated. A wide number of papers describing characteristics, proposing new features, business applications, and so on, regarding of NoSQL database engines [8,9,11].

Studies like [4], research performance characteristics, and mechanisms, to evaluate the advantages of NoSQL scalability over relational. The proposed work differs from these contributions by focusing on the comparison and analysis of performance execution time quantitatively.

The principle that NoSQL databases operate is on BASE (i.e., Basically Available, Soft State, and Eventually Consistent), which offers high availability, but low consistency [1,3,13]. On the other hand, ACID (i.e., Atomic, Consistent, Isolated, and Durable) is used to represent relational databases, in which transactions and committed data are always consistent. For both cases, ACID and BASE, come from CAP theorem (i.e., Consistency, Availability, and Partition

Tolerance) [14]. Following CAP theorem, only two out of three guarantees can be provided. If data consistency is crucial, relational databases should be used. If working with distributed data, consistency is more important and harder to achieve. NoSQL can be separated into four main categories, each oriented to different optimization's [10], such as:

- Key-value, where all data is stored as a pair of a key and a value, as happens with "hash tables". Data access is performed by searching for the corresponding key.
- Document, as in XML or JSON, data is stored in documents [16], where the format can variate.
- Column family, data is stored as a set of rows and columns. Columns are grouped according to relations, and often retrieved together.
- Graph, all data is represented as a graph with interlinked elements. For instance, social networks or maps.

In this research paper the testes and evaluated databases have the following type(s):

- Redis, Memcached, and Voldemort are key-value based.
- MongoDB and OrientDB, are document store based. Where OrientDB is an hybrid, also widely used for graphs.
- Cassandra and HBase, are column family based.

3   Experimental Setup

In order to evaluate the performance of the tested NoSQL databases, YCSB benchmark was used [4]. Two components set YCSB benchmark. First, a data generator, second, performance testing, i.e., read and insert operations. The tested workloads are defined by certain parameters that comprehend: percentage of reads; percentage or updates; percentage of scans; percentage of inserts; the number of operations; and the number of records. By default, there are predefined workloads, such as:

- Workload A, 50% reads and 50% updates;
- Workload B, 95% reads and 5% updates;
- Workload C, 100% reads;
- Workload D, 95% reads and 5% inserts;
- Workload E, 95% scan and 5% insert;
- Workload F, 50% reads and 50% read-modify-write;
- Workload H, 100% write;

In this paper, the main focus is to compare the execution speed of getting and put operations. For that purpose, workload A, C, and workload H, consisting of 100% write.

In order to perform the workload testes over the databases, 6.000.000 randomly generated records were used, each with ten fields of 100 bytes for the key registry. In total, around 1 kb per each record. While the number of stored records was varying, 5.000 requests per second, to the testing database were being performed.

Other benchmarks such as TPC-H or SSB, could be used to test performance. However, this benchmarks, are more suitable for decision support. YCSB was chosen, due to its simplicity, and to test the two principal operations that are performed over NoSQL databases, get and put.

Tests were performed in a single node, with Ubuntu 18.10, i5 processor, 3.4 GHz, 16 GB memory, and 1 TB HDD.

4   Performance Evaluation

The next sections present the execution times, when using YCSB workloads A, C, H. Section 4.1 shows the performance evaluation using workload A. Section 4.2, uses workload C. Section 4.3, uses workload H. Finally in Sect. 4.4, the global performance evaluation of all tested databases are compared.

4.1   Workload A

Workload A, consists of 50% reads and 50% updates, on top of 6.000.000 records. Figure 1, shows the obtained results for each tested database.

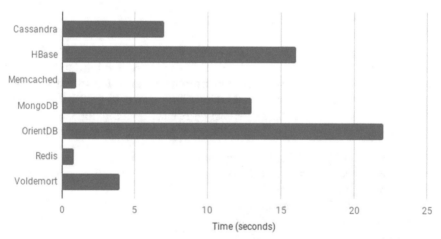

Fig. 1. Workload A

From Fig. 1, Redis and Memcached database, key-value store, achieves good performance. Redis and Memcached, both use volatile memory, to store and retrieve data, allowing this way lower execution times. Voldemort shows slower performance due to the combination of in-memory caching with the storage system.

Column-family databases, such as Cassandra, present better performance tan HBase. Overall, the worst performance was from the Document Store database, OrientDB, this happens because, in OrientDB, records must be read from disk.

4.2 Workload C

Workload C, consists of 5.000 random reads, on top of 6.000.000 records. Figure 2, shows the obtained results for each tested database.

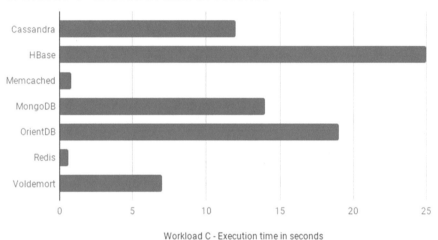

Fig. 2. Workload C

Concerning HBase and OrientDB, these have the lowest execution times when performing read operations. The reason is that different parts of the records are stored in different disk files, resulting in higher overheads. Nevertheless, HBase is optimized for updates as it will be shown later with workload H. OrientDB, stores data in the disk, and does not load it into memory; therefore it presents the second worst result. Since Redis and Memcached are designed to operate in volatile memory (but very fast), reading data is just a matter of key-value mapping in memory. Despite Voldemort being slower than Redis, and Memcached, it shows better performance than other tested databases.

4.3 Workload H

Workload H performance is shown in Fig. 3. This workload is based on 5.000 updates over 6.000.000 records.

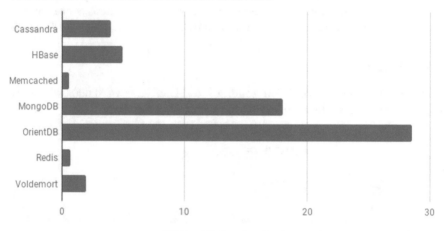

Workload H - Execution time in seconds

Fig. 3. Workload H

Workload H uses 5.000 write operations. This workload allows to verify the performance of the database when executing, read and update operations. Column databases, like Cassandra and HBase, are more optimized for updates, as they load large volumes of records into memory, the number of operations performed on disk is reduced. This performance is increased.

Document store database, OrientDB, shows the highest performance time (higher is worst), performing slower than MongoDB. The main reason for this difference is the way records are kept on disk, rather than on memory. MongoDB poor results occur because of the locking mechanisms in place when performing updates, leading to an increase in execution time.

Finally, key-value database, operate mainly in volatile memory. Records are directly mapped from memory; therefore, performance increases substantially.

4.4   Global Evaluation

This section, Fig. 4, shows how the tested databases handle a mix of all workloads at once.

In general, results show that the tested in-memory NoSQL database, Redis, Memcache, and Voldemort have the best performance of all. However, in order to assure in-memory performance, it sacrifices data consistency in case of failures.

Column-oriented databases, such as HBase and Cassandra, show good performance during update operations. Although, in general, they show themselves to be quite slow when compared with key-value approaches.

Document store databases, OrientDB, MongoDB, in general, comparing with all the rest, present the slowest performance.

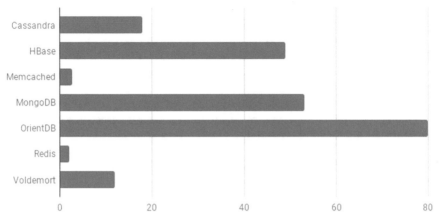

Fig. 4. Workload combination

5   Conclusions and Future Work

NoSQL popularity dealing with huge amounts of data, for storage and process, has increased substantially in the last decade, boosting the development of such systems.

Different NoSQL databases types, provide different features each with its performance. Performance evaluation, for reference purposes, is relevant for deciding which database is more relevant according to applications necessities.

This paper provides a comparative study of different NoSQL databases performance. First, the most popular NoSQL databases, like, Cassandra and HBase, which are column-family based. Second, MongoDB and OrientDB, which are document-store based. Finally, Redis and Memcached, are a key-value in-memory database.

In order to test the performance of the database, YCSB benchmark was used with several workloads. Results show that the in-memory database, Redis and Memcached, are extremely fast, loading, updating and reading data. Nevertheless, a big limitation must be accounted, and the used memory is volatile, more expensive and relatively limited. Voldemort combines key-valued approach with in-memory and file-system for data persistence, however, it comes with a performance cost.

With the worst performance are the document store databases, OrientDB. The use of virtual machine environment, regarding computing resources, is more demanding, therefore implies larger overheads.

HBase and Cassandra, are optimized for sequential reading and writing, reducing the accesses to disk, as well as, number of operations by temporally storing records in memory.

MongoDB shows relatively good performance when dealing with large volumes of records. Performance issues occur from locking mechanisms to keep data consistent.

In conclusion, two groups of databases can be created. First, optimized for reading operations (MongoDB, Redis, Memcached, OrientDB, Voldemort). Second, optimized for updates/writes (Cassandra, HBase).

As related future work research, besides testing other non-relational databases, there will be applied parallelism mechanisms, to test performance in terms of response time, and scalability. More research directions related with this study, comprehend the testing of NoSQL over the cloud.

Acknowledgements. "This article is a result of the CityAction project CENTRO-01-0247-FEDER-017711, supported by Centro Portugal Regional Operational Program (CENTRO 2020), under the Portugal 2020 Partnership Agreement, through the European Regional Development Fund (ERDF), and also financed by national funds through FCT Fundação para a Ciência e Tecnologia, I.P., under the project UID/Multi/04016/2016. Furthermore, we would like to thank the Instituto Politécnico de Viseu for their support."

References

1. Carro, M.: NoSQL databases (2014). arXiv preprint arXiv:1401.2101
2. Chang, F., Dean, J., Ghemawat, S., Hsieh, W.C., Wallach, D.A., Burrows, M., Chandra, T., Fikes, A., Gruber, R.E.: Bigtable: a distributed storage system for structured data. ACM Trans. Comput. Syst. (TOCS) **26**(2), 4 (2008)
3. Cook, J.D.: Acid versus base for database transactions. Cook, J.D., Blog (2009)
4. Cooper, B.F., Silberstein, A., Tam, E., Ramakrishnan, R., Sears, R.: Benchmarking cloud serving systems with YCSB. In Proceedings of the 1st ACM Symposium on Cloud Computing, pp. 143–154. ACM (2010)
5. DeCandia, G., Hastorun, D., Jampani, M., Kakulapati, G., Lakshman, A., Pilchin, A., Sivasubramanian, S., Vosshall, P., Vogels, W.: Dynamo: amazon's highly available key-value store. In: ACM SIGOPS operating systems review, vol. 41, pp. 205–220. ACM (2007)
6. Elbushra, M.M., Lindström, J.: Eventual consistent databases: state of the art. Open J. Databases (OJDB) **1**(1), 26–41 (2014)
7. Gajendran, S.K.: A survey on NoSQL databases. University of Illinois (2012)
8. Han, J., Haihong, E., Le, G., Du, J.: Survey on NoSQL database. In: 2011 6th International Conference on Pervasive Computing and Applications (ICPCA), pp. 363–366. IEEE (2011)
9. Hecht, R., Jablonski, S.: NoSQL evaluation: a use case oriented survey. In: 2011 International Conference on Cloud and Service Computing (CSC), pp. 336–341. IEEE (2011)
10. Indrawan-Santiago, M.: Database research: are we at a crossroad? reflection on NoSQL. In: 2012 15th International Conference on Network-Based Information Systems (NBiS), pp. 45–51. IEEE (2012)
11. Leavitt, N.: Will NoSQL databases live up to their promise? Computer **43**(2), 12–14 (2010)

12. Moniruzzaman, A., Hossain, S.A.: NoSQL database: new era of databases for big data analytics-classification, characteristics and comparison (2013). arXiv preprint arXiv:1307.0191
13. Pritchett, D.: Base: an acid alternative. Queue **6**(3), 48–55 (2008)
14. Simon, S.: Brewer's Cap Theorem. CS341 Distributed Information Systems, University of Basel (HS2012) (2000)
15. Stonebraker, M.: SQL databases v. NoSQL databases. Commun. ACM **53**(4), 10–11 (2010)
16. Zhang, H., Tompa, F.W.: Querying xml documents by dynamic shredding. In: Proceedings of the 2004 ACM Symposium on Document Engineering, pp. 21–30. ACM (2004)

A New Model for Evaluation of Human Resources: Case Study of Catering Industry

João Paulo Pereira[1](✉) ⓘ, Natalya Efanova[2], and Ivan Slesarenko[2]

[1] UNIAG (Applied Management Research Unit),
Instituto Politécnico de Bragança,
Campus Sta Apolónia, 5300-253 Bragança, Portugal
jprp@ipb.pt
[2] Kuban State Agrarian University, Krasnodar, Russian Federation
efanova.nv@gmail.com, one.concealed.light@gmail.com

Abstract. This paper presents a competency model that would improve a personnel assessment process in the catering business. The research showed that existing appraisal methods, such as KPI and 360-degree feedback method, are difficult to apply when evaluating the performance of non-managerial employees (waiters). Moreover, test method comes with several shortcomings. The proposed competency model allows for combining various assessment results and forming an employee competency profile. This profile stores information about the professional level and skills of that employee. A computer system for testing the model was developed. The test results obtained were used as data sources for the employee profile. Experiments were conducted using this computer system and details provided by a pizza chain. Results from these experiments confirmed that the model is adequate. Experiments identified events that need to be taken into account when using the model. It is necessary integrity and reliability of the source data and understand that other effects that can influence the performance appraisal results.

Keywords: Competency model · Performance appraisal · Performance level · Catering industry · Employee profile · Computer program · Experiment

1 Introduction

In today's world, employee appraisal is an important duty of the HR department of any major organization.

With the development of catering business and spread of business chains, choosing the right performance appraisal method for a large number of employees has become a pressing issue. So much attention now needs to be devoted not only to managerial staff but also to non-managerial employees. A company's revenue and number of customers directly depend on the performance of these lower-level employees.

Various evaluation methods are used depending on the structure and capabilities of that particular company. The pros and cons of the two most popular performance evaluation methods today will be reviewed. The review is centered around a competency-based approach. Presently, this approach is the most popular – it evaluates not only the professional but also the personal qualities of an employee.

© Springer Nature Switzerland AG 2019
Á. Rocha et al. (Eds.): WorldCIST'19 2019, AISC 930, pp. 612–621, 2019.
https://doi.org/10.1007/978-3-030-16181-1_58

The 360-degree feedback method has a good track record and is widely used in performance appraisal. Under this method, an employee's performance is evaluated based on the feedback from his working environment – supervisors, subordinates and colleagues. They give performance appraisal scores. The arithmetic average of these scores is then calculated and a competency schedule is built.

This method has several advantages. Firstly, it gives a comprehensive opinion not just on the manager but also on each employee [1]. Secondly, with the 360-degree feedback method, you can identify not only the qualities of an employee, but also the practicality of these qualities. Thirdly, each employee of the company can be evaluated in numerical terms irrespective of his or her post. Implementing a 360-degree appraisal system on the levels of distributive, procedural, and interactional justice in organizations may cause positive effect [2].

However, this method has major shortcomings [3, 4]. The evaluation is subjective in nature. Besides, it is near impossible to evaluate the specific performance of an employee. These drawbacks make the effectiveness of the 360-degree feedback method in the catering industry disputable. For example, the ability of a waiter to serve a customer or to apply sales methods may not always be noticeable to his colleagues and superiors. The same applies when that waiter lacks some skills in this area. One more problem is negative reaction of employees on negative feedback that may cause prejudice of facilitator [5]. Because of this, the 360-degree feedback method can only be an addition to any already existing performance appraisal technique.

Another equally popular performance appraisal method is the Key Performance Indicators (KPI) method, which introduces a balanced scorecard.

With the KPI method, employees' performance is aligned with the company's strategic and operational objectives. KPI is used in enterprises to measure the general achievements of the company, the individual departments and employees themselves. This method also allows to create an incentive program.

KPIs can be the interface between scheduling and control which will be used as a strategy for maximizing the plant performance [6]. Also KPIs monitoring can be used for decision support [7].

Organizations often have their own system of performance indicators and the formulas for calculating them. That is why there is presently no uniform standard for applying the KPI method [8]. Companies are not willing to share their achievements with business rivals. There are little or no publicly available information on the best practices of using the KPI method. To apply this technique in the catering business, one would need to take a number of actions aimed at determining the performance indicators of departments and employees, and the formulas for calculating them. This could lead to costs.

Owing to these pitfalls, the methods described above are most often used to evaluate managerial personnel. Evaluating 600 waiters using the 360 or KPI method is much harder than evaluating just 60 managers and supervisors. Therefore, in evaluating non-managerial staff, one needs to choose methods that are simpler to implement.

At the moment, tests represent such a simple method. Implementing a test does not require major expenses. Where resources are limited, one would only need to create the test, upload it to a free testing platform and provide access to it for employees. The total score gained by the staff will be their performance assessment. With all the simplicity, testing still has a number of significant shortcomings [9, 10]:

1. It checks only theoretical knowledge. A person's ability to put his theoretical knowledge into practice is hard to assess.
2. Test answers can be memorized. Moreover, an employee can decide to cheat during the test. These undermine the reliability of result obtained.
3. Since only the final grade is most often taken into account, the manager may not always get a complete picture of the abilities of an employee.

These disadvantages are critical in the catering industry – a waiter's ability to provide quality service to a customer determines whether that customer will visit again or not in the future. This ability cannot be evaluated through a test. Therefore, in order to evaluate such ability, mystery shoppers are recruited. This requires additional resources. In addition, if attention is paid only to the total score earned by the employee, testing will not be able to determine areas where the employee is stronger or weaker. This complicates the process of distributing employees in shifts and restaurants. Here, the growth/decline dynamics of certain skills is not taken into account.

Thus, choosing a suitable appraisal method for non-managerial employees becomes a challenge. Testing alone cannot solve this problem due to its number of shortcomings. Also, there is no standard employee competency card because internal standards vary from restaurant to restaurant.

Nowadays competency approach is perspective direction of researches [11, 12]. This research proposes a competency model for performance appraisal. The model makes it possible to strengthen the efficiency of the existing test-based appraisal system. The authors have tried to eliminate the third shortcoming and determine how negatively the second shortcoming affects results. They have also attempted to find a correlation between theoretical knowledge test results and the ability to put the theoretical knowledge into practice.

Thus, the paper does not aim at dismantling the already established performance appraisal system at an enterprise. The purpose of the research is to design the so-called superstructure. This will help not only to determine the current position of an employee, but also to track the dynamics and draw appropriate conclusions about his ability to learn and refine himself.

2   Materials and Methods

In this section, the proposed competency model for evaluation of non-managerial employees in catering enterprises is considered in more detail.

This model represents a weighted graph:

$$G = (V, R), \tag{1}$$

where $V = <P, Q, C>$ is a set of vertices of the graph:

$P = \{p_i\}$ – positions, $i = 1..N_P$;
$Q = \{q_j\}$ – competencies, $j = 1..N_Q$;
$C = \{c_k\}$ – competence components; $k = 1..N_K$;
$R = \{\{r_{ij}\} \cup \{r_{jk}\}\}$ – a set of edges describing the connections between the vertices.

For example, the weight of edge r_{jk} between competence q_j and its component c_k determines what proportion of competence level L_{qj} makes up component level L_{ck}. Thus, competence level L_{qj} is calculated based on the levels of its N_{qj} components via the formula:

$$L_{qj} = \sum_{k=1}^{N_{qj}} \left(L_{ck} \cdot r_{jk} \right) \qquad (2)$$

Position level L_{pi} is calculated based on the levels of its N_{pi} competencies using the formula:

$$L_{pi} = \sum_{j=1}^{N_{pi}} \left(L_{qi} \cdot r_{ij} \right) \qquad (3)$$

These formulas are not applied when calculating the levels for leaf vertices.
Thus, the graph can be provisionally divided into three layers:

1. Layer of positions held by employees.
2. Layer of competencies that must be possessed by employees holding certain positions.
3. Layer of components of competencies, which are knowledge, skills and abilities of employees.

Some vertices of the competency layer may be leaves and have no connection with the components layer. This situation usually occurs where it is impractical to divide a competence into components.

To ensure flexibility of the proposed competency graph model, the "here and now" assembly principle is contemplated. This means storing data only on leaf vertices. A complete graph is formed on the basis of formulas (2) and (3). Due to this, the graph can be changed relative to the layers without triggering serious consequences.

An example of a fragment of the model is presented in Fig. 1.

The diagram is interpreted as follows: A waiter should be able to sell the company's products, which implies having a knowledge of sales and sales methods, as well as being able to offer various menu items. In addition, there are other competencies that a waiter should possess. At the same time, some of the competencies have cross-links to components, or may also apply to another position.

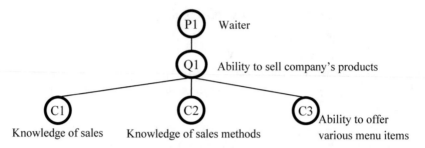

Fig. 1. A fragment of a waiter competency model

The proposed graph competency model underlies the formation of an employee profile – a special structure for storing information about the employee himself, his professional and personal qualities, and the dynamics of changes in these qualities. In total, there are three elements of the profile:

4. Personal data (full name, place of work, position, etc.).
5. Information on level of competence.
6. History of changes in the level of competence.

The profile can be formed using various sources.

So, an employee profile contains both static (1) and dynamic (2 and 3) elements. Of interest in this study are the dynamic elements. This paper will focus on them.

Using the profile, one can determine the current level of employee competencies and track his professional growth. Preserving the history of changes in the level of competence will help not only to monitor the dynamics but also to determine the employee's response to changes in external factors influencing competencies, for example, when the company updates its standards.

In order to test the competency model developed, a software system was designed to track the professional growth of employees at a restaurant chain. Employee profile is a key element of the software system. Test results serve as data sources in the study. Having a clear structure of results and being able to relate them to competence allows for automatic processing.

The system can fill in the database, which includes importing data from sources, editing employee profiles, and obtaining results of analysis of competency change dynamics. Analysis of the dynamics of changes in competencies involves evaluating the nature of changes in the level of each competence over a selected period of time. It enables one to automate not only the process of identifying problem-plagued employees, but also the process of pinpointing their specific shortcomings. This can help in optimization when planning on how to eliminate the shortcomings.

Test results are used to form employee profile as follows. In the process of compiling a test, each of the questions is assigned a competence or its component, thereby creating a correspondence table for questions and competencies. In the course of processing the test results, the automated system counts the number of correct answers A for an employee and the total number of questions T for each competency or component. The level of competence or component L_{leaf} is the ratio of the number of correct answers A to the total number of questions T:

$$L_{leaf} = \frac{A}{T} \tag{4}$$

Formula (4) can only be used to calculate the levels of leave vertices.

After the levels are defined, they are saved in the database, while the previous profile state is saved in the profile history.

The software system is implemented in the training and development department of a pizza restaurant chain, whose employees are trainers. Their responsibility includes training and monitoring employees, as well as solving tasks aimed at developing the company. Trainers are tasked with compiling tests, scheduling visits by mystery shoppers, and processing the results obtained.

The competency model is tested at five restaurants of a pizza restaurant chain, which has 30 restaurants and over 300 waiters. It was decided to conduct the test, which involves checking whether in the formed employee competency model, there is a correlation between practical skills determined by a set of performance indicators and the theoretical knowledge identified during testing.

Two experiments are conducted. The first experiment is to determine whether the test results and the results obtained from the mystery shopper match. The second experiment is aimed at determining whether there is a correlation between the waiter's average check (bill) and his knowledge of sales methods.

During the first experiment, the level of competence associated with the standard of service is compared with the level of ability to serve customers appropriately (as determined by the mystery shopper). Since newly recruited employees are not always able to stick to instructions due to lack of experience, experimental data includes only levels of employees who have at least than two month of experience in the restaurant.

In the second experiment, the competency dynamics associated with sales methods is compared with the dynamics of the relative average check. The average check is a financial indicator of a waiter's performance, which may be influenced by his ability to sell various menu items. Since this average check depends on many external factors, the authors decided to choose the relative average check instead of the absolute average check to be compared with the competency dynamics. The relative value is calculated via the formula:

$$AOV_{ir} = \frac{AOV_{ia}}{AOV_c} \tag{5}$$

where AOV_{ia} is the waiter's average check, and AOV_c is the restaurant's average check.

Also taken into account is the total deviation AOV_d of the waiter's average check from the restaurant's average check. It is calculated via the formula:

$$AOV_d = AOV_{ir} - 1 \qquad (6)$$

This is due to the fact that in some cases, the relative value decreases if the waiter's average check has decreased, and in other cases, if the restaurant's average check has increased, which can occur thanks to improvement in the skills of those waiters that are lagging behind.

3   Results

The following charts show results obtained from the experiments.

3.1   Comparison of Competency Dynamics with the Mystery Shopper

The principle of constructing a petal diagram representing the experimental results is as follows. Each axis represents the details of a particular employee. Two points are placed on each axis: one denotes the level of competence in the profile, while the other is the level of competence in the results obtained from the mystery customer.

Figure 2 shows the performance of employees that have been with the pizza chain for at least two months.

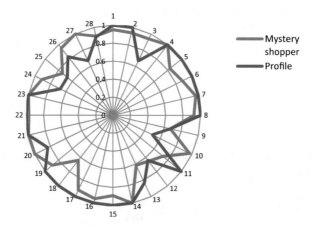

Fig. 2. Comparison of evaluation results (experienced employees) obtained from mystery shoppers with the profiles of these employees

In most cases there was minor (less than 0.1) difference between profile data and information obtained from mystery shoppers. This suggests that there is correlation between compared indicators. But there were about one third cases with major difference. It could occur due to uncontrollable events like full restaurant load, cheating during the test or obtaining a warning about mystery shopper coming.

3.2 Comparison of Competency Dynamics with the Average Check Dynamics

Figure 3 presents results obtained from comparing the average check dynamics with the employee competency dynamics. A total of 50 employees participated in the sample, but cases, where there were no competence dynamics, were excluded from the results. This was done because such cases related to analysis of the impact of the static state of competencies on a check (bill).

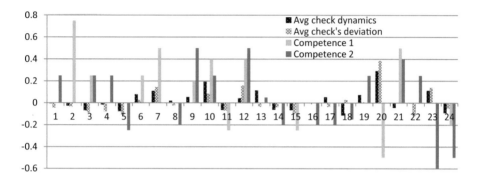

Fig. 3. Comparison of the competency dynamics with the average check dynamics

Now, with regards to schedule creation, the Y-axis shows the values of employee details, while the X-axis represents the numbers of employees participating in the experiment. There are four bars in each row. The first two relate to the average check dynamics: one is the dynamics of the relative average check, the other is the total deviation from the standard (which is equal to 1). The remaining bars represent the dynamics of competencies that are associated with sales methods and are checked during testing.

A fall in the relative average check is negative if its total deviation is less than zero, meaning that the waiter's indicator falls below his colleagues' average indicator.

Comparison of the competency dynamics with the average check dynamics (separately for each competence), revealed that they differed in one-third of all cases. But at the same time, a concurrent rise or fall in the two competencies was accompanied by a similar average check dynamics. Exceptions were cases 3 and 21. For case 21, the fall in the relative average check could be as a result of an increase in the average check of the restaurant as a whole, since the average check of this employee did not fall below the restaurant's average check. When examined individually, case 3 showed that the waiter was unprofessional in his official duties but still delivered good test results, just to avoid losing his pay rise. This was due to loss of motivation.

The results of this experiment show that there is a correlation between the waiter's average check dynamics and the dynamics of his knowledge of sales methods. However, like in the first experiment, there were deviations due to uncontrollable events.

4 Discussion

Experiments conducted in the process of testing the competency model confirmed there is correlation between employees' theoretical knowledge, which were checked via testing, and the practical skills and abilities determined using various performance indicators. This lends weight to the claim that the model is adequate and consistent. However, there are a number of circumstances that must be regarded when using the model.

The first is the "purity" or reliability of the source data. The model assumes that the original data is 100% pure, which is quite difficult to achieve in reality. The main obstacle is that a considerable proportion of employees tended to cheat during test in order to mask their lack of certain knowledge. This cheating attitude was partly attributed to the employees' desire to gain the minimum required test score. It is based on this minimum bar that decisions regarding the employee are made. Cheating during tests leads to unreliable employee profile data. This erases any correlation between the profile and the real state of things. It is a major problem, whose solution lies beyond the scope of this study.

The second circumstance is uncontrollable situations. As experiments showed, evaluation of an employee through tests is not always consistent with his other performance indicators. Here, there are cases where cheating was the reason for such deviation. For example, when there are too many customers at the restaurant at a time, which is a uncontrollable for the restaurant – priority shifts from quality to speed. During such peak hours, the restaurant tries to ensure that the customer does not leave displeased and with empty stomach. The waiter's average check, in turn, depends not only on knowing the sales methods but also, for example, on the number of orders attended to by the waiter and the proportion of business lunches in these orders. Business lunches are typically cheap. It is therefore a uncontrollable for the waiter whose customers are mainly ordering business lunches – the waiter's financial performance will be relatively lower. Here, the average bill of a business-lunch customer is usually lower than the average restaurant bill.

When taking uncontrollable into account, the range of profile formation sources should be expanded, ensuring flexibility in customizing the model. This would improve accuracy in determining competence levels.

There are three promising directions towards which the model can be developed.

7. Creating a shift profile. Based on waiters' profiles, it is possible to create compensating pairs (in which the shortcomings of one employee are compensated by the advantages of the other) and mentor-mentee pairs. On the basis of these pairs, an effective shift profile can be designed.
8. Forecasting the waiter's financial performance. Since some competencies are connected with a waiter's financial performance, changing such competencies may affect the restaurant's profit. Tracking the dynamics of these competencies in a timely manner would make it possible to predict a possible drop in profits from a particular employee due to a drop in his skills.
9. Determining employee potential. Tracking the dynamics of changes in competence levels helps in determining how quickly an employee learns and how he responds to

certain changes in his surrounding work environment (for example, when the menu is updated). In turn, this will tell whether the employee can work in extreme conditions (when the restaurant is working under peak demand especially during the holiday season) and during uncontrollable events. It will also tell whether the employee is ready for promotion or transfer to another structural division.

In conclusion, it can be claimed that the use of the developed model can really improve the efficiency of the existing test-based performance appraisal system. When solving the problems identified in the course of the research, the model can also serve as basis for a comprehensive performance appraisal methodology for the catering industry.

Acknowledgments. UNIAG, R&D unit funded by the FCT – Portuguese Foundation for the Development of Science and Technology, Ministry of Science, Technology and Higher Education. Project no. UID/GES/4752/2019.

References

1. Popova, N.V.: Features of the personnel assessment method "360°". Sbornik nauchnyh trudov, pp. 383–387 (2018). (In Russian)
2. Karkoulian, S., Assaker, G., Hallak, R.: An empirical study of 360-degree feedback, organizational justice, and firm sustainability. J. Bus. Res. **69**(5), 1862–1867 (2016)
3. Carson, M.: Saying it like it isn't: the pros and cons of 360-degree feedback. Bus. Horiz. **49**(5), 395–402 (2006)
4. Jackson, E.: The 7 Reasons Why 360 Degree Feedback Programs Fail (2012). https://www.forbes.com/sites/ericjackson/2012/08/17/the-7-reasons-why-360-degree-feedback-programs-fail
5. Brett, J.F., Atwater, L.E.: 360° feedback: accuracy, reactions, and perceptions of usefulness. J. Appl. Psychol. **86**(5), 930–942 (2001)
6. Bauer, M., Lucke, M., Johnsson, C., Harjunkoski, I., Schlake, J.C.: KPIs as the interface between scheduling and control. IFAC-PapersOnLine **49**(7), 687–692 (2016)
7. Pérez-Álvarez, J.M., Pérez-Álvarez, J.M., Maté, A., Gómez-López, M.T., Trujillo, J.: Tactical business-process-decision support based on KPIs monitoring and validation. Comput. Ind. **102**, 23–39 (2018)
8. Malysheva, M.A.: KPI: advantages and disadvantages of implementing. Informaciya kak dvigatel nauchnogo progressa: sbornik statey Mezhdunarodnoy nauchno - prakticheskoy konferencii, pp. 134–137 (2017). (In Russian)
9. Nagaeva, I.A.: The organization of electronic testing: advantages and disadvantages. On-line J. "Naukovedenie" **5**(18), 119 (2013). In Russian
10. Loban, N.E., Dongak, H.G.: The problem of testing personnel in the enterprise. Ekonomika I socium **5–1**(36), 806–809 (2017). In Russian
11. Boyatzis, R.E.: Competencies in the 21st century. J. Manage. Dev. **27**(1), 5–12 (2008)
12. Baylina, P., Barros, C., Fonte, C., Alves, S., Rocha, Á.: Healthcare workers: occupational health promotion and patient safety. J. Med. Syst. **42**(9), 159 (2018)

A Governing Framework for Data-Driven Small Organizations in Colombia

Diana Heredia-Vizcaíno[1]([⊠]) [iD] and Wilson Nieto[2] [iD]

[1] Universidad Simón Bolívar, Barranquilla, Colombia
dianahv@unisimonbolivar.edu.co
[2] Universidad del Norte, Barranquilla, Colombia
wnieto@uninorte.edu.co

Abstract. Historically, data has been seen as the result of business transactions, with a little value beyond them, but they were not treated as a valuable shared asset. From the early 1990s, business decisions and processes started to be driven increasingly by data and data analysis. Further investment in data management was the approach taken to tackle the increasing volume, velocity, and variety of data, such as complex data repositories, data warehouses, and complex information systems. Data needed to be shared between multiple systems, which gave origin to the master data management. The first attempts of Data Management and Data Governance were failed, owing that, they were conducted by Information Technology division; today, Data Governance is still an under-researched area.

In Colombia, very few organizations have started to implement Data Governance programs. There are IT Governance initiatives, within which there may be some particular linings for Data Quality or Metadata Management, as well as for Information Security. In the year 2014, the *Ministerio de Tecnologías de la Información* (MinTIC, a national government organization) designed a guide for Data Governance, which must be followed by every public institution, as part of the national policies such as *Gobierno Digital Colombia*. Small organizations, generally, know little or nothing about this initiative, but are aware of the importance of exploiting the data to obtain useful knowledge for the development of their businesses.

Keywords: Data governance · Data driven organizations

1 Introduction

Small businesses, even more so in the Colombian environment, do not apply mechanisms for the use of their data, in such a way that they effectively support their decision-making processes. Data governance can provide an organization with a methodology to comprehensively manage their data, but it is necessary to adapt it to the particular characteristics of those small.

This document shows the result of a research work, in which an initial review of the state of the art of data governance models was carried out; from it, an appropriate framework for small organizations is proposed.

© Springer Nature Switzerland AG 2019
Á. Rocha et al. (Eds.): WorldCIST'19 2019, AISC 930, pp. 622–629, 2019.
https://doi.org/10.1007/978-3-030-16181-1_59

2   Data Governance

2.1   Basic Concepts

Data governance is defined as "the exercise of decision-making and authority for data-relates matters" [3] and enables the organization to integrate and manage the data and applications, the processes, people, tools and the policy formulation [4]. It needs to be connected with the corporate governance and the IT governance [5]. Data governance, must ensure the data attributes: quality, accessibility, availability, security, among others [6]. It requires a framework, which contains the strategies and methods to support the information management programs [7]. Data Governance must be a constant program and must establish metrics that ensure the effective and efficient use of data and information in enabling the organization to achieve its goals [8, 9]. Organizations that have given important guidelines to data governance are, among others, ISO 38500, Data Management Association (DAMA), Data Governance Institute (DGI), IBM Data Governance Council, Information Systems Audit and Control Association (ISACA) [10].

The DGI's data governance framework [3] was created, principally, to help to create a clear mission, maintain the scope and focus, establish accountabilities and define measurable achievements in data issues, among others. DGI establishes seven universal goals for data governance programs: Enable better decision-making, reduce operational friction, protect the needs of data stakeholders, train management and staff to adopt common approaches to data issues, reduce costs and increase effectiveness, ensure transparency of processes and build standard and repeatable processes. DGI also defines six focus areas: Policy, Standards, and Strategy; Data Quality; Privacy, Compliance, and Security; Architecture and Integration; Data warehouse and Business Intelligence; Management Support.

A particular Data Governance Framework can be defined for one or more of these focus areas.

2.2   Data Governance Frameworks: State of the Art

A researching for models or frameworks of data governance shows that several of them try to propose generic models or adaptable to any type of organization, however there are some initiatives designed for small companies. Other proposals take into account the implications of the large volumes of data currently available and the availability of cloud computing to manage data.

Any of these works and their principal contributions are:

Khatri and Brown [11] provide a framework for Data Governance with five data decision domains: Data Principles, Data Quality, Metadata, Data Access, Data Lifecycle. For each of them, authors suggest potential roles of accountability.

Yulfitri [12] proposes an Operational Data Governance Model based on DMBOK (Data Management Body of Knowledge), which focuses on three functions: Data Governance, Metadata Management and Data Quality Management; the model was designed with the level of the management of the current organizational structure.

Yamada and Peran [13] present an Analytics Governance Framework to help organizations to transform their business by maximizing the usage of data (big data) and analytics. Authors propose the building of a synchronized roadmap for each analytics project, in which data, infrastructure and skill development must be synchronized with other projects. The governance board must coordinate all roadmaps according the business priorities and then, creating an Analytic Roadmap.

Nwabude and others [14] focus in the need of a particular Data Governance Framework for small businesses, attending their differences with large ones: Lack of resources, time, technology or expertise to research in new business lines and innovation. Also, small businesses, generally, have small organizations and the decision making is done for a unique person. Authors consider that this type of framework must be easy to implement, low cost (Implementation, maintenance, and training), consider the internal business structure, have fewer roles, and require no IT staff or expertise.

Eber and others [15] propose a flexible data governance model, which contains roles, decision areas and responsibilities. They propose five roles: Executive Sponsor, Data Quality Board, Chief Steward, Business Data Steward and Technical Data Steward. The model has three business perspectives: Strategy, Organization and Information System.

Al- Riuthe and Benkhelifa [16] present a conceptual framework for cloud data governance based on Analytic Theory. It consists of five phases, each with its components: Initial phase, where requirements are established; Design phase, to design cloud data governance functions and the contextual integration and alignment; Deploy phase, where are implemented the governance processes; Monitor phase, ensures all data in the environment meets the governance rules; Sustain phase, enables the maintenance of the governance in the time.

Kim and Cho [7] highlight that general data governance framework doesn't work with big data, due to most sources, confidence and authority on data are uncertain. Authors define the big data quality attributes as timeliness, trustfulness, meaningfulness and sufficiency, then the framework proposed is based on them. This framework has four important aspects: Objective, Strategy, Components, IT Infrastructure (which include the big data infrastructure) and Audit and Control.

Zhan and others [10] consider that Big Data has three attributes: Data source, data processing and data application, thus big data governance is a complex activity with environment, social and technology aspects. This Big Data Governance model includes Governance objectives, the Top-level design, Governance objects and methods, the internal and external environments and contributing factors.

Al-Ruithe and others [17] Identifies six key dimension for Traditional (Non-cloud) Data Governance Programs: Data Governance Function, Data Governance Structure, Organizational, Technical, Environmental, Measuring and Monitoring tools. In addition, authors identify the dimensions for Cloud Data Governance Programs: Cloud Deployment Model, Service Delivery Model, Cloud Actors, Service Level Agreement, plus the dimensions in traditional model.

In addition, for the Colombian organizational environment, any initiatives were found:

Osorio and others [18] present a Data Governance Plan designed for higher education institutions in Colombia, focus on data quality. The definitive plan proposed by authors contains three dimensions: accuracy, consistency and completeness, and four components: Structure (Vision, goals, work team, roles and responsibilities, data quality policies), Processes (Supervision and monitoring, Problem resolution), Strategies (Projects and services for data governance support) and Communications (Methods and accountable for the communication process).

Anaya [19] proposal is a Governance plan for open data in territorial public entities in Colombia. Author shows first the process for formulating this plan: Diagnosis of currently published open data and interviews with IT Offices in Metropolitan Area of Bucaramanga (Colombia), Open Data Government Plan design and Strategy design for future implementations. This plan is focused in the publication of open data and has four components: Policies (Defines vision and legal framework), Institutionalization (Defines roles and responsibilities, Supervision and monitoring), Problem resolution and Open data management processes.

Those works evidence that Data Governance Models are designed, generally, for a unique focus and for the specific organization characteristics, such as business model, internal structure, and information needs.

3 Proposed Data Governance Framework for Small Organizations

The proposed Data Governance Framework for small organizations is based on DGI model, but taking into account that these types of companies have limited human, financial and technological resources. The framework has the activities in the six focus areas given for DGI, defines five roles for Data Governance Organization, a Data Governance Maturity model, and a Workflow for its implementation. A specific program does not need to implement the six focus areas, since the previous diagnosis must indicate which ones are priority and, from there, design the most appropriate program.

3.1 Roles in the Data Governance Framework

Roles refer to the work or tasks that people perform within data governance programs. One person can assume different roles, according to the needs and abilities. Proposed roles are:

1. Executive Sponsor (ES): Provides sponsorship, strategic direction, funding and oversight. Must be an executive or senior master.
2. Data Governance Board (DGB): Defines the data governance framework and controls its implementation. Must be integrated for Chief Steward, Business units and Technical stewards.
3. Chief Steward (CS): Puts the board's decision into practice, enforces the adoption of standards and helps establish metrics and targets. Must be a senior management with data management background.
4. Business data steward (BDS): Details standards and policies for her/his area. Must be a professional from business unit.

5. Technical data steward (TDS): Provides standardized data elements definitions and formats, source systems details and data flow between systems. Must be a professional from IT department.

Each focus area has particular activities and responsibilities, each of them must have accountable roles.

3.2   Workflow for Data Governance Implementation

The implementation of a Data Governance program must be cyclic, due to it is a permanent process while organization needs to work with data, as showed in Fig. 1.

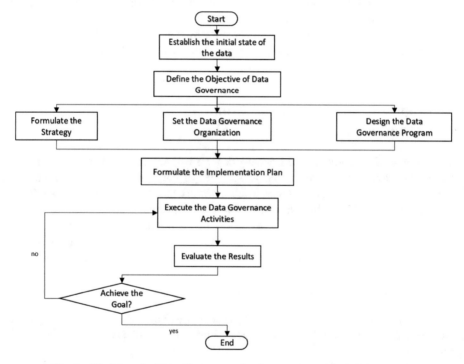

Fig. 1. Workflow for Data Governance implementation (Adapted from [10])

The process begins with a diagnosis of the current situation of the data, using the Data Maturity Model, adapted from [20, 21]; from it, the objective to be achieved in Data Governance is defined, the DG Program and its implementation plan is designed. Then, that plan is executed and periodically evaluated in order to adjust the program. This process is cyclic and must be repeated until the objective be achieved.

The Data Maturity Model has six levels (0–5), everyone with three dimensions or points of view:

- People: It refers to organization structure for Data Governance, roles definition, profiles and skills that are required.
- Policy: It is about development, auditing and enforcement of data rules, standards and best practices.
- Architecture: It contains the technological components necessary to support the data lifecycle

And five aspects or variables of DG that are evaluated:

- Formalization: Evaluates existence, definition and implementation.
- Metadata: Describes other data and IT assets by relating essential business and technical information. Facilitates the consistent understanding of the characteristics and usage of data.
- Data Stewardship: The formalization of accountability for the definition, usage and quality standards of specific data assets within a defined organizational scope.
- Master Data: Business-critical data that is highly shared across the organization. Master data is often codified data, data describing the structure of the organization or key data entities.
- Data Quality: The continuous process for defining the parameters for specifying acceptable levels of data quality to meet business needs, and for ensuring that data quality meets these levels.

The Maturity levels are defined based on each dimension and each variable:

Level 0 – Incomplete: the processes of Data Governance do not exist or are informal. Roles are not assigned. Architecture is basic and does not support the DG activities.

Level 1 – Executed: Some processes are implemented, a few DG roles are assigned and Architecture support some of the DG activities.

Level 2 – Managed: The formal DG policies and organization are defined and some roles are assigned, architecture is defined and some basic layers are developed.

Level 3 – Established: DG policies are implemented and roles are mostly assigned. The architecture is developed.

Level 4 – Predictable: Roles totally assigned, policies implemented and architecture supports effectively the business and DG activities.

Level 5 – Optimized: The DG policies, organization and architecture are totally defined, implemented, measured and in constant improvement.

The goal of an execution cycle of the framework is to reach the next level of maturity, from the diagnosis; a cycle can be applied to a single variable or dimension, or several of them.

4 Conclusions

In the Colombian business environment, the small ones are the majority. They can make better use of their data and those available from multiple sources, in order to innovate in products and processes and make better decisions. For this, these organizations must implement Data Governance programs, but appropriate to their needs and availability of resources. These programs must be flexible, easy to implement (in short cycles), scalable and low cost.

References

1. Al-Ruithe, M., Benkhelifa, E., Hameed, K.: A systematic literature review of data governance and cloud data governance. Pers. Ubiquit. Comput. **22**, 1–21 (2018)
2. Mejía Jaramillo, M.: Arquitectura TI Colombia. http://www.mintic.gov.co/arquitecturati/630/w3-article-9258.html. Accessed: Apr 2018
3. The Data Governance Institute, The Data Governance Framework. http://www.datagovernance.com/wp-content/uploads/2014/11/dgi_framework.pdf. Accessed Feb 2018
4. DeStefano, R., Tao, L., Gai, K.: Improving data governance in large organizations through ontology and linked data. In: IEEE 3rd International Conference on Cyber Security and Cloud Computing, Beijing (2016)
5. Tallon, P.: Corporate governance of big data: perspectives on value. Risk Cost Comput. **46**(6), 32–38 (2013)
6. Castillo, L., Raymundo, C., Domínguez, F.: Information architecture model for the successful data governance initiative in the Peruvian higher education sector. In: IEEE XXIV International Conference on Electronics, Electrical Engineering and Computing (INTERCON), Cusco (2017)
7. Kim, H., Cho, J.: Data governance framework for big data implementation with a case of Korea. In: IEEE 6th International Congress on Big Data, Honolulu (2017)
8. Wróbel, A., Komnata, K., Rudek, K.: IBM data governance solutions. In: International Conference on Behavioral, Economic, Socio-cultural Computing (BESC), Krakow (2017)
9. Bhansali, N.: The Role of Data Governance in an Organization, Data Governance: Creating Value from Information Assets, p. 268. Auerbach Publications (2013)
10. Zhan, S., Gao, H., Yang, L., Song, J.: Research on big data governance based on actor-network theory and petri nets. In: Proceedings of the 2017 IEEE 21st International Conference on Computer Supported Cooperative Work in Design, Wellington (2017)
11. Khatri, V., Brown, C.: Designing data governance. Commun. ACM **53**(1), 148–152 (2010)
12. Yulfitri, A.: Modeling operational model of data governance in government. In: International Conference on Information Technology Systems and Innovation (ICITSI), Bandung (2016)
13. Yamada, A., Peran, M.: Governance framework for enterprise analytics and data. In: IEEE International Conference on Big Data (BIGDATA), Boston (2017)
14. Nwabude, C., Begg, C., McRobbie, G.: Data governance in small businesses – why small business framework should be different. In: 3rd International Conference on Business, Management and Governance, Las Vegas (2014)
15. Eber, K., Otto, B., Österle, H.: One size does not fit all—a contingency approach to data governance. ACM J. Data Inf. Qual. **1**(1), 1–27 (2009)

16. Al- Riuthe, M., Benkhelifa, E.: A conceptual framework for cloud data governance-driven decision making. In: International Conference on the Frontiers and Advances in Data Science (FADS), Xian (2017)
17. Al-Ruithe, M., Benkhelifa, E., Hameed, K.: Key dimensions for cloud data governance. In: IEEE 4th International Conference on Future Internet of Things and Cloud, Vienna (2016)
18. Osorio, M., Guerrero, C., González-Zabala, M.: La gobernabilidad de datos como apoyo en la gestión de datos de instituciones de educación superior, Revista Espacios, 38(51), 11–23 (2017)
19. Anaya, B.: Plan de Gobernabilidad de Datos Abiertos Para Entidades Públicas Territoriales del Área Metropolitana de Bucaramanga, Bucaramanga (2017)
20. Stanford University: Data Governance at Stanford: A Data Governance Overview. http://web.stanford.edu/dept/pres-provost/irds/dg/files/DG-News001.pdf. Accessed Mar 2017
21. Stanford University: Data Governance Maturity Model - Guiding Questions for each Component-Dimension, https://web.stanford.edu/dept/pres-provost/irds/dg/files/StanfordDataGovernanceMaturityModel.pdf. Accessed Mar 2018
22. IBM Corporation: IBM Data Governance Council Maturity Model, IBM Corp. (2007)

Critical Success Factors for Corporate Data Quality Management

Ana Lucas[✉]

ADVANCE/CSG ISEG, Lisbon, Portugal
ana.lucas@iseg.ulisboa.pt

Abstract. Currently data are often considered the most vital asset of an organization; they need to be of good quality in order to properly sustain the
organization's competitive advantage and support operational and decision
support systems. The identification and rating or ranking of the critical success
factors (CSFs) for data quality management is still understudied, although it is
very significant for organizations that intend to implement an adequate management of data quality. This work is an exploratory study on that subject. It
identifies and rates 19 CSFs, through a qualitative research methodology, using
a focus group, followed by a Delphi study. The results show that the top three
factors are Data Governance, Management Commitment and Leadership and
Continuous Data Quality Management Improvement.

Keywords: Critical success factors · Data quality · Data quality management ·
Qualitative research · Focus group · Delphi study

1 Introduction

Data quality management is an issue of growing importance to academic and professional communities. Today, there is great concern for the quality of corporate data, as
data of poor quality mean inaccurate information, which in turn leads to waste of
resources, poor decision making and damage to the organization, particularly with
regard to regulatory compliance and the relationships with its customers [1]. Con-
comitantly, digital enterprise transformation and the increasing use of big data- particularly for predictive analysis, machine learning and data mining - have led to the
explosive growth of data in almost every industry and business area, further under-
scoring the increasing importance of data quality [2, 3]. The concept of big data was
first defined in 2001 by Gartner [4]; it comprised 3Vs - Volume, Velocity and Variety,
to which IBM added in 2012 another V -Veracity [5], meaning that data should be
trustworthy and accurate [6]. According to a survey conducted by CrowdFlower and
reported by [7], cleaning and organizing data account for 60% of the work of data
scientists[1].

---

[1] The title was coined in 2008 by D.J. Patil and Je Hammerbacher. Data scientists are high-ranking
professionals with the training and curiosity to make discoveries in the world of big data [50].

© Springer Nature Switzerland AG 2019
Á. Rocha et al. (Eds.): WorldCIST'19 2019, AISC 930, pp. 630–644, 2019.
https://doi.org/10.1007/978-3-030-16181-1_60

To obtain an idea of poor data quality costs, The Data Warehousing Institute calculated that data quality issues cost U.S. businesses about USD 600 billion a year [8]. Similarly, a study conducted in 2016 by Royal Mail shows that the poor quality of customer contact data costs, on average, 5.9% of the annual revenue to UK companies [9]. The costs of poor data quality have been classified by English [10] into three categories:

- Process failure costs, which occur because processes are not properly performed due to poor quality data, such as costs related to incorrectly delivered or undeliverable mail due to inaccurate mailing addresses;
- Data scrap and rework, such as costs related to resending mail or scrapping of defective data and their processing, to achieve the desired quality levels;
- Opportunity costs, due to lost and missed revenues. For example, due to the low accuracy of customer addresses related to "loyalty cards", some of these card owners do not seem to be reached in advertising campaigns, which leads to lower revenues [1]. Another example is a customer loss due to incorrect billing.

Although data and information mean slightly different things, for reasons of simplicity the terms are often used interchangeably [11, 12]. Data quality research, which began in the late eighties of the last century, mainly focused on computer science issues, such as developing techniques for querying multiple data sources and building large data warehouses [13]. Later on, in the early 1990s, MIT established its Total Data Quality Management (TDQM) program and developed the TDQM framework, which extends to the domain of data the Total Quality Management framework for quality improvement in the manufacturing domain [13]. Nowadays, research in data quality operates in two major scientific fields: Computer Science (CS) and Management Information Systems (MIS) [13]. Our study falls clearly in the latter.

The main objective of this research study is to find the critical success factors for data quality management, including their relative importance to the process. Given that quality of unstructured data is still an under-researched area, it will not be considered in this study. The study approaches the topic using two qualitative methods – a focus group and a Delphi study.

The research questions can be summarized as follows:

- What are the critical success factors (CSF) for data quality management?
- What is the relative importance of the various CSF?

2   Literature Review

This section reviews the literature about Corporate Data Quality Management and Critical Success Factors.

2.1   The Data Quality Concept

Data quality (DQ) can be defined as "fitness for use by data consumers" [14] p. 6. This definition takes the consumer viewpoint of quality in line with the general literature on quality. In addition, the concept of "data quality dimension" can be defined as a set of data quality attributes that represent a single aspect or concept of data quality [14].

Examples of data quality dimensions are accuracy, timeliness, completeness and consistency. "High-quality data should be "intrinsically good, contextually appropriate for the task, clearly represented, and accessible to the data consumer" [14] p. 6.

Three main scientific approaches and one ISO standard identify and define universal (domain independent) dimensions for data quality:

- Theoretical [15];
- Empirical [14];
- Intuitive [16];
- ISO Standard [17].

These approaches refer to both the data in extension, i.e., their values [10, 14–17], and in intention, i.e., their models or database schemas and business rules that apply to the data [16, 10]. Moreover, some authors [14, 16–10] also consider data presentation dimensions, contextual data quality dimensions, as well as system dependent dimensions, for instance data security and accessibility dimensions [14, 17]. Other dimensions such as compliance, efficiency, traceability, portability and recoverability are recognized by ISO [17].

2.2 Data Quality Management and Data Governance

According to ISO [18] p. 4, data quality management is a set of "coordinated activities to direct and control an organization with regard to data quality".

To be effective, data quality management should transcend the activities of fixing non-quality data to prevent data quality problems by managing data over its life cycle to meet the information needs of its stakeholders. In addition, DQM needs to break down the stove pipes that separate data across business units and create collaboration between business and IT functions to address both organizational and technical perspectives. This process requires a profound cultural change, demanding leadership, authority, control and allocation of resources. Thus, it entails governance, specifically data governance (DG) [19]. DG can be defined as "the exercise of authority, control and shared decision making (planning, monitoring and enforcement) over the management of data assets" [20] p. 38. Nevertheless, DG does not equal DQM, either with regard to the entity that makes the decisions, or to their scope. DG is the responsibility of the board of directors and executive management, and it is more focused on the corporate environment and strategic directions, including areas beyond DQM, like Data Security, Privacy and Information Life-Cycle Management [21]. With data governance, organizations are able to implement corporate-wide accountabilities for DQM, encompassing professionals from both business and IT units [22].

2.3 Critical Success Factors

The concept of Critical Success Factors (CSFs) was introduced by D. Ronald Daniel in 1961 [23] and further popularized by John F. Rockart [24]. According to [24] p. 85

Critical success factors are, for any business, the limited number of areas in which results, if they are satisfactory, will ensure successful competitive performance for the organization.

Rockart categorized the CSFs in two groups: monitoring and building. The CSFs in the former group serve, as the name implies, to monitor the current results, and the CSFs in the latter group are oriented towards building for the future [24].

2.4 Critical Success Factors for Data Quality Management

The literature on CSFs for data quality management is scarce. Previous studies deal with their identification [11, 25–27], or with their rating or ranking [25, 26, 28]. Some papers rate CSFs in different industries [29] or in different sized organizations [30], or group CSFs into categories [31] while others only rate the social, cultural and organizational CSFs [32]. In view of this situation, it was decided to start the literature review by CSFs for Total Quality Management (TQM), for it is a much more mature subject. In addition, data quality management can be considered a sub-area of the TQM, because there is an analogy between the quality aspects between the manufacturing of products and that of information [33].

For the literature review on CSFs for TQM it was decided to review papers that are themselves literature reviews [34–37] and the criteria underlying two important quality awards, one for the USA, Malcolm Baldrige National Quality Award [38], and the other for Europe, EFQM Excellence Model [39] and the ISO Quality Management Principles [40].

Table 1 summarizes a literature review on CSFs for DQM, where the first column corresponds to a synthesis of CSFs for TQM.

Table 1. Critical success factors for data quality management

Nb	Critical success factors	CSF for TQM	Baškarada and Koronios [11]	Xu, Koronios and Brown [19]	Xu [10]	Akpon-Ebiyomare, Chiemeke, and Egbokhare [15]
1	Management commitment and leadership	Management commitment and leadership	Information quality management governance	Top management/Middle management commitment to DQ	Top management/Middle management commitment to DQ	Management commitment & support
2	Organization for quality	Organization for quality		Organizational structure	Appropriate organizational structure/Establishing DQ manager position	
3	Training	Training and learning	Training	Education and training	Education and training	Training and communication
4	Information quality requirements management	Product/Service Design	Information quality requirements management			
5	Supplier partnership	Supplier partnership		Data supplier quality management	Data supplier quality management	Information supplier quality management
6	Information product lifecycle management	Process management	Information product lifecycle management			
7	Information quality assessment/ monitoring	Quality data and reporting	Information quality assessment/ monitoring		Measurement and reporting	

(*continued*)

Table 1. (*continued*)

Nb	Critical success factors	CSF for TQM	Baškarada and Koronios [11]	Xu, Koronios and Brown [19]	Xu [10]	Akpon-Ebiyomare, Chiemeke, and Egbokhare [15]
8	Teamwork	Employee involvement		Teamwork	Teamwork	Teamwork between different departments and within departments
9	Customer focus and satisfaction	Customer focus and satisfaction	Information quality requirements management	Customer focus	User focus	Customer focus/User involvement
10	Strategic Data Quality Planning	Strategic quality planning				
11	Continuous information quality management improvement	Operations	Continuuous information quality management improvement			
12	Culture and communication	Culture and communication	Organizational culture of focusing on DQ		Organizational culture of focusing on DQ/Clear DQ vision for entire organization	Training and communication
13	Continuous improvement	Continuous improvement	Continuous information quality improvement	Continuous improvement	Continuous improvement	Continuous improvement
14	Information architecture management		Information architecture management			
15	Information security management		Information security management			
16	Storage management		Storage management			
17	Risk management		Information quality risk management	Risk management	Risk management	
18	Physical environment			Physical environment	Physical environment	Conducive physical environment/ continuous power supply
19	Understanding of the systems and DQ			Understanding of the systems and DQ		Understanding the system and importance of DQ
20	Personnel competency			Personnel competency	Personnel competency	
21	DQ policies and standards			DQ policies and standards	DQ policies and standards	DQ policies and standards
22	DQ controls/Input Controls			DQ controls/Input Controls	DQ controls/Input Controls/Internal controls	DQ controls & improvement/Input controls/Internal controls
23	Nature of the IS			Nature of the IS	Nature of the AIS	Nature of IS
24	Employee relations			Employee relations	Effective employee relations	Employee personnel relations
25	Management of changes			Management of changes	Management of changes	Change management
26	Audit and reviews			Audit and reviews	Audit and reviews	
27	Evaluate cost/benefit trade-offs		Continuous improvement	Evaluate cost/benefit trade-offs	Evaluate cost/benefit trade-offs	

3 Data Collection and Methods

For this study, a Focus Group followed by a Delphi Study was chosen. The focus group research method is useful when information systems researchers are interested in new topics to be studied through exploratory research [41]. By contrast, Delphi is a method for structuring a group communication process that enables a group of individuals as a whole to deal with a complex problem [42]. The method allows for the analysis of qualitative data, and, instead of random sampling, it uses a number of experts. In the context of a Delphi study, an expert is a specialist in the field of knowledge within which the study is developed.

A Delphi study consists of a series of questionnaires, each one corresponding to a round. The rounds continue until a consensus is reached or it is found that one is not possible.

The Q-methodology was developed by [43], cited by [44], and it provides grounds for the systematic study of subjectivity. The distinctive feature of the Q-Sort technique, a component of Q-methodology, is that panel members must order the questions provided under a predefined distribution, usually approximately normal [44].

Qualitative studies follow the process of analytical generalization [45], where:

Analytic generalization may be defined as a two-step process. The first involves a conceptual claim whereby investigators show how their study's findings are likely to inform a particular set of concepts, theoretical constructs or hypothesized sequence of events. (…). The second step involves applying the same concepts or theoretical constructs to implicate other similar situations.
In [45], p. 105]

3.1 Focus Group

On the 28th of June 2018, a Focus Group was held in a meeting room of Lisbon School of Economics and Management (ISEG). The duration of the meeting was 1 h and 16 min and the 6 attendees were recognized academics and practitioners. The focus group discussion was recorded, with the participants' permission.

The participants were previously informed of the focus group objectives, which sought consensus in the answers to the following three questions:

– Is there any CSF that should not be considered as such?
– Are there some CSFs missing? Which ones?
– Are there some CSFs that can be merged?

The focus group reached a consensus that "physical environment", "storage management", "employee relations" and "nature of IS" should be considered contingency factors, not CSFs. It was decided to add "data governance", defined as "set of key actions to ensure data compliance with organizational strategies", to the list of CSFs. The group also decided to include "organization for quality" and "teamwork" in the CSF "data governance", thus adapting its definition. The CSF "strategic data quality planning", considered as "data quality should be in line with the strategic plan", was included in the CSF "management commitment and leadership".

In addition, the participants reached a consensus to include the CSFs "continuous improvement" and "DQ controls/input controls" in the CSF "continuous information quality management improvement", defined as "set of actions there must be taken to improve data quality".

Some CSFs were renamed: the CSF "understanding the information systems and DQ" was renamed "understanding the information systems and the relevance of DQ", the CSF "risk management" changed to "DQ risk management", and "customer focus and satisfaction" changed to "focus on data customer satisfaction", considering two types of customers: clients (external customers) and the so-called users, who are internal customers. The definition of DQ-KPIs was removed from CSF "Continuous Information Quality Management Improvement" and included in "Information Quality Assessment/Monitoring".

The following text identifies and defines the CSFs that resulted from the Focus Group:

- Management Commitment and Leadership — Top management must form a sound foundation for clear values and data quality policies and provide the corresponding resources. Companies must integrate data quality into the organizational strategy to achieve consistent and lasting excellence [35];
- Data Governance – This includes a set of key actions to ensure data compliance with organizational strategies. It defines a suitable organizational structure to produce high quality information. It should define responsibilities for the DQ: identify the owners and custodians; appoint data stewards and a data champion; appoint an expert or a group of experts as DQ managers [28]. It should promote teamwork between business and IT people, as a key to improving data quality;
- Training - Training in data quality concepts, methods and tools is a precondition for employee involvement and empowerment [35]. Relevant training needs should be identified and documented, and training workshops should be conducted regularly. In addition to formal training, mentoring programs should ensure on-the-job professional development [11];
- Data Quality Requirements Management – It is important to identify all the key stakeholders and collect and model their requirements [11];
- Supplier Partnership - "Data supplier quality management means to have an effective data quality management relationship with raw data suppliers, which has two important parts: 1. To have agreement about the acceptable level of quality of raw data to be supplied, such as the requirements of availability, timeliness, accuracy, and completeness; 2. To provide regular data quality reports and technical assistance to data suppliers" [29] p. 291;
- Data Product Lifecycle Management - Managing information as a product as well as effectively managing the information processes (life-cycles of critical information products) is important for effective data quality management. One of the aspects of this CSF includes identifying and documenting the data flow within the organization as well as between the organization and any external parties (i.e., information product supply chain management) [11]. Clarity of process ownership (process owners), boundaries, and steps must be established [34];

- Data Quality Assessment/Monitoring - Before any DQ improvements can be attempted, the current state of DQ first needs to be assessed both qualitatively and quantitatively [11]. Profiling tools can be used to access most of the data quality dimensions. Qualitative and quantitative DQM metrics or Key Performance Indicators (DQ-KPIs) should be defined, and then used to continuously monitor the effectiveness of organizational DQM efforts [11]. At certain time intervals DQ should be monitored using the same data profiling tools. In addition, compliance with policies and standards should be monitored;
- Focus on Data Customer Satisfaction – This entails focusing on data customers' needs and their quality requirements. It should enable active participation from data customers to ensure and improve data quality [27]. "Data customer" can refer to the client (external customer) and the internal customer;
- Continuous Data Quality Management Improvement - Continuous DQM improvement deals with using Key Performance Indicators (DQM-KPIs) to continuously monitor the effectiveness of organizational DQM efforts [11]. There is a need for continuous and consistent data quality improvement, materialized as a set of actions that must be taken to improve data quality, such as input validations [27];
- Culture and Communication – This involves encouragement of an organization-wide culture committed to data quality improvement [37]. Communication is viewed as a two-way process with feedback channels available. Communication is seen as an ongoing process, taking into account ways of strengthening concepts in the future [36];
- Data Architecture Management - Architecture of the IS ecosystem or its geography is relevant to the type of DQ initiative. The architecture of the IS ecosystem should be described, namely the flows of information should be depicted, and the ownership of the data in each system should be identified;
- Data Security Management - Access security is a key DQ dimension and data security management requires an organization to have effective access controls in place. The controls must ensure that all users are appropriately authenticated as well as authorized with the least set of privileges they require. For instance, IS developers should not have access to the production environment. Furthermore, audit trails (logs of users' activities on the IS) should be analyzed (e.g., for exceptions) and periodically reviewed [11];
- DQ Risk Management - Risk management can be defined as the awareness of and the level of commitment to the reduction of the consequences of poor DQ [27]. DQ risks to business objectives (including financial risks, reputation risks, regulatory risks, etc.) should be diagnosed, documented, analyzed, classified, prioritized and mitigated/controlled. Effective DQ Risk Management should allow organizations to focus their DQM efforts on the most critical information products, thus, increasing DQM efficiency and effectiveness [11];
- Understanding of the IS and the relevance of DQ – It is important to understand how the information systems work (technical competence) and IT personnel an data customers need to understand the importance of data quality [27];
- Personnel Competence - The competence of personnel responsible for DQ is particularly important. For instance, employees should be exceptionally skilled and informed in both technical and business areas. [27];

- DQ Policies and Standards – The organization should have data quality policies and standards that are simple, relevant and consistent. There are two main components: 1. Establishing appropriate and specific data quality policies and standards; 2. Implementing/enforcing policies and standards [27];
- Management of Changes – DQ requirements, which can be internal or external, should be included and consistently updated in the process of management of changes. Internal changes include structural changes, such as organizational restructuring as well as micro changes, such as the change of an attribute domain. External changes include things such as government regulations, technology, economy, and market changes [27];
- Audit and Reviews – It is important to have independent internal and external regular data quality audits and reviews to ensure appropriate controls are in place [27];
- Evaluating Cost/Benefit Trade-offs – Before any process improvements are made, it is critical to estimate the costs associated with poor DQ and corresponding improvement initiatives, as well as any potential benefits or cost savings that may result from any process improvements [11].

3.2   Delphi Study

In order to refine the results obtained in the Focus Group and rate the CSFs, a Delphi online questionnaire was administered to a group of experts. The Delphi questionnaire consisted of the 19 CSFs described above. They were presented to the participants mostly in alphabetical order, with their corresponding descriptions. Unfortunately the CSF "Data Governance" contained a space at the beginning so it was presented in first place, and that was the only exception to the alphabetical order. The initial version of the questionnaire was submitted to two data quality experts to check its readability.

First Round

For the first round, which ran from September 10 to 26, 2018, 94 experts were selected from data quality professionals acting in various industries all over the European Union (89%) and academics knowledgeable on the subject (11%). 56 experts answered, which corresponds to a response rate of 60%.

As a result of the first round, the factors were ordered by the sum of the points assigned by each panel member, with 1 point assigned to the most important factor and 19 points to the least important. In this round experts were also asked to suggest factors that were not represented on the list submitted to them. Two new CSFs were proposed, which were not considered because they were included in CSFs presented in the initial list. The result of the first round is presented in the first column of Table 2.

In order to evaluate the consensus among the participants, the Kendall's W coefficient of agreement was used. Its value was 0.195, significant at 0.000, which, according to [46], indicates very weak agreement. For that reason, it was decided to launch a second round.

Second Round

The resultant order of CSFs from the first round was sent to all the respondents, and they were asked to respond to a new round, given the very weak consensus obtained. The second round took place from 3 to 21 October 2018 and only the 56 experts who responded to the first round were invited to participate. The result of the second round is presented in the second column of Table 2.

Table 2. Results of the Delphi Study

CSF numbers ordered by first round	CSF numbers ordered by second round	CSF numbers ordered by third round	CSF descriptions ordered by third round	Position in third round
1	1	1	Data Governance	1
3	14	14	Management Commitment and Leadership	2
7	3	3	Continuous Data Quality Management Improvement	3
14	5	5	Data Architecture Management	4
5	8	4	Culture and Communication	5
8	4	8	Data Quality Policies and Standards	6
4	7	7	Data Quality Assessment/Monitoring	7
9	9	9	Data Quality Requirements Management	8
11	16	13	Focus on Data Customer Satisfaction	9
16	13	6	Data Product Lifecycle Management	10
13	11	16	Personnel Competency	11
10	15	15	Management of Changes	12
6	2	11	Data Security Management	13
2	6	10	Data Quality Risk Management	14
18	18	2	Audit and Reviews	15
15	10	18	Training	16
19	19	19	Understanding of the IS and the relevance of Data Quality	17
12	12	12	Evaluating Cost/Benefit Trade-offs	18
17	17	17	Supplier Partnership	19

The second round was answered by 29 experts, which corresponds to a response rate of 52%, lower than in the first round. The Kendall's W coefficient of agreement was 0.317, significant at 0.000, which according to [46] indicates weak agreement. The Spearmen's rank correlation coefficient between the results of the first and the second rounds [47] was 0.674 significant at 0.01 level, which indicates a good positive correlation.

The CSF Data Governance held the first place in both rounds (obtaining 99 points versus 300 in the first round), and the last three CSFs also held the same positions in the two rounds. This result reflects the theory associated with the Q-Sort which states that the panel members are more confident on the most and least important issues [48].

Considering the weak consensus obtained and despite fearing a sharp drop in the number of responses, it was decided to launch a third round.

Third Round

The order of CSFs resulting from the second round was sent to all the respondents of that round and they were asked to respond to a new round, given the still weak consensus obtained. The third round took place from November, 18, 2018 to December, 7, 2018 and only the 29 experts who responded to the second round were invited.

The detailed results of the third round are presented in Table 3. Table 2 compares its results with those of rounds 1 and 2.

Table 3. Detailed results of the third round

Position	CSF	Sum of points	Average	Variance	Standard deviation
1	Data Governance	55	3,44	18,13	4,26
2	Management Commitment and Leadership	74	4,63	9,18	3,03
3	Continuous Data Quality Management Improvement	79	4,94	8,06	2,84
4	Data Architecture Management	109	6,81	17,63	4,2
5	Culture and Communication	111	6,94	13,53	3,68
6	Data Quality Policies and Standards	112	7	14,27	3,78
7	Data Quality Assessment/Monitoring	126	7,88	23,05	4,8
8	Data Quality Requirements Management	155	9,69	11,56	3,4
9	Focus on Data Customer Satisfaction	159	9,94	26,46	5,14
10	Data Product Lifecycle Management	164	10,25	26,07	5,11
11	Personnel Competency	179	11,19	28,16	5,31
12	Management of Changes	179	11,19	15,9	3,99
13	Data Security Management	190	11,88	18,52	4,3
14	Data Quality Risk Management	203	12,69	10,36	3,22
15	Audit and Reviews	206	12,88	42,65	6,53
16	Training	209	13,06	13,66	3,7
17	Understanding of the IS and the relevance of Data Quality	214	13,38	25,85	5,08
18	Evaluate Cost/Benefit Trade-offs	237	14,81	10,83	3,29
19	Supplier Partnership	279	17,44	5,46	2,34

The third round was answered by 18 experts, which corresponds to a response rate of 62%, higher than in the first and second rounds. The higher response rate was the result of insistence with the experts.

The Kendall's W coefficient of agreement was 0.433, significant at 0.000, which according to [46] indicates weak to moderate agreement. The Spearmen's rank correlation coefficient between the results of the second and the third rounds [47] was 0.256 significant at 0.01 level, indicating weak correlation.

It was intriguing that this coefficient dropped from 0.674 (good positive correlation) between the first and the second rounds to 0.256 (weak positive correlation) between the second and third rounds. One possible explanation for the result was that the experts might have tried to respond according to the summary of the second round. To verify if this was the case, some of them were contacted, and they confirmed the hypothesis. According to [42] respondents are sensitive to feedback from the scores of the entire group and tend to move towards the perceived consensus.

The CSF Data Governance held the first place in all rounds (obtaining 55 points in the third round versus 99 in the second round) and the last three CSFs (Understanding of the IS and the Relevance of Data Quality, Evaluating Cost/Benefit Trade-offs and Supplier Partnership) also retained the same last positions in the three rounds. It should be noted that the first eight CSFs remained in the top eight positions in the three rounds, although in some cases they changed their order.

4   Discussion

The first 10 CSFs that were found are in line with results from previous research [11, 25–28, 32]. The consistent and clear ranking of Data Governance in first place (see Tables 2 and 3) is noteworthy, for previously it was only identified in [11, 28]. This result is probably due to the increasing importance of data to organizations, particularly in the context of Big Data Management and Business Analytics, which have given rise to the implementation of Data Governance and the function of Chief Data Officer [49], among other new functions. It should also be noted that the first three CSFs - Data Governance, Management Commitment and Leadership and Continuous Data Quality Management Improvement - clearly stand out from the rest.

The CSFs in positions 6 and 8 - respectively Data Quality Policies and Standards and Data Quality Requirements Management - are the least frequently identified in previous studies, having only been recognized in one single work. In contrast, the CSFs 2, 3 and 10 - respectively, Management Commitment and Leadership, Continuous Data Quality Management Improvement and Data Product Lifecycle Management - are the most widely reported, having been identified in five of the six papers analyzed.

5   Limitations

This study suffers from the limitations of the qualitative research methodology used. As a result, it only allows analytical generalization, generalization to the theory, and not generalization to the population.

On the other hand, it is an exploratory study about a poorly studied subject, whose objective was to understand the phenomenon and to explore the two research questions:

- What are the critical success factors (CSF) for data quality management?
- What is the relative importance of the various CSF?

6 Future Work

Based on the results of this work, a new explanatory study, using a multiple case study as a research method, will be carried out in different industries to confirm, or disconfirm, the results obtained in [29] that the importance of some CSFs depends on the type of industry.

Acknowledgments. I gratefully acknowledge financial support from FCT- Fundação para a Ciencia e Tecnologia (Portugal), national funding through research grant UID/SOC/04521/2019.

I thank the University of Minho and particularly Professor Delfina Sá Soares, for making the e-Delphi tool available and for her kind support.

I sincerely thank Professor Ann Henshall for her diligent proofreading of my English.

References

1. Lucas, A.: Corporate data quality management in context. In: Proceedings of the 15th International Conference on Information Quality, ICIQ 2010, pp. 1–19 (2010)
2. Gao, J., Xie, C., Tao, C.: Big data validation and quality assurance – issues, challenges, and needs. In: 2016 IEEE Symposium on Service-Oriented System Engineering Big, pp. 433–441 (2016)
3. Jin, X., Wah, B.W., Cheng, X., Wang, Y.: Significance and challenges of big data. Big Data Res. **2**(2), 59–64 (2015)
4. Laney, D.: "Gartner," 3D Data Management: Controlling Data Volume, Velocity, and Variety (2001). http://blogs.gartner.com/doug-laney/files/2012/01/ad949-3D-Data-Management-Controlling-%0AData-Volume-Velocity-and-Variety.pdf%0A. Accessed 31 Dec 2018
5. Experian Data Quality: The data quality benchmark report (2015)
6. IBM: The path to data veracity (2018)
7. Press, G.: Cleaning Big Data: Most Time-Consuming, Least Enjoyable Data Science Task, Survey Says, Forbes (2016). https://www.forbes.com/sites/gilpress/2016/03/23/data-preparation-most-time-consuming-least-enjoyable-data-science-task-survey-says/#7903c3fa6f63. Accessed 31 Dec 2018
8. The Data Warehousing Institute [TDWI]: Data quality and the Bottom Line: Achieving Business Success through a Commitment to High Quality Data (2002)
9. Royal Mail, D.: How better customer data drives marketing performance and business growth (2016)
10. English, L.P.: Improving Data warehouse and Business Information Quality. Wiley, New York (1999)
11. Baškarada, S., Koronios, A.: A critical success factor framework for information quality management. Inf. Syst. Manag. **31**(4), 276–295 (2014)

12. Pipino, L.L., Lee, Y.W., Wang, R.Y.: Data quality assessment. Commun. ACM **45**(4), 211 (2002)
13. Madnick, S.E., Wang, R.Y., Lee, Y.W., Zhu, H.: Overview and Framework for Data and Information Quality Research, 1(1) 2:1–2:22 (2009)
14. Wang, R.Y., Strong, D.M.: Beyond accuracy: what data quality means to data consumers. J. Manag. Inf. Syst. **12**(4), 5–34 (1993)
15. Wand, Y., Wang, R.Y.: Anchoring data quality dimensions in ontological foundations. Commun. ACM **39**(11), 86–95 (1996)
16. Redman, T.C.: Data Quality for the Information Age. Artech House Inc., Norwood (1996)
17. ISO: ISO/IEC 25012, vol. 2008 (2008)
18. ISO: ISO standard 8000 part (2) version (1) - Data Quality : Vocabulary, vol. 2017 (2017)
19. Lucas, A.: Corporate data quality management - from theory to practice. In: 5a Conferência Ibérica de Sistemas e Tecnologias da Informação (2010)
20. DAMA: The DAMA Dictionary of Data Management (2008)
21. IBM Data Governance Council: The IBM Data Governance Council Maturity Model : Building a roadmap for effective data governance (2007)
22. Wende, K.: A model for data governance – organising accountabilities for data quality management. In: 18th Australasian Conference on Information Systems, pp. 417–425 (2007)
23. Daniel, D.: Management information crisis. Harv. Bus. Rev. **39**(5), 111–121 (1961)
24. Rockart, J.F.: Chief executives define their own data needs. Harv. Bus. Rev. **39**(5), 81–94 (1979)
25. Akpon-Ebiyomare, D., Chiemeke, S.C., Egbokhare, F.A.: A study of the critical success factors influencing data quality in Nigerian higher institutions. Afr. J. Comput. ICT **5**(2), 45–50 (2012)
26. Xu, H.: What are the most important factors for accounting information quality and their impact on AIS data quality outcomes? ACM J. Data Inf. Qual. **5**(4), 14:1–14:22 (2015)
27. Xu, H., Koronios, A., Brown, N.: Managing data quality in accounting information systems. In: IT-based management: challenges and solutions, pp. 277–299 (2003)
28. da C. dos Santos, M.P.: Fatores Críticos de Sucesso na Gestão da Qualidade dos Dados (2015)
29. Xu, H., Lu, D.: The critical success factors for data qualty in accounting information system - different industries' perspective. Issues Inf. Syst. **4**, 762–768 (2003)
30. Xu, H.: Would Organization Size Matter for Data Quality. In: Proceedings of the Eighth International Conference on Information Quality (ICIQ 2003), pp. 365–379 (2003)
31. Xu, H.: Factor analysis of critical success factors for data quality. In: Proceedings of the Nineteenth Americas Conference on Information Systems, pp. 1–6 (2013)
32. Williams, T.L., Becker, D.K., Robinson, C., Redman, T.C., Talburt, J.R.: Measuring Sociocultural Factors of Success in Information Quality Projects. In: Proceedings of 20th International Conference on Information Quality, pp. 48–69 (2015)
33. Wang, R.Y.: A product perspective on total data quality management. Commun. ACM **41**(2), 58–66 (1998)
34. Saraph, J.V., Benson, P.G., Schroeder, R.G.: An instrument for measuring the critical success factors of quality management. Decis. Sci. **20**(4), 810–829 (1989)
35. Hietschold, N., Reinhardt, R., Gurtner, S.: Measuring critical success factors of TQM implementation successfully – a systematic literature review. Int. J. Prod. Res. **52**(July), 1–19 (2014)
36. Porter, L.J., Parker, A.J.: Total quality management—the critical success factors. Total Qual. Manag. **4**(1), 13–23 (1993)
37. Black, S.A., Porter, L.J.: Identification of the critical factors of TQM. Decis. Sci. **27**(1), 1–21 (1996)

38. NIST: Baldrige Performance Excellence Framework (2017)
39. EFQM: An Overview of the EFQM Excelence Model (2013)
40. International Standards Organization: ISO 9001:2015 (2015)
41. Belanger, F.: Theorizing in information systems research using focus groups. Australas. J. Inf. Syst. **17**(2), 109–135 (2012)
42. Linstone, H.A., Turoff, M.: The Delphi Method - Techniques and Applications (2002)
43. Stephenson, W.: The Study of Behavior: Q-technique and its Methodology (1953)
44. Brown, S.R.: A primer on Q methodology. Operant Subj. **16**(3/4), 91–138 (1993)
45. Yin, R.K.: Qualitative Research from Start to Finish (2016)
46. Schmidt, R.C.: Managing Delphi surveys using nonparametric statistical techniques. Decis. Sci. **28**(3), 763–774 (1997)
47. Santos, L., Amaral, L.: Estudos Delphi com Q-Sort sobre a web – A sua utilização em Sistemas de Informação. In: Proceedings of the Conferência da Associação Portuguesa dos Sistemas de Informação (2004)
48. L. D. dos Santos, "Factores Determinantes do Sucesso de Adopção e Difusão de Serviços de Informação Online em Sistemas de Gestão de Ciência e Tecnologia," 2004
49. DAMA: DAMA - DMBOK Data Management Body of Knowledge, 2nd Edn. (2017)
50. Davenport, T.H., Patil, D.J.: Data scientist: the sexiest job of the 21st century. Harv. Bus. Rev. **90**(5), 70–77 (2012)

Solutions for Data Quality in GIS and VGI: A Systematic Literature Review

Gabriel Medeiros$^{(\boxtimes)}$ and Maristela Holanda

Department of Computer Science (CIC),
University of Brasilia (UnB), Brasilia, DF, Brazil
`gabriel.medeiros93@gmail.com`, `mholanda@unb.br`

Abstract. In Geographic Information Systems (GIS), and more specifically in Volunteered Geographic Information (VGI), users can actively participate in the processes of inclusion, change and exclusion of data. Therefore, quality becomes a central issue in these types of application, with the objective of verifying the correctness and consistency of the information, for example. In this context, this paper provides a Systematic Literature Review (SLR) of the literature on data quality in GIS and VGI, in order to analyze several academic papers related to the topic, examining what quality dimensions are being analyzed and what types of solution are being proposed. In addition, this paper intends to make explicit what themes are recurrent in papers related to the area of data quality and what are the main challenges to be explored by the researchers. For the accomplishment of this work, a research protocol was implemented, which involved following the definition of a group of research questions, used for the subsequent selection and extraction of the data.

Keywords: Systematic review · Data quality · GIS · VGI ·
Quality dimensions.

1 Introduction

According to Dueker and Kjerne [1], a Geographic Information System (GIS) is an aggregation of hardware, software, data, people, organizations and institutional arrangements for collecting, storing, analyzing and disseminating information about areas of the earth. In recent years, other types of geographical systems have also emerged, which have facilitated the active participation of its users.

In the year 2007, the term Volunteered Geographic Information (VGI) was developed with the purpose of describing computer systems in which a large number of private citizens are engaged in the creation or editing of geographic information [2]. An example of a successful VGI is the collaborative mapping tool OpenStreetMap (OSM), which surpassed 4 million registered users in 2017[1].

---

[1] https://wiki.openstreetmap.org/wiki/Stats [Accessed on November 15, 2018].

© Springer Nature Switzerland AG 2019
Á. Rocha et al. (Eds.): WorldCIST'19 2019, AISC 930, pp. 645–654, 2019.
https://doi.org/10.1007/978-3-030-16181-1_61

Although collaborative geographic information systems have contributed to the access of geographical information from many parts of the world, it has also brought some challenges related to data quality and reliability [3].

In order to seek solutions to the issue of data quality, many authors have used the term "quality dimension" to refer to parameters such as completeness, logical consistency and accuracy [4–6]. In this way, each dimension captures a specific aspect included under the general umbrella of data quality [4].

The objective of this paper is to present a Systematic Literature Review (SLR), analyzing what the scientific community has researched in the area of data quality in GIS and VGI, using the approach of [7]. As such, the goal of this literature review is to provide an overview of the research reported in the field and identify possible issues that existing literature is not addressing adequately. This work is proposed to function as a snapshot of the research in the field, identifying what quality dimensions are being analyzed and classifying the research by type of solution.

The paper is structured in the following sections: Sect. 2, in which some related works are presented; Sect. 3, in which the systematic literature review is shown; Sect. 4, in which the results of the SLR are presented; and finally, the conclusion is presented in Sect. 5.

2   Related Works

In recent years, a considerable number of researchers have already presented some works about the quality of data in GIS and VGI, including systematic literature reviews. For example, in [8] some research papers were reviewed and summarized in a detailed report on measures for each spatial data quality element. Also, the difficulties in using authoritative datasets were briefly presented and new proposed quality indicators were discussed, as recorded through the literature review.

In [9] an SLR was carried out to discover current research on Crowd-sourced Geographic Information (CGI), which is an umbrella term that encompasses both "active/conscious" and "passive/unconscious" georeferenced information [10]. The SLR aimed at providing an overview of the characteristics of the existing methods for researchers and developers of crowdsourcing-based platforms.

Furthermore, [11] reviewed various quality measures and indicators for selected types of VGI and existing quality assessment methods, presenting as an outcome a classification of VGI with current methods utilized to assess the quality of this type of system. Through these findings, data mining was introduced as an additional approach for quality handling in VGI.

The systematic review in this paper seeks to analyze which quality dimensions are being studied by the scientific community and also the main solutions proposed for the issue of data quality in GIS and VGI.

3 Systematic Literature Review (SLR)

A SLR consists of creating an analytic methodology to identify, select, evaluate and synthesize the main scientific research that allows the elaboration of a more objective bibliographic review, using a rigorous review protocol [12]. In this paper, the applied protocol is based on the strategies of [7]. According to these authors, the use of a review protocol is essential to guarantee that the literature review will be systematic and to avoid personal opinions on the part of the research authors [13].

The main objective of a systematic review is to answer a series of research questions, which will be described in this paper in Sect. 3.1. Furthermore, Sects. 3.2 and 3.3 define the paper literature selection/extraction strategy including the list of resource libraries, the search queries and inclusion/exclusion criteria.

3.1 Research Questions

The intention of this literature review was to answer the following research questions (RQ):

- RQ1: What quality dimensions are being analyzed in the field of Geographic Information Systems?
- RQ2: What type of solution is being proposed in relation to data quality in Geographic Information Systems in general?
- RQ3: What type of solution is being proposed in relation to data quality specifically in Volunteered Geographic Information Systems, as such the collaborative mapping tool OpenStreetMap?

The review protocol was divided into two phases: data selection and data extraction. These two steps will be described in the next subsections.

3.2 Data Selection

The searches were realized in three different scientific libraries, chosen because of their significant relevance in the field of Information Systems:

1. The ACM Digital Library[2]
2. IEEE Xplore[3]
3. Scopus (Elsevier)[4]

In order to perform the searches into the scientific sources, two search strings were elaborated, considering data quality in Geographic Information Systems in general (in order to answer RQ1 and RQ2), and considering especially data

---

[2] http://dl.acm.org/ [Accessed on August 15, 2018].
[3] http://ieeexplore.ieee.org [Accessed on August 15, 2018].
[4] https://www.scopus.com/ [Accessed on August 15, 2018].

quality in Volunteered Geographic Information Systems, such as the collaborative mapping tool OpenStreetMap (in order to answer RQ3). Thus, the search strings elaborated were as follows:

(i) ("Geographic Information System") AND (QUALITY) AND (PARAMETERS OR ASSESSMENT OR INDICATORS OR METRICS OR STUDY OR ANALYSIS)

(ii) ((VGI OR OPENSTREETMAP OR "Volunteered Geographic Information") AND (QUALITY) AND (PARAMETERS OR ASSESSMENT OR INDICATORS OR METRICS OR STUDY OR ANALYSIS))

Furthermore, the selected literature body collected should commit to a set of inclusion criteria:

- Papers should be available for download in the previously defined digital databases;
- Papers published since 2007 were preferred. However, classic sources (books or papers with classic concepts or definitions) were also considered, even with publication prior to the year 2007;
- Preference was given to papers with more than 3 citations;
- The paper should explicitly mention the issue of data quality in Geographic Information Systems.

On the other hand, the following exclusion criteria were applied:

- Papers published before the year 2007 that are not relevant classic sources should not be considered;
- Duplicate works, which have been published in more than one database, should not be considered;
- Papers with less than 3 citations published before 2017 should not be considered;
- Papers which were identified as outwith the scope or theme of this work should not be considered.

3.3   Data Extraction

Applying the two search strings, 1,192 papers were extracted from the libraries. After applying the inclusion/exclusion criteria, 1,146 papers were rejected and only 46 papers were found relevant. Figure 1 shows the distribution of papers encountered by source. According to the figure, 12% of the papers were found in the ACM database, 16% in the IEEE database and 72% in the Scopus database.

During the data extraction phase, the abstract and introduction of the papers considered relevant were read in order to extract their most important information, priority being given to those which answered the pre-defined research questions. After reading the abstract and the introduction of all 46 papers, it was ascertained that a further 16 papers were unrelated to the scope of this work and therefore they were discarded. Thus, at the end of the systematic review, we obtained a total of 30 papers relevant to the proposed theme. In order to assist the classification process of the papers, the StArt[5] tool was used.

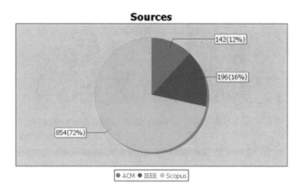

Fig. 1. Distribution of papers found in the scientific sources.

4 Results

Answering the research question RQ1, it was observed that the quality dimension most explored by researchers in recent years has been the accuracy, being addressed in all the 30 relevant papers (100% of the total). The completeness was explored in 13 papers (43% of the total) and the logical consistency was explored in 12 papers (40% of the total). Some papers addressed more than one dimension and for that reason, the total exceeds 100%. The large number of academic works exploring the accuracy can be easily explained, since this dimension is responsible for describing the correctness or incorrectness of the objects in the database, functioning as an essential aspect for the quality of the data. These results can be seen in Fig. 2.

In relation to the research question RQ2, the following results were obtained: 21 papers proposed the elaboration of a framework or algorithm for the improvement of data quality; 19 papers used location-based services; 11 papers made a comparison to satellite images or used image processing techniques; 9 papers structured their work in the form of a report, without any kind of practical implementation; 8 papers conducted analysis based on road networks; 7 papers made use of machine learning techniques; 6 papers made use of metadata; 4 papers made use of interpolation techniques; 2 papers realized a kind of ranking of users' reliability; 2 papers made use of human computation techniques. These results can be observed in Table 1, noting that some papers addressed more than one type of solution.

---

[5] http://lapes.dc.ufscar.br/tools/start_tool [Accessed on August 15, 2018].

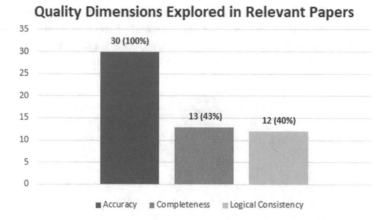

Fig. 2. Quality dimensions explored in relevant papers.

Table 1. Solutions proposed in relation to data quality in GIS

Type of solution proposed	Papers	Total
Framework / Algorithm implementation	[22–42]	21
Location based services	[14–16, 20–22, 24–29, 35–39, 41, 43]	19
Comparison to satellite images / Image processing	[14, 16, 17, 19, 21–26, 30]	11
Report	[14–21, 43]	9
Road networks analysis	[21, 22, 24, 25, 27, 32, 34, 42]	8
Machine learning techniques	[21, 23, 32, 39–42]	7
Use of metadata	[15, 19, 20, 28, 31, 43]	6
Interpolation techniques	[17, 23, 26, 33]	4
Ranking of users' reliability	[31, 35]	2
Human computation techniques	[28, 35]	2

Finally, answering the research question RQ3, the following results were obtained: 12 papers used location-based services; 11 papers proposed the elaboration of a framework or algorithm for the improvement of data quality; 11 papers made a comparison to satellite images or used image processing techniques; 8 papers structured their work in the form of a report, without any kind of practical implementation; 6 papers made use of metadata; 4 papers conducted an analysis based on road networks; 4 papers made use of machine learning techniques; 2 papers realized a kind of ranking of users' reliability; 2 papers made use of human computation techniques; no paper was found dealing with interpolation techniques. These results can be observed in Table 2.

Following this systematic review, other important results were obtained. In many of the academic works read, the great heterogeneity in data quality was

Table 2. Solutions proposed in relation to data quality in VGI

Type of solution proposed	Papers	Total
Location based services	$[14\text{--}16, 20\text{--}22, 28, 35, 37\text{--}39, 43]$	12
Framework / Algorithm implementation	$[22, 28, 30, 31, 34, 35, 37\text{--}40, 42]$	11
Comparison to satellite images / Image Processing	$[14, 16, 17, 19, 21\text{--}26, 30]$	11
Report	$[14\text{--}16, 18\text{--}21, 43]$	8
Use of metadata	$[15, 19, 20, 28, 31, 43]$	6
Road networks analysis	$[21, 22, 34, 42]$	4
Machine learning techniques	$[21, 39, 40, 42]$	4
Ranking of users' reliability	$[31, 35]$	2
Human computation techniques	$[28, 35]$	2
Interpolation techniques	-	0

discussed because of the discrepancy in the degree of knowledge among users of this type of system. This was discussed, for example, in the works [35, 40].

Moreover, the issue of routing is fundamental in GIS. To look for the smallest possible route between distances it is essential that the objects have a good level of accuracy. In this context, [25] proposed an automatic road network generation algorithm that takes vehicle tracking data in the form of trajectories as input and produces a road network graph; and [27] measured the distance between road networks in GIS.

5   Conclusion

This work describes a systematic literature review held in the field of data quality in GIS and VGI. The purpose of this paper was to provide an overview of the field and verify the quality of data present in these types of system. As a result, we found and analyzed 30 relevant papers from a gross total of 1,192 papers, extracted from a list of prestigious scientific libraries.

From this work, it can be observed that the field of data quality in GIS and VGI is still expanding and there is much to be explored, especially with regard to the quality dimension of accuracy. In addition, there is a great amount of written work involving a framework or algorithm implementation. This fact evidences the importance of practical work involving the issue of data quality in collaborative systems. Thus, with the aim of increasing data quality specifically in VGI, it is intended to conduct a further investigation in order to carry out more work with a practical bias.

References

1. Dueker, K.J., Kjerne, D.: Multipurpose Cadastre: Terms and Definitions. Falls Church VA, American Society for Photogrammetry and Remote Sensing and American Congress on Surveying and Mapping (1989)
2. Goodchild, M.F.: Citizens as sensors: the world of volunteered geography. GeoJournal **69**(4), 211–221 (2007)
3. Flanagin, A.J., Metzger, M.J.: The credibility of volunteered geographic information. GeoJournal **72**, 137–148 (2008)
4. Batini, C., Scannapieco, M.: Data and Information Qualty - Dimensions, Principles and Techniques, 1st edn. Springer, Switzerland (2016)
5. Firmani, D., Mecella, M., Scannapieco, M., Batin, C.: On the meaningfulness of 'big data quality'. Data Sci. Eng. **1**(1), 6–20 (2016). (Invited Paper)
6. Francisco, M.M.C., Alves-Souza, S.N., Campos, E.G.L, De Souza, L.S.: Total data quality management and total information quality management applied to costumer relationship management. In: ICIME 2017: 2017 9th International Conference on Information Management and Engineering (2017)
7. Kitchenham, B., Charters, S.: Guidelines for performing systematic literature reviews in software engineering. Technical report. EBSE 2007-001, Keele University and Durham University (2007)
8. Antoniou, V., Skopeliti, A.: Measures and indicators of VGI quality: an overview. ISPRS Annals of the Photogrammetry, Remote Sensing and Spatial Information Sciences, Volume II-3/W5 (2015)
9. Dregossi, L.C., Albuquerque, J.P., Rocha, R.S., Zipf, A.: A framework of quality assessment methods for crowdsourced geographic information: a systematic literature review. In: 14th International Conference on Information Systems for Crisis Response and Management, At Albi, France (2017)
10. See, L., Mooney, P., Foody, G., Bastin, L., Comber, A., Estima, J., Rutzinger, M.: Crowdsourcing, citizen science or volunteered geographic information? the current state of crowdsourced geographic information. In: ISPRS International (2016)
11. Senaratne, H., Mobasheri, A., Ali, A.L., Capineri, C., Haklay, M.: A review of volunteered geographic information quality assessment methods. Int. J. Geogr. Inf. Sci. **31**(1), 139–167 (2017)
12. Camacho, L.A.G., Souza, S.N.A.: Social network data to alleviate cold-start in recommender system: a systematic review. Inf. Process. Manag. **54**, 529–544 (2018)
13. Manikas, K., Hansen, K.M.: Software ecosystems - a systematic literature review. J. Syst. Softw. **86**, 1294–1306 (2013)
14. Ahlers, D.: Assessment of the accuracy of GeoNames gazetteer data. In: Proceedings of the 7th Workshop on Geographic Information Retrieval (GIR 2013), pp. 74–81. ACM, New York (2013)
15. Estima, J., Painho, M.: Exploratory analysis of OpenStreetMap for land use classification. In: Proceedings of the Second ACM SIGSPATIAL International Workshop on Crowdsourced and Volunteered Geographic Information (GEOCROWD 2013), pp. 39–46. ACM, New York (2013)
16. Foody, G.M., Boyd, D.S.: Using volunteered data in land cover map validation: mapping West African forests. IEEE J. Sel. Top. Appl. Earth Obs. Remote Sens. **6**(3), 1305–1312 (2013)
17. Chiang, Y., Leyk, S., Nazari, N.H., Moghaddam, S., Tan, T.X.: Assessing the impact of graphical quality on automatic text recognition in digital maps. Comput. Geosci. **93**, 21–35 (2016)

18. Demetriou, D.: Uncertainty of OpenStreetMap data for the road network in Cyprus. In: Fourth International Conference on Remote Sensing and Geoinformation of the Environment, Paphos, Cyprus (2016)
19. Mooney, P., Corcoran, P., Winstanley, A.C.: Towards quality metrics for OpenStreetMap. In: Proceedings of the 18th SIGSPATIAL International Conference on Advances in Geographic Information Systems (GIS 2010), pp. 514–517. ACM, New York (2010)
20. Mooney, P., Corcoran, P., Sun, H., Yan, L.: Citizen generated spatial data and information: risks and opportunities. In: 2012 International Conference on Industrial Control and Electronics Engineering (2012)
21. Sehra, S.S., Singh, J., Rai, H.S.: A systematic study of OpenStreetMap data quality assessment. In: 11th International Conference on Information Technology: New Generations (2014)
22. Biagioni, J., Eriksson, J.: Map inference in the face of noise and disparity. In: Proceedings of the 20th International Conference on Advances in Geographic Information Systems (SIGSPATIAL 2012), pp. 79–88. ACM, New York (2012)
23. Helmholz, P., Büschenfeldc, T., Breitkopf, U., Müller, S., Rottensteiner, F.: Multitemporal quality assessment of grassland and cropland objects of a topographic dataset. In: The International Archives of the Photogrammetry, pp. 67–72 (2012)
24. Hoffman, S., Brenner, C.: Quality assessment of automatically generated feature maps for future driver assistance systems. In: Proceedings of the 17th ACM SIGSPATIAL International Conference on Advances in Geographic Information Systems (GIS 2009), pp. 500–503. ACM, New York (2009)
25. Karagiorgou, S., Pfoser, D.: On vehicle tracking data-based road network generation. In: Proceedings of the 17th ACM Sigspatial International Conference on Advances in Geographic Information Systems, pp. 89–98 (2012)
26. Zhang, C., Fraser, C.S.: Automated registration of high-resolution satellite images. Photogram. Rec. **22**(117), 75–87 (2007)
27. Ahmed, M., Fasy, B.T., Wenk, C.: Local persistent homology based distance between maps. In: Proceedings of the 22nd ACM SIGSPATIAL International Conference on Advances in Geographic Information Systems (SIGSPATIAL 2014), pp. 43–52. ACM, New York (2014)
28. Celino, I.: Human computation VGI provenance: semantic web-based representation and publishing. IEEE Trans. Geosci. Remote Sens. **51**(11), 5137–5144 (2013)
29. Che, Y. Chiew, K., Hong, X., He, Q.: SALS: semantics-aware location sharing based on cloaking zone in mobile social networks. In: Proceedings of the First ACM SIGSPATIAL International Workshop on Mobile Geographic Information Systems (MobiGIS 2012), pp. 49–56. ACM, New York (2012)
30. Subbiah, G., Alam, A., Khan, L., Thuraisingham, B.: Geospatial data qualities as web services performance metrics. In: Proceedings of the 15th International Symposium on Advances in Geographic Information Systems ACM GIS (2007)
31. Camara, J.H.S., Vegi, L.F.M., Pereira, R.O., Geöcze, Z.A., Lisboa-Filho, J.: ClickOnMap: a platform for development of volunteered geographic information systems. In: 12th Iberian Conference on Information Systems and Technologies (CISTI), pp. 1–6 (2017)
32. Delling, D., Goldberg, A.V., Goldszmidt, M., Krumm, J., Talwar, K., Werneck, R.F.: Navigation made personal: inferring driving preferences from GPS traces. In: Proceedings of the 23rd SIGSPATIAL International Conference on Advances in Geographic Information Systems (SIGSPATIAL 2015). ACM, New York (2015). Article 31, 9 p

33. Galarus, D.E., Angryk, R.A.: A SMART Approach to Quality Assessment of Site-Based Spatio-Temporal Data. In: IEEE International Conference on Big Data, pp. 2636–2645 (2016)

34. Wang, M., Li, Q., Hu, Q., Zhou, M.: Quality analysis of open street map data. international archives of the photogrammetry, remote sensing and spatial information sciences. In: 8th International Symposium on Spatial Data Quality, Hong Kong, Volume XL-2/W1 (2013)

35. Karam, R., Melchiori, M.: A Crowdsourcing-based framework for improving geospatial open data. In: IEEE International Conference on Systems, Man, and Cybernetics (2013)

36. Xu, T., Cai, Y.: Location anonymity in continuous location-based services. In: Proceedings of the 15th Annual ACM International Symposium on Advances in Geographic Information Systems (GIS 2007). ACM, New York (2007). Article 39, 8 p

37. Ye, M., Janowicz, K., Mülligann, C., Lee, W.: What you are is when you are: the temporal dimension of feature types in location-based social networks. In: Proceedings of the 19th ACM SIGSPATIAL International Conference on Advances in Geographic Information Systems (GIS 2011), pp. 102–111. ACM, New York (2011)

38. Zhou, M., Hu, Q., Wang, M.: A quality analysis and uncertainty modeling approach for crowd-sourcing location check-in data. In: 8th International Symposium on Spatial Data Quality (2013)

39. Zheng, Y., Fen, X., Xie, X., Peng, S., Fu, J.: Detecting nearly duplicated records in location datasets. In: Proceedings of the 18th ACM Sigspatial International Conference on Advances in Geographic Information Systems, pp. 137–143 (2010)

40. Ali, A.L., Schmid, F., Al-Salman, R., Kauppinen, T.: Ambiguity and plausibility: managing classification quality in volunteered geographic information. In: Proceedings of the 22nd ACM SIGSPATIAL International Conference on Advances in Geographic Information Systems, pp. 143–152 (2014)

41. Hassan, A., Jones, R., Diaz, F.: A case study of using geographic cues to predict query news intent. In: Proceedings of the 17th ACM Sigspatial International Conference on Advances in Geographic Information Systems, pp. 33–41 (2009)

42. Jilani, M., Corcoran, P., Bertolotto, M.: Automated highway tag assessment of OpenStreetMap road networks. In: Proceedings of the 22nd ACM SIGSPATIAL International Conference on Advances in Geographic Information Systems (SIGSPATIAL 2014), pp. 449–452. ACM, New York (2014)

43. Langley, S.A., Messina, J.P., Moore, N.: Using meta-quality to assess the utility of volunteered geographic information for Science. Int. J. Health Geogr. **16**(1), 40 (2017)

Use of the Lean Methodology to Reduce Truck Repair Time: A Case Study

Alexander Börger$^{(\boxtimes)}$, Javiera Alfaro, and Priscila León

Department of Industry and Business, University of Atacama, Copiapó, Chile
alexander.borger@uda.cl,
{javiera.alfaro, priscila.leon}@alumnos.uda.cl

Abstract. The freight transport is considered as a vital system for today's society because it is the vehicle that mudas the productive activity of the country. However, as a consequence of inadequate management and improperly performed maintenance work, a greater number of hours is required for the repair of trucks. This work aims to improve truck repair times by applying Lean methodology. The research methodology involves the selection of 16 trucks considered part of the problem, the identification of the most frequent failures of these trucks, and the causes associated with these failures, defining a set of 23 countermeasures to eliminate or reduce the effects of the failures. To carry out these actions, a Gantt chart and a schedule of activities are prepared to monitor compliance. Based on the development of this proposal, it is expected that operability of the truck fleet will increase by 20% in order to improve customer satisfaction. Therefore, it is necessary to review periodically during the year to make the necessary adjustments.

Keywords: LEAN · Reduction · Maintenance

1 Introduction

The freight transport is considered a vital system for today's society [1] because it is essentially the vehicle that moves the activity of the country. Based on this, a transport company in Copiapo city has established as a rule that the Mean Time To Repair (MTTR) is 24 h of work, that is, a maximum of 3 working days.

However, between February and July 2017, only 62% of the trucks entered the workshop were repaired within the period established by the company, which was due to a deficiency in the administration and execution of maintenance that does not add value. In view of this, LEAN manufacturing is one of the best-known systems to improve productivity [2].

The core benefits of Lean (flow, minimizing waste, etc.) have become the paradigm for manufacturing operations [3]. This is due to the fact that when implementing Lean, the complete production can be increased, increasing the output product and reducing the input factors such as lead time, time, manpower, raw material and defect products [2]. The objective of the study is to apply Lean methodology in order to improve equipment repair times, specifically to reduce them to a maximum of 24 working hours, that is, a maximum of 3 business days.

© Springer Nature Switzerland AG 2019
Á. Rocha et al. (Eds.): WorldCIST'19 2019, AISC 930, pp. 655–665, 2019.
https://doi.org/10.1007/978-3-030-16181-1_62

Notwithstanding, it is expected that at least 82% of the trucks that access the Copiapó branch offices may be available to operate in a period less than or equal to 24 working hours. Therefore, the gap that is defined for the project is that it increases the operation of the trucks by 20% in order to comply with expectations. This 20% is considered the first goal because continuous improvement with Lean is measured using increases or improvements made gradually in the short term.

2   Background

"Lean is an English word that can be translated as "fat-free, scarce, slender", but applied to a productive system means "agile, flexible", that is, capable of adapting to the needs of the client" and is "the pursuit of an improvement of the manufacturing system through the elimination of waste" [4].

The Lean concept, evolved from Japanese industries especially from Toyota [5] in the automotive sector after the Second World War [6] in their effort to reduce cost [7]. The introduction of lean changed the market in the development of the car industry that was pioneered by Toyota Production System (TPS) [7].

The term Lean is an approach practiced to eliminate waste in a production that consumes resources, but creates no value to customers [8], that is, it seeks to maximize the value of the product by reducing waste.

Lean tools that is most commonly apply include Kaizen, Kanban, Just in Time (JIT), Supply Chain Management (SCM), Total Quality Management (TQM), Total Productive Maintenance (TPM), The Seven Waste Concept (MUDA), y the 5S workplace methodology [9].

3   Related Work

The Lean methodology is described in the literature as a general solution to the effective management of operations [10]. In their article, the authors review the theoretical and practical aspects of the Lean methodology in order to show the fundamental role it plays in organizations related to increased production and competitiveness.

The LEAN model has been applied in different productive sectors, for example, in the mining industrial sector, experiences such as those made by Klippel and others are presented [11] who propose to apply the Lean methodology as a new management method in the mineral extraction industry. The authors also show the application of the methodology to 2 practical cases: Fluorite and Amethyst. In these they obtained an increase in the productivity and an improvement in the quality of life of the workers.

In the health-care sector, the Lean methodology has been used on 3 main areas, defining the value from the patient's point of view, mapping value streams and eliminating waste in an attempt to create a continuous flow [12].

Regarding the reduction of times, several works have been found that use the LEAN methodology for time reduction. Ganesan [13] apply the concepts of LEAN to reduce the manufacturing times of monobloc pumps. The results show that the total waiting time was reduced from 26.3 to 24.9 days, with the reduction rate from 16.66 to 25%.

In the work of Lingam et al. [14], some LEAN tools were used in order to reduce the time of making shirts. With this a saving of 20% in the cycle times was achieved.

Likewise, Siva et al. [15] manages to improve maintenance cycle times in the aviation industry through the use of LEAN tools for the elimination of seedlings or waste.

Finally, Saravan et al. [16], through LEAN's Single Minutes Exchange of Die (SMED) method, seek to reduce the loss of time in the exchange of molds or dies, in order to reduce the change in configuration time Current from 34.94 min to less than 10. As a result, they achieved a reduction of 67.72% over time.

4   Research Methodology

This case study is based on records corresponding to the first semester of 2017, specifically February through July, contributed by personnel in maintenance and operations. In addition, of the total fleet owned by the company, only trucks (including silos and hoppers) that enter and leave the workshop located at the Copiapó branch are considered and, given the requirements of those companies for which this transport company provides services and compliance with internal regulations, only those with a maximum age of 5 years are included.

On the other hand, it is worth noting that the personnel assigned to carry out the maintenance activities make repairs only during the 8-h workday even when the trucks remain in circuit 24 h a day. It is also necessary to emphasize that the platforms of the mechanical workshop are intended for various types of maintenance; for this study, only the corrective type is considered. Repairs with a preventive purpose are exempt from analysis. Finally, the Deming Cycle, or PDCA, is used to solve the problem since it is the basic model for carrying out continuous improvement. The methodology used in the investigation is shown in Fig. 1.

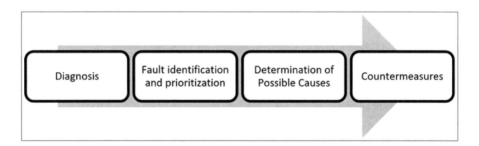

Fig. 1. Research methodology

The tasks performed in the methodology are described below:

- Diagnosis: Documentation associated with the company, work orders, and fleet control are reviewed in order to extract relevant information about the fleet and the processes that are carried out.

- Identification and Prioritization of Faults and Mudas: Based on the information obtained, the mudas (waste) present in the maintenance activities are identified. Likewise, the number and types of failures for which the trucks are subjected to maintenance during the semester under study are identified. Then, the mudas and failures are prioritized according to an 80%–20% relationship (Pareto's Law).
- Determination of Possible Causes: Through the Ishikawa diagram and the rule of 5 Whys are the possible causes of delays in maintenance times determined.
- Countermeasures: Finally, to reduce the effect of possible causes of delayed maintenance times, a set of countermeasures is offered based on brainstorming and consultation with experts in the area.

5 Results

Of a total of 77 tracts, only 16 exceeded the average repair time of 24 working hours and did not exceed the maximum age of 5 years. These 16 were considered for this study.

5.1 Fault Assessment

The type of failures for each of the 16 tracts from February through July is shown in Table 1, in order to identify the most frequent types of failures and to organize and prioritize the obtained data, an 80%–20% relationship was carried out. In Table 1, the most relevant types of failures that reached 80% as a whole are shown.

Table 1. Pareto development to obtain types of failures.

Causes	Frequency	%	Accumulated	Accumulated %	80–20
Ext E Int Structure	137	25%	137	25%	80%
Tires	134	25%	271	50%	80%
Electric	104	19%	375	59%	80%
Breaks	53	10%	428	78%	80%
Lubrication	44	8%	472	86%	80%
Engine	33	6%	505	92%	80%
Suspension	19	3%	524	96%	80%
Transmission	14	3%	538	99%	80%
Hydraulic	8	1%	546	100%	80%
Total	546	100%			

5.2 Analysis of Causes

For this study, the segmentation of possible causes has been made for Business Resources. The number of possible causes, for each of the business resources (6M) are:

Environment: 8; Methods: 8; Measurement: 6, Machines: 6; Materials: 11 y Peoples: 14, obtaining a total of 53 possible causes

These causes are ranked according to their Priority Indicator (PI), which arises as a result of the product (multiplication) of 2 criteria, Frequency (F) and Importance (I), to which a score is assigned (1: low, 2: medium, 3: high). The causes, the type of business resource, and the decision criteria are shown in Table 2.

Table 2. Matrix of prioritization of possible causes

No	Possible causes	Type of business resource	F	I	PI
1	Climb and descend the ladder frequently.	Environment	5	5	25
2	There is no standardization	Methods	5	5	25
3	Need for a program to evaluate work done	Measurement	5	5	25
4	Lack of repair history updated daily	Measurement	5	5	25
5	Old machines	Machines	5	5	25
6	Unexcused absences	People	5	5	25
7	Routes traveled by tracts	Machines	5	5	25
8	Missing a daily record of progress	Measurement	5	5	25
9	Absence of machinery to remove tires	Machines	5	5	25
10	There is only one stock of spare parts requested	Materials	5	5	25
11	Low quality of certain spare parts	Materials	5	5	25
12	Incomplete staffing	People	5	5	25
13	Lack of competencies	People	5	5	25
14	Various simultaneous tasks	People	5	5	25
15	Performing activities not related to their job	People	5	5	25
16	Imprecision of drivers to detect faults	People	5	5	25
17	Unplanned workshop	Environment	5	5	25
18	Disorder in the area	Environment	5	5	25
19	Limited number of tools	Materials	5	5	25
20	Lot of older staff	People	5	5	25

There are 20 possible causes with the maximum score (25), for which counter-measures were sought to reduce their effect. Therefore, as shown in Table 2, when deploying the methodology, a total of 26 root causes to be treated were obtained.

5.3   Countermeasures

The actions or concrete activities that seek to eliminate, lessen, impact, or diminish the effect of each of the root causes shown in Fig. 2 and also of the changes found by using Pareto's Law have been determined by brainstorming and by consulting experts in the area. Table 3 lists the countermeasures associated with the 26 root causes and Table 4 lists the 17 previously reported mudas.

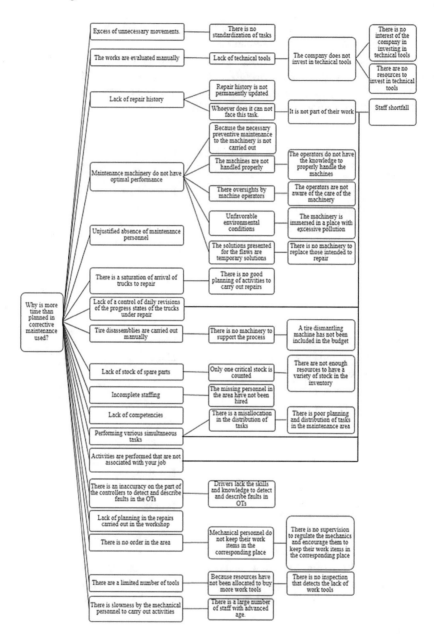

Fig. 2. Diagram cause - effect applied to the organization.

Table 3. Countermeasures associated with root causes

No	Description	Countermeasures
1	Lack of standardization of tasks	1. Operational standards
2	Absence of standardization of tasks	2. 5S 3. Standard layout 4. Visual control 5. Value Stream Mapping (VSM)
3	There is no interest on the part of the company in acquiring technical tools	1. Performance dialogs 2. Managing for Daily Improvement (MDI)
4	There is no inspection to detect deficiencies of tools	1. Assign a person to cover pending activities
5	A person assigned to update reparations history has not been hired	
6	Machines destined to carry out repairs are not given the necessary preventive maintenance	1. Carry out planning that contemplates the preventive maintenance of the machines destined to carry out repairs 2. Autonomous maintenance
7	The operators do not have the knowledge of correct handling for the machines destined to repair	1. Train and promote LEAN behavior 2. Roles and Agendas
8	The operators are not aware of the care of machinery intended to repair	1. Autonomous maintenance 2. Matrix of skills 3. Roles and Agendas
9	The machinery destined to repair is immersed in a place with excessive pollution	1. 5s's
10	Immediate solution is sought to repair the machines intended to carry out the maintenance and continue the operation	1. Train and promote LEAN behavior 2. ANDON 3. Joidka 4. Autonomous maintenance 5. Single Minutes Exchange of Die (SMED)
11	There is no one who executes the activity of control of daily revisions of trucks and advances	1. Assign a person to cover pending activities 2. Implement status sheet by day and send report to all involved
12	Unjustified absence of maintenance personnel	1. Train and promote LEAN behavior

(*continued*)

Table 3. (*continued*)

No	Description	Countermeasures
13	There is no good planning of repair activities when there is a truck arrival concentration	1. Implement sampling process times and define productivity metrics (time studies) 2. Roles and Agendas 3. 5s's 4. Task Time
14	There is no machinery that replaces the machines intended to repair	1. Evaluate and designate resources according to the most urgent needs of the maintenance area
15	A budget for the purchase of tire removal machines has not been contemplated	
16	There are not enough resources to have a variety of stock in the inventory	
17	The missing person in the area has not been hired	
18	There is a large number of staff with advanced age	
19	Lack of staff for each specialty	
20	Not enough staff has been hired to cover all maintenance tasks	
21	There is a lack of resources to invest in technical tools	1. Evaluate and designate resources according to the most urgent needs of the maintenance area
22	Bad planning in the distribution of tasks in the maintenance area	1. MDI 2. Value Stream Mapping (VSM) 3. Maintenance cells
23	Lack of planning in the repairs carried out in the workshop	1. Operational standards (SW) 2. MDI 3. Roles and Agendas 4. VSM
24	There is no supervision to regulate the mechanics and encourage them to keep their work items in the corresponding place	1. 5S's 2. Assign a person to cover pending activities
25	Drivers do not have skills and knowledge to detect and describe faults in the OT's	1. Training and Training 2. Matrix of skills
26	Lack of competencies	

Table 4. Countermeasures associated with mudas

No	Description	Countermeasures
1	Generally, more work must be done than required and specified in OTs	1. Check list 2. Training
2	Climbing and descending stairs on a frequent basis for inquiries and requirements to the head of maintenance	1. Standard layout
3	Drivers are not specific when detecting equipment malfunctions	1. Training 2. Matrix of Skills
4	Lack of competences	
5	Staff performs work in parallel that are not corresponding to their specialty	1. Evaluate and designate resources 2. Roles and Agendas
6	Unnecessary movements while running a maintenance	1. Implement sampling process times and define productivity metrics (time studies) 2. Implement status sheet per day and send a report to all involved
7	Work done improperly that must be re-executed while repairing the tract	1. Management or Visual Control 2. One point lesson (OPL) 3. Standard work (SW)
8	Bureaucratic procedure for documents	1. Incorporate use of digital system and not manual sheets
9	Stacking OT's in office	1. 5S's
10	Work done with lack of thoroughness	1. Training and Training 2. Standard work (SW)
11	Trucks in standby mode due to lack of spare parts	1. Evaluate and designate resources 2. Carry out a task planning
12	Some parts are of low quality	1. Evaluate and designate resources
13	Ambiguity in interpretation of OT's by manual writing	1. Incorporate use of digital system and not manual sheets 2. Train in the use of electronic spreadsheets to streamline the process and reduce ambiguities
14	Wait for some to be used	1. Perform task planning
15	Existence of spare parts for trucks that no longer belong to the company	2. Manage the destination of spare parts without use

6   Conclusions

The aim of this study is to reduce the time it takes for trucks to undergo maintenance in order to increase the operation of these trucks by 20%. For this purpose, the Lean methodology is used as a management system whose main focus is the elimination of all waste, thus reducing repair times in order to deliver greater customer satisfaction, improve quality, and reduce costs.

The results obtained show that the delay in the maintenance of the truck tracts is due to mudas or waste, detected mainly in the areas of inventory and defects and reprocessing. To these mudas, the constant failures experienced by trucks due to various causes associated with business resources (The 6M) are added. To this end, a set of countermeasures was proposed that seek to reduce the effect of maintenance delays.

In this sense, the transport company is guided to follow the Lean philosophy and thus be able to proactively search for problems and their solutions, minimize activities that do not generate value, optimize the use of typically scarce resources, and implement continuous improvement, reducing costs and increasing productivity while applying useful tools and gaining the commitment of the organization.

This is why the great importance of having a Lean Philosophy is concluded because it is a management system that guides when operating a business since its main focus is the elimination of all waste allowing to reduce the time for deliver greater customer satisfaction, improving quality and reducing costs.

References

1. Engström, R.: The roads' role in the freight transport system. Transp. Res. Procedia **14**, 1443–1452 (2016)
2. Ng, T.C., Ghobakhloo, M.: What determines Lean manufacturing implementation. A CB-SEM model. Economies **6**, 9 (2018)
3. Womack, J.P., Jones, D.T.: Lean Thinking: Banish Waste and Create Wealth in Your Organisation, p. 397. Simon and Shuster, New York (1996)
4. Carreras, M.R., García, J.L.S.: Lean Manufacturing. La evidencia de una necesidad. Ediciones Díaz de Santos, Madrid (2010)
5. Sundar, R., Balaji, A.N., Kumar, R.S.: A review on Lean manufacturing implementation techniques. Procedia Eng. **97**, 1875–1885 (2014)
6. AlManei, M., Salonitis, K., Xu, Y.: Lean implementation frameworks: the challenges for SMEs. Procedia CIRP **63**, 750–755 (2017)
7. Wahab, A.N.A., Mukhtar, M., Sulaiman, R.: A conceptual model of Lean manufacturing dimensions. Procedia Technol. **11**, 1292–1298 (2013)
8. Anvari, A., Ismail, Y., Hojjati, S.M.H.: A study on total quality management and Lean manufacturing: through Lean thinking approach. World Appl. Sci. J. **12**(9), 1585–1596 (2011)
9. Herron, C., Braiden, P.M.: A methodology for developing sustainable quantifiable productivity improvement in manufacturing companies. Int. J. Prod. Econ. **104**(1), 143–153 (2006)
10. Kariuki, B.M., Mburu, D.K.: Role of Lean manufacturing on organization competitiveness. Ind. Eng. Lett. **3**(10), 81–82 (2013)
11. Klippel, A.F., Petter, C.O., Antunes Jr., J.A.V.: Lean management implementation in mining industries. Dyna **75**(154), 81–89 (2008)
12. Poksinska, B.: The current state of Lean implementation in health care: literature review. Qual. Manag. Healthcare **19**(4), 319–329 (2010)
13. Ganesan, K., Prasad, M.M., Suresh, R.K.: Lead time reduction through Lean technique in an monobloc (SWJ1HP) pump industry. Appl. Mech. Mater. **592**, 2671–2676 (2014)

14. Lingam, D., Ganesh, K.S., Ganesh, K.: Cycle time reduction for t-shirt manufacturing in a textile industry using Lean tools. In: IEEE Sponsored 2nd International Conference on Innovations in Information, Embedded and Communication Systems, pp. 2–7. IEEE, Coimbatore (2015)

15. Siva, R., Purusothaman, M., Jegathish, Y.: Process improvement by cycle time reduction through Lean methodology. In: IOP Conference Series: Materials Science and Engineering, vol. 197, no. 1, p. 012064 (2017)

16. Saravanan, V., Nallusamy, S., Balaji, K.: Lead time reduction through execution of Lean tool for productivity enhancement in small scale industries. Int. J. Eng. Res. Afr. **34**, 116–127 (2018)

Proposal to Avoid Issues in the DevOps Implementation: A Systematic Literature Review

Mirna Muñoz, Mario Negrete[✉], and Jezreel Mejía

Parque Quantum, Ciudad del conocimiento Avenida Lasec andador Galileo Galilei manzana 3 lote 7, 98160 Zacatecas, Zac., Mexico
{mirna.munoz,mario.negrete,jmejia}@cimat.com.mx

Abstract. Nowadays many software companies and startups try to do software development faster and with better quality. Companies have tried to readjust their process with new techniques and practices in order to achieve these goals for example, getting continuous automation using DevOps practices or some companies tried to switch their habits getting software as faster as they can. Companies commonly use DevOps with agile methodologies such as SCRUM, XP, and lean. However most of the time they do not ensure that implementation of those practice in a correct way neither the use of models nor standards. This paper presents the results of a performed systematic review protocol focused on getting information about the real state of implementation of a DevOps approach regarding the use of methodologies, frameworks and standards. The obtained results allow us to identify challenges and barriers that companies have when they try to implement a DevOps approach.

Keywords: DevOps · Approach · Software development · Agile practices · ISO/IEC 29110

1 Introduction

Some companies are offering internet-based services such as Facebook, IBM, Atlassian among others. They are deploying software functionality to customers almost every day to provide features to their needs [1]. However software development is changing fast and continuously; continuous changes are bringing concepts such as continuous deployment which has opportunities of automation and challenges that affect habits for companies. These new practices have been tried to be met by companies handled new concepts like testing automation or even have tried to find a way to solve the continuous changes of customer requirements [2]. DevOps (Development and Operations) is a concept introduced for many authors, referring not only of software practices but also of management. DevOps is a set of practices that intent to reduce the time between making a change in the system and delivering it in the baseline or production with high quality. Besides, DevOps practices have influence on: how to organize teams

© Springer Nature Switzerland AG 2019
Á. Rocha et al. (Eds.): WorldCIST'19 2019, AISC 930, pp. 666–677, 2019.
https://doi.org/10.1007/978-3-030-16181-1_63

to build systems and to structure systems [3]. DevOps has different practices to satisfy its function, which is improving quality and application speed such as (1) TreatOps (Treatment Operations) which involves operators in requirements stage, (2) making developers more responsible for incidents, (3) reinforcing the deployment process to reduce the time to diagnose and repair any error, (4) continuous deployment to reduce the time between committing a code to a repository and then to deploy it, developing infrastructure code to ensure high quality in every deployment [3]. However, these type of practices are used based on different perspectives in many organizations due to companies do not apply DevOps in a correct way, or perhaps, they do not have a guidance to implement DevOps correctly. Most of the time the reason to implement a DevOps approach is for pressure by other companies, or because a company just wants to be at the forefront.

DevOps is also confused with Continuous Delivery (CD) because some companies sell this idea, however CD is just one activity within DevOps process. DevOps is not just a group of tools working together to reduce release time. It also contributes to eliminate wait times caused by manual activities among the team [4]. This research aims to set the state of the art of DevOps approach and its relationship with their implementation under agile methodologies or models to identify which problems or barriers make difficult to implement a DevOps approach in companies. The rest of the paper is structured as follows: Sect. 2 presents a Systematic Literature Review (SLR), Sect. 3 shows analysis of the data obtained and Sect. 4 presents a discussion of findings and future work.

2   Systematic Literature Review

A Systematic Literature Review (SLR) is a protocol that allows to identify, evaluate and interpret all available research relevant to a particular research question, topic area, or phenomenon of interest. SLR has already been proven in several studies that have helped in establishing and focusing results with respect to a research question in software engineering field such as in [5–7].

The SLR is based on three main phases: planning the review, conducting the review, and reporting the results [5].

2.1   Review Planning

Planning is the first phase of SLR. In this phase, research objectives must be defined as well as how the review will be performed. It includes: identify the need to perform the SLR, define the research questions, create the search string and select the data sources.

2.1.1   Identify the Need to Perform the SLR

Companies have adopted agile practices called DevOps, which promises to optimize the development process using practices such as continuous integration, continuous deployment, monitoring, testing, continuous release with high quality. According to Virmani [8], DevOps is trying to bridge developers-operations gap and at the same time covering all aspects among team; that help in the whole process such as speedy, optimized and high quality software delivery. However companies said that it is complicated to implement a DevOps process in its current environment because it is a big cultural shift, a key impediment for successfully adopting DevOps is insufficient communication or lacking a clear DevOps definitions causing as much problems as benefits [A4]. The systematic review is performed in order to know the current status of DevOps.

2.1.2 Define the Research Questions

Three research questions were set: (RQ1) Which agile methodologies are used in DevOps environment to develop software? (RQ2) Which models and standards are used in DevOps environment to develop software? and; (RQ3) Which are the main problem or barriers to setup a DevOps environment?

2.1.3 Create the Search String

After establishing the research questions, a set of keywords were selected. As Table 1 shows, synonyms and terms associated to the keywords were listed. After selecting the keywords, logical connectors were used such as 'AND' and 'OR' to create the search string. As result, this research establish 2 search strings to cover the whole topic.

Table 1. Keywords and search strings

Questions	Keywords	Synonyms or related terms	Search strings
1, 2, 3	DevOps		(S1) DevOps AND development AND software AND environment AND software engineering, AND (methodology OR framework)
1, 2, 3	Environment		
2, 3	Problems	Barriers, challenges	
2, 3	Model	Standard, methodology	
2, 3	Software development	Framework, software engineering, software deployment	(S2) DevOps AND development AND software AND environment AND software engineering AND (model OR standard OR methodology)

2.1.4 Select the Data Sources

The following list contains the data sources that were selected as sources to perform the defined strings. All of them are focused on the software engineering area such as ACM Digital Library, SpringerLink, Elseiver Science, IEEE Xplorer, Software Engineering Institute (SEI), Web of Science.

2.2 Conducting the Review

The next phase of the SLR is conducting the review, which focuses on collecting a set of studies and select the primary studies. It includes the following activities: establish the inclusion and exclusion criteria, define selection of primary studies quality assurance criteria and data extraction.

2.2.1 Establish the Inclusion and Exclusion Criteria

The first part of conducting the review consists on setting the inclusion and exclusion criteria.

Inclusion Criteria:

1. Studies in english, french or spanish languages.
2. Studies published between 2014 and 2018.
3. Studies containing at least 3 keywords on title, abstract, keywords.
4. Studies containing DevOps implementation at industry (case studies).
5. Studies showing results of DevOps implementation with agile methodologies.
6. Studies containing DevOps theory and application in industry environment.

Exclusion Criteria:

1. Studies repeated in more than one digital library.
2. Studies that do not contain information about the inclusion criteria 4 or 5 or 6.

2.2.2 Primary Studies Selection

The selection process consisted of seven steps as follows: (1) using the search string, adapting search string to selected digital libraries; (2) filter studies considering the first 3 inclusion criteria (language, year and keywords); (3) read titles and abstracts to identify relevant studies; (4) apply the remaining criteria and if it's necessary reading introduction, conclusion and methods; (5) select the primary studies; (6) filter the results of the second search string according to the first search string to remove duplicate studies; (7) apply the quality criteria to ensure the quality of studies. Figure 1 shows the implementation of the steps and the number of primary studies obtained and Fig. 2 in the same way but with a result filtered by the second search string. As Fig. 1 shows the results of S1 performed got 11 studies that met all the inclusion and exclusion criteria. In the Fig. 2 primary studies were selected. These studies were analyzed and used to answer the research questions of this research work. The list of primary studies of both of these search strings are shown in Appendix A.

2.2.3   Primary Studies Quality Assurance

Quality criteria was used to ensure the quality and fidelity of the studies. The quality criteria were established in two questions: (1) Is the study mainly focused on the DevOps application with model or standard or an agile methodology? (2) Does the study focus on case studies where DevOps has been implemented or has been completely proposed? These questions were applied in every primary study to guarantee their quality of this research.

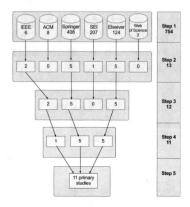

Fig. 1. Primary studies collected by selection process in first search.

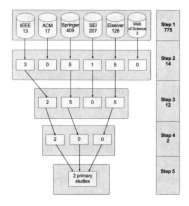

Fig. 2. Primary studies collected by selection process in second search.

2.2.4   Data Extraction

The information obtained from the primary studies was registered in a template developed in a spreadsheet editor (Google Sheets), which contains the following data: id, title, author, year, keywords, data source, objective, problem, strategy, validation, findings, targeted phase or process, proposal type (method, methodology or framework).

3   Results

This section is about the analysis of main results.

3.1   Data Obtained

Table 2 shows the coverage of the primary studies regarding the research questions defined. Subsequently, the data was analyzed by research question. Research questions were answered as follows

Table 2. Tracking of research questions results and primary studies

ID PS	Research questions		
	1	2	3
1	X		X
2	X	X	
3	X		
4			X
5			X
6			X
7			X
8			
9	X		
10	X		X
11			X
12	X		X
13			

(1) Which agile methodologies are used in DevOps environment to develop software?

Table 3 lists the agile methodologies used by companies that have implemented a DevOps approach to develop software. These primary studies showed case studies about companies that implement an agile environment configured and its results after setting up a DevOps approach. Besides, complementary practices are included to the agile process. DevOps was implemented in most of the time after using SCRUM as main methodology due to companies had already its process defined. However, they recommend to test possible options such as DSDM, XP and tools like kanban to use along a DevOps approach.

(2) Which models and standards are used in DevOps environment to develop software?

Only one primary study [A2] mentions information about DevOps approach using standards. The paper proposes and recommends the use of ITIL and

Table 3. Results obtained about agile methodologies used in DevOps environment

ID primary study	Methodologies	Tools
1	SCRUM	
2	SCRUM	
3	SCRUM	
9	SCRUM	
10	LEAN	KANBAN
12	SCRUM	

OWASP's Software Assurance Maturity Model (OWASP SAMM). Its DevOps set of processes based on the methodology OWASP SAMM to formulate and implement a strategy for software security on risk management. ITIL elements once applied by the organization for establishing integration with the organization's strategy, delivering value, and maintaining a minimum level of competency.

Moreover, ITIL within DevOps approach enables the operation of IT infrastructure, using qualitative and quantitative indicators to generate information for decision-making based on best practices.

(3) Which are the main problem or barriers to setup an environment DevOps?

Results obtained were classified in three categories: processes, guidance and team because these problems or barriers were similar in some aspects. The processes category was identified because the introduction of a new paradigm into the current process, which is not defined, causes problems to setup a DevOps approach building immature systems. The following list presents the main problems or barriers in processes that the companies had when they used DevOps.

– *Immature systems.* This means that there is not a control to follow system versions [A4].
– *Processes are not defined.* Companies do not have a defined process to follow, it causes problems in management and software development. Operational routines are not being established prior to deployment [A4].
– *Problems to instance, adapt and use tools and automation.* Introducing a new paradigm in the current process of the companies could be difficult because they have been operating on its own way and to implement new practices without help might be frustrating for the team [A5].
– *Line of business.* Companies which develop software such as embedded software, game development have different perspective and process [A7].

The guidance category was identified because contains problems related to misconception of DevOps causing problems to implement DevOps practices such as testing, continuous deployment, incorporate operators and lack of management because there is not a defined guide. The following list presents the main problems or barriers when companies tried to implement DevOps without a guidance.

- *Lack of experience implementing testing.* Companies have problems when they tried to implement testing causing unsatisfactory test environments [A1, A4].
- *Lack of guidance.* Companies have problems when they tried to setup a DevOps environment because there's not a guidance to consult for an advice [A1, A4, A11, A12].
- *Operators inclusion.* Operators are not being involved in requirements specification [A4].
- *Overload information.* New knowledge to learn (Ops working with Dev). Deployment Risk. Companies do not trust at all in DevOps implementation because they already have a established process [A5].
- *Project Management.* Part of the team has excess of freedom because anyone knows what need to do in the new defined process [A5].

The team category was identified because cross-team collaboration usually is disabled. Friction among team cause severity problems in short time. The following list has the main problems or barriers among team, when companies tried to implement DevOps.

- *Changing habits.* Change the traditional practice not only with Dev and Ops but management as stakeholders too [A1].
- *Poor communication among team.* There is not defined way to communicate among team causing problems in follow instructions or having an evidence about data, progress of work [A1, A4].
- *Deep-seated company culture.* Companies do not want to change the current process to avoid risks. It is almost inevitable to change the current process or team habits and do not get any claim about it [A4].
- *Team frustration.* Chaos and frustration because team tried to implement a DevOps approach [A1, A10].
- *Uncoordinated activities.* Management of software development in companies have a seated process which defines roles, activities, so that trying to adopt new practices to get an approach to DevOps causing uncoordinated activities among team [A4].
- *Operations and developers interests.* Operations engineers care about infrastructure and developers care about deploy code faster [A4].
- *Lack of management.* Mix responsibilities Dev and Ops is a common problem when companies have a DevOps approach in the current process induce friction among team [A5, A6].

4   Discussion

In this research paper the benefits and barriers in the implementation of DevOps were identified and classified in three categories: processes, guidance and team. Some of the main barriers identified are a lack of an agreement or document regarding processes that specifies a way to establish communication channels among the team and between team and stakeholders because a DevOps approach focused only on the software implementation; a lack of knowledge of how to integrate testing environments with continuous integration. A lack of DevOps information, so that it is with other concepts like continuous integration. Below, the problems and barriers detected in the systematic literature review are inspected and an analysis is made of the main problems of each category.

The main problem in processes category is the lack of the definition of process in the company. In Canada was identified a company that is specializing in the integration of interactive systems, communication and security, which had followed the ISO/IEC 29110 to solve the problem that they have. They kept a record of their processes based on employees' experience as well as recognized practices. This approach wasn't possible to replicate and get consistent deliverables in the project. Also they do not have templates or checklists to manage projects. They got a defined process after implementing the ISO/IEC 29110, which defines the roles needed to produce and review the project deliverables and how to reduce project risks, working in systematic and disciplined way, getting better quality and communication among team using standard [14].

One of the main problems in guidance category is deployment risk, because companies do not trust at all in implement DevOps practices without a guidance in their current process. In Canada, another company benefited with the implementation of ISO/IEC 29110 was a cash management IT department, of large financial institution that had a problem to know the status of specific requests. They had incidents when a change was placed in production. The development process had been painful and the documentation produced was not very useful. After using the ISO/IEC 29110 to prevent risks and uncoordinated activities in order to reduce the deployment risk. They defined a process covering the gaps helping to assure the quality in the stage of production [13]. ISO/IEC 29110 has a set of deployment packages which have been developed to define guidelines and explain in more detail the processes to adapt tools getting a future automation in some areas like test procedure [13].

One of the main problems in the team category is that companies do not want to change their established practices because imply a risk and they expect to get results fast. After applying these changes in the current process, team suffer frustration trying to do activities that do not correspond according their role, causing friction among team. In Peru was identified a company which followed the ISO/IEC 29110 to solve problems that they have among team like how to know who is responsible of software component. After implementing the ISO/IEC 29110 in its current process, each team member knew the tasks that they were responsible for executing and implemented the required software components, avoiding freedom or lack of management, taking into account the architectural design of the system. The results of this project demonstrate that the company, using ISO/IEC 29110, can quickly reach a high level of quality not only in its software development projects but in team members performance too [12]. These type of companies have those problems mentioned above frequently. Related to the identified problems/barriers ISO/IEC 29110 provides a set of proven practices that enable companies to plan and execute a project reducing rework [12]. Proven software engineering practices are needed to develop quality products that allow them to improve their operation and processes, the standard ISO/IEC 29110 arises to cover this need, it includes guidelines and technical reports focused on this companies to increase the quality of their products and services [9]. The main features of ISO/IEC 29110 are:

- *Profiles.* ISO/IEC 29110 has 4 profiles according their objectives. These profiles are entry, basic, intermediate, advance.
- *Process area.* ISO/IEC 29110 has 2 process area, it depends on the profile but project management process and the software implementation process are defined in entry and basic profile to carry software project with good practices.
- *Adoption.* Companies can adapt independently of the development approach or methodology used the ISO/IEC 29110 in the current process.
- *ISO/IEC 29110 elements.* ISO/IEC 29110 provides a set of process elements that facilitate its adoption, such as objectives, tasks, roles and work products [10].

There is not a specific guidance to implement a DevOps approach. It is a lack of support its implementation. The need to bridge this gap using the standard ISO/IEC 29110 was identified due to this standard has profiles to be applied in different companies according to size and demand.

5 Conclusions and Future Work

DevOps has several opportunities as well as barriers or problems to follow software development process, most of them are in project management. A SLR was performed to establish the state of the art of DevOps focused on three ideas: (1) Methodologies used along DevOps; (2) Standards applied in a DevOps approach (3) Barriers or problems to set up a DevOps environment. After applying the selection process using two search strings, only 13 studies were selected as "primary studies". Based on the results obtained, DevOps is relatively new in its implementation of environment. DevOps is based on lean so it's common to hear about companies using agile methodologies such as SCRUM, XP and DevOps simultaneously. Most of the barriers or problems obtained in the primary studies are related to a lack of management. In addition, most of the cases, it was discovered that there are companies that do not want to take a risk to change the process because they have a established culture and do not want to learn or adapt different practices. Besides, small companies usually do not have expertise to search for and adapt process improvement best practices [11]. Companies that have had similar problems to solve them, have opted to implement the ISO/IEC 29110. As a result of this research, future work to develop a guideline to implement a DevOps approach with ISO/IEC 29110 to solve gaps discovered after this SLR.

Appendix A: Primary Studies

A1 F. Elberzhager and T. Arif: Software Quality. Complexity and Challenges of Software Engineering in Emerging Technologies, vol. 269, pp. 33–44 (2017).

A2 M. Muñoz and O. Díaz.: Engineering and Management of Data Centers (2017).

A3 L. Ellen. Lwakatare. B, Pasi. Kuvaja, and Markku. Oivo: Product Focused Software Process Improvement, vol. 10027, pp. 399–415 (2016).

A4 L. Ringu-kalliosaari. B and M. Simo: Product-Focused Software Process Improvement, vol. 10027, pp. 590–597 (2016).

A5 K. Nybom. B, J. Smeds, and I. Porres: Agile Processes, in Software Engineering, and Extreme Programming, vol. 251, pp. 131–143 (2016).

A6 B. S. Farroha and D. L. Farroha: A framework for managing mission needs, compliance, and trust in the DevOps environment, Proc. IEEE Mil. Commun. Conf. MILCOM, pp. 288–293 (2014).

A7 T. Laukkarinen, K. Kuusinen, and T. Mikkonen: DevOps in regulated software development: Case medical devices, Proc. IEEE/ACM 39th Int. Conf. Softw. Eng. New Ideas Emerg. Results Track, ICSE-NIER 2017, pp. 15–18 (2017).

A8 R. Bierwolf, P. Frijns, and P. van Kemenade: Project management in a dynamic environment: Balancing stakeholders, IEEE Eur. Technol. Eng. Manag. Summit, pp. 1–6 (2017).

A9 B. Fitzgerald and K. J. Stol: Continuous software engineering: A roadmap and agenda, J. Syst. Softw., vol. 123, pp. 176–189 (2017).

A10 L. Chen: Continuous Delivery: Overcoming adoption challenges. J. Syst. Softw., vol. 128, pp. 72–86 (2017).

A11 R. Colomo-Palacios, E. Fernandes, P. Soto-Acosta, and X. Larrucea. (2018): A case analysis of enabling continuous software deployment through knowledge management, Int. J. Inf. Manage., vol. 40, no. November, pp. 186–189 (2017).

A12 S. Makinen et al.: Improving the delivery cycle: A multiple-case study of the toolchains in Finnish software intensive enterprises, Inf. Softw. Technol., vol. 80, pp. 1339–1351 (2016).

A13 V. Gupta, P. K. Kapur, and D. Kumar: Modeling and measuring attributes influencing DevOps implementation in an enterprise using structural equation modeling, Inf. Softw. Technol., vol. 92, pp. 75–91 (2017).

References

1. Claps, G.G., Berntsson Svensson, R., Aurum, A.: On the journey to continuous deployment: technical and social challenges along the way. Inf. Softw. Technol. **57**(1), 21–31 (2015)

2. Lassenius, C., Dingsøyr, T., Paasivaara, M.: Agile processes in software engineering, and extreme programming: 16th international conference, XP 2015, Helsinki, Finland, 25–29 May 2015 proceedings. In: Lecture Notes in Business Information Processing, vol. 212, pp. 212–217 (2015)

3. Bass, L., Weber, I., Zhu, L.: DevOps: a software architects perspective. In: SEI. Addison Wesley, New York (2015)
4. IBM Corporation: DevOps: the IBM approach, pp. 1–12 (2014)
5. Kitchenham, B.: Procedures for performing systematic reviews, joint technical report. Keele University Technical Report TR/SE-0401 (2004). ISSN 1353-7776
6. Pedreira, O., Piattini, M., Luaces, R.M., Brisaboa, R.N.: A systematic review of software process tailoring. ACM SIGSOFT Softw. **32**(3), 1 (2007)
7. Miramontes, J., Munoz, M., Calvo-Manzano, J.A.: Trends and Applications in Software Engineering, vol. 405 (2016)
8. Virmani, M.: Understanding DevOps and bridging the gap from continuous integration to continuous delivery. In: 5th International Conference on the Innovative Computing Technology (INTECH 2015), pp. 78–82 (2015)
9. Munoz, M., Mejía, J., Laporte, C.Y.: Implementación del Estándar ISO/IEC 29110 en Centros de Desarrollo de Software de Universidades Mexicanas: Experiencia del Estado de Zacatecas, no. 2 (2018)
10. Laporte, C., Muñoz, M., Geranon, B.: The education of students about ISO/IEC 29110 software engineering standards and their implementations in very small entities. In: IEEE Canada International Humanitarian Technology Conference (IHTC), pp. 94–98 (2017)
11. Laporte, C., Connor, R.: Software process improvement standards and guides for very small organizations: an overview of eight implementations. CrossTalk J. Defense Softw. Eng. **3**, 23–27 (2017)
12. Laporte, C., García, L., Bruggman, M.: Implementation and Certification of ISO/IEC 29110 in an IT Startup in Peru (2015)
13. Laporte, C., O'Connor, R.: A multi-case study analysis of software process improvement in very small companies using ISO/IEC 29110. In: EuroSPI 2016. CCIS, vol. 633, pp. 30–44 (2016)
14. Laporte, C., Tremblay, N., Menaceur, J., Poliquin, D.: Implementing the new ISO/IEC 29110 systems engineering process standard in a small public transportation company. In: EuroSPI 2016. CCIS, vol. 633, pp. 15–29 (2016)

Role of Green HRM Practices in Employees' Pro-environmental IT Practices

Adedapo Oluwaseyi Ojo$^{(\boxtimes)}$ (iD) and Murali Raman

Multimedia University, 63100 Cyberjaya, Selangor, Malaysia
Ojo.adedapo@mmu.edu.my

Abstract. Green human resource management (GHRM) involves the alignment of the firm's HRM practices to environmental management system. However, the specific impact of GHRM practices on employees' pro-environmental behaviours, most especially in the information technology (IT) domain has not been fully clarified. Therefore, this study posits GHRM practices like recruitment and selection, training and development, compensation and reward, performance management, and empowerment and participation as determinants of employees' pro-environmental IT practices. Based on the responses from 68 HR managers and 333 IT professionals in Malaysia, the results of path model analysis confirm the significant impact of green training and development, and empowerment and participation. Furthermore, the implications of the findings are discussed, and suggestions offered on the deployment of GHRM practices in stimulating employees' pro-environmental IT behaviour.

Keywords: GHRM practices · Pro-environmental behaviour ·
Sustainable IT practices · Environmental management system

1 Introduction

Sustainable environmental performance has become an essential part of corporate strategy, thereby resulting in the rapid adoption of the environmental management system (EMS) like ISO 14001 certification. The EMS is a regulatory framework that enables an organisation to document the procedures and policies guiding the environmental impact of its operations [1]. However, the implementation of EMS does not necessarily translate into sustainable environmental performance [2]. The implementation of EMS in an organisation is considered as a change initiative which is likely to be resisted by the employees, who are expected to adopt new workplace practices. Recent empirical investigations of the manufacturing processes reported higher environmental performance for firms that integrate their EMS with HRM practices [1, 3, 4]. This integration has resulted in the deployment of GHRM practices. Through GHRM an organisation develops the relevant policies and practices to encourage the sustainable use of resources by its employees. Although, studies have acknowledged that behavioural change at the individual level is critical to environmental performance, yet, few empirical studies have clarified the impact of GHRM practices on such change [1, 2]. For example, Zibarras and Coan [2] highlighted the GHRM practices for facilitating

© Springer Nature Switzerland AG 2019
Á. Rocha et al. (Eds.): WorldCIST'19 2019, AISC 930, pp. 678–688, 2019.
https://doi.org/10.1007/978-3-030-16181-1_64

employees' pro-environmental behaviour but did not investigate the extent to which these practices relate to the employee's behaviour. Therefore, further studies are needed to clarify the specific GHRM practices underlying employees' pro-environmental behaviour.

Studies on the adoption of green IT practices have acknowledged the lack of employees' engagement as the main barrier to sustainable IT performance in the workplace [5, 6]. According to Jenkin et al. [5], a significant portion of carbon emission and energy wastage from IT operations can be associated with human attitude and behaviour. To this end, subsequent studies have examined the factors of employee's attitude and behaviour towards green IT practices, but the impact of GHRM practices on behavioural change has been overlooked [7–9]. Thus, the present study aims to investigate the impact of GHRM practices in stimulating employees' engagement in sustainable workplace IT practices.

This paper is structured as follows: Sect. 2 explains the theoretical background and hypotheses; Sect. 3 describes the methodology and measurement; Sect. 4 is the results of data analyses; Sect. 5 discusses the implications of findings and conclusion, with suggestions for future research.

2 Theoretical Background

GHRM entails the deployment of sustainable HR practices to enhance processes, reduce resource wastage, and align products, services, and procedures to stimulate employees' pro-environmental behaviour. Although many organisations have acknowledged the strategic implications of GHRM in the sustainability agenda, however, most environmental initiatives have not generated the expected behavioural responses from the employees [1, 4]. According to Jabbour and Santos [10], the alignment between HRM practices and EMS is an essential factor for better organisational environmental performance. The benefit of EMS is most significant for organisations that engage her employees in the implementation [2]. The organisations that encourage employees pro-environmental behaviour attain higher productivity, which in turn result in sustainable competitive advantage [11, 12].

The attainment of sustainable environmental performance depends on the alignment between HR practices and corporate strategy. According to the resource-based view, the bundle of HRM practices is a form of internal resource, which can enable a firm to develop the essential employees' skills, knowledge, and attitude to support the execution of strategic objectives [13]. Therefore, successful intervention to enable employees' pro-environmental behaviour require the deployment of specific HRM practices. However, most studies on pro-environmental behaviour have been outside the workplace [14]. The few studies on the workplace have prioritized the impact of organisational green initiatives and leadership on employees' behaviour [15, 16].

GHRM practices have been reported to have significant impact on the firms' environmental performance, but the specific impact of these practices have been less examined [14, 16]. According to [11] "employers need to understand how employees make decisions about whether to participate in organisational roles and activities." Employees behaviours towards organisation's green initiative could be stimulated through the in-role or extra-role actions [14]. The in-role behaviours are stated in the job descriptions and employees are expected to demonstrate such actions and are rewarded for doing so. However, the extra-role behaviours are the discretionary actions performed by the employees to enhance environmental performance. Within the IT domain, these include actions that are taking to reduce energy wastage, such as turning off computers after use or encouraging colleagues to adopt Green IT practices. Unlike the in-role behaviour, these practices are not specified in the job descriptions, and they are neither required nor rewarded. However, they constitute the pro-environmental behaviour, which is deeply rooted in the organisational citizenship behaviour (OCB) framework [14].

Sequel to the above, we argue that the impact of GHRM practices on organisation's environmental performance manifests through practices that encourage employees' pro-environmental behaviour in the work practices. The research framework as shown in Fig. 1.

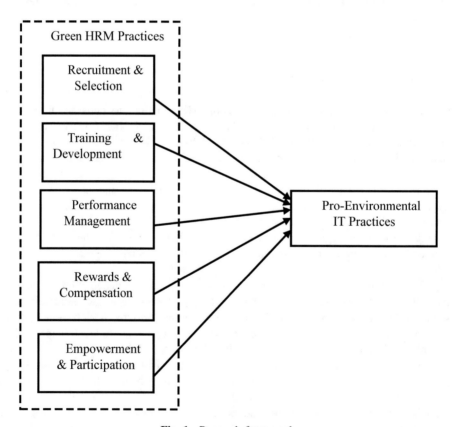

Fig. 1. Research framework

2.1   Green Recruitment and Selection

Organisation's commitment towards sustainable environmental work practice is reflected through the recruitment and selection process that focuses on the hiring of environmentally conscious employees [12]. Such employees are attracted to organisations that have invested in developing their environmental reputations and images [1]. Organisations can communicate their sustainability agenda to prospective candidates through various touchpoints including websites and other publications [17]. Guerci et al. [17] found that environmentally conscious employees are attracted to organisation that prioritize environmentally sustainable work practices. Besides, the importance of sustainable work practices is emphasised in job description that focuses on environmental values [18]. The employees' perceive environmentally friendly values as the accepted norms in the organisation, thereby shaping their pro-environmental behaviours. Based on the above, the following hypothesis is suggested:

> H1. Green recruitment and selection influence employee's engagement in pro environmental IT practices.

2.2   Green Training and Development

An organisation's transition to sustainable environmental work practices can be considered as a change initiative, which requires adequate training and re-orientation of the employees. Through training, employees are exposed to the essential environmental management standards which enable them to gain awareness of pro-environmental work practices and stimulate their interest in such practices [2, 19]. The impact of green training can also be assessed in the creation and nurturing of green consciousness among the employees [20]. Training on green practices enable the employees to acquire the relevant knowledge and capability to support their organisation's environmental initiatives, which can in turn nurtures the culture of sustainable practices. Therefore, training programmes that are aligned with closing the environmental knowledge gaps can be effective in stimulating employees' pro-environmental behaviour. Accordingly, environmental training should incorporate hands-on activities that can enhance the employees' awareness and knowledge of environmental initiatives. Employees should be trained on how to conduct green analysis of workspace and equipped with the personal skills to support green initiatives [12]. Therefore, the following hypothesis is suggested:

> H2. Green training and development influence employee's engagement in pro-environmental IT practices.

2.3   Green Performance Management

The adoption of organisation metrics for the assessment of workplace practices can contribute to the sustainable management of the resource consumption and production activities [1, 11]. Environmental management metrics can also be extended to assess and track employees' engagement in pro-environmental practices [19]. Through these metrics, employees are encouraged to participate and provided the relevant feedback on their level of involvement in sustainable practices. By incorporating environmental metrics into the job tasks and providing timely feedbacks, the employees are able to adjust their behaviour in accordance to the metrics used in assessing their performance. In line with the above, it is hypothesized that:

H3. Green performance management influence employee's engagement in pro-environmental IT practices.

2.4   Green Reward and Compensation

Employees' engagement in sustainability initiative can be enhanced through the introduction of the appropriate reward and compensation system [10]. Reward system has a more significant impact on environmental performance, when it is focused on encouraging pro-environmental behaviour [21]. An effective reward system is designed to reflect the management's commitment to environmental performance through the reinforcement of employees' pro-environmental behaviours [12]. Pro-environmental behaviour can be reinforced by linking employees' engagement in green practices with rewards like promotion, career advancement and other monetary and non-monetary incentives. However, the application of negative reinforcement can inhibit pro-environmental behaviour [1]. In attempt to avoid punishment, employees might conceal environmental problem or completely withdraw from sustainable practice that are prone to failure.

H4. Green reward and compensation influence employee's engagement in pro-environmental IT practices.

2.5   Green Empowerment and Participation

Employee's empowerment involves the management's delegation of authority to the employees on environmental issues [21]. Employees are empowered when they are able to contribute in shaping the organisation's environmental objectives and exercise autonomy in taking actions on environmental issues. By participating in shaping the organisation's environmental agenda, the employees assume social responsibility over their actions. Several studies have validated the significant impact of employees' empowerment on environmental performance [1, 22, 23]. Nevertheless, the complexity of most environmental problems requires the collective participation of empowered individuals with diverse expertise. Therefore, empowered employees are more likely to collaborate with others, through their engagement in pro-environmental behaviour to support the firm's environmental performance. Consistent with these research, the following hypotheses are suggested.

H5. Green empowerment and participation influence employee's engagement in pro-environmental IT practices.

3   Research Method

3.1   Sample and Procedures

The data were collected from the HR managers, and IT professionals in ISO14001 certified MSC status companies in Malaysia. The HR managers completed the questions on the GHRM practices and the IT professionals answered the questions on pro environmental IT behaviour. The MSC status companies are those which develop or use information and communication technologies in the production and delivery of their products, services, and process. The two-stage sampling technique was followed in the selection of the respondents. At the first stage, the researcher contacted the HR managers of MSC status companies on the list of ISO14001 certified companies from the SIRIM QAS international directory. After sending three reminders, the consent to participate was obtained from the HR managers of eighty eight (88) companies. At the second stage, two different sets of questionnaires were distributed to the HR managers and IT professionals in the 88 companies. The HR managers were asked to complete the questionnaires on the GHRM practices, while the IT professionals completed the questionnaires on pro environmental IT behaviour. A total of 440 questionnaires were distributed to IT professionals, and 355 were returned as completed. After excluding unmatched and incomplete cases, we collated 68 matched questionnaires, which were used for the final analysis. The response rates were 73.91% for the HR managers and 80.68% for the IT professionals.

3.2   Measurement

All the variables were measured with scales adapted from the literature to suit the context of the present study. The multidimensional Green HRM practices of training and development, recruitment and selection, performance management, reward and compensation, and empowerment and participation were measured with items adapted from Masri and Jaaron [1] and Jabbour [24]. The pro-environmental IT behaviour was measured with seven items adapted from Paillé et al. [14]. All the items were based on the 5-point Likert scale ranging from 1 for "strongly disagree" to 5 for strongly "strongly agree".

4   Result

4.1   Measurement Model

The data analysis was conducted using the PLS-SEM technique. This involves the simultaneous evaluation of the measurement and structural model within a single, inclusive and systematic procedure. For the measurement model, the uni-dimensionality of each latent construct was assessed using the composite reliability (CR) with a threshold value of 0.70. As shown in Table 1, the CR values for all the constructs exceeded 0.70, and all the factor loadings are greater than 0.60. Therefore, the measurement model satisfies the conditions for internal consistency reliability [25].

Table 1. Results of the measurement model

Variable	Cronbach Alpha	CR	AVE	# of items	Factor loading
Green selection and recruitment	0.880	0.912	0.675	5	0.754–0.830
Green training and development	0.850	0.893	0.628	5	0.685–0.846
Green performance management	0.880	0.912	0.674	5	0.767–0.866
Green reward and compensation	0.826	0.896	0.743	3	0.835–0.903
Green employee empowerment	0.831	0.881	0.598	5	0.675–0.805
Pro-environmental IT practices	0.966	0.972	0.833	7	0.885–0.949

The average variance extracted for all constructs exceeded the threshold value 0.50, thereby indicating high convergent validity [26]. Furthermore, the discriminant validity was assessed by comparing the square root of the AVE for each construct with the pair of correlations between other constructs [26]. As shown in Table 2, this study satisfies the conditions for the discriminant validity because the square root of AVE for each construct (i.e., the diagonal elements) is greater than the correlations of the construct with other constructs (i.e., the off-diagonal elements).

Table 2. Results of discriminant validity

Variables	1	2	3	4	5	6
1. Green Empowerment & Participation (GEP)	**0.773**					
2. Green Recruitment & Selection (GRS)	0.745	**0.822**				
3. Green Training & Development (GTD)	0.458	0.624	**0.793**			
4. Green Performance Management (GPM)	0.570	0.665	0.669	**0.821**		
5. Green Reward & Compensation (GRC)	0.624	0.637	0.416	0.446	**0.862**	
6. Pro-Environmental IT practices (PIT)	0.582	0.535	0.550	0.525	0.412	**0.913**

Following Henseler et al. [27], the heterotrait-monotrait ratio of correlations (HTMT) was examined to ascertain strong discriminant validity. This is established when all the correlation values are smaller than 0.9. As shown in Table 3, all the correlation values are lower than the threshold value of 0.9. Therefore, the measurement model satisfies the condition for a strong discriminant validity.

4.2   Structural Model

The hypothesized relationships were tested using the significance levels of the path coefficient (i.e., beta) and the coefficient of determination (R^2) [28]. As shown in

Table 3. Heterotrait-monotrait ratio (HTMT)

Variables	GEP	PIT	GRS	GTD	GPM
PIT	0.640				
GRS	0.780	0.555			
GTD	0.560	0.599	0.712		
GPM	0.679	0.549	0.755	0.776	
GRC	0.750	0.459	0.753	0.501	0.539

Table 4, green training and development ($\beta = 0.311$, $p < 0.05$), and empowerment ($\beta = 0.399$, $p < 0.01$) are significant predictors of pro-environmental IT practices. While the other GHRM practices like recruitment and selection, reward and compensation, and performance management were not significantly related to pro-environmental IT practices. Therefore, H2 and H5 were supported, but H1, H3, and H4 were not supported.

The R^2 value of 0.445 indicates that 44.5% of the variance in pro-environmental IT practices can be explained by green training and development, and green employees' empowerment. Besides, we found moderate effect sizes for training and development, and empowerment and participation, with the f^2 values of 0.114, and 0.192, respectively. Moreover, the predictive capability of the model was assessed based on the Stone-Geisser's, Q^2 value [29]. Following Hair et al. [25], the Q^2 value of 0.327 for pro-environmental IT practices suggests that the model has a good predictive relevance.

Table 4. Results of hypotheses testing

H	Relationships	Beta	t-value	p-value	Decision	f^2
H1	GRS → PIT	0.032	0.199	0.421	Not Supported	0.001
H2	GTD → PIT	0.314	1.945	0.026	Supported	0.114
H3	GPM → PIT	0.107	0.702	0.241	Not Supported	0.009
H4	GRC → PIT	0.006	0.053	0.479	Not Supported	0.001
H5	GEP → PIT	0.397	2.324	0.010	Supported	0.192

5   Discussion and Implications

This study examines the specific impact of GHRM practices on employees' pro-environmental IT practices. The results support the significant impact of green training and development, and green empowerment and participation on the employees' engagement in pro-environmental IT practices. However, GHRM practices like recruitment and selection, performance management, and reward and compensation were found not to be associated with employees' pro-environmental IT practices. The lack of support for these hypotheses can be partly explained through the role of social exchange relationships in stimulating discretionary environmental behaviour [30–32]. Unlike the economic exchange relationship, which are characterized by enforceable contractual arrangements, social exchange relationship entails socio-emotional benefits that are long-term [32]. The deployment of recruitment and selection, performance management, and reward and compensation in stimulating pro-environmental IT practices align with the economic exchange relationship, because these practices focus on the organisation's control over the outcome. Therefore, employees are more likely to engage in pro-environmental IT practices when they are assured that such actions are required and rewarded by the organisation. Whereas the GHRM practices of training and development, and empowerment and participation are likely to be perceived by the employees as the organisation's appreciation of their need for advancement and contributions. Thus, when an organisation directs positive and beneficial actions towards the employees, it creates the obligations for the employees to reciprocate through desirable work-related behaviours [31].

Secondly, this study offers a response to the recent calls for further study on the organisational antecedents of employees' pro-environment behaviour [7, 33]. The prevalence of studies on the organisational antecedents of GIT practices and performance have overlooked the significant role of the employees in the sustainability agenda. And the few studies on employees' adoption of GIT have only examined the impact of individuals' cognition and motivation on pro-GIT behaviour [8, 9, 34, 35]. Therefore, our study addresses the need for more interdisciplinary investigation of organisation's environmental management initiative from the perspective of an emerging economy [3, 36].

6 Conclusion

This study draws on literature from GHRM and GIT domains to examine the impact of GHRM practices on employees' pro-environmental IT practices. The results confirm the significant impacts of training and development, and empowerment and participation on pro-environmental IT practices. Nevertheless, the findings from this study should be interpreted considering the inherent limitations. The data were collected from the HR managers and IT personnel in Malaysia's based ICT organisations. Future study are implored to adopt this model in other context like the manufacturing and aviation, to ascertain the extent to which the findings can be generalized. The impacts of other individual and organisational factors like leadership support and culture on GHRM practices are also important issues that could be addressed in subsequent studies. Besides future research is also needed to clarify the implication of GHRM practices in enabling employees' pro-environmental behaviours towards sustainable firm's environmental performance.

Acknowledgments. This research was funded by Fundamental Research Grant Scheme [FRGS FRGS/1/2016/SS03/MMU/03/1], Ministry of Higher Education Malaysia.

References

1. Masri, H.A., Jaaron, A.A.M.: Assessing green human resources management practices in Palestinian manufacturing context: an empirical study. J. Clean. Prod. **143**, 474–489 (2017)
2. Zibarras, L.D., Coan, P.: HRM practices used to promote pro-environmental behavior: a UK survey. Int. J. Hum. Resour. Manag. **26**, 2121–2142 (2015)
3. Renwick, D.W.S., Jabbour, C.J.C., Muller-Camen, M., Redman, T., Wilkinson, A.: Contemporary developments in Green (environmental) HRM scholarship. Int. J. Hum. Resour. Manag. **27**, 114–128 (2016)
4. Jabbour, C.J.C., Santos, F.C.A.: Relationships between human resource dimensions and environmental management in companies: proposal of a model. J. Clean. Prod. **16**, 51–58 (2008)
5. Jenkin, T.A., Webster, J., McShane, L.: An agenda for "Green" information technology and systems research. Inf. Organ. **21**, 17–40 (2011)
6. Gholami, R., Sulaiman, A.B., Ramayah, T., Molla, A.: Senior managers' perception on green Information Systems (IS) adoption and environmental performance: results from a field survey. Inf. Manag. **50**, 431–438 (2013)
7. Loeser, F., Recker, J., vom Brocke, J., Molla, A., Zarnekow, R.: How IT executives create organizational benefits by translating environmental strategies into Green IS initiatives. Inf. Syst. J. **27**, 503–553 (2017)
8. Molla, A., Abareshi, A., Cooper, V.: Green IT beliefs and pro-environmental IT practices among IT professionals. Inf. Technol. People **27**, 129–154 (2014)
9. Ojo, A.O., Raman, M., Vijayakumar, R.: Cognitive determinants of IT professional belief and attitude towards green IT. In: Advances in Intelligent Systems and Computing, pp. 1006–1015 (2018)
10. Jabbour, C.J.C., Santos, F.C.A.: The central role of human resource management in the search for sustainable organizations. Int. J. Hum. Resour. Manag. **19**, 2133–2154 (2008)
11. Jackson, S.E., Seo, J.: The greening of strategic HRM scholarship. Organ. Manag. J. **7**, 278–290 (2010)
12. Renwick, D.W.S., Redman, T., Maguire, S.: Green human resource management: a review and research agenda. Int. J. Manag. Rev. **15**, 1–14 (2013)
13. Bowen, D.E., Ostroff, C.: Understanding HRM-firm performance linkages: the role of the "strength" of the HRM system. Acad. Manag. Rev. **29**, 203–221 (2004)
14. Paillé, P., Chen, Y., Boiral, O., Jin, J.: The impact of human resource management on environmental performance: an employee-level study. J. Bus. Ethics **121**, 451–466 (2014)
15. Robertson, J.L., Barling, J.: Greening organizations through leaders' influence on employees' pro-environmental behaviors. J. Organ. Behav. **34**, 176–194 (2013)
16. Dumont, J., Shen, J., Deng, X.: Effects of green HRM practices on employee workplace green behavior: the role of psychological green climate and employee green values. Hum. Resour. Manage. **56**, 613–627 (2017)
17. Guerci, M., Longoni, A., Luzzini, D.: Translating stakeholder pressures into environmental performance – the mediating role of green HRM practices. Int. J. Hum. Resour. Manag. **27**, 262–289 (2016)
18. Wehrmeyer, W.: Greening people: human resources and environmental management. Greenleaf, Sheffield (1996)
19. Govindarajulu, N., Daily, B.F.: Motivating employees for environmental improvement. Ind. Manag. Data Syst. **104**, 364–372 (2004)
20. Opatha, H.H.D.N.P., Arulrajah, A.A.: Green human resource management: simplified general reflections. Int. Bus. Res. **7**, 101 (2014)

21. Daily, B.F., Huang, S.: Achieving sustainability through attention to human resource factors in environmental management. Int. J. Oper. Prod. Manag. **21**, 1539–1552 (2001)

22. Del Brío, J.Á., Fernández, E., Junquera, B.: Management and employee involvement in achieving an environmental action-based competitive advantage: an empirical study. Int. J. Hum. Resour. Manag. **18**, 491–522 (2007)

23. Rothenberg, S.: Knowledge content and worker participation in environmental management at NUMMI. J. Manag. Stud. **40**, 1783–1802 (2003)

24. Jabbour, C.J.C.: How green are HRM practices, organizational culture, learning and teamwork? A Brazilian study. Ind. Commer. Train. **43**, 98–105 (2011)

25. Hair, J., Hollingsworth, C.L., Randolph, A.B., Chong, A.Y.L.: An updated and expanded assessment of PLS-SEM in information systems research. Ind. Manag. Data Syst. **117**, 442–458 (2017)

26. Fornell, C., Larcker, D.F.: Evaluating structural equation models with unobservable variables and measurements error. J. Mark. Res. **18**, 39–50 (1981)

27. Modeling, S.E., Henseler, J., Ringle, C.M., Sarstedt, M.: A new criterion for assessing discriminant validity in variance-based a new criterion for assessing discriminant validity in variance-based structural equation modeling (2015)

28. Hair, J.F., Ringle, C.M., Sarstedt, M.: Partial least squares structural equation modeling: rigorous applications, better results and higher acceptance. Long Range Plann. **46**, 1–12 (2013)

29. Henseler, J., Ringle, C.M., Sinkovics, R.R.: The use of partial least squares path modeling in international marketing. Adv. Int. Mark. **20**, 277–319 (2009)

30. Nishii, L.H., Lepak, D.P., Schneider, B.: Employee attributions of the "Why" of HR practices: their effects on employee attitudes and behaviors, and customer satisfaction. Pers. Psychol. **61**, 503–545 (2008)

31. Gould-Williams, J., Davies, F.: Using social exchange theory to predict the effects of HRM practice on employee outcomes: an analysis of public sector workers. Public Manag. Rev. **7**, 1–24 (2005)

32. Aryee, S., Budhwar, P.S., Chen, Z.X.: Trust as a mediator of the relationship between organizational justice and work outcomes: test of a social exchange model. J. Organ. Behav. **23**, 267–285 (2002)

33. Asadi, S., Hussin, A.R.C., Dahlan, H.M.: Organizational research in the field of Green IT: a systematic literature review from 2007 to 2016. Telemat. Inform. **34**(7), 1191–1249 (2017)

34. Mishra, D., Akman, I., Mishra, A.: Theory of reasoned action application for green information technology acceptance. Comput. Hum. Behav. **36**, 29–40 (2014)

35. Dao, V., Langella, I., Carbo, J.: From green to sustainability: information technology and an integrated sustainability framework. J. Strateg. Inf. Syst. **20**, 63–79 (2011)

36. Jackson, S.E., Renwick, D.W.S., Jabbour, C.J.C., Muller-Camen, M.: State-of-the-art and future directions for green human resource management. Ger. J. Res. Hum. Resour. Manag. **25**, 99–116 (2011)

Innovation Trends for Smart Factories: A Literature Review

Maria José Sousa[1,2(✉)], Rui Cruz[1], Álvaro Rocha[2], and Miguel Sousa[3]

[1] Universidade Europeia, Lisbon, Portugal
mjdcsousa@gmail.com
[2] University of Coimbra, Coimbra, Portugal
[3] University of Essex, Colchester, UK

Abstract. The purpose of this article is to analyze the different dimensions of innovation in order to create a smart factory, including emerging technologies and their impacts on the operations management process, and besides, the digital competencies needed to cope with technological and organizational changes. The main findings of the systematic literature review were Artificial Intelligence, Big data, 3D Printing, Robotization, Internet of Things, and Augmented Reality. The results can help factories to be prepared for the digital transformation process and become a smart factory to meet the challenges of an ever-changing economy.

Keywords: Smart Factories · Innovation · Trends · Digital technologies · Digital economy

1 Introduction

Digital technologies drive for several large-scale transformations in the production processes and in operations management of industries helping them to become Smart Factories. The recent studies tried to analyses the manufacturing operations' strategic alignment and responsiveness to market and the need for customization and an increase in performance. One of the answers can be the analysis of relationships between technology and supply chain integration, innovative capabilities, skills (Sousa and Wilks 2018) and the manufacturing performance.

This literature review tries to undercover some of this topic and the technology add value in the various stages of the new product development and on the production along the supply chain and the operations management processes.

An important topic framed by the robotization, the artificial intelligence, and the augmented reality is the analysis of the impact of AM (addictive manufacturing – 3DP) technology and the challenges and barriers posed in the development and deployment of 3DP including the research of the obstacles that resist mass-scale applications of this technology. The analysis of the conceptual benefits of the implementation of AM is also an important issue for the industry.

© Springer Nature Switzerland AG 2019
Á. Rocha et al. (Eds.): WorldCIST'19 2019, AISC 930, pp. 689–698, 2019.
https://doi.org/10.1007/978-3-030-16181-1_65

This paper begins by understanding what innovation is, what for it can be used and how organizations can become more innovative. For this purpose, it includes a systematic literature review of new production technologies and processes.

The literature review intent to understand theories of innovation, namely the technological and organizational innovation, regarding research and development, new processes and materials in industrial processes and their importance for the increase of productivity and competitiveness.

2 Literature Review

2.1 Methodological Approach

A systematic search of online scientific databases using b-on, a scientific information research tool, was conducted in the middle of November 2018. The search was made using several queries, containing the keywords "Innovation," "Additive Manufacturing"; and "Technology." The articles selected for this systematic review are presented in Table 1, containing information regarding the bibliographical reference of the publication, the primary attributes researched and the key findings.

Table 1. Articles selection criteria

Criteria	Number of articles
Operations management	5,132,279
No expanders	680,040
Academic reviews (peer review and in English)	427,696
Since 2015	107,535
7 main journals with more than 1000 articles	14,618
Main Keywords: Innovation; Additive Manufacturing; Technology	217
Other Keywords: Product development; New product development; Operations management research	37
Final	**27 articles**

2.2 Results of the Paper's Search

The total of articles eligible according the criteria defined are 27 (3 eliminated because did not fit the purpose of the search and 7 were duplicate), from the following journals: European Journal of Operational Research, International Journal of Production Economics, International Journal of Production Research, Operations Management Research, and Production and operations management (Table 2).

Table 2. Number of articles per journal

Journal/Impact factor	Number of articles
European Journal of Operational Research, 4.06	2
International Journal of Production Economics, 4.34	8
International Journal of Production Research, 2.55	11
Operations Management Research, 1.10	1
Production and operations management, 2.02	5
Total	**27**

The technological innovation attribute is the one with the most articles (14 or 52%), followed by the attribute organizational innovation (13 or 48% of the articles) as presented in Table 2 and Graph 1:

Graph 1. Types of innovation

From the literature review and according to the types of innovation: organizational and technological, it was possible to identify the most recent and vital articles (according to the criteria of this research) studying the new operation management processes and also the technologies to make the production processes more innovative, efficient and productive, as showed in Table 3.

Table 3. New research on smart factory technologies by type of innovation

Type of Innovation/Journal/Article/Author(s)
Organizational Innovation
European Journal of Operational Research
Cooperation royalty contract design in research and development alliances: help vs. knowledge-sharing
Yu et al. (2018)
International Journal of Production Economics
Exploring the managerial dilemmas encountered by advanced analytical equipment providers in developing service-led growth strategies
Raja et al. (2017)

(*continued*)

Table 3. (*continued*)

Type of Innovation/Journal/Article/Author(s)
Knowledge sharing dynamics in service suppliers' involvement for servitization of manufacturing companies
Ayala et al. (2017)
The impact of strategic alignment and responsiveness to market on manufacturing firm's performance
Sardana et al. (2016)
International Journal of Production Research
Collaborative networks: a systematic review and multi-level framework
Durugbo (2016)
Inter-firm partnerships–strategic alliances in the pharmaceutical industry
Yoon et al. (2018)
New product development in new ventures: the quest for resources
Bolumole et al. (2015)
Strategic alliance formation and the effects on the performance of manufacturing enterprises from supply chain perspective
Yang et al. (2015)
The impact of supply chain relationships and integration on innovative capabilities and manufacturing performance: the perspective of rapidly developing countries
Adebanjo et al. (2018)
Through entrepreneurs' eyes: the Fab-spaces constellation
Mortara and Parisot (2016)
Production and operations management
Creativity and risk taking aren't rational: behavioural operations in MOT
Loch (2017)
How excessive stage time reduction in NPD negatively impacts market value
Bendoly and Chao (2016)
Making the best idea better: the role of the idea pool structure
Erat (2017)
The impact of contracts and competition on upstream innovation in a supply chain
Wang and Shin (2015)
Technological Innovation
European Journal of Operational Research
Innovation and technology diffusion in competitive supply chains
Aydin and Parker (2018)
International Journal of Production Economics
Economic implications of 3D printing: market structure models in light of additive manufacturing revisited
Weller et al. (2015)
Green product development and environmental performance: investigating the role of government regulations
Hafezi and Zolfagharinia (2018)

(*continued*)

Table 3. (*continued*)

Type of Innovation/Journal/Article/Author(s)
The barriers to the progression of additive manufacture: perspectives from the UK industry
Thomas-Seale et al. (2018)
The impact of 3D Printing Technology on the supply chain: manufacturing and legal perspectives
Chan et al. (2018)
The perceived value of additively manufactured digital spare parts in industry: an empirical investigation
Chekurov et al. (2018)
International Journal of Production Research
3D printing technology and its impact on Chinese manufacturing
Long et al. (2017)
Additive manufacturing management: a review and future research agenda
Khorram Niaki and Nonino (2017)
Leveraging prototypes to generate value in the concept-to-production process: a qualitative study of the automotive industry
Elverum and Welo (2016)
Technology alignment and business strategy: a performance measurement and Dynamic Capability perspective
McAdam et al. (2017)
The best of times and the worst of times: empirical operations and supply chain management research
Melnyk et al. (2018)
Operations Management Research
The direct digital manufacturing (r) evolution: definition of a research agenda
Holmström et al. (2016)
Production and operations management
The role of project and organizational context in managing high-tech R&D projects
Chandrasekaran et al. (2015)

2.3 Innovation Concept

The innovation theory literature gives the idea that innovations occur mostly within the national system of innovation (Freeman 1987; Lundvall 1992; Nelson 1993; Edquist 1997). However, another perspective was studied by organizational academics in innovation in organizational microsystems (Van de Ven 1986; Aldrich and Fiol 1994; Van de Ven et al. 1999; den Hertog and Huizenga 2000; Raja et al. 2017; Ayala et al. 2017): literature shows that the concept of innovation is very complicated, which makes it difficult to have a single definition. The Green Book on Innovation from the European Commission (1996) defines innovation as "the successful production, assimilation, and exploration of something new." Mulgan and Albury (2003) made their contribution to the concept pointing out the importance of the innovation implementation results: "new processes, products, services and methods of delivery which result in significant improvements in outcomes efficiency, effectiveness or quality."

Leadbeater (2003) exposes the complexity of the concept including the interactive and social dimensions: he argues that "the process of innovation is lengthy, interactive and social; many people with different talents, skills and resources have to come together." The literature assumes various categorizations of innovation: OECD (2002) structures the concept around three areas: (i) the renewal and broadening of the range of and services and associated markets; (ii) the creation of production, procurement and distribution methods; (iii) the introduction of changes to management, work organization and workers' qualifications. Baker and Sinkula (2002) typology also differentiates three types of innovation: (i) process, (ii) product/service, (iii) strategy/business. Process innovation (i.e., work organization, new internal procedures, policies and organizational forms) and the strategic and new business models (i.e., new missions, objectives, and strategies) are called organizational innovation.

2.4   Technological Innovation

Today the profusion and advancement of technologies (Wang and Shin 2015) is enormous and occurs at high speed redefining world economies and, in a more micro-analysis, companies and how they are managed, produce and interact with the market. The most important contemporary technologies (Aydin and Parker 2018) have artificial intelligence (AI) incorporated, which is machines' ability to think as human beings - to have the power to learn, reason, perceive and decide rationally and intelligently. AI applications are numerous at the enterprise level in their digital transformation process, for example, virtual assistants or chatbots.

Also, big data and analytics technologies bring with it the ability to use new tools, architectures and methodologies to analyze new types of information (Holmström et al. 2016; Melnyk et al. 2018), such as sensors, audio, and video for which traditional information management platforms do not respond.

New advanced analytical tools are required that automatically identify business behaviors and forecasts, integrating analytical models into the business processes of companies, which play a central role in the design of business strategies, with a significant impact on the design of organizations where new functions emerge (Adebanjo et al. 2018).

Another critical technology for companies is the Internet of Things (IoT) as a network of billions of digitally connected devices that collect data and communicate with each other. Its application to the various sectors of activity is an added value, as it allows greater efficiency and efficiency of the processes. The application of IoT is vast and intricate to define boundaries, from the monitoring of the production process to identify problems with impact on the final quality of the products, allowing in real time to activate corrective actions, improve the efficiency of machines and other activities (Chandrasekaran et al. 2015; Holmström et al. 2016).

Moreover, Virtual Reality (VR) which replaces visual reality with a digital reality, while augmented reality (AR) overlaps digital elements with physical reality and both accelerates the learning of new competencies, the immediate resolution of tasks and the visual understanding of processes - contributing to greater effectiveness in the decision making. In addition robots and drones are capable of performing the work of a human through a programmed process - robotic process automation (Aydin and Parker 2018).

Furthermore, the application of drones is one of the emblematic elements of industry 4.0 and has revolutionized the way of monitoring equipment, locations, and specific situations. The adoption of drones in the industry allows greater precision and agility in the inspection of large equipment, obtaining evidence and data that help the decision making (McAdam et al. 2017).

3D Printing (or additive manufacturing) (Chan et al. 2018) is also redefining the way the production processes, equipment's, and materials are used to produce 3-dimensional object.

3 Smart Factory Model Proposal

Innovation is crucial to organizations particularly in encouraging the creation of new products and services, and in the implementation of new practices and processes. In this context, two elements need to be managed together: people and technology.

The following model proposes a set of technologies that should be nuclear to any innovative operations management process in order to create a Smart Factory (Fig. 1):

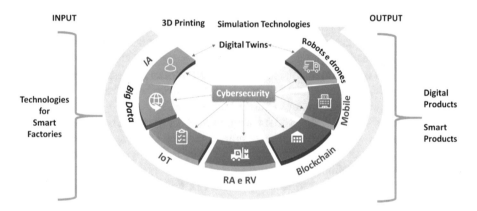

Fig. 1. Model proposal for Smart Factories

Digital manufacturing is on an improvement trajectory based on emergent technologies. Organizations are still integrating Robots, Drones, IoT and 3DP in the supply chain; AI, Big data, AR and VR are fundamental for product conception and design; and production cost models begin to use blockchain technologies.

Manufacturing systems are facing strategic challenges like open-source innovation and new digital production systems. Also some barriers, to a total integration of emergent technologies in operations management and production processes, can be identified in the literature: education, cost, design, software, materials, traceability, machine constraints, in-process monitoring, mechanical properties, repeatability, scalability, validation, standards, quality, inspection, tolerances, and finishing.

The main characteristics of a Smart Factory are: (a) *interoperability*, enables all technologies and other devices essential to the production process to communicate with one another; (b) *virtualization*, enables to simulate or create a virtual copy of the real world and the smart factory is able use simulations and real-time monitoring; (c) *decentralization*, control isn't limited to just a single computing unit being more flexible and optimized in solving problems; (d) *real-time capability*, data is monitored in real time so the smart factory notice and even predict anomalies.

This research tried to create an analytical framework to current challenges faced by industrial organizations to turn into Smart Factories, and the results reveal a portfolio of technologies that can support innovation and research in order to face economic current and future contexts and competitiveness.

4   Conclusions

This article does a literature review about the innovation process of organizations and analyzes the articles under the organizational innovation process. The proposed model can be used as a roadmap in order to operationalize the strategy of the organizations, focusing on the adoption of technologies in operations management and in the production process to transform the industries in Smart Factories.

The main expected impacts of the increasingly widespread use of technology are: reduction of labor costs, greater flexibility and shorter delivery times of products on the market, reduction of the performance of dangerous tasks by people (i.e. manual labor automation), productivity growth, higher quality products; new challenges in terms of employment and education and the nature of work (robotics growth), analysis of large amounts of data, creation of new products and services, and changes in the way companies and other organizations structure themselves.

Digital will shape the future and impact on societies, organizations and people and education-learning systems need to be rethought to respond to the needs arising from this context and the constant change brought about by technological advancement, and that is continually redefining economies.

As main limitations we can state the number of articles analyzed, expanding the search to other databases and introducing specific technologies as keywords during the search would help to have a more wider analysis of the application of the technologies to industry. As future research on this theme it is a goal to have empirical data and several case studies of Smart Factories already implemented.

References

Adebanjo, D., Teh, P.L., Ahmed, P.K.: The impact of supply chain relationships and integration on innovative capabilities and manufacturing performance: the perspective of rapidly developing countries. Int. J. Prod. Res. **56**(4), 1708–1721 (2018)

Aldrich, H., Fiol, C.M.: Fools rush in? The institutional context of industry creation. Acad. Manag. Rev. **19**(4), 645–670 (1994)

Ayala, N.F., Paslauski, C.A., Ghezzi, A., Frank, A.G.: Knowledge sharing dynamics in service suppliers' involvement for servitization of manufacturing companies. Int. J. Prod. Econ. **193**, 538–553 (2017)

Aydin, A., Parker, R.P.: Innovation and technology diffusion in competitive supply chains. Eur. J. Oper. Res. **265**(3), 1102–1114 (2018)

Baker, W.E., Sinkula, J.M.: Market orientation, learning orientation and product innovation: delving into the organization's black box. J. Mark. Focused Manag. **5**(1), 5–23 (2002)

Bendoly, E., Chao, R.O.: How excessive stage time reduction in NPD negatively impacts market value. Prod. Oper. Manag. **25**(5), 812–832 (2016)

Bolumole, Y.A., Calantone, R.J., Di Benedetto, C.A., Melnyk, S.A.: New product development in new ventures: the quest for resources. Int. J. Prod. Res. **53**(8), 2506–2523 (2015)

Chan, H.K., Griffin, J., Lim, J.J., Zeng, F., Chiu, A.S.: The impact of 3D printing technology on the supply chain: manufacturing and legal perspectives. Int. J. Prod. Econ. **205**, 156–162 (2018)

Chandrasekaran, A., Linderman, K., Schroeder, R.: The role of project and organizational context in managing high-tech R&D projects. Prod. Oper. Manag. **24**(4), 560–586 (2015)

Chekurov, S., Metsä-Kortelainen, S., Salmi, M., Roda, I., Jussila, A.: The perceived value of additively manufactured digital spare parts in industry: an empirical investigation. Int. J. Prod. Econ. **205**, 87–97 (2018)

den Hertog, J.F., Huizenga, E.: The Knowledge Enterprise. Implementation of Intelligent Business Strategies. Imperial College Press, London (2000)

Durugbo, C.: Collaborative networks: a systematic review and multi-level framework. Int. J. Prod. Res. **54**(12), 3749–3776 (2016)

Edquist, C. (ed.): Systems of Innovation; Technologies, Institutions, and Organizations. Pinter, London and Washington (1997)

Elverum, C.W., Welo, T.: Leveraging prototypes to generate value in the concept-to-production process: a qualitative study of the automotive industry. Int. J. Prod. Res. **54**(10), 3006–3018 (2016)

Erat, S.: Making the best idea better: the role of idea pool structure. Prod. Oper. Manag. **26**(10), 1946–1959 (2017)

Freeman, C.: Technology Policy and Economic Performance: Lessons from Japan. Pinter, London (1987)

Hafezi, M., Zolfagharinia, H.: Green product development and environmental performance: investigating the role of government regulations. Int. J. Prod. Econ. **204**, 395–410 (2018)

Holmström, J., Holweg, M., Khajavi, S.H., Partanen, J.: The direct digital manufacturing (r) evolution: definition of a research agenda. Oper. Manag. Res. **9**(1–2), 1–10 (2016)

Khorram Niaki, M., Nonino, F.: Additive manufacturing management: a review and future research agenda. Int. J. Prod. Res. **55**(5), 1419–1439 (2017)

Leadbeater, C.: Open innovation in public services. The Adaptive State - Strategies for personalising the public realm. In: Bentley, T., Wilsdon, J. (eds.) Demos, pp. 37–49 (2003). http://www.demos.co.uk/files/HPAPft.pdf

Loch, C.H.: Creativity and risk taking aren't rational: behavioural operations in MOT. Prod. Oper. Manag. **26**(4), 591–604 (2017)

Long, Y., Pan, J., Zhang, Q., Hao, Y.: 3D printing technology and its impact on Chinese manufacturing. Int. J. Prod. Res. **55**(5), 1488–1497 (2017)

Lundvall, B.A. (ed.): National Systems of Innovation: Towards a Theory of Innovation and Interactive Learning. Pinter, London (1992)

McAdam, R., Bititci, U., Galbraith, B.: Technology alignment and business strategy: a performance measurement and dynamic capability perspective. Int. J. Prod. Res. **55**(23), 7168–7186 (2017)

Melnyk, S.A., Flynn, B.B., Awaysheh, A.: The best of times and the worst of times: empirical operations and supply chain management research. Int. J. Prod. Res. **56**(1–2), 164–192 (2018)

Mortara, L., Parisot, N.G.: Through entrepreneurs' eyes: the Fab-spaces constellation. Int. J. Prod. Res. **54**(23), 7158–7180 (2016)

Mulgan, G., Albury, D.: Innovation in the public sector, Strategy Unit, Cabinet Office (2003)

Nelson, R.R. (ed.): National Innovation Systems: A Comparative Analysis. Oxford University Press, Oxford (1993)

Raja, J.Z., Frandsen, T., Mouritsen, J.: Exploring the managerial dilemmas encountered by advanced analytical equipment providers in developing service-led growth strategies. Int. J. Prod. Econ. **192**, 120–132 (2017)

Sardana, D., Terziovski, M., Gupta, N.: The impact of strategic alignment and responsiveness to market on manufacturing firm's performance. Int. J. Prod. Econ. **177**, 131–138 (2016)

Sousa, M.J., Wilks, D.: Sustainable skills for the world of work in the digital age. Syst. Res. Behav. Sci. **35**(4), 399–405 (2018)

Thomas-Seale, L.E.J., Kirkman-Brown, J.C., Attallah, M.M., Espino, D.M., Shepherd, D.E.T.: The barriers to the progression of additive manufacture: perspectives from UK industry. Int. J. Prod. Econ. **198**, 104–118 (2018)

Van de Ven, A.H.: Central problems in the management of innovation. Manage. Sci. **32**(5), 590–607 (1986)

Van de Ven, A.H., Policy, D.E., Garud, R., Venkataraman, S.: The Innovation Journey. Oxford University Press, Oxford (1999)

Wang, J., Shin, H.: The impact of contracts and competition on upstream innovation in a supply chain. Prod. Oper. Manag. **24**(1), 134–146 (2015)

Weller, C., Kleer, R., Piller, F.T.: Economic implications of 3D printing: market structure models in light of additive manufacturing revisited. Int. J. Prod. Econ. **164**, 43–56 (2015)

Yang, J., Lai, K.H., Wang, J., Rauniar, R., Xie, H.: Strategic alliance formation and the effects on the performance of manufacturing enterprises from supply chain perspective. Int. J. Prod. Res. **53**(13), 3856–3870 (2015)

Yoon, J., Rosales, C., Talluri, S.: Inter-firm partnerships–strategic alliances in the pharmaceutical industry. Int. J. Prod. Res. **56**(1–2), 862–881 (2018)

Yu, X., Lan, Y., Zhao, R.: Cooperation royalty contract design in research and development alliances: help vs. knowledge-sharing. Eur. J. Oper. Res. **268**(2), 740–754 (2018)

Open Government Data in Kingdom of Bahrain: Towards an Effective Implementation Framework

Abdull-Kareem Katbi$^{(\boxtimes)}$ and Jaflah Al-Ammary

College of Information Technology,
University of Bahrain, Zallaq, Kingdom of Bahrain
Abdulkareem.katbi@khuh.org.bh, jafuob@gmail.com

Abstract. Governments around the world have realized the importance of Open Government Data - OGD as new paradigm shift in government that focuses on making governments more service oriented, transparent and competent. However, the situation of OGD initiative in Kingdom of Bahrain is not promising as reflected by number of assessments that measure the implementation and progress of OGD worldwide. Therefore, the current research in progress aims at investing the local situation regarding supplying OGD to the public by determine the motivations and impediments that affect government agencies' behavior towards publishing OGD and assess the level of OGD Maturity Level in the government agencies participating in publishing open data. Second, it will investigate the local situation regarding consuming and reusing OGD in Kingdom of Bahrain by assess the level of citizen awareness towards OGD, determine citizen requirements of OGD and identify the key challenges and obstacles in using/reusing OGD. Based on the initial assessment of the available OGD measurement tools and reviewing the literature review, it have been noticed that the OGD assessment framework for Kingdom of Bahrain needs to consider both supply and demand sides, assessing OGD initiative at a local/Government Agency Level and consider quality issues of OGD: Portal, Data and Meta-data. Therefore, two questionnaires were developed: one will assess the level of OGD Maturity level (supply side), while the second will be used to investigate the demand side.

Keywords: Open data government · Electronic government · Supply · Demand · Digital society · Kingdom of Bahrain

1 Introduction

Nowadays, advancements in Information Technologies (IT) have changed the way in which information is collected and disseminated and affected both governments and civilian society's altogether. Civilian societies started experiencing a dramatic shift in the way of obtaining and using information and data. This shift led to the creation of what's known as digital society. Accessing, using and sharing information became an integral part of citizen's lives. In response to the changes, governments realized the importance of coping with this new paradigm shift of digital society and started to look

© Springer Nature Switzerland AG 2019
Á. Rocha et al. (Eds.): WorldCIST'19 2019, AISC 930, pp. 699–715, 2019.
https://doi.org/10.1007/978-3-030-16181-1_66

for ways to improve the services provided to and the ways of interacting with citizens. This led governments to look for and adopt citizen-centric approaches. Such approaches require governments to become more transparent, efficient, citizen oriented and innovative. To fulfill those requirements, governments have to be open. One effective and essential way to make governments open is through embracing an Open Government Data (OGD) initiative. Governments around the world have realized the importance of using such technologies to improve and provide new services to the community. Such approaches promise greater benefits not only to the community but also to the government as well. While e-government initiatives have proven to be successful in delivering valued services to the community, they provide limited interaction and empowerment to citizens. Thus, for the governments to be more efficient and citizen centric they have to be more open than ever before. One noticeable way of making governments open is by adopting an OGD initiative.

The concept of OGD is based on the premise that governments are the largest collector and producer of public data. It is considered as an essential step that governments should undertake in order to improve the services and relationships with citizens [1]. Moreover, it is also a main requirement for proceeding towards Gov 3.0 [2] - a new paradigm shift in government that focuses on making governments more service oriented, transparent and competent [1]. The benefit of open data does not lie on keeping it secret or selling it to others. Rather, it lies on keeping it open and available to the public [3] which could help in increasing transparency, social involvement, providing new and improved services, improving trust in government, facilitating and improving the life of citizens etc. [4]. Open data is defined as "publicly available data that can be universally and readily accessed, used, and redistributed free of charge. Open data is released in ways that protect private, personal, or proprietary information. It is structured for usability and computability" [5]. Janssen et al. [4] defined OGD as "non-privacy-restricted and non- confidential data which is produced with public money and is made available without any restrictions on its usage or distribution". Another definition from Zuiderwijk [6] states that "Open Government Data are structured, machine-readable and machine-actionable data that governments and publicly-funded research organizations actively publish on the internet for public reuse and that can be accessed without restrictions and used without payment." OGD is basically open data that is collected and published by government entities [7]. For government data to be qualified as open, it should conform to the principles of open data. In this sense, several organizations have attempted to develop guiding principles for open data. For example, OGD should be complete, primary, timely, accessible, machine-process-able, non-discriminatory, non-proprietary, license free, online and free, permanent, trusted, documented, safe to open etc. [8–10]. Currently, there are around 115 counties who implemented an OGD initiative [11]. The level of implementation and maturity of OGD varies greatly among those countries. According to Open Data Barometer [11], the top performers are UK, Canada, France, USA and Korea. Unfortunately, Kingdom of Bahrain isn't in the list of top performers, instead it positioned to be in the nascent stage of OGD readiness, implementation and impact [11, 12]. This raises the question of why Kingdom of Bahrain isn't performing well in the area of OGD.

Unfortunately, there is also a lack of academic studies that investigate the situation of OGD in Kingdom of Bahrain. Only very few articles found focusing on OGD in GCC countries and none of them studied the situation in Kingdom of Bahrain thoroughly [12–15]. Moreover, OGD entails both publishing data by governments (supply) and using the data by the community (demand). The majority of studies found focusing only on supply side and ignoring an important aspect of OGD: Demand [7, 15]. This raised many calls to address the demand side of OGD. Saxena [12] stressed the point that there is a clear academic gap in the area of OGD in GCC. Several recommendations were proposed by Saxena [12] such as assessing the users' perceptions and attitudes towards of OGD, assessing the impacts of OGD on governments and citizens etc. The lack of academic research resulted in a knowledge gap that must be covered properly. The absence of any evidence that OGD had any kind of desirable positive impacts on the country should not be omitted. Therefore, it is necessary to conduct a thorough research that will investigate the various aspects of OGD initiative with the aim of determining the impediments and recommending solutions that will foster and make OGD a successful initiative. Therefore, the current research in progress is aimed at highlighting the key motivators and importance behind conducting a thorough assessment of OGD initiative in Kingdom of Bahrain. Thus, the main objectives of this research are first to investigate the local situation regarding supplying open government data to the public by determine the motivations and impediments that affect government agencies' behavior towards publishing OGD and assess the level of OGD Maturity Level in each of the government agencies participating in publishing open data. Second, it will investigate the local situation regarding consuming and reusing OGD in Bahrain by assess the level of citizen awareness towards OGD, determine citizen requirements of OGD and identify the key challenges and obstacles in using/reusing OGD data.

2 Why Open Government Data for Kingdom of Bahrain

Since 2011, Kingdom of Bahrain became one of many countries in the world that embraced the idea of opening up its government data to the public as a way to increase government transparency, realize economic gains, improve citizen engagement and foster the wheel of creating a culture of innovative and participative society [7]. However, for such an initiative to succeed, a number of preconditions have to be met such as: implementing a technical infrastructure capable of supporting open data, providing a consistent supply of "quality" data that address the needs of the public society, increasing the awareness of citizens towards using OGD, providing protections against data privacy and security issues, etc. The more efforts devoted to improve OGD, The more benefits are expected to be gained.

The situation of Open Government Data initiative in kingdom of Bahrain is not promising as reflected by number of assessments that measure the implementation and progress of OGD worldwide. For Example: WORLD WIDE WEB FOUNDTION [11] surveyed OGD initiatives around the word in 2015 and Positioned Kingdom of Bahrain in the 57th place among 92 positions covering the surveyed 122 countries. Specifically, Bahrain scored 18 out of 100 in its global open data Barometer index. This score is unfortunately below the average which is 36.5/100. More importantly, the foundation stated there is a lack of any evidence that OGD in Bahrain has contributed to any positive impact to the social, economic or political domains. Recently, WORLD WIDE WEB FOUNDTION [11] resurveyed the countries. The results indicate that very little efforts have been made to improve the position of Kingdom of Bahrain with respect to open government data. Another evaluation done by Open Knowledge Network [16] in 2015 that ranked Kingdom of Bahrain 78 out of 122 evaluated countries. Their global open data index revealed that only 25% of the common surveyed 13 datasets are open.

Additionally, the literature confirms that there is a clear lack of research in the area of OGD in Bahrain. To the best of my knowledge, no previous research has been found specifically focusing and thoroughly studying OGD in Bahrain. Only very few studies [12–15] found investigating some aspects of OGD in GCC countries. Unfortunately, all of those studies confirm the fact that Bahrain is still in the nascent stage of OGD development.

Moreover, currently there is an academic and government movement towards embracing and implementing nation-wide smart city initiative. In this context, Open Government data is considered to be of a high value and a key enabler for the success of smart city initiative [17]. By opening up government data, a common and standard pool of shared data will exist. Such pool of open data will not only benefit the society and government, but also will enable different smart city projects to cooperate and integrate with each other using the easily available high quality open data. Thus, OGD will act as a common integrator and enabler for the various smart city projects. So, unless we investigate and solve the issues related to OGD, smart city initiative will not be successfully attained. With respect to the economic vision of Kingdom of Bahrain 2030, OGD is a necessity. A direct impact on the economy and social life of citizens can be realized once the OGD initiative is improved. Countries who implemented a mature OGD initiative have gained 1% direct increase in GDP, as reported by [18, 19]. Academic professionals, universities and institutions in Bahrain require OGD to enable them conduct their research by using OGD as a valuable secondary source of Data. Businesses also have the same demand for OGD to enable them create new and innovative applications that will better serve the community.

On the other hand, the need for assessing and evaluating OGD initiatives in the Kingdom of Bahrain has been necessitated by the literature review and the academic works. There exist few studies that shed the light on OGD initiative in Kingdom of Bahrain. Elbadawi [14] examined the state of OGD in three GCC countries: Bahrain, Saudi Arabia and UAE. His examination was based on case study that examined the technical characteristics of the OGD portals as well as an interview government officials in the mentioned countries. The Findings revealed that OGD portals does not match the desired standards. Several challenges also found in his study such as: lack of cooperation between government agencies caused by cultural barriers, lack of legislations pertaining to OGD publishing and use, uncertainties among government agencies regarding the value of publishing OGD. Alromaih et al. [15] examined the technical maturity of four GCC countries: Saudi Arabia, Oman, UAE and Bahrain. The examined features include: data linkage, category browsing, data representation, APIs etc. Their findings revealed that both UAE and Oman found to have more mature OGD portal than Bahrain and Saudi. Recently, Saxena [12] conducted a study to assess the nature and scope of OGD in all GCC counties. Her study is based on a qualitative approach to evaluate the OGD portals. The results of her study indicates that all of GCC counties are in the initial stage of OGD maturity, there is an obvious lack of citizen participation from the perspective of portal feedback features, the data sets of the portals are not exhaustive and does not cover all aspects and services. The progress of OGD implementation and use is very slow. Thus, it is obvious that OGD initiative in Bahrain requires a lot of attention.

3   Driver and Inhibitors for OGD

One effective way to motive governments towards embracing and continually improving their open government data initiative is through the identification of benefits that will be gained as a result of implementing the initiative. With this respect, the literature presented ample amount of searchers and practitioners who demonstrated the benefits of OGD [7, 20, 21]. Three main benefits have been believed to be at the center motivating governments to embrace OGD, namely: Increased transparency, releasing social and economic value and participatory governance [7]. Martin and Begany [20] studied the benefits of opening specific type of government data (health data) to the public. Such benefits include: more efficient public health operations, improved healthcare delivery and health literacy, reaching new audiences etc. Hardy and Maurushat [22] mentioned three main benefits of OGD: improving the effectiveness and efficiency of government policy and services, Improving transparency and accountability, and promoting democratic participation. Thirty two benefits of OGD and classified those under three categories: political and social benefits (increased transparency and the empowerment of the public, etc.), economic benefits (stimulated competitiveness, contribution towards the improvement of products and service, reduced government spending) and technical and operational benefits (ability to reuse data, easier access to data, and easier discovery of data) [4, 21]. OGD can be used to empower citizens, promote equal access to data, create new and innovative social services, provide direct and indirect improvement to the economy, improve government and citizen decision making [4, 21]. Other benefits include reducing government costs, providing new knowledge from combined data sources and patterns in large data volumes and enabling self-empowerment and personalized services to the citizens [1]. OGD is also regarded as a key enabler for smart city initiatives [23, 24]. Indeed, OGD is considered as an essential requirement and main enabler for government evolution towards U-Government-Gov 3.0 [1, 2].

Despite the promised benefits, many impediments exist and have to be taken into account while undertaking an OGD initiative. Data providers – government agencies- and data users both citizens and business community are facing many obstacles towards adopting or using OGD which could prevent them from reaching their full potential of OGD initiatives. Zuiderwijk et al. [25] reviewed relevant literature, conducted interviews and workshops to provide a detailed review of socio-technical data use impediments that influence the open data initiative from the viewpoint of OGD users. Eight broad categories of impediments were identified, namely: availability and access, find-ability, usability, understandability, quality, linking and combining data, comparability and compatibility and metadata impediments. Janssen et al. [4] identified a list of adoption barriers for OGD and categorized them under the following categories: institutional task complexity, use and participation, legislation, information quality and technical barriers. Zuiderwijk et al. [25] also examined the open data process (lifecycle) and identified barriers related to the use and publication of open data grouped under the different processes of OGD lifecycle. Conradie and Choenni [26] took a different perspective and studied the barriers that affect local governments in the process of releasing OGD. Some of the identified issues include: government agencies fear of false conclusions that could be interpreted by users from the released data, unknown data ownership and unknown data locations, financial effects of OGD release, etc. Attard et al. [7] argued that the challenges of OGD usually vary depends on the specific domain, but mostly are related to the technical issues. Examples of challenges identified include: data formats, data ambiguity, data discoverability, data representation, overlapping scope and public participation. Magalhães and Roseira [27] explored the barriers that private companies encounter when using OGD. Examples of the identified barriers include: Data accessibility and find-ability issues, Data quality and metadata issues, etc. Brugger et al. [28] examined the main barriers affecting the usage of OGD by a specific user group, namely media, political parties, associations and NGOs. Some of the barriers identified are lack of skills and knowledge within the organizations that prevent them from using OGD, existing culture that discourage OGD use and limited budget, etc. Chorley [29] studied a specific governmental group (Records management team) within the local context of England healthcare to identify the challenges they face as a result of implementing the OGD processes. Major challenges include lack of formal leadership and direction, lack of common and comprehensible understanding of what OGD is, lack of consistent metadata for the records, etc. Other impediment categories include legislation, information quality, technical, cultural, organizational and financial barriers [20, 22, 26, 30]. In conclusion, it is important to review and determine the barriers and challenges related to OGD implementation, adoption and use as it enable the governments and decision makers to anticipate, plan and take care of the possible issues that hider the successful implantation of the initiative.

4   OGD Assessment Tools and Approach

The above mentioned impediments require governments to proactively address them from the early stages of the initiative in order to ensure a successful implementation at the later stages. In this sense, it is essential to identify the success factors that could facilitate addressing the impediments more effectively. Zuiderwijk et al. [21] identified 64 success factors pertaining to open data publication and use. The most important success factors were found related to legislation, regulations and license. Their study highlighted the importance of considering the context of a particular country while implementing OGD initiative as it have a direct influence on the criticality of the factors identified. With respect to publishing data, several authors provided guidelines intended to streamline the process and overcome the challenges [31].

Once governments started to implement OGD initiative, they should look for assessment tools to evaluate their adherence to the basic principles of OGD, ensure that the impediments are being considered and the initiative is being directed towards achieving the benefits. By reviewing the literature, it is confirmed that various assessment tools exist and they vary greatly in several perspectives. Some assessment tools focus on specific agencies within a country [32, 33], whole country [34]; a group of countries [35] or worldwide [36]. Moreover, the focus of assessment tools also vary. For example, some tools assess the capabilities/readiness of OGD initiative [37]. Other tools assess the implementation of the initiative [37–40] or the impact of the initiative [41]. There exist few assessment tools that have a broader theme and assess all three aspects of OGD initiative: readiness, implementation and impact [11,42]. The aspects of OGD initiatives being evaluated also vary. Some studies focused on the data, functionality and features of OGD portals [15]. Other studies evaluated stakeholder participation [43], initiative maturity [44], and citizen engagement and feed-back [45]. Both quantitative and qualitative approaches have been used in the previous studies. Thus, there is no universally applicable assessment tool and the selection of a tool depends on the purpose of the assessment, scope, area of investigation and other considerations [7, 46]. Susha et al. [46] noted that it is beneficial to use more than one assessment tool to get a wider understanding of the initiative being implemented. Moreover, they mentioned that assessments should be carried out across different levels i.e.: country, government agencies, etc. [46]. This is also supported by a recent study by Donker and Loenen [47], in which they investigated the open data ecosystem and found that the majority of assessment tools focus on one or two aspects of the eco-system while paying little or no attention to other aspects. For example, they demonstrated that the data ecosystem consists of 5 main constructs: Data (Supply) Users (Demand), access networks, policy and standards. Unfortunately, the majority of assessments were found to be focusing on supply side. Thus, it is also important to use not only one assessment tool but also to use or develop other assessment tools that focus on the demand side of OGD as well.

It has been mentioned above that there are vital amount of benefits and advantages that motivate governments worldwide to embrace an OGD initiative. However, the challenges and obstacles impose major concerns regarding how to achieve a successful implementation and gain fruitful results. Hence, it is necessary to utilize diagnostic mechanisms to assess the current situation and recommend corrective actions for continuous improvement. Such assessments have to occur at the various levels and different stages of the OGD ecosystem. Literature have presented sufficient number of studies evaluating the various aspects of the OGD initiatives taking into account different perspectives and using different measurements.

4.1   Rationale Behind Using OGD Assessment Tools

OGD initiatives usually require the involvement and commitment of various stake-holders, considerable investments as well as efforts to implement, monitor and improve the initiatives. Such large and complex project requires continuous monitoring and periodical measurement in order to understand the weaknesses and strengths of the initiative, identify and prioritize the areas for improvement and ultimately assisting in creating an impactful positive effect on the society as whole. An OGD assessment tool are necessarily for successful embedding of OGD initiative in society and improving the uptake of released data by the various stakeholders [47]. It used to stimulate use, reuse of OGD, and determine the actions required to improve the quality and usability aspects of published data and possibly strengthening the OGD ecosystem to deliver social, economic and political impacts [48]. There exist different purposes, reasons and motivations for OGD assessment tools. For example OGD assessment can be used to benchmark countries regarding the implementation and use of OGD, thus helping governments and citizens to understand where the country is located among other countries [48]. In addition, OGD assessment can be used for benchmarking specific agencies within the country, thus obtaining a closer and detailed look at the actual performance of each agency with respect to the quality of published datasets, the progress made and the effects gained from implementation etc. [48]. On the other hand, such assessment can be used to identify the social and commercial opportunities that can be reaped by businesses and civic society and identify ways to encourage the generation of public values (financial, political, social, strategic, quality of life and stewardship values) [46, 49]. Finally OGD assessment can act as a strong reasoning mechanism to derive support for more investment in OGD initiative as well as coordinate and prioritize the efforts to be made to make OGD a successful initiative [46, 48].

4.2 Categories of OGD Assessment Tools

Due to the variety of OGD initiatives proposed and implemented worldwide, the tools and frameworks used to assess them vary greatly as well. This variation could be reasoned partially to the different purposes and reasons that tools intend to cover. Another reason is related to the nature of OGD and its life cycle that encompass large and varied number of elements. Each assessment tool tries to uncover certain aspects of the larger eco-system of OGD. By doing so, assessment tools could evaluate same or different aspects of OGD, Thus certain overlapping in their coverage could be noticed.

A common way to look at OGD assessment tools is to examine what aspects they evaluate. Verhulst [50] proposed looking at four main dimensions of OGD assessment tools, namely: context/environment, data, use and impact. The Context/environment refers to the context within which OGD initiative is implemented which could be at a national level or at a particular sector or agency level. Data considers the nature and quality of OGD datasets by looking at the technical, legal and social openness, in addition to the important core categories of OGD and the overall quality issues of the released data. However, use considers the usage of data by different stakeholders taking into account the type of users, their purposes and the activities being undertaken in the process of using data. The impact, instead, is identifying the benefits realized when consuming and utilizing open government data. Such benefits could be related to social, economic, environment and political/governance dimensions [50]. A similar but simplified way was proposed by Shekhar and Padmanabhan [48] in which they categorize OGD assessment tools according to whether they measure capabilities, implementation or impact. In this case, capabilities or readiness investigates the preconditions necessarily for OGD initiative to thrive. Important aspects include legal, social, organizational and economical contexts as well as existence of political will and technical capacity necessarily for effective launching and implementation of OGD initiative. Implementation deals with the actual implementation of OGD policies in a daily basis. It covers all issues related to the use and reuse of data including data quality, accessibility, availability, use purposes, activities being followed and internal management issues. Impact which is the most difficult aspect of OGD initiative to be measured, identifying the various benefits accrued by the utilization of OGD, including social environmental economic and political benefits [48].

Dawes et al. [49, 51] categorize the assessment tools based on the assumption that OGD initiatives are sociotechnical phenomena that covers a range of dynamically interacting aspects of organizational, human, material and technological nature. Thus, all of the assessment tools should cover certain factors that deal with social and technical considerations. By following this theory, OGD initiatives could fall under one of the following perspectives: data-Oriented approaches (the supply side), use and User oriented approaches (the demand side), program Oriented approaches (structural considerations), scorecard and Impact approaches (emphasis on results) and network and ecosystem approaches [49, 51]. Susha et al. [46] demonstrated other elements that could be used to classify the assessment tools such as: frequency of assessment, scope (focus), scale, source, rationale, collection methods etc.

The above presented categories and classification though they differ to a certain extent in their description and focus, can be mapped too each other. For example, both data-oriented and program oriented approaches (Dawes et al. 2015) reflect to a certain extend the characteristics of the Implementation approaches proposed by Shekhar and Padmanabhan [48]. The holistic view of impact and scorecard approaches can be mapped to the three elements of capabilities, implementation and impact. However, user and use oriented approaches are tied more to the impact and partially to the implementation. Additionally, all of the above-mentioned approaches emphasize the measurement of quality related aspects of OGD (data, implementation, data oriented and program oriented).

4.3 Current Landscape of Assessment Approaches

The assessment approaches can be broadly divided in to two categories: (i) assessment approaches utilized by practitioners and organizations (ii) assessment approaches proposed and used by academics. The first category comprises all the assessment approaches used by OGD advocates, government and non-government organizations as well as OGD industry leaders and practitioners. Based on earlier work done by Verhulst [50], the authors Shekhar and Padmanabhan [48] conducted a detailed review of all assessment tools and frameworks that are used by practitioners to measure OGD initiatives. In their study, nineteen assessment frameworks were reviewed and compared across several dimensions. Below is a summary of the most common and known frameworks followed by a comparison table adopted from their work. The Global Open Data Index (GODI) is a benchmark used by Open knowledge Network to evaluate the openness of datasets according to the open definition. Thus, it focuses only on data and does not consider other aspects such as context or impact. Fifteen OGD datasets where chosen for annual evaluation based on the belief that they are the most useful for the public. The evaluation process is crowd sourced but are subjected to experts review for additional validity. Examples of datasets include: national statistics, procurements, company register, land ownership, locations etc. Open Data Readiness Assessment (ODRA) is a diagnostic and planning tool developed by the word bank. It aims at exclusively evaluating the readiness of countries or specific agencies with regard to the adoption of OGD initiative. Thus, it analyzes how well the government is prepared before the actual lunching of OGD initiative. Their assessment aims to provide an action plan for OGD initiative launching and act as a basis for establishing fruitful and continuous dialogue between the various OGD stakeholders. This tool mostly focuses on developing countries and evaluates aspects such as: senior leadership, policy/legal framework, institutional structures, government data management policies, demand for data, civic engagement, etc. A more comprehensive tool that takes a wider look at several aspects including readiness, implementation and impact is the Open Data Barometer (ODB) developed by World Wide Web foundation and Open Data institute. This annual ranking benchmark of countries worldwide utilizes in-depth methodology that combines contextual data, technical assessments, expert surveys and secondary data to provide country ranking comparable with others worldwide. It examines the readiness of governments, society as well as businesses and entrepreneurs. Also, it evaluates the availability and openness of datasets and the various possible social political and economic impacts of OGD initiatives. While Open data Barometer evaluates countries worldwide, European Public sector information (PSI) Scoreboard instead evaluates EU countries (subject to PSI directive) regarding the implementation and Impact of OGD initiatives. Their assessment framework focuses on the following aspects: Implementation of PSI directive, practice of use and reuse, data formats, pricing, implementation of PSI directive by local agencies, the existence or absence of events and activities pertaining to OGD initiative and exclusive arrangements. The results were derived from crowd sourced and expert's survey of the key dimensions of PSI policies and practices.

UN E-Government survey focuses on how governments utilize Information technologies to provide access and inclusion for all. Though it's not primarily an OGD assessment tool, it provides some indicators that assess certain parts of OGD initiative namely data sharing, data openness and transparency. Both desk research and structured survey were utilized to answer and evaluate all of the dimensions across the surveyed UN member states and countries. While the above-mentioned approaches focus on countries or governments, other approaches have a narrower field of study. For example, Open Data 500 aims to assess the value of open data in organizations. It studies the use and impact of open data by individual organizations across several countries. It holds the assumption that new approaches for understanding the various impacts of OGD can't be realized unless the use of data is mapped within and across countries. Thus, commercial activities based on OGD should clearly demonstrate the use of and have links to the source data sets. Table 1 provides a general description regarding the assessment approaches. Table 1 lists and compares the various assessment tools proposed and used by practitioners. The classification of assessments were based on capabilities, implementation, impact that are proposed by Shekhar and Padmanabhan [48].

Table 1. OGD assessment approach (utilized by practitioners and organization)

Measurement	Coverage		Focus					Classification			Methods						
	Local	Countries	Government	Specific sector	individual organization	Datasets	portal	Capabilities	Implementation	impact	Desk research	Survey	Secondary Data	Self-assessment	automated assessment	crowdsourced	Expert assessment/review
Open Data Index and Census		v				v		v								v	v
Open Data Inventory		v	v	v				v					v				v
Open Data Readiness Assessment (ODRA)		v	v					v			v		v				v
Open Data Barometer		v	v					v	v	v			v				v
European PSI Scoreboard		v	v						v	v						v	v
Open Data 500		v	v		v					v	v	v					
Open Data Certificate		v				v		v						v			
Maturity Model and Pathway	v				v			v	v					v			
UN E-Government Survey		v	v				v	v			v	v					
Common Assessment Framework		v						v	v	v							
Open Data Impact Map		v			v					v			v				v
OUR Data Index (OECD)		v	v					v					v				
Right to Information Ratings		v	v					v			v						v
Capgenimi Open Data Economy		v	v					v	v		v						v
Open Data Monitor		v	v				v	v								v	

The second category of OGD assessment frameworks belongs to the academic studies that propose and/or assess OGD initiatives. With this respect, research scholars usually assess OGD initiative by either adopting a wider look shedding the light on many aspects of OGD at the same time, such approaches include maturity approaches, ecosystem and general approaches. The second category of scholars usually look at narrower and sometimes much focused parts of OGD initiative. This category comprises mainly the studies that evaluate specific aspects of OGD quality. A review of literature reveals an ample amount of studies focusing on quality issues pertaining to OGD. Another type of studies that have a narrow focus are those that examines the behaviors, intensions and preferences of stakeholders involving in OGD initiative. Although those studies are not directly related to the evaluation of progress of OGD initiative, they deal with important aspects that have effect on the successful implementation and adoption of OGD initiatives. Veljković et al. [52] proposed a comprehensive model to assess open government from the perspective of Open data. Their model consists of sources (open data, user involvement), Indicators (Basic data set, openness, transparency, participation, collaboration) and Results (E-government openness index and Maturity). The focus of this model was on evaluating the aspects of OGD that would ultimately improve the degree of government openness and its ability to implement new concepts and innovative ideas. Maturity was determined to be a function of Indicators' results (E-gov openness index) overtime. Susha et al. [46]

reviewed four other maturity models that are related to open data. While each model emphasizes different unit of analysis, all of them represent the progress of OGD through different stages that require more effort and dedication in order to move from lower to advanced stages.

5   Research Method

The literature reviewed revealed that the majority of OGD assessment frameworks that were carried out by industry practitioners have a broader focus which is the whole country as shown in Table 1. While such assessments are useful in gauging the general situation of OGD, it lacks the specific identification of which governmental agencies of the country are performing better than the others with respect to OGD publication and use. Hence, it's important to have a micro investigation that evaluates the different governmental agencies with respect to OGD publication and use. Besides, it has been noticed that the majority of assessment approaches focusing on the supply side of the OGD initiative and in most cases are ignoring the demand side [46]. This raised the importance of adopting an ecosystem approach when developing an OGD evaluation metrics [46, 47, 51, 53, 54]. Such approach promotes investigating many aspects of OGD ecosystem including OGD supply, demand and etc. With regard to OGD quality assessments, it has been observed that most of the proposed assessment tools measure OGD quality from a very basic and simple perspective (i.e. adherence to certain principles concerning formats of published data) and ignoring other important quality aspects of portal, data and metadata. For example, open data index and open data barometer assesses the implementation of OGD (could be considered as quality of the published data) based on: weather the data is open to the public, accessible, machine-readable, and updated frequently. Such quality attributes are considered as an essential prerequisite for a quality dataset but does not guarantee a quality datasets [55, 56]. Thus, there is clear lack of a comprehensive OGD quality measurement framework that takes into account the various quality issues related to portal, data and Metadata across different levels of granularity. Yet, it is of a vital importance for any OGD evaluation to consider the quality aspects of the published data.

Based on the initial assessment and the observations mentioned above, we can define the main requirements of OGD assessment framework for Kingdom of Bahrain needs to:

- Consider both supply and demand.
- Assess OGD initiative at a local/Government Agency Level.
- Consider quality issues of OGD: Portal, Data and Meta-data.

To be able to provide such assessment and achieve the main objectives of the research, quantitative approach was adopted. Two questionnaires were developed and used. The first questionnaire will assess the level of OGD Maturity level in each of the government agencies participating in publishing open data (supply side). The second questionnaire will attempt to gauge the level of citizen awareness and obtain an idea about the possible challenges and requirements of citizens (demand side). In both questionnaires, perception on drivers and challenges of OGD will be explored and identified. Both surveys will be based on validated survey items from the literature with possible modifications to fit the local context of Bahrain.

Two different populations are included in the current study. The first population covered all residents in the Kingdom of Bahrain, while the second population covered all government entities. Each population has a different nature and characteristics, and then different approaches were therefore used to identify the sample size of each population. Thus, cluster sampling was adopted for the first population to cover the whole residents from the main four governorates in Kingdom of Bahrain followed by a probabilistic random sampling method approach to select the sample size of residents as 2400. However, for government, a probabilistic stratified sampling method was followed in which the government population was divided into smaller government entities. Then, a representative sample size was calculated using the sampling size equation from each sector and the sample sizes for the government was identified to be 257 from 52 government entities in Kingdom of Bahrain.

6   Conclusion

This research in progress intends to make OGD a successful initiative in Kingdom of Bahrain by analyzing the various aspects of the initiative and providing solutions to the exiting issues pertaining to the implementation and use of OGD. By doing so, the research will have a number of valuable contributions relating to the community, practice and science. OGD promise to deliver an ample amount of benefits to the society and government as well. Such benefits include increasing government transparency, social involvement, providing new and improved services, improving trust in government, facilitating and improving the life of citizens etc. [4]. Those benefits can't be attained without having a mature OGD initiative. The current research will help the government and society to gain the promised benefits by recommending an effective implementation framework that enable OGD initiative to progress towards becoming a mature initiative.

From a practical perspective, this research will help government agencies understand their strengths and weaknesses with regard to publishing and using open data. By determining the level of OGD maturity for each selected government agency, the current and future actions required to improve OGD initiative will be known. Thus, a clear roadmap will be provided to guide the implementation and reuse process. Besides the maturity, identifying the barriers and motivators of government agencies behavior towards publishing OGD and the actual demands of citizens could give valuable insights to the authorities. Such insights will enable the authorities to formulate new regulations pertaining to publishing and using OGD. Moreover, the research will contribute to the government's vision of becoming a world-leading center for innovation and growth by exploiting the opportunities created by a mature OGD initiative. Businesses and entrepreneurs will also benefit from improving the OGD initiative.

It will enable them to create new and innovative applications based on OGD. Businesses will utilize the available OGD to make wise business decisions and identify the promising areas for future investments. Economic gains will be noticed also once OGD became mature.

From a science perspective, this research will contribute to the body of knowledge relating to OGD in several ways. The majority of OGD research were based on studding various aspects of the supply side of OGD while ignoring the actual needs of the citizens [7, 15]. This led researchers such Saxena [12], Susha et al. [46], Wirtz et al. [57], Ohemeng and Ofosu-Adarkwa [58] to raise many calls for studying the demand side of OGD. The current research will attempt to answer such calls by examining both supply and demand perspectives of OGD. Such examination is valuable to the literature as most of the current studies focus merely on only one aspect of OGD at a time [7, 47]. Thus, a comprehensive examination would lead to better insights that can uncover issues that were previously unknown. This will adds to the knowledge pool and enable the research community to get better insights regarding OGD assessment and evaluation. Moreover, the research will attempt to develop a framework that guides to an effective implementation of OGD. The framework would be valuable not only to the GCC and Arab countries but also to any country that wishes to implement and improve the progress of OGD initiative. Such framework would be a valuable element to the literature as well. Unfortunately, to the best of our Knowledge, there is no study found in the literature focusing specifically on the OGD Initiative in Kingdom of Bahrain. Thus, this is an obvious gap that have to be covered. The proposed research will attempt to address the various calls of researchers mentioned above and tries to gauge OGD maturity form several perspectives. This will help in developing a comprehensive context-related framework that will guide towards an effective implementation of OGD in Kingdom of Bahrain.

References

1. Al-Khouri, A.: Open data: a paradigm shift in the heart of government. J. Public Adm. Gov. 4(3), 217 (2014)
2. Nam, T.: Challenges and concerns of open government: a case of government 3.0 in Korea. Soc. Sci. Comput. Rev. 33(5), 556–570 (2014)
3. Jung, K., Park, H.: A semantic (TRIZ) network analysis of South Korea's "Open Public Data" policy. Gov. Inf. Q. 32(3), 353–358 (2015)
4. Janssen, M., Charalabidis, Y., Zuiderwijk, A.: Benefits, adoption barriers and myths of open data and open government. Inf. Syst. Manag. 29(4), 258–268 (2012)
5. Young, A., Verhulst, S.: The Global Impact of Open Data Key Findings from Detailed Case Studies Around the World, 1st edn., p. 5. O'Reilly Media Inc., Sebastopol (2016). http://www.oreilly.com/data/free/the-global-impact-of-open-data.csp. Accessed 17 May 2017
6. Zuiderwijk, A.: Open data infrastructures: the design of an infrastructure to enhance the coordination of open data use. Ph.D. thesis. Tehnische Universiteit Delft (2015)
7. Attard, J., Orlandi, F., Scerri, S., Auer, S.: A systematic review of open government data initiatives. Gov. Inf. Q. 32(4), 399–418 (2015)
8. Opengovdata.org: The 8 Principles of Open Government Data (OpenGovData.org) (2007). https://opengovdata.org/. Accessed 18 May 2017

9. International Open Data Charter: G8 Open Data Charter - International Open Data Charter (2013). http://opendatacharter.net/resource/g8-open-data-charter/. Accessed 18 May 2017
10. Tauberer, J.: Open Government Data: The Book, 2nd edn. (2014). https://opengovdata.io/2014/8-principles/. Accessed 18 May 2017
11. Open Data Barometer: Global Report (2017). http://opendatabarometer.org/3rdEdition/report/. Accessed 23 May 2017
12. Saxena, S.: Open Public Data (OPD) and the Gulf Cooperation Council (GCC): challenges and prospects. Contemp. Arab Aff. 1–13 (2017)
13. Alanazi, J., Chatfield, A.: Sharing government-owned data with the public: a cross-country analysis of open data practice in the middle east. In: In Proceedings of Americas Conference on Information Systems (2012). http://aisel.aisnet.org/cgi/viewcontent.cgi?article=1150&context=amcis2012. Accessed 13 May 2017
14. Elbadawi, I.: The state of open government data in GCC countries. In: Proceedings of the 12th European Conference on eGovernment, pp. 193–200. Institute of Public Governance and Management, Barcelona (2012)
15. Alromaih, N., Albassam, H., Al-Khalifa, H.: A proposed checklist for the technical maturity of open government data: an application on GCC countries. In: Proceedings of the 18th International Conference on Information Integration and Web-Based Applications and Services, pp. 494–499. ACM, Singapore (2016)
16. Open Data Index: Open Data Index - Open Knowledge (2017). http://2015.index.okfn.org/place/bahrain/. Accessed 1 May 2017
17. Abella, A., Ortiz-de-Urbina-Criado, M., De-Pablos-Heredero, C.: A model for the analysis of data-driven innovation and value generation in smart cities' ecosystems. Cities **64**, 47–53 (2017)
18. Smartdubai.ae: Smart Dubai (2017). http://www.smartdubai.ae/dubai_data.php. Accessed 13 May 2017
19. Srinivasan, P.: The GovLab Index: Open Data - 2016 Edition - The Governance Lab @ NYU (2017). The Governance Lab @ NYU. http://thegovlab.org/govlab-index-on-open-data-2016-edition/. Accessed 13 May 2017
20. Martin, E., Begany, G.: Opening government health data to the public: benefits, challenges, and lessons learned from early innovators. J. Am. Med. Inform. Assoc. **24**(2), 345–351 (2016)
21. Zuiderwijk, A., Susha, I., Charalabidis, Y., Parycek, P., Janssen, M.: Open data disclosure and use: critical factors from a case study. In: Proceedings of the International Conference for E-Democracy and Open Government. Donau-Universität Krems, pp. 197–208 (2015)
22. Hardy, K., Maurushat, A.: Opening up government data for Big Data analysis and public benefit. Comput. Law Secur. Rev. **33**(1), 32–33 (2017)
23. Pereira, G., Macadar, M., Luciano, E., Testa, M.: Delivering public value through open government data initiatives in a Smart City context. Inf. Syst. Frontiers **19**(2), 213–229 (2016)
24. Gagliardi, D., Schina, L., Sarcinella, M., Mangialardi, G., Niglia, F., Corallo, A.: Information and communication technologies and public participation: interactive maps and value added for citizens. Gov. Inf. Q. **34**(1), 153–166 (2017)
25. Zuiderwijk, A., Janssen, M., Choenni, S., Meijer, R., Alibaks, R.: Socio-technical impediments of Open Data. Electron. J. e-Gov. **10**(2), 156–172 (2012)
26. Conradie, P., Choenni, S.: On the barriers for local government releasing open data. Gov. Inf. Q. **31**, S10–S17 (2014)
27. Magalhaes, G., Roseira, C., Strover, S.: Open government data intermediaries: a terminology framework. In: Proceedings of the 7th International Conference on Theory and Practice of Electronic Governance. ACM (2013)

28. Brugger, J., Fraefel, M., Riedl, R., Fehr, H., Schoeneck, D., Weissbrod, C.S.: Current barriers to open government data use and visualization by political intermediaries. In: Conference for E-Democracy and Open Government (CEDEM), Krems, pp. 219–229 (2016). https://doi.org/10.1109/cedem.2016.18

29. Chorly, K.: The challenges presented to records management by open government data in the public sector in England: a case study. Rec. Manag. J. **27**(2), 149–158 (2017)

30. Martin, C.: Barriers to the open government data agenda: taking a multi-level perspective. Policy Internet **6**(3), 217–240 (2014)

31. de Rosnay, M.D., Janssen, K.: Legal and institutional challenges for opening data across public sectors: towards common policy solutions. J. Theor. Appl. Electron. Commer. Res. **9** (3), 1–14 (2014)

32. Verhulst, S., Noveck, B., Caplan, R., Brown, K., Paz, C.: The open data era in health and social care: a blueprint for the National Health Service (NHS England) to develop a research and learning programme for the open data era in health and social care, 1st edn. The GOVLAB - New York University, New York (2014). http://thegovlab.org/nhs/. Accessed 23 May 2017

33. Dodds, L., Newman, A.: A guide to the open data maturity model: assessing your data publishing and use, 1st edn. Open Data Institute, London (2015). https://theodi.org/guides/maturity-model. Accessed 23 May 2017

34. Solar, M., Concha, G., Meijueiro, L.: A model to assess open government data in public agencies. In: 11th IFIP WG 8.5 International Conference on Electronic Government, pp. 210–221. Springer, Kristiansand (2012)

35. Carrara, W., Nieuwenhuis, M., Vollers, H.: Open Data Maturity in Europe 2016: Insights into the European State of Play, 1st edn. Capgemini Consulting (2016). https://www.europeandataportal.eu/sites/default/files/edp_landscaping_insight_report_n2_2016.pdf. Accessed 23 May 2017

36. Máchová, R., Lnenicka, M.: Evaluating the quality of open data portals on the national level. J. Theor. Appl. Electron. Commer. Res. **12**(1), 21–41 (2017). http://www.scielo.cl/pdf/jtaer/v12n1/art03.pdf. Accessed 5 May 2017

37. worldbank.org: Readiness Assessment Tool (2017). http://opendatatoolkit.worldbank.org/en/odra.html. Accessed 23 May 2017

38. Open Data Certificate: ODI Open Data Certificate (2017). https://certificates.theodi.org/en/. Accessed 23 May 2017

39. Global Open Data Index: Place Overview - Global Open Data Index (2017). https://index.okfn.org/. Accessed 23 May 2017

40. UN E-Government Knowledge Database: UN E-Government Survey 2016 (2017). https://publicadministration.un.org/egovkb/en-us/Reports/UN-E-Government-Survey-2016. Accessed 23 May 2017

41. Open Data 500: Open Data 500 (2017). http://www.opendata500.com/. Accessed 23 May 2017

42. Caplan, R., Davies, T., Wadud, A., Verhulst, S., Alonso, J., Farhan, H.: Towards common methods for assessing open data: workshop report & draft framework, 1st edn. Open Data Research Network, New York (2014). http://opendataresearch.org/sites/default/files/posts/Common%20Assessment%20Workshop%20Report.pdf. Accessed 23 May 2017

43. Sayogo, D., Pardo, T., Cook, M.: A framework for benchmarking open government data efforts. In: 47th Hawaii International Conference on System Sciences (HICSS). IEEE, Waikoloa (2014). https://doi.org/10.1109/HICSS.2014.240. Accessed 23 May 2017

44. Yang, T., Lo, J., Wang, H., Shiang, J.: Open data development and value-added government information: case studies of Taiwan E-government. In: Proceedings of the 7th International Conference on Theory and Practice of Electronic Governance, pp. 238–241. ACM, New York (2013). https://doi.org/10.1145/2591888.2591932. Accessed 23 May 2017
45. Sieber, R., Johnson, P.: Civic open data at a crossroads: dominant models and current challenges. Gov. Inf. Q. **32**(3), 308–315 (2015)
46. Susha, I., Zuiderwijk, A., Janssen, M., Gronlund, A.: Benchmarks for evaluating the progress of open data adoption: usage, limitations, and lessons learned. Soc. Sci. Comput. Rev. **33**(5), 613–630 (2014). http://journals.sagepub.com/doi/abs/10.1177/0894439314560852. Accessed 7 Apr. 2017
47. Donker, F., Loenen, B.: How to assess the success of the open data ecosystem? Int. J. Digit. Earth **10**(3), 284–306 (2016). http://www.tandfonline.com/doi/pdf/10.1080/17538947.2016.1224938?needAccess=true. Accessed 9 Apr. 2017
48. Shekhar, S., Padmanabhan, V.: How to apply assessment tools for shaping successful public sector open data initiatives. Open Data Institute, London (2016)
49. Dawes, S.; Vidiasova, L., Trutnev, D.: Approach to assessing open government data program: comparison of common traits and differences at global context. In: EGOSE 2015 Processing of the 2nd International Conference on Electronic Governance and Open Society: Challenges in Eurasia, Russian Federation, 24–25 November 2015
50. Verhulst, S.: How to advance open data research: towards an understanding of demand, users and key data. WWW Foundation press release, GovLab, posted on 15 September 2016 (2016)
51. Dawes, S.S., Vidiasova, L., Parkhimovich, O.: Planning and designing open government data programs: an ecosystem approach. Gov. Inf. Q. **33**, 15–27 (2016)
52. Veljković, N., Bogdanovic-Dinic, S., Stoimenov, L.: Benchmarking open government: an open data perspective. Gov. Inf. Q. **31**(2), 278–290 (2014)
53. Styrin, E., Luna-Reyes, L., Harrison, T.: Open data ecosystems: an international comparison. Transforming Gov. People Process Policy **11**(1), 132–156 (2017). https://doi.org/10.1108/TG-01-2017-0006
54. Srimuang, C., Cooharojananone, N., Tanlamai, U., Chandrachai, A.: Open government data assessment model: an indicator development in Thailand. In: Proceedings of the 2017 19th International Conference on Advanced Communication Technology, pp. 341–347 (2017)
55. Belhiah, M., Bounabat, B.: A user-centered model for assessing and improving open government data quality. In: MIT International Conference on Information Quality, UA Little Rock, USA, 6–7 October 2017
56. Vetro, A., Canova, L., Torchiano, M., Minotas, C.O., Iemma, R., Morando, F.: Open data quality measurement framework: definition and application to Open Government Data. Gov. Inf. Q. (2016). http://dx.doi.org/10.1016/j.giq.2016.02.001
57. Wirtz, B., Weyerer, J., Rsch, M.: Citizen and open government: an empirical analysis of antecedents of open government data. Int. J. Public Adm. 1–13 (2017)
58. Ohemeng, F., Ofosu-Adarkwa, K.: One way traffic: the open data initiative project and the need for an effective demand side initiative in Ghana. Gov. Inf. Q. **32**(4), 419–428 (2015)

Modelling Reporting Delays in a Multilevel Structured Surveillance System - Application to Portuguese HIV-AIDS Data

Alexandra Oliveira[1,2(✉)], Humberta Amorim[1], Rita Gaio[3,4],
and Luís Paulo Reis[2,5]

[1] School of Health, Polytechnic Institute of Porto, Porto, Portugal
aao@ess.ipp.pt
[2] LIACC - Artificial Intelligence and Computer Science Laboratory, Porto, Portugal
[3] DM - FCUP - Department of Mathematics,
Faculty of Sciences of the University of Porto, Porto, Portugal
argaio@fc.up.pt
[4] CMUP - Center of Mathematics of the University of Porto, Porto, Portugal
[5] DEI - FEUP - Department of Informatics Engineering,
Faculty of Engineering of the University of Porto, Porto, Portugal
lpreis@fe.up.pt

Abstract. In a deeply interconnected world of people and goods, infectious diseases constitute a serious threat. An active vigilance is required. The collection of adequate data is vital and coordinated by surveillance systems. It is widely-acknowledged that every case-reporting system has some degree of under-reporting and reporting delay in particular in HIV-AIDS Portuguese Surveillance System. To better understand the processes generating the reporting delays, which is an administrative process, it was used a flexible continuous time fully parametric survival analysis approach. It was taken into consideration the hierarchical administrative and organizational structure of the system as well as the relevant changes in the procedures throughout the time. The best multilevel structure to represent reporting delays in continuous time is the model where the individuals are nested into Reporting Entities (20.24% of the variance) which are nested into Type of services (8% of the variance) with the log-normal distribution.

Keywords: Reporting delay · HIV-AIDS · Continuous outcome ·
Multilevel modelling · Random effects modelling · Surveillance system

1 Introduction

In a deeply interconnected world of people and goods, infectious diseases constitute a serious threat. An active vigilance "for signs of an outbreak, rapid recognition of its presence, diagnosis of its microbial cause" is required [1]. It is

© Springer Nature Switzerland AG 2019
Á. Rocha et al. (Eds.): WorldCIST'19 2019, AISC 930, pp. 716–726, 2019.
https://doi.org/10.1007/978-3-030-16181-1_67

also necessary to identify modes of transmission and at-risk population groups, and to define public measures for target prevention [2,3]. The collection of adequate data is vital, to evaluate the burden of the disease and the impact of prevention and control programmes, aiming at leading to effective and efficient responses.

Surveillance systems are responsible for the collection of these data and relay on processes and individuals thereby can be found to differ substantially according to the disease, health condition and from country [3–5].

It is widely-acknowledged that every case-reporting system has some degree of under-reporting and reporting delay. Timely report may enable public health authorities to take fast and effective actions to prevent outbreaks by reducing disease transmission in a population [6–8].

Reporting delays is defined as the time between the diagnosis (by the physician and/or the laboratory) and its report to the national surveillance system and may depend on a number of factors such as the patient's recognition of symptoms; the patient's acquisition of medical care; the administrative tasks involving the reporting process by the health care provider or the laboratory to the local, region, or state public health authority; the volume of cases identified in the geographic region; periods of variable employment levels; computer system down-time for maintenance, upgrades, or new application development; organizational structure of the system, government rules and regulations and data processing routines, such as data validation or error checking [4,9]. Under-reporting can be defined as a reposting-delay with infinity length.

A particular case of a surveillance system is that for the Human Immunodeficiency Virus (HIV)-Acquired Immunodeficiency Syndrome (AIDS). This infection has several particularities reflecting the special transmission patterns, long asymptomatic or with mild symptoms latency period of the infection, lack of affordable treatment and cure, high case fatality rates, and the social stigma associated with it which makes its surveillance a particular difficult task [10].

Similarly to other European Countries, the Portuguese HIV-AIDS surveillance system suffers from under-reporting and reporting delay affecting timeliness and data quality. Since reporting is an administrative process the main goal of this paper is to model the reporting delay and under-reporting distributions considering the hierarchical administrative and organizational structure of the system as well as the relevant changes of the procedures throughout the time.

In Sect. 2 it is presented an overview the Portuguese HIV-AIDS surveillance system and its implementation; in the Sect. 3 it is presented the related work concerning reporting delay modelling; in the Sect. 4 it is presented the methodology and main results and finally in Sect. 5 it is presented the main conclusion.

2   The Portuguese HIV-AIDS Surveillance System and National Health System

The Portuguese HIV-AIDS is a confidential case-based reporting system where the report was performed voluntarily by physicians until 2005. Since then, it is mandatory to report all cases in any of the stages of the infection (asymptomatic, AIDS related complex and AIDS) and all the progressions including death, identifying the probable source of transmission (Heterosexual, Men Who have Sex (MSM) with Men and Injecting Drug Users (IDU)) and always assuring patient's confidentiality. Until 2011, the filled forms were sent by mail to Centro de Vigilância Epidemiológica das Doenças Transmissíveis (CVEDT), who collected, manually updated the data set, maintained, stored, analysed, produced and disseminated bi-annual reports [8].

Over the years, the surveillance procedure has suffered some changes that may have altered the quality of the reports. We point out the following:

1. in 1993, tuberculosis was included as an AIDS defining disease;
2. in 1996, Highly Active Anti-Retroviral Therapy was introduced;
3. in 2005, the notification forms were re-structured;
4. in 2009, CVEDT was re-structured [8].

Portuguese health care providers play an important role in care and support for HIV-AIDS patients but are also one of the principal components of the reporting system since are the primary source of health information. They work in a variety of settings of the health care system: hospitals, local health units and grouped health centres. Grouped health centres provides primary health care to the local communities, hospitals provide specialize secondary health care services.

These health providers' system are managed by the Central Administration of the Health System (ACSS) and by the five Regional Health Administration (RHA) boards implemented in mainland North, Centre, Lisbon and Tagus Valley, Alentejo and the Algarve regions (Fig. 1).

The RHAs are responsible for the regional implementation of national health policy objectives and for coordinating all levels of health care. They work in accordance with principles and directives issued in regional plans and by the ACSS. In the autonomous regions of Azores and Madeira, health policy followed the same general constitutional principles of the National Health Service, but are implemented locally by regional governments who retained full administrative flexibility [11] (Fig. 2).

The hospital network (the number of hospitals, their location and typology) should be understood as an integrated system of health care, thought and organized in a coherent way, based on principles of rationality and efficiency [12]. The organization of the network respects national geographic diversity, equity in distribution and access to health services, different levels of intervention, primary health care, hospital care and continued care.

Portugal has adopted a "gate-keeping" system; this means that, in general, when seeking for health evaluation, a patient must first go to a general practitioner (called a family doctor) and, if necessary, the primary care unit will send the patient to other levels of care, e.g. a hospital, where the specific care will be provided [13]. So, when a patient is diagnosed with HIV by a primary health care centre, usually it is referenced to see a specialist in the hospital. These special visits can also be required directly in the hospital.

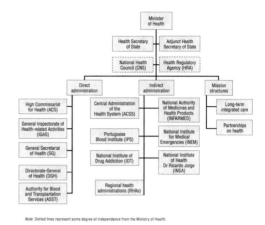

Fig. 1. Portuguese health regions

Fig. 2. Organizational chart of the Ministry of Health from [11]

The National Institute of Drug Addiction (IDT), that until 2011 were independent from RHAs, promoted the reduction of both legal and illegal drugs consumption, as well as the prevention and treatment of drug addictions.

"The management responsibilities of these boards were a mix of strategic management of population health, supervision and control of hospitals, and centralized direct management responsibilities for primary care services [11]".

3   Related Work

Most of the approaches addressing the reporting delay focus on the joint modelling of the reporting delay and incidence of HIV-AIDS. In this setting, the two major approaches are the reverse time hazard and the discrete outcome models.

The reverse time hazards approach focus on the reporting delay distribution and allow the visualization of the hazard function in continuous time. Moreover, the analysis of the effect of the covariates is done in a natural fashion. The long delay is taken into consideration and so the entire delay distribution is analysed [14–16]. The main weakness of this approach is that, in general, it implies data transformation on the time axis and the interpretation of the results is more difficult.

The discrete outcome (typical considering quarters) models focus on incidence and accommodate right truncation without transforming the data. They can be implemented with standard statistical packages being the most common approach to these problems [9, 17–22]. Usually Poisson and Multinomial assumptions are involved, which may yield biased inferences if data fails to meet the posited distributional assumptions.

Recently, separate approaches are being considered to provide more flexibility and complexity to the reporting delay and incidence sub-models. Each model can even be fitted in different samples and, like the discrete outcome approach, it may need parametric assumptions.

Previous studies on HIV-AIDS Portuguese Surveillance data pointed that a large majority of cases are reported within $[0, 3]$ months masking the reporting delay information [23–26]. To better understand the processes generating the reporting delays, it was used a flexible continuous time fully parametric survival analysis approach.

The parametric survival analysis is an appropriate method because reporting delay can be viewed as a 'time from diagnosis to report' avoiding the need to, somehow arbitrarily, classify the delays into discrete categories. Creating categories implies that no matter whether a person is diagnosed at the beginning or at the end of a time unit, a same reporting delay proportion applies [27]. Moreover, allows a flexible representation of the reporting delay distribution at moderately long delays and accommodates the observed extremely long tails (underreporting), which are a striking feature of reporting delay data [16, 23–26].

4   Methods and Results

To determine the appropriate multilevel, nested structure a preliminary was evaluated which distribution should be specified.

It was considered that the data collected before 1992 had not enough quality for performing any model, so the data were truncated at this time stamp and was fitted generalized linear model with the common survival analysis distributions: normal, log-normal, Weibull and Gamma fitted using the iteratively reweighted least squares (IWLS) method. These models are not nested since they assume different distributions for the response, which makes direct comparison using AIC and BIC problematic. But, the collection of findings in the analysis of the estimates of the coefficients, their standard errors as well as residual diagnostics, scientific context and interpretation let to the choice of the log-normal distribution for the response.

In order to determine the best multilevel/hierarchical structure, a full theoretical model was considered. The first approach was developed based on the contextual framework of this problem. Namely, individuals (Population) are diagnosed in a health service institution that gets the case reported (Reporting Entity), which in turn belongs to a certain type of service, such as Hospital care, primary care, additive behaviours... (Type of Service). These entities are under an Administrative responsibility, which in turn follows therapeutic, administrative rules and regulations defined by Scientific Communities and Governments indexed to a time period. This framework and the observations in each level are represented on Fig. 3.

This context can be expressed as the following multilevel structure

Individual Level 1 - Population (i = 1, ..., 27461): $log(Y_{ijklm}) = \beta_{0jklm} + \epsilon_{ijklm}$

Group Level 2 - Reporting Entity (j = 1, ..., 593): $\beta_{0jklm} = \gamma_{00klm} + \nu_{jklm}$

Group Level 3 - Type of service (k = 1, ..., 100): $\gamma_{00klm} = \theta_{000lm} + \upsilon_{klm}$

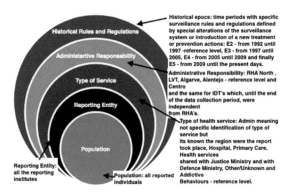

Fig. 3. Conceptual representation of the nested structure

Group Level 4 - Administrative Responsibility (l = 1, ..., 38): $\theta_{000lm} = \psi_{0000m} + \varpi_{lm}$

Group Level 5 - Historical and Context Epocs (m = 1, ..., 4): $\psi_{0000m} = \varsigma_{00000} + \varphi_m$ where, $\epsilon_{ijklm} \sim N(0, \sigma)$, $\nu_{jklm} \sim N(0, \sigma_1)$, $\upsilon_{klm} \sim N(0, \sigma_2)$, $\varpi_{lm} \sim N(0, \sigma_3)$, $\varphi_m \sim N(0, \sigma_4)$.

The model was fitted with the restricted maximum likelihood (REML) whose results are represented on Table 1. Note that, σ is much higher than the other standard deviations and there is evidence that σ_3 and σ_4 are not significant due to the curvilinear form of Fig. 4.

Table 1. Full multilevel model

Random effects	Variance	Std. Dev.	Std. Dev. 2.5% C.I. limit	Std. Dev. 97.5% C.I. limit
Level 2 - σ_1	0.36	0.6	0.45	0.56
Level 3 - σ_2	0.25	0.50	0.30	0.55
Level 4 - σ_3	0.002	0.042	0.00	0.32
Level 5 - σ_4	0.17	0.41	0.13	0.80
Residual - σ	1.18	1.09	0.93	0.94
Fixed effects	Estimate	Std. Error		
Intercept	4.68	0.1636		

The profile zeta plot ζ for the fixed - effect parameter (Intercept) term is slightly over-dispersed relative to the normal distribution. While for σ, σ_1 and σ_2 have a good normal approximation while the standard deviations for σ_3 and σ_4, are skewed. The skewness for σ_4 is worse than that for σ_3, making the estimate of σ_4 less precise than of σ_3, in both absolute and relative senses (Fig. 4). For an absolute comparison we compare the widths of the confidence intervals

$(\sigma_3 \in [0.00; 0.32]$ and $\sigma_4 \in [0.13; 0.80])$ for these parameters. Clearly, the lack of precision of these estimates is a consequence of only having 4 distinct levels of the Epocs factor.

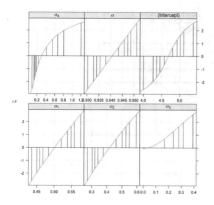

Fig. 4. Profile zeta plot of the parameters in full null hierarchical model.

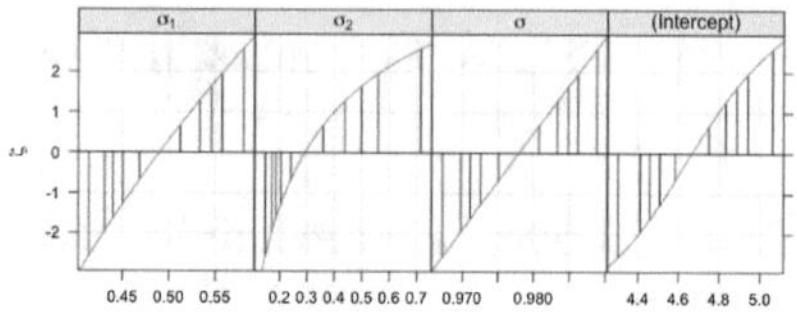

Fig. 5. Profile zeta plot of the parameters in the minimal multilevel model

We proceed to fit four null nested models in order to validate the minimal multilevel structure.

Model A - random effect on the Reporting Entity

$$log(Y_{ij}) = \beta_0 + u_{0j} + \epsilon_{ij} \quad \epsilon_{ij} \sim N \quad and \quad u_{0j} \sim N \tag{1}$$

$j = 1, \ldots, 349$ (Number of reporting entities) and $i = 1, \ldots, 26461$ (Number of individuals).

Model B - random effect by Entity nested within the Type of Service

$$log(Y_{ijk}) = \beta_0 + u_{00k} + \nu_{0jk} + \epsilon_{ijk} \tag{2}$$

with $\epsilon_{ijk} \sim N$ and $u_{00k} \sim N$ and $\nu_{0jk} \sim N$ $j = 1, \ldots, 349$ (Number of reporting entities), $k = 1, \ldots, 8$ (Number of type of services) and $i = 1, \ldots, 26461$ (Number of individuals).

Model C - random effect by Reporting Entity, nested within the Type of Service which are nested in Administrative Responsibility

$$log(Y_{ijkl}) = \beta_0 + u_{000l} + \nu_{00kl} + \theta_{0jkl} + \epsilon_{ijkl} \tag{3}$$

with $\epsilon_{ijkl} \sim N$, $u_{000l} \sim N$, $\nu_{00kl} \sim N$ and $\theta_{0jkl} \sim N$ $j = 1, \ldots, 349$ (Number of reporting entities), $k = 1, \ldots, 8$ (Number of type of services), $l = 1, \ldots, 10$ and $i = 1, \ldots, 26461$ (Number of individuals).

Model D: random effect by Reporting Entity, nested within the Type of Service which are nested in Administrative Responsibility, nested within Historical Rules and Regulations

$$log(Y_{ijklm}) = \beta_0 + u_{0000m} + \nu_{000lm} + \theta_{00klm} + \varphi_{0jklm} + \epsilon_{ijkl} \tag{4}$$

with $\epsilon_{ijkl} \sim N$, $u_{0000m} \sim N$, $\nu_{000lm} \sim N$ and $\theta_{00klm} \sim N$, $\varphi_{0jklm} \sim N$ $j = 1, \ldots, 349$ (Number of reporting entities), $k = 1, \ldots, 8$ (Number of type of services), $l = 1, \ldots, 10$ (Number of ARS), $m = 1, .., 5$ (Number of Historical Epocs) and $i = 1, \ldots, 26461$ (Number of individuals).

The interclass correlation associated to the group level Reporting Entity of model A is approximately equal to 29.67%. Thus, 29.67% of the variance of the reporting delay is at the Reporting Entity group level. Considering the model B, 20.24% of the variance of the reporting delay is at the Reporting Entity and 8% is at the Type of Services. For the model C, 18.37% of the variance of the reporting delay is at the Reporting Entity, 10.50% is at the Type of services group level and 1.42% is at the Administrative Responsibility group level. For the model D, 18.15% of the variance is at the Reporting Entity level, 12.89% at the Type of Services level, 0.09% at the Administrative Responsibility level and 8.66% at the Historical Rules and Regulations level (Table 2). Since these models are intercept - only models that do not contains no explanatory variables, the residuals variances represent unexplained error variance.

Table 2. Null models I.C.C.

	Model A		Model B		Model C		Model D	
	Std	ICC	Std	ICC	Std	ICC	Std	ICC
Reporting Entity	0.537	29.67%	0.359	20.24%	0.336	18.37%	0.357	18.15%
Type of Services			0.142	8.00%	0.192	10.50%	0.253	12.89%
Administrative Responsibility					0.026	1.42%	0.002	0.09%
Historical Rules and Regulations							0.170	8.66%
Residuals	1.273		1.273		1.273		1.183	

We found that the model B was the best using the Likelihood Ratio tests. The tests statistics are presented on Table 3. It can be seen that B improve A, but C does not improve B. So, the multilevel structure represented by model B

Table 3. Models Evaluation

Models	Df	AIC	BIC	logLik	Deviance	Chisq	Chi Df	
A	3	80188	80213	−40091	80182			
A–B	4	77085	77118	−38538	77077	3105.7	1	<0.001
B–C	5	77100	77141	−38545	77090	0.001	1	1.000
C–D	6	75359	75409	−37674	75347	1742.5	1	<0.001

was select. The profile zeta plot for the constant intercept (β_0) for the model B, which it was called the minimal null model has almost symmetric intervals. The profile zeta plot for σ, σ_1 appears to be good normal approximation and although the plot σ_2 is not a straight line, the bias is very mild (Fig. 5).

5   Conclusions

Since the majority of cases were reported in the reporting delay group [0, 3] months and some fluctuations of the percentage of this pattern were observed, it was decided to model the delays in continuous time and to use a fully parametric approach. This allows the model to assess the under-reporting considered as a particular case of reporting delay with infinite length. It was fitted an explicit model for the reporting delay (measured in number of days), using commonly distributions from survival analysis parametric models: normal, lognormal, exponential, Weibull and gamma allowing a flexible representation of the reporting delay distribution at moderately long delays and accommodating the observed extremely long tails, which are a striking feature of these data. Moreover, if the cases are reported in batches, the multinomial assumption does not capture the pattern and if the discretized time intervals are large this can yield very imprecise estimates for recent HIV-AIDS incidence.

Within this approach, it was also studied the possibility of creating a model where the patients are nested in health care facilities, which are grouped according to its type (hospital, primary care services, ...), which in turn are grouped in administrative regions and government by historical policies. Considering that the simpler the model the better, the full model was prune to minimal null possible model which was composed by the individual nested into Reporting Entities which are nested into Type of services.

This model states that 20.24% of the variance of the reporting delay is at the Reporting Entity and 8% is at the Type of Services.

Acknowledgements. Luís Paulo Reis and Alexandra Oliveira were partially founded by the European Regional Development Fund through the programme COMPETE by FCT (Portugal) in the scope of the project PEst-UID/CEC/00027/2015 and QVida+: Estimação Contínua de Qualidade de Vida para Auxílio Eficaz à Decisão Clínica, NORTE010247FEDER003446, supported by Norte Portugal Regional Operational Programme (NORTE 2020), under the PORTUGAL 2020 Partnership Agreement.

Rita Gaio was partially supported by CMUP (UID/MAT/00144/2019), which is funded by FCT with national (MCTES) and European structural funds through the programs FEDER, under the partnership agreement PT2020.

The authors are grateful to Professor Henrique Barros, a former coordinator of the Coordenação Nacional para a Infecção VIH/SIDA, for the HIV/AIDS Portuguese data.

References

1. Choffnes, E., Sparling, P., Hamburg, M., Lemon, S., Mack, A., et al.: Global Infectious Disease Surveillance and Detection: Assessing the Challenges-Finding Solutions, Workshop Summary. National Academies Press (2007)
2. World Health Organization Communicable disease surveillance and response systems: guide to monitoring and evaluating. World Health Organization, Lyon (2006)
3. European Centre for Disease Control and Prevention: Data quality monitoring and surveillance system evaluation. ECDC, Stockholm (2014)
4. Jajosky, R., Groseclose, S.: Evaluation of reporting timeliness of public health surveillance systems for infectious diseases. BMC Pub. Health **4**, 29 (2004)
5. Doyle, T., Glynn, M., Groseclose, S.: Completeness of notifiable infectious disease reporting in the United States: an analytical literature review. Am. J. Epidemiol. **155**, 866–874 (2002)
6. Marinovic, A., Swaan, C., van Steenbergen, J., Kretzschmar, M.: Quantifying reporting timeliness to improve outbreak control. Emerg. Infect. Dis. **21**, 209–216 (2015)
7. Reijn, E., Swaan, C., Kretzschmar, M., van Steenbergen, J.: Analysis of timeliness of infectious disease reporting in the Netherlands. BMC Pub. Health **11**, 409 (2011)
8. Mauch, S.: Situational Assessment of the HIV/AIDS Notification System - A Portuguese Experience. National Coordination For HIV Infection (2009)
9. Pagano, M., Tu, X., De Gruttola, V., MaWhinney, S.: Regression analysis of censored and truncated data: estimating reporting-delay distributions and AIDS incidence from surveillance data. Biometrics **50**, 1203–1214 (1994)
10. World Health Organization: WHO report on global surveillance of epidemic-prone infectious diseases (2000)
11. Barros, P., Machado, S., Simões, J.: Portugal: health system review. Health Syst. Transition **13**, 1–156 (2011)
12. Ministério da Saúde: A Organização Interna e a Governação dos Hospitais (2010)
13. Paulo, A.: SNS: Caracterização e desafios. GPEARI-MFAP. Lisboa, Setembro, 16 p. (2010)
14. Brookmeyer, R., Liao, J.: Statistical modelling of the AIDS epidemic for forecasting health care needs. Biometrics **46**, 1151–1163 (1990)
15. Kalbfleisch, J., Lawless, J.: Regression models for right truncated data with applications to AIDS incubation times and reporting lags. Statistica Sinica **1**, 19–32 (1991)
16. Noufaily, A., Ghebremichael-Weldeselassie, Y., Enki, D., Garthwaite, P., Andrews, N., Charlett, A., Farrington, P.: Modelling reporting delays for outbreak detection in infectious disease data. J. Roy. Stat. Soc.: Ser. A (Stat. Soc.) **178**, 205–222 (2015)
17. Harris, J.: Reporting delays and the incidence of AIDS. J. Am. Stat. Assoc. **85**, 915–924 (1990)
18. Barnard, J., Meng, X.: Applications of multiple imputation in medical studies: from AIDS to NHANES. Stat. Methods Med. Res. **8**, 17–36 (1999)
19. Green, T.: Using surveillance data to monitor trends in the AIDS epidemic. Stat. Med. **17**, 143–154 (1998)
20. Amaral, J., Pereira, E., Paixão, M.: Data and projections of HIV/AIDS cases in Portugal: an unstoppable epidemic? J. Appl. Stat. **32**, 127–140 (2005)
21. Bellocco, R., Marschner, I.: Joint analysis of HIV and AIDS surveillance data in back-calculation. Stat. Med. **19**, 297–311 (2000)

22. Midthune, D., Fay, M., Clegg, L., Feuer, E.: Modeling reporting delays and reporting corrections in cancer registry data. J. Am. Stat. Assoc. **100**, 61–70 (2005)
23. Oliveira, A., Costa, J., Gaio, R.: VIH/SIDA Estimação de probabilidades de atraso e sub-notificaçõ. In: Giovani Silva, L.G.P.O. (ed.) Programa de Resumos dos Encontros de Biometria 2013 (2013)
24. Oliveira, A., Costa, J., Gaio, R.: The incidence of AIDS in Portugal adjusted for reporting delay and underreporting. In: 9th Iberian Conference on Information Systems and Technologies (CISTI), pp. 1–5. IEEE (2014)
25. Oliveira, A., Faria, B., Gaio, R., Reis, L.: Data mining in HIV-AIDS surveillance system. J. Med. Syst. **41**, 51 (2017)
26. Oliveira, A., Gaio, R., da Costa, J., Reis, L.: An approach for assessing the distribution of reporting delay in Portuguese AIDS data. In: New Advances in Information Systems and Technologies, pp. 641–649. Springer (2016)
27. Cui, J., Kaldor, J.: Changing pattern of delays in reporting AIDS diagnoses in Australia. Aust. N. Z. J. Public Health **22**, 432–435 (1998)

Study of a Successful ERP Implementation Using an Extended Information Systems Success Model in Cameroon Universities: Case of CUCA

Chris Emmanuel Tchatchouang Wanko[1],
Jean Robert Kala Kamdjoug[1(✉)], and Samuel Fosso Wamba[2]

[1] FSSG, GRIAGES, Université Catholique d'Afrique Centrale,
Yaoundé, Cameroun
chrisemmanuelt@gmail.com, jrkala@gmail.com
[2] Toulouse Business School, Université Fédérale de Toulouse Midi-Pyrénées,
20 Boulevard Lascrosses, 31068 Toulouse, France
s.fosso-wamba@tbs-education.fr

Abstract. Nowadays, the performance of organizations including universities depends on the alignment of their educational strategy with the evolution of technology. For increased efficiency and improved productivity in their operations, universities around the world have been implementing and boosting ITs. This paper seeks to study a case of successful implementation of ERP system in the higher education sector in Cameroon. A model was based on the information system success model (ISSM) by Delone and Mclean (updated in 2003) from which we extracted only the relevant constructs according to our context. Results of the study showed that ISSM is well suited for studying a successful ERP implementation for universities, particularly in the Cameroonian environment. They also indicated that the feeling of belonging to an organization, work satisfaction and the perception of technological innovation are strongly influenced by the use of a university ERP system and the satisfaction that a user can draw from this use.

Keywords: ISSM · Delone & Mclean · Job satisfaction ·
Organization commitment · Innovation process performance

1 Introduction

ERP (Enterprise resource planning) refers to a complete and integrated software solution [1]. It is composed of a set of modules that seek to integrate all the departments and functions of an organization into a single computer system that can meet different specific needs in the area of production, sales, human resources, finance, etc. [2]. ERP tools have the potential to enable organizations to have a more effective and efficient management of their internal resources (human resources, material resources, financial resources, information resources, etc.) [3]. These different attributes make enterprise resource planning (ERP) become a powerful IT tool and driver of performance

© Springer Nature Switzerland AG 2019
Á. Rocha et al. (Eds.): WorldCIST'19 2019, AISC 930, pp. 727–737, 2019.
https://doi.org/10.1007/978-3-030-16181-1_68

improvement in organizations across the world [4]. African countries are also involved this global trend and many of them are steadily embarking on the acquisition and implementation of ERP in their national organizations, especially in universities [5, 6]

Despite the strong development of ERP, non-profit structures or organizations such as universities started adopting it only in the 2000s [7]. For example, higher education ERP, like traditional enterprise ERP, connects and unifies all organization's computer systems for materiel management, inventory, finance, and general administration [8]. Though the major ERP publishers have not really engaged in designing a university-specific ERP, all ERP platforms have content with common functionalities for all organizations as well as some web client interfaces and object-oriented systems [9]. The IT literature on successful implementation of ERP is quite abundant, but few studies are dedicated to ERP for higher education institutions and especially to ERP implementation in African countries. The main objective of this study is to examine the enabling factors for a successful implementation of this software technology in Cameroon universities. We will dwell on the following research question to guide our study:

May of the extended and updated version of Delone and Mc Lean's information systems success model [10] be adequately used to study the successful implementation of an ERP in the Cameroon's higher education system?

In order to answer this research question, we have developed and tested a research model based on the extension of Delone and Mclean's information systems success model (ISSM) that was completed by these constructs: *job satisfaction* (JS), *organization commitment* (OC), and *innovation process performance* (IPP), all from the extant relevant literature. The data that were analyzed to answer our research question were collected with ERP users in Cameroonian universities.

2   Theoretical Background

Our study is based on the research model presented in Fig. 1. It is inspired by the Delone and Mclean's information systems success model as well as by the ERP Information System literature.

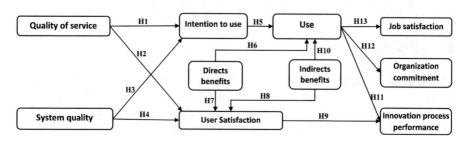

Fig. 1. Conceptual research model

ISSM was developed in 1992 by Delone and McLean [11] and was updated ten years later, in 2003 [10]. These two versions of ISSM aim to anticipate the effects of information technology on the performance of an organization by taking into account the individual performance of the organization's members [10]. The original version of ISSM measures six constructs, namely: the *quality of service* (QS), the *quality of the information* (IQ), *use* (U), *user satisfaction* (US), the *individual impact* (II) and the *organizational impact* (OI) [11]. The update made in 2003 added the *system quality* (SQ), *the intention to use* (IU), which were both largely inspired by TAM (Theory of Acceptance Model) and TRA (Theory of reasoned action). The ISSM also used impact constructs such as *net profit* or *benefits* of the system (NB) [10]. The ISSM establishes meaningful causal relationships between each of the variables quality of service and system quality and each of the other user satisfaction variables and intention to use [10]. It also supports that net profits are major drivers or causes of both use and user satisfaction [10].

In our study, we built a model with the variables *Quality of Service* (QS), *Use* (U), *User Satisfaction* (US), *System Quality* (SQ), *Intention to Use* (IU), *Net Benefits* (NB) which were divided into *direct benefits* (DB) and *indirect benefits* (IB) all of which were from the ISSM. To the relevant ISSM constructs that aimed to measure ERP impact in our context, we add these constructs: *job satisfaction* (JS), *organization commitment* (OC), and the *innovation process performance* (IPP). *Job satisfaction* is a pleasant emotional state resulting from the evaluation of one's work as reaching or facilitating the achievement of one's professional values and goals [12]. *Organization commitment* refers to an affective or emotional attachment of an individual to an organization so that it is strongly engaged, identified with the organization and perceived as a participant in the organization in one way or another [13]. The *innovation process performance* refers to standard costs, efficiency and speed [14].

2.1 Quality of Service (QS)

There are two relevant trends concerning the quality of service: (1) the trends from the American school, which is defended by Parasuraman, Zeithaml and Berry, and which defines the quality of service as the direct characteristics of the service offered [15]; and (2) the trend from the European school, which is supported by Christian Gréinroos and for which quality of service encompasses the service structure and its quality components [15]. In this paper, like in ISSM [10], we follow the American approach to measure Quality of service as system's characteristics and reactivity when it is required. In fact, Quality of service is crucial for the measurement of IS success. In this case, it refers to the ability of the information system to meet the different user needs [11]. So we formulate the following hypotheses:

H1: QS has a significant positive effect on IU.
H2: QS has a significant positive effect on US.

2.2   System Quality (SQ)

System quality refers to the technical quality of the system and its use. As stipulated by Delone, William H and McLean, System quality positively influences the intention to use the system and the user satisfaction of the system [10]. This construct is concerned with two elements: (1) the ability of the system to consistently perform user demands and security concerns, and (2) the system reactivity to perform a given task [16]. Then we formulate the following hypotheses:

H3: SQ has a significant positive effect on IU.
H4: SQ has a significant positive effect on US.

2.3   Intention to Use (IU)

The intention to use is defined as the strength of an individual's intention to adopt and use a technology generally in the future [17–19]. In addition, it refers to the frequency and manner in which an individual intends to use technology in general [20]. This is an indication of an individual's willingness to engage in immediate behavior [21]. It is widely used in the literature to predict the adoption and use of a technology by an individual based on the attitude towards use. Then we formulate the following hypothesis:

H5: IU has a significant positive effect on Use.

2.4   Net Benefits (DB, IB)

Net benefits, which is divided into Direct Benefits (DB) and Indirect Benefits (IB), are a measure of first-rate success because they reflect the balance of positive and negative impacts of a technology and its use [10]. Net benefits are an idealized understandable measure of the sum of all the past and expected benefits, minus all the past and future costs [22]. The net benefits are seen by the different stakeholders of an information system in different ways, these net benefits are perceived differently depending on whether we are an individual, a group of individuals, and the management of an organization or the society herself [22]. So we formulate the following hypotheses:

H6: Direct Benefits have a significant positive effect on Use.
H7: Direct Benefits have a significant positive effect on User Satisfaction.
H8: Indirect Benefits have a significant positive effect on User Satisfaction.
H10: Indirect Benefits have a significant positive effect on Use.

2.5   User Satisfaction (US)

User satisfaction is a subjective evaluation of the pleasant or unpleasant emotion aroused in an individual as a result of using a system. It is closer to the individual's perception of the benefits of using the system [22]. User satisfaction is generally substituted in the mind of the user for the usefulness of the system in decision-making, or for objective determinants like the effectiveness of an information system [23].

In our study we use this last approach to measure the user satisfaction. Therefore, we formulate the following hypothesis:

H9: User Satisfaction has a significant positive effect on Innovation Process Performance.

2.6 Use (U)

Use refers to the extent of system practice, which would imply a consumption of resources such as human effort. It can be measured in hours of work, in hours of reports analysis, in use frequency, in users' number, or simply in binary variable: use/non-use [22]. For some authors, the use of the system is a behavior that should be included not as a success variable in a causal model, but rather in a process model [22]. Others argue that the success measurement is multi-dimensional, taking into account the use extension, the nature, the quality and the relevance of the system [24]. In this study, we measure of the use in terms of frequency of use. Then, we formulate the following hypotheses:

H11: Use has a significant positive effect on Innovation Process Performance.
H12: Use has a significant positive effect on Organizational Commitment.
H13: Use has a significant positive effect on Job Satisfaction.

3 Methodology

The population of this study is made of the users of any ERP system in the university environment in Cameroon. The use of this type of technology is not wide-spread in the country, and that is why we mainly focus on the Catholic University of Central Africa (CUCA), which has recently integrated ERP in its academic management. The data collection process was conducted in two phases. The first phase was a pre-test that involved twelve people with varied profiles (06 students from various departments of the CUCA, a lawyer, two computer scientists, three professionals in Information Systems). This phase allowed us to refine and improve the stability of our model. The second phase one was the actual data collection from a significant sample of our population.

The data was collected from 175 CUCA (Catholic University of Central Africa) ERP users over 800 by means of a web questionnaire. The data collection activity took place in October 2018. Participants, who were all volunteers, were able to click on the ERP link to access the form and fill it out. A total of 178 answers were obtained from the web platform, and 175 of them were exploited. The two figures accounted respectively for 22.25% of our population and 98.31% of usable responses.

We used a total of thirty nine (39) items from the literature to measure our ten (10) constructs composed of seven (7) independent variables and three (3) dependent variables. Our constructs were adapted from ISSM constructs, which were measured by means of a seven-point Likert scale. The structural equation modeling on the SmartPLS software (version 3.2.7) was used to analyze various aspects of the model.

This application is recognized as which suitable for this sample size [25]. The method used was therefore a partial least squares approach. We used four main criteria to evaluate our model: Cronbach's Alpha, Rho_A, composite reliability, average extracted variance (AVE). The minimum thresholds required for these criteria are respectively 0.7, 0.7, 0.7 and 0.5 [26, 27].

4   Results and Discussion

Table 1 presents the external loads, all of which are higher than 0.7 as recommended by Hair et al. [26]. It also presents the validity of our model. It clearly appears that the Cronbach's Alpha, Rho_A and composite reliability are greater than 0.7 and that each of the AVE values greater than 0.5, which is one of the thresholds suggested by [26, 27].

Table 1. Reliability and validity of the construct

Construct	Items	Loadings	Cronbach's Alpha	Rho_A	Composite reliability	AVE
DB	DB1	0,917	0,897	0,903	0,928	0,764
	DB2	0,854				
	DB3	0,886				
	DB4	0,839				
IB	IB1	0,839	0,908	0,916	0,935	0,784
	IB2	0,910				
	IB3	0,924				
	IB4	0,867				
IPP	IPP1	0,775	0,841	0,843	0,894	0,678
	IPP2	0,824				
	IPP3	0,830				
	IPP4	0,861				
IU	IU1	0,962	0,969	0,969	0,977	0,915
	IU2	0,945				
	IU3	0,967				
	IU4	0,952				
JS	JS1	0,786	0,869	0,879	0,911	0,719
	JS2	0,906				
	JS3	0,850				
	JS4	0,845				
OC	OC1	0,884	0,908	0,949	0,941	0,841
	OC2	0,934				
	OC3	0,933				
QS	QS1	0,864	0,921	0,928	0,944	0,809
	QS2	0,920				
	QS3	0,917				

(*continued*)

Table 1. (*continued*)

Construct	Items	Loadings	Cronbach's Alpha	Rho_A	Composite reliability	AVE
	QS4	0,895				
SQ	SQ1	0,852	0,889	0,890	0,923	0,751
	SQ2	0,891				
	SQ3	0,843				
	SQ4	0,880				
US	SU1	0,831	0,878	0,885	0,916	0,732
	SU2	0,882				
	SU3	0,883				
	SU4	0,825				
U	U1	0,823	0,891	0,898	0,925	0,754
	U2	0,873				
	U3	0,853				
	U4	0,922				

Table 2 presents the correlations between the constructs. We can see that the square roots of the AVE on the diagonal are indeed higher than all the values of inter-correlations as suggested by Hair et al. [26, 28].

Table 2. Discriminant validity

	DB	IB	IPP	IU	JS	OC	QS	SQ	U	US
DB	**0,874**									
IB	0,702	**0,885**								
IPP	0,674	0,634	**0,823**							
IU	0,525	0,505	0,543	**0,956**						
JS	0,678	0,605	0,610	0,419	**0,848**					
OC	0,654	0,571	0,569	0,430	0,613	**0,917**				
QS	0,610	0,509	0,584	0,538	0,448	0,373	**0,899**			
SQ	0,712	0,651	0,752	0,587	0,548	0,555	0,614	**0,867**		
U	0,476	0,486	0,459	0,390	0,515	0,449	0,333	0,337	**0,868**	
US	0,581	0,590	0,563	0,518	0,589	0,519	0,518	0,551	0,529	**0,856**

Table 3 presents the results of our structural model. We can observe that all normalized path coefficients from the ISSM constructs are positive and significant at 0.05 level at least. Thus, the part of our model extracted from ISSM is relevant to our study. In fact, our study confirms almost all the assumptions of the extracted part of ISSM, except for the hypothesis H4.

According to the results obtained, the extensions that we made on the ISSM, namely the hypotheses (H9, H11, H12, H13), are all supported with the level 0.001.

Table 3. Results of analysis

		Original Sample (O)	Sample Mean (M)	Standard Deviation (STDEV)	T Statistics (|O/STDEV|)	P Values	Significance level		Hypotheses
H1	QS -> IU	0,286	0,285	0,094	3,044	0,002	99%	***	Supported
H2	QS -> US	0,189	0,195	0,091	2,085	0,038	95%	**	Supported
H3	SQ > IU	0,411	0,405	0,078	5,241	0,000	99,90%	****	Supported
H4	SQ -> US	0,120	0,115	0,092	1,307	0,192	No		Not supported
H5	IU -> U	0,145	0,148	0,073	1,984	0,048	95%	**	Supported
H6	DB -> U	0,217	0,222	0,092	2,354	0,019	95%	**	Supported
H7	DB -> US	0,174	0,171	0,100	1,738	0,083	90%	*	Supported
H8	IB -> US	0,293	0,294	0,098	3,002	0,003	99%	***	Supported
H9	US -> IPP	0,444	0,448	0,088	5,028	0,000	99,90%	****	Supported
H10	IB -> U	0,261	0,254	0,099	2,636	0,009	99%	***	Supported
H11	U -> IPP	0,224	0,223	0,067	3,363	0,001	99,90%	****	Supported
H12	U -> OC	0,449	0,451	0,053	8,423	0,000	99,90%	****	Supported
H13	U -> JS	0,515	0,517	0,056	9,268	0,000	99,90%	****	Supported

****P < 0.001; ***P < 0.01; **P < 0.05; *P < 0.1

t > 1,65: 90%; t > 1,96: 95%; t > 2,57: 99%; t > 3,29: 99,9%

5 Conclusion and Future Research Directions

In this study, a number of constructs were extracted from ISSM and completed with some impact measurements in order to measure the implementation success of ERP system in an African university, and in this case the CUCA. In fact, our study should certainly contribute to enriching the literature on information systems or information technologies. Therefore, university decision-makers could leverage this study's results and findings to operate a successful implementation of ERP systems in their institutions. This is all the more important as the ERP literature remains very scarce in sub-Saharan Africa and especially in Cameroon, as compared to developed countries.

It has been concluded that the extracted part of ISSM is relevant for our study, and that the set of fundamental relationships established in the model are supported, with the exception of the quality of the system which is not significantly positive related to user satisfaction. The rejection of this hypothesis, although surprising, may be due to the users being forced to adopt the technology and not necessarily to the user perception of criteria like quality, utility, etc. For example, in a context where the great part of the communication between actors must be done through a specific system, the users would tend to minimize the system defects and tend to valorize the benefits that they can shoot there. This could explain the fact that our work reveals that the quality of the system does not have a significant influence on the use of the system. Then, future studies might consider the effect of non-constraint system on the satisfaction of users.

Our study also revealed that the use of an ERP system in an organization can reinforce users' sense of belonging to the organization and improve the organization's place in their mind. Indeed, with customize system, user tends to be more efficient and satisfy. Then he seems that organization management pays attention to his contribution in the organization functioning. Hence, this would improve the image and his feeling towards his organization. The use of an ERP system increases a user's value of his work within an organization. We were able to realize that the use of the system had a significant positive influence on job satisfaction. Together with the user's satisfaction for using ERP, this system constitutes a real innovation in the user's way of doing things. Furthermore, such a system in an African university would speed up management processes, improve communication between university actors. Thus, it is in the interest of Cameroonian universities and those of Africa as a whole to integrate ERP in order to would benefit from its many advantages: increased efficiency and profitability; high service quality; interoperability with other systems; user-friendliness, etc. Additionally, ERP system can be relevant for the development of staff membership, for self-confidence and for improved productivity.

Despite the actual quality of our study's results, we believe future studies should consider investigating this topic for a longer period of time in order to seize all the facets of ERP impact on university activities; they may also involve more survey participants (respondents) that will increase the representativeness of the sampled population. Furthermore, the factors that influence the adoption and use of ERP systems in institutions of higher learning may be further examined using for example TAM or UTAUT. Finally, we think it would be important to adopt a research method that would include a qualitative study, so as to improve the understanding of the ERP phenomenon in the academic context.

References

1. Klaus, H., Rosemann, M., Gable, G.G.: What is ERP? Inf. Syst. Frontiers **2**(2), 141–162 (2000)
2. Botta-Genoulaz, V., Millet, P.-A.: An investigation into the use of ERP systems in the service sector. Int. J. Prod. Econ. **99**(1–2), 202–221 (2006)
3. Nah, F.-H., Lau, J.L.-S., Kuang, J.: Critical factors for successful implementation of enterprise systems. Bus. Process Manag. J. **7**(3), 285–296 (2001)
4. Hallé, M.-F., Renaud, J., Ruiz, A.: Progiciels de gestion intégrée: expériences d'implantation dans cinq entreprises Québécoises. Logistique & Management **13**(2), 31–43 (2005)
5. Mahanga, K.M., Seymour, L.F.: Enterprise resource planning teaching challenges faced by lecturers in Africa. In: Proceedings of the 9th IDIA Conference, IDIA2015, Nungwi, Zanzibar (2015)
6. Worou, D.: Impact des pratiques de gestion des ressources humaines sur l'acceptation de l'ERP dans les entreprises en Afrique: cas de deux entreprises en Afrique de l'Ouest. In: Tidjani, B., Kamdem, E. (eds.) Gérer les ressources humaines en Afrique: entre processus sociaux et pratiques organisationnelles, Caen, Éditions Management et Société, pp. 121–144 (2010)
7. Althonayan, M.: Evaluating stakeholders performance of ERP systems in Saudi Arabia higher education. Brunel University, School of Information Systems, Computing and Mathematics (2013)
8. Pollock, N., Cornford, J.: Implications of enterprise resource planning systems for universities: an analysis of benefits and risks (2005)
9. Awad, H.: One ERP system for twenty five universities an empirical investigation for development ERP private cloud: Kingdom of Saudi Arabia universities case. J. Inf. Eng. Appl. **4**, 77–81 (2014)
10. Delone, W.H., McLean, E.R.: The DeLone and McLean model of information systems success: a ten-year update. J. Manag. Inf. Syst. **19**(4), 9–30 (2003)
11. DeLone, W.H., McLean, E.R.: Information systems success: the quest for the dependent variable. Inf. Syst. Res. **3**(1), 60–95 (1992)
12. Locke, E.A.: What is job satisfaction? Organ. Behav. Hum. Perform. **4**(4), 309–336 (1969)
13. Kanter, R.M.: Commitment and social organization: a study of commitment mechanisms in Utopian communities. Am. Sociol. Rev. **35**, 499–517 (1968)
14. Hsueh, J.-T., Lin, N.-P., Li, H.-C.: The effects of network embeddedness on service innovation performance. Serv. Ind. J. **30**(10), 1723–1736 (2010)
15. Binani, K.: La perception de la qualité de service rendue par le personnel des institutions financières au Québec (2013)
16. Lupo, G.: Evaluating e-Justice: the design of an assessment framework for e-Justice systems (2016)
17. Taylor, S., Todd, P.A.: Understanding information technology usage: a test of competing models. Inf. Syst. Res. **6**(2), 144–176 (1995)
18. Venkatesh, V., Brown, S.A.: A longitudinal investigation of personal computers in homes: adoption determinants and emerging challenges. MIS Q. **25**, 71–102 (2001)
19. Venkatesh, V., et al.: User acceptance of information technology: toward a unified view. MIS Q. **27**, 425–478 (2003)

20. Gupta, B., Dasgupta, S., Gupta, A.: Adoption of ICT in a government organization in a developing country: an empirical study. J. Strateg. Inf. Syst. **17**(2), 140–154 (2008)
21. Ajzen, I.: The theory of planned behavior. Organ. Behav. Hum. Decis. Process. **50**(2), 179–211 (1991)
22. Seddon, P.B.: A respecification and extension of the DeLone and McLean model of IS success. Inf. Syst. Res. **8**(3), 240–253 (1997)
23. Ives, B., Olson, M.H., Baroudi, J.J.: The measurement of user information satisfaction. Commun. ACM **26**(10), 785–793 (1983)
24. DeLone, W.H., McLean, E.R.: Information systems success revisited. In: Proceedings of the 35th Annual Hawaii International Conference on System Sciences (HICSS 2002). IEEE (2002)
25. Ringle, C.M., Wende, S., Becker, J.-M.: SmartPLS 3. SmartPLS GmbH, Boenningstedt (2015)
26. Hair, J., Hult, G., Ringle, C., Sarstedt, M.: A Primer on Partial Least Squares Structural Equation Modeling (PLS-SEM). Sage Publication Inc., Thousand Oaks (2014)
27. Nunnally, J.C., Bernstein, I.: Psychometric Theory. McGraw-Hill Series in Psychology, vol. 3. McGraw-Hill, New York (1994)
28. Fornell, C., Larcker, D.F.: Structural equation models with unobservable variables and measurement error: algebra and statistics. J. Mark. Res. **18**, 382–388 (1981)

Conceptual Approach for an Extension to a Mushroom Farm Distributed Process Control System: IoT and Blockchain

Frederico Branco[1,2], Fernando Moreira[3,4(✉)], José Martins[1,2], Manuel Au-Yong-Oliveira[5], and Ramiro Gonçalves[1,2]

[1] University of Trás-os-Montes and Alto Douro, Vila Real, Portugal
{fbranco,jmartins,ramiro}@utad.pt
[2] INESC TEC and UTAD, Vila Real, Portugal
[3] REMIT, IJP, University Portucalense, Porto, Portugal
fmoreira@upt.pt
[4] IEETA, University of Aveiro, Aveiro, Portugal
[5] GOVCOPP, Department of Economics, Management,
Industrial Engineering and Tourism, University of Aveiro, Aveiro, Portugal
mao@ua.pt

Abstract. Collecting, storing, integrating and transforming data, together with the problem of security and privacy, are topics that present great challenges for society. The needs of the industry in general and agri-food in particular in the mushroom production sector, due to their specificities, requires the adoption of emerging technologies to make them more productive and more competitive in this global market. In this type of industry it is important, and essential, the control of the environmental variables of the production areas and the way they are presented, because the information associated with these variables would provide an important complement to the established production control system. In this paper we propose a conceptual approach for an extension to a mushroom farm distributed process control system with IoT and blockchain integration that not only allows to collect distributed data on the environmental indicators inherent to the mushrooms production, but also complement the already existent production control system, which is extremely important for the overall success of the farm management information system used by the group managers.

Keywords: Blockchain · IoT · Precision agriculture · Information system · Architecture

1 Introduction

The world is undergoing a transformation because of increasing data dependence [1]. This dependence is directly related to how we interact with each other using existing technology with the aim of achieving a better understanding of the world while improving citizens' living conditions. According to Shrier et al. [2], this transformation occurs in all social systems through several services, including road traffic, health,

© Springer Nature Switzerland AG 2019
Á. Rocha et al. (Eds.): WorldCIST'19 2019, AISC 930, pp. 738–747, 2019.
https://doi.org/10.1007/978-3-030-16181-1_69

governance, logistics, marketing, security, agriculture, among others [3]. However, problems such as the (voluntary or involuntary) loss of information on the part of the providers of these services require the creation of new security models.

In innovation accelerating technologies, the Internet of Things (IoT) offers a unique opportunity in the transformation of many industries [4, 5], including the agri-food sector. Traditionally this sector has low investment in information and communication technologies (ICT) [6]. In IoT it is possible to include sensors, actuators, navigation systems, cloud-based data services, among others, which provide a wide variety of data to "feed" decision support tools [7].

However, citizens today, as a matter of principle, rely on the information provided by financial and government entities; nevertheless, it is possible to question whether the information provided by both the referred entities and other external entities (e.g., companies that provide IoT, companies that provide cloud storage services) has not undergone any type of alteration. This is difficult to clarify by considering the centralized nature of architecture. Thus, this problem leads to the need to verify whether the information has been modified or not. One of the technologies that have emerged recently and with great potential to meet this need is the blockchain.

A great benefit of the blockchain is being a decentralized information technology that can be applicable in many situations (e.g., cryptocurrency, financial assets, among others). The decentralized nature of the blockchain makes it a technology of equality for all entities in the world, human and machine alike [8]. The blockchain is currently a relevant research topic, with investments of several billion dollars [9].

After analysing all the new challenges associated with the agri-food industry, and particularly with the mushroom production sector, it was possible to perceive that, even with the existing technology [10], there is still an important gap in what concerns the control of the environmental variables of the production areas and the way they are presented; the information associated with these variables would provide an important complement to the established production control system.

One of the technologies that can bring an important contribution to the development of a distributed security system that responds to all the identified needs comes in the form of the IoT devices integrated with blockchain. These devices are composed by a multitude of other small autonomous devices distributed throughout the physical environments; their role is to monitor the existing environmental conditions in a cooperative manner and send the collected data through blockchain to headquarters. Additionally, mushroom production is distributed across multiple sites, with the need to integrate this information into a Mushroom Farm Distributed Process Control System [11].

In the present paper we would like to present a conceptual approach for an extension to a mushroom farm distributed process control system with IoT and blockchain integration that not only allows to collect distributed data on the environmental indicators inherent to the mushrooms production, but also complement the already existent production control system, which is extremely important for the overall success of the farm management information system used by the group managers.

The presented paper is divided in five section, with the first one being the introduction. In the second section, the background and motivation are discussed, and the IoT and blockchain integration is the main goal of the third section. The last two sections consist of the conceptual approach and conclusions.

2 Background and Motivation

2.1 IoT

The IoT comprises a broad range of technologies, ranging from Radio Frequency Identification (RFID) to Wireless Sensors Networks (WSN) required to communicate over the Internet [9]. An IoT device can take on various forms (e.g., wearable) as well as a broad set of applications covering different areas and needs of society. Due to the increasing number of IoT devices, Đurić [12] predicts that the number of devices connected will increase 150% by 2020 [13].

IoT presents a fully connected world where things are capable of communicating collected data as well as interacting with each other. This capability enables real-world digital representation through which intelligent applications can be developed in a variety of industries. We can characterize IoT applications because they require the power to operate for long periods of time, require connectivity and generate large amounts of data. However, it also presents a large number of challenges, including memory limitations, processing capacity, communication networks and limited energy supply, among others.

The reliability of the data generated by the IoT devices can be achieved through a distributed service that ensures that they are not changed if all the nodes that participate in the validation and storage of the data have means to verify that it has not been changed.

2.2 Blockchain

Satoshi Nakamoto, in 2008 [14], introduced two concepts with a strong impact on the financial markets. Bitcoin was the first concept presented, a cryptocurrency that emphasizes its value with the absence of a centralized authority; the security of the coin is maintained collectively through a decentralized Peer-to-Peer (P2P) network of actors. The second, which has achieved even greater success than the cryptocurrency, is the blockchain.

The blockchain is based on the reliability of transaction verification by a group of unreliable actors. It provides an auditable, secure, transparent, immutable and distributed ledger. All the records stored in the blockchain can be accessed openly and can be checked and grouped by entity at any time. The protocol structures the information in block chain, each block being responsible for storing a set of transactions performed at a given time. Each block refers to the previous block, constituting a string, and adopts a hash-based distributed consensus algorithm Proof-of-Work (PoW).

To support blockchain operation and maintenance, peers must provide the following functionalities: routing, storage, wallet services, and mining [14]. There may be different types of nodes in the network depending on the features provided. The propagation of transactions in the network is guaranteed through the routing function in the P2P network. Storing the string on the node is guaranteed by the storage feature. The security of the keys, which allow the accomplishment of transactions of coins, is assured by the wallet services. The creation of new blocks is the responsibility of the mining function.

Due to the high complexity of the blockchain its implementation presents itself as a huge challenge [15]. To ensure that technology is useful and profitable, all actors must ensure proper implementation and adoption. This is a technology that is still taking the first steps, so there are no consolidated implementation standards yet.

2.3   Motivation

Precision agriculture (PA) is based on an abundant set of techniques and methodologies aimed at the optimization of agricultural crops and providing maximum economic efficiency. The PA instruments allow for a rational use of resources like energy, fertilizers and water, thus decreasing the inherent environmental impact [16].

The PA involves the use of electronic technology to collect large amounts of data from production areas, data that is going to be used in production management activities. One of the most critical issues associated with PA concerns the complexity inherent to the analysis and comprehension of the collected data. This is true due to the criticality associated with the evaluation of the production environment, the specific culture and inherent variability in the referred data. The implemented AP systems must be able to offer solid and effective strategies for the management of the entire production process variability [17].

The main activity of Sousacamp Group is the production of fresh Agaricus bisporus mushrooms. The markets where the Group companies operate are mostly located on the Iberian Peninsula, France, Netherlands and North Africa, and their success strongly depends on the production process optimization, which is already highly industrialized and incorporates several complex technologies. Additionally, the Group's strategy focuses on investment in IS/IT as a vehicle for efficiency gains and competitiveness in respect to its direct competitors [11, 18].

In the analysed scientific literature there were no studies that described the implementation of IoT and blockchain integration to PA with the specific focus on mushroom production. In order to address this gap we propose a production monitoring system based on IoT and blockchain integration that meets the following requirements: (1) Develop a distributed monitoring system that amplifies the collection of environmental-related data created by the existent process control system (PCS); (2) Create a distributed IoT and blockchain integration (by all production units) to monitor various production-related indicators (temperature, relative humidity, CO_2 concentration); (3) Develop a decision support system based on data visualization techniques and methodologies, which delivers to the user a set of intuitive dashboards with concentrated information on the production process; (4) In addition, implement an alert functionality responsible for monitoring all critical indicators.

3   IoT and Blockchain Integration

In recent years, according to Díaz et al. [9], the need to analyse and process data in real time by IoT technology was greatly facilitated by the massification of cloud computing technologies. IoT's growth has created new opportunities by providing mechanisms for acquiring and sharing information. Still, one of the biggest weaknesses of this technology is lack of trust.

Centralized architectures such as those used by the most popular cloud storage applications (e.g., Dropbox, Google Drive, and so on) are in a growth phase due to their characteristics in storing various types of files (documents, photos, videos, music, etc.); however, according to Crosby, et al. [19–21], they are faced with challenges in areas such as security, privacy and data control.

A centralized database is more susceptible to hackers, corruption or failures [22]. The main question posed by Crosby et al. [19] is the need to rely on a third party regarding the confidentiality of the data. Confidence is one of the main consequences of decentralization, since there is no need to assess the reliability of the intermediary or other network participants [23]. Thus, decentralization is an important property to take into consideration when choosing an alternative technology.

The blockchain technology, due to its characteristics, allows the verification of any information tampering, thus increasing its validity. Deletion of records kept collectively is impractical and verified for each single transaction that is accessible to participants through public or private distributed ledgers [19]. Participants can view ledgers and analyse transactions. This feature provides transparency [22], while ensuring anonymity by preserving records through the use of cryptography.

IoT technology, because of its centralized model in which a broker or hub controls the interaction between devices, raises questions about security and privacy as well. The issue has to do with the need for some devices to exchange data independently of one another. This problem leads to the application of blockchain technology to facilitate the implementation of decentralized IoT platforms, such as the safe and reliable exchange of data as well as record keeping. In this architecture, the blockchain serves as a general ledger, maintaining a reliable record of all messages exchanged between devices in a decentralized IoT topology.

Thus, it is possible to recognize the potential of the blockchain to overcome most of the problems of using IoT. According to Malviya [24], the use of the blockchain can complement the IoT technology with respect to scalability, reliability, privacy, traceability and information security.

The improvements that this integration can bring include, among others: (i) decentralization and scalability [25]; (ii) identity [26]; (iii) autonomy [27]; (iv) reliability [28]; (v) security [29]; (vi) service market [30]; (vii) secure code implementation [25].

As a conclusion, the functionalities of blockchain technology can contribute in large scale to the development of IoT technology. However, as these technologies are at an early stage, there are still many research challenges.

4   Approach Proposed

In this section we would like to present a distributed sensing system proposal that not only allows to collect data on the environmental indicators inherent to the mushroom production, but also complements the already existent production control system.

The production of mushrooms is of extreme importance to the Sousacamp Group, since it constitutes the Group's most important revenue source. In order to maintain a sustainable level of performance, the Group decided that it needed to minimize production-related problems by monitoring the environmental conditions of production areas (distributed over multiple units, as shown in Fig. 1), and with this controlling the various stages of the production process and ensuring that the appropriate guidelines and metrics are applied [10, 11, 18].

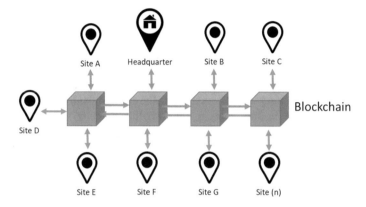

Fig. 1. Units connected to blockchain

IoT devices distributed and integrated with blockchain network contribute in a substantial manner to an effective monitoring of the mushrooms production rooms' environmental conditions, giving extra information to production managers.

The massive collection of data will provide a basis for the desired visual analysis that is composed by dashboards presenting readings of multiple IoT devices distributed in a graphical representation, in a given time interval and using different types of filters.

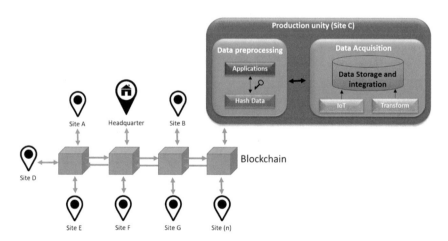

Fig. 2. Integration of information collected from IoT devices and other devices and connect to block-chain

Additionally, the proposed IoT devices distributed for all nine production units can also be used to detect power failures and possible problems associated with electrical equipment. The proposed architecture for distributed collection in all production units, as can be seen in Fig. 2, consists of a unit that integrates the collection of information from IoT devices and other devices ("Transform"), consolidating all the information collected in the "Data Storage and Integration" that is then passed to the "Data pre-processing" module. This module has as main functionality to handle the first processing of the data so that when they are sent to the blockchain network, they are already in the appropriate format and can serve the decision-making modules without the need to perform any pre-processing. For instance, one can verify that a given equipment is not functioning in full conditions if the environmental indicators values of the room are outside the desired range. As one should expect, the sooner the equipment failures are detected, the easier it will be to fix the problem, thus reducing the risk of production losses.

If a malfunction is detected in its initial stage, one can prevent a crop to be completely lost, mitigating the cost of completely stopping production. Due to this, when an anomaly is detected within the production rooms, the proposed system will send an alert message to all production managers prompting them to take the necessary actions. Through the proposed system, when a given erroneous situation is solved, the production managers also receive a notification indicating that conditions are normalized, and production can continue.

The proposed system architecture (Fig. 3) is subdivided into three large blocks: (1) the IoT devices network with pre-processing and data storage and integration, (2) the back-end and front-end and (3) the blockchain network. The IoT devices are dedicated to the task of obtaining the necessary data to feed the modules of the knowledge generator system and are composed by both back-end and front-end. All modules are responsible for establishing communication with blockchain network, storing the IoT data samples, analysing and providing graphic dashboards to the end-user.

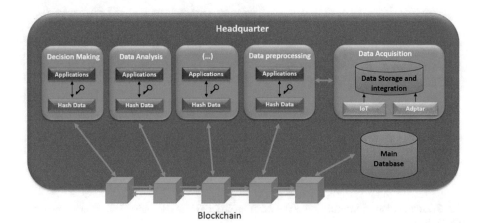

Fig. 3. General architecture

The back-end block (Fig. 3) is composed of a set of components ("Decision Making"; "Data Analysis", "Data pre-processing", etc.), and each one is composed in its turn of several applications. The components are connected to a blockchain network including the main database.

In a simple scenario, the components representing "Data Pre-processing" and "Data Acquisition" and their relation to the blockchain network can be described as follows: (i) the "Data Acquisition" module sends the data to the module "Data Pre-processing" through an IoT device; (ii) The data is processed in the "Data Pre-processing" module responsible for creating the final Hash Data (HD) containing the key and block required for storage on the blockchain shared network situated in each of the production units and in the headquarters; (iii) The generated block contains HD and sends to the blockchain network.

One component, which is responsible for saving all HD and transactions in the Main Database, first checks all blocks in the blockchain network. The other components then use similar mechanisms and processes to send data to the system. The system is largely constituted by the blockchain network, which functions as a distributed network for the system.

In the blockchain network, each node after winning the consensus competition creates a new block and assigns all related HD in the block based on timestamps, indicating the block creation time. The developed blocks have two parts (header containing meta information and a body part containing all verified data that are stored in the form of a hash); this is accomplished using a specific algorithm, for example the SHA256 [31]. After the block has been generated, it is transmitted to the blockchain network, where all nodes continue to search. Each node can check a block in the blockchain network according to the defined specification (key); if the key is invalid, it is sent to other neighbouring nodes for data sharing. This means that the blocks are checked by all nodes. The nodes trust each other, which creates an advantage by promoting all decentralized nodes reaching consensus on the validity of the data. In the latter part, several pieces of information are packaged, which serve as an important determent for the block in the blockchain network.

The front-end block is composed of multiple dashboards, integrated into a collaborative system, which allows the end-user to consult different graphics and apply filters that represent several relevant dimensions, such as: the year, room and week of production. Additionally, zoom/drill down features are also provided, as well as the ability to export data in table format or graphics for external use. Finally, when user logs in, a set of relevant warnings are displayed, for example, temperature warnings or alerts of lack of communication of devices.

5 Conclusions

When perceiving the relevance of the agri-food industry and the inherent need to achieve greater production efficiency and overall control, the research team underwent a thorough analysis on what technologies might fit the identified industry needs and allow for the development of a sustainable and focused solution.

In order to respond to the enumerated challenges, we propose in this manuscript a conceptual approach for an extension to a mushroom farm distributed process control system with IoT and blockchain integration. This conceptual approach, still in a design stage, allows the collection, storage and processing of data to be used in a distributed and scalable, immutable, transparent, auditable and essentially secure manner. However, as these technologies are at an early stage, there are still many research challenges. This proposed approach not only allows to collect distributed data on the environmental indicators inherent to the mushrooms production, but also complements the already existent production control system, which is extremely important for the overall success of the farm management information system used by the group managers.

Future work will specify an information systems architecture for the mushroom production sector that contemplates the vision of the current proposed model, and its emerging advantages, with the critical information subsystems for this sector of activity, i.e., the production control systems, collaborative and document management, enterprise resource planning/customer relationship management, quality management, integration of services with customers and suppliers, maintenance management, legal and social responsibility and, not least, an integrated system of decision support.

References

1. Estanislau Ferreira, J., Costa Pinto, F.G., dos Santos, S.C.: Estudo de Mapeamento Sistemático sobre as Tendências e Desafios do Blockchain. Revista Eletrônica de Gestão Organizacional **15**, 108–117 (2017)
2. Shrier, D., Wu, W., Pentland, A.: Blockchain & infrastructure (identity, data security). Massachusetts Institute of Technology-Connection Science, p. 1 (2016)
3. Moreira, F., Oliveira, M., Gonçalves, R., Costa, C.: Transformação Digital: Oportunidades e ameaças para uma competitividade mais inteli-gente. Sílabas & Desafios, Lisboa (2017)
4. Da Xu, L., He, W., Li, S.: Internet of things in industries: a survey. IEEE Trans. Ind. Inform. **10**, 2233–2243 (2014)
5. Vermesan, O., Friess, P.: Building the Hyperconnected Society–IoT Research and Innovation Value Chains, Ecosystems and Markets. ISBN: 978-87-93237-99-5. River Publishers, Gistrup (2015)
6. EIP-Agri Focus Group: Mainstreaming Precision Farming (2015)
7. Al-Fuqaha, A., Guizani, M., Mohammadi, M., Aledhari, M., Ayyash, M.: Internet of things: a survey on enabling technologies, protocols, and applications. IEEE Commun. Surv. Tutorials **17**, 2347–2376 (2015)
8. Swan, M.: Blockchain thinking: the brain as a dac (decentralized autonomous organization). In: Texas Bitcoin Conference, pp. 27–29, Chicago (2015)
9. Díaz, M., Martín, C., Rubio, B.: State-of-the-art, challenges, and open issues in the integration of Internet of things and cloud computing. J. Netw. Comput. Appl. **67**, 99–117 (2016)
10. Branco, F., Gonçalves, R., Martins, J., Cota, M.P.: Decision Support System for the Agrifood Sector–The Sousacamp Group Case. New Contributions in Information Systems and Technologies, pp. 553–563. Springer (2015)

11. Branco, F.A.D.S.: Uma proposta de arquitetura de sistema de informação para as empresas agroalimentares do setor de produção de cogumelos: o caso Grupo Sousacamp (2014)
12. Đurić, B.O.: Organisational metamodel for large-scale multi-agent systems: first steps towards modelling organisation dynamics (2017)
13. Rivera, J., van der Meulen, R.: Forecast alert: internet of things—endpoints and associated services, worldwide. Gartner (2016)
14. Antonopoulos, A.M.: Mastering Bitcoin: Unlocking Digital Cryptocurrencies. O'Reilly Media, Inc. (2014)
15. Iansiti, M., Lakhani, K.R.: The truth about blockchain. Harvard Bus. Rev. **95**, 118–127 (2017)
16. Mulla, D.J.: Twenty five years of remote sensing in precision agriculture: key advances and remaining knowledge gaps. Biosyst. Eng. **114**, 358–371 (2013)
17. Zhang, C., Kovacs, J.M.: The application of small unmanned aerial systems for precision agriculture: a review. Precision Agric. **13**, 693–712 (2012)
18. Branco, F., Martins, J., Gonçalves, R.: Das Tecnologias e Sistemas de Informação à Proposta Tecnológica de um Sistema de Informação Para a Agroindústria: O Grupo Sousacamp. RISTI-Revista Ibérica de Sistemas e Tecnologias de Informação, pp. 18–32 (2016)
19. Crosby, M., Pattanayak, P., Verma, S., Kalyanaraman, V.: Blockchain technology: beyond bitcoin. Appl. Innov. **2**, 6–10 (2016)
20. Lopes, I.M., Sá-Soares, F.D.: Information security policies: a content analysis. In: PACIS-The Pacific Asia Conference on Information Systems (2012)
21. Lopes, I.M., Oliveira, P.: Evaluation of the adoption of an information systems security policy. In: 2015 10th Iberian Conference on Information Systems and Technologies (CISTI), pp. 1–6. IEEE (2015)
22. Tian, F.: An agri-food supply chain traceability system for China based on RFID & blockchain technology. In: 2016 13th International Conference on Service Systems and Service Management (ICSSSM), pp. 1–6. IEEE (2016)
23. Nofer, M., Gomber, P., Hinz, O., Schiereck, D.: Blockchain. Bus. Inf. Syst. Eng. **59**(3), 183–187 (2017). https://doi.org/10.1007/s12599-017-0467-3
24. Malviya, H.: How Blockchain will Defend IOT. https://ssrn.com/abstract=2883711. Accessed 10 Oct 2018
25. Veena, P., Panikkar, S., Nair, S., Brody, P.: Empowering the edge-practical insights on a decentralized internet of things. IBM Institute for Business Value 17 (2015)
26. Gan, S.: An IoT Simulator in NS3 and a Key-Based Authentication Architecture for IoT Devices using Blockchain. Indian Institute of Technology Kanpur (2017)
27. Block Chain of Things. https://www.blockchainofthings.com/. Accessed 10 Oct 2018
28. Modum. https://modum.io/. Accessed 10 Oct 2018
29. Prisco, G.: Slock. it to Introduce Smart Locks Linked to Smart Ethereum Contracts, Decentralize the Sharing Economy. Bitcoin Magazine, November 2015. https://bitcoinmagazine.com/articles/sloc-it-to-introduce-smart-locs-lined-to-smart-ethereum-contractsdecentralize-the-sharing-economy-1446746719. Accessed 20 May 2016
30. LO3ENERGY. https://lo3energy.com/. Accessed 20 Oct 2018
31. Courtois, N.T., Grajek, M., Naik, R.: Optimizing sha256 in bitcoin mining. In: International Conference on Cryptography and Security Systems, pp. 131–144. Springer (2014)

Improving Automatic BPMN Layouting by Experimentally Evaluating User Preferences

Tobias Scholz and Daniel Lübke[✉]

FG Software Engineering, Leibniz University Hanover,
Welfengarten 1, 30167 Hannover, Germany
daniel.luebke@inf.uni-hannover.de

Abstract. BPMN Process Models are the basis for implementing and optimizing business processes in the era of digitization. Good layouts of BPMN diagrams help readers to better understand them. In order to improve the quality of automatic layouting, we need to identify subjective and objective characteristics of diagram layouts. While research into objective criteria has been ongoing, we want to evaluate subjective layout preferences by model readers. Therefore, we conducted an experiment, which let participants select the preferred layout out of two layout variants. The experiment yields significant findings for subjective layout preferences, e.g., with regards to spacing of elements, layout options after gateways, and the routing of message flows. The findings help modelers to produce better process models and we incorporated them in an automatic layout tool.

Keywords: BPMN · Automatic layout · Experiment · Element size · Element placement · Edge routing · Boundary event placement

1 Motivation

Business processes are continuously evolving. Especially in digitization projects, processes are analyzed, optimized and implemented in software.in order to improve their efficiency. Processes have to be presented in a unified and consistent manner to allow for easier analysis and optimization. BPMN was created to tackle these problems and allows the modeling and presentation of business processes with a standardized notation. Syntactical rules are defined to guarantee the correct usage of the notation elements; each for itself and within the context of the model. The layout of the associated diagram, however, has to be created manually by the modeler. Rules considering the position and order of the elements have been defined to a certain extent but are mainly focused on the semantics of the corresponding elements and the definitions of the BPMN specification [11].

Layout principles helping to create BPMN diagrams, which are easily comprehensible and visually pleasing, have not been researched as much. However,

© Springer Nature Switzerland AG 2019
Á. Rocha et al. (Eds.): WorldCIST'19 2019, AISC 930, pp. 748–757, 2019.
https://doi.org/10.1007/978-3-030-16181-1_70

it has been examined that the layout can improve the intelligibility of a diagram [9]. Specified layout principles are thus necessary to develop an automatic layout algorithm. Part of our research project was to find out which layout principles are preferred by model readers. The resulting experiment, its results and their interpretation are described in this paper.

This paper is structured as follows: Related work is discussed in the next section. Section 3.1 presents the experiment design. The results are presented in Sect. 4 and are interpreted in Sect. 5. Section 6 presents the take-aways for automatic BPMN layouting before we finally conclude and give ideas for future research.

2   Related Work

Kitzman et al. developed a grid-based layout algorithm for BPMN models [5]. The generator specifies the ordering of BPMN elements relative to each other (e.g., next to, below, ...). The grid needs to be transformed into absolute coordinates. The describing paper does not give the values used by the generator nor does it justifies any absolute positions empirically.

Specific layout criteria for BPMN diagrams have already been defined in the works of Effinger et al. [4] which were mostly covered by Kitzmann et al.'s algorithm. Criteria like keeping the workflow direction or minimizing overlapping elements, for example, were already fulfilled through the grid-based approach. Furthermore, Effinger et al. developed an algorithm, which globally optimizes the placement of BPMN elements. However, elements are placed without a strict relation to each other. Therefore, the example diagrams do not have an as clear layout direction as the diagrams layouted by the Kitzmann et al.'s algorithm.

Most of the aforementioned layout criteria were researched in-depth by Purchase et al. and refined into specific metrics to measure the presence of those criteria in graph drawing algorithms [8]. For example, this includes the minimization of edge bends. Furthermore, it has been examined which of those criteria are actually preferred for domain-specific graphs like UML diagrams [7]. Siebenhaller and Kaufmann used those criteria to develop a new layout approach for activity diagrams [10]. Their work is based on a combination of the TSM approach of Tamassian et al. [14] and the Sugiyama approach [13].

Störrle conducted several experiments in his series "On the Impact of Layout Quality to Understanding UML Diagrams" [12]. Most notably he found that diagrams are profiting more of a good layout in their early lifecycles. He also hypothesizes that layout is benefiting more from the simultaneous use of many different principles; every single one in isolation, however, has not as much impact on the end result.

3   Experimental Design

3.1   Goals, Hypothesis and Variables

Nick and Tautz [6] demonstrated that the extended use of the Goal-Question-Metric (GQM) approach [1] in guiding research can be very helpful. We follow it and started by defining our goal as

> *Understand* the *User Preferences*
> with regard to the *Layout of BPMN 2.0 Diagrams*
> from the viewpoint of a *Model Reader.*

From this goal we derived our research questions numbered RQ1 to RQ7:

RQ1: Which horizontal spacing, i.e., the distance between two BPMN elements, is preferred? Automatic layout tools need to place elements not only relatively to each other but also calculate absolute positions. Therefore, we need to evaluate which horizontal spacing between elements is preferred. We created three diagrams with small, medium and large spacings, which correspond to 25%, 50% and 70% of an activity's width.

RQ2: Which vertical spacing, i.e., the distance between two BPMN elements, is preferred? Similarly, we evaluate the vertical spacing with three options small, medium, and large, which correspond to 12.5%, 25%, and 50% of an activity's height.

RQ3: What relative size for subprocess contents compared to main process elements is preferred? The activities within a subprocess can have the same size and spacing as the rest of the diagram. In order to highlight the hierarchical structure, activites could be made smaller. The freed space can be used to make the diagram more compact. Therefore, want to evaluate preferences for these three diagram options (same sizes and spacings as rest of process, smaller-sized activities, and totally smaller-sized subprocesses).

RQ4: What placements of BPMN elements after gateways are preferred? A BPMN gateway represents a decision point. Multiple sequence flows start from the gateway and form different branches in the diagram. The branches can be symmetrical, i.e., one branch is higher than the gateway and the other is lower, or one branch could extend on the level of the gateway while the other branch is offset above or below. In addition, branches can contain the same number of activities or have different sizes. We want to evaluate preferences of branch placement after gateways for these cases.

RQ5: What placements of pools are preferred? BPMN diagrams can use collapsed and uncollapsed pools. Collapsed pools hide the control flow with the modeled party. Collapsed pools usually exchange messages represented with message flows with other pools. We want to evaluate whether there are preferences on placing collapsed pools above all other pools or whether they can be distributed on top and below an uncollapsed pool.

RQ6: What routing strategy of message flows is preferred? Message flows represent messages between different pools. Because the process in the pools is usually differently complex, message flows require more sophisticated routing, i.e., they need to be bent or the process models need ot be changed in order to make space for them. We want to evaluate preferences for different options on routing message flows.

RQ7: What placements of boundary events are preferred by model readers? BPMN defines boundary events, which are placed on the edge of an activity. The standard allows to attach events at all places on the border. We want to evaluate, whether there is a preference for the bottom border, which is often used in examples and books (e.g., [11]) or whether the placement can occur on the top and bottom, especially when many boundary events are attached.

The metrics for all questions is the subjective preference between diagrams exposing the different layout options as shown in Fig. 1.

3.2   Objects and Participants

The experiment was conducted as an online test. Participants could test themselves for their "BPMN Modeler Type" by choosing the diagrams, which they liked best. For this they had to choose the best out of two model layouts: Diagrams were syntactically the same but differed in one aesthetic characteristic. All diagrams are shown in Fig. 1. The concrete pairs of diagrams the participants had to choose between are listed in Table 1. After a participant had answered all questions a summary of his/her layout preferences was shown, i.e., the BPMN Modeler Type was described.

We advertised the BPMN quiz on various social media platforms (Facebook & Xing) and at the university's student forums. Furthermore, we sent out invitations to personal contacts and asked for snowballing the invitation to colleagues.

The quiz is programmed to log every user choice to one CSV file per user. Each choice consists of a unique session ID, the question to be answered, and the user's answer. This means that a trace of user decisions is available for statistical analysis. Because the data is collected in parallel to the quiz no further instrumentation is required.

3.3   Analysis Procedure

The data will be aggregated and the number for each of the two answers for every question will be counted. This means that $x_q <= n$ out of n participants will have opted for answer x for a given question q and in turn $y_q = n - x_q$ participants will have opted for answer y.

If we code all participants, who chose answer x, as 0 and the others as 1 and sum up their answers, we will get a mean value of $m_q = y_q/n$ for a given question q.

If the difference in layout is insignificant, i.e, the users randomly choose one of the answers, the null hypothesis is $x_q = y_q$. In this case $m_q = 0.5$ holds. The

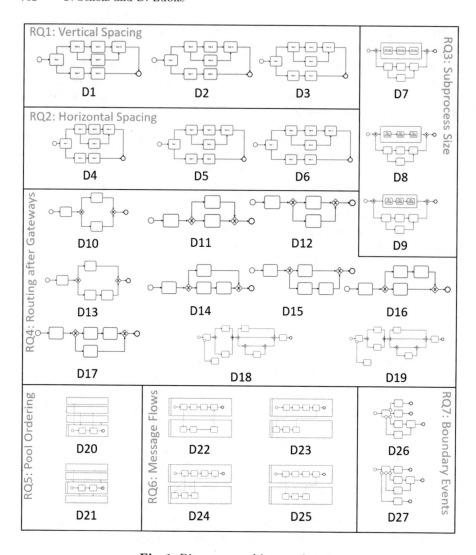

Fig. 1. Diagrams used in experiment

expected standard deviation for the null hypothesis is 0.5. We will use a z-test with a 5% alpha-level to test whether the actual measurements for each question differ significantly from this null hypothesis. The effect size is calculated using Cohen's d.

3.4 Evaluation of Validity

In order to eliminate learning effects the order of all questions was randomized. In addition, to avoid bias due to the order of the presented options, the order of the two answers was randomized for every question as well.

However, our control over both the experiment participants is limited. This applies to both the sampling of the participants as well as the way how seriously they conduct the quiz. The sample might be not representative of BPMN modelers and participants might just click through the quiz out of curiosity.

4 Analysis

121 people participated in our experiment. We excluded the data for (a) duplicate IP addresses (which seemed to come from bots) and (b) incomplete data sets. This left us with 70 records.

Table 1. Statistics summary for experiment questions

	Option A	Option B	Count A	Count B	p-value	Eff/d
Q1	D1	D2	35	35	1.00000	0.00
Q2	D1	D3	25	45	0.01683 (*)	0.29 (+)
Q3	D2	D3	31	39	0.33898	0.11
Q4	D4	D5	21	49	0.00082 (***)	0.40 (+)
Q5	D4	D6	28	42	0.09426	0.20
Q6	D5	D6	37	33	0.63259	0.06
Q7	D7	D8	48	22	0.00189 (**)	0.37 (+)
Q8	D7	D9	37	33	0.63259	0.06
Q9	D8	D9	31	39	0.33898	0.11
Q10	D10	D11	51	19	0.00013 (***)	0.46 (+)
Q11	D10	D12	46	24	0.00855 (**)	0.31 (+)
Q12	D11	D12	31	39	0.33898	0.11
Q13	D13	D14	37	33	0.63259	0.06
Q14	D13	D15	27	43	0.05583	0.23 (+)
Q15	D14	D16	46	24	0.00855 (**)	0.31 (+)
Q16	D15	D17	52	18	0.00005 (***)	0.49 (+)
Q17	D18	D19	19	51	0.00013 (***)	0.46 (+)
Q18	D20	D21	40	30	0.23200	0.14
Q19	D22	D23	57	13	0.00000 (***)	0.63 (++)
Q20	D23	D24	45	25	0.01683 (*)	0.29 (+)
Q21	D22	D25	50	20	0.00034 (***)	0.43 (+)
Q22	D23	D25	17	53	0.00002 (***)	0.51 (++)
Q23	D26	D27	27	43	0.05583	0.23 (+)

We used the z-test for statistically determining whether we could reject the null hypothesis that no differences in user preferences exist for each pair of diagrams. All gathered counts, as well as p-values and the effect sizes are shown in Table 1. p-values below 0.05 are deemed a significant finding (*). p-values below 0.01 are marked with (**), and below 0.001 are marked with (***). Effects above 0.8 are deemed a large effect size (+++), above 0.5 a medium effect size (++), and above 0.2 a small effect size (+).

In total we found 2 *-significant differences (Q2 & Q20) below 0.05, 3 **-significant differences (Q7, Q11, Q15), and 7 ***-differences (Q4, Q10, Q16+Q17, Q19, Q21+Q22). For all these significant differences, 2 had a medium effect size (Q19, Q22) and the rest had a small effect size. Two differences had a small effect size (Q14 & Q23) but gained only a nearly significant p-value.

In addition to our measurements one participant emailed us feedback and told us that his/her preference for positioning of elements after a split is context-dependent: if the probability of the branches is the same, he/she prefers the symmetric split. Otherwise, the most probable branch should extend on the same level as the gateway.

5   Interpretation

5.1   Evaluation of Results and Implications

Overall the number of usable data points (n=70) is relatively small. However, due to the experiment design, many of the significant findings have a very low p-value. Overall, we were positively surprised by the high number of significant findings of subjective layout preferences. The findings are explained in more detail in the following.

RQ1: Which vertical spacing, i.e., the distance between two BPMN elements is preferred? Considering the results for diagrams with varied vertical spacings, not all cross comparisons yield significant results. However, all in all, large vertical spacings are overall preferred by readers. They are significantly preferred over small spacings (Q2) and are non-significantly preferred over medium spacings (Q3).

RQ2: Which horizontal spacing, i.e., the distance between two BPMN elements is preferred? As with vertical spacings not all cross comparisons yield significant results, medium horizontal spacings are overall preferred by readers. They are significantly preferred over small spacings (Q4) and are non-significantly preferred over diagrams with large spacings (Q6).

RQ3: What relative size for subprocess contents compared to main process elements is preferred? Experiment participants favored that the subprocess has the same layout (size and spacings of activities) as the embedding process. Significant differences were found compared to adjusting the activity sizes (Q7), although the preference compared to subprocesses that are smaller altogether is only minimal (Q8).

RQ4: What placements of BPMN elements after gateways are preferred? The preference of element placements after a gateway seems to be dependent on the characteristics of the branches: In questions Q10 to Q12 only one activity was in each branch. In this case the symmetrical split, i.e., one branch continues above the gateway and one below the gateway, is significantly preferred over the other variants. However, when one branch consists of two activities and the other one only of one activity, preferences change: Depending on whether the second branch is above or below the gateway, it is preferred that the longer branch is on the level of the gateway (Q15) or below the gateway (Q16). Comparing symmetric splits to non-symmetric splits yielded no significant results (Q13/14). Space after gateways should be minimized and not tried to be made consistent with other gateways (Q17).

RQ5: What placements of pools are preferred? No significant difference was found between the two offered layout variants by placing all collapsed pools on top or distributed on top and on the bottom (Q18). Therefore, no subjective preference can be evaluated.

RQ6: What routing of message flows is preferred? All questions regarding preferences for routing the message flows yielded significant results. In general participants preferred less bends in the message flows. The variants in which activities in all pools were distributed so that message flows ran straight were preferred (Q19 & Q21). If not given this option, it is preferred that the left-most message flow is not bent (Q20 & Q22). All in all, existing literature, which says that edge bends are to be avoided, was fully confirmed.

RQ8: What placements of boundary events are preferred by model readers? The result of question Q23 is nearly significant ($p = 0.056 > 0.05$) in favor of using only the bottom edge of an activity to place boundary events but it has a small effect size ($d = 0.23$). Further research is required in order to judge whether this layout is significantly better or not, although our results indicate that this is probable. For choosing an option for automatic layouts, it is therefore currently sensible to place boundary events only at the bottom.

5.2   Limitations of Study

This experiment exclusively focused on the layout of BPMN diagrams. This also means that no business context was given for the BPMN models and users simply voted for the best aesthetics but not necessarily for the most understandable model. However, "[i]n general, researchers associate aesthetics with readability, and readability with understanding" [2].

Because no business context is present, some results may be refined. One angle is the suggestion by one practitioner for layouting gateways differently based on the probability of branches. Context could also influence placing of pools, e.g., importance or time of involvement in process orchestration might be a context factor for layouting different pools.

All diagrams were layouted left to right, i.e., horizontally. Some results – especially vertical and horizontal gaps – are therefore probably not generalizable to vertical layouts.

6   Improvements for BPMN Layout Algorithms

Our goal was to improve the current state of automatic BPMN layouting. As such, we wanted to incorporate our findings into a generator tool. As the basis we chose the algorithm of Kitzmann et al. [5] because it generates diagrams that look more like human-modeled BPMN. The algorithm by Effinger [3] tries to excel on objective metrics but does not necessarily place consequitive activities in a row in order to avoid edge crossings.

Kitzmann et al.'s BPMN layout algorithm follows a local strategy: Activities are first sorted and then placed activity by activity into a grid. The grid is expanded as required during this process. Finally, the grid is optimized by removing cells, which are not required.

The grid and the containing BPMN elements are then written back as a valid BPMN model following the OMG's standardized BPMN XML format. This grid-structure allows easy implementation of the findings: When calculating absolute positions from the relative grid, the horizontal and verticals spacings can be easily considered.

We also added support for the layout of boundary events. Although there was only a near-significant finding for the placements along the bottom of the activity, this was the most preferred option. Consequently, the layout tool is implemented to support the layout boundary events at the bottom edge of an activity. The Java implementation of the BPMN layouter is freely available on GitHub[1].

7   Conclusions and Outlook

Within this paper we described an experiment, which aims for evaluating subjective layout diagram choices. The preferred constructs have been implemented in an automatic BPMN layout tool, which is freely available. However, the findings can also guide modelers to create better process diagrams, who manually layout their models.

The findings were that (a) medium horizontal spacing and large vertical spacing were preferred by the participants, (b) sub-processes should be layouted with the same sizes and spacings as the rest of the process, (c) splits after gateways should be symmetric if the number of activities are the same in both branches and should otherwise contain the branch with the most activities in the same level as the gateway, and (d) message flows should have no bends even if this means to add additional spacings between the activities in the pools.

However, while we could find many significant preferences, more research questions emerged: Especially the influence of the semantics of the modeled process could have an influence on the layout being used. Also we found a nearly significant finding for placing boundary events, which should be explored further in future research. Our experiment did not dive deeply enough into the placement and order of pools. Positioning those is a further research possiblity worth

---

[1] https://github.com/t0b1z/BPMNLayouter.

pursuing. Furthermore, our experiment used small, simple diagrams in order to control for layout choices better. However, the experiment could be extended to cover larger diagrams and see what preference patterns emerge.

We encourage everyone to use our BPMN layouter tool and explore more improvements for automatic layouts of BPMN processes!

References

1. Basili, V.R.: Applying the goal/question/metric paradigm in the experience factory. In: Software Quality Assurance and Measurement: A Worldwide Perspective, pp. 21 44 (1993)
2. Bennett, C., Ryall, J., Spalteholz, L., Gooch, A.: The aesthetics of graph visualization. In: Proceedings of Computational Aesthetics in Graphics, Visualization, and Imaging, pp. 57–64 (2007)
3. Effinger, P.: Business process model and notation: third international workshop, BPMN 2011, Lucerne, Switzerland. In: Proceedings, chap. Layout Patterns with BPMN Semantics, pp. 130–135. Springer, Heidelberg, 21–22 November 2011. https://doi.org/10.1007/978-3-642-25160-3_11
4. Effinger, P., Jogsch, N., Seiz, S.: On a study of layout aesthetics for business process models using BPMN. In: International Workshop on Business Process Modeling Notation, pp. 31–45. Springer (2010)
5. Kitzmann, I., König, C., Lübke, D., Singer, L.: A simple algorithm for automatic layout of BPMN processes. In: 2009 IEEE Conference on Commerce and Enterprise Computing, pp. 391–398. IEEE (2009)
6. Nick, M., Tautz, C.: Practical evaluation of an organizational memory using the goal-question-metric technique. In: German Conference on Knowledge-Based Systems, pp. 138–147. Springer (1999)
7. Purchase, H.C., Allder, J., Carrington, D.: Graph layout aesthetics in UML diagrams: user preferences. J. Graph Algorithms Appl. **6**(3), 255–279 (2002)
8. Purchase, H.C., Carrington, D., Allder, J.A.: Experimenting with aesthetics-based graph layout. In: Anderson, M., Cheng, P., Haarslev, V. (eds.) Theory and Application of Diagrams: First International Conference, Diagrams 2000 Edinburgh, Scotland, UK, 1–3 September 2000. Proceedings, pp. 498–501. Springer, Heidelberg (2000). https://doi.org/10.1007/3-540-44590-0_46
9. Sharif, B., Maletic, J.I.: An eye tracking study on the effects of layout in understanding the role of design patterns. In: 2010 IEEE International Conference on Software Maintenance (ICSM), pp. 1–10. IEEE (2010)
10. Siebenhaller, M., Kaufmann, M.: Drawing activity diagrams. In: Proceedings of the 2006 ACM Symposium on Software Visualization, pp. 159–160. ACM (2006)
11. Silver, B., Richard, B.: BPMN Method and Style, vol. 2. Cody-Cassidy Press, Aptos (2009)
12. Störrle, H.: Model-driven engineering languages and systems. In: 17th International Conference, MODELS 2014, Valencia, Spain, 28 September – 3 October 2014. Proceedings, chap. On the Impact of Layout Quality to Understanding UML Diagrams: Size Matters, pp. 518–534. Springer, Cham (2014). https://doi.org/10.1007/978-3-319-11653-2_32
13. Sugiyama, K., Tagawa, S., Toda, M.: Methods for visual understanding of hierarchical system structures. IEEE Trans. Syst. Man Cybern. **11**(2), 109–125 (1981)
14. Tamassia, R., Di Battista, G., Batini, C.: Automatic graph drawing and readability of diagrams. IEEE Trans. Syst. Man Cybern. **18**(1), 61–79 (1988)

Drone Based DSM Reconstruction for Flood Simulations in Small Areas: A Pilot Study

P. Rinaldi[1,3], I. Larrabide[1,2], and J. P. D'Amato[1,2(✉)]

[1] Univ. Nac. Del Centro, UNICEN, Tandil, Argentina
`juan.damato@gmail.com`
[2] National Scientific and Technical Research Council,
CONICET, Buenos Aires, Argentina
[3] Buenos Aires Province Research Council, CICPBA, Buenos Aires, Argentina

Abstract. Satellite based Digital Elevation Models of free availability have been the standard for 2D flood simulation models of large areas. But for small urban areas where high definition and up to date details are needed a different approach must be taken. Digital Surface Models generated with photogrammetry software based on Unmanned Aerial Vehicles imagery is a low cost and high quality solution applicable to this problem. Combined with distributed flood simulation software can give highly detailed results of rain events in urbanized landscapes. In this study, a complete processing workflow is presented to analyze the response to different rain conditions in a complex small area. All the phases are presented, including topographic survey, DEM correction and flood modeling. The solution was applied to a complex environment of semi-urbanized area reproducing even little waterlogged ground depressions.

Keywords: DSM · UAV · Flood simulation

1 Introduction

Satellite build based Digital Elevation Models (DEM) are the most widely applied data input for 2D flood simulation models in medium and large areas. DEM or DSM (Digital Surface Model) production techniques generally imply a trade-off between cost, precision, spatial cover and resolution [1]. Most of the free available DEMs, like SRTM or ASTER provide good precision, almost worldwide cover and cell size in the order of 10 meters. But this resolution makes them impractical for small areas, due to the impossibility to represent urban zones with adequate level of detail [2], where a simple wall or channel can change the water course in a flood event. Furthermore, these DEMs are based on satellite missions flown over a decade ago, not showing the latest structure of buildings and constructions.

Today, the use of Unmanned Aerial Vehicles (UAV) and digital photogrammetry has evolved drastically as a versatile and low cost technology for small to medium scale terrain mapping technology [3, 4]. UAV Photogrammetry software is composed by essentially two software tools, pre-planned mission control software like MAP, Drone Deploy [5], or the Pix4D Capture [6] used in this work and post-acquisition image processing software like Open Drone Map [6].

© Springer Nature Switzerland AG 2019
Á. Rocha et al. (Eds.): WorldCIST'19 2019, AISC 930, pp. 758–764, 2019.
https://doi.org/10.1007/978-3-030-16181-1_71

UAV generated DEMs can reach centimeter level resolutions easily [2, 5] but they require corrections before been used for flood simulation. There are still substantial limitations to the efficient 3D reconstruction of areas with dense vegetation and the inability to properly reconstruct the submerged zone of channels [7].

In large plains like Buenos Aires Province in Argentina, flooding is a critical problem. For large productive lands and rural areas, very accurate simulations can be achieved using satellite DEMs [8, 9], but a different approach is needed for urban zones. Even more, as the cities keep growing in a poorly planned manner an efficient tool for analyzing this phenomenon and prevent mayor losses is needed.

This project, intends to work towards a complete software platform to deal with such problem that integrates both DSM generation and flood simulation. The flood simulation is based on an original development named Aqua. Aqua automaton [10] is a hydrological model based in the cellular automata paradigm that simulates the natural surface water flow by tracking local water stocks, book-keeping precipitation, infiltration and intercellular flows. Aqua can easily detect waterlogged areas and natural courses starting from a rain event hyetograph and a highly detailed DSM. The model was originally tested in rural areas of Buenos Aires Province in Argentina [11] using a DEM produced by means of radar interferometry from an ERS Tandem mission in 1997 satellite images. Recently Aqua was applied for urban flood simulations over a 90 m SRTM DEM [12].

In this paper, first results obtained by the entire process from UAV flying to water runoff during an intense rain event is presented giving special details on surface DSM generation. The proposed solution is based on open-source GIS and integrated to a parallel implementation of Aqua automata.

2 Materials and Methods

2.1 DSM Generation

The University Campus was chosen as the study area initially for the proximity, which made it easy to fly the UAV, plus allowing a local (walking) survey of different terrain characteristics (waterlogging, channels, obstructions, etc.). This region also presents a nice blend of urban and rural landscape with buildings, roads, cars as well as variety of vegetation with high trees, bushes and grass, making it a real challenge to test and setup the proposed methodology. The campus is located on top of a hill making it easier to set up boundary conditions, since it does not receive external contributions from neighboring regions during intense rains.

The drone used for this project is a DJI Phantom 4 Advance quadcopter (DJI, Shenzhen, China). It can fly up to 30 min in light wind conditions. It has a 20-megapixel CMOS sensor digital camera stabilized with a three axis gimbal and a GPS to automatically geo-tagging images during acquisition.

Pix4DCapture (Pix4D S.A., Lausanne, Switzerland) was used for the flying plan, it is a free drone flight planning app for optimal mapping and 3D modeling data running over an Android platform [6]. The app allows programming a flight plan offline, taking into account the specific drone limitations as maximum flight time. The plan is finally loaded on the UAV, which is afterwards flown autonomously.

Flying parameters were set to a Grid mission, at 100 m altitude with an overlap of 70% between images at max speed with vertical camera alignment. Total flight time was 23 min and 51 s taking into account the return to the starting point and resulting in 584 photographs in total.

OpenDroneMap is an Open Source toolkit for processing aerial imagery [13], this is a collaborative project in the early stage of development. The software accepts aerial georeferenced images and ground control points, delivering a dense point cloud, orthophotos and digital surface models as shown in Fig. 1. All processes are computed using central processing unit (CPU), graphical processing unit (GPU) processing is not supported yet.

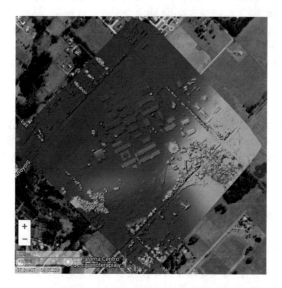

Fig. 1. Campus DSM layer over satellite google images on WebODM.

The original DJI photos with a resolution of 5472 × 3648 pixels are resized to 3072 × 2048 due to ensure the solution can be satisfactory computed in reasonable time with the available hardware, which is limited in memory (16 Gb) and CPU capacity (Intel i7). Parameter *min_num_features* significantly improves the quality of the reconstruction at the expense of longer computational times. Finding the optimal parameters values required trial and error, mainly because the software is running on a desktop computer which can easily result in filling-up system memory if too many features or too big images are used. Final values different from default were: *min_num_features* = 12.000,   *rezise_to* = 2048,   *dem_resolution* = 10 cm,   *dsm* = enable.

2.2 DEM Correction

The Digital Surface Model obtained from ODM is a 2D matrix with the relative mean height of each cell (in this case a 60 × 60 cm cell) showing the actual height of grass, roads, buildings and trees. Still, for an accurate flood runoff simulation proper underground channel simulation, like presented in Fig. 2, is required.

Fig. 2. Underground channels in campus.

Starting from a vector layer of the channels given by campus maintenance office a *streamburning* process was run [14]. A binary raster layer with the DSM dimensions is created using QGis to identify channels cells. Then, the algorithm applies two different corrections to DSM's "channel" cells: obstructions elimination, and height decrease. The obstruction elimination traverses every channel correcting any height increase between neighbor cells downward the channel. Then, all channel cells are reduced by a given value, usually the estimated channel mean depth.

The method is implemented as part of the flood simulation software, the algorithm was developed to force channels in flat areas where DSM resolution was not enough to resolve a channel, but can still be used to identify buildings or other structures, cutting a line in the building where the channel passes underneath as shown in Fig. 3.

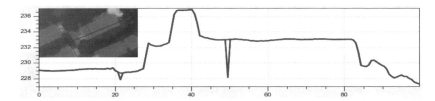

Fig. 3. Detail of some buildings with burned channels and the corresponding profile. The blue line corresponds to the burned DSM profile and the red line to the original one.

2.3 Flood Simulation Model

The AQUA automaton [10] is a flood simulation model based on simple empirical rules following a Cellular Automata (CA) paradigm over a highly detailed DSM. In AQUA, the surface state of each cell is determined by a scalar representing the water level. The model works by creating floodgates connecting cells groups that are opened and closed in turns, allowing the water to flow driven by the elevation differences between cells. In addition, water mass sources and sinks can be associated to each cell, which represents infiltration, precipitation and the contributions of canals or rivers. The automaton was originally developed for extra-urban areas but it was successfully applied to urban flood modeling with minor adjustments [12].

Being AQUA an implicit algorithm running over a regular grid it falls into the embarrassingly parallel category, making it highly suitable for GPU platforms. For this work, the simulation model was implemented using NVidia CUDA 10 (NVidia, Santa Clara, CA, USA), using one parallel thread for each basic partition and 9 sequential kernel calls for each time step. Near two orders of magnitude of speed-up where reached over single-threaded CPU code using an NVidia TITAN XP graphics processor. The computation time needed to simulate a one day event over a 2 Million cell DSM is less than a minute. This GPU hardware has been supported by the NVidia Grant research program.

Relaxation parameter α on AQUA model should be calibrated against field measurements or numerical simulations with physically based models like HEC-RAS. As a first approach we estimated the parameters starting from the ones on the original publication given the similarities of the study zones slopes and vegetation. The main difference stands for the cell size change from 80 m in original work to 0.6 m in the present one making a variation of 1/133 in Δx.

So, a similar scale change must be made to Δt what gives the model an adequate approximation for precipitation and infiltration rates in relation to water surface movement. Infiltration parameters also were estimated from the original study, assuming high level of soil saturation due to previous rainfall events.

3   Results

A particular intense rain event from September 28–29 2018 was simulated for the first approach, the hyetograph was registered by the weather station installed on campus but no precise measurements were found of the rain runoff results. Being the working place of the authors qualitative comparisons could be done and pictures were taken from some of the campus waterlogged areas immediately after the event.

Figure 4 shows the water height raster layer generated by the AQUA model after rain simulation, overlapped to the campus orthomosaic. Some of the waterlogged areas are highlighted and compared to the event real pictures.

Very good agreement was found in the highlighted zones in Fig. 4 and other little ground depressions correctly reconstructed in the DSM. It can be seen how the simulation model correctly conducts the water runoff which turns some of the small depressions into important puddles while others dry out.

The simulation also registered some waterlogged areas that didn´t existed in reality. This was caused mainly for the tree representation in de DSM reconstruction. In the DSM trees are represented as a solid structure with the size of the crown projected towards the ground preventing underneath water flow.

Fig. 4. Pictures of real event in comparison with simulation.

4   Discussion

A complete workflow for flooding studies in small areas is presented, based on free or open source software solutions and low cost hardware achieving promising results. The detail level of the DSM and the correct simulation of water runoff open new horizons to rain management and architectural planning for this kind of areas. Little flooded areas are detected easily by the simulation model and also the naturally created channels. Although transitory runoff simulations seems to be correct, further measuring of real events will be required to accurately calibrate the flooding model. The use of GPU computing reduces greatly the computation time, allowing simulations of future events in minutes once precipitation forecast is known.

Future works should incorporate GCPs for global DSM accuracy and solution scaling and also the tree identification and removal is been studied.

Acknowledgments. The author would like to express their appreciation to NVidia Grant Research Program for the GPU Hardware used in this project.

References

1. Robinson, N., Regetz, J., Guralnick, R.P.: EarthEnv-DEM90: a nearly-global, void-free, multi-scale smoothed, 90 m digital elevation model from fused ASTER and SRTM data. ISPRS J. Photogram. Remote Sens. **87**, 57–67 (2014)
2. Feng, Q., Liu, J., Gong, J.: Urban flood mapping based on unmanned aerial vehicle remote sensing and random forest classifier-a case of yuyao. China. Water **7**, 1437–1455 (2015)
3. Nex, F., Remondino, F.: UAV for 3D mapping applications: a review. Appl. Geomatics **6**(1), 1–15 (2014)
4. Vasić, D., Ninkov, T., Bulatović, V., Sušić, Z., Marković, M.: Terrain mapping by applying unmanned aerial vehicle and lidar system for the purpose of designing in Serbia. In: FIG, Ingeo 2014, 6th International Conference on Engineering Surveying, pp. 217–222 (2014)
5. Sharan Kumar, N., Ismail, M.A.M., Sukor, N.S.A., Cheang, W.: Method for the visualization of landform by mapping using low altitude UAV application. In: IOPScience (eds.) IOP Conference Series: Materials Science and Engineering, vol. 352, conference 1, pp. 012–032. IOP Publishing (2018)
6. Burdziakowski, P.: Evaluation of open drone map toolkit for geodetic grade aerial drone mapping – case study. In: International Multidisciplinary Scientific GeoConference Surveying Geology and Mining Ecology Management, SGEMAt: Albena, Bulgaria, vol. 17(23), pp. 101–110 (2017)
7. Langhammer, J., Bernsteinová, J., Miřijovský, J.: Building a high-precision 2D hydrody-namic flood model using UAV photogrammetry and sensor network monitoring. Water **9**(11), 861 (2017)
8. Tosi, L., Kruse, E.E., Braga1, F., Carol, E.S., Carretero, S.C., Pousa, J.L., Rizzetto, F., Teatini, P.: Hydro-morphologic setting of the Samborombon Bay (Argentina) at the end of the 21st century. Nat. Hazards Earth Syst. Sci. **13**(3), 523–534 (2013)
9. Patro, S., Chatterjee, C., Mohanty, S., Singh, R., Raghuwanshi, N.S.: Flood inundation modeling using MIKE FLOOD and remote sensing data. J. Indian Soc. Remote Sens. **37**, 107–118 (2009)
10. Rinaldi, P.R., Dalponte, D.D., Venere, M.J., Clausse, A.: Cellular automata algorithm for simulation of surface flows in large plains. Simul. Model. Pract. Theor. **15**(3), 315–327 (2007)
11. Dalponte, D., Rinaldi, P., Cazenave, G., Usunoff, E., Vives, L., Varni, M., Venere, M., Clausse, A.: A validated fast algorithm for simulation of flooding events in plains. Hydrol. Process. **21**, 1115–1124 (2007)
12. La Macchia, M.L.: Modelización y análisis espacial del drenaje urbano de la ciudad de Tandil mediante TIG's. II Jornadas Nacionales de Ambiente "Seguimos Comprometidos". Tandil, Argentina. ISBN: 978-950-658-359-0 (2014)
13. Dakota, B., Fitzsimmons, S., Toffanin, P.: Open Drone Map. www.opendronemap.org. Accessed 21 Nov 2018
14. Lindsay, J.B.: The practice of DEM stream burning revisited. Earth Surf. Proc. Land. **41**(5), 658–668 (2016)

Web Application for Management of Scientific Conferences

João Bioco[1]([✉]) and Alvaro Rocha[2]

[1] Instituto de Telecomunicações, Dep. de Informática, Universidade da Beira Interior,
6200-001 Covilhã, Portugal
`joao.bioco@ubi.pt`
[2] Departamento de Engenharia Informática, Universidade de Coimbra,
3030-290 Coimbra, Portugal

Abstract. A scientific conference is an event with a great importance within scientific community, fundamentally due the results obtained from it. Researchers of various research centers, laboratories, and universities have been meeting periodically according to their specific areas, in order to present their studies and thus contribute for the development of science. The preparation of a conference is not a simple activity; there are several processes taking into account to promote the success of a conference. Thus it is very important develop and implement software applications capable of doing the management of scientific conferences from planning to the end of the conference. Although there are several conference management systems, it is always necessary to have applications capable of increasingly facilitating and improving the activities carried out by all those involved in a scientific conference. In this paper we propose a web application to support the management of scientific conferences. Innovative features that are not evident in existing systems are implemented in this web application.

Keywords: Conferences management system · Conference · Author · Chair · Reviewer · Submission · Paper

1 Introduction

A scientific conference is an event organized and directed to academic and scientific community. This event has been generally based on presentations and discussion of studies, discoveries, novelties that have great value and interest to the scientific community.

Researchers usually participate at least in one conference per year, in their areas of interests. There are lot of players involved in conference tasks. Among them we highlight conference chairs, scientific committee, general chair, chair of conference publication and authors. In order to the organization of conference be succeed, a process must be properly defined. The process of organizing a conference consists of the following main phases: submission of papers, review,

© Springer Nature Switzerland AG 2019
Á. Rocha et al. (Eds.): WorldCIST'19 2019, AISC 930, pp. 765–774, 2019.
https://doi.org/10.1007/978-3-030-16181-1_72

discussion about review of papers, notification of acceptance, submission of final version and registration of authors. Advancement of technology affecting every perspective of our life, scientific and technical conferences grows in number and quality, increasing also the number of papers submitted, that implies the increase of reviewers in scientific committees. Because of that it will become impractical to schedule face-to-face meetings between the various members of the scientific committee.

In general scientific conferences are constituted by several oral presentations ranging from 10–20 min each, following a period of discussion. In addition to these presentations, there are also another activities such as discussion panels, workshops, keynote speakers, summers school, etc. Every paper presented in a conference must be submitted and approved by a scientific committee.

Being an event of great importance to the scientific community and with several activities and sessions, it is urgent to manage it carefully in order to reach the objectives for which it was proposed. Therefore, systems capable of manage the processes of a scientific conference have been of great interest for universities and institutions that organize and manage these activities.

This paper proposes a web-based conference management system focused on the main tasks of a scientific conference, such as (1) paper submission, (2) registration of reviewers and editors, (3) automatic assignment of papers for reviewers, and (4) review papers.

The rest of the paper is organized as follows: In Sect. 2 some conference management systems are addressed; In Sect. 3 the system requirements are presented; In the Sect. 4 the proposed system architecture is described, followed by the concluding section.

2   Related Work

In this section we describe some conference management systems and we do a comparative analysis of existing solutions taking into account some comparison vectors previously defined. We have chosen for this list of conference management systems those systems that are widely referenced and used by several international conferences.

OpenConf [6] is an conference management system developed with PHP, MySQL and Apache. Openconf has three editions: (1) Community Edition, (2) Plus Edition, and (3) Professional Edition. The main features of OpenConf are as follow: online submission, upload files, assignment of reviewers, conflict detection, notification by email, committee discussion, online proceedings, customized forms, plagiarism check and Copyright. OpenConf is available to be installed locally, and also as software as a service (SaaS).

EasyChair [7] is one of the most used conference management system. It was developed in order to help the organizers to deal with the complexity of some processes. EasyChair currently has three types of licenses: (1) free, (2) professional and (3) executive. The main features of easychair are as follows: paper submission, assignment of articles according to reviewers preference, list of last events, file upload, notification by email, online discussion and conference program.

Open Conference System (OCS) [5] is a conference management system licensed under the GNU General Public License v2. Their users are allowed to share and modify it, according to their needs and also according to license terms. OCS allows users create the conference website, online submission, update papers, make the proceedings available, integrate online discussion after finishing the conference.

IAPR Commence [2] is a web-based software management system used to manage technical conferences. IAPR Commence is free license and open source for scientific community. In order to use IAPR Commence, users have to send an email to the project manager, just so he knows which organizations are using the source code. IAPR Commence provides online submission, assignment of reviewers, email or acceptance letter, and organization of conference proceedings. Conference Management System (COMS) [8] is a web-based system that includes planning and detailed creation of an event. COMS has been developed to manipulate events with thousands of users and submissions. The main features are: author and reviewers registration, online submission, payment management, detailed planning of the event, notification by email, creation of the book of abstracts.

ConfMaster [3] is a conference management system based on LAMP (Linux, Apache, MySQL and PHP). The system supports all process of planning and preparation of a scientific conference. ConfMaster is a scalable system. For this reason is possible to organize scientific conferences with high numbers of submissions. ConfMaster covers requirements such as papers submission, assignment of reviewers, including payment with credit card.

PaperCept [1] is a conference management system developed and maintained by PaperCept, Inc. It is a web-based system where all active conference are available. In addition to the registration of authors and reviewers, and the submission of articles and submission of evaluation by reviewers, PaperCept also has as one of the main features the management of online payments.

In order to compare existing conference management systems solutions, and also taking into account the project that we intend to implement, some keys features were considered (see Table 1). According to Table 1 it is possible to verify a great similarity between the systems described above with respect to its features. All systems implements the features that we consider essential in a conference management system. OpenConf and EasyChair are the systems with the most functionality implemented.

However, there are features that have not been implemented by the systems described in Sect. 2. These features are: (1) choose the language of the paper during papers submission, (2) choose the languages that a reviewer can read the papers (during reviewers registration), (3) generate invitation letter for visa purposes after the paper is accepted, (3) generate invitation letter for scientific committee members for visa purposes, (4) assignment of papers to reviewers based on the language, and (5) gather feedback at the end of the conference.

Table 1. Comparison between existing solutions.

Features	OpenConf	EasyChair	OCS	IAPR	COMS	ConfMaster.net	PaperCept
1. Server SaaS	✓	✓			✓	✓	✓
2. Online submission	✓	✓	✓	✓	✓	✓	✓
3. Editing articles	✓	✓	✓	✓	✓	✓	✓
4. Upload files	✓	✓	✓	✓	✓	✓	✓
5. Assignment of reviewers	✓	✓	✓	✓	✓	✓	✓
6. Email notification	✓	✓	✓	✓	✓		✓
7. Export data	✓	✓				✓	✓
8. User account recovery	✓	✓	✓	✓	✓	✓	✓
9. Generation of the conference program	✓	✓	✓	✓	✓	✓	✓
10. Generation of conference report	✓	✓			✓		
11. Online discussion	✓	✓	✓				
12. Reviewers registration	✓	✓	✓	✓	✓	✓	✓
13. Plagiarism check	✓						
14. Choose the language of the paper							
15. Choose the languages by reviewers							
16. Generation of invitation letter for visa purposes							
17. Assign articles to reviewers based on language							
18. Satellite events							
19. Gather feedback at the end of the conference					✓		
20. Quantity of conferences using the system	5600+	68230	–	36	735	1000+	–

In this sense, our web-based conference management systems will focus on these features that are not implemented in the conference management systems described in this section.

3   Proposed System Architecture

In this web application we choose multi-tier architecture because it guarantees greater efficiency, security and scalability. On the other hand, multi-tier architecture makes a clear separation between the layers, and thus any change in one layer does not affect the others [10]. Multi-tier architecture allows you to distribute team elements to each layer and work at same time without interference (Fig. 1).

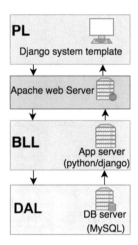

Fig. 1. General system architecture

Data Access Layer (DAL). DAL contains all aspects related to the data. In python-django framework [4] the data access layer is defined in a file named models.py. In this file we write the classes, their attributes and their methods. Each class defined in the file corresponds to a table in the database. DAL communicate with business logic layer (BLL) and the database. In oder to communicate to the database, django uses a file called settings.py where the connection data are provided [9].

Business Logic Layer (BLL). BLL is responsible for receiving the http requests from the presentation layer (PL), communicating with the database and after that, sending a response to the presentation layer. All business logic has been developed in a file called views.py.

Presentation Layer (PL). PL is responsible for visualizing data, defining how data should be presented in web pages and in another types of documents (e.g. pdf file). In presentation layer we used django system template, with Bootstrap (front-end framework), HTML and JavaScript.

3.1 How the Application Works?

Figure 2 describes the operation of the application: (1) the browser sends an http request to the web server (apache); (2) the web server forwards the request to the application server; (3) the URL setting contained in the urls.py file selects the view according to the url specified in the request; (4) the view communicates with the database via models.py, renders the html or other format using templates (loads the static files) and returns the http response to the web server; (5) finally, the web server provides the desired page to the browser.

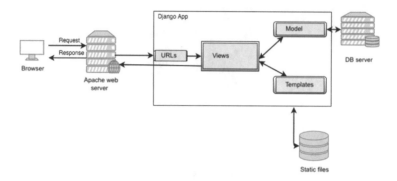

Fig. 2. Specific Django architecture

3.2 Security

Cross-site Scripting (XSS). Django prevents XSS (mostly JavaScript) in web pages viewed by users. To deal with these XSS attacks, the entire Django template by default automatically escapes potentially harmful HTML special characters. For example, if an attacker enters special HTML characters (such as double quotation marks) into a field in an application form, this content will be converted to hard-coded HTML (& quot in the case of double quotation marks), making it impossible to inject any scripts into the application. Another mechanism used to prevent XSS attacks was the validation of the input fields of the HTML forms using the Django forms.

SQL Injection. In order to prevent the application against SQL injection attacks, the queryset was used. Queryset is a database abstraction API that allows you to query without using pure SQL code. Due to some limitations of the queryset API, we also used queries with pure SQL code, however preventing the introduction of parameters that the user can control.

Authentication. The application uses users and groups to limit access to the system to authenticated users with a certain profile. Each user of the application has a username and password that allows him to access the application after a previous registration. Initially, the application validates the user's credentials, confirming that the user with the username + password exists and is active; If is active, it checks the group to which the user belongs and then the user has access to the application according to the group to which it belongs.

3.3   Architecture in the Cloud (Amazon Web Service)

We used aws (amazon web service) to host the application and make it available as an online service (SaaS). Figure 3 shows the architecture using the aws services.

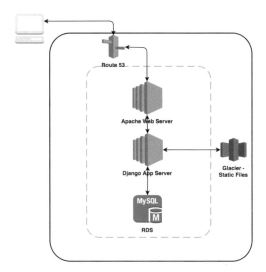

Fig. 3. Architecture in cloud (Amazon web service)

- **Amazon Route 53**: in order to register and manage application domains.
- **Apache web/Django App Server**: we use Amazon EC2, virtual computer with scalable capabilities, designed to be scalable and easy to use.
- **Amazon Glacier**: to store static files (css, javascript, pdf, media, templates).

3.4   Data Modeling

In this section we present the physical data model of our database, and a brief description of the database tables. The database has 17 tables as we can see in the physical data model (Fig. 4) (Table 2).

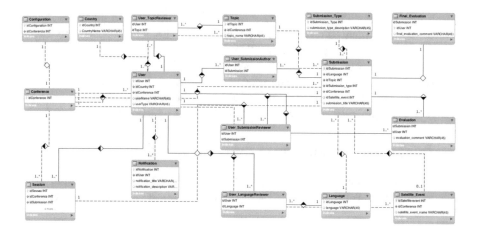

Fig. 4. Database - physical data model

3.5 The Web Application

The application was developed in three modules (management, review and submission). The management model is used by chair of the conference, review module is used by reviewers and submission module is used by authors. In order to use the app, users must be logged in. There are three possible profile in the application: organizer (chair), reviewer and author. The chair has full access to the application and can perform both the reviewer's and author's features. According to the users profile the application shows the corresponding module (management, review or author module of the application). In the management module, home screen displays some general statistics for a conference, such as: number of submissions, submissions with reviewers assigned, articles evaluated, articles accepted, number of reviewers, reviewers with articles assigned, number of organizers, and also shows the last submission made at the conference, the last reviewer and last organizer registered at the conference.

The assignment of papers to reviewers can be done manually or automatically. Manually, the assignment of papers is made to each reviewer individually; and automatically, the assignment of papers is made to all reviewers at same time.

Reviewers download the papers assigned to them, and they insert the evaluation of a paper. Reviewers can edit or remove the evaluation of the paper if necessary. Personal information can also be edited.

An author can submit more than one paper to a conference and can visualize and edit these submissions. After the evaluation (review) process, our system allows authors to see all the comments made by reviewers and the final decision of the chair. After the paper is accepted the acceptance certification and the invitation letter is available for the author.

Table 2. Description of the database tables

N°	Table	Description
1	Country	Record all countries
2	Language	Record languages
3	Submission_Type	Record all submission types of a conference. For example full paper, short paper, poster paper
4	Conference	Records all conferences
5	Satellite_Event	Records all satellite events of a particular conference
6	User	Records all user of the application (authors, reviewers, chairs)
7	Configuration	Records the configuration of all conferences registered in the system
8	Topic	Records all topics belonging to a particular conference
9	Submission	Records the articles belonging to a conference or to a particular satellite event of a conference
10	Session	Records all sessions of a given conference
11	Evaluation	Records all evaluations of papers in a particular conference
12	Final_Evaluation	Records the final evaluation by organizers
13	Notification	Records all notifications created for a given conference
14	User_SubmissionAuthor	Records authors associated with submitted articles
15	User_SubmissionReviewer	Records reviewers with assigned articles
16	User_TopicReviewer	Records reviewers associated with the topics chosen
17	User_LanguageReviewer	Records the reviewers associated with the chosen languages

We have implemented almost all the previously defined features of our system. However, there are still remaining some features such as: generate the conference program, generate the conference report, close submissions, online discussion, plagiarism check, email notification, invitation of reviewers, and gather feedback at the end of the conference.

4 Conclusion

We developed a web application for management of scientific conferences that brings some features that differ from the others (e.g. (1) choose the language of the paper during papers submission, (2) choose the languages that a reviewer can read the papers (during reviewers registration), (3) generate of invitation letter for visa purposes after an paper is accepted, (3) generate invitation letter for scientific committee members for visa purposes, (4) assignment of papers to reviewers based on the language, and (5) gather feedback at the end of the conference). This web application is divided into three modules: management, review and submission. These three modules are accessed from a web browser. In the implementation phase of the project, architecture design, database design and modeling, back-end and front-end development are included. We implement and test all "must" requirements, and thus the application perform all main tasks without any difficult. For future work we intend do implement the remaining features of the application. We also intend to implement the application in a conference in order to evaluate its features in a real environment and collect necessary feed-backs to continuously improve the application in terms of robustness, performance and security.

Acknowledgments. We thank Associação Ibérica de Sistemas e Tecnologias de Informação (AIST) for the support in this project.

References

1. http://ifac.papercept.net/conferences/scripts/start.pl. Accessed 25 Oct 2018
2. Commence conference system. https://iaprcommence.sourceforge.io/. Accessed 22 Oct 2018
3. Confmaster.net—conference management system. https://confmaster.net/. Accessed 25 Oct 2018
4. General django documentation. https://docs.djangoproject.com/en/2.1/. Accessed 10 Nov 2018
5. Open conference systems. https://pkp.sfu.ca/ocs/. Accessed 22 Oct 2018
6. Peer-review, abstract and conference management. https://www.openconf.com/. Accessed 20 Oct 2018
7. The world for scientists. https://www.easychair.com/. Accessed 20 Oct 2018
8. Dienstleistungen: Coms - conference management software. https://www.conference-service.com/. Accessed 3 Nov 2018
9. Ravindran, A.: Django Design Patterns and Best Practices. Packt Publishing Ltd., Birmingham (2015)
10. Schuldt, H.: Multi-tier architecture. In: Encyclopedia of database systems, pp. 1862–1865. Springer (2009)

Localizing Inconsistencies into Software Process Models at a Conceptual Level

Noureddine Kerzazi[✉]

ENSIAS School of Engineering, Mohammed V University in Rabat,
Avenue Mohammed Ben Abdallah Regragui, Madinat Al Irfane, Agdal, BP 713,
Rabat, Morocco
n.kerzazi@um5s.net.ma

Abstract. Software process modeling aims to provide an abstract description of roles, activities, and artifacts used to lead the development and maintenance of software projects. The designed process models must be correct syntactically and consistent semantically in order to improve teams' productivity and enable developers to achieve product quality goals. To address those challenges, we introduce an approach to ensure the correctness and consistency of designed process models at a conceptual level. The proposed approach is based on: *(1)* a syntactic verification of correctness at the meta-model level and *(2)* a semantic validation based on a rules' engine seeking consistency at the operational level. Using this approach, software development teams can define new validation rules to constrain the semantic of their software processes. We implemented the approach and evaluate its effectiveness through two case studies. The results have shown that we are able to support process modelers identifying inconsistencies at a conceptual level. Most importantly, we further discuss the reusability of semantic validation rules produced by experts for an effective analysis of process models.

Keywords: Software process modeling · Exception handling · Process models consistency · Rules engine

1 Introduction

Software process modeling consists in organizing and visualizing concepts (*e.g.,* ≺ *Roles, Activities, WorkProducts* ≻) and their relationships, with sufficient details, with the aim to provide explicit guidance for executing a software development project [4]. Process models provide constructs useful for representing and reasoning about the various aspects of a process so that it may be analyzed and improved. It is acknowledged that an abstract conceptualization helps to share a common understanding of software process activities. However, having conceptual views of process models does not necessarily mean that these models are consistent and defects free [18]. To the best of our knowledge, very little research studies have focused on developing techniques to check the validity of conceptual models [5,20]. Yet the correctness of these models is essential to ensure

© Springer Nature Switzerland AG 2019
Á. Rocha et al. (Eds.): WorldCIST'19 2019, AISC 930, pp. 775–788, 2019.
https://doi.org/10.1007/978-3-030-16181-1_73

correct guidance of software engineers involved in the projects. By verification we mean the process of checking whether a conceptual model is correctly represented according to the *Software & Systems Process Engineering Meta-model* (SPEM) requirements [22].

Models validation procedures are needed by process engineers to reduce the time, cost, and risk [2] associated with designing and running inconsistent processes. In fact, the empirical verification of the correctness of a software process requires between 50 to 200 man days [19], for gathering the data, structuring and processing the data, assessing the outcome of the process (in terms of products' quality, productivity), measuring the process deviation (using the so-called Δ-approach), and finally performing root cause analysis to identify the origin of issues. Hence, to reduce this verification time, tools are needed to identify inconsistencies directly in software processes (instead of in the outcomes), in order to help process designers in developing high-quality process models quickly.

1.1   Problem Statement

Current verification approaches: empirical [19], early analysis [24], metrics [3], formal verification [8], and simulation [25] do not yet support software process designers in an optimal way [8,21], especially for detecting inconsistencies [9] early (*i.e.*, at the conceptual level). Even though process modeling has been used extensively over the past three decades, surprisingly we know little about how to verify a process model at a conceptual level [18]; in particular, it is not clear which characteristics should be considered when checking whether or not a process model contains flaws. Among the many aspects that should be verified in a software process (*e.g.*, usability [13], quality [18], communication [1], cognitive integration [15]), the semantic validity is the most important.

1.2   Objective

In this paper, we aim to support process designers with an approach and a tool for the verification of correctness and consistency of process models at the conceptual level. Also, we seek to find an efficient way of reusing verification experiences, especially for recurrent semantic inconsistencies.

1.3   Contributions

The paper makes the following three contributions. First, we propose a method for a syntactic verification of process models using an ontological approach acting at the meta-model level. Second, we introduce a semantic verification technique using an easy-to-use declarative rule engine in XML format, aiming to visualize the origin of semantic flaws according to predefined rules. Third, we show, through two case studies performed in both academic and industrial contexts, that our proposed approach can provide a reusable knowledge base of verification rules, by allowing import/export of these rules in an XML format.

1.4   Outline

The remainder of this paper is organized as follows: Sect. 2 presents the architectural background underpinning the DSL4SPM tool on top of which we implement our approach. Section 3 introduces our method for localizing syntactic and semantic inconsistencies within process models, at the conceptual level. Section 4 discusses the evaluation of our approach through two case studies. Section 5 summarizes related work. Finally, Sect. 6 presents concluding remarks and points out the future work.

2   Background

What does it mean to say that a process model conforms to the principles of Software Process Engineering Meta-model (SPEM)? Process Model refers to a conceptual view of interrelated $\prec$ *Roles, Activities, WorkProducts* $\succ$; and what does a defect mean in the context of process modeling? In this work, we advocate two types of process models' inconsistencies. The first refers to the meta-model level including domain classes and their relationships, whereas the second refers to the usage level according to a set of predefined rules.

2.1   Architecture of DSL4SPM Tool

Figure 1 depicts the layered architecture supporting our DSL4SPM tool [14] dedicated to software process modeling. Our activity-oriented modeling perspective has been built with respect to *Software & Systems Process Engineering Meta-Model* (SPEM 2.0), a defacto standard from the OMG [22]. **Layer 1** relates to the technical infrastructure such as Frameworks, serialization, and data persistence. **Layer 2** refers to the translation of the SPEM Meta-model as specified in [22], into our own meta-model, which then has been enriched with other perspectives of modeling (*e.g.,* Stochastic Simulation, Knowledge representation [12], CMMI alignment [11]). The relationships between concepts at the meta-level were formally represented based on our predefined ontology. The ontological approach allows us a formal representation of concepts and inference on model concepts. **Layer 3** constitutes the conceptual view such as provided to process designers (*e.g.,* Toolbox of process components, modeling area, Exceptions handling, Repositories for content re-usability, etc.). Finally, **Layer 4** embodies the verification and validation mechanism which we present in this work. the simulation engine, along with other features.

A functional representation of DSL4SPM tool can be seen in Fig. 6 (Sect. 3.2). This paper focuses on layers 2 and 4. In layer 2, we define a formalism based on an ontological approach in order to formally represent relationships between SPEM elements, as illustrated in Fig. 2. This dynamic perspective is a cornerstone of the conceptual approach behind DSL4SPM. In Layer 4, we provide a rule engine for semantic validation. Process designers, users of our tool, can request an on-demand semantic validation according to the rules they have defined as shown in Fig. 6-(b).

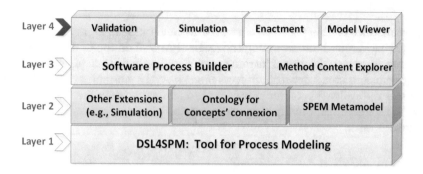

Fig. 1. Layered architecture of DSL4SPM tool

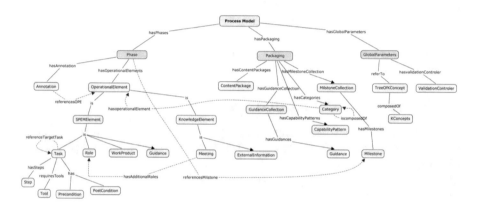

Fig. 2. Ontological view of the meta-model

2.2 Ontological View of the Meta-model

According to Gruber [7], an ontology is a collection of key concepts and their inter-relationships collectively providing an abstract view of an application domain. We used this ontological approach to formalize relationships between the elements of a process model at design. As depicted in Fig. 2, a design element can relate to many other elements through multiple links which improve the expressiveness in modeling and allows the consolidation of multiple views of process models. For instance, we can generate an activity-oriented view based on activity relationship as well as an artifacts-oriented view based on another kind of relationships such as Roles or Tasks.

By doing so, we are able to support process designers in constraining the types of connectors depending on the type of nodes, source and target. For example, a designer cannot instantiate a direct link between a *Role* and a *WorkProduct*, because it is specified that *WorkProducts* are the inputs or outputs of tasks and accordingly can only be related to *Tasks*.

3 Localizing Inconsistencies

In this section, we describe how we implement our approach of syntactic and semantic verification. Section 3.1 describes the approach for syntactic verification, while Sect. 3.2 our approach for semantic validation.

Upon activation of the rule engine, within DSL4SPM tool, according to these set of rules, performs semantic verifications on the process model that is being designed (Sect. 3.3). The identification and visualization of the origin of semantic flaws including the recommendation on how to fix the flaws are returned (Sect. 3.4). We conclude the section with a discussion on the quality of the conceptual verification [18] and the performance and scalability of our rules engine (Sect. 3.5).

3.1 Syntax Verification

Figure 3 shows an example from our meta-model (left-hand side) which represents domain classes and relationships as described originally in SPEM [22] and enriched by adding other concerns of process modeling. For instance, we have a domain class called *"Task"* and its attributed relationships linking this domain class with other classes *Step*, *Role*, *WorkProduct*, and itself (a Task can be linked to other *Tasks*). Unlike UML[1], we have attributed relationships which embed semantics in order to better qualify the meanings of each relationship. We used an ontological approach, as described in Sect. 2.2, to build the perspective of interactions between concepts of process models. It is worth noticing that DSL4SPM is a multi-perspective modeling tool, which means that we can have more than one relationship between two process elements. This enables us to address more than one view.

We have implemented and enriched the semantics provided by SPEM in a previous work [14]. The validation of our proper semantic at the meta-model level was done using First Order Logic, as seen in Fig. 4. We used a first-order logic as a representational language [6], which serves two motivations: *(1)* define axioms constraining object structures. For example, as shown in Fig. 3, a relationship between two nodes can have only one semantic sense (directed connection), which implies a relation between a source and a target; and *(2)* properties for each possible relationship between two nodes. For instance, if a relationship between two nodes has the *'Inverse'* property, then the other sense of the relationship has also a meaning specified by another kind of relationship (*i.e.*, other semantics).

---

[1] SPEM has been designed as a UML Profile and thus inherits the poor concepts of relationships between concepts to represent effectively a specific domain. We have mitigated this drawback in previous work by adding the concept of attributed relationships between domain classes to be able to address enriched semantics according to multidimensional concerns.

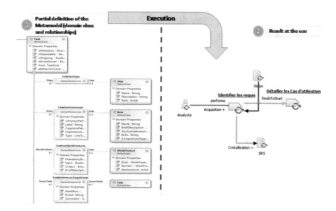

Fig. 3. (1)- Partial conceptual model; (2)- Modeling example

The right-hand side of Fig. 3-(2) shows an example of the kind of design that can be carried out at the operational level using our DSL4SPM tool.

3.2 Rules for Semantic Validation

Before checking the semantic validity of a designed process model, designers must add or import semantic rules, which are stored in XML format, and link this repository of rules to the process to be verified. Figure 6-(b) shows the form of rules settings, severity, and export/import of features. An internal rule engine loads the rules and extracts the semantic information of interest, as well as the severity of each rule (Error, Warning, or Information Message). We made the choice to store semantic rules in XML format to be able to articulate and share verification experiences and to gather a body of knowledge on how experts perform verification checking for complex process models. It's worth noting that, in an academic context, the approach has been successfully used by instructors to evaluate the students' work effectively and automatically. We provide a wizard, within DSL4SPM tool, to support building semantic rules.

3.3 Semantic Verification

After the repository of semantic rules has been populated by either adding new rules or by importing a set of rules from an XML file, the designer can activate semantic verification on the process model at hand. DSL4SPM tool loads the rules using our internal engine and is ready to run semantic checks. Figure 5 illustrates an example of internal implementation of rules. When a designer asks for verification, DSL4SPM run all rules on each process element and each relationship instantiated in the process model. The results are then returned in a specific form.

Axiom	First-order	Description
Axiom 1	$\forall xy[xRy \rightarrow \sim(yRx)]$	A relationship between two nodes has only one semantic sense.
Axiom 2	$\forall x.(Rc(x) \vee Ra(x))$	A relationship between two nodes can only be a classifier or association.
Axiom 3	$\forall r.(Ra(r) \rightarrow \exists y.MA(r,y))$	An association relationship can embed one or more attributes.
Axiom 4	$\forall r.(Rc(r) \rightarrow \neg\exists y.MA(r,y))$	A classification relationship cannot contain any attribute. Where Rc is the predicate for a classification relationship and MA a predicate for a meta attribute.
Axiom5	$\forall x,y,r.(Es(x) \wedge Ec(y) \wedge Ra(r) \rightarrow notationG(r))$	The graphical notation of links depends on both source and target nodes as well as on the direction of the relationship.

Property	Description	Example
Inverse	A property may have a corresponding inverse property	The property 'TaskHasSteps' has its inverse property 'is-PartOfTask'.
Functional	If a relationship has the property functional then it must exist at least one instance of this relationship between two nodes	$Role_n$ is responsible for $Artifact_m$.
Inverse Functional	If a property is inverse functional, it means that the inverse property is also functional	if $Artifact_m$ is produced by $Role_a$ and $Role_b$ is responsible for $Artifact_m$, then $Role_a$ and $Role_b$ are the same.
Transitive	If a property P is transitive, and that property relates node A to B, and node B to C, then node A is related to C via property P. $\forall x.y.z.(R(x,y) \wedge R(y,z)R(x,z)$	If $Artifact_m$ is related to a $Task_n$, which in turn should be related to a $Role_a$, then we can infer that $Artifact_m$ is related to the $Role_a$.
Symmetric	P is symmetric if the link from A to B, means also a relationship from B to A. All paths are 2-way.	askHasPerformer is symmetric, but the relation TaskHasSteps is not.
Antisymmetric	If P is Antisymmetric, then if A is linked to B then B cannot be linked to A	$Step_n$ is a partOfTask $Task_m$. The relation partOfTask cannot be used to link $Task_m$ to $Step_n$.
Reflexive	A property P is reflexive when that property must relate element A to itself	Work Product can refer to itself using the relationship 'isDeliverable'.
Irreflexive	Cannot link an element to itself	$Task_a$ cannot be preceeded ('isPrecededBy') by itself.

Fig. 4. Semantic representation with first-order logic

Figure 6-(c) shows the form returning semantic inconsistencies identified in the designed process model according to predefined verification rules. One can see how the tool visualizes the list of inconsistencies according to their level of severity based on the same view provided for developers. The designer can double-click on each line in Fig. 6-(c) to highlight the element of the process model responsible for the flaw. As one can see, the tool provides also useful descriptions of each identified flaw helping to fix them.

3.4   Performance

To support designers and to provide initial experience with some kind of rules, we provide a set of examples rules. Typically, the example set considers basic principles and contains ten modeling rules and ten operational rules. Figure 6-(b) summarizes some of them. Verifying semantics for a variety of different real word process models and returning identified flaws is fast, typically taking less than 50 ms. However, further analysis of reported experiences will optimize the structure of rules by *(1)* organizing them according to relevant perspectives of modeling which will help to focus on specific concerns, and *(2)* adding hierarchical rules allowing us to increase the efficiency of semantic verification.

```
 9  □namespace POLYMTL.Process_Modeling_DSL1.CustomCode.Coherence
10  {
11      /// This class is used to enable certain rules which cannot be runned during the initial
12      /// load because it causes errors.
13      [RuleOn(typeof(Diagram), FireTime = TimeToFire.TopLevelCommit)]
          1 reference
14      public sealed class Diagram_Loaded_Rule : AddRule {
          13 references
15          public override void ElementAdded(ElementAddedEventArgs e) {
16              base.ElementAdded(e);
17              // Enables the slack (or float) calculation rules.
18              e.ModelElement.Store.RuleManager.EnableRule(typeof(Task_Slack_ChangeRule));
19              e.ModelElement.Store.RuleManager.EnableRule(typeof(Task_Slack_AddRule));
20              e.ModelElement.Store.RuleManager.EnableRule(typeof(Task_Slack_DeleteRule));
21          }
```

Fig. 5. An example of internal rule implementation

4 Related Work

Different approaches have been proposed to identify flaws in process models. Besides traditional empirical analysis of the process outcomes [19], we distinguish between approaches based on early analysis [10,17,24], Metrics [3], formal verification [8], and simulation [25]. The empirical analysis is a posteriori approach that use historical data of projects to analyze the overall performance of process models and potential defects. For instance, Mendez et al. [19] carried out an analysis of 9 empirical studies of real software development processes aiming for methodological pattern-based guidelines for an empirical qualitative analysis of process models. Although both quantitative and qualitative analysis are drastically efficient when carried out by experts, they are not very cost-effective and are time-consuming and dependent on reliable data. We believe that the early verification of software process models at a conceptual level would provide team process management with support which would make their improvement of tasks easier.

Approaches for early analysis are generally based on an abstract view (source code or descriptions) of the process model in order to identify patterns of flaws. Osterweil and Wise [24] implemented procedures written in their Little-JIL process definition language aiming to analyze Scrum Process models and to determine the extent to which this process satisfies its Agile requirements. Authors demonstrated that they can find issues related to Scrum iterations. In the same vein, Hurtado et al. [10] presented a visual approach for the verification of software process models which are implemented using the EPF Composer tool. Authors proposed an approach supported by a tool to identify a set of common error patterns. Lerner et al. [17] tried to introduce patterns of exception handling in the field of process modeling in the same way as programming languages. Authors argued that software processes are software too [23]. They implement their exception handling mechanisms in Little-JIL tool. However, Little-JIL is not compliant with the SPEM specification and consequently, the semantics that they try to validate are proper to the Little-JIL tool.

Canfora et al. [3] explored the maintainability of process models aiming to understand which metrics can be used as maintainability indicators. Authors focus on three sub-characteristics: *Analyzability*, *Understandability*, and *Modifiability*.

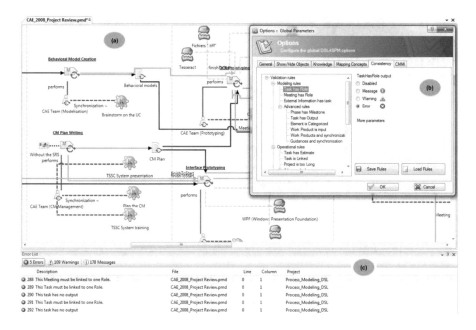

Fig. 6. Localizing inconsistencies within process models

We were more interested in Analyzable metrics which refers to easiness in discovering errors or deficiencies. In our work, we were inspired by these metrics as a key starting point in order to provide a framework for users, at the meta-model level, so that they can define their validation rules. Furthermore, we visualize different concerns about process models and provide immediate feedback on where the errors are located and how to fix them.

Formal verification and simulation approaches are suitable for detecting defects but they are often too costly and time-consuming, as stated by Law and McComas [16]. These approaches require resources for building new dedicated models, using specific semantics aiming at representing the process according to specific attributes. The research reported in this paper concentrates on supporting designers by identifying and highlighting dynamically and visually the origin of flaws. Our proposed approach does not need to rebuild or maintain specific models for simulation. For instance, we store some anti-patterns, in XML format, for designing semantically valid models. By doing so, we are able to provide recommendations on how to fix the identified semantic flaws.

5   Evaluation

To be useful, a process modeling tool should support process designers in identifying inconsistencies and help them building a reusable knowledge base of validation rules (i.e., what to look for to assess the correctness of process models).

Thus, we considered two research questions aiming to gain evidence about this statement:

- (RQ1) Can process designers be supported for identifying effectively defects in process model?
- (RQ2) Can process designers improve their experience by reusing validation rules?

Two studies have been conducted in order to evaluate our approach for processes' validation. The first study takes place in an academic context, specifically within a laboratory session of a software engineering course. While the second study was carried out in an industrial context aiming to improve a subprocess of software releases for a North American company producing a complex online banking system used across 192 countries.

5.1 Studies Methods

Study 1- The study implies a pool of 34 candidates' senior undergraduate students, paired into 17 teams along with the involvement of 3 instructors. All the participants were enrolled in a mandatory course of software process engineering and have learned the core area components of process modeling. The session of laboratory lasts 3 h which was expected to be more than enough. Students were tasked to diagnose a process model and then produce validation rules using DSL4SPM tool. The process model diagnosed by students was well-known and straightforward. At the end of the laboratory session, students must submit to the instructors an XML file of rules generated from DSL4SPM tool. In order to assess the students' work, instructors use the validation file to raise and show all defects found by the students.

Study 2- The study takes place in a North American company. The primary goal was the diagnose and analyze the current software release process as part of a broader project of software process improvement. Our process validation aims at *(i)* assessing the quality of the software release process and *(ii)* identifying anomalies (Activity workflows blueprints) in the process model and *(iii)* providing hints on how to fix those anomalies.

5.2 Data Analysis

For the academic study, we counted the number of defects found by each group of students regarding $\prec$ *Roles - Tasks - Artifacts* $\succ$. A descriptive statistical analysis was performed and data were compared to the mean values. For the industrial study, we were more interested in the number of semantic defects and coordination issues since the purpose was to improve the subprocess of software release carried out by a relatively new role called DevOp. We can classify these inconsistencies into four broad groups: *(1) Missing Output Artifacts, (2) Missing Guidance; (3) Needs for Coordination, (4) Needs for Communication.* This turned out to be a simple, yet meaningful and comprehensive set of categories.

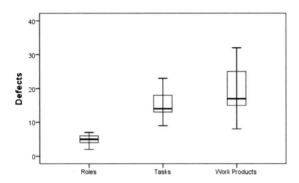

Fig. 7. Defects localized by students.

5.3   Results

RQ1. Can process designers be supported for identifying effectively defects in process model? - To evaluate our first research question, we considered how many defects participants were able to identify correctly.

Study 1- Within a laboratory session which last 3 h, our approach supported students in identifying five inconsistencies related to Roles (mean $= 5.11, \pm 1.57$); fifteen inconsistencies related to Tasks (mean $= 15.35, \pm 4.04$); and nineteen inconsistencies related to Work products (mean $= 19.29, \pm 7.15$). Despite the low amount of training, the students were able to use the system and to localize defects. Figure 7 summaries the defects found in the academic context.

One surprising observation is the negative correlation between defects found for Roles and those found for Work Products. More the students' attention is focusing on finding defects for "Roles", fewer defects they found for "Tasks". We consider that this finding shows strong support for the usefulness of our method.

It is worth noticing that the tool was useful for students in their learning approach regarding the design of consistent process models. Also, the tool makes it easier for instructors to evaluate the students' works considering rules such as a checklist for assessment.

Study 2- The main defects found in the software release process are about missing links and guidance as shown in Fig. 8. Several problems were identified directly, where others required further analysis and inferring. For instance, missing guidance about the branching structure of the source code to lead parallel development was missed as well as the checklist used before pushing new code to the production environment. Problems related to coordination required an overall view and inferring of the process model. For example, the release engineer should coordinate with a DBA to sync pushing new packages with DB scripts to avoid system crashes.

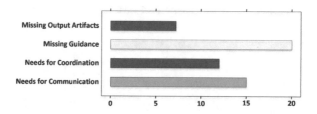

Fig. 8. Semantic defects of the software release process.

***RQ2. Can process designers improve their experience by reusing vali-
dation rules?*** - To evaluate our second research question, the instructors were
able to evaluate the work of students using the XML file of rules they submitted. Moreover, we have proven with the industrial case study that the effort of
reviewing a process model can be reused as best practice recommendation in
future modeling projects. For instance, we have learned how to detect coordination issues and how to anticipate communications needs by analysis interrelated
tasks.

6   Threats to Validity

The academic study was not representative enough for real software processes.
Our sample size (34 students) did not provide us with enough data for a solid
comparative statistical analysis. Additionally, as the laboratory session was tailored towards educational purpose, the students may have had to make less effort
in defining new rules. Time constraint made it difficult for students to understand the specification of the process to be diagnosed and then create validation
rules.

Nevertheless, the laboratory instructors were able to evaluate student's work
using a common XML file of rules. This gives us a piece of strong evidence about
our second research question (reusability of rules). We mitigate the representativeness threat of the academic study by an industrial case study, assuming that
it was more robust and effective.

7   Conclusion

What makes a software process modeling tool effective at defects detection? To
be effective, a software process model should be inspected and validated by localizing inconsistencies. To this end, we have introduced an approach supported by
a tool aiming at helping designers performing semantic and syntactic verifications of designed process models compliant with SPEM 2.0. We have shown how
the syntactic verification can be performed at the meta-model level, whereas
semantic verification can be based on an articulated rule engine. Verification
rules are defined by designers and stored in XML format for sharing and reusing

purposes. Finally, we have demonstrated the usefulness of our approach and the effectiveness of the tool through two case studies.

The first case study, carried out in an academic context, showed that students were able to activate verification rules and check whether a relatively small software process model (IPSW6) contained semantic flaws. Using the tool, the students were able to localize efficiently the origin of these flaws and use recommendations provided by the tool to fix them. This academic case study allows us to get insights on the reusability of verification experiences and how to build and share knowledge about rules. The second case study, which was performed in an industrial context, allowed us to identify several inconsistencies within a software release process as well as improvement opportunities, some of them were fixed according to the tool's recommendations, while others needed articulated and deep changes in the process model.

The outcomes of these two studies have a number of practical implications on the field of process modeling. First, practitioners can evaluate the consistency of their process models at a conceptual level in a cost-effective manner; in terms of time and effort required to avoid the effects of missing or poorly implementing software processes. Second, the verification rules can be reused to evaluate other process models or can be shared with other process designers, which enhance the body of knowledge. In future works, we plan to investigate two aspects: first, we shall refine the rules store by adding hierarchical rules and then propose a structured way to separate concerns for a guided semantic verification. Second, we plan to explore the mapping of our verification mechanism with a formal verification technique based on model checking. This formal method should enhance the expressiveness of our semantics validation rules.

References

1. Aranda, J., Ernst, N., Horkoff, J., Easterbrook, S.: A framework for empirical evaluation of model comprehensibility. In: Proceedings of the International Workshop on Modeling in Software Engineering, MISE 2007, p. 7, Washington, DC, USA. IEEE Computer Society (2007)
2. Boehm, B.W.: Software Engineering Economics, 1st edn. Prentice Hall PTR, Upper Saddle River (1981)
3. Canfora, G., García, F., Piattini, M., Ruiz, F., Visaggio, C.A.: A family of experiments to validate metrics for software process models. J. Syst. Softw. **77**(2), 113–129 (2005)
4. Curtis, B., Kellner, M.I., Over, J.: Process modeling. Commun. ACM **35**(9), 75–90 (1992)
5. Davies, I., Green, P., Rosemann, M., Indulska, M., Gallo, S.: How do practitioners use conceptual modeling in practice? Data Knowl. Eng. **58**(3), 358–380 (2006)
6. Fitting, M.: First-order Logic and Automated Theorem Proving, 2nd edn. Springer-Verlag New York Inc., Secaucus (1996)
7. Gruber, T.R.: Toward principles for the design of ontologies used for knowledge sharing. Int. J. Hum.-Comput. Stud. **43**(5–6), 907–928 (1995)
8. Gruhn, V.: Validation and verification of software process models. In: Proceedings of the European Symposium on Software Development Environments and CASE Technology, pp. 271–286, New York, USA. Springer-Verlag Inc. (1991)

9. Hungerford, B., Hevner, A., Collins, R.: Reviewing software diagrams: a cognitive study. IEEE Trans. Softw. Eng. **30**(2), 82–96 (2004)
10. Hurtado Alegría, J.A., Bastarrica, M.C., Bergel, A.: Analyzing software process models with avispa. In: Proceedings of the 2011 International Conference on Software and Systems Process, ICSSP 2011, pp. 23–32 (2011)
11. Kerzazi, N.: Conceptual alignment between spem-based processes and CMMI. In: 2015 10th International Conference on Intelligent Systems: Theories and Applications (SITA), pp. 1–9, October 2015
12. Kerzazi, N., ElBouzidi, D.: On the discontinuity of knowledge flows between software teams coworkers. In: 5th International Symposium ISKO-Maghreb Knowledge Organization in the perspective of Digital Humanities, pp. 136–145 (2015)
13. Kerzazi, N., Lavallee, M.: Inquiry on usability of two software process modeling systems using ISO/IEC 9241. In: 2011 24th Canadian Conference on Electrical and Computer Engineering (CCECE), pp. 773–776, May 2011
14. Kerzazi, N., Robillard, P.: Multi-perspective software process modeling. In: 2010 Eighth ACIS International Conference on Software Engineering Research, Management and Applications (SERA), pp. 85–92, May 2010
15. Kim, J., Hahn, J., Hahn, H.: How do we understand a system with (so) many diagrams? Cognitive integration processes in diagrammatic reasoning. Info. Sys. Res. **11**(3), 284–303 (2000)
16. Law, A.M.: How to build valid and credible simulation models. In: Proceedings of the 37th Conference on Winter Simulation, WSC 2005, pp. 24–32. Winter Simulation Conference (2005)
17. Lerner, B., Christov, S., Osterweil, L., Bendraou, R., Kannengiesser, U., Wise, A.: Exception handling patterns for process modeling. IEEE Trans. Softw. Eng. **36**(2), 162–183 (2010)
18. Lindland, O., Sindre, G., Solvberg, A.: Understanding quality in conceptual modeling. IEEE Softw. **11**(2), 42–49 (1994)
19. Mendez Fernandez, D., Penzenstadler, B., Kuhrmann, M.: Pattern-based guideline to empirically analyse software development processes. In: 16th International Conference on Evaluation Assessment in Software Engineering (EASE 2012), pp. 136–145 (2012)
20. Mendling, J., Neumann, G., van der Aalst, W.: Understanding the occurrence of errors in process models based on metrics. In: On the Move to Meaningful Internet Systems 2007: CoopIS, DOA, ODBASE, GADA, and IS, vol. 4803, pp. 113–130. Springer (2007)
21. Mendling, J., Verbeek, H.M.W., van Dongen, B.F., van der Aalst, W.M.P., Neumann, G.: Detection and prediction of errors in EPCs of the SAP reference model. Data Knowl. Eng. **64**(1), 312–329 (2008)
22. OMG. Software process engineering metamodel spem 2.0 omg. Technical report ptc/08-04-01, Object Managemente Group, April 2008
23. Osterweil, L.: Software processes are software too. In: Proceedings of the 9th International Conference on Software Engineering, ICSE 1987, pp. 2–13, Los Alamitos, CA, USA. IEEE Computer Society Press (1987)
24. Osterweil, L., Wise, A.: Using process definitions to support reasoning about satisfaction of process requirements. In: Proceedings of the 2010 International Conference on New Modeling Concepts for Today's Software Processes: Software Process, ICSP 2010, pp. 2–13. Springer-Verlag, Heidelberg (2010)
25. Zhang, H., Jeffery, R., Houston, D., Huang, L., Zhu, L.: Impact of process simulation on software practice: an initial report. In: Proceedings of the 33rd International Confefernce on Software Engineering, ICSE 2011, pp. 1046–1056 (2011)

Software and Systems Modeling

On the Practicality of Subspace Tracking in Information Systems

Noor Ahmed[1]([⊠]) and Gregory Hasseler[2]

[1] AFRL/RI, Rome, NY 13441, USA
norman.ahmed@us.af.mil
[2] PAR Government Systems Corporation, Rome, NY 13440, USA
gregory.hasseler@partech.com

Abstract. Modeling and characterizing information systems' observation data (i.e., logs) is fundamental for proper system configuration, security analysis, and monitoring system status. Due to the underlying dynamics of such systems, observations can be viewed as high–dimensional, time–varying, multivariate data. One broad class for concisely modeling systems with such data points is low–rank modeling where the observations manifest themselves in a lower-dimensional subspace. Subspace Tracking plays an important role in many applications, such as signal processing, image tracking and recognition, and machine learning. However, it is not well understood which tracker is suitable for a given information system in a practical setting. In this paper, we present a comprehensive comparative analysis of three state-of-the-art low–rank modeling approaches; *GROUSE*, *PETRELS*, and *RankMin*. These algorithms will be compared in terms of their convergence and stability, parameter sensitivity, and robustness in dealing with missing data for synthetic and real information systems data sets, and then summarize our findings.

Keywords: Information systems · Subspace tracking · *GROUSE* · *PETRELS* · *RankMin*

1 Introduction

Many real world information systems continuously observe and log real-time system behavior for the health and status of the system. Enterprise web application servers in data centers, for example, observe each systems' component usage (e.g. CPU, memory, and network traffic) to accommodate for load balancing, performance tuning, and security auditing. Information systems for critical infrastructures like electric power grids or water systems, for instance, observe data that's critical for the safety of the system such as; the current flow within the grid, the flow and pressure of the water in each pipe and concentration of the chemical levels in the water. Due to the underlying dynamics of such systems, these observations can be viewed as high–dimensional, time–varying, multivariate data. One broad class for concisely modeling these systems is low–rank

This is a U.S. government work and not under copyright protection in the U.S.;
foreign copyright protection may apply 2019
Á. Rocha et al. (Eds.): WorldCIST'19 2019, AISC 930, pp. 791–800, 2019.
https://doi.org/10.1007/978-3-030-16181-1_74

modeling where the observations manifest themselves in a lower-dimensional sub-space. Principal Component Analysis and Subspace Tracking are the essential building blocks for exploring and understanding these observation data.

Typically, data from information systems has many attributes that can be observed at each time tick, t. Each observation may be viewed as a high–dimensional vector. In other words, if a system has n attributes, then for each observation x, we have $x_t \in \mathbb{R}^n$. However, in practice, there may be times when not all attributes are observed or recorded at a particular time, due to cir-cumstances like system or network failure. As a result, we know which system attribute observation x is missing in the log data at any given time t. Therefore, each observation x_t is accompanied by a sampling vector, p_t, $p_t \in \{0, 1\}^n$, where entries of 1 correspond with positions in which an attribute was observed and with 0s otherwise; thereby, generating a matrix with missing entries.

Such unpredictable system properties that can change over time pose a greater challenge for applying low–rank modeling techniques, especially, for naive low–rank modeling approaches like *Singular Value Decomposition (SVD)* in their default formulations. *SVD* is intended for a full observed static matrix. However, using techniques like sliding window analysis can overcome this limitation but poses the challenge for appropriately setting the window size. In order to cap-ture the evolving dynamics of complex systems, low–rank modeling techniques for information systems must be online and able to cope with partial observations.

The three widely studied online subspace tracking algorithms we consider evaluating in this paper are *GROUSE* [1], *PETRELS* [2], and *RankMin* [3]. The behaviors of these algorithms are not sufficiently explored with regard to each other and this knowledge gap can adversely affect deciding the best one for exploring and understanding information systems' log data; thus, is the focus of this work. *GROUSE, PETRELS*, and many others are described in detail in a recent review article on subspace tracking [9], including benchmarks of their performance in different criterion, such as memory and computational complex-ity, and convergence. Furthermore, a high-dimensional analysis of *GROUSE, PETRELS*, and *Oja* [4] recently reported in [5].

Previous analysis of these three trackers were primarily theoretical and not comparative, impeding the ability to select a suitable tracker for a given applica-tion. As such, a comprehensive comparative analysis among these trackers with real and synthetic information systems' data sets is conducted in this paper. We have organized the paper as follows: we give a brief background of the trackers in relation to each other in Sect. 2, followed by their comparative analysis and the summary of our findings in Sect. 3, and conclude our work in Sect. 4.

2   Background

Online Subspace Tracking plays a central role in the signal processing domain [6]. In many of these applications, the data set includes unobserved data entry caused by a malfunctioning system or delay in the network within the observa-tion time interval. As such, information systems observation data points (i.e., log

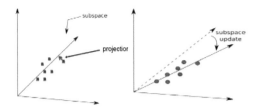

Fig. 1. Illustration of projecting the current observation into the subspace estimate (left) and the subspace updates using the projection on the (right).

entries) exhibit an embedding of low-dimensional structure in a high-dimensional manifold. It is of great significance to concisely model and structure this information for inference and learning with small storage constraint and computational complexity. In this work, we focus on the comparative analysis of state-of-the-art subspace tracking algorithms to aid in selection for a given information system. We'll consider further extending this analysis with regard to inference and learning for the selected algorithm in the future work.

Grassmannian Rank-One Update Subspace Estimation *(GROUSE)* [1], Parallel Estimation and Tracking by REcursive Least Squares *(PETRELS)* [2], and Rank Minimization *(RankMin)* [3] are the three well-known state-of-the-art subspace tracking algorithms for effectively modeling and characterizing incomplete or partial observation data. In general, subspace tracking consists of two generalized steps. The first step is to find the projection of the current observation into the subspace estimate, as depicted in Fig. 1 (left). This step can be formalized by Eq. (1), where a_t is the projection of the current observation, x_t, onto the subspace estimate, $U \in \mathbb{R}^{nxr}$, and r is the rank of the model. The second step is to find an updated subspace using the projection computed in step one, as depicted in Fig. 1 (right). As a result, the traditional subspace tracking fails to deal with partial observations as formalized in Eq. (1).

$$f_t(U) = \min_{a_t} \|U a_t - x_t\|_2^2 \tag{1}$$

Hence, these trackers deal with partial observations by computing both the projection of the current observation into the subspace and subspace update using only the rows in the observation corresponding to the sampled attributes. Therein, the objective function shown in Eq. (1) is updated to reflect the changes as shown in Eq. (2),

$$f_t(U) = \min_{a_t} \|P_t(U a_t - x_t)\|_2^2 \tag{2}$$

where P_t is a zero matrix with the sampling vector, p_t, along its diagonal. It is important to note that there is a theoretical bound as to how many data points may be missing from an observation for these trackers to work effectively. Balzano et al. [1] show this theoretical bound to be $r \log r$, where r is the rank of the complex system being modeled. While existing state–of–the–art subspace trackers fundamentally exploit the structure of the observed data, they do not

Fig. 2. Illustration of the projection and the residual to compute the gradient. The gray surface notionally depicting the Grassmannian manifold, and the dot representing the current location on the manifold (ergo) of the current subspace estimate, and the arrow is the computed gradient shown on the left, and the projection of the gradient onto the tangent space of the manifold depicted the rectangular boxes with the additional arrow depicting the projection of the gradient onto this tangent space shown on the center. On the right show the project from the tangent space back onto the manifold with the updated location on the manifold depicted on the dot.

exploit the structure in the projections of the observations into the subspace. We postulate that this structure can be used to improve the ability to track with overly sparse observations. As such, these trackers differ from one another in how they compute their subspace estimates, discussed next.

2.1   GROUSE

Grassmannian Rank-One Update Subspace Estimation *(GROUSE)* is an efficient online algorithm for subspace identification and tracking when data vectors and the matrix are highly incomplete. It adopts basic linear algebraic manipulations at each iteration and updates can be performed in linear time of the dimension of the subspace. *GROUSE* computes its subspace update by performing gradient descent along the Grassmannian manifold with descent constrained to only orthogonal subspaces. The Grassmannian manifold is the set of all linear subspaces of given dimension and rank. The algorithm is derived by analyzing incremental gradient descent on the Grassmannian manifold of subspaces with a slight modification. As illustrated in Fig. 2, the gradient descent of *GROUSE* is performed following three steps: (1) subspace update is to compute the gradient (left), (2) project the gradient onto the tangent space of the manifold (center), and (3) project from the tangent space back onto the manifold (right). *GROUSE* consumes a single parameter, referred to as of `step size`. The `step size` is used in the above steps and effectively limits the size of the descent along the manifold.

2.2   PETRELS

Parallel Estimation and Tracking by REcursive Least Squares (*PETRELS*) first identifies the underlying low-dimensional subspace and then reconstructs the missing entries via least-squares estimation, if required. Subspace identification is performed via a recursive procedure for each row of the subspace matrix in parallel with discounting for previous observations. The subspace update strategy employed by *PETRELS* varies greatly from that of *GROUSE*. *PETRELS'* subspace update strategy considers all past projections. However, it does not do so with equal importance, as shown in Eq. (3), where λ represents *PETRELS'* only input parameter, known as the `forgetting factor`.

$$U_n = \arg \min_U \sum_{t=1}^{n} \lambda^{n-t} f_t(U) \tag{3}$$

PETRELS employs a recursive least squares technique to solve this equation. This approach allows rows of the subspace to be updated simultaneously, thereby, making it trivial to parallelize an implementation of *PETRELS*. It is important to note that since the update subspace is solved as a least squares problem, as a result, the orthogonality constraint imposed in *GROUSE* no longer holds.

GROUSE performs first-order incremental gradient descent on the Euclidean space and the Grassmannian respectively, whereas *PETRELS* can be interpreted as a second-order stochastic gradient descent scheme. These algorithms have been shown to be highly effective in practice but their performance depends on the careful choice of algorithmic parameters, such as the step size for *GROUSE* and the discount parameter for *PETRELS*.

2.3 RankMin

RankMin is identical to *PETRELS*, except that it incorporates regularization terms into each step of the subspace tracking. This is to cope with inherent numerical stability issues found in the recursive least squares-based algorithm that *PETRELS* employs. Thus, the incorporation of regularization terms changes the first subspace tracking step into that shown in Eq. (4),

$$f_t(U) = \min_{a_t} \| P_t(Ua_t - x_t) \|_2^2 + \beta_t \|a\|_2^2 \tag{4}$$

$$U_n = \arg \min_U \sum_{t=1}^{n} \lambda^{n-t} f_t(U) + \beta_t \|U\|_F^2 \tag{5}$$

where β is the `regularization` parameter. As such, the subspace update step of *PETRELS* changes to the one shown in Eq. (5).

3 Comparative Analysis

The behavior of these three tracking algorithms is not well understood with regards to one another. This adversely affects leveraging them in practice and it is for this reason that comparative analysis is critical. Figure 3 below depicts the distribution and the projection of the subspace used in our analysis. The three criteria of evaluating the model are; *Convergence and Stability, Parameter Sensitivity, and Robustness of Missing Data*. In this section, we first discuss the evaluation framework and settings followed by comparative analysis and the rationale behind each criterion. Finally, we summarize our findings.

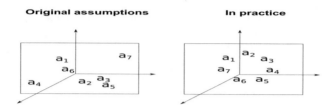

Fig. 3. A notional depiction of assumptions made about the distribution of subspace projections by existing subspace trackers (left) and a notional depiction of structure in how the projections are distributed (right).

3.1   Evaluation Data and Settings

To facilitate the analysis of these trackers, we developed a framework and implemented the three algorithms with Python programming language. In our analysis, we consider both synthetic data and real data from application servers or middleware systems in a private cloud cluster and EPANET [7]. We excluded the evaluations of application servers in our analysis due to space limitations. The real data used in this analysis was gathered from the EPANET water distribution piping simulator. This simulated data has many attributes including flow of water in each pipe, pressure at each node in the distribution network, height of water in each tank, type and concentration of chemicals throughout the network, water age, water source, and tracing. The real data has an ambient dimension of 166 and an apparent effective rank of 4. The synthetic data used in this analysis had an ambient dimension of 100 and an ambient dimension of 10. Observations were generated by projecting a random vector into a random subspace that has been rotated by a constant for each observation.

Since the true subspace was known for only synthetic data, error had to be calculated differently depending on whether the real or synthetic data set was used. The error calculation used for synthetic data is shown in Eq. (6) left and for the real data in Eq. (6) right.

$$E = \frac{\|I - \tilde{U}\tilde{U}^2\|_2}{\|U\|_2} \qquad\qquad E = \frac{\|\tilde{x} - x\|_2}{\|x\|_2} \qquad\qquad (6)$$

3.2   Convergence and Stability

The analysis of the convergence and the stability of the trackers attempts to answer the following questions: (1) *Do the trackers converge to subspaces?* (2) *How quickly do the trackers converge to subspaces?* (3) *Are the trackers able to maintain a subspace?* Although the convergence of *GROUSE* is reported in [8] and recently compared with five other trackers in [9], this criterion explores the selected three algorithms with regard to each other for a given data set.

In this analysis, all tracker parameters were set empirically and then held constant. The *RankMin*'s regularization parameter was automatically set by that tracker's implementation. Observation were sampled at 70%, and the experiment was conducted on both synthetic and real data sets.

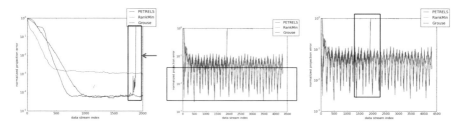

Fig. 4. Stability on synthetically generated data.

Fig. 5. The rectangle box depicts GROUSE difficulties of converging on real data

Fig. 6. The rectangle box depicts PETRELS stability issues on the real dataset

All of the trackers were able to converge to subspaces on synthetic data as shown in Fig. 4. On real data, we again observed that all of the trackers were able to converge to subspaces as shown in Fig. 5, however, *GROUSE* undulated a bit, suggesting that it experienced difficulties in descending to a single subspace on the manifold. We also observed that *PETRELS* suffered from stability issues on real data, as shown in rectangle box in Fig. 6. As a result, we dropped it from the remaining two analysis.

3.3   Parameter Sensitivity

We analyzed the trackers in terms of their sensitivities to their input parameters. For instance, if a tracker is overly sensitive to an input parameter, then its applicability to information systems in highly dynamic environment could be limited, especially, if the parameter is hard to set. For this experiment, free tracker parameters were again set empirically and then held constant. *RankMin*'s regularization parameter was automatically set, observations were sampled at 70%, and the experiment was conducted on both synthetic and real data sets.

For the `forgetting` parameter of *RankMin*, we observed that it achieved the lowest error when its parameter was set to a value between 0.99 and 0.91 on real data, as shown in Fig. 7. In both cases, while the `forgetting` parameter remained influential, it did not show itself to be overly sensitive. We observed the *step size* parameter of *GROUSE* to be quite sensitive when considering synthetic data as shown in Fig. 8, albeit quite a bit less sensitive on real data shown in Fig. 9.

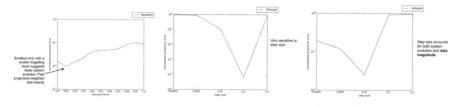

Fig. 7. Performance of RankMin on real data as a function of its `forgetting` parameter

Fig. 8. Performance of GROUSE on synthetic data as a function of its `step size` parameter.

Fig. 9. Performance of GROUSE on real data as a function of its `step size` parameter.

In Summary, *GROUSE* significantly outperformed *RankMin* on synthetic data when a good setting was found for `step size`, but no such setting could be found when it came to real data. One possible explanation to this could be that the `step size` parameter is responsible for tuning the rate of system evolution and the magnitude of attributes within the observations contained in the data. Having to tune for both of these factors, particularly, when the magnitude of the attributes cannot be known a priori, can significantly complicate finding an acceptable setting for `step size` parameter.

3.4   Robustness to Missing Data

This analysis is to understand the trackers in terms of their robustness in dealing with missing data. As noted earlier, log data from information systems are more likely to represent observations with missing data entries, due to system failure, for instance. Another consideration is that a tracker could be more robust to missing data but be slower than a tracker that is less robust. If fully sampled observations were available in a particular application, it would then probably become more desirable to use the fastest of the trackers instead of the tracker most robust to missing data. For this evaluation, free tracker parameters were empirically set and then held constant. *RankMin*'s parameter was automatically set. Sampling rate was varied from 10% to 100%, and the experiment was conducted on synthetic and real data sets.

On synthetic data, we observed the performance of *RankMin* and *GROUSE* to be similar under very aggressive subsampling, when only 10%–20% of the attribute values are available in the observations. However, as shown in Fig. 10, *GROUSE* significantly outperformed *RankMin* as the subsampling became less aggressive. In particular, when sampling between 40% and 90%, *GROUSE* achieved a projection error lower than that of *RankMin* by an order of magnitude. Not until fully sampling the data did the trackers again have similar performance. It is worth nothing, though, that once sampling rates were greater than 20%, error of less than 10^{-3} was consistently achieved by both trackers. In summary, the performance of each tracker was similar across all tested sampling rates as shown in Fig. 11. As for the real data, the differences became less exaggerated as shown in Fig. 12.

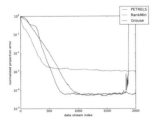

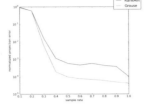

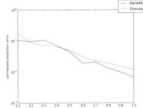

Fig. 10. Results of how trackers performed over time on synthetically generated data

Fig. 11. Tracker performance on synthetic data as a function of missing data.

Fig. 12. Tracker performance on real data as a function of missing data.

3.5 Summary

In our analysis, the *PETRELS* tracker was found to suffer from numeric stability problems on both synthetic and real data sets. This is likely caused by the recursive least squares procedure integrated into its algorithm. *RankMin* and *GROUSE* were found to not suffer from such stability problems. In terms of parameter sensitivity, *RankMin* was found to be quite easy to use, unlike *GROUSE*. *GROUSE* could not only be quite a bit harder to use but it could potentially be entirely unusable, as in a case where the data magnitude varies widely or cannot be known a priori. Despite its higher parameter sensitivity, *GROUSE* could occasionally outperform *RankMin* on synthetic data. It is also important to note that in addition to occasionally outperforming *RankMin*, the subspace estimate provided by *GROUSE* is always already an orthogonal matrix, which could be important in allowing some potential applications to avoid performing even more costly computations.

Generally, the subspace trackers considered in this work are limited in that partial observations they are consuming must contain at least $r \log r$ observed attributes, where r is the rank of the complex system being observed. With this lower bound honored, there are strong theoretical guarantees on the ability of the underlying subspace to be recovered and tracked by these algorithms. For such, the fundamental question is *Is it possible to move this bound even lower?*. Finally, the work conducted under this task led to the question of *"Can tracking with overly sparse observations be improved below the existing theoretical bounds of a minimum of rlogr attributes observed in each observation?"*

4 Conclusions

A comprehensive comparative analysis of *GROUSE*, *PETRELS*, and *RankMin* is discussed. To facilitate the analysis, a modular and extensible experimentation framework was designed and implemented the algorithms, and finally summarized our findings. We consider further experiments in tracking real-time log entries for inference and learning.

Acknowledgments. Authors would like to thank Matthew Paulini for his copy editing and proof reading. The implementation of the algorithms and the experiments was developed by the second author while he was employed at AFRL/RI. First author would like to sincerely thank Jim Hanna, Rick Metzger, Dr. Mark Linderman, Steven Farr and Lt. Col. Scott Cunningham at AFRL/RI for their continuous support and guidance.

References

1. Balzano, L., Nowak, R.D., Recht, B.: Online identification and tracking of subspaces from highly incomplete information. In: 48th IEEE Annual Allerton Conference in Communication, Control, and Computing, pp. 704–711 (2010)
2. Chi, Y., Eldar, C., Calderbank, R.: PETRELS: parallel subspace estimation and tracking by recursive least squares from partial observations. IEEE Trans. Signal Process. **61**(23), 5947–5959 (2013)
3. Mardani, M., Mateos, G., Giannakis, G.B.: Rank minimization for subspace tracking from incomplete data. In: ICASSP (2013)
4. Oja, E.: Simplified neuron model as a principal component analyzer. J. Math. Biol. **15**(3), 267–273 (1982)
5. Wang, C., Eldar, Y.C., Lu, Y.M.: Subspace estimation from incomplete observations: a high-dimensional analysis. In: Proceedings of the Signal Processing with Adaptive Structured Representatives (SPARS) Workshop, Lisbon, Portugal, 5–8 June 2017 (2017)
6. Delmas, J.-P.: Subspace Tracking for Signal Processing. Adaptive Signal Processing: Next Generation Solutions, pp. 211–270. Wiley-IEEE Press (2010). ISBN 978-0-470-19517-8
7. EPANET. https://www.epa.gov/water-research/epanet. Accessed Nov 2018
8. Zhang, D., Balzano, L.: Convergence of a Grassmannian gradient descent algorithm for subspace estimation from undersampled data (2016). arXiv:1610.00199
9. Balzano, L., Chi, Y., Lu, Y.M.: Streaming PCA and subspace tracking: the missing data case. Proc. IEEE (2018). arXiv:1806.04609

A Model-Driven Approach for the Integration of Hardware Nodes in the IoT

Darwin Alulema[1,2]([✉]), Javier Criado[2], and Luis Iribarne[2]

[1] Universidad de las Fuerzas Armadas ESPE, Sangolquí, Ecuador
doalulema@sespe.edu.ec
[2] Applied Computing Group, University of Almería, Almería, Spain
{javi.criado,luis.iribarne}@ual.es

Abstract. We currently live in continuous interaction with people and things, giving rise to the era of the Internet of Things (IoT). This has led the creation of new applications in diverse fields such as asset and stock tracking, transportation, electricity grids, industry automation, smart homes, agriculture or sports, among others. However, the growing number of platforms and the growing variety of end devices make application development a difficult task that requires a lot of time. A technology currently being used to solve such problems is modeling, because it can enhance the reuse of different elements to simplify developers' work. Model-Driven Engineering (MDE) suggests a development process based on model making and transformation. For this reason, we propose a solution based on models to generate code automatically. Specifically, we focus on a Domain-Specific Language (DSL), a graphical editor and a Model to Text (M2T) transformation for hardware-node code generation. The proposed methodology automates the development process, allowing developers not to have an in-depth knowledge of all hardware and software platforms. To demonstrate this approach, a scenario for a smart home (with different sensors and actuators) has been designed, as well as an application for mobile devices, which allows system to monitor and control the scenario.

Keywords: Model-Driven Engineering (MDE) ·
Domain Specific Language (DSL) · Internet of Things (IoT) ·
Smart home

1 Introduction

The concept of Internet of Things (IoT) introduces the vision of a global network of objects that interact with people [9]. This interaction allows objects to be seen by applications as means to be aware of the surrounding environment [9]. The knowledge offered by interconnected objects allows the creation of a global network where it is possible to develop new applications and scenarios, such as smart cities, health, transportation, and industry monitoring, among others [6]. IoT has grown significantly in recent years due to the increase of smartphones,

© Springer Nature Switzerland AG 2019
Á. Rocha et al. (Eds.): WorldCIST'19 2019, AISC 930, pp. 801–811, 2019.
https://doi.org/10.1007/978-3-030-16181-1_75

tablets, smart TVs and smart speakers, among other devices, coming onto the market [3]. In fact, it is expected that billions of devices play an important role in the future network, bringing data from the physical world to the world of digital content and services [1]. By including sensors, these devices allow to continuously collect data from their environment and people, without interfering with daily activities [15].

Creating IoT applications in the real world can be a challenge, even for experienced developers. This is mainly due to hardware and software heterogeneity [5]. For this reason, low-level programming skills related to the hardware platform are required, as well as a deep knowledge of the application field and the business model requirements [13]. To show the complexity of the scenarios that may arise in the IoT, let us provide an example to use as a case study in this text. The proposed scenario consists of a house and a garden, which are shown in Fig. 1. Inside the house there are a bulb, a temperature sensor, a smoke (CO_2) sensor, a speaker, a contact switch, a humidity sensor, a fan, a thermostat and a humidity sensor coordinated by a controller that manages the Internet connection. The garden has a soil humidity sensor and an irrigation system. Information generated by the sensors and used by actuators is stored in a database on an Internet server. To control and monitor the Smart Home system, the user, who owns the house, has an application he opens on his smartphone to access the database and update the values associated with the house sensors and actuators. Users receive information about ambient temperature, humidity, smoke in atmosphere, seismic vibrations, garden soil humidity, and the status of the contact switch in their smartphone. They can also enable or disable the irrigation system, fan, heating, light bulb, speaker and light.

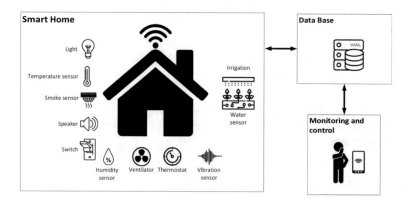

Fig. 1. Heterogeneous hardware systems in a Smart Home scenario.

The scenario used as a case study shows the viability of using cyber-physical components for homes. However, these components can be very heterogeneous, and their implementation may get really difficult. This diversity makes it difficult to develop new approaches, since developers must have extensive hardware and software knowledge to adapt systems to new scenarios. For this reason, herein we propose a methodology to develop IoT hardware nodes adaptable to new scenarios. Developers will be able to incorporate specific sensors and actuators for each scenario. The rest of this article is organized as follows. Section 2 reviews some related work. Section 3 presents a Domain-Specific Language (DSL), a graphical editor and a Model to Text (M2T) transformation for hardware nodes code generation. Section 4 shows the feasibility of the approach through a test scenario. Finally, future work and conclusions are drawn in Sect. 5.

2   Related Work

In literature, there are different types of scenarios in which the IoT is present. One of these are the health-care services, such as that proposed in [15], in which authors suggest a system to monitor physical activity in real time in a smart home. This solution uses multi-task learning, and dictionaries and reasoning learning based on rules to observe and quantify changes in readings of the sensors installed in domestic environments to carry out a continuous monitoring of residents' behavior and to detect any abnormal event for early medical assistance. The main difference with our approach is the use of Model-Driven Engineering (MDE) for application design. Our approach, by using generic models, can be used in different domains.

In [4] authors suggest an intelligent home using cyber-physical systems to monitor and measure physical activities. In this case, authors propose a component-based architecture to incorporate heterogeneous cyber-physical systems. Each device can be managed as an encapsulated component within the IoT concept. The approach enables interoperability by applying a homogeneous component representation providing communication functions via web sockets and implementing gateways. In contrast, our proposal models the cybernetic components at a higher level of abstraction, allowing them to be used in a wide variety of IoT applications. The use of MDE in a heterogeneous environment, such as the IoT, allows accelerating the development process because it standardizes the parameters of the cybernetic components. In this way, other developers can create new applications. In addition, our approach uses M2T transformations, which makes it possible to automate the software design process for hardware components.

IoT Industry or Industry 4.0 is another field in constant development because it allows process automation. In [9] authors propose an architecture based on five layers: (a) Detection, (b) Database, (c) Network, (d) Data-response, and (e) User layer. Authors test the architecture in a metamodel for social networking integration with Industry 4.0. A Raspberry Pi, a cloud storage server and a mobile device to control production processes online are used for implementation. A difference with our approach is the development of a graphical editor that allows application design visually, and the use of the Arduino platform [17], which is more focused on hardware.

Due to the different application scenarios for the IoT, underlying technologies have also evolved. A technology that has benefited from this development are Wireless Sensor Networks (WSN). For this reason, in [11] authors propose a modeling platform for a development architecture and WSN networks analysis by code generation. The proposed platform consists of three modeling languages to describe specific architecture views of a WSN sensor network: (a) Software architecture modeling language, (b) Node modeling language, and (c) Environment modeling language. The main difference with our approach is the use of the 802.11 protocol (WiFi), enabling hardware nodes control and monitoring to be conducted remotely because the nodes are treated as another element in the services. Another technology powered by the IoT are hardware platforms operating systems. In this context, in [5] authors propose a model-based tool to develop and configure applications for Contiki OS, but specifically for the network protocol stack. Unlike this project, our approach uses drivers not running operating systems, and includes sensors and actuators.

In [9] authors conducted a study to determine the lines of research for modeling and automatic code generation for wireless sensor network applications. Their results show the prevalence of proposals for generating code based on MDE approaches, highlighting their benefits and suitable features for code generation. They also mention the relevance of smartphones for new IoT applications, because they include several communication protocols and have a large number of sensors. For that matter, as proposed in [3], the author presents a device-independent architecture separating the applications from the devices and enabling application development. To demonstrate the approach, the author introduces a scenario to implement the service-oriented architecture for mobile devices to access resources. Furthermore, crowdsourcing is used to determine application performance in relation to delay to access the services. The main difference with our approach is the use of MDE to develop hardware nodes modeling that interconnect with other applications using a services architecture.

Another approach where MDE is used is [7], here authors propose a DSL to create software to interconnect any object to an IoT platform. The approach uses a graphical editor to create objects that can be interconnected to a platform. The architecture is divided into three layers: (a) Object definition, (b) Service objects generation, and (c) Objects. The main difference of our proposal is code generation for a development board, used to create a hardware node with WiFi connectivity that can be used in any kind of scenario. Another project that deals with creating hardware nodes with MDE is [18], where authors propose a system for generating Arduino code for children as a tool for learning by creating small programs for Arduino. Created programs can control small electronic modules, so the design process is simplified. Unlike our proposal that generates code for more complex components oriented to a web-services connection.

As shown in referenced projects, there is great interest in IoT; however, not many approaches using MDE for application design have been developed yet. For this reason, our proposal contributes to the state-of-the-art of IoT because of the following main reasons:

(a) A modeling language for circuit description of IoT hardware node is defined.

(b) A methodology for modeling and code generation using a graphical editor and an M2T transformation is provided.

(c) A mechanism for nodes to have connectivity to a service-based architecture and be treated as a software component is proposed.

3   Proposed Methodology

This section describes the methodology proposed for application design with hardware nodes according to MDE.

3.1   A Metamodel for Hardware Nodes

Eclipse Modeling Framework (EMF) [19] was used for model development. Figure 2 shows the metamodel that describes the services and hardware nodes built on a development platform (*e.g.*, Arduino and Raspberry Pi). These nodes also have connectivity (*e.g.*, USB, serial, Bluetooth, WiFi and Ethernet) to be linked with other nearby devices or services. The platform also allows to control multiple types of analog or digital components (*i.e.*, sensors and actuators). However, the proposed metamodel is at a higher level of abstraction, so many details of the interfaces of ports and components are not considered. At this level, ports have been represented as inputs and outputs, and components as sensors and actuators. This representation allowed to simplify the process of connection of the components and the controller.

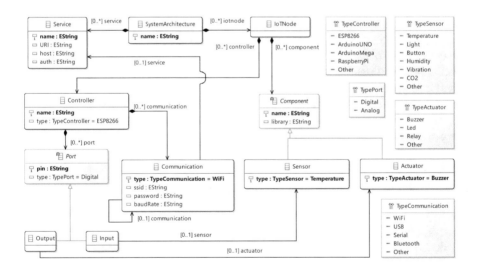

Fig. 2. A metamodel that connects some IoT hardware nodes.

A graphical editor based on the previous metamodel was created for hardware node design. Figure 3 shows a screenshot of the editor and the View Specification

Model (VSM) setup file, which has been developed with Sirius [21]. The VSM allows configuring classes and their relations, it also allows specifying which objects are displayed and how they are shown. Diagram was used for metamodel configuration, which includes: nodes (service), containers (IoT node), subnodes (controller, sensor and actuator), edge nodes (communication and ports), and tool options to create system elements and the connections between them.

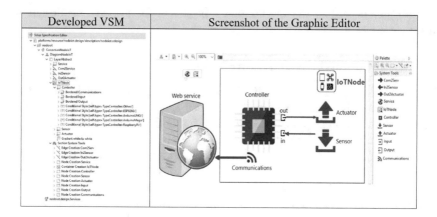

Fig. 3. Screenshot of the tool for developing the Graphical Editor.

The graphical editor has two sections: (a) Canvas, to display and edit the diagram, and (b) Tool palette, which includes the components to draw and edit the visual representation. The palette also allows to define the service that will be connected to the circuit and the circuit to be implemented. The components available in the palette are: (a) Controller, displays the development board, (b) Communications, establishes the communication protocol that will be used for communication, (c) Input and Output, elements of the development board used to connect sensors or actuators, (d) Sensors, allow selecting the sensors that will be connected to the board, and (e) Actuators, allow selecting the actuators that will be connected to the board.

3.2   M2T Transformation

Code generation takes place after defining the metamodel. M2T transformation process should be implemented to achieve this goal, Acceleo was used in order to do so [16]. In this case, according to the model of the scenario, transformation takes place for the development board. To define transformation rules, the Arduino platform [12] was considered due to the simplicity of use, low cost and community creation [14]. The structure of Arduino programs is divided into three sections: (a) Variable declaration, (b) setup() function, where all the initial parameters for the board are initialized, and (c) loop() function, where software that will run continuously is programmed.

ESP8266 is the specific development board model for testing. This board can receive and send information through the Internet [2]. The main features that make the ESP8266 particularly attractive for IoT applications are [8]: (a) Compliance with the IEEE 802.11 standard, (b) WPA/WPA2 security protocols, (c) 32-bit and 160 MHz Tensilica L106 RISC processor, (d) Internal 36 kB SRAM, and (e) 4 MB external SPI Flash. Firebase Cloud System was also used [20], which is a combination of various Google services, including instant messaging, user authentication, real-time database, storage and hosting, among others [10].

Arduino projects generated by the tool consist of three sections: (a) Global variable declaration and library calling, (b) Initial project setup, and (c) Execution loop. Library (e.g., DHT.h, ESP8266WiFi.h and FirebaseArduino.h) calling takes place according to this structure, the declaration of constants associated with sensors and actuators, and the controller board pin port, which can be analog or digital. Then the **setup()** function, which declares the port operation mode, is configured, the DHT11 sensor is initialized, WiFi-network connection is configured, is connected to Firebase and the variable identifier is created. Finally, in the **loop()** function instructions for the sensors and actuators are created and associated with a service variable. Thus the driver behaves like a gateway, sending and receiving data with the service.

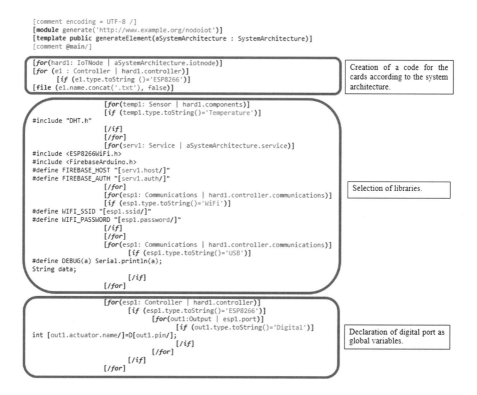

Fig. 4. M2T transformation with Acceleo.

Figure 4 shows a code snippet implemented in Acceleo for Arduino code generation. First, this figure shows the source code files creation according to the board that may exist in the design. Then, the selection of the libraries that will be used takes place. Finally, we see the global variable declaration associated with sensors or actuators.

4   A Case Study Scenario

To demonstrate the functionality of our approach, a scenario shown in Fig. 5 was designed and discussed in the introduction. This scenario has two boards: (a) the first controls two LEDs, one DHT11 sensor measuring ambient humidity and temperature, one CO_2 MQ7 sensor, one buzzer and one switch, and (b) the second board controls one FC28 soil humidity sensor, one LED, one switch, one SW520D vibration sensor, and four relays. The graphical editor is used for architecture design and setup parameters (WiFi network, database, pin number and port type) are entered. With the system architecture in place, source code (.ino) with M2T transformation is created.

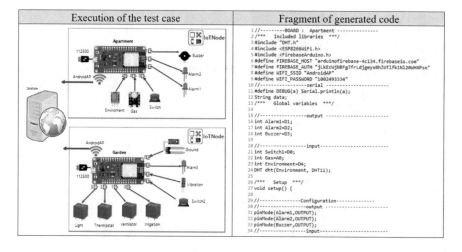

Fig. 5. System architecture for Smart Home and code generated for the first board.

Before the etching process, it is necessary to complete the circuit, which must be consistent with the settings configured using the graphical editor. In order to control the system, an Android application that allows managing actuators and displaying sensor values in real time was developed. Thus the sensor data and the actuator status can be displayed on the mobile device side. Figure 6 shows some screen-shots of this app built.

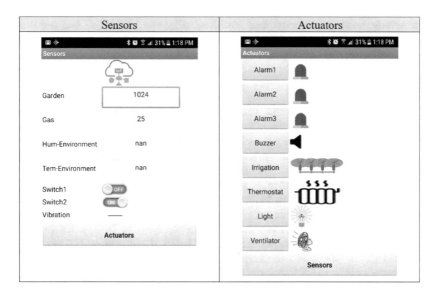

Fig. 6. Screenshots of the Android App.

5   Conclusions and Future Work

This article has proposed a graphical editor and an M2T transformation for code generation based on MDE to simplify hardware node design for the IoT. The approach allows to standardize applications description representing them as models, enabling developers to define new architectures without deep knowledge of hardware and software for IoT systems design. Created nodes can include several sensors and actuators, which may be associated with an Internet service.

The DSL is one of the software tools of the proposed methodology and is used in M2T transformation. To transform the model, the Arduino platform was used because it offers a number of benefits: (a) it is free and expandable, (b) has a large community, (c) multiplatform programming environment, (d) simple programming language, (e) low cost, and (f) reusable. In addition, some real devices such as a temperature sensor, a smoke (CO_2) sensor, a contact switch, an environmental humidity sensor and a soil humidity sensor were used. Several indicator lights (LEDs), a speaker and four relays were also used to control a fan, a thermostat, an irrigation system and a bulb. To define the specific model for the scenario, we used the proposed graphical editor, which allows configuring the hardware nodes used for the application graphically.

As future work, some secondary lines are still undone: (a) expanding the methodology for the automatic code generation for other hardware platforms, (b) expanding the metamodel to include instructions for control, loops and operations, among others, (c) automating the methodology to automatically generate mobile or web applications, and (d) performance testing.

Acknowledgments. This work has been funded by the EU ERDF and the Spanish Ministry MINECO under the AEI Projects TIN2013-41576-R and TIN2017-83964-R.

References

1. Atzori, L., Iera, A., Morabito, G.: From "smart objects" to "social objects": the next evolutionary step of the IoT. IEEE Com. Mag. 97–105 (2014)
2. Badamasi, Y.: The working principle of an Arduino. In: 11th International Conference on Electronics, Computer and Computation (ICECCO), pp. 1–4. IEEE (2014)
3. Chmielewski, J.: Device-independent architecture for ubiquitous applications. Pers. Ubiquit. Comput. **18**, 481–488 (2014)
4. Criado, J., Asencio, J., Padilla, N., Iribarne, L.: Integrating cyber-physical systems in a component-based approach for smart homes. Sensors **18**(7), 2156 (2018)
5. Gomes, T., Lopes, P., Alves, J., Mestre, P., Cabral, J., Monteiro, J., Tavares, A.: A modeling domain-specific language for IoT-enabled operating systems. In: Annual Conference of the IEEE Industrial Electronics Society, pp. 3945–3950. IEEE (2017)
6. Gonçalo, M., Garcia, N., Pombo, N.: A survey on IoT: architectures, elements, applications, QoS, platforms and security concepts. In: Advances in Mobile Cloud Computing and Big Data in the 5G Era, vol. 22. Springer (2016)
7. Gonzalez, C., Pascual, J., Nuñez, E., García, V.: Midgar: domain-specific language to generate smart objects for an Internet of Things platform. In: Conference on Innovative Mobile and Internet Services in Ubiquitous Computing, pp. 352–357. IEEE (2014)
8. Mesquita, J., Guimares, D., Pereira, C., Santos, F., Almeida, L.: Assessing the ESP8266 WiFi module for the Internet of Things. In: International Conference on Emerging Technologies and Factory Automation (ETFA), pp. 784–791. IEEE (2018)
9. Rodríguez, J., Cueva, J., Montenegro, C., Granados, J., González, R.: Metamodel for integration of Internet of Things, social networks, the cloud and industry 4.0. J. Ambient Intell. Human. Comput. **9**(3), 709–723 (2017)
10. Li, W., Yen, C., Lin, Y., Tung, S., Huang, S.: JustIoT Internet of Things based on the firebase real time database. In: International Conference on Smart Manufacturing, Industrial and Logistics Engineering (SMILE), pp. 43–47. IEEE (2018)
11. Malavolta, I., Mostarda, L., Muccini, H., Ever, E., Doddapaneni, K., Gemikonakli, O.: A4WSN: an architecture driven modelling platform for analysing and developing WSNs. Softw. Syst. Model. 1–21 (2018)
12. Saha, S., Majumdar, A.: A review of Arduino board's, Lilypad's and Arduino shields. In: Devices for Integrated Circuit (DevIC), pp. 307–310. IEEE (2017)
13. Teixeira, S., Alves, B., Gonçalves, J., Filho, P., Rossetto, S.: Modeling and automatic code generation for wireless sensor network applications using model driven or business process approaches - a systematic mapping study. Syst. Softw. **132**, 50–71 (2017)
14. Torroja, Y., López, A., Portilla, J., Riesgo, T: A serial port based debugging tool to improve learning with Arduino. In: Conference on Design of Circuits and Integrated Systems (DCIS), pp. 1–3 (2015)
15. Yao, L., Sheng, Q., Benatallah, B., Dustdar, S., Wang, X., Shemshadi, A., Kanhere, S.: WITS: an IoT endowed computational framework for activity recognition in personalized smart homes. Computing **4**, 369–385 (2018)

16. Acceleo — The Eclipse Foundation (2018). https://www.eclipse.org/acceleo/
17. Arduino (2018). https://www.arduino.cc/
18. Arduino Designer (2018). https://github.com/mbats/arduino/wiki
19. Eclipse Modeling Project (2018). https://www.eclipse.org/modeling/emf/
20. Firebase (2018). https://firebase.google.com/
21. Sirius: The easiest way to get your own Modeling Tool (2018). https://www.eclipse.org/sirius/

Multiphase CFD Simulation of Photogrammetry 3D Model for UAV Crop Spraying

Héctor Guillermo Parra[1]([✉]), Victor Daniel Angulo Morales[2]([✉]),
and Elvis Eduardo Gaona Garcia[2]([✉])

[1] Facultad de Ingeniería, Fundación Universitaria Colombo Germana,
Bogotá, Colombia
hector.parra@colombogermana.edu.co

[2] Doctorado en Ingeniería, Universidad Distrital Francisco José de Caldas,
Bogotá, Colombia
vdangulom@correo.udistrital.edu.co, egaona@udistrital.edu.co

Abstract. In this paper, Computational Fluid Dynamics (CFD) is used
to perform a transient analysis of physical variables associated with aerodynamic and water behavior an unmanned aerial vehicle (UAV) for crop
spraying. The simulation is used to observe vortex, pressure, and turbulence kinetic energy (TKE) and particle volume with post-processing.
These variables facilitate the aerodynamic and fluid evaluation in aerial
devices. In the research, images of a plant were captured and the 3D
model was obtained through digital photogrammetry techniques. CFD
studies are currently being conducted to improve efficiency in a large
number of devices such as wind turbines, aircraft and, as in our case,
unmanned aerial vehicles in precision agriculture. It was observed how
the aerodynamic and water fluid behavior varies in time and the simulation show an impact on the coverage area in the 3D model of the
plant.

Keywords: UAV · Computational fluid dynamics ·
Precision agriculture · k–ε model

1 Introduction

Aerial spraying is a type of engineering in which the prediction of tracking fluids
is one of it's main problems. In 2018 an analysis of computational fluid dynamics
(CFD), techniques was performed to predict the trajectories of droplets in the lift
off of an airplane [1]. Computational fluid dynamics is a field of fluid mechanics
that is based on the development of numerical analysis and algorithms that seek
to solve and analyze problems related to fluid flow. In general terms, it can be
described as the analysis of a volume of interest that is known in the literature
by a computational domain [2]. Once this domain is established, the fluid within
the volume is analyzed in samples, which is called mesh.

© Springer Nature Switzerland AG 2019
Á. Rocha et al. (Eds.): WorldCIST'19 2019, AISC 930, pp. 812–822, 2019.
https://doi.org/10.1007/978-3-030-16181-1_76

In 2016, a CFD simulation research of nozzles for aerial vehicle fumigation through the Tun Hussein Onn University in Malaysia, concludes that a CFD simulation is useful to improve the quality of materials and reduce the drift of the fumigation with UAV [3]. Others research has shown the impact of simulation with CFD techniques, in the design of components and nozzles to optimize inputs in the area of precision agriculture [4–7].

The unmanned aerial vehicles (UAV), began their development in military practice, to then perform functions in civil environments. These vehicles have also been used to measure mineral exploration. By the year 2004, 2% of the UAVs that were in the United States corresponded to civil operation [8], but thanks to the development and impact of these technologies in different fields of science, have allowed growing up. The report of the International Association of Unmanned Vehicles (AUVSI) on the economic impact of the integration of Unmanned aerial vehicles in the United States shows that the benefit will be 82 billion dollars and more than 100,000 jobs created [9].

The purpose of this article is to create a CFD simulation of a UAVs that are equipped with nozzles to fumigate crops in precision agriculture applications. With a rotating domain (blade), a domain that corresponds to a fumigation nozzle and a 3D model of a plant, which can be compared with mechanical and physical modifications for future simulations, which ease the study of structural behavior, turbulence analysis and fluid coverage area.

Six simulations were carried out to show results. The first three simulations were carried out assuming that the plant was a sphere-shaped object and modifying the location of the fumigation nozzle in three positions. The other three simulations were made by taking a 3D model of a real potato plant, this model was generated from photogrammetry techniques.

The simulation was developed with the phases of the MEF method and the turbulence model k–ε [10]: CAD design, meshing, a configuration of physical variables, simulation, post-processing and an analysis of contour results and coverage area.

Turbulence Model k–ε. The model k-ε, It is the most commonly used and is generally defined as a standard model of turbulence for CFD simulation. The model calculates the effective viscosity and the equations of the kinetic energy and the dissipation of the turbulence to simulate. The Eq. 1 is the effective viscosity, as it is the viscosity of a fluid under specific conditions.

$$\mu_{eff} = \mu + \mu_t, \mu_t = \frac{C_\mu \, \rho k^2}{\varepsilon} \tag{1}$$

Effective viscosity: μ_{eff}, viscosity of the fluid: μ, turbulent viscosity: μ_{t}, density of the fluid: ρ, the constant C_μ, kinetic energy: k, and turbulence dissipation ε.

The values of **k** and ε of the turbulence model come directly from the kinetic energy differential transport mathematical model, Eq. 2, and the turbulence dissipation rate, Eq. 3.

$$\frac{\partial(\rho k)}{\partial t}+\frac{\partial}{\partial x_j}\left(\rho U_j k\right)=\frac{\partial}{\partial x_j}\left(\left(\mu+\frac{\mu_t}{\sigma_k}\right)\frac{\partial k}{\partial x_j}\right)+P_k-\rho\varepsilon+P_{kb} \qquad (2)$$

$$\frac{\partial(\rho\varepsilon)}{\partial t}+\frac{\partial}{\partial x_j}\left(\rho U_j\varepsilon\right)=\frac{\partial}{\partial x_j}\left(\left(\mu+\frac{\mu_t}{\sigma_\varepsilon}\right)\frac{\partial\varepsilon}{\partial x_j}\right)+\frac{\varepsilon}{k}(C_{\varepsilon1}P_k-C_{\varepsilon2}\rho\varepsilon+C_{\varepsilon1}P_{\varepsilon b}) \qquad (3)$$

The magnitude of velocity U and $\sigma_\varepsilon, C_{\varepsilon1}, C_{\varepsilon2}$ are constants of turbulence model k–ε, σ_k is constant, and finally in the turbulence model is calculated with P_k, which is the production of turbulence due to viscous forces, Eq. 4.

$$P_k=\mu_t\left(\frac{\partial U_i}{\partial x_j}+\frac{\partial U_j}{\partial x_i}\right)\frac{\partial U_i}{\partial x_j}-\frac{2}{3}\left(\frac{\partial U_k}{\partial x_k}\right)\left(3\mu_t\frac{\partial U_k}{\partial x_k}+\rho k\right) \qquad (4)$$

P_{kb} y $P_{\varepsilon b}$ represent the production of turbulent influence by flotation forces, and are defined by the Eqs. 5 and 6.

$$P_{kb}=-\frac{\mu}{\rho\sigma_\rho}\,g_i\frac{\partial\rho}{\partial x_i} \qquad (5)$$

$$P_{\varepsilon b}=C_3\mathrm{max}\left(0,\ P_{kb}\right)\sin\theta \qquad (6)$$

The dissipation coefficient $C_3=1$, with Schmidt turbulence number $\sigma_\rho=1$, the gravity vector g_i and the angle between speed and gravity vector θ.

2   Data and Methods

2.1   Crop Data

The information was captured on April 7, 2018. The optical sensor was installed on a base 2 m above ground level to capture information, to obtain images in the visible spectrum. The images were captured from 12:01 pm until 12:27 pm, local time Colombia (GMT-5). This ensured the necessary lighting conditions to minimize the effects of shadows on the crop. A total of 238 images were captured. The optical sensor measures 6.16×4.62 mm with a resolution in pixels of 4608×3456, the pixel size is 1.34 um.

(i) **Study area.** The study area is a potato crop located in the Cundinamarca zone. This zone is located in the western part of the savannah of Bogotá, which is part of the Cundiboyacense high plateau at an altitude of 2600 m above sea level. The potato crop comprises an area of about 1/4 ha. The altitude values of this area range from 2650 m to 2678 m above sea level. This area was photographed with an optical Canon A2300 sensor.

(ii) Data processing. At this stage, the photogrammetric adjustment of the images captured with the optical sensor was made, the software used to make this adjustment is Agisoft Photoscan. This software searches for similar points by analyzing all the captured images. The algorithm used to perform this photogrammetric adjustment is the SIFT adjustment function for artificial vision, which also uses binary descriptors to facilitate the work of the processor and to match the key points of each image with its adjacent ones. With the alignment of the images, the coinciding points are calculated in their XYZ coordinates and with each point, an interpolation is made to form an irregular triangular mesh, Fig. 1a, and a dense cloud points, Fig. 1b, with positioning attributes and color value captured by the optical sensor.

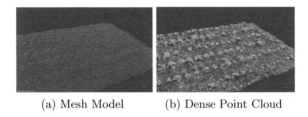

(a) Mesh Model   (b) Dense Point Cloud

Fig. 1. Photogrammetric adjustment for the crop.

(iii) Data post-processing. At this stage, the processing of the generated mesh information was performed. The objective of this stage is to segment and perform smoothing processes to the current information to have data that can be simulated through CFD techniques. Since the data structure that can be simulated with CFD techniques corresponds to a solid object, the dense cloud is filtered to obtain which surface objects can be closed to generate a solid object. To achieve this condition, three candidate plants were obtained to be simulated, Fig. 2. The element that is chosen to perform the simulation, is chosen by the number of faces and vertices, in the Table 1, the main attributes of the models.

Table 1. Attributes of candidate plants models.

	Selected plants		
Attribute	Plant 1	Plant 2	Plant 3
Dense point cloud-Points:	2.662.184	2.460.443	2.099.783
Mesh - Faces:	177.105	163.669	139.709
Mesh - Vertices:	88.731	82.006	69.986
Mesh - Vertex colors:	3 Bands, uint8	3 Bands, uint8	3 Bands, uint8

At the next stage the mesh is processed with three filters: (i) Decimate Filter, (ii) Smooth Filter, (iii) Close holes Filter, Fig. 3. This is done to have a mesh of

(a) Plant1-PointCloud   (b) Plant2-PointCloud   (c) Plant3-PointCloud

(d) Plant1-Mesh Model   (e) Plant2-Mesh Model   (f) Plant3-Mesh Model

Fig. 2. Dense point cloud and mesh model of selected plants

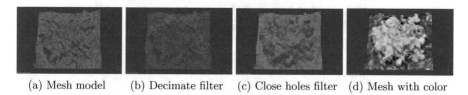

(a) Mesh model   (b) Decimate filter   (c) Close holes filter   (d) Mesh with color

Fig. 3. Filters

the object that can be simulated by CFD methods. Once the filtering process is finished, there is a solid object with 38,716 faces and 19,539 vertices, maintaining a structural distribution of the original object, Fig. 3d.

2.2   CFD Simulation

To define the size of the finite element, it is necessary to pre-simulate CFD, with several element sizes or mesh convergence analysis. A medium size in the simulations of the element on the surface of the UAV was established, due to the behavior of the values of the force of the surface in the plant. To analyze the vortex, it is common to generate a mesh with influence on the structure of the UAV that causes an increase in the elements, and have a finer mesh near the object to be simulated. The CFD simulation has the following steps: (i) CAD design and assembly design, (ii) CFD meshing, (iii) simulation parameters and (iv) design of experiments.

(i) CAD design. Using a CAD design tool, the UAV is assembled to analyze with CFD, in order to achieve adequate simulation results, the body of the UAV was covered with a cylinder to avoid the construction of a mesh of smaller size of elements, which allows a faster simulation, Fig. 4a. Also, the design of the nozzle is made to simulate a constant flow of water of 10 m/s, Fig. 4b. Figure 4c, the CAD design and assembly is represented, parts of the UAV are removed from the CAD design and only one of the rotation domain is simulated. The parts of the drone were replaced by cylinders to ease the mesh, additionally in order to obtain a successful CFD simulation only one rotor with variations of the position of the nozzle keeping the physical dimensions of the blades was placed.

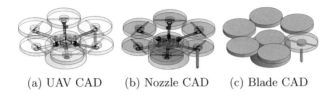

(a) UAV CAD   (b) Nozzle CAD   (c) Blade CAD

Fig. 4. Design and Assembly CAD - UAV and nozzle

(ii) CFD meshing. Figure 5, shows the mesh generated by ICEM of the ANSYS software, which is used to perform the simulations without mesh options. generating a basic rotating domain mesh, Fig. 5a, and two mesh of basic static domains, Fig. 5b and c, the mesh has a total of 653,487 nodes and 3,719,512 finite elements approximately by simulation.

(a) Domain mesh   (b) Sphere domain   (c) 3D model domain

Fig. 5. Meshing of static and rotating domains ICEM

(iii) Simulation parameters. Figure 6, shows the configuration of the edge surfaces of the domain in "opening" configuration, with the surface of the base or floor in "wall" configuration, the output vectors in the nozzle are also shown. To perform the multi-phase or DPM simulation, the following parameters were defined: 2000 rpm for the rotation speed of the blade, 10 m/s the speed of the fluid, 25 °C the temperature of the fluids water and air, 10 m/s of gravity, turbulence model k–ε and initial pressure in the domains is 1 atm. The regions and surfaces of the rotor and the plant were configured as a "wall" and the simulation time was 1 s with time intervals of 0.1 s.

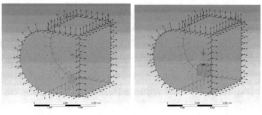

(a) Sphere parameters   (b) 3D model parameters

Fig. 6. Simulation parameters

(iv) Experiment design. The simulation of CFD was done with two static domains with three variations of the position of the nozzle, a total of 6 combinations or simulations as shown in the Fig. 7. In the first static domain a sphere is located two meters from the center of mass of the UAV and centered with the rotating domain. In the second static domain the 3D reconstructed model of the potato plant is added on the upper surface of a 1000 mm cube of edge at a distance of 1 m from the center of mass of the UAV. C is the position of the nozzle with regard to the center of the rotor, and r is the radius of the rotor.

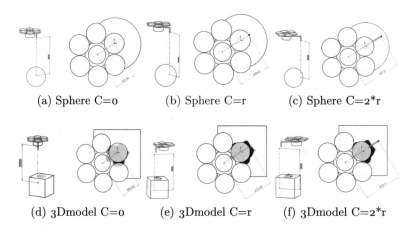

(a) Sphere C=0   (b) Sphere C=r   (c) Sphere C=2*r

(d) 3Dmodel C=0   (e) 3Dmodel C=r   (f) 3Dmodel C=2*r

Fig. 7. Nozzle position in each simulation

To perform the simulations, the nozzle was located at a distance c, in three positions; (i) c = center of the rotating domain, (ii) c = radius of the rotating domain, or the edge of the blade and (iii) c = twice the radius of the rotating domain.

3   Results

Figure 8, show a water distribution of the simulations, the contours of greater coverage and density of water in the regions of the plant when the nozzle is in the center of the rotor, Figs. 8a and d.

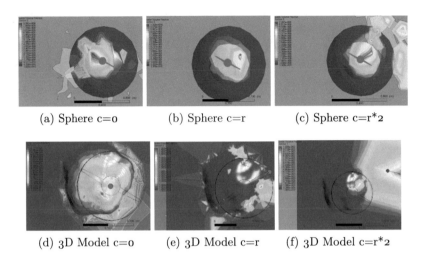

(a) Sphere c=0   (b) Sphere c=r   (c) Sphere c=r*2

(d) 3D Model c=0   (e) 3D Model c=r   (f) 3D Model c=r*2

Fig. 8. Contour plane of volume fraction for experiment designs

Figure 9, show the isometric simulation view, the water particles and as the particles vary by the effect of blade or the rotation domain and the position of the nozzle in each of the simulations. Due to the fact the 3D model of the plant is centered in the rotation domain. It observes a smaller coverage with the displacement of the nozzle, nevertheless, it is observed that in the cases where the blade is at a double distance from the radius, Fig. 9c, there is an effect on the particles, where the liquid can penetrate the lower structure of the plant. Table 2, shows the distance measurements from the center of the rotor to the maximum limit of the contour of the amount of water fluid in the object's surface (Sphere or 3D Model), also shows the diameter of the maximum contour of the volume fraction = 1 of the water fluid in millimeters.

Table 2. Maximum contour diameter of volume fraction.

Attribute	Experiments					
	Expe. 1	Expe. 2	Expe. 3	Expe. 4	Expe. 5	Expe. 6
Distance[a] (mm)	160	200	20	5	140	600
Diameter[b] (mm)	60	40	80	20	10	60

[a]Distance from rotor center to maximum contour of water volume fraction (mm).
[b]Diameter of maximum contour of volume fraction = 1 of water in plant (mm).

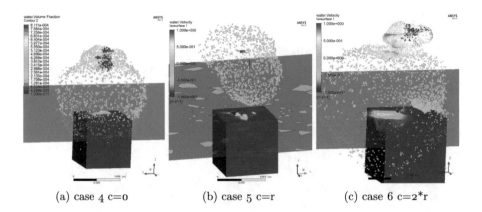

| (a) case 4 c=o | (b) case 5 c=r | (c) case 6 c=2*r |

Fig. 9. Contour plane of volume fraction for 4–6 experiment designs

Table 3, shows the diameters of the maximum spray lobe in meters and the boundary diameter of the volume fraction when it is 0.002. Figure 10, shows the simulation of the water particles and the air flow lines and their projection in the 3D model of the plant. The dynamic behavior of these two fluids is similar,

Table 3. Diameter of maximum contour of volume fraction.

Attribute	Experiments					
	Expe. 1	Expe. 2	Expe. 3	Expe. 4	Expe. 5	Expe. 6
Diameter A[a](m)	1.7	1.5	2.0	1.7	1.5	2.0
Diameter B[b](mm)	300	250	250	250	50	600

[a]Maximum diameter of spray lobe (m).
[b]Contour diameter of volume fraction = 0.002, water fluid in plant (mm).

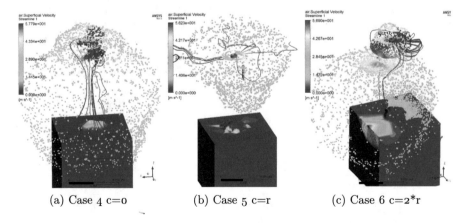

| (a) Case 4 c=o | (b) Case 5 c=r | (c) Case 6 c=2*r |

Fig. 10. Contour plane of air velocity for 4–6 experiment designs

however, due to the conical shape of the nozzle, the water particles present greater dispersion. It also shows greater speed and wind in the area near the rotating domain.

4 Conclusions

Multiphase simulations with CFX solver allows the behavior of the dispersed fluid on the surface of the 3D models of the crops surrounded by air, to be observed allowing the best location of the nozzle in fumigation systems with UAVs to be evaluated. The simulations, show better water fluid coverage in cases where the nozzle is in the center of the rotating domain.

The methods of reconstruction of 3D models with photogrammetry techniques are suitable for simulations of dynamic fluids, however, the great number of details of the geometry of the plant, is necessary to make filters that allow obtaining smoother meshes, so that they can combine with the ICEM.

Due to the partially obtained results, we suggest the future development of a 3D modeling methodology with photogrammetry techniques, to improve the spatial resolution, in order to perform a simulation closer to the real behavior and evaluate different finite element sizes for CFD simulation with CFX solver.

The study method proposed to simulate the behavior of UAVs for crop fumigation, allows to improve of analysis and design of irrigation systems in crops to efficiently use liquids such as fungicides or fertilizers.

The generation of iso-surface in 3D with the CFX post processing, allows the observation of fluid dynamics in 3D, this process has a high computational cost, this is the cause of having simulations of 1 s. We suggest future works use high performance computers and clusters to increase the simulation time.

Acknowledgment. This work was supported by GITUD Universidad Distrital Francisco José de Caldas research group & Fundación Universitaria Colombo Germana.

References

1. Zhang, B., Tang, Q., Chen, L.p., Zhang, R.r., Xu, M.: Numerical simulation of spray drift and deposition from a crop spraying aircraft using a CFD approach. Biosyst. Eng. **166**, 184–199 (2018)
2. Wilcox, D.C., et al.: Turbulence modeling for CFD, vol. 2. DCW industries La Canada, CA (1998)
3. Omar, Z., Qiang, K.Y., Mohd, S., Rosly, N.: CFD simulation of aerial crop spraying. In: IOP Conference Series: Materials Science and Engineering, vol. 160, p. 012028. IOP Publishing (2016)
4. Endalew, A.M., Debaer, C., Rutten, N., Vercammen, J., Delele, M.A., Ramon, H., Nicolaï, B., Verboven, P.: Modelling pesticide flow and deposition from air-assisted orchard spraying in orchards: a new integrated CFD approach. Agric. Forest Meteorol. **150**(10), 1383–1392 (2010)
5. Tsay, J.R., Liang, L.S., Lu, L.H.: Evaluation of an air-assisted boom spraying system under a no-canopy condition using CFD simulation. Trans. ASAE **47**(6), 1887 (2004)

6. Melese, E., Debaer, C., Rutten, N., Vercammen, J., Delele, M.A., Ramon, H., Nicolaï, B., Verboven, P., et al.: A new integrated CFD modelling approach towards air-assisted orchard spraying. part i. model development and effect of wind speed and direction on sprayer airflow. Comput. Electron. Agric. **71**(2), 128–136 (2010)
7. Delele, M.A., Jaeken, P., Debaer, C., Baetens, K., Endalew, A.M., Ramon, H., Nicolaï, B., Verboven, P.: CFD prototyping of an air-assisted orchard sprayer aimed at drift reduction. Comput. Electron. Agric. **55**(1), 16–27 (2007)
8. Newcome, L.R.: Unmanned aviation: a brief history of unmanned aerial vehicles. American Institute of Aeronautics and Astronautics (2004)
9. Jenkins, D., Vasigh, B.: The economic impact of unmanned aircraft systems integration in the United States. Association for Unmanned Vehicle Systems International (AUVSI) (2013)
10. CFX, A., et al.: Ansys CFX-solver theory guide. ANSYS Inc. (2009)

Towards a Taxonomy of Software Maintainability Predictors

Sara Elmidaoui[⊠], Laila Cheikhi, and Ali Idri

Software Project Management Team (SPM), ENSIAS, Mohamed V University,
Rabat, Morocco
sarah.elmidaoui@gmail.com, laila.cheikhi@gmail.com,
ali.idri@um5.ac.ma

Abstract. Software maintainability prediction has gained more attention in the last decade. Several studies have conducted empirical studies to look for models to predict software maintainability more accurately. In a previous work, a systematic mapping study (SMS) was performed in this context and a set of metrics used as predictors of software maintainability has been identified. But, unfortunately those metrics are not organized in a structured way. Moreover, some authors may raise the same metric with the same meaning, but with different wordings. Hence, it becomes a necessity to unify all the metrics in a taxonomy that will help researchers build maintainability models in an easy way. The proposed taxonomy is 3 levels with categories, subcategories and metrics. We expect that the use of this taxonomy by researchers can help us identify other options to both propose a more useful taxonomy and to perform its evaluation.

Keywords: Maintainability · Metrics · Taxonomy · Predictors · Classification

1 Introduction

Nowadays, maintainability is considered an important quality characteristics of the software product. This characteristics is defined by the ISO 25010 as the "degree of effectiveness and efficiency with which a product or system can be modified by the intended maintainers" [1]. Research in software engineering have found that failures of software are often caused by its unsatisfied users' requirements, leading to an increasing cost of software maintenance [2–4]. In this perspective, many software product maintainability prediction (SPMP) models have been proposed in the literature in order to successfully predict the maintainability of the software and therefore contribute to reduce the maintenance cost. To empirically validate these models, many metrics or factors have been used as predictors. These metrics can help identifying potential problems in the software structure [5], and directs management by guiding managerial and technical decisions [6]. In a previous work [7], a systematic mapping study (SMS) has been conducted and a set of SPMP empirical studies has been selected. From these studies, various metrics or factors used as predictors have been identified such as: Chidamber and Kemerer metrics [8], Li and Henry metrics [9], McCabe cyclomatic complexity metrics [10], web based application metrics, database metrics, Lorenz and Kidd [11] metrics, etc. But, unfortunately those metrics are not

© Springer Nature Switzerland AG 2019
Á. Rocha et al. (Eds.): WorldCIST'19 2019, AISC 930, pp. 823–832, 2019.
https://doi.org/10.1007/978-3-030-16181-1_77

organized in a structured way. Moreover, some authors may raise the same metric with the same meaning, but with different wordings. This non-categorization and different wordings may cause ambiguities for researchers. Hence, it becomes a necessity to unify all the metrics in a taxonomy. This will help researchers build SPMP techniques in an easy and structured way. Besides, from the literature, three studies have been conducted to compile metrics related to software maintainability by [12–14], but this study is different from them in three aspects. Firstly, the current study is carried out based on empirical studies on SPMP identified in the SMS. Secondly, in this study, the taxonomy includes metrics about different types of software applications and not restricted to object oriented ones. Thirdly and most importantly, none of the three studies is as comprehensive as the current one in which more than 80 studies published from the year 2000 till date are investigated. The purpose of this study is to propose a way to unify the metrics wordings and categorize them into a structured way by proposing a taxonomy with 3 levels with categories, subcategories and metrics.

This study is structured as follows: Sect. 2 provides the classification methodology. Section 3 provides the proposed taxonomy of software maintainability predictors. Section 4 provides the discussion and the conclusion of this study, and gives directions for future work.

2   Classification Methodology

Metrics used as predictors for maintainability can be gathered at different stages of the software lifecycle. Authors in [15, 16] have found that the metrics used as predictors in SPMP models are mainly source code and design metrics as well as factors related to software process. Moreover, the SMS [16] have also identified the use of some quality attributes as predictors of maintainability. Thus, the maintainability predictors can be therefore grouped according to these categories, which are: design metrics, source code size metrics, quality attributes metrics, and process metrics. These categories constitute the first level of the proposed taxonomy. Each category includes subcategories in the second level and examples of metrics are also provided in the third level. The metrics are included in their corresponding categories according to their definitions founded in selected studies in the SMS previously performed (Table A in Appendix lists these studies with their corresponding Identifiers), their purposes and the measured attributes.

3   Proposed Taxonomy of Software Maintainability Predictors

The taxonomy includes four main categories subdivided in subcategories described as follows; with their corresponding metrics. Table 1 provides these metrics and their supported SPMP studies from Table A in Appendix.

Table 1. Software maintainability predictors' categories, subcategories and metrics

Category	Sub-category	Metrics	ID study
Design metrics	Coupling	Data Abstraction Coupling (DAC), Response For a Class (RFC), Message Passing Coupling (MPC), Coupling between objects (CBO), Coupling factor (COF), Average response per class (SRFC), Class Coupling (CC), OCMEC of class (OCMEC), CBO accounts for import coupling (CBO_IUB), CBO accounts for export coupling (CBO_U), Data Abstraction Coupling 1 (DAC1), Data Abstraction Coupling 2 (DAC2), Efferent Coupling (Ce), Afferent coupling (Ca), Web Data Coupling (Wdata), Web Control Coupling (WContr), Entropy Coupling (EntCoup) Control coupling, Data Coupling, Coupling between methods (CBM), Inheritance Coupling (IC), Coupling through inheritance (CTI)	S2, S6, S8, S9, S14, S16, S19, S21, S23, S24, S26, S28, S29, S31, S32, S33, S34, S35, S37, S38, S41, S42, S43, S44, S45, S46, S47, S48, S49, S52, S53, S54, S55, S56, S57, S58, S59, S60, S61, S62, S63, S64, S65, S66, S68, S69, S70, S71, S72, S73, S75, S77
	Cohesion	Lack of Cohesion in Methods (LCOM), Lack of Cohesion Among Methods of a Class (LCOM3), Average lack of cohesion on methods per class (SLCOM), Cohesion (Coh), Cohesion Among Methods in a Class (CAMC), Tight Class Cohesion (TCC), Loose Class Cohesion (LCC), Low-level Design Similarity-based Class Cohesion (LSCC), Class Cohesion Metric (SCOM), Path Connectivity Class Cohesion (PCCC), The average strength of the attributes (OL2), Entropy Cohesion (EntCoh), Information Flow based Cohesion (ICH), Degree of cohesion-direct (DCd), Degree of cohesion-indirect (Dci), Class cohesion (CC), Class cohesion metric (SCOM), Normalized Hamming distance (NHD), Scaled normalized Hamming distance (SNH), Weighted cyclomatic complexity (WCC)	S2, S6, S8, S9, S14, S16, S19, S21, S23, S24, S26, S28, S29, S31, S32, S33, S35, S37, S38, S42, S43, S44, S45, S47, S46, S48 S49, S52, S53, S54, S55, S56, S57, S58, S59, S60, S61, S62, S64, S65, S66, S68, S69, S70, S71, S72, S73, S75, S77
	Inheritance	Depth of Inheritance Tree (DIT), Number of Children (NOC), Average depth of inheritance tree per class (SDIT), Average number of children per class (SNOC), Method inheritance factor (MIF), Attribute inheritance factor (AIF), Maximum depth of inheritance trees (PDIT), Maximum number of Depth of Inheritance Tree (MaxDIT), Class-to-Leaf Depth (CLD), Maximum DIT (MaxDIT), Average inheritance depth (AID) Number of children in the hierarchy (NOCC), Specialization ratio (SRatio), Reuse ratio (RRatio), Attribute inheritance factor (AIF), Method inheritance factor (MIF), Number of attributes inherited in hierarchy (NAIH), Number of methods inherited in hierarchy (NMIH)	S2, S6, S8, S9, S14, S16, S19, S21, S23, S24, S26, S28, S29, S31, S32, S33, S35, S37, S38, S41, S42, S43, S44, S45, S47, S52, S54, S55, S56, S57, S58, S59, S60, S61, S62, S64, S65, S66, S68, S69, S70, S71, S72, S73, S75, S77
	Complexity	Weighted Methods per Class (WMC), Average McCabe Cyclomatic Complexity, McCabe's cyclomatic complexity, Bandwidth, Average weighted methods per class (SWMC), Maximum Operation Complexity (Ocmax), Attribute Complexity (AR), Total Cyclomatic Complexity, Web Page Cyclomatic Complexity, Average method complexity (AMC), McCabe's VG complexity (MVG), Cyclomatic complexity (Ocavg), Average method complexity (AMC)	S1, S2, S3, S4, S5, S6, S7, S8, S9, S10, S12, S14, S15, S16, S19, S20, S21, S22, S23, S24, S26, S28, S29, S31, S32, S33, S34, S35, S37, S38, S39, S41, S42, S43, S44, S45, S47, S48, S49, S51, S52, S54, S55, S56, S57, S58, S59, S60, S61, S62, S64, S65, S66, S69, S70, S71, S72, S73, S75, S76, S77

(continued)

Table 1. (*continued*)

Category	Sub-category			Metrics	ID study
	Encapsulation			Attribute Hiding factor (AHF), Method hiding factor (MHF)	S8, S16, S33, S64
	Polymorphism			Polymorphism factor (POF)	S8, S16, S62
Class diagram	Classes			Number of classes (NC, NCLASS, NCL, NOA), Number of base classes (NIB)	S3, S4, S12, S22, S47, S51, S53, S59, S65, S73, S77
	Attributes			Number of attributes/Class Size Attributes (NA, CSA), Average number of attributes-unweighted (ANAUW), Average number of attributes-weighted (ANAW), Number of Attributes Added (NAA), Number of Attributes Inherited (NAIC)	
	Methods			Number of methods/class size operation (NM, NMETH, NOM, CSO), Average number of methods-unweighted (ANMUW), Average number of methods-weighted (ANMW), Number of parameters, Number of foreign methods accessed, Number of local methods accessed, Number of Operations Inherited (NOIC), Number of Operations Overridden (NOOC), Number of Operations added (NOAC), Number of Public methods (NPM), Number of evaluated methods, Average number of methods per class (CSO), Number of protected methods (NPROM), Number of private methods (NPRM), Number of defined methods (NMD)	
	Relationship	Association		Number of Associations (NASSOC), Average number of association relationships (ANAsso)	
		Aggregation		Number of Aggregations (NAGG), Number of Aggregations hierarchies (NAggH), Average number of aggregation relationships (ANAgg), Maximum number of aggregation hierarchies (MaxHAgg)	
		Generalization		Number of Generalisations(NGen), Number of Generalisation hierarchies (NGenH), Average number of generalization relationships	
Sequence diagram	Scenarios			Number of scenarios (NOS)	S4
	Messages			Weighted messages between objects (WMBO), Average number of return messages (ANRM), Average number of the directly dispatched messages (ANDM), Average number of the elements in the transitive closure of the directly dispatched messages (ANET)	
	Conditions			Average number of condition messages (ANCM)	
Quality attributes	Related ISO standard			Stability, Changeability, Analyzability, Testability, Adaptability, Modifiability, Understandability	S11, S17, S18, S25, S40, S50, S67, S78, S79, S80, S81, S82

(*continued*)

Table 1. (*continued*)

Category	Sub-category	Metrics	ID study
	Related to source code	Readability of source code, Document quality, Understandability of software	
	Related to software design	Simplicity, Accessibility, Fix structure, Workspace, Standardization/Modularization/Reciprocation, Adjustment degree, Assembly/Disassembly, Identification, origin of UML diagrams, level of detail of UML diagram, Methods, Models	
Source code size metrics	Size	Line of Code (Size1, LOC), Source lines of code (SLOC),Total logical lines of code for the whole system (TLLOC), Logical lines of code (LLOC), Average number of LLOC for the evaluated methods, Toral Lines of Code (TLOC), JavaDoc lines of Code (JLoc), The number of executable lines (rather than comments or whitespace), Comments Lines of Code (CLOC), Number of Commands (Command), Comment Ratio (CR), Size of the change impact set (Impact), Halsted Bugs (B), Halsted Volume (V), Halsted Effort (E), Halsted Difficulty metric (D), Halsetad Length (N), Halstead's Nh, Halstead Vocabulary (n), Jensen's Nf, Physical source lines of code gives (SLOP-P), Blank lines of code (BLOC), The summation of both code and comments lines (C&SLOC), Comment lines of code (CLOC), Count of comment words (CWORD), number of the header comments (HCLOC), Number of words (HCWORD), Number of classes in hierarchy (NCH), Number of attributes in hierarchy (NAH), Number of methods in hierarchy (NMH), Number of statements (NOS), Comment lines before class/method/function (CLB)	S2, S3, S5, S6, S7, S8, S9, S10, S12, S14, S15, S16, S19, S20, S21, S22, S23, S24, S26, S29, S31, S32, S33, S34, S35, S36, S38, S39, S41, S42, S43, S44, S45, S46, S47, S48, S51, S52, S53, S54, S56, S57, S58, S59, S60, S61, S62, S65, S66, S68, S69, S70, S71, S73, S74, S76, S77
	WA Size	Total Web Page, WebObject, Server Scripts, Client Scripts, Page Code Size, Total Data, Total number of data references, Number of server Page (NServP), Number of client pages (NClienP), Number of web pages = server Pages + client pages (NWebP), Number of Form Pages (NFrmP), Number of Form Elements (NFrmE), Number of Client scripts components (NCsrC), Number of server scripts components (NSscrC)	
Process metrics	Effort Time	Effort, Time-scale metrics, Fault and change metrics	S1

Design category is related to software design properties of the software. This category includes metrics related to coupling which provides an indication about the association or interdependency among modules, cohesion which provides an indication of the degree to which the methods and attributes of a class bind together, inheritance which consists of sharing attributes and operations amongst classes based on a hierarchical relationship, complexity of program control flow obtained by calculating the number of loops and branches in a component, encapsulation which is the wrapping up of data and functions into a single unit, polymorphism which means the ability to take more than one form, class diagram which represents the static structure of the system described in terms of classes, attributes, methods and relationship, and sequence diagram that shows the interaction of the system with its actors in terms of scenarios, messages and conditions.

Quality attributes category is related to quality aspects of the software. It includes quality characteristics related to the ISO 9126 and 25010 standards on software product quality such as analyzability, modifiability, modularity, etc., quality attributes related to source code such as readability of source code, and document quality, and those related to software design, such as simplicity, accessibility, fix structure, etc.

Source code size category is related to the way used to quantify the structural elements of the software [17]. This category includes the size measured by counting the number of lines of code and its different alternatives such as; total logical lines of code (TLLOC), etc. as well as the size of web application (WA) such as number of server Page (NServP), etc.

Process category is related to software process management factors, which are derived from the activities and tasks that make up the design process [18]. This category includes metrics related to effort and time, such as; time-scale and effort metrics.

4   Discussion and Conclusion

The main goal of this study is to present and categorize the metrics used as predictors of maintainability in SPMP empirical studies. To achieve this purpose, a set of metrics has been identified from the empirical studies collected in the SMS previously performed. Although these metrics are widely used, it was noticed that they are not grouped in a structured way, nor the metrics are unified. For instance, some metrics have more than one abbreviation such as NC, NCLASS, and NCL for number of classes, while others have the same abbreviation but with different meanings such as CC, which can be Class Coupling, Class Cohesion or Clone Coverage. To tackle these problems, a taxonomy is proposed with 3 levels. Four main categories were proposed in the first level: design metrics, source code metrics, quality attributes and process metrics, which are subdivided into subcategories in the second level, to which some metrics are provided in the third level. From Table 1, it can be noticed that the design metrics were the most widely used in 65 studies, followed by source code size metrics in 57 studies, quality attributes in 12 studies, and process metrics in one study. Besides, this taxonomy can help the researchers in their choice of metrics to include in the empirical studies about SPMP models since it is not software application types dependent. For instance, coupling subcategory includes object-oriented metrics such as data abstraction coupling (DAC), etc. and web application metrics such as web data coupling (Wdata), etc. Moreover, source code size category includes the size measured in terms of LOC, number of executable lines, etc. in object oriented project, and in web application in terms of number of server page or client pages, etc. Therefore, we do not claim that this is an exhaustive list of predictors used in SPMP studies, nor do we claim that we have chosen the best possible categorization. As a future work, we intend to validate the proposed taxonomy by experts using a survey to prove that the proposed categorization is valid and reliable.

Appendix

Table A. Empirical SPMP studies

ID	Author	Year	Title	ID	Author	Year	Title
S1	S. Muthanna et al.	2000	A maintainability model for industrial software systems using design level metrics	S21	M. O. Elish et al.	2009	Application of treenet in Predicting Object-Oriented Software Maintainability: A Comparative Study
S2	M.M.T Thwin et al.	2003	Application of Neural Networks for Estimating Software Maintainability Using OO Metrics	S22	S. Rizvi et al.	2010	Maintainability Estimation Model for Object-Oriented Software in Design Phase (MEMOOD)
S3	M. Genero et al.	2003	Building UML class diagram maintainability prediction models based on early metrics.	S23	A.Kaur et al.	2010	Soft computing approaches for prediction of software maintenance effort
S4	M.Kiewkayna et al.	2004	A methodology for constructing maintainability model of object-oriented design	S24	C. Jin et al.	2010	Applications of support vector mathine and unsupervised learning for predicting maintainability using OO metrics
S5	G.A.D. Lucca et al.	2004	Towards the definition of a maintainability model for Web applications	S25	L. CAI et al.	2010	Evaluating software maintainability using fuzzy entropy theory
S6	C.V. Koten et al.	2005	An application of Bayesian network for predicting object-oriented software maintainability	S26	S. O. Olatunji et al.	2010	Extreme Learning Machine as Maintainability Prediction model for Object Oriented Software Systems
S7	J.H. Hayes et al.	2005	Maintainability prediction: a regression analysis of measures of evolving systems	S27	P. Dhankhar et al.	2011	Maintainability Prediction for Object Oriented Software
S8	S.C. Misra	2005	Modeling design/coding factors that drive maintainability of software systems	S28	S.K. Dubey et al.	2012	A Fuzzy Approach for Evaluation of Maintainability of Object Oriented Software System
S9	Y. Zhou et al.	2006	Predicting object-oriented software maintainability using multivariate adaptive regression splines	S29	S. K. Dubey et al.	2012	Maintainability Prediction of Object-oriented Software System by Multilayer Perceptron Model
S10	X. Jin et al.	2006	Locality Preserving Projection on Source Code Metrics for Improved Software Maintainability	S11	S.S. Dahiya et al.	2007	Use of genetic algorithm for software maintainability metrics conditioning
S32	Al-Jamimi et al.	2012	Prediction of software maintainability using fuzzy logic	S31	Sharawat et al.	2012	Software maintainability prediction using neural networks
S12	M. Genero et al.	2007	Building measure-based prediction models for UML class diagram maintainability	S33	R. Malhotra et al.	2012	Software maintainability prediction using machine learning algorithms
S13	K. Shibata et al.	2007	Quantifying Software Maintainability Based on a Fault-Detection/Correction Model	S34	T. Bakota et al.	2012	A cost model based on software maintainability

(continued)

Table A. (*continued*)

S14	K.K.Aggarwal et al.	2008	Application of artificial neural network for predicting maintainability using object-oriented metrics	S35	Y. Dash et al.	2012	Maintainability Prediction of Object Oriented Software System by Using Artificial Neural Network Approach
S15	Y. Thian et al.	2008	AODE for source code metrics for improved software maintainability.	S36	P. Hegedűs et al.	2012	Towards Building Method Level Maintainability Models Based on Expert Evaluations
S16	Y. Zhou et al.	2008	Predicting the maintainability of open source software using design metrics	S74	G. Szoke et al.	2017	Empirical study on refactoring large-scale industrial systems and its effects on maintainability
S17	YU, Haiquan et al.	2009	An application of case-based reasoning to predict structure maintainability	S38	H. Aljamaan et al.	2013	An ensemble of computational intelligence models for software maintenance effort prediction
S18	A. Sharma et al.	2009	Predicting Maintainability of Component-Based systems using Fuzzy Logic	S39	P. Hegedűs et al.	2013	A Drill-Down Approach for Measuring Maintainability at Source Code Element Level
S19	W. Li-jin et al.	2009	Predicting object-oriented software maintainability using projection pursuit regression	S40	X.L. Hao et al.	2013	Research on software maintainability evaluation based on fuzzy integral
S20	H. Mittal et al.	2009	Software maintainability assessment based on fuzzy logic technique	S41	F. Ye et al.	2013	A new software maintainability evaluation model based on multiple classifiers combination
S42	M. A. Ahmed et al.	2013	Machine learning approaches for predicting software maintainability: a fuzzy-based transparent model	S61	L. Kumar et al.	2016	Maintainability prediction of web service using support vector machine with various kernel methods
S43	S.O. Olatunji et al.	2013	Sensitivity-based linear learning method and extreme learning machines compared for software maintainability prediction of OO software systems	S82	G. Scanniello et al.	2018	Do software models based on the UML aid in source-code comprehensibility? Aggregating evidence from 12 controlled experiments
S44	A. Kaur et al.	2013	Statistical Comparison of Modeling Methods for Software Maintainability Prediction	S63	S. Almugrin et al.	2016	Using indirect coupling metrics to predict package maintainability and testability
S45	A. Mehra et al.	2013	Maintainability Evaluation of Object-Oriented Software System Using Clustering Techniques	S64	S. Tarwani et al.	2016	Sequencing of refactoring techniques by Greedy algorithm for maximizing maintainability
S46	J. Al Dallal.	2013	Object-oriented class maintainability prediction using internal quality attributes	S65	L. Kumar et al.	2017	Empirical Analysis on Effectiveness of Source Code Metrics for Predicting Change-Proneness
S47	R. Malhotra et al.	2014	Application of group method of data handling model for software maintainability prediction using OO systems	S66	K. Gupta et al.	2017	Evaluation of instance-based feature subset selection algorithm for maintainability prediction
S48	A. Kaur et al.	2014	A proposed new model for maintainability index of open source software	S67	S. Kundu et al.	2017	Maintainability assessment for software by using a hybrid fuzzy multi-criteria analysis approach
S49	A. Kaur et al.	2014	Software maintainability prediction by data mining of software code metrics	S68	B.R. Reddy et al.	2017	Performance of Maintainability Index prediction models: a feature selection based study
S71	L. Kumar et al.	2017	Using source code metrics and multivariate adaptive regression splines to predict maintainability of service oriented software	S69	L. Kumar et al.	2017	Software maintainability prediction using hybrid neural network and fuzzy logic approach with parallel computing concept
S51	G. Laxmi et al.	2014	Maintainability Measurement Model for Object Oriented Design	S70	L. Kumar et al.	2017	The impact of feature selection on maintainability prediction of service-oriented applications
S52	R. Malhotra et al.	2014	A metric suite for predicting software maintainability in data intensive applications	S50	A. Pratap et al.	2014	Estimation of software maintainability using fuzzy logic technique
S62	S. Tarwani et al.	2016	Predicting Maintainability of Open Source Software using Gene Expression Programming and Bad Smells	S72	L. Kumar et al.	2017	Using source code metrics to predict change-prone web services: A case-study on ebay services

(*continued*)

Table A. (*continued*)

S54	M. O. Elish et al.	2015	Three empirical studies on predicting software maintainability using ensemble methods	S73	R. Malhotra et al.	2017	Prediction & Assessment of Change Prone Classes Using Statistical & Machine Learning Techniques
S55	L. Kumar et al.	2015	Neuro- Genetic Approach for Predicting Maintainability Using Chidamber and Kemerer Software Metrics Suite	S37	D. Chandra	2012	Support vector approach by using radial kernel function for prediction of software maintenance effort on the basis of multivariate approach
S56	S.O. Olatunji et al.	2015	Type-2 fuzzy logic based prediction model of object oriented software maintainability	S75	Y. Gokul et al.	2017	An authoritative method using fuzzy logic to evaluate maintainability index & utilizability of software
S57	A. K. Soni et al.	2015	Validating the Effectiveness of Object-Oriented Metrics for Predicting Maintainability	S76	P. Hegedűs et al.	2018	Empirical evaluation of software maintainability based on a manually validated refactoring dataset
S58	L. Kumar et al.	2016	Hybrid functional link artificial neural network approach for predicting maintainability of object-oriented software	S53	S. Misra et al.	2014	Framework for maintainability measurement of web application for efficient knowledge-sharing on campus intranet
S59	A. Jain et al.	2016	An empirical investigation of evolutionary algorithm for software maintainability prediction	S78	G. Scanniello et al.	2014	On the impact of UML analysis models on source-code comprehensibility and modifiability
S60	A. Chug et al.	2016	Benchmarking framework for maintainability prediction of open source software using object oriented metrics	S79	A.M. Fernándz-Sáez et al.	2015	Are Forward Designed or Reverse-Engineered UML diagrams more helpful for code maintenance?: A family of experiments
S80	G. Scanniello et al.	2015	Studying the Effect of UML-Based Models on Source-Code Comprehensibility	S81	A.M. Fernándz-Sáez et al.	2016	Does the level of detail of UML diagrams affect the maintainability of source code?: a family of experiments
S77	L. Kumar et al.	2018	A comparative study of different source code metrics and machine learning algorithms for predicting change proneness of object oriented systems				

References

1. ISO: Systems and Software Engineering, Systems and Software Quality Requirements and Evaluation, System and Software Quality Models. ISO/IEC 25010. Geneva (Switzerland): International Organization for Standardization, p. 34 (2010)
2. Glass, R.L., Noiseux, R.A.: Software Maintenance Guidebook. Prentice Hall, Englewood Cliffs (1981)
3. Jones, C.: Assessment and Control of Software Risks. Prentice Hall, Englewood Cliffs (1994)
4. Pigoski, T.M.: Practical Software Maintenance: Best Practices for Managing Your Software Investment. Wiley, New York (1996)
5. Bandi, R.K., Vaishnavi, V.K., Turk, D.E.: Predicting maintenance performance using object-oriented design complexity metrics. IEEE Trans. Soft. Eng. **21**(1) (2003)
6. Srinivasan, K.P., Devi, T.: A novel software metrics and software coding measurement in software engineering. Int. J. Adv. Res. Comput. Sci. Softw. Eng. **4**(1), 303–308 (2014)
7. Elmidaoui, S., Cheikhi, L., Idri, A.: Empirical Studies on Software Product Maintainability Prediction: A Systematic Mapping Study, In Process
8. Chidamber, S.R., Kemerer, C.F.: A metrics suite for object-oriented design. IEEE Trans. Softw. Eng. **20**(6), 476–493 (1994)
9. Li, W., Henry, S.: Object-oriented metrics that predict maintainability. J. Syst. Softw. **23**(2), 111–122 (1993)
10. McCabe, T.J.: A complexity measure. IEEE Trans. Softw. Eng. **2** (1976)
11. Lorenz, M., Kidd, J.: Object-Oriented Software Metrics. Prentice Hall, USA (1994)

12. Saraiva, J., Soares, S., Castor, F.: Towards a catalog of object-oriented software maintainability metrics. In: International Workshop Emerging Trends in Software Metrics, pp. 84–87 (2013)

13. Abreu, F.B., Carapua, R.: Candidate metrics for object-oriented software within a taxonomy framework. J. Syst. Softw. **26**(1), 87–96 (1994)

14. Archer, C., Stinson, M.: Object Oriented Software Measure, Technical report CMU/SEI-95-TR-002, ESC-TR-95-002 (1995)

15. Riaz, M., et al.: A systematic review of software maintainability prediction and metrics. In: International Symposium on Empirical Software Engineering and Measurement, pp. 367–377 (2009)

16. Saraiva, J. et al.: Aspect-oriented software maintenance metrics: a systematic mapping study. In: International Conference on Evaluation Assessment in Software Engineering, pp. 253–262 (2012)

17. Misra, S., Egoeze, F.: Framework for maintainability measurement of web application for efficient knowledge-sharing on campus intranet. In: International Conference on Computational Science and Its Applications, pp. 649–662. Springer (2014)

18. Muthanna, S. et al.: A maintainability model for industrial software systems using design level metrics. In: Conference on Reverse Engineering, pp. 248–256 (2000)

A BPMN Extension for Business Process Outsourcing to the Cloud

Karim Zarour[1,2(✉)], Djamel Benmerzoug[1], Nawal Guermouche[2],
and Khalil Drira[2]

[1] LIRE Laboratory, University of Constantine 2 - Abdelhamid Mehri,
Constantine, Algeria
{zarour.karim,djamel.benmerzoug}@univ-constantine2.dz
[2] LAAS-CNRS, University of Toulouse, Toulouse, France
{nguermou,khalil}@laas.fr

Abstract. The trend of business process outsourcing (BPO) has been intensi-fied with the advent of Cloud computing that brings a new way of paying and consuming resources. Nevertheless, ensuring the compliance of outsourced business processes (BPs), protecting sensitive data, and making the right choices of offers, remain among the most important challenges in BPO to the Cloud. In this context, it is crucial to clearly specify the awaited requirements when modelling BPs. BPMN (Business Process Model and Notation) is the most used standard for BP modelling. However, it is a generic language and does not support BPO characteristics. This paper proposes a BPMN extension (called BPOMN) that allows specifying the requirements of BP activities in terms of security, compliance, cost, and performance. This enables specifying and con-sidering all relevant information for effective Cloud BPO.

Keywords: Business process outsourcing · Cloud computing ·
BPMN extension · Security · Compliance · Cost · Performance

1 Introduction

Given the harsh competition that companies face today, many of them opt for business process outsourcing (BPO), which is seen among the most beneficial strategies, especially for small and middle-size companies. BPO has been intensified with the advent of Cloud computing that has attracted companies with its new economic model of pay-per-use and its high elasticity in resource consumption. In fact, the Gartner research firm [1] estimates the BP market in the Cloud at $40.8 billion, which exceeds the market of other service models.

However, the Cloud may endanger the security of outsourced BPs by exposing them to different threats such as data breach and denial of service. In order to avoid being sanctioned, organizations are also required to be well aware of the increasing number of laws and standards imposed on their BPs, especially those operating in heavily regulated sectors like the financial or medical sector. Furthermore, companies are struggling to find the most suited deals for their outsourced BPs, owing to the

© Springer Nature Switzerland AG 2019
Á. Rocha et al. (Eds.): WorldCIST'19 2019, AISC 930, pp. 833–843, 2019.
https://doi.org/10.1007/978-3-030-16181-1_78

proliferation of Cloud providers that invade the market with services of various qualities and at different prices.

Given the previous factors, a company must identify accurately the requirements of BP activities in order to decide for each of them, whether it should be outsourced or kept locally, and secondly to select the most suitable Cloud offers for the outsourced activities. In this context, it is necessary to specify clearly the awaited requirements when modelling BPs. Several works (e.g., [2–5]) have proposed extensions of the BPMN standard to consider for example the performance criterion [6] or the security criterion [7]. However, there is a lack for Cloud-based BPO criteria specifications.

In this paper, we propose BPOMN (Business Process Outsourcing Model Notation), an extension of the BPMN language by defining new artefacts to be able to specify the requirements of BP activities in terms of security, compliance, cost, and performance. We chose BPMN since it is a very expressive language that is defined by the Object Management Group (OMG) and approved as an ISO standard. Moreover, BPMN provides an extension mechanism and it is supported by a wide range of modelling tools like Bizagi and MS Visio. The extension BPOMN aims to help organizations in outsourcing their BPs by assigning to each activity, the Cloud offer that best meets its requirements and therefore to obtain an optimal configuration for outsourced BPs. In order to provide a valid extension, we rely on the official documentation of the BPMN language and we use the two formats defined in the extension mechanism, namely, the MOF-based meta-model as well as the XML Schema. For the visual aspect, we propose new graphical elements that we add to the predefined artefacts.

The business process management (BPM) aims to improve processes in a continuous way by handling their iterative lifecycle, which includes the modelling, automation, execution, and optimization phases [8]. BPOMN is intended for the modelling phase and complements the business intelligence (BI) approach that we have proposed in [9]. Indeed, the latter contributes in the optimization phase by exploiting the execution history of outsourced BPs to evaluate the Cloud offers and the activities confided to them.

The remainder of this paper is organized as follows. Section 2 addresses some related work. In Sect. 3, we explain and justify the choice of each outsourcing criteria. Section 4 describes BPOMN and demonstrates its feasibility through an illustrative example and modelling tools. Finally, Sect. 5 concludes this paper and provides directions for future work.

2   Related Work

There exist many works in the literature that extend the BPMN language. We have classified and compared in Table 1 a set of fairly recent extensions that seem closer to our work, that is, those that extend BPMN for specifying BP requirements.

Table 1. Presentation and comparison of existing work on BPMN extension

Work	Aim of the BPMN extension	Domain
[10]	Specifying the outsource-ability characteristics of BPs	BPO
[4]	Specifying security requirements (e.g., attack harm detection, integrity, privacy, access control)	Security
[7]	Modelling security concepts (like authorization, encrypted message and harm protection) in BPMN-based healthcare processes	
[3]	Specifying privacy requirements and capturing data leakage	
[2]	Handling security risks by aligning BPMN constructs to the IS security risk management (ISSRM) domain model	Risks
[11]	Management of security risks based on the alignment of BPMN constructs to the ISSRM concepts	
[12]	Representing audit-relevant concepts into BP models for improving process audits	Compliance
[13]	Representing legal constraints and business rules through artefacts	
[5]	Facilitating access and interpretation of accurate cost information at the activity level	Other non-functional requirements
[6]	Specifying performance and reliability properties in BPMN models	
[14]	Extending BPMN with new elements for the performance and cost analysis of BPs	

Based on existing BPMN extensions, we have found that regarding the involved domains:

- A large number of studies are concerned with security in general or by targeting a particular sector such as security in healthcare. Their main objective is to secure the BPs during the modelling phase but they do not consider Cloud-related threats.
- There exists some work that extend BPMN for risk management and others for compliance but not both at once.
- There are work that extend BPMN for specifying other non-functional requirements such as cost, performance, and reliability. However, they do not consider the Cloud context, which has its own characteristics like pay-per-use and resource allocation.
- Finally, we found only one work in [10] that proposes a BPMN extension for BPO to the Cloud. Unfortunately, it does not take into account the compliance, which is a very important criterion. Moreover, we found a problem in the graphical representation. Indeed, the authors propose to add icons inside the rectangle representing BP activities (in the four corners of the rectangle) which leaves no place for specifying the type of activity with the icons predefined by BPMN (e.g., manual task, user task, service task, etc.)

As regards the extension mechanism, we have retained the following points:

- Despite XML Schema specifies an interchange format for BPMN models, existing work rarely use it to define their extension as has been done by [6, 15]. In fact, none of the BPMN extensions cited in Table 1 used XML Schema.
- Some works propose new graphic elements while others reuse existing elements such as artefacts.
- Most work overlook the BPMN extension mechanism by extending their own meta-model and not the one specified by the OMG. This hampers comprehensibility and comparability of BPMN extensions and impedes their straightforward integration in modelling tools [15].

3 Identified BPO Criteria

There are two major steps when extending a standard, particularly BPMN [16]: (1) identifying the requirements of the targeted domain, and (2) proposing the corresponding BPMN extensions. In this section, we first present and explain the considered criteria and their importance in Cloud based BPO. Then, in Sect. 4, we present BPOMN, a BPMN extension for Cloud-based BPO.

Organizations should examine several criteria in order to successfully outsource their BPs to the Cloud. Currently, the focus is shifting towards problems of security risks and compliance [17]. Hence, in our work, we approach the criteria from the perspective of Governance, Risk and Compliance (GRC) that is defined as a framework to govern an organization through strategic decision-making by taking into account the incurred risks and compliance [18]. In fact, the decisions to make in BPO are strategic for an organization and therefore they are part of its governance. Moreover, given the overgrowth of Cloud providers that propose services at different prices and with different quality levels, we have decided to combine the criteria of cost and performance with those considered in GRC, namely risks and compliance. In what follows, we explain and justify the choice of each criterion:

- **Security:** The main goals of security consist in ensuring confidentiality, integrity, availability, non-repudiation, and authenticity (CIANA) [19]. These types of security can be endangered because the outsourcing of BPs exposes them to different Cloud threats that are regularly published and ranked by the Cloud Security Alliance (CSA). The CSA also indicates in its reports the types of security affected by each Cloud threat [19] as well as the security mechanisms implemented by each Cloud provider.
- **Compliance:** The organizations have to ensure that their outsourced BPs comply with regulations, contracts, laws, standards, etc. Compliance is a significant criterion for organizations since it allows them to maintain a good reputation and avoid fines that may outweigh the cost benefits of Cloud. Ensuring compliance becomes more challenging when BPs are under the control of Cloud providers. Indeed, BP activities may be spread, replicated, and shifted across different data centers that follow laws and regulations of the country in which they are located [17].

Furthermore, the location of the data center may remain unknown since several Cloud providers hide it to prevent cyber-attacks [17].

- **Cost:** Saving money is a much-awaited profit when outsourcing BPs to the Cloud [10]. Nonetheless, the Cloud market is flooded with a growing number of offers, some of which are cheap for storage services and expensive for computation while others propose the contrary. Therefore, it becomes very difficult for an organization to choose the adequate offers for their BPs.
- **Performance:** As for the cost, there exists an increasing number of Cloud offers that propose similar services but at different performance levels. Furthermore, the activities that make up time-constrained BPs should be assigned to Cloud offers with enough calculation capacity to respect the imposed deadlines.

4   A BPMN Extension for Specifying BP Requirements

In the previous section, we identified the criteria that should be considered when outsourcing BPs to a Cloud environment. In this section, we present the extension and demonstrate its feasibility through an example and modeling tools.

4.1   BPOMN Representation

This part describes the proposed extension based on the MOF meta-model and XML schema as well as the corresponding graphic elements.

Extension Based on MOF Meta-Model
The extensible meta-model defined in [20] allows to be still BPMN-compliant and it consists of four elements: Extension, ExtensionDefinition, ExtensionAttributeDefinition, and ExtensionAttributeValue. Figure 1 presents the proposed class diagram corresponding to the MOF-based meta-model. The BPO criteria class represents the ExtensionDefinition class that contains the extension attributes: security, compliance, cost, and performance.

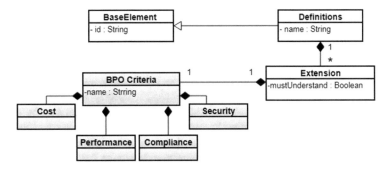

Fig. 1. The proposed BPMN extension based on MOF meta-model

Extension Based on XML Schema

To specify an interchange format for BPMN models, we present in Listing 1 an XML Schema for BPOMN. We adopt the same structure as that defined in [20] and used in most BPMN extensions like [6, 8, 15]. However, unlike them, we use the constructor 'all' instead of 'sequence' to nest the ExtensionAttributeDefinition elements in ExtensionDefinition, since the appearance order of the attributes is not important in our case.

We also mention via comments the correspondences with the elements of the MOF meta-model as indicated in [21].

Listing 1. An XML Schema definition for the proposed BPMN extension

```
<?xml version= "1.0" encoding="UTF-8" ?>
<xsd:schema xmlns:xsd="http://www.w3.org/2001/XMLSchema">
<xsd:element name="BPOcriteria" type="BPOcriteriaType"/><!-- ExtensionDefinition -->
<xsd:complexType name="BPOcriteriaType">
 <xsd:all>
  <xsd:element name="security" minOccurs="0"> <!-- ExtensionAttributeDefinition -->
   <xsd:complexType>
    <xsd:sequence>
     <xsd:element name="SecuRequirement" maxOccurs="5" type="SecuRequirementType"/>
    </xsd:sequence> </xsd:complexType> </xsd:element>
   <xsd:element name="compliance" minOccurs="0"/> <!-- ExtensionAttributeDefinition-->
    <xsd:complexType>
     <xsd:sequence>
      <xsd:element name="regulation" type="xsd:string" maxOccurs="unbounded"/>
      <xsd:element name="standard" type="xsd:string" maxOccurs="unbounded"/>
     </xsd:sequence> </xsd:complexType> </xsd:element>
    <xsd:element name="cost" minOccurs="0">    <!-- ExtensionAttributeDefinition -->
     <xsd:complexType>
      <xsd:simpleContent>
       <xsd:extension base="xsd:decimal">
        <xsd:attribute name="currency" type="xsd:string" use="required"/>
       </xsd:extension> </xsd:simpleContent> </xsd:complexType> </xsd:element>
     <xsd:element  name="performance"  type="xsd:duration"  minOccurs="0"  />  <!--
ExtensionAtributeDefinition -->
  </xsd:all> </xsd:complexType>
</xsd:schema>
```

New Graphical Elements

The BPMN specification does not provide any graphical notation for the representation of extensions. However, new graphic forms can be added as artefacts [20]. Thus, we have proposed a set of graphical elements presented in Fig. 2 that allows us to annotate BP activities. For providing further details about activity requirements, we use the 'Text annotation' artefact. Thus, for security, we can specify precisely the types of security (CIANA) that are required. Regarding compliance, we can list the rules and standards that must be respected. Finally, for performance and cost, we can indicate the runtime that should not to be exceeded and the amount of estimated cost. Seen that BPOMN applies to the activity element, we consider that the requirements of an activity also depend on the objects it manipulates.

Fig. 2. The graphical elements of BPOMN

4.2 Illustrative Example

In order to illustrate the use of BPOMN, we give a simplified example (see Fig. 3) on the patient's medical visit to a healthcare institution. Furthermore, we consider in this example that the health organization desiring to migrate to the Cloud is in the USA. Hence, it must comply with the regulations imposed in that country. For representing participants, we have defined two pools: Patient and Healthcare institution. The second one is divided into three lanes: secretariat, doctor, and infirmary. Concerning the requirements that we have annotated for decision-making in BPO to the Cloud, we specified that:

- The activity 'check patient record' requires confidentiality. Indeed, when the secretary verifies the existence of a record, she has to see only the administrative data but not the medical data that must remain confidential. This activity has also to comply with Health Insurance Portability and Accountability Act (HIPAA), which imposes rules on access rights and protection of medical data.
- The creation of the patient record has to comply with the norm ISO 18308:2011, which recommends a certain structure for the electronic health record.
- The activity 'consult medical history' requires two types of security: availability and integrity. Indeed, the doctor must always have at his disposal complete and reliable medical data. This activity is also exigent in terms of performance, as the data to be downloaded can be very voluminous (e.g., medical imaging).
- When recording the consultation report and new medical data, it is necessary to respect the international health regulations (IHR) that imposes on 196 countries the declaration of certain epidemiological diseases. This activity can also be very expensive if the new data require a large storage space.
- If the patient wants to pay by card, the secretary has to verify that the provided card meets the Payment Card Industry Data Security Standard (PCI DSS).

From the process model modelled through BPOMN, we can determine an appropriate Cloud offer for each activity. For instance, activities to be performed by the secretary and that are involved in the management of patient's records should be assigned to Cloud providers that respect the medical confidentiality and the record structuring standards. As for the physician' activities to be performed during the consultation, they should be entrusted to providers that guarantee the integrity and availability of medical data and that propose a low storage costs.

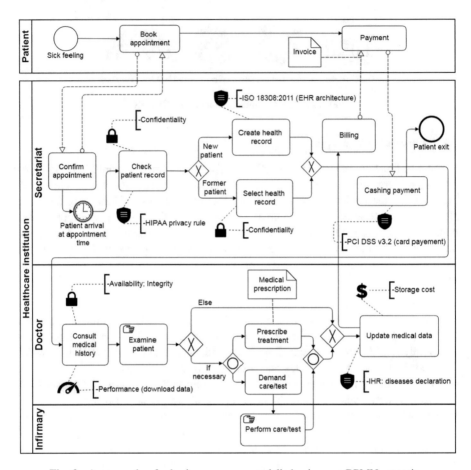

Fig. 3. An example of a business process modelled using our BPMN extension

4.3 Implementation

Modellers or modelling tools are free to add as many artefacts as necessary but without changing the shapes of existing elements [20]. Therefore, we have defined new artefacts and added them to the palette of BIZAGI BPM Modeler, which is a well-known tool for process modelling as it has more than 500 000 active users. We have also integrated the BPOMN extension into the Microsoft Visio tool by defining the four artifacts as stencils as shown in the left screenshot of the Fig. 4. As for tools that do not offer the possibility to add new artefacts like Activiti and jBPM, we define custom tasks bearing the BPO criteria icons, namely cost, compliance, performance, and cost.

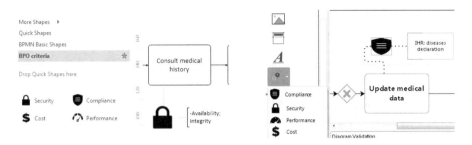

Fig. 4. Integration of BPOMN elements in MS Visio (left) and Bizagi (right) tools

5   Conclusion and Future Work

The Cloud computing allures companies that opt for a BPO strategy to be more competitive. Indeed, it ensures them a high elasticity by adjusting the resource consumption according to the BP demand. The Cloud also allows companies to save money by paying for only what they consume.

Despite these advantages, there are many worrisome concerns in BPO to the Cloud that are mainly related to compliance violation, security issues, and loss of control over BPs. In fact, the organizations wanting to outsource their BPs are afraid to expose their critical activities and sensitive data to Cloud threats. Moreover, it becomes very difficult for organizations to choose the adequate Cloud solutions given the growing number of offers available on the market and the diversity of prices and service qualities. Furthermore, companies should not entrust all the activities of a given process to the same Cloud provider. On the contrary, they have to spread their BPs across several Cloud providers to take advantage from the strengths of each of them.

Hence, BPO to the Cloud may be very beneficial for organizations on condition that they make the right choices by investigating thoroughly the characteristics of their BPs as well as the incurred risks. In this context, we have proposed in this paper BPOMN, an extension of the BPMN language that enables organizations to specify clearly and precisely the requirements of their BPs in terms of security, compliance, cost, and performance. BPOMN was developed on the basis of two representations provided by the BPMN extension mechanism, namely, the MOF meta-model and the XML Schema. These two representations were supplemented by new graphical elements that we integrated into a modelling tool.

For future work, we envisage to consider the criteria of cost and time of data transfer between BP activities. Indeed, the data generated by an activity may be needed to execute another and the data centers of Cloud providers are geographically dispersed. Therefore, activities that exchange a large amount of data should be assigned to the same provider. It would also be interesting to consider technical aspects such as data formats, platforms, and programming languages to avoid incompatibility with the heterogeneous Cloud providers. Finally, we plan for an automatic deployment by taking into account both activity requirements and Cloud offer features.

References

1. Gartner, Inc. (2017). http://www.gartner.com/newsroom/id/3616417
2. Matulevicius, R.: Security risk-oriented BPMN. In: Fundamentals of Secure System Modelling, pp. 63–76 (2017)
3. Pullonen, P., Matulevicius, R., Bogdanov, D.: PE-BPMN: privacy enhanced business process model and notation. In: Proceedings of the International Conference on Business Process Management (BPM), Barcelona, Spain, pp. 40–56 (2017)
4. Rodriguez, A., Fernandez-Medina, E., Piattini, M.: A BPMN extension for the modeling of security requirements in business processes. Trans. Inf. Syst. **90**(4), 745–752 (2007)
5. Thabet, D., Ghannouchi, S., Ben Ghezala, H.: A general solution for business process model extension with cost perspective based on process mining. In: Proceedings of the 11th International Conference on Software Engineering Advances (ICSEA), Seville, Spain, pp. 238–247 (2016)
6. Bocciarelli, P., D'Ambrogio, A.: A BPMN extension for modeling non functional properties of business processes. In: Proceedings of the Symposium on Theory of Modeling & Simulation, Boston, USA, pp. 160–168 (2011)
7. Sang, K.S., Zhou, B.: BPMN security extensions for healthcare process. In: Proceedings of the IEEE International Conference on Computer and Information Technology; Pervasive Intelligence and Computing, Liverpool, UK, pp. 2340–2345 (2015)
8. Workflow Management Coalition (WfMC), What is BPM ? https://www.wfmc.org/what-is-bpmn. Accessed 20 Nov 2017
9. Zarour, K., Benmerzoug, D.: Multicriteria-based analysis and evaluation of business processes executed in multi-cloud environment. In: Proceedings of the Computational Methods in Systems and Software (CoMeSySo), pp. 315–327 (2017)
10. Rekik, M., Boukadi, K., Ben-Abdallah, H.: Specifying business process outsourcing requirements. In: Proceedings of the 10th International Conference on Software Technologies (ICSOFT), Alsace, France, pp. 175–190 (2015)
11. Altuhhova, O., Matulevicius, R., Ahmed, N.: An extension of business process model and notation for security risk management. Int. J. Inf. Syst. Model. Des. **13**(4), 93–113 (2013)
12. Mueller-Wickop, N., Schultz, M.: Modelling concepts for process audits - empirically grounded extension of BPMN. In: Proceedings of the 21st European Conference on Information Systems (ECIS), Utrecht, Netherlands, pp. 194 (2013)
13. Goldner, S., Papproth, A.: Extending the BPMN syntax for requirements management. In: Business Process Model and Notation, pp. 142–147 (2011)
14. Lodhi, A., Koppen, V., Saake, G.: An extension of BPMN meta-model for evaluation of business processes. Sci. J. Riga Tech. Univ. **43**(1), 27–34 (2011)
15. Braun, R., Schlieter, H., Burwitz, M., Esswein. W.: BPMN4CP: design and implementation of a bpmn extension for clinical pathways. In: Proceedings of the IEEE International Conference on Bioinformatics and Biomedicine (BIBM), Belfast, UK, pp. 9–16 (2014)
16. Yousfi, A., Bauer, C., Saidi, R., Dey, A.: uBPMN: a BPMN extension for modeling ubiquitous business processes. Inf. Softw. Technol. **74**, 55–68 (2016)
17. Martens, B., Teuteberg, F.: Risk and compliance management for cloud computing services: designing a reference model. In: Proceedings of the 17th Americas Conference on Information Systems Americas, Detroit, USA, pp. 1–10 (2011)
18. Papazafeiropoulou, A., Spanaki, K.: Understanding governance, risk and compliance information systems (GRC IS): the experts view. J. Res. Innov. **18**(6), 1251–1263 (2016)
19. CSA. The notorious nine: Cloud Computing Top Threats in 2013. https://cloudsecurityalliance. org/download/the-notorious-nine-cloud-computing-top-threats-in-2013/

20. Object Management Group (OMG). Business Process Model and Notation (BPMN) Version 2.0.2 (2013). http://www.omg.org/spec/BPMN/2.0.2/
21. Stroppi, L.J.R., Chiotti, O., Villarreal, P.D.: Extending BPMN 2.0: method and tool support. In: Proceedings of the 3rd International Workshop on the Business Process Model and Notation (BPMN), Lucene, Switzerland, pp. 59–73 (2011)

A Computational Modeling Based on Trigonometric Cubic B-Spline Functions for the Approximate Solution of a Second Order Partial Integro-Differential Equation

Arshed Ali[1(✉)], Kamil Khan[1], Fazal Haq[2], and Syed Inayat Ali Shah[1]

[1] Department of Mathematics, Islamia College University, Peshawar, Pakistan
arshad.ali@icp.edu.pk
[2] Department of Maths/Stats/Computer Science, The University of Agriculture, Peshawar, Pakistan

Abstract. In this paper, the trigonometric cubic B-spline collocation method is extended for the solution of a second order partial integro-differential equations with a weakly singular kernel. The method is obtained by discretization of time derivative using backward finite difference formula while trigonometric cubic B-spline functions are used to approximate the spatial derivative. The scheme is validated through two benchmark test problems. Accuracy of the present approach is assessed in terms of L_∞, L_2 error norms and pointwise error. Better accuracy is obtained and the results are compared with quasi wavelet method (QWM), quintic B-spline collocation method (QBCM) and sinc-collocation method using Linsolve Package (SMLP).

Keywords: Trigonometric cubic B-spline functions · Collocation method · Partial integro-differential equation · Weakly singular kernel

1 Introduction

Partial Integro-Differential Equations (PIDEs) are usually used to model many phenomena of engineering and science phenomena such as heat conduction, electricity swaptions, viscoelasticity, chemical kinetics, nuclear reactor dynamics, population dynamics, biological models, fluid dynamics mathematical biology and financial mathematics [1–16].

Various techniques have been presented for the numerical solution of PIDEs including finite element method [5], finite-difference methods [6, 7], sinc collocation methods [8, 16], spectral method [9], quasi-wavelet methods [9, 10], B-spline methods [11–15], and Legendre multiwavelet collocation method [17]. In this paper we construct a collocation method using trigonometric cubic B-spline functions for the solution of the following second order PIDE with a weakly singular kernel, and this equation plays an important role in the modeling of various physical phenomena [10, 14, 16]:

© Springer Nature Switzerland AG 2019
Á. Rocha et al. (Eds.): WorldCIST'19 2019, AISC 930, pp. 844–854, 2019.
https://doi.org/10.1007/978-3-030-16181-1_79

$$\frac{\partial u(\zeta,\tau)}{\partial \tau} = \int_0^\tau (\tau - v)^{-\alpha} u_{\zeta\zeta}(\zeta, v) dv + f(\zeta, \tau), \zeta \in [a,\ b], \tau > 0, \tag{1}$$

with initial condition

$$u(\zeta, 0) = \sigma(\zeta), \zeta \in [a,\ b], \tag{2}$$

and boundary conditions

$$u(a, \tau) = \rho_0(\tau), u(b, \tau) = \rho_1(\tau), \tau \geq 0, \tag{3}$$

where $0 \leq \alpha \leq 1$.

Due to special features of trigonometric cubic B-spline functions such as diagonal structure of the system matrix, shape control and its analysis (see [18, 25] and the references therein), several methods based on trigonometric cubic B-spline functions are recently developed for the solution of different problems including hyperbolic problems [18], wave equation [19], non-classical diffusion problem [20], second-order hyperbolic telegraph equation [21], non-linear two-point boundary value problem [22], Hunter Saxton equation [23], time fractional diffusion-wave equation [24], Burgers' equations [25, 26], and Fisher's reaction-diffusion equation [27].

The rest of the paper is outlined as follows: In Sect. 2, we extend trigonometric cubic B-spline collocation method for the solution of the problem defined in (1)–(3). Section 3 is devoted to numerical tests and results. Section 4 concludes work the paper.

2 Construction of the Trigonometric Cubic B-Spline Collocation Method

Consider the problem (1)–(3). The interval $[a,\ b]$ is partitioned into N subintervals of equal length h by the nodes $\zeta_j, j = 0, 1, 2, \cdots, N$ such that $a = \zeta_0 < \zeta_1 < \cdots < \zeta_{N-1} < \zeta_N = b$ and $h = \frac{b-a}{N}$.

The trigonometric cubic B-spline functions $B_j(\zeta), j = -1, 0, 1, \cdots, N+1$ at these nodes are defined as [18, 25]

$$B_j(\zeta) = \frac{1}{\omega} \begin{cases} \beta^3(\zeta_{j-2}), \zeta \in [\zeta_{j-2}, \zeta_{j-1}], \\ \beta(\zeta_{j-2})(\beta(\zeta_{j-2})\gamma(\zeta_j) + \beta(\zeta_{j-1})\gamma(\zeta_{j+1})) + \beta^2(\zeta_{j-1})\gamma(\zeta_{j+2}), \zeta \in [\zeta_{j-1}, \zeta_j], \\ \beta(\zeta_{j-2})\gamma^2(\zeta_{j+1}) + \gamma(\zeta_{j+2})(\beta(\zeta_{j-1})\gamma(\zeta_{j+1}) + \beta(\zeta_j)\gamma(\zeta_{j+2})), \zeta \in [\zeta_j, \zeta_{j+1}], \\ \gamma^3(\zeta_{j+2}), \zeta \in [\zeta_{j+1}, \zeta_{j+2}], \end{cases} \tag{4}$$

where $\beta(\zeta_j) = \sin\left(\frac{\zeta - \zeta_j}{2}\right)$, $\gamma(\zeta_j) = \sin\left(\frac{\zeta_j - \zeta}{2}\right)$, $\omega = \sin\left(\frac{h}{2}\right)\sin(h)\sin\left(\frac{3h}{2}\right)$.

The set of trigonometric cubic B-spline functions $\{B_{-1}, B_0, B_1, B_2, \cdots, B_{N+1}\}$ forms a basis for the functions over the interval $[a, b]$. Let $U(\zeta, \tau)$ be an approximation to the exact solution $u(\zeta, \tau)$, then

$$U(\zeta, \tau) = \sum_{j=-1}^{N+1} C_j(\tau) B_j(\zeta), \tag{5}$$

where $C_j(\tau)$ are unknown time dependent quantities. The nodal values U_j, U_j' and U_j'' at the grid points ζ_j are obtained from Eqs. (4) and (5) in the following form:

$$\begin{cases} U_j = \delta_1 C_{j-1} + \delta_2 C_j + \delta_1 C_{j+1}, \\ U_j' = \delta_3 C_{j-1} + \delta_4 C_{j+1}, \\ U_j'' = \delta_5 C_{j-1} + \delta_6 C_j + \delta_5 C_{j+1}, \end{cases} \tag{6}$$

where dashes represent derivatives with respect to the space variable and $\delta_1 = \dfrac{\sin^2\left(\frac{h}{2}\right)}{\sin(h)\sin\left(\frac{3h}{2}\right)}$,

$\delta_2 = \dfrac{2}{1+2\cos(h)}$, $\delta_3 = -\dfrac{3}{4\sin\left(\frac{3h}{2}\right)}$, $\delta_4 = \dfrac{3}{4\sin\left(\frac{3h}{2}\right)}$, $\delta_5 = \dfrac{3(1+3\cos(h))}{16\left(2\cos\left(\frac{h}{2}\right) + \cos\left(\frac{3h}{2}\right)\right)\sin^2\left(\frac{h}{2}\right)}$ and

$\delta_6 = -\dfrac{3\cot^2\left(\frac{h}{2}\right)}{2+4\cos(h)}$.

To develop the trigonometric cubic B-spline collocation technique for the problem (1)–(3), we proceed as follows:

The time level is denoted by $\tau^m = m\Delta\tau, m = 1, 2, \ldots, M$, where $\Delta\tau$ is the increment in time. The time derivative $u_\tau(\zeta, \tau^{m+1})$ in Eq. (1) is approximated by backward difference formula [12] as follows:

$$u_\tau\left(\zeta, \tau^{m+1}\right) \approx \frac{u^{m+1}(\zeta) - u^m(\zeta)}{\Delta\tau}, \quad m \geq 1. \tag{7}$$

Substituting Eq. (7) in Eq. (1), we get

$$\frac{u(\zeta, \tau^{m+1}) - u(\zeta, \tau^m)}{\Delta t} = \int_0^{\tau^{m+1}} (\tau^{m+1} - v)^{-\alpha} u_{\zeta\zeta}(\zeta, v) dv + f\left(\zeta, \tau^{m+1}\right). \tag{8}$$

The integral part in Eq. (8) is evaluated as follows [12, 14]:

$$\int_0^{\tau^{m+1}} (\tau^{m+1} - v)^{-\alpha} u_{\zeta\zeta}(\zeta, v) dv = \int_0^{\tau^{m+1}} v^{-\alpha} u_{\zeta\zeta}(\zeta, \tau^{m+1} - v) dv$$

$$= \sum_{j=0}^m \int_{\tau^j}^{\tau^{j+1}} v^{-\alpha} u_{\zeta\zeta}(\zeta, \tau^{m+1} - v) dv \tag{9}$$

$$\approx \sum_{j=0}^m u_{\zeta\zeta}(\zeta, \tau^{m-j+1}) \int_{\tau^j}^{\tau^{j+1}} v^{-\alpha} dv.$$

Substituting Eq. (9) in Eq. (8), we obtain

$$\frac{u(\zeta, \tau^{m+1}) - u(\zeta, \tau^m)}{\Delta \tau} = \frac{(\Delta \tau)^{1-\alpha}}{1-\alpha} \sum_{j=0}^{m} u_{\zeta\zeta}(\zeta, \tau^{m-j+1}) \left[(j+1)^{1-\alpha} - j^{1-\alpha}\right] + f(\zeta, \tau^{m+1}).$$

$$(10)$$

Equation (10) can be rearranged as

$$u^{m+1}(\zeta) - \frac{(\Delta \tau)^{2-\alpha}}{1-\alpha} u_{\zeta\zeta}^{m+1}(\zeta) = u^m(\zeta) + \frac{(\Delta \tau)^{2-\alpha}}{1-\alpha} \sum_{j=1}^{m} b_j u_{\zeta\zeta}^{m-j+1}(\zeta) + \Delta \tau f^{m+1}(\zeta), m \geq 1,$$

$$(11)$$

where $u^{m+1}(\zeta) = u(\zeta, \tau^{m+1})$, $f^{m+1}(\zeta) = f(\zeta, \tau^{m+1})$ and $b_j = (j+1)^{1-\alpha} - j^{1-\alpha}$, $j = 1, 2, 3, \ldots$

To develop the general scheme, use Eq. (6) at $\zeta = \zeta_j, j = 0, 1, 2, \ldots, N$, and substituting the approximate values of $u^{m+1}(\zeta)$ and $u_{\zeta\zeta}^{m+1}(\zeta)$ in Eq. (11), we get

$$E_1 C_{j-1}^{m+1} + E_2 C_j^{m+1} + E_3 C_{j+1}^{m+1} = F_j, \qquad (12)$$

where $E_1 = \delta_1 - \frac{(\Delta \tau)^{2-\alpha}}{1-\alpha} \delta_5$, $E_2 = \delta_2 - \frac{(\Delta \tau)^{2-\alpha}}{1-\alpha} \delta_6$, $E_3 = \delta_1 - \frac{(\Delta \tau)^{2-\alpha}}{1-\alpha} \delta_5$, and

$F_j = u^m(\zeta_j) + \frac{(\Delta \tau)^{2-\alpha}}{1-\alpha} \sum_{j=1}^{m} b_j u_{\zeta\zeta}^{m-j+1}(\zeta_j) + \Delta \tau f^{m+1}(\zeta_j)$.

Equation (12) represents a system of $N+1$ equations in $N+3$ unknowns $C_j^{m+1}, j = -1, 0, \ldots, N+1$. In order to get a unique solution; the parameters $C_{-1}^{m+1}, C_{N+1}^{m+1}$ can be eliminated using the boundary conditions (3) and (6). The resulting linear system can be solved by a tridiagonal solver successively once initial time solution is obtained from Eqs. (2) and (3). Finally the approximate solution will be obtained from Eq. (5). The scheme obtained in this section can be implemented in the following algorithm:

Algorithm

The algorithm works in the following manner:

1. Choose $N+1$ equal spaced nodes from the domain set $[a, b]$.
2. Choose the parameter $\Delta \tau$.
3. Compute the initial solution u^0 from Eq. (2).
4. The parameters $\left\{C_j^{m+1}\right\}$ are calculated from Eq. (12) after eliminating $C_{-1}^{m+1}, C_{N+1}^{m+1}$ using (3) and (6).
5. The approximate solution u^{m+1} at the successive time levels is obtained from step 4 and Eq. (5)

3 Numerical Tests

We provide the following examples for validation of the present technique (12) for the solution of the problem defined in Eqs. (1) and (3). The examples are taken from the references [10, 14, 16] with $\alpha = \frac{1}{2}$ in order to compare the results of the present method with the methods QWM, QBCM and SMLP. In all the figures, the symbols x and t represent the space variable ζ and time variable τ respectively. All the computations are carried using Processor Core i3 2.40 GHz and 2 GB RAM.

Example 1: We examine the performance of the proposed method in solving the PIDE (1)–(3) with the given weakly singular kernel. Consider Eq. (1), with the following initial and boundary conditions: $u(\zeta, 0) = \sin(\pi\zeta)$ and $u(0, \tau) = u(1, \tau) = 0$.

$f(\zeta, \tau)$ is to be chosen so that $u(\zeta, \tau) = \sin(\pi\zeta) - \frac{4\tau^{3/2}}{3\sqrt{\pi}}\sin(2\pi\zeta)$ is exact solution of Eqs. (1)–(3). Numerical results are obtained using the parameter values $N = 10, 20$, $\Delta\tau = 10^{-5}, 10^{-6}$ and $M = 50, 150, 250, 350, 450$ and the results are provided in Tables 1, 2, 3 and 4 along with the results obtained by quintic B-spline collocation method and quasi-wavelet method given in [10, 14]. From these tables it is clear that the results obtained by the present method show better accuracy as compared to quintic B-spline method and quasi wavelet method given in [10, 14]. Figure 1 shows the trigonometric cubic B-spline and exact solution at $M = 1000$ whereas Fig. 2 represents the error graph at $M = 1000$. Figure 3 shows solutions at various time levels corresponding to $N = 20, \Delta\tau = 10^{-6}$ over the time interval [0, 0.001].

Table 1. Error norms for $N = 10, \Delta\tau = 10^{-5}$ for Example-1

M	Trigonometric Cubic B-spline L_∞	L_2	QBCM [14] L_∞	QWM [10] L_∞
50	8.93×10^{-6}	5.96×10^{-6}	1.18×10^{-4}	1.58×10^{-3}
150	4.42×10^{-5}	3.09×10^{-5}	6.75×10^{-4}	7.89×10^{-3}
250	9.45×10^{-5}	6.65×10^{-5}	1.40×10^{-3}	1.61×10^{-2}
350	1.56×10^{-4}	1.10×10^{-4}	2.51×10^{-3}	2.53×10^{-2}
450	2.28×10^{-4}	1.60×10^{-4}	3.70×10^{-3}	3.46×10^{-2}

Table 2. Error norms for $N = 20, \Delta\tau = 10^{-5}$ for Example-1

M	Trigonometric Cubic B-spline L_∞	L_2	QBCM [14] L_∞	QWM [10] L_∞
50	2.82×10^{-7}	1.88×10^{-7}	3.75×10^{-5}	5.04×10^{-5}
150	1.40×10^{-6}	9.77×10^{-6}	2.13×10^{-5}	2.60×10^{-4}
250	2.99×10^{-6}	2.10×10^{-6}	4.72×10^{-5}	5.57×10^{-4}
350	4.94×10^{-6}	3.48×10^{-6}	7.94×10^{-5}	9.21×10^{-4}
450	7.20×10^{-6}	5.08×10^{-6}	1.16×10^{-4}	1.34×10^{-3}

Table 3. Error norms for $N = 10, \Delta\tau = 10^{-6}$ for Example-1

Trigonometric Cubic B-spline		QBCM [14]	QWM [10]
M	L_∞	L_2	L_∞	L_∞
50	8.51×10^{-6}	6.00×10^{-6}	1.18×10^{-4}	4.93×10^{-4}
150	4.21×10^{-5}	3.09×10^{-5}	6.71×10^{-4}	2.52×10^{-3}
250	8.99×10^{-5}	6.65×10^{-5}	1.48×10^{-3}	5.36×10^{-3}
350	1.49×10^{-4}	1.10×10^{-4}	2.40×10^{-3}	8.76×10^{-3}
450	2.16×10^{-4}	1.60×10^{-4}	3.67×10^{-3}	1.26×10^{-2}

Table 4. Error norms for $N = 20, \Delta\tau = 10^{-6}$ for Example-1

Trigonometric Cubic B-spline		QBCM [14]	QWM [10]
M	L_∞	L_2	L_∞	L_∞
50	2.69×10^{-7}	1.88×10^{-7}	3.75×10^{-6}	1.56×10^{-5}
150	1.33×10^{-6}	9.77×10^{-6}	2.12×10^{-5}	8.04×10^{-5}
250	2.84×10^{-6}	2.10×10^{-6}	4.68×10^{-5}	1.72×10^{-4}
350	4.70×10^{-6}	3.48×10^{-6}	7.87×10^{-5}	2.85×10^{-4}
450	6.84×10^{-6}	5.08×10^{-6}	1.15×10^{-4}	4.16×10^{-4}

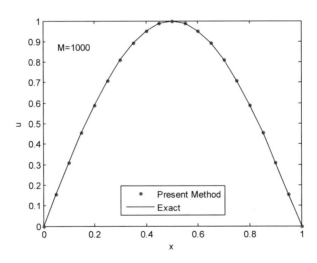

Fig. 1. Trigonometric cubic B-spline and exact solution for $N = 20, \Delta\tau = 10^{-6}$.

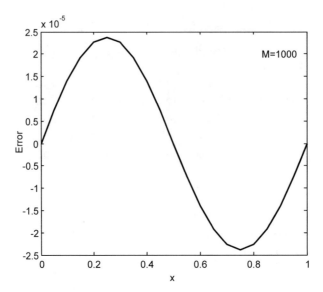

Fig. 2. Error Plot for $N = 20, \Delta\tau = 10^{-6}$ for Example-1

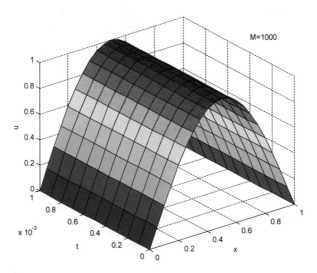

Fig. 3. Trigonometric cubic B-spline solutions at various time levels for $N = 20, \Delta\tau = 10^{-6}$.

Example 2: In this example we consider Eq. (1)–(3), with the following initial and boundary conditions: $u(\zeta, 0) = \sin(\pi\zeta), 0 \leq \zeta \leq 1, u(0, \tau) = u(1, \tau) = 0, \tau > 0$.

Here $f(\zeta, \tau) = 0$. The exact solution of Eq. (1)–(3) is given by [10] $u(\zeta, \tau) = \sum_{j=0}^{\infty} (-1)^j \Gamma\left(\frac{3}{2}j + 1\right)^{-1} \left(\pi^{\frac{5}{2}} \tau^{\frac{3}{2}}\right)^j \sin(\pi\zeta)$. Numerical results are obtained using the parameter values $N = 10$, $\Delta\tau = 10^{-2}, 10^{-3}, 10^{-4}$ and $\tau = 1$ and the results are reported in Table 5 along with the results given in [16]. From this table it is clear that the solutions obtained by the present method show better accuracy as compared to the SMLP method given in [16]. Figure 4 shows the trigonometric cubic B-spline and exact solution at $\tau = 1$ whereas Fig. 5 represents the error in solution at $\tau = 1$. (Table 6)

Table 5. Pointwise absolute error using $N = 10, \tau = 1$ for Example-2

$\Delta\tau = 10^{-2}$			$\Delta\tau = 10^{-3}$			$\Delta\tau = 10^{-4}$		
ζ	Present method	SMLP [16]	Present method	SMLP [16]	Present method	SMLP [16]		
0.1	3.54×10^{-5}	1.9×10^{-3}	5.90×10^{-6}	4.1×10^{-4}	6.56×10^{-6}	2.6×10^{-4}		
0.2	6.74×10^{-5}	3.4×10^{-3}	1.12×10^{-5}	6.1×10^{-4}	1.25×10^{-5}	3.1×10^{-4}		
0.3	9.28×10^{-5}	4.4×10^{-3}	1.54×10^{-5}	5.3×10^{-4}	1.72×10^{-5}	1.3×10^{-4}		
0.4	1.09×10^{-4}	5.1×10^{-3}	1.81×10^{-5}	4.9×10^{-4}	2.02×10^{-5}	1.1×10^{-5}		
0.5	1.15×10^{-4}	5.3×10^{-3}	1.91×10^{-5}	4.9×10^{-4}	2.12×10^{-5}	1.7×10^{-5}		
0.6	1.09×10^{-4}	5.1×10^{-3}	1.81×10^{-5}	4.9×10^{-4}	2.02×10^{-5}	1.1×10^{-5}		
0.7	9.28×10^{-5}	4.4×10^{-3}	1.54×10^{-5}	5.3×10^{-4}	1.72×10^{-5}	1.3×10^{-4}		
0.8	6.74×10^{-5}	3.4×10^{-3}	1.12×10^{-5}	6.1×10^{-4}	1.25×10^{-5}	3.1×10^{-4}		
0.9	3.54×10^{-5}	1.9×10^{-3}	5.90×10^{-6}	4.1×10^{-4}	6.56×10^{-6}	2.6×10^{-4}		

Table 6. Error norms $\Delta\tau = 10^{-4}, \tau = 0.01$ for Example 2

N	L_∞	L_2
10	1.74×10^{-4}	3.84×10^{-4}
20	1.13×10^{-4}	3.23×10^{-4}
40	9.88×10^{-5}	3.98×10^{-4}
80	9.50×10^{-5}	5.41×10^{-4}
160	9.41×10^{-5}	7.58×10^{-4}
320	9.38×10^{-5}	1.07×10^{-4}

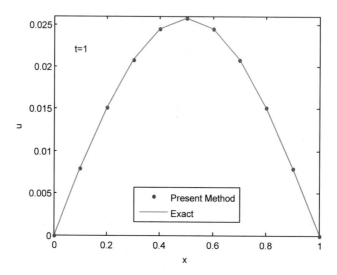

Fig. 4. Trigonometric cubic B-spline and exact solution for $N = 10, \Delta\tau = 10^{-4}$.

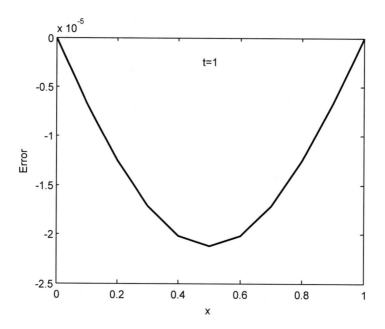

Fig. 5. Error Plot for $N = 10, \Delta\tau = 10^{-4}$ for Example-2

4 Conclusion

In this paper, we extended the trigonometric cubic B-spline collocation scheme to solve a second order partial integro-differential equation with a weakly singular kernel. From the numerical simulation, it is found that the present method provides accurate results, and better accuracy is obtained as compared to QBCM, QWM and SMLP. Thus the proposed method is economical, valid and simple to implement for solving the second order partial integro-differential equation with a weakly singular kernel. However, the numerical technique provides higher accuracy for small time step size.

References

1. Gurtin, M.E., Pipkin, A.C.: A general theory of heat conduction with finite wave speed. Arch. Ration. Mech. Anal. **31**, 113–126 (1968)
2. Miller, R.K.: An integro-differential equation for rigid heat conductors with memory. J. Math. Anal. Appl. **66**, 313–332 (1978)
3. Lodge, A.S., Renardy, M., Nohel, J.A.: Viscoelasticity and rheology. Academic Press, New York (1985)
4. Ortega, J.M., Davis, S.H., Rosemblat, S., Kath, W.L.: Bifurcation with memory. SIAM J. Appl. Math. **46**, 171–188 (1986)
5. Chen, C., Thome, V., Wahlbin, L.: Finite element approximation of a parabolic integro-differential equation with a weakly singular kernel. Math. Comput. **58**, 587–602 (1992)
6. Tang, T.: A finite difference scheme for partial integro-differential equations with a weakly singular kernel. Appl. Numer. Math. **11**(4), 309–319 (1993)
7. Dehghan, M.: Solution of a partial integro-differential equation arising from viscoelasticity. Int. J. Comp. Math. **83**(1), 123–129 (2006)
8. Zarebnia, M.: Sinc numerical solution for the Volterra integro-differential equation. Comm. Nonlinear Sci. Num. Simul. **15**(3), 700–706 (2010)
9. Fakhar-Izadi, F., Dehghan, M.: The spectral methods for parabolic Volterra integro-differential equations. J. Comput. Appl. Math. **235**(14), 4032–4046 (2011)
10. Long, W.T., Xu, D., Zeng, X.Y.: Quasi wavelet based numerical method for a class of partial integro-differential equation. Appl. Math. Comput. **218**, 11842–11850 (2012)
11. Yang, X., Xu, D., Zhang, H.: Crank-Nicolson/quasi-wavelets method for solving fourth order partial integro-differential equation with a weakly singular kernel. J. Comput. Phys. **234**, 317–329 (2013)
12. Zhang, H., Han, X., Yang, X.: Quintic B-spline collocation method for fourth order partial integro-differential equations with a weakly singular kernel. Appl. Math. Comput. **219**, 6565–6575 (2013)
13. Ali, A., Ahmad, S., Shah, S.I.A., Haq, F.I.: A quartic B-spline collocation technique for the solution of partial integro-differential equations with a weakly singular kernel. Sci. Int. **27**(5), 3971–3976 (2015)
14. Ahmad, S., Ali, A., Shah, S.I.A., Haq, F.I.: A computational algorithm for the solution of second order partial integro-differential equations with a weakly singular kernel using quintic B-spline collocation method. SURJ. **47**(4), 709–712 (2015)
15. Ali, A., Ahmad, S., Shah, S.I.A., Fazal-i-Haq: A computational technique for the solution of parabolic type integro-differential equation with a weakly singular kernel. SURJ. **48**(1), 71–74 (2016)

16. Fahim, A., Araghi, M.A.F., Rashidinia, J., Jalalvand, M.: Numerical solution of Volterra partial integro-differential equations based on sinc-collocation method. Adv. Dif. Equ. **2017**, 362 (2017)

17. Aziz, I., Khan, I.: Numerical Solution of Partial Integrodifferential Equations of Diffusion Type. Math. Prob. Eng. **2017**, 11 (2017). Article ID 2853679

18. Abbas, M., Majid, A.A., Ismail, AIMd, Rashid, A.: The application of cubic trigonometric B-spline to the numerical solution of the hyperbolic problems. Appl. Math. Comput. **239**, 74–88 (2014)

19. Zin, S.M., Majid, A.A., Ismail, AIMd, Abbas, M.: Cubic trigonometric B-spline approach to numerical solution of wave equation. Int. J. Math. Comput. Sci. **8**(10), 1302–1306 (2014)

20. Abbas, M., Majid, A.A., Ismail, AIMd, Rashid, A.: Numerical method using cubic trigonometric B-Spline technique for nonclassical diffusion problems. Abstract Appl. Anal. **2014**, 11 (2014). Article ID 849682

21. Nazir, T., Abbas, M., Yaseen, M.: Numerical solution of second-order hyperbolic telegraph equation via new cubic trigonometric B-splines approach. Cogent Math. **4**, 1382061 (2015)

22. Heilat, A.S., Ismail, AIMd: Hybrid cubic b-spline method for solving Non-linear two-point boundary value problems. Int. J. Pure Appl. Math. **110**(2), 369–381 (2016)

23. Hashmi, M.S., Awais, M., Waheed, A., Ali, Q.: Numerical treatment of Hunter Saxton equation using cubic trigonometric B-spline collocation method. AIP Adv. **7**, 095124 (2017). https://doi.org/10.1063/1.4996740

24. Yaseen, M., Abbas, M., Nazir, T., Bale, D.: A finite difference scheme based on cubic trigonometric B-splines for a time fractional diffusion-wave equation. Adv. Differ. Equ. **2017**, 274 (2017)

25. Dag, I., Hepson, O.E., Kaçmaz, O.: The trigonometric cubic B-spline algorithm for burgers' equation. Int. J. Nonlin. Sci. **24**(2), 120–128 (2017)

26. Arora, G., Joshi, V.: A computational approach using modified trigonometric cubic B-spline for numerical solution of Burgers' equation in one and two dimensions. Alex. Eng. J. **7**(2), 1087–1098 (2018)

27. Tamsir, M., Dhiman, N., Srivastava, V.K.: Cubic trigonometric B-spline differential quadrature method for numerical treatment of Fisher's reaction-diffusion equations. Alex. Eng. J. **7**(3), 2019–2026 (2018)

Addressing Fine-Grained Variability in User-Centered Software Product Lines: A Case Study on Dashboards

Andrea Vázquez-Ingelmo[1]([⊠]) [iD], Francisco J. García-Peñalvo[1] [iD],
and Roberto Therón[1,2] [iD]

[1] GRIAL Research Group, Computer Sciences Department, Research Institute
for Educational Sciences, University of Salamanca, Salamanca, Spain
{andreavazquez, Fgarcia, theron}@usal.es
[2] VisUSAL Research Group, University of Salamanca, Salamanca, Spain

Abstract. Software product lines provide a theoretical framework to generate and customize products by studying the target domain and by capturing the commonalities among the potential products of the family. This domain knowledge is subsequently used to implement a series of configurable core assets that will be systematically reused to obtain products with different features to match particular user requirements. Some kind of interactive systems, like dashboards, require special attention as their features are very fine-grained. Having the capacity of configuring a dashboard product to match particular user requirements can improve the utility of these products by providing the support to users to reach useful insights, in addition to a decrease in the development time and an increase in maintainability. Several techniques for implementing features and variability points in the context of SPLs are available, and it is important to choose the right one to exploit the SPL paradigm benefits to the maximum. This work addresses the materialization of fine-grained variability in SPL through code templates and macros, framed in the particular domain of dashboards.

Keywords: Software product lines · SPL · Granularity · User interfaces · Dashboards · Customization

1 Introduction

Software product lines (SPLs) address the systematic development of software assets for building families of products that share a specific domain [1, 2]. By reutilizing, configuring and composing these software assets, the time-to-market of new derived products decreases, in addition to an increase in requirements traceability, customization levels, flexibility, maintainability and of course, productivity.

However, implementing and introducing an SPL is not a straightforward job. The domain in which the SPL will be framed must be thoroughly studied to extract significant features and capture the commonalities among the potential products that could be developed through this paradigm. Planning the development of highly configurable software components allows the delay of design decisions, enhancing flexibility

© Springer Nature Switzerland AG 2019
Á. Rocha et al. (Eds.): WorldCIST'19 2019, AISC 930, pp. 855–864, 2019.
https://doi.org/10.1007/978-3-030-16181-1_80

regarding the materialization of dynamic or even new requirements. These delayed design decisions are the so-called variability points [3].

The study of the target domain is the first step regarding an SPL design process, but the implementation of the identified variability points within the core assets of the product family remains a crucial and a critical challenge for this paradigm to succeed.

There are several techniques to materialize variability points, and the desired granularity of the SPL features is a relevant factor to choose the right method referring to the ability to modify the products behavior or their underlying functionality. In addition to the desired granularity level, the target domain of the SPL is also a key factor regarding the choice of the implementation technique.

For instance, user-centered tools require high levels of customization, both at functional and at visual design level. Developing these type of tools need further efforts on the requirements elicitation processes, in order to fully understand the final users' necessities and to provide them with helpful interfaces. Customizing user interfaces within the SPL paradigm context, however, is still a complex task, yet requiring semi-automatic or completely manual processes [4]. A large number of possible user profiles (and their associated particular requirements) could make the automatic derivation of interactive systems chaotic regarding its possible features, hampering the evolution and maintainability of the product line. The main issue regarding these interaction-intensive systems is the fine-grained nature of their features: a slight modification on interaction patterns, interface layout, color palette, etc. could be crucial regarding the final perceived usability of a generated product.

A particular case of these interactive systems is dashboards. These tools aim at helping users to reach useful insights about datasets, facilitating the discovery of unusual patterns or significant data points. The potential of dashboards resides in their ability to present information at-a-glance, supporting complex procedures like decision-making processes, communication, and learning, etc. [5]. A lot of profiles could be involved in these procedures though, being difficult to provide a common and general dashboard useful for each of them. That is why the SPL paradigm can ease the development of customized dashboards by reutilizing its different components (i.e., visualizations, controls, filters, interaction patterns, etc.), instead of implementing a single dashboard for each data domain or user involved. However, dashboards need fine-grained variability to provide powerful customizations and to support particular configurations for different user profiles, helping them to reach their own goals regarding data exploration and data explanation.

The remainder of this paper is structured as follows. Section 2 is an overview of a set of available methods for implementing variability within SPLs. Section 3 analyzes the particular case of the dashboards domain regarding the granularity of its features, presenting the case study in which an experimental framework for generating dashboards has been developed in Sect. 4. Finally, Sect. 5 discusses the achieved results regarding granularity, and Sect. 6 presents the conclusions of this work.

2   Variability Mechanisms

There exist different techniques to implement variability points in SPLs. It is important to choose wisely given the requirements of the product line itself (i.e., the complexity of the software to develop, its number of features, their granularity requirements, etc.). Generally, at the code level, the variability points that correspond to a specific feature will be spread across different source files [6]. That is why separating concerns at the implementation level is essential to avoid the variability points to be scattered, as this feature dispersion would decrease code understandability and maintainability. Implementing each feature in individual code modules can help with this separation of concerns [6], but it is difficult to achieve fine-grained variability through this approach. A balance between code understandability and granularity should be devised to choose both a maintainable and highly customizable SPL.

This section will briefly describe different methods that are potentially suitable to the dashboards' domain given their particular features, although there are more approaches to implement variability in SPLs that can be consulted in [6].

2.1   Conditional Compilation

Conditional compilation uses preprocessor directives to inject or remove code fragments from the final product source code. This method allows the achievement of any level of feature granularity due to the possibility of inserting these directives at any point of the code, even at expression or function signature level [7]. Also, although pretended to the C language, preprocessor directives can be used for any language and arbitrary transformations [8]. The main drawback of this approach is the decrease of code readability and understandability as interweaving and nesting these preprocessor directives makes the code maintainability a tedious task [9].

2.2   Frames

Frame technology is based on entities (frames) that are assembled to compose final source code files. Frames use preprocessor-like directives to insert or replace code and to set parameters [6]. An example of a variability implementation method based on frame technology is the XML-based Variant Configuration Language (XVCL) [10]. Through this approach, only the necessary code is introduced in concrete components by specifying frames that contain the code and directives associated with different features and variants. XVCL is independent of the programming language and can handle variability at any granularity level [11].

2.3   Template Engines

Template engines allow the parameterization and inclusion/exclusion of code fragments through different directives. If the template engine allows the definition of macros, features can be refactored into different code fragments encapsulated through these elements, improving the code organization and enabling variability at any level of granularity. Templating engines can also be language-independent, providing a powerful tool for generating any type of source file [12] by using programming directives such as loops and conditions.

2.4 Aspect-Oriented Programming

Aspect-Oriented Programming (AOP) allows the implementation of crosscutting concerns through the definition of aspects, centralizing features that need to be present in different source files through unique entities (aspects) thus improving code understandability and maintainability by avoiding scattered features and "tangled" code [13].

AOP is a popular method to materialize variability points in SPLs due to the possibility of modifying the system behavior at certain points, namely join points [14–16]. However, AOP could lack fine-grained variability (i.e., variability at sentence, expression or signature level) and particular frameworks or language extensions are necessary to implement aspects in certain programming languages.

3   The Dashboards' Domain

Regarding the present work target domain (i.e., dashboards), the chosen implementation technique was to use a template engine. The decision was made due to the fine granularity that can be achieved through this method, which is necessary to materialize even the slightest variability on the visualization components. Another factor for choosing this technique lies in the straightforward way of implementing variability regarding the products' features and its language-independent nature.

Framing technology could also be a potential solution within the dashboards' domain, but the decision of designing a DSL to wrap the features at a higher level made the use of code templates a more suitable solution, providing complete freedom to define the syntax of the DSL (specification x-frames are based on a fixed syntax [11], which could result in lack of flexibility for this work's approach) as the directives within the templates can be fully parameterized.

The selected template engine was Jinja2 (http://jinja.pocoo.org/docs/2.10/) given its rich API and powerful features such as the possibility of defining macros, importing them, defining custom filters and tags in addition to its available basic directives (loops, conditions, etc.).

4   Results of the Case Study

As it has been aforementioned, a DSL has been designed along with the SPL to abstract and ease the application engineering process. This DSL binds the feature model with the implementation method at code-level [17], enabling the specification of features through XML technology. Designing a DSL not only eases the configuration of variants but also improves the traceability of features through the different SPL paradigm phases (and opens up the possibility of combining the SPL paradigm with model-driven development [16]).

For this case study, it is necessary to provide a configurable SPL that enable automatic generation of dashboards with different features. These features involve a variety of potential requirements: from the modification of the dashboard layout (i.e., including or removing whole visualization components) to the modification of a particular interaction pattern to manage to zoom on visualizations, for example. To achieve the desired levels of granularity and to support the DSL for automating the application engineering phase, a template engine (Jinja2) was selected, as indicated in Sect. 3.

Templating resembles conditional compilation, as their underlying behavior based in programming directives is very similar. The main benefit of templates is that they support these directives and macros in a more sophisticated manner.

As presented in Sect. 2, the main drawback of conditional compilation is the scatter of concerns and features, decreasing code maintainability and readability. One of the benefits of using a powerful template engine like Jinja2 is the possibility of clustering the necessary code fragments that compose a certain feature in sets of macros. This approach improves maintainability, as the code fragments in charge of the features will be contained and organized in associated files.

The practical approach followed to apply this implementation method is exemplified in the remainder of this section.

Figure 1 shows a high-level view of the feature model for the dashboard product line developed for the Spanish Observatory for University Employability and Employment (OEEU, https://oeeu.org) [18, 19] to allow users to explore and reach insights about the data collected by this organization [20–25]. The generated dashboard can have different pages, each one composed of different visualizations and data filters. At this high-level view, features are coarse-grained; whole components can be included or removed from the final generated dashboard.

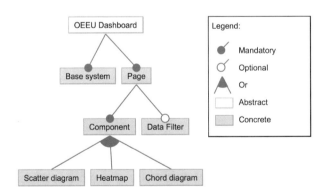

Fig. 1. High-level view of the dashboard SPL's feature diagram.

Low-level features (i.e., leaf nodes of the feature diagram) require fine-grained granularity within the dashboard domain, as these features concern minor visual, functional or interaction characteristics. Figure 2 shows low-level features for a scatter diagram component about the possible functionalities related to its data and behavior.

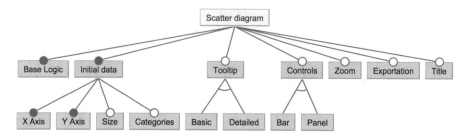

Fig. 2. A snippet of the feature diagram showing lower-level features regarding a scatter diagram component. Some of these features (e.g., the "controls" feature) have their own subsequent features to provide higher customization levels regarding the visualization's functional and information requirements.

To materialize these features at code-level, each feature is arranged in its own file and each file is composed with a set of macros Fig. 3. This set of macros contains the required code fragments associated with an SPL feature.

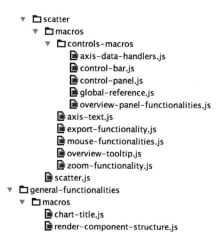

Fig. 3. Example of the code templates organization.

The macros calls are executed within the base logic of the component (in this case the "scatter.js" file contains the basic logic for the scatter diagram, which is mandatory and common for all possible product derivations, as specified in the feature diagram).

The macros themselves are affected by the conditional directives in charge of adapting the code giving particular configurations. This means that the base code will only contain the macro calls, delegating the condition check to the macros and making the code cleaner. By using this approach at the implementation level, concerns are not continuously scattered through the code as it could happen with pure preprocessor directives Fig. 4.

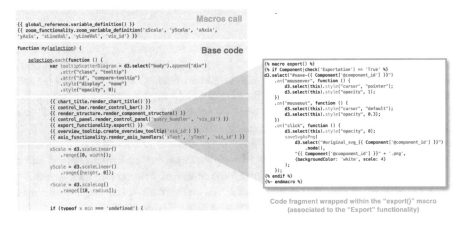

Fig. 4. A snippet of the scatter diagram's JavaScript code. The base code (highlighted in blue) contains macro calls (highlighted in green). If the condition wrapped within the macro is matched, the associated code is injected (i.e., the associated feature will be supported).

Through the DSL and the code templates, a custom code generator can build personalized dashboards that meet the specified requirements automatically Fig. 5.

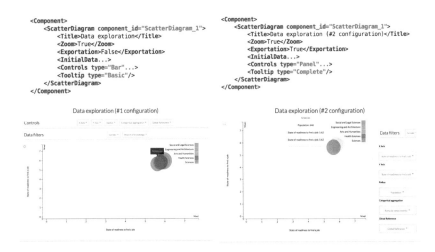

Fig. 5. Two different scatter diagram configurations achieved through the DSL (on top). As it can be seen, the tooltip type, for example, provides different behaviors when interacting with the visualization elements. Also, the layout of the whole visualization can be modified

5 Discussion

SPLs have proved to be a powerful paradigm for managing particular sets of requirements in an efficient and maintainable way. However, these requirements could need different granularity levels, as some important features could be coarse-grained while others could be fine-grained. Choosing the right implementation technique is a complex task because several factors must be taken into account: the levels of granularity, the understandability, and maintainability of the code, the viability of the technique, etc. This work addresses fine-grained granularity in a SPL of dashboards. Dashboards are key tools for reaching of insights regarding particular datasets and to support decision-making processes. Having the power of customizing their features at fine-grained level could be highly valuable, as dashboards usually ask to be user-tailored to provide useful support for particular and individual goals.

In the presented case study, a DSL has been designed for abstracting the configuration process. The use of this DSL to feed a code generator has been one of the determining factors to choose a template engine as the implementation method of the SPL's variability points. Although this approach still lacks powerful maintainability levels, it maintains a proper requirements' traceability by arranging features in a variety of macro definitions. Using XVCL [10] could have been another solution to manage these fine-grained features, but the decision of wrapping the SPL specification through a DSL asked for a more flexible and customizable method such as a template engine. What is more, a combination of the AOP paradigm with the templating method could be highly beneficial providing both customizations regarding directives and a better technique to manage crosscutting concerns (an issue that a template engine could not solve straightforwardly). Also, the approach asks for a method to address data heterogeneity in order to visualize data from any kind of source. However, although presenting these caveats, the results are promising and prove that a robust template engine could be a beneficial method to materialize fine-grained variability within the SPL paradigm context.

Regarding the application on the dashboard domain, having a dashboard SPL could address several problems related to individual personalization, meeting particular requirements. This approach could provide tailored dashboards efficiently after an in-depth elicitation of requirements without consuming many resources, avoiding overwhelming configuration processes delegated to end-users themselves [26].

6 Conclusions

Dashboards are sophisticated tools that require fine-grained features to offer valuable user experiences to their target users. A template-based approach to implement variability points at code level has been applied to an SPL of dashboards.

Creating an SPL of dashboards is not a straightforward task, as different variability dimensions are involved (variability regarding visual design, functionality, layout, data sources, etc.). Using a template engine to implement the core assets of the SPL can address the mentioned fine-grained variability and increase the traceability of features.

This SPL paradigm application to the dashboards' domain opens up different research paths, such as experimenting with different fine-grained configurations to find the best configuration for a particular user profile (A/B testing [23, 27]) or applying machine learning or knowledge bases [28] to provide potentially suitable configurations automatically given certain contexts or user characteristics. Also, developing an automatic link between the feature diagram and the DSL, as well between the DSL and the code templates' directives could further improve maintainability and traceability.

Acknowledgements. This research work has been supported by the Spanish *Ministry of Education and Vocational Training* under an FPU fellowship (FPU17/03276). This work has been partially funded by the Spanish Government Ministry of Economy and Competitiveness throughout the DEFINES project (Ref. TIN2016-80172-R) and the PROVIDEDH project, funded within the CHIST-ERA Programme under the national grant agreement: PCIN-2017-064 (MINECO, Spain).

References

1. Clements, P., Northrop, L.: Software Product Lines. Addison-Wesley, Boston (2002)
2. Pohl, K., Böckle, G., Linden, van Der Linden, F.J.: Software Product Line Engineering: Foundations, Principles and Techniques. Springer, New York (2005)
3. Van Gurp, J., Bosch, J., Svahnberg, M.: On the notion of variability in software product lines. In: 2001. Proceedings. Working IEEE/IFIP Conference on Software Architecture, pp. 45–54. IEEE (2001)
4. Pleuss, A., Hauptmann, B., Keunecke, M., Botterweck, G.: A case study on variability in user interfaces. In: Proceedings of the 16th International Software Product Line Conference, vol. 1, pp. 6–10. ACM (2012)
5. Sarikaya, A., Correll, M., Bartram, L., Tory, M., Fisher, D.: What do we talk about when we talk about dashboards? IEEE Trans. Visual. Comput. Graph (2018)
6. Gacek, C., Anastasopoules, M.: Implementing product line variabilities. In: ACM SIGSOFT Software Engineering Notes, pp. 109–117. ACM (2001)
7. Liebig, J., Apel, S., Lengauer, C., Kästner, C., Schulze, M.: An analysis of the variability in forty preprocessor-based software product lines. In: Proceedings of the 32nd ACM/IEEE International Conference on Software Engineering, vol. 1, pp. 105–114. ACM (2010)
8. Favre, J.-M.: Preprocessors from an abstract point of view. In: Proceedings of the Third Working Conference on Reverse Engineering 1996, pp. 287–296. IEEE (1996)
9. Spencer, H., Collyer, G.: #ifdef considered harmful, or portability experience with C News (1992)
10. Jarzabek, S., Bassett, P., Zhang, H., Zhang, W.: XVCL: XML-based variant configuration language. In: Proceedings of the 25th International Conference on Software Engineering, pp. 810–811. IEEE Computer Society (2003)
11. Zhang, H., Jarzabek, S., Swe, S.M.: XVCL approach to separating concerns in product family assets. In: International Symposium on Generative and Component-Based Software Engineering, pp. 36–47. Springer, Heidelberg (2001)
12. Cisco Blogs. https://blogs.cisco.com/developer/network-configuration-template
13. Kiczales, G., Lamping, J., Mendhekar, A., Maeda, C., Lopes, C., Loingtier, J.-M., Irwin, J.: Aspect-oriented programming. In: European Conference on Object-Oriented Programming, pp. 220–242. Springer, Heidelberg (1997)

14. Waku, G.M., Rubira, C.M., Tizzei, L.P.: A case study using AOP and components to build software product lines in android platform. In: 41st Euromicro Conference on Software Engineering and Advanced Applications (SEAA), pp. 418–421. IEEE (2015)
15. Heo, S.-h., Choi, E.M.: Representation of variability in software product line using aspect-oriented programming. In: Fourth International Conference on Software Engineering Research, Management and Applications, 2006, pp. 66–73. IEEE (2006)
16. Voelter, M., Groher, I.: Product line implementation using aspect-oriented and model-driven software development. In: 11th International Software Product Line Conference, SPLC 2007, pp. 233–242. IEEE (2007)
17. Voelter, M., Visser, E.: Product line engineering using domain-specific languages. In: 15th International Software Product Line Conference (SPLC), pp. 70–79. IEEE (2011)
18. Michavila, F., Martínez, J.M., Martín-González, M., García-Peñalvo, F.J., Cruz-Benito, J., Vázquez-Ingelmo, A.: Barómetro de empleabilidad y empleo universitarios. Edición Máster 2017. Observatorio de Empleabilidad y Empleo Universitarios, Madrid, España (2018)
19. Michavila, F., Martínez, J.M., Martín-González, M., García-Peñalvo, F.J., Cruz-Benito, J.: Barómetro de empleabilidad y empleo de los universitarios en España, 2015 (Primer informe de resultados). Observatorio de Empleabilidad y Empleo Universitarios, Madrid (2016)
20. Vázquez-Ingelmo, A., Cruz-Benito, J., García-Peñalvo, F.J., Martín-González, M.: Scaffolding the OEEU's data-driven ecosystem to analyze the employability of spanish graduates. In: García-Peñalvo, F.J. (ed.) Global Implications of Emerging Technology Trends, pp. 236–255. IGI Global, Hershey (2018)
21. García-Peñalvo, F.J., Cruz-Benito, J., Martín-González, M., Vázquez-Ingelmo, A., Sánchez-Prieto, J.C., Therón, R.: Proposing a machine learning approach to analyze and predict employment and its factors. Int. J. Interact. Multimedia Artif. Intell. 5(2), 39–45 (2018)
22. Vázquez-Ingelmo, A., García-Peñalvo, F.J., Therón, R.: Generation of customized dashboards through software product line paradigms to analyse university employment and employability data. Learning Analytics Summer Institute Spain 2018 – LASI-SPAIN 2018, León, Spain (2018)
23. Cruz-Benito, J., Vázquez-Ingelmo, A., Sánchez-Prieto, J.C., Therón, R., García-Peñalvo, F. J., Martín-González, M.: Enabling adaptability in web forms based on user characteristics detection through A/B testing and machine learning. IEEE Access 6, 2251–2265 (2018)
24. Vázquez-Ingelmo, A., García-Peñalvo, F.J., Therón, R.: Domain engineering for generating dashboards to analyze employment and employability in the academic context. In: 6th International Conference on Technological Ecosystems for Enhancing Multiculturality, Salamanca, Spain (2018)
25. Vázquez-Ingelmo, A., García-Peñalvo, F.J., Therón, R.: Application of domain engineering to generate customized information dashboards. In: International Conference on Learning and Collaboration Technologies, pp. 518–529. Springer, Switzerland (2018)
26. Elias, M., Bezerianos, A.: Exploration views: understanding dashboard creation and customization for visualization novices. In: IFIP Conference on Human-Computer Interaction, pp. 274–291. Springer, Heidelberg (2011)
27. Kakas, A.C.: A/B Testing (2017)
28. Moritz, D., Wang, C., Nelson, G.L., Lin, H., Smith, A.M., Howe, B., Heer, J.: Formalizing visualization design knowledge as constraints: actionable and extensible models in Draco. IEEE Trans. Visual. Comput. Graph. 25, 438–448 (2019)

Estimate of Discharge of Lithium-Ion Batteries

Erick Frota da Costa[1]($\boxtimes$), Darielson Araújo de Souza[1],
Miquéias Silva Araújo[1], Vandilberto Pereira Pinto[1],
Artur Melo Peixoto[1], and Erivaldo Pinheiro da Costa Júnior[2]

[1] Federal University of Ceará – UFC, Electrical Department Campus Sobral,
Sobral, CE, Brazil
erickfrotac@gmail.com
[2] Federal University of Rio Grande do Norte – UFRN, Natal, RN, Brazil

Abstract. The present work has the objective to estimate the curves of discharge of Lithium-Ion batteries. Several methods were tested for comparative analysis. The used database set of batteries were made available by NASA. All of the estimates had a good dash beyond the expected estimate. We used a Artificial Neural Network, Least Squares, Kalman Filter with least squares, ARX e ARMAX models. The evaluations of each method were done according to the computational cost and the coefficient of determination.

Keywords: Estimate · Battery discharge · Lithium-Ion batteries ·
NASA dataset

1 Introduction

The earliest military uses of unmanned aircraft date from the backward century, but it was from the 1980s that they began to draw more attention - from simple "toys" to real weapons when used by the Israeli Air Force against the Air Force Syria in 1982.

As soon as they became more easily available devices, the drones were first implemented for tasks that could help the public safety sectors - they were used by firefighters to put out fires in a region that was difficult to access, or too risky for rescue workers. Most unmanned aerial devices use batteries, as well as other vehicles and devices. Monitoring the batteries can be a rather complex task, as it depends on their usability.

Batteries are devices that transform chemical energy into electrical energy by means of electrochemical reactions. Lithium-ion batteries represent the "state of the art" in energy conversion systems. The advantages of this technology are: the highest energy density (Wh/g) and the low weight. This combination is great for use in portable devices [1, 2].

There are many surveys to estimate battery variables or behaviors, in the work of [3] it has the objective of verifying which analytical model is more accurate for the prediction of the life of Lithium Polymer (LiPo). However, the article [4] evaluates some methods that estimate the discharge voltage of Li-Po batteries in UAVs.

The objective of this work is to perform an estimation of the discharge voltage of Lition Ion batteries. Estimates are made by 4 methods: Non-recursive least squares,

© Springer Nature Switzerland AG 2019
Á. Rocha et al. (Eds.): WorldCIST'19 2019, AISC 930, pp. 865–874, 2019.
https://doi.org/10.1007/978-3-030-16181-1_81

Kalman Filter + LSM, ELM neural network and an ARX model. The method will be evaluated by the coefficient of determination (R^2). The set of data used belongs to NASA's repository, they are batteries with random currents.

2 Battery Datasets

Lithium-Ion batteries are types of rechargeable batteries widely used in portable electronic equipment. The set of data used in this research is called the "Randomized Battery Usage Data Set" which is of Lions Ions and is on the NASA website [5]. The process of data acquisition is very similar to the classics, discharging and loading the tensions. In these tests random currents were placed every minute in the discharge process and the currents varied from 0.5A, 1.0A, 1.5A, 2.0A, 2.5A, 3.0A, 3.5A, 4.0A, 4.5A, 5.0A.

The database has several sets of batteries with the names of RW. In this work, only one battery will be used in RW13, in which it will be estimated the discharge curves of the first, middle and last cycles. Figure 1 shows the first discharge cycle of the RW13 battery, voltage with respect to time. It also shows each discharge cycle after the fully charged battery. This behavior of tension with respect to time always decreases with each discharge.

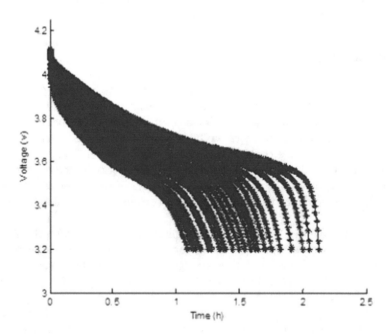

Fig. 1. Discharge curve.

3 Methodology

All methodologies use the coefficient of determination to evaluate the estimates. Five methods were used for an accurate comparative analysis. The methods were evaluated with the same database, in the pre-processing stage, the shuffling and a separation for training and for the tests were done.

3.1 Least Squares Method

Ordinary Least Squares (OLS) is a mathematical optimization technique that seeks to find the best fit for a set of data by minimizing the sum of the squares of the differences between the estimated value and the observed data (such differences are called residues). The least squares technique fits into the General Multiple Linear Regression.

The least squares method models the relationship between two variables by fitting an equation to the observed data. Equation 1 shows the adjustment function.

$$y = a_0 + a_1x + a_2x^2 + \ldots + a_mx^m + \in \tag{1}$$

Where $a_0 \ldots a_m$ are polynomial coefficients, $x \ldots x^m$: is the predicted value of the independent variable, $\in$: variable that includes all residual factors plus possible measurement errors.

According to [6], the least squares criterion weighs all errors equally, and this corresponds to the assumption that all measures have the same precision.

3.2 MMQ + FK

For the hybridization of the Kalman filter methods with least squares, the two parallel ones were implemented, but the least squares estimator equation is updated as a function of the current estimator added with the Kalman gain product with Kalman error. The equation of the least squares estimator in the hybrid system can be presented in Eq. (2).

$$\theta = \theta + K\varepsilon \tag{2}$$

3.2.1 Kalman Filter

The Kalman filter is a set of mathematical equations that constitutes an efficient recursive estimation process, since the quadratic error is minimized. By observing the variable called "observation variable" another variable, not observable, called "state variable" can be estimated efficiently. Past states, the present state, and even predicted future states can be estimated.

In order to use the Kalman filter to estimate the complete state of a process given only a sequence of noisy observations, it is necessary to model the process according to the Kalman filter framework. This means specifying the following matrices: Fk, the state transition model; Hk, the model of observation; Qk, the covariance of process noise; Rk, the noise covariance of observation; and, occasionally, Bk, the model of the control inputs, for each step of time k. The model for the Kalman filter assumes that the real state at time k is obtained through the state in time $(k - 1)$ according to Eq. (3).

$$x_k = F_k x_{k-1} + B_k u_k + w_k \tag{3}$$

and:

F_k: Is the state transition model, applied in the previous state.

B_k: Is the control inputs model, applied to the input control vector.

w_k: Is the process noise, assumed to be sampled from a normal multivariate zero mean distribution and covariance.

At time k, an observation (or measurement) z_k of the real state x_k is made according to Eq. (4).

$$z_k = H_k x_k + v_k \tag{4}$$

Where Hk is the observation model, which maps the real state space in the observed state space, and vk is the observation noise, assumed to be a Gaussian white noise of zero mean and covariance Rk. Figure 2 presents the steps of the Kalman filter algorithm.

$$
\begin{array}{c}
\textbf{Prediction} \\[4pt]
\hat{X}_{k|k-1} = F_k \, \hat{X}_{k-1|k-1} + B_k u_k \\[4pt]
P_{k|k-1} = F_k P_{k-1|k-1} F_k^T + Q_k \\[10pt]
\textbf{Update} \\[4pt]
\tilde{Y}_k = z_k - H_k \, \hat{X}_{k|k-1} \\[4pt]
S_k = H_k P_{k|k-1} H_k^T + R_k \\[4pt]
K_k = P_{k|k-1} H_k^T S_k^{-1} \\[4pt]
\hat{X}_{k|k} = \hat{X}_{k|k-1} + K_k \tilde{Y}_k \\[4pt]
P_{k|k} = (I - K_k H_k) P_{k|k-1}
\end{array}
$$

Fig. 2. Kalman filter algorithm

3.3   ELM ANN

Second [7] Artificial Neural Networks (ANNs) are computer models implemented by software capable of processing data similar to the human brain, simulating learning and generalization. An ANN is defined by one or more layers of fundamental constructs called neurons with weighted interconnections between each. In the present work an ANN is used with the Extreme Learning Machine training (ELM). Figure 3 presents the architecture of the ELM neural network.

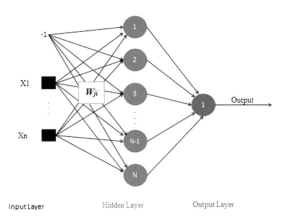

Fig. 3. ELM ANN architecture

The ELM ANN has several advantages over the others as: the input weights and layer bias hidden are chosen randomly, another advantage is the weights of the output layer are determined analytically [8]. In Fig. 4 the steps of the ELM ANN algorithm are shown.

---

begin

Step 1: Randomly select values for Wji and bias weights bi, i = 1, ..., N;

Step 2: Calculate the output matrix H of the hidden layer.

Step 3: Calculate the output weights $\beta = H^{\dagger}T$

end.

Fig. 4. ELM algorithm

3.4   ARX and ARMAX Model

ARX models (AutoRegressive with external inputs) and ARMAX (AutoRegressive with moving average and external inputs) are commonly used in the theory of dynamical system identification. Second [6] systems identification are models that approximate real cofactors.

These models are easy to use and generally have good accuracy depending on the system. The ARX model can be obtained from the system to be idealized, taking $C(q) = D(q) = F(q) = 1$ and $A(q)$ and $B(q)$, arbitrary polynomials, resulting in Eq. (5).

$$A(q)y(k) = B(q)u(k) + v(k) \tag{5}$$

The ARMAX model can be obtained from the system to be idealized, taking $D(q) = F(q) = 1$ and $A(q)$, $B(q)$, $C(q)$ arbitrary polynomials, resulting in Eq. (6).

$$A(q)y(k) = B(q)u(k) + C(q)v(k) \tag{6}$$

4   Results

The methods estimated the discharge curves of the NASA RW13 batteries. The methods used were Least Squares, Kalman filter + Least Squares, ELM ANN and ARX and ARMAX models. The comparative analyzes were made according to the R^2 which is the coefficient of determination, it verifies the quality of the estimate compared with the actual data, its results vary from 0 to 1, the closer to 1 the better the estimate.

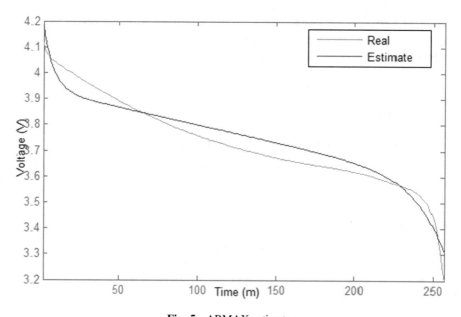

Fig. 5. ARMAX estimate

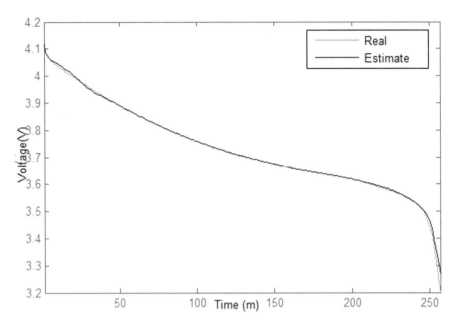

Fig. 6. ARX estimate

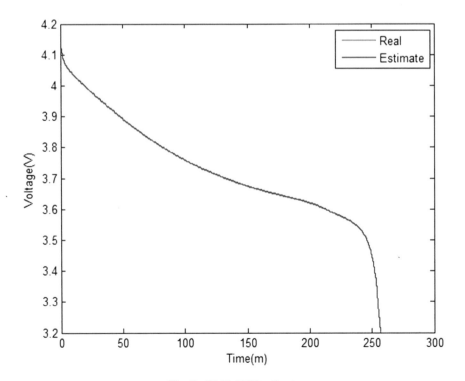

Fig. 7. ELM ANN estimate

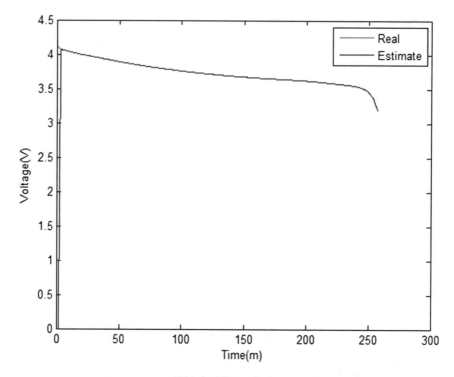

Fig. 8. LSM estimate

The comparative analyzes of all methods are shown in Table 1, where the best esti-mates of each proposed measure are found and evaluated by R^2 and computational cost. When doing the comparative analysis almost all computational costs were very close.

Table 1. Shows all the estimate being evaluated by the coefficient of determination.

Models	R^2	Computational cost
LSM	99.340	0.0000021 s
KF + LSM	98.075	0.0000021 s
ANN ELM	99.541	0.0000019 s
ARX	94.950	0.0000021 s
ARMAX	72.610	0.0000021 s

The ELM neural network obtained a better estimate than the others, it obtained its best R^2 score of **99.541%** or **0.99541**. The architecture used is very simple, only a hidden layer with 5 neurons and a learning step of 0.9.

The method that obtained the worst result was the ARMAX model, which is due to the fact that the model requires carefully controlled experiments on the operating range of the process. Figures 5, 6, 7, 8 and 9 present the estimates of the discharge curves of the techniques, ARMAX, ARX, ELM ANN, LSM and LSM + KF.

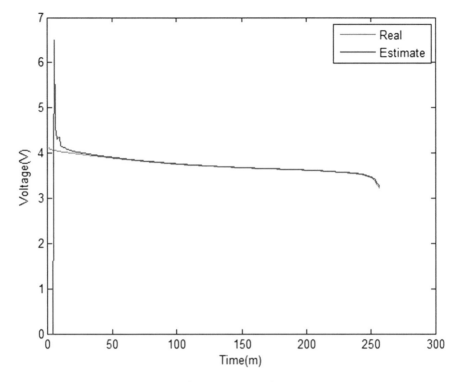

Fig. 9. LSM + KF estimate

5   Conclusion

The work presented satisfactory results, so all methods were able to estimate the discharge curves of Lithium-Ion batteries. The ELM neural network achieved a better result compared to the other methods. The ELM got the R^2 equal 99,541 and computational cost of 0.0000019 s. The other models also had good results with the exception of ARMAX.

The work proposal intends to estimate other aspects of the batteries, such as capacity, remaining life and health among other behaviors that cause uncertainty of lithium ion batteries.

References

1. Linden, D.: Handbook of Batteries. MC Graw-Hill, New York (1995)
2. Mahan, B.H.: Química, um curso universitário. Edgard Blücher Ltda, São Paulo (1980)
3. Julia, G.Z., Douglas, J.B.F., Airam, T.Z., Romcy, S., Paulo, S.S.: Análise Comparativa de Modelos Analíticos para a Predição do Tempo de Vida de Baterias sob o mesmo Cenário de Simulação considerando Correntes de Descargas Variáveis. In: Proceeding Series of the Brazilian Society of Computational and Applied Mathematics (2018)

4. Souza, D.A., Pinto, V.P., Nascimento, L.B.P., Torres, J.L.O., Gomes, J.P.P., Junior, J.J.M.S. A., Almeida, R.N.C.: Battery Discharge forecast applied in Unmanned Aerial Vehicle. Przeglad Elektrotechniczny (2016)
5. Hertz, J., Palmer, R.G., Krogh, A.S.: Introdução à teoria da computação neural. Perseus Books, New York City (1990)
6. Coelho, A.A.R., Coelho, L.S.: Identificação de sistemas Dinâmicos Lineares, Editora UFSC, 2º Edição (2015)
7. Chansanroj, K., Petrovic, J., Ibric, S., Betz, E G.: Drug release control and system understanding of sucrose esters matrix tablets by artificial neural network, Eur. J. Pharm. Sci. (2011)
8. Huang, G.-B., Zhu, Q.-Y., Siew, E.C.-K.: Extreme learning machine: theory and applications. Neurocomputing (2006)

Study of the Number Recognition Algorithms Efficiency After a Reduction of the Characteristic Space Using Typical Testors

Kuntur Muenala[1], Julio Ibarra-Fiallo[1(✉)], and Monserrate Intriago-Pazmiño[2]

[1] Colegio de Ciencias e Ingenierías,
Universidad San Francisco de Quito, Cumbayá, Ecuador
kmuenala@gmail.com, jibarra@usfq.edu.ec
[2] Departamento de Informática y Ciencias de la Computación,
Escuela Politécnica Nacional, Quito, Ecuador
monserrate.intriago@epn.edu.ec

Abstract. The present work studies an application of typical testors to reduce the space of characteristics used as inputs in statistical models to recognize a manuscript digit from its corresponding binary image. The fundamentals of typical testors are described, and the use of logistic and multilogistic regression models in the predictions. Images from the public database of the Semeion Research Center of Sciences of Communication are used. Adequate prediction results were achieved in a reduced space of characteristics.

Keywords: Typical testor · Number recognition algorithm ·
Stochastic algorithms · Logistic regression

1 Introduction

A testor is a collection of characteristics that discriminate descriptions of objects belonging to different classes, and it is minimal in the partial order determined by the collection of these sets. Typical testors play a very important role in problems of supervised pattern recognition, by using the combinatorial logic approach. In patterns recognition problems, the typical testors has several applications in variables selection problems, text mining, and many others. In the combinatorial logic approach, the supervised pattern recognition information can be reduced to a matrix, which is called the M matrix. Typical testors are searched among all possible subsets of columns of this M array, these columns keep information about the characteristics of each object [1]. There are many applications of typical testors. Hence, the interest in this theory, its applications, and its computational optimization are still valid in the last years [2–4]. In the present work, we study the performance measured in efficiency of two recognition algorithms comparing their performance using a large set of characteristics with their

© Springer Nature Switzerland AG 2019
Á. Rocha et al. (Eds.): WorldCIST'19 2019, AISC 930, pp. 875–885, 2019.
https://doi.org/10.1007/978-3-030-16181-1_82

performance providing a reduced set obtained with typical testors. In the next section, the main theoretical foundations of typical testors and the two logistic regression methods used to calculate manuscript digits are described. The third section, presents the results and analysis of this study, in which the success of the methods is verified using the reduced set of characteristics obtained with typical testors. Finally, some conclusions are presented.

2   Methods

2.1   Typical Testor

Let U be a collection of objects, these objects are described by a set of n characteristics and are grouped into l classes. It compares characteristic to characteristic of each objects belonging to different classes, in this way the matrix is obtained $M = [m_{ij}]_{p \times n}$ where $m_{ij} \in \{0, 1\}$ and p is the number of pairs. $m_{ij} = 1$ means that the objects of the pair denoted by i are similar in the characteristic j, while $m_{ij} = 0$ indicates that the objects of the pair denoted by i are different. This comparison between characteristics of each pair of objects forms the matrix M of similarity. Therefore, $m_{ij} = 0$ for similarity between characteristics of each pair of objects and $m_{ij} = 1$ for different characteristics of each pair of objects, they form the M matrix of dissimilarity. In the present work we will work with the M matrix of dissimilarity.

Let $I = \{i_1, ..., i_p\}$ the set of the rows of M and $J = \{j_1, ..., j_n\}$ the set of the columns (characteristics) of M. Let $T \subseteq J$ and $M_{/T}$ get the matrix obtained from M by eliminating all the columns that do not belong to the set T, that is $M_{/T}$ is the matrix associated to the set T to represent it as a matrix and not as a row of columns, the matrix $M_{/T}$ has dimensions $p \times q$, where p is the number of comparison pairs and q is the number of elements of the set J.

Definition 1: A set $T = j_{k_1}, ..., j_{k_s} \subseteq J$ is a testor of M if there is no row of zeroes in $M_{/T}$, where $\{j_{k_s}\}$ are sub index of $\{j_n\}$.

Definition 2: The characteristic $j_{k_r} \in T$ is typical with respect to T and M if $\exists h, h \in \{1, ..., p\}$ such that $a_{i_h j_{k_r}} = 1$ and for $s > 1$ $a_{i_h j_{k_t}} = 0, \forall t, t \in \{1, ..., s\}$ $t \neq r$.

Definition 3: A set T has the property of typicality with respect to an array M if all the characteristics in T are typical with respect to T and M.

Definition 4: A set $T = j_{k_1}, ..., j_{k_s} \subseteq J$ is called typical testor of M if it is a testor and has the property of typicality with regarding M.

Proposition 1: A set $T = j_{k_1}, ..., j_{k_s} \subseteq J$ has the property of typicality with respect to the M array if and only if it can be obtain an identity matrix in $M_{/T}$, eliminating and exchanging some rows.

Let a and b two rows of M.

Definition 5: We say that a is less than b ($a < b$) if $\forall i \ a_i \leq b_i$ and $\exists j$ such that $a_j \neq b_j$, where $\{a_i\}$ and $\{b_i\}$ are elements of the rows a and b respectively, and i and j are indexes of the number of columns of M, $i = \{1, .., q\}$.

Definition 6: a is a basic row of M if there is no other row less than a in M.

Definition 7: The basic matrix of M is the matrix M' that only contains all the basic rows of M.

    The next proposition is a characterization of the basic matrix.

Proposition 2: M' is a basic matrix if and only if for two rows a and b any, $a, b \in M'$ there are two columns i and j such that $a_i = b_j = 1$ and $a_j = b_i = 0$. It is said that all the basic rows of a basic matrix are unmatched rows.

    Given a A matrix, we denote as $\Psi^*(A)$ the set of all the typical A testers.

Proposition 3: $\Psi^*(M) = \Psi^*(M')$.

    According to Proposition 3, it is convenient to find the matrix M' and then calculate the set $\Psi^*(M')$. Because M' has less or equal number of rows than M, that helps in the efficiency of the algorithms to find the set of all typical testers of M.

2.2   Logistic Regression

A regression model is a mathematical model that seeks to determine the relationship between a dependent variable or response with respect to other independent or explanatory variables [6]. The dependent variable (response) is represented as Y, and the independent variable (explanatory) as X. There are several regression models which have many applications in different academic areas such as the social sciences to biology. In this work, we only study the logistic regression and the multilogistic regression. In some situations of the regression models, the answer variable y_i has only two possible outcomes 0 or 1. For example, if the blood pressure is high or low, if the cancer in a person continues to develop or not, and others more examples [7]. In these cases the variable y_i is assigned one of the values 0 or 1, as a yes or no, to predict the probability p_i of the result on the database of one or several x_i. The logistic regression model is given by the function:

$$p_i = E(y_i) = \frac{1}{1 + e^{-(\beta_0 + \beta_1 x_{1,i} + ... + \beta_n x_{n,i})}} \tag{1}$$

The model can be linearized by means of the transformation:

$$\ln\left(\frac{p_i}{1 - p_i}\right) = \beta_0 + \beta_1 x_{1,i} + ... + \beta_n x_{n,i} \tag{2}$$

This transformation is called *logit*.

    The coefficients $\beta_0, ..., \beta_n$ are estimators calculated by the maximum likelihood method [7].

2.3   Multi-logistic Regression

Multilogistic regression is also called multinomial logistic regression, because the dependent variable is a nominal type with more than two categories (polytomy) as a response and it is also a multivariate extension of the binary logistic regression presented above [8].

The multilogistic regression model is given by the following functions, let $Z_{ij} = \beta_{0,j} + \beta_{1,j}x_{i,1} + ... + \beta_{n,j}x_{i,n}$

$$p_{ij} = \frac{e^{Z_{ij}}}{1 + \sum_{k=1}^{g-1} e^{Z_{ik}}} = P[Y_i = j|x_1, x_2, ..., x_p]; j = 1, 2, .., g - 1 \tag{3}$$

$$p_{ig} = \frac{1}{1 + \sum_{k=1}^{g-1} e^{Z_{ik}}} = 1 - \sum_{j=1}^{g-1} p_{ij} \tag{4}$$

Where p_{ij} is the probability of the individual i that belongs to the category j, p_{ig} is the probability of the individual i that is part of the category g, the category g is referred to as a reference category, of the variable with multinomial distribution Y [9]. In the same way, the model can be linearized by:

$$\ln\left(\frac{p_{ij}}{p_{ig}}\right) = \beta_{0,j} + \beta_{1,j}x_{i,1} + ... + \beta_{n,j}x_{i,n} = Z_{ij} \tag{5}$$

This transformation is called *logit*. The coefficients $\beta_{0,j}, \beta_{1,j}, ..., \beta_{n,j}$, are estimators of the multilogistic regression, these estimators are calculated by the method of maximum likelihood [9].

2.4   Computation of β Estimators

In this study, Matlab software is used to compute the coefficients β of each regression. In Matlab, the functions of logistic and multilogistic regression are already predetermined, and what is intended is to implement them in the study. For logistic regression the function *glmfit* (Generalized linear model regression) is used. The function *glmfit* returns a vector b of $(p+1) \times 1$, where p is the number of independent variables (predict), b is the vector of the estimation coefficients β, X are the independent variables (predictors), X is a matrix of $n \times p$, n is the number of events and *and* are the dependent variables (answer), *and* is a vector of $n \times 1$. *distr* is the distribution of the dependent variables: 'binomial', 'gamma', and others. *param1* are parameters set in matlab as 'link', 'estdisp', and others. *val1* is the assignment of the model, this can be 'identity', 'log', 'logit', and others [10]. To establish the logistic regression model in the *glmfit* function, the 'binomial' distribution, the 'link' parameter and the 'logit' value are established. To see other types of linear regression models, review the Matlab documentation for this function.

For multilogistic regression is used the function *mnrfit* (Multinomial logistic regression). The *mnrfit* function returns a B matrix of $(p+1) \times (q-1)$, where p is the number of independent variables (predict) and q the number of response categories that has the model, B is the matrix of the estimates of the coefficients for the multilogistic regression of the nominal responses in Y on the predictors X. Each column of B are the coefficients β of a category with respect to the reference category [10].

2.5   Computation of All Typical Testors

In present study, a sample of data obtained from Semeion Handwritten Digit Data Set (*semeion.data*) [5] is used. This sample is a compilation of 1593 handwritten digits, with the help of approximately 80 people, digitized each sample in a box rectangular 16×16 gray-scale pixels of 256 values. Then, each pixel of each image was assigned a boolean value $(1/0)$ according to the image. Each person wrote the digits from 0 to 9 twice on a paper. The first time they wrote it in a normal way trying to write the digit with precision, the second time they wrote it quickly losing the precision [5]. An example of each digit is presented in the Fig. 1. In the sample *semeion.data* we have a boolean matrix 1593×266 of which each row represents a different image, and the first 256 columns represent the pixels of each image of 16×16 pixels, and the last 10 columns classify the type of digit from 0 to 9 that represent the images. Therefore, only the sub-matrix 1593×256 is used for the analysis of each image, besides each digit in the images turns out to be a class of the sample, therefore in the whole sample there are 10 different classes, each one respecting to a different digit. It was coded in Julia language [11].

Fig. 1. Sample of the digits from 0 to 9

It is needed to compare each image of the sample with the other images of the sample, this comparison is made with the pixel (characteristic) of each image with respect to the same pixel (characteristic) of another image. In the case of the data matrix you have, each value of the same column is analysed in different rows. This matrix is called the matrix M. $M = [m_{ij}]$, where $m_{ij} \in \{0,1\}$, i is the comparison number of two rows of the matrix M, j is the number of the column, in this case the pixel (characteristic), which is compared to one image to another, an example is shown in the Fig. 2.

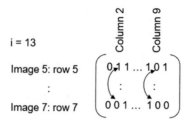

Fig. 2. Comparison between pixels of different images

In our study we will choose the M dissimilarity matrix, defined in the Typical Testors section. As an example in the Fig. 2 we have $m_{13\ 2} = 1$ and $m_{13\ 9} = 0$, where $i = 13$ is the comparison of the image 5 and 7 of the sample. In the same way, using Julia, an algorithm is created to extract the M matrix of dissimilarity, when comparing pixel (characteristic) to pixel (characteristic) of each pair of images (rows) belonging to different classes.

The algorithm specifies that each class has the same number of objects, this does not necessarily have to be that way. However, due to the efficiency of the algorithm, the same number of objects in each class is chosen. By the variable *Digit* it is easy to see how many images (objects) you have of each digit (class), this is represented in the following matrix:

$$\begin{bmatrix} 0 & 1 & 2 & 3 & 4 & 5 & 6 & 7 & 8 & 9 \\ 161 & 162 & 159 & 159 & 161 & 159 & 161 & 158 & 155 & 158 \end{bmatrix}$$

To facilitate the analysis, 150 objects of each class will be chosen, therefore a sample of 1500 images and a matrix of 1500×256 will be obtained. The result of the matrix M, given by the algorithm with the chosen sample matrix, is a Boolean matrix of 1012500×256 rows and columns, the number of rows indicates that you have 1012500 comparisons of different kinds of objects of the 1500 images in the sample. Once the M matrix is obtained, it has information of dissimilarity of all the images of different classes of the sample. Now it is necessary to reduce the M matrix to its basic matrix M'. To obtain the basic M matrix, two algorithms are used in Julia's language. The first algorithm is responsible for eliminating all rows zero of the M matrix.

The second algorithm eliminates all rows that are not basic, and it contains all the minor or basic rows. It is based on proposition 2 of the typical testors section. Because it is more efficient to find non-comparable rows than one row greater than another. By reducing the M matrix to its basic matrix M' with the algorithms shown, you get an array of 1011965×256 rows and columns, obtaining a reduction of 535 rows. This matrix contains necessary information of the dissimilarity between images of different classes. Therefore, all the sample typical testors are calculated since the matrix is already in reduced form. For this calculation, the algorithms Sparsity-based Hypergraph Dualization (SHD) are used to find the *minimal hitting set*, and they are equivalent to the typical testors in the theory of Hypergraphs [12].

Once $\Psi\{M'\}$ has been calculated for the set of all typical M' testors, the typical testor that has the most frequent characteristics within the set $\Psi\{M'\}$ are chosen. However, for ease of calculation is the matrix M of only two digits 0 and 1 with 150 images each, this matrix will be called M_1 and its basic matrix M'_1, in this way it will be chosen $T_1 \in \Psi\{M'_1\}$ to be used in the logistic regression model. For the multilogistic regression we will use a $T_2 \in \Psi\{M'_2\}$, where M'_2 is the basic matrix of M_2 and this is the comparison of the digits 0, 1, 2, 3, 4, with 50 images each and M'_3 is the basic matrix of the comparison of the digits 5, 6, 7, 8, 9 with 50 images each.

3   Results and Discussion

For the case of logistic regression, we have chosen 100 images of digit 0, and 100 images of digit 1. Leaving the other images out of the sample for the prediction model. The function *glmfit* is implemented in Matlab, assigning the case that the answer variable $y_i = 1$ for the images of digit 1 and $y_i = 0$ for the images of digit 0. We obtain the vector b of the coefficients β of the regression model, this gives 257 different betas, where the β_0 is the independent coefficient of the independent variables X. Then, the process is repeated choosing a typical testor previously calculated, that is to say the model is not repeated with all the pixels of the images but with the pixels assigned by the typical testor T_1. A candidate as typical testor of cardinality 14 is:

$$T_1 = \{j_{247}, \ j_{249}, \ j_{228}, \ j_{231}, \ j_{245}, \ j_{129}, \ j_{200}, \ j_{243}, \ j_{236}, \ j_{242}, \ j_{135},$$
$$j_{156}, \ j_{207}, \ j_{237}\}$$

Where the elements j_k are columns of the matrix M', and have the information that discriminate the different objects of several classes. Therefore, these columns will be the most significant pixels (characteristics) of the images (objects) when comparing the images of the digit 0 with the images of the digit 1. The same program as before is chosen choosing the columns of the chosen testors in the previous sample. Fifteen different β are obtained instead of the previous 257. With the calculated betas, the Eq. 1 is applied, in the different models obtained, to predict the probability of comparing new images with respect to the sample. Current, for this comparison is chosen an image that does not belong to the first 100 images of each digit chosen for the calculation of the betas, this image is chosen in random way. The Table 1 presents the respective comparisons.

In the case of multilogistic regression, it is assigned to the $y_i = 0$ if the image is the digit 0, $y_i = 1$ if the image is 1, $y_i = 2$ if the image is 2 and so respectively. In the Matlab model, always chooses the last category as a reference category. Due to limitations in computer materials, the calculation speed is agile, dividing the multilogistic regression into two parts: a model for the digits from 0 to 4 and another model for the ones from 5 to 9. Therefore, only 5 categories will be calculated in each model. We also choose 50 images of each of them for the sample. In this way, we obtain the B matrix of (257×4) of the coefficients

Table 1. Comparison of logistic regression using typical testors

	Images	
	Digit 0	Digit 1
Without typical testor: p	0	1
With typical testor: p	0.0051	0.9866

of the multilogistic regression model, where the first column is the β_0 of each $logit\left(\dfrac{p_{ij}}{p_{ig}}\right)$, respectively and each column contains the betas of the respective $logit\left(\dfrac{p_{ij}}{p_{ig}}\right)$. Now, same procedure is performed for multilogistic regression with typical testors, previously calculated, two with different cardinalities are chosen for a comparison between them. Cardinalites 24 and 21 respectively, where the j_k are columns of matrix M', and has significant information that discriminate the digits 0, 1, 2, 3, 4.

Table 2. Comparison of the first multilogistic regression model using typical testors

	Images					
		Digit 0	Digit 1	Digit 2	Digit 3	Digit 4
All features	$P(y_i = 0) = p_0$	1	0	0	0	0
	$P(y_i = 1) = p_1$	0	1	0	0	0
	$P(y_i = 2) = p_2$	0	0	1	0	0
	$P(y_i = 3) = p_3$	0	0	0	1	0
	$P(y_i = 4) = p_4$	0	0	0	0	1
Typical testor characteristics T_2	$P(y_i = 0) = p_0$	1	0	0	0	0
	$P(y_i = 1) = p_1$	0	0.9961	0.0038	0	0.0001
	$P(y_i = 2) = p_2$	0	0.1239	0.8662	0.006	0.0040
	$P(y_i = 3) = p_3$	0	0.0577	0.0031	0.9392	0
	$P(y_i = 4) = p_4$	0	0.0048	0	0.0001	0.9951
Typical testor characteristics T_3	$P(y_i = 0) = p_0$	1	0	0	0	0
	$P(y_i = 1) = p_1$	0	0.7195	0.0002	0.2510	0.0292
	$P(y_i = 2) = p_2$	0	0.149	0.8508	0.0002	0
	$P(y_i = 3) = p_3$	0.0215	0	0.0082	0.9703	0.0001
	$P(y_i = 4) = p_4$	0	0	0	0.0004	0.9996

We obtain a matrix BT of 25×4 coefficients β's from multilogistic model, when characteristics come from typical testor T_2. For T_3, model has a 22×4 matrix. Once the β's are obtained, the Eqs. 3 and 4 are applied, to obtain the multilogistic regressions. Table 2 summarizes comparison of performance between testors T_2 and T_3.

$$T_2 = \{\hat{j}_{247}, \hat{j}_{249}, \hat{j}_{228}, \hat{j}_{231}, \hat{j}_{245}, \hat{j}_{129}, \hat{j}_{200}, \hat{j}_{237}, \hat{j}_{83}, \hat{j}_{93}, \hat{j}_{207}, \hat{j}_{191}, \hat{j}_{250},$$
$$\hat{j}_{189}, \hat{j}_{241}, \hat{j}_{140}, \hat{j}_{252}, \hat{j}_{121}, \hat{j}_{256}, \hat{j}_{98}, \hat{j}_{244}, \hat{j}_{227}, \hat{j}_{192}, \hat{j}_{229}\}$$

$$T_3 = \{\hat{j}_{247}, \hat{j}_{249}, \hat{j}_{228}, \hat{j}_{231}, \hat{j}_{245}, \hat{j}_{129}, \hat{j}_{200}, \hat{j}_{237}, \hat{j}_{83}, \hat{j}_{93}, \hat{j}_{207}, \hat{j}_{191}, \hat{j}_{250},$$
$$\hat{j}_{189}, \hat{j}_{241}, \hat{j}_{140}, \hat{j}_{252}, \hat{j}_{114}, \hat{j}_{58}, \hat{j}_{160}, \hat{j}_{159}\}$$

$$T_4 = \{\hat{j}_{251}, \hat{j}_{252}, \hat{j}_{247}, \hat{j}_{157}, \hat{j}_{215}, \hat{j}_{249}, \hat{j}_{219}, \hat{j}_{244}, \hat{j}_{227}, \hat{j}_{245}, \hat{j}_{242}, \hat{j}_{241}, \hat{j}_{243}, \hat{j}_{197}, \hat{j}_{288},$$
$$\hat{j}_{174}, \hat{j}_{56}, \hat{j}_{70}, \hat{j}_{182}, \hat{j}_{202}, \hat{j}_{127}, \hat{j}_{240}, \hat{j}_{148}, \hat{j}_{223}, \hat{j}_{248}, \hat{j}_{187}, \hat{j}_{239}, \hat{j}_{207}, \hat{j}_{205}, \hat{j}_{176}, \hat{j}_{80},$$
$$\hat{j}_{58}, \hat{j}_{186}, \hat{j}_{112}, \hat{j}_{96}\}$$

$$T_5 = \{\hat{j}_{251}, \hat{j}_{252}, \hat{j}_{247}, \hat{j}_{157}, \hat{j}_{215}, \hat{j}_{249}, \hat{j}_{219}, \hat{j}_{244}, \hat{j}_{227}, \hat{j}_{245}, \hat{j}_{242}, \hat{j}_{241}, \hat{j}_{250}, \hat{j}_{84}, \hat{j}_{74},$$
$$\hat{j}_{90}, \hat{j}_{3}, \hat{j}_{80}, \hat{j}_{216}, \hat{j}_{254}, \hat{j}_{246}, \hat{j}_{54}, \hat{j}_{186}, \hat{j}_{186}, \hat{j}_{156}, \hat{j}_{220}, \hat{j}_{144}, \hat{j}_{61}\}$$

In same way, a second set of multilogistic regressions are implemented for recognizing digits from 5 to 9. Table 3 summarizes this comparison.

Table 3. Comparison of the second multilogistic regression using typical testors

	Images					
		Digit 5	Digit 6	Digit 7	Digit 8	Digit 9
All features	$P(y_i = 5) = p_0$	1	0	0	0	0
	$P(y_i = 6) = p_1$	0	1	0	0	0
	$P(y_i = 7) = p_2$	0	0	1	0	0
	$P(y_i = 8) = p_3$	0	0	0	1	0
	$P(y_i = 9) = p_4$	0	0	0	0	1
Typical testor characteristics T_4	$P(y_i = 5) = p_0$	0.9176	0	0	0.0816	0.0007
	$P(y_i = 6) = p_1$	0.0901	0.9090	0	0.0008	0.0001
	$P(y_i = 7) = p_2$	0.0002	0.0210	0.9780	0	0.0007
	$P(y_i = 8) = p_3$	0	0.0006	0	0.9781	0.0214
	$P(y_i = 9) = p_4$	0	0	0	0.0602	0.9398
Typical testor characteristics T_5	$P(y_i = 5) = p_0$	0.9852	0	0	0.0141	0.0008
	$P(y_i = 6) = p_1$	0.4330	0.4304	0	0.1310	0.0056
	$P(y_i = 7) = p_2$	0	0	0.6510	0.0595	0.2895
	$P(y_i = 8) = p_3$	0	0	0	0.8578	0.1422
	$P(y_i = 9) = p_4$	0.0002	0	0.0027	0.0729	0.9242

In Tables 1 and 2, approximation obtained from testors is very good, so the theory of typical testors is checked. In Table 3, about prediction obtained from testor T_5, there is a fairly large error in predicting digit 6, however, when using T_4, error is minimal. A typical testor that provides better performance, should be choosen.

4   Conclusions

Set $\Psi\{M'\}$ provides a domain where we can search a reduced set of characteristics in problems of hard classification. This work has established a technique that allows making predictions, with very acceptable performance, using only around between 10 and 20% of information provided in a binary image. The main advantage is that reduction of characteristics space, implies a reduction of complexity for predictor model. This brings us closer to the behavior of the brain, when it comes to reconstructing pattern information. We worked with a database of written digits, including ideal writings and other handwritten digits of difficult recognition. To distinguish complex images more features are required. Part of the errors found in our study, may be because probability model get confused with those complex images, the search for better sets of characteristics becomes interesting, since the cardinality of set $\Psi\{M'\}$ is enormous, we do not know yet, if the errors can be reduced, just looking for a better typical testor.

References

1. Alba-Cabrera, E., Godoy-Calderon, S., Ibarra-Fiallo, J.: Generating synthetic test matrices as a benchmark for the computational behavior of typical testor-finding algorithms. Pattern Recogn. Lett. **80**, 46–51 (2016). https://doi.org/10.1016/j.patrec.2016.04.020
2. Sanchez-Diaz, G., Diaz-Sanchez, G., Mora-Gonzalez, M., Piza-Davila, I., Aguirre-Salado, C.A., Huerta-Cuellar, G., Reyes-Cardenas, O., Cardenas-Tristan, A.: Generación de matrices para evaluar el desempeño de estrategias de búsqueda de testores típicos. Avances en Ciencias e Ingenierías **2**, A30–A35 (2010). https://doi.org/10.1016/j.patrec.2013.11.006
3. Lazo-Cortés, M.S., Martínez-Trinidad, J.F., Carrasco-Ochoa, J.A., Sanchez-Diaz, G.: On the relation between rough set reducts and typical testors. Inf. Sci. **294**, 152–163 (2015). https://doi.org/10.1016/j.ins.2014.09.045
4. Piza-Davila, I., Sanchez-Diaz, G., Lazo-Cortes, M.S., Rizo-Dominguez, L.: A CUDA-based hill-climbing algorithm to find irreducible testors from a training matrix. Pattern Recogn. Lett. **95**, 22–28 (2017). https://doi.org/10.1016/j.patrec.2017.05.026
5. Semeion Research Center of Sciences of Communication via Sersale 117, 00128. Semeion Handwritten Digit Data Set, Rome. http://archive.ics.uci.edu/ml/datasets/semeion+handwritten+digit
6. Roldan, P.: Economipedia, Modelo de regresión (2015). http://economipedia.com/definiciones/modelo-de-regresion.html
7. Rencher, A.C., Schaalje, G.: Linear Models in Statistics. Wiley, Provo (2008)

8. Pando, V., Martín, R.S.: Regresión Logística Multinomial. Sociedad Española de Ciencias Forestales **18**, 323–327 (2004). https://dialnet.unirioja.es/servlet/articulo?codigo=2981898. ISSN 1575-2410

9. Osorio, D., Ospina, J., Lenis, D.: Planteamiento del modelo logístico multinomial a través de la función canónica de enlace de la familia exponencial. Heuristica **16**, 105–115 (2009)

10. Matlab: MathWorks, documentation, generalized linear regression (2006). https://la.mathworks.com/help/stats/glmfit.html

11. Bezanson, J., Karpinski, S., Shah, V., Edelman, A.: Manual the Julia language (2013). https://docs.julialang.org/en/stable/stdlib/io-network/#Text-I/O-1

12. Uno, T.: Program codes and instances for hypergraph dualization (minimal hitting set enumeration). http://research.nii.ac.jp/~uno/dualization.html

Patterns of Ambiguity in Textual Requirements Specification

David Šenkýř$^{(\boxtimes)}$ ⓘ and Petr Kroha ⓘ

Faculty of Information Technology, Czech Technical University in Prague,
Prague, Czech Republic
{david.senkyr,petr.kroha}@fit.cvut.cz

Abstract. In this contribution, we investigate the ambiguity problem in textual requirements specifications. We focused on the structural ambiguity and extracted some patterns to indicate this kind of ambiguity. We show that the standard methods of linguistics are not enough in some cases, and we describe a class of ambiguity caused by coreference that needs an underlying domain model or a knowledge base to be solved. Part of our implemented solution is a cooperation of our tool TEMOS with the Prolog inference machine working with facts and rules acquired from OCL conditions of the domain model.

Keywords: Requirements specification · Text processing · Ambiguity · Coreference · Domain model · OCL rules · Prolog inference machine

1 Introduction

Textual formulated requirements specifications are necessary as a base of communication between the customer, the user, the domain expert, and the analyst. Unfortunately, any text suffers by ambiguity, incompleteness, and inconsistency.

In this paper, we focus on the problem of ambiguity that is caused by the phenomenon that natural language is inherently ambiguous. Requirements specifications have to be unambiguous because, in general, any ambiguous requirement is not verifiable, and there is a risk that programmers implemented the meaning that was not intended by the specification.

Information retrieval operates with a process called *word sense disambiguation* (WSD) [3]. It has to determine the correct sense of the given word in the context it was used in the text. The goal is to identify a unique meaning of each word. Text mining operates with a more complex version of the ambiguity problem in text documents that can be called *sentence sense disambiguation*. The goal is to identify a unique meaning of each sentence.

In our tool TEMOS [5], we construct and maintain a *glossary* that defines and explains the meaning of words used in a textual requirements specification. The word meaning is derived from an ontological database in Web [5], and the

© Springer Nature Switzerland AG 2019
Á. Rocha et al. (Eds.): WorldCIST'19 2019, AISC 930, pp. 886–895, 2019.
https://doi.org/10.1007/978-3-030-16181-1_83

glossary is discussed by the customer, the user, the domain expert, and the analyst during the analysis of requirements. In this contribution, we explain why it is not enough for our purpose, and how we solved it.

In common, the ambiguity of sentences is given by *structural ambiguity* that was investigated by linguists. We have constructed the corresponding patterns and described them in Sect. 3.

Practically, we found that in most cases, ambiguity is combined with the coreference problem in textual requirement specifications. Coreference resolution has to find all expressions that refer to the same entity in a text. In textual requirements specification, it often happens that a pronoun in a sentence refers to a specific noun from the previous clause. If there is more than one noun as a possible candidate, we need to decide which one of them is the right one. We propose a new method to solve this problem because we have found that the text layer used in computational linguistics for solving the problem of ambiguity is not strong enough for our purpose. The roots of ambiguity in our textual requirements specifications concern not only the linguistic part of requirements but also the semantics of the underlying model as we show in Sect. 4.

Our paper is structured as follows. In Sect. 2, we discuss the problems of ambiguity and the related works. We present the introduced *patterns* of structural ambiguity in Sect. 3 and our solution of *semantic sentence ambiguity* in Sect. 4. Our implementation, used data, experiments, and results are described in Sect. 5. Finally, we conclude in Sect. 6.

2   Related Work Concerning Problem of Ambiguity

We targeted the issue of ambiguity in textual requirements specification from two perspectives – the *ambiguity of words* (also called the *lexical ambiguity*) and the *ambiguity of sentences* (also called the *structural ambiguity* or the *syntactic ambiguity*). We explain the difference in the following subsections.

2.1   Ambiguity of Words

Natural languages are inherently ambiguous. Ambiguity arises whenever a word or an expression (concept) can be interpreted in more than one way. If the expression is a single word, we speak about *word disambiguation* to resolve it (discussed by computational linguists). The user, the customer, and the domain expert understand the domain and know that two or more different words have the same meaning (synonyms), or what a meaning of a homonym is the right one in the given context. But the analyst does not know that and can model them as unique entities. It happens that project teams are places in different countries where different terminology has been used.

We handle the word disambiguation with the help of a *glossary* supported by thesaurus that is freely available on the Internet. As a result, questions are generated by asking stakeholders whether they agree with the equivalence found [5]. Finally, the result is stored in the *glossary again*.

Wang et al. [7] presented the different approach via *ranking method* to deal with a problem of *overloaded* (it refers to different semantic meanings) and *synonymous concepts* (several different concepts are used interchangeably to refer to the same semantic meaning).

2.2 Ambiguity of Sentences

Ambiguity of sentences is a more complex problem because the parsing structure of a sentence is not unique in all cases, i.e., identical sentence text segments (composed from the same words – even if these words would have unique meaning), can have a different meaning that depends on their position in the parsing tree of the sentence.

This problem has been discussed in details by computational linguists as the problem of *structural ambiguity*, e.g., in [4]. The complete analysis is given, e.g., in [4] and [2], and it is out of the scope of our paper. Our contribution is that we have developed a set of patterns that indicate possible ambiguities in textual requirements, and in some cases offer possible solutions. Gleich et al. [1] presented similar patterns based on the *part of speech tagging*. Our work differs in that we also use *dependency parsing* in our patterns and we focus on other use cases.

Actually, the semantics of the world is very complex, so we do not think it is possible to solve the ambiguity problem automatically in all cases. Some of the ambiguities can be indicated by tools of computational linguistics as will be shown in Sect. 3. Some other can not, and they are shown in Sect. 4.

3 Patterns of Structural Ambiguity

In this contribution, we focus on the ambiguity of sentences that should be solved by *patterns*. The candidates for structural ambiguity patterns can be derived from the following linguistic features: *attachment ambiguity* and *analytical ambiguity* [2].

Legend: Within the following examples of patterns, we use Penn Treebank [6] *part of speech* and *dependency* tags. In some cases, there is possible to have multiple part of speech candidates. We use the pipe symbol (|) in these situations. The meaning of this symbol is a logical OR.

Often, the subjects or the objects are represented by a composition of several nouns. Therefore, we introduce a shortened notation **NN*** means that at least one noun is required and there should also be more nouns composited via *compound* relation.

3.1 Patterns of Attachment Ambiguity

There is more than one node to which a particular syntactic constituent can be attached [2]. The nature of the constituent is clear, but it is not clear where to put it. Below, we attach selected examples with sample patterns that help to reveal ambiguity.

Pattern #1 (Prepositional Phrase Modifier): A prepositional phrase may either modify a verb or an immediately preceding noun phrase, e.g., *"He saw the man with field glass"*.

VB —dobj→ NN* —prep→ IN —pobj→ NN*
"SEE"|"HEAR" NOUN₁ "WITH" NOUN₂

– *Example of generated warning*: ambiguity — **"VERB NOUN₁** via **NOUN₂"** vs. **"VERB NOUN₁** that has/have **NOUN₂"**

Pattern #2 (Preposition Phrase): A prepositional phrase may have more than one noun phrases available to attach it to, e.g., *"The door near to stairs with the Members Only sign ..."* [2].

NN* —advmod→ IN —prep→ IN —pobj→ NN* —prep→ IN
NOUN₁ ADPOSITION ADPOSITION NOUN₂ "WITH"

– *Example of generated warning*: (?) **"with"** — ambiguity between **NOUN₁** and **NOUN₂**

Pattern #3 (Relative Clause): Relative clauses – there may be a relation to more than one noun of the main clause, e.g., *"The door near to stairs that had the Members Only sign ..."* [2].

NN* —advmod→ IN —prep→ IN —pobj→ NN* —relcl— JJ ←nsubj— VB
NOUN₁ ADPOSITION ADPOSITION NOUN₂ "THAT" VERB

– *Example of generated warning*: (?) **"that"** — ambiguity between **NOUN₁** and **NOUN₂**

Pattern #4 (Subsentence Attachment): When a sentence contains a subsentence, both may contain places for the attachment of a prepositional phrase or adverb, e.g., *"He said that she had done it yesterday"* [2].

PRN|NN ←nsubj— VB —ccomp— PRN|NN —nsubj— IN —mark→ VB —npadvmod→ NN
PRONOUN|NOUN₁ VERB PRONOUN|NOUN₂ "THAT" VERB NOUN₃

– *Example of generated warning*: (?) **NOUN₃** — ambiguity between **"PRONOUN|NOUN₁ ..."** and **"PRONOUN|NOUN₂ ..."**

Pattern #5 (Adverbial Position (1)): An adverbial placed between two clauses can be attached to the verb of either [2], e.g., *"The lady you met now and then came to visit us"*.

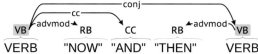

VB —advmod→ RB —cc— CC —conj— RB —advmod→ VB
VERB "NOW" "AND" "THEN" VERB

– *Example of generated warning*: a possible **"now and then"** ambiguity — do you mean **"sometimes"** or not?

Pattern #6 (Adverbial Position (2)): Adverb placement ambiguity [2], e.g., *"A secretary can type quickly written reports"*. The pattern of this example is a little bit complicated because of a bad annotated word *written* as a verb instead of an adjective as shown in the following parsing result.

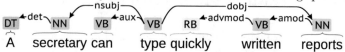

The same issue is also discussed in the next section. When we replace the word *written* to *handwritten*, then the word *handwritten* is correctly recognized as an adjective. Based on this observation, we created the corresponding pattern.

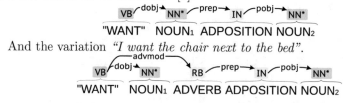

– *Example of generated warning*: please consider a possible ambiguity of the adverb **ADV**.

3.2 Patterns of Analytical Ambiguity

There is more than one possible syntactic analysis, and the nature of the constituent is itself in doubt [2].

Patterns #7 and #8 (Reduced Restrictive Relative Clause (1) and (2)): Reducing a pronoun of the restrictive relative clause, e.g., *"I want the box on the table"* can have the meaning "I want the box that is on the table" or "I want the box to be on the table" [2].

VB ⌒dobj→ NN* ⌒prep→ IN ⌒pobj→ NN*
"WANT" NOUN₁ ADPOSITION NOUN₂

And the variation *"I want the chair next to the bed"*.

VB ⌒dobj→ NN* RB ⌒prep→ IN ⌒pobj→ NN*
"WANT" NOUN₁ ADVERB ADPOSITION NOUN₂

– *Example of generated warning*: **NOUN₁** — that is/are **"ADPOSITION ..."** or to be **"ADPOSITION ..."**?

Pattern #9 (Present Participle vs. Adjective): Distinguishing a present participle from an adjective, e.g., *"We are writing a letter"* and *"Pen and pencils are writing implements"* [2].

VB ←aux— VBG ⌒dobj→ NN*
"BE" VERB (present participle) NOUN

– *Example of generated warning*: please consider a possible ambiguity of the word **VERB**

Pattern #10 (Participle): The participle could be attached to the object either as a reduced restrictive relative clause or as a verb complement, e.g., *"The manager approached the boy smoking a cigar"* or *"The manager caught the boy smoking a cigar"*.

$$VERB_1 \quad NOUN \; VERB_2 \text{ (participle)}$$

– *Example of generated warning*: please consider a possible ambiguity of the word **VERB**$_2$

3.3   Result and Discussion

According to the rules described above, we defined and implemented a set of structural ambiguity patterns that may indicate a suspicion of structural ambiguity in the given textual requirement. Users obtain a *generated warning* that the analyzed sentence may have more that one meaning.

The question is how often such cases of structural ambiguity occur in textual requirements. The texts discussed in [4] and [2] are taken from novels, newspapers, jokes, etc. Textual requirements have specific features, e.g., they do not contain a description of feelings, direct or indirect speech, etc. Their sentences are not intentionally constructed to obtain structural ambiguities like it is in case of jokes [4]. On the other hand, we are also able to find examples of coreferential ambiguity in specifications as shown in Fig. 1[1].

In requirements specifications, we expect that the text interpretation semantically corresponds to the *domain model*, i.e., the text makes sense in the context of the domain. Systems based only on parsing that we discussed in Sect. 2.1 and Sect. 3, do not have access to the knowledge stored in the domain model required to make an interpretation properly.

We construct the model (in the form of *UML class diagram*) from the textual requirements specification, but in this model, some interpretations do not make sense. To solve this problem, we need a domain model and the part of the problem model that has already been built. Of course, at the starting point, the problem model is not available. Unfortunately, the domain model is often not available, too.

The only practical possibility is to build both these models incrementally and iterative during textual requirements processing through discussions with stakeholders. Then suspicions of ambiguity can be generated by comparing the textual requirements specification and the model derived.

---

[1] https://docplayer.net/24435423-System-requirements-specification-e-voting-authored-by-seth-appleman-patrick-coffey-david-kelley-cliff-yip.html.

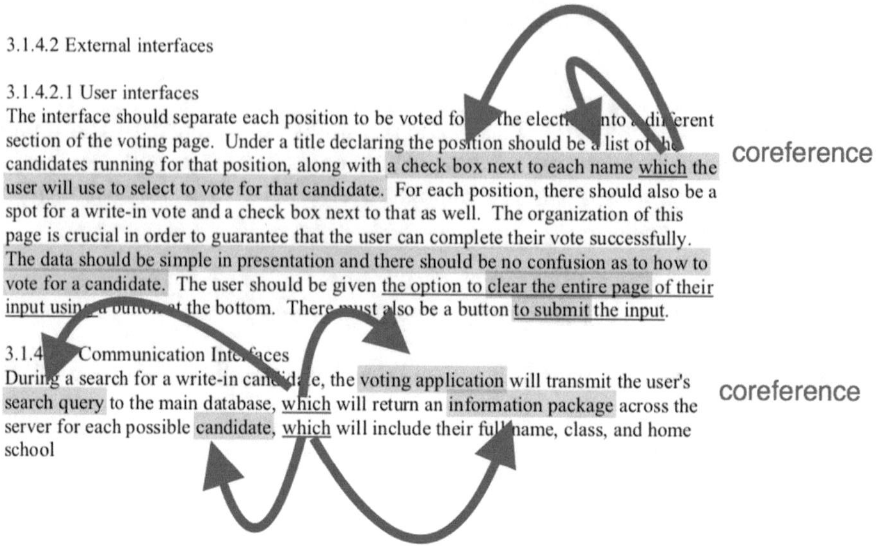

3.1.4.2 External interfaces

3.1.4.2.1 User interfaces
The interface should separate each position to be voted for the election into a different section of the voting page. Under a title declaring the position should be a list of the candidates running for that position, along with a check box next to each name which the user will use to select to vote for that candidate. For each position, there should also be a spot for a write-in vote and a check box next to that as well. The organization of this page is crucial in order to guarantee that the user can complete their vote successfully. The data should be simple in presentation and there should be no confusion as to how to vote for a candidate. The user should be given the option to clear the entire page of their input using a button at the bottom. There must also be a button to submit the input.

coreference

3.1.4.2.2 Communication Interfaces
During a search for a write-in candidate, the voting application will transmit the user's search query to the main database, which will return an information package across the server for each possible candidate, which will include their full name, class, and home school

coreference

Fig. 1. Example of coreferential ambiguity

4   Problems of Semantic Sentence Ambiguity and Coreference

We argue that not all ambiguity problems can be solved by methods of computational linguistics. In [4], this problem is briefly mentioned (but not solved) in comments, and it is illustrated by the following example.

4.1   Linguistic Approach Completed by Knowledge Base

1. *John murdered Bill. His funeral was held on Monday.* — "His" refers to Bill.
2. *John murdered Bill. His trial was held on Monday.* — "His" refers to John.

The knowledge from the real world, i.e., from its *domain model*, is necessary to disambiguate the co-referencing "His" that should refer to a subject or an object from the previous sentence. The possibility, we used to find a solution, is a knowledge base written in *Prolog*. In the case above, we had two nouns in the role of *subject* and *object*. The textual sentence is a source of the following facts:

```
subject(john). object(bill). murder(john, bill).
```

The knowledge base has to obtain the following rules:

```
murder(X, Y) :- subject(X), object(Y).
funeral(X) :- murder(_, X).        trial(X) :- murder(X, _).
```

Prolog answers what we expect:

```
?-trial(Who).                      ?-funeral(Who).
Who=john.                          Who=bill.
```

4.2 Model Approach

In the previous section, we use only the linguistic information. In this section, we include the information coming from the constructed model. This suggests a two-phase processing. In the first step, we use the linguistic information coming from the text, and we build the skeleton of the model. In the second phase, we analyze the text with regard to the obtained model. Because of that, we can find that two classes have the same attribute as you can see in this example: *"Our delivery contains product XYZ-123 in container X-50. Its weight is 50 kg."*

In the model, there are three classes: *Delivery*, *Product*, and *Container*. Delivery consists of *Product* and *Container*. All these classes have an attribute *Weight*. The weight of 50 kg may concern any of them.

However, maybe that the Product XYZ-123 is a radioactive isotope. In this case, its weight will not be 50 kg because, e.g., 235-uranium has 48 kg as its critical mass, so 50 kg would mean a nuclear explosion. It is surely obvious for all stakeholders, but it is a question, how obvious it is for the programmer.

4.3 Specific Attribute Values Distinguish the Coreference

In textual formulations, there may be a context (e.g., given by a coreference), in which the relation between an entity and the mentioned property represented by an attribute or its value is ambiguous. These ambiguities cannot be solved by syntactic means (like part of speech tagging) or by statistical means (like statistic sample data about co-occurrence).

To introduce the core of the problem, we present the following example sentence: *In the picture, there is Lady Diana and her grandmother, as she was <number> years old.* We can see that changing the age number affects the coreference meaning of the pronoun "she". If there are some *OCL expressions* associated with these classes, we can use, e.g., *USE tool*[2] (UML-based Specification Environment) to check which coreference option is consistent with the domain model.

Let's start with the UML model:

```
model HUMAN                          association mother_of between
                                       HUMAN_BEING[1..*] role child
class HUMAN_BEING                      MOTHER[0..1] role mother
attributes                           end
  age : Integer
end                                  association grandmother_of between
                                       HUMAN_BEING[1..*] role grandchild
class MOTHER < HUMAN_BEING             GRANDMOTHER[0..2] role grandmother
end                                  end

class GRANDMOTHER < MOTHER
end
```

---

[2] http://www.db.informatik.uni-bremen.de/projects/USE-2.3.1.

We continue with the *OCL expressions* to support the previous UML model with the constraints.

```
context HUMAN_BEING inv: self.age >= 0

context MOTHER inv MotherAge:
                self.child->forAll(y|y.age < self.age - 12)
context HUMAN_BEING inv MotherExclude: self.mother->excludes(self)

context GRANDMOTHER inv GrandmotherAge:
                self.child->forAll(y|y.age < self.age - 24)
context HUMAN_BEING inv GrandmotherExclude:
                self.grandmother->excludes(self)
context HUMAN_BEING inv GrandmotherExclude2:
                self.grandmother->excludes(mother)
```

Using this simple model, we can resolve the ambiguity for number = 18 of our example, i.e., that a grandmother cannot have 6 years, and the pronoun "she" is co-referencing Lady Diana.

In the case of number = 96, we can resolve the ambiguity too, because if Lady Diana were 96, her mother had to have at least 110 years and her grandmother at least 124 years, which is a possibility not contained in our realistic domain model.

In the case of number = 32, our domain model will not help us to solve the ambiguity shown above, because both ladies can be 32 years old. In such a case, we have to generate a remark concerning the found ambiguity and to generate a question for the analyst.

From the linguistic analysis, we have the information that Lady Diana is a woman (pronouns "she" and "her"), so we identify her as an object of the class WOMAN from our domain model. We associate the found word "grandmother" with the class GRANDMOTHER in our domain model.

We are able to check defined *OCL expressions* with the mentioned *USE tool*.

5   Implementation and Experiments

We implemented the module for solving ambiguity as a part of our tool called TEMOS (Textual Modelling System). TEMOS is written in Python based on *Spacy* NLP framework. We use *SWI-Prolog* runtime that serves solutions for constraints written in Prolog generated by our class diagram generator from textual requirements specification. *SWI-Prolog* is running like a command line client, and it is capable of solving constraints inserted like a text file. The output is also a text file that we parse back to our structure. Based on this response, we are able to generate questions for the user.

We tested our TEMOS tool on selection of freely available specifications from the Internet. Below, in Table 1, we presented the list of the frequency of recognized patterns from Sect. 3. We display only the patterns that have been recognized at least once.

Table 1. Frequency of patterns

Software Requirements Specification\Pattern No.	2	3	4	9	10
E-Voting Systems[a]	0	1	0	0	7
Restaurant[b]	2	3	0	0	21
Amazing Lunch Indicator[c]	2	1	0	1	3
Online National Election Voting[d]	1	3	1	0	4

[a]https://pdfs.semanticscholar.org/318f/989bc774c9c3e907c470ebb3b1016672f-679.pdf
[b]https://kungfumas.files.wordpress.com/2017/09/099.pdf
[c]http://www.cse.chalmers.se/~feldt/courses/reqeng/examples/srs_example_2010_group2.pdf
[d]https://senior.ceng.metu.edu.tr/2011/iteam4/documents/srs-iTeam4.pdf

6   Conclusions

Using our patterns defined above, we are able to generate the warning messages that there are some sentence ambiguities in the textual requirement specifications. In some cases, we can recommend the solution, in other cases, the semantics is too complex.

Our proposed method can be used in the very first phase of analysis to start a discussion with stakeholders about possible ambiguity of texts. It can be important especially in the case of geographically distributed software development.

References

1. Gleich, B., Creighton, O., Kof, L.: Ambiguity detection: towards a tool explaining ambiguity sources. In: Requirements Engineering: Foundation for Software Quality, pp. 218–232. Springer, Heidelberg (2010)
2. Hirst, G.: Semantic Interpretation and the Resolution of Ambiguity. Cambridge University Press, New York (1987)
3. Kwong, O.Y.: New Perspectives on Computational and Cognitive Strategies for Word Sense Disambiguation. Springer Publishing Company, Incorporated, New York (2013)
4. Oaks, D.D.: Structural Ambiguity in English, 1st edn. Continuum International Publishing Group, London (2010)
5. Šenkýř, D., Kroha, P.: Patterns in textual requirements specification. In: 2018 13th International Conference on Software Technologies (2018)
6. Taylor, A., Marcus, M., Santorini, B.: The Penn Treebank: An Overview, pp. 5–22. Springer, Dordrecht (2003)
7. Wang, Y., Winbladh, K., Fang, H.: Automatic detection of ambiguous terminology for software requirements. In: Natural Language Processing and Information Systems, pp. 25–37. Springer, Heidelberg (2013)

3D Simulator Based on SimTwo
to Evaluate Algorithms
in Micromouse Competition

Lucas Eckert[1], Luis Piardi[2], José Lima[2,5($\boxtimes$)], Paulo Costa[3,5],
Antonio Valente[4,5], and Alberto Nakano[1]

[1] Federal University of Technology - Paraná, Toledo, Brazil
`lucas.eck95@gmail.com, nakano@utfpr.edu.br`
[2] Research Centre in Digitalization and Intelligent Robotics (CeDRI),
IPB Bragança, Bragança, Portugal
`{piardi,jllima}@ipb.pt`
[3] Faculty of Engineering of University of Porto (FEUP), Porto, Portugal
`paco@fe.up.pt`
[4] Universidade de Trás-os-Montes e Alto Douro (UTAD), Vila Real, Portugal
`avalente@utad.pt`
[5] INESC-TEC, Centre for Robotics in Industry and Intelligent Systems,
Porto, Portugal

Abstract. Robotics competitions are increasing in complexity and number challenging the researchers, roboticists and enthusiastic to address the robot applications. One of the well-known competition is the micromouse where the fastest mobile robot to solve a maze is the winner. There are several topics addressed in this competition such as robot prototyping, control, electronics, path planning, optimization, among others. A simulation can be used to speed-up the development and testing algorithms but faces the gap between the reality in the dynamics behaviour. In this paper, an open source realistic simulator tool is presented where the dynamics of the robot, the slippage of the wheels, the friction and the 3D visualization can be found. The complete simulator with the robot model and an example is available that allow the users to test, implement and change all the environment. The presented results validate the proposed simulator.

Keywords: Mobile robot · Simulation environment ·
Micromouse competition

1 Introduction

Robotics competitions are an excellent way to foment research and to attract students to technological areas [1]. The robotics competitions present problems that can be used as benchmark, in order to evaluate and to compare the performances of different approaches [4]. Several of these implementations can be

© Springer Nature Switzerland AG 2019
Á. Rocha et al. (Eds.): WorldCIST'19 2019, AISC 930, pp. 896–903, 2019.
https://doi.org/10.1007/978-3-030-16181-1_84

utilized in industrial environments. The micromouse competition present a problem in mobile robots movements. The robot must be able to navigate, self-localize and map an unknown environment. When the map is completed it must calculate and go trough the best path to the goal in the lowest time possible.

The development of robots is a process that involves several areas of engineering, such as programming, electronics and mechanics, among others. In this sense, advanced technological materials and design methods are needed, which implies that the development of new robotic solutions is often an expensive practice [11]. Due to such a financial factor the simulation has established itself as an important tool in the field of mobile robotics. The simulation provides an efficient way to fulfill test and develop robotic solutions in a virtual environment without materials and/or personal risks.

This paper describes a novel application including the model and implementation of the micromouse robot in the SimTwo simulator(a 3D realistic open source robot simulator) [10]. This application described in the present work, has the objective to provide a dynamic and faithful simulation environment for the competition micromouse, where the competitors can interact with the mobile robot evaluating its strategies of control, mapping and path planning in different mazes (capability of import classic competitions), thus contributing to competitiveness among participating teams.

An outline of this paper is as follows. In Sect. 2, the state of the art will present the competition and the main simulators adopted by the mobile robotics community. In Sect. 3 is described the procedure for using different maps in the simulator. In Sect. 4 the model of the simulator is presented. Simulated results are presented in Sects. 5 and 6 concludes the paper.

2   State of Art

Competitions and robotic challenges have a positive impact on educational learning [9], introduction to new technologies, teamwork [5] and even develop solutions to real challenges in industry. In this context, some competitions deserve to be highlighted, such as RoboCup, with a worldwide repercussion, where robots up to 40 kg meet in a robotic football match [6]. Another interesting competition is Robot@Factory inspired by the challenges that an autonomous robot will have to face during its use in a factory [13], and other diverse competitions.

Micromouse is one of the most popular competitions among mobile robotics researchers. The competition is based on a small autonomous mobile robot, called micromouse. This robot need navigate through an unknown and arbitrary maze to locate the center. The objective of the competition is to find the center of the maze in as short a time as possible. The robot that makes the fastest timed run from the start to the center is declared the winner of the tournament [14].

To solve the maze, algorithm-based approaches aim to optimize the process of mapping and planning the best trajectory to the center of the maze. The widely disseminated algorithms are left wall follower, right wall follower and Floodfill algorithm. As reported in [12], the first two algorithms (left/right wall follower) are not efficient for most scenarios and can result in infinite loops without ever finding the center of the maze. The Floodfill algorithm has the ability to solve a wide variety of mazes. This algorithm assigns weights for each cell as a function of the distance of the cell with the objective, that is, the closer to the center cell the smaller the value assigned to the current cell [8]. In order to optimize the original algorithm, the work described in [3] adds an expression for optimizing situations with walls still unknown by the process of exploration of the algorithm, assigning a value of uncertainty to calculate the values assigned to each cell of the labyrinth, resulting in a shorter time to solve the labyrinth.

For the purpose of preparing the teams for the competitions, there are online simulators to test and validate the algorithms. However, these simulators are limited to 2D environments, as is the case of work presented in [2]. Although it is an essential tool for the preparation of competitions, the current simulators do not represent with fidelity the challenges encountered in a real competition, since it disregards the dimensions of the robot, the friction, dynamic effect, possible slippage, models for motor control, algorithm for the location and analysis of sensor data, and the uncertainty involved in a real environment. This work demonstrates the development of a simulation environment for the micromouse competition that takes into account these relevant elements, where each team can interact and implement their codes to perform tests and validations in a complete virtual environment with all the features present in competition.

3   Map Generation

In order to represent the classic competition with greater fidelity, the simulated maze has the same characteristics of a real environment. The maze, depicted in Fig. 1, is composed of multiples of an 18 cm × 18 cm unit square and comprises 16 × 16 unit squares. The walls are 5 cm high and 1.2 cm width. The outside wall encloses the entire maze, the start position (S) of micromouse robot is located at

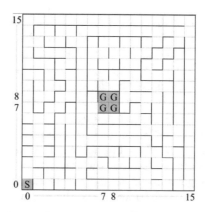

Fig. 1. Micromouse maze

left down corner (easily modified), and the goal (G) is located at the four center cells. The start square is bounded on three sides by walls.

To validate and test the algorithms developed for the micromouse, the simulator has the capacity to represent more than 450 different classic labyrinths (with valid start and goal), in a simple and fast way to change between them, where many have already been used in past competitions.

Originally, the labyrinths are encoded in a text file to facilitate their visualization (credits to [7]). A map converter has been developed that extracts the information from the text file and encodes it into an "obstacle" file in XML which is interpreted by the simulator. The maze generator is shown in Fig. 2a, where it has a brief description of the required procedure and an button "Generate Maze" that when pressed opens a tab (directory containing the text files) to select the desired maze as presented in Fig. 2b. After finishing the maze generation it is possible to observe the 3D simulation environment with the selected classic maze, as in Fig. 2c. This project can be found at https://github.com/J24a/SimTwoScenes.

(a)   (b)   (c)

Fig. 2. Generate maze: (a) Encoder application. (b) Maze files directory. (c) Micromouse simulator 3D.

4   SimTwo Robot Model

The construction of a physically simulated robot is an option that allows to reduce the cost of the implementation and realization of experiments guaranteeing a greater security, however it is almost as arduous as the construction of a real robot. SimTwo is a realistic simulation software suitable for the design and development of solutions for several types of robots. This application is based on a multiple document interface where the scenario and the robot's body and dynamics are implemented on XML language meanwhile the robots control is implemented on Pascal programming language.

SimTwo provides a platform where several robots can be simulated at the same time. Each robot is defined by various solids (cuboid, cylinder and sphere) interconnected through joints (slider, socket and hinge) and sensors that detect information from the robot and the environment. The dynamics realism is obtained by simulating each body and electric motor numerically using its

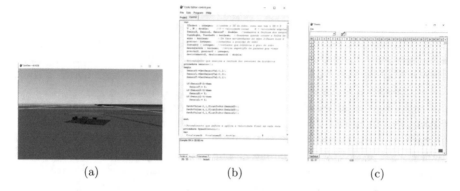

Fig. 3. Simulation environment: (a) Micromouse robot. (b) Code editor. (c) Sheet displaying the explored map (1-walls 2-free pass).

physical characteristics: shape, mass and moments of inertia, surface friction and elasticity [4]. The SimTwo simulator can be downloaded at https://paginas. fe.up.pt/~paco/wiki/index.php?n=Main.SimTwo.

The micromouse robot is assembled with the dimensions presented at the Table 1. The model is based on two motors, located at the lateral extremities of the base, that are connected into wheels 2 mm away from the edge. An omnidirectional wheel is positioned in the back of the base allowing the robot to freely move. Three distance sensors are located in the front with angles of 45, 90 and 135 degrees. The developed robot is presented on Fig. 3a, but this configuration can be altered according to the user's need.

Table 1. Simulated robot measures

Robot description	Dimension	Unit
Width	0.096	m
Length	0.120	m
Wheel diameter	0.032	m
Wheel thickness	0.008	m
Robot mass	1.25	kg

The robot's control is defined in the "code editor" (Fig. 3b). This editor offers an IDE for high-level programming based in Pascal language. The control script is divided into two main procedures, Initialize and Control. The Initialize procedure is executed once the code starts setting the initial configuration of the variables. Meanwhile the control procedure repeats itself every 40 ms executing the functions responsible for the control of the robot. The resulting robot movements obtained of such functions can be visualized in the main window.

In order to navigate without collision and map the maze environment the micromouse requires a constant interaction with the environment. Such interaction is realized by sensors that measures the distance between the robot laterals and frontal walls. To obtain the value of the sensor distance the function *GetSensorValue(IDrobot, IDsensor)* is used. This function is applied to each sensor returning the respective measured distance in millimeters.

The robots movements is based on the motors actuation. To control the speed and direction of the robot two variables are implemented, V and W. The V variable stores the linear velocity of the wheels, controlling the final speed of the robot, while the variable W keep the angular velocity, controlling the direction of the robot. The angular velocity is related to the distances measured by the lateral sensors, in order to keep the robot always centered in the current cell of the maze. The commands to set the final speed on each motor are shown in Algorithm 1.

Algorithm 1. Motor's Speed Control

1: **procedure** SPEEDCONTROL(V, W)   ▷ angular and linear speed
2:     $FinalSpeedRight \leftarrow V - W$
3:     $FinalSpeedLeft \leftarrow V + W$
4:     $SetAxisSpeedRef(IDrobot, IDmotor, FinalSpeedRight)$
5:     $SetAxisSpeedRef(IDrobot, IDmotor, FinalSpeedLeft)$

Another requirement to the micromouse implementation is the capacity of self-locate in the environment. This is implemented utilizing odometry sensors. The measured values of odometry per cycle(40 ms) can be obtained using the command *GetAxisOdo(IDrobot, IDmotor)*, based on this value is possible to obtain the current speed and position of the robot.

5   Results

SimTwo provides a realistic environment allowing the configuration of several robotics and environmental characteristics, such as the motor settings and environment friction. This improves the users experience since non-idealities errors present in the real robot implementation are experienced along the simulation.

The main error present on the real development of the micromouse is due to non-systematic errors present on the odometer measure, in this case caused by wheel slippage. To check this error and minimize it the simulator provides the robots position in real time making possible to compare the data from the odometry and the real position/velocity. A comparison between these factors is presented on Fig. 4

To facilitate the visualization of results and possible errors in real time, such as mapping and sensors values, the simulator utilize a "spreadsheet" window, presented on Fig. 3c. This window can exhibit all variables values and also can be configured to generate a 2D visualization of the map.

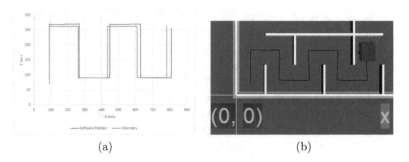

(a)　　　　　　　　　　　　　　　　　(b)

Fig. 4. Comparative between odometry and real robot movimentation: (a) Graphic. (b) Traveled route.

6   Conclusion and Future Work

A complete micromouse simulation environment was addressed in this paper. It is composed by a labyrinth generator that allows the use of available text file maze. The dynamic simulator of the robot implements the concepts as control, sliding, localization, decision and path planning that can be tested in a fast way, increasing the competition. All tools are available online and can be downloaded to speed-up the development of the micromouse software module. Results emphasize the approach and show the realistic dynamic behaviour of the model that can also be adapted to different projects such as the half size competition.

As future work, encoders, time of flight distance sensors and motor model can be tuned for more accuracy. It is important to emphasize that the integration of the hardware-in-the-loop is being developed where user can program and control the simulator robot in a real microcontroller connected to the simulator by Ethernet or USB.

Acknowledgement. This work is financed by the ERDF – European Regional Development Fund through the Operational Programme for Competitiveness and Internationalisation - COMPETE 2020 Programme within project POCI-01-0145-FEDER-006961, and by National Funds through the FCT – Fundação para a Ciência e a Tecnologia (Portuguese Foundation for Science and Technology) as part of project UID/EEA/50014/2013.

References

1. Almeida, L.B., Azevedo, J., Cardeira, C., Costa, P., Fonseca, P., Lima, P., Ribeiro, A.F., Santos, V.: Mobile robot competitions: fostering advances in research, development and education in robotics (2000)
2. Cai, J., Wu, J., Huo, M., Huang, J.: A micromouse maze sovling simulator. In: 2010 2nd International Conference on Future Computer and Communication (ICFCC), vol. 3, p. V3–686. IEEE (2010)

3. Cai, Z., Ye, L., Yang, A.: Floodfill maze solving with expected toll of penetrating unknown walls for micromouse. In: 2012 IEEE 14th International Conference on High Performance Computing and Communication 2012 IEEE 9th International Conference on Embedded Software and Systems (HPCC-ICESS), pp. 1428–1433. IEEE (2012)

4. Gonçalves, J., Lima, J., Costa, P.J., and Moreira, A.P.: Modeling and simulation of the EMG30 geared motor with encoder resorting to SimTwo: the official robot@ factory simulator. In: Advances in Sustainable and Competitive Manufacturing Systems, pp. 307–314. Springer (2013)

5. Kandlhofer, M., Steinbauer, G.: Evaluating the impact of educational robotics on pupils' technical-and social-skills and science related attitudes. Robot. Auton. Syst. **75**, 679–685 (2016)

6. Kitano, H., Asada, M., Kuniyoshi, Y., Noda, I., and Osawa, E.: Robocup: the robot world cup initiative. In: Proceedings of the first international conference on Autonomous agents, pp. 340–347. ACM (1997)

7. Micromouseonline: Github repository with maze files to micromouse (2018). https://github.com/micromouseonline/mazefiles

8. Mishra, S., Bande, P.: Maze solving algorithms for micro mouse. In: IEEE International Conference on Signal Image Technology and Internet Based Systems, 2008, SITIS 2008, pp. 86–93. IEEE (2008)

9. Nugent, G., Barker, B., Grandgenett, N., Welch, G.: Robotics camps, clubs, and competitions: results from a us robotics project. Robot. Auton. Syst. **75**, 686–691 (2016)

10. Paulo, C., José, G., José, L., Paulo, M.: Simtwo realistic simulator: a tool for the development and validation of robot software. Theory Appl. Math. Comput. Sci. **1**(1), 17–33 (2011)

11. Pinho, T., Moreira, A.P., Boaventura-Cunha, J.: Framework using ROS and SimTwo simulator for realistic test of mobile robot controllers. In: CONTROLO'2014–Proceedings of the 11th Portuguese Conference on Automatic Control, pp. 751–759. Springer (2015)

12. Sharma, M., Robeonics, K.: Algorithms for micro-mouse. In: 2009 International Conference on Future Computer and Communication, pp.581–585. IEEE (2009)

13. Sileshi, B.G., Oliver, J., Toledo, R., Gonçalves, J., Costa, P.: Particle filter slam on FPGA: a case study on robot@ factory competition. In: Robot 2015: Second Iberian Robotics Conference, pp. 411–423. Springer (2016)

14. Tondr, A., Drew, H.: The Inception of Chedda: A Detailed Design and Analysis of Micromouse. University of Nevada, Las Vegas (2004)

Towards Forest Fire Prevention and Combat Through Citizen Science

João Bioco$^{(\boxtimes)}$ and Paulo Fazendeiro

Instituto de Telecomunicações, Dep. de Informática,
Universidade da Beira Interior, 6200-001 Covilhã, Portugal
`joao.bioco@ubi.pt`

Abstract. Involving the community (volunteers) in citizen science projects is a good way to address and prevent a lot of societal concerns. The participation of volunteers has been quit frequent in citizen science projects; making them a fundamental key for the success of these projects. Volunteers participate in citizen science projects by collecting and processing data that can be used for various purposes such as educational, scientific, preservation of biodiversity, decision-making, etc. In forest fire prevention, participation of citizens in collecting and processing data could help significantly in decision-making related to forest fire prevention and mitigation. Mobile phones can be the tool of choice for collecting data due to not only to its wide availability and powerful communication features but also to its embedded sensing capabilities such as GPS location, camera and microphone.

This study is concerning to the development of a mobile-based citizen science project that allows volunteers to report fire-prone area, the occurrence of fire, and area where fire has occurred; then these information are used by firefighters and specialists for decision-making. In a scenario of fire occurrence, volunteers can take a picture of the place where the fire is occurring, upload to the mobile application, and send GPS location of the place; then the application notifies the firefighters and helps them allocate the needed resources to combat the fire based on the information sent by the volunteers.

Keywords: Citizen science · Forest fire · Prevention ·
Mobile application

1 Introduction

Nowadays citizen science projects are promoting more and more a great interaction between scientists and citizens. It is a good opportunity for collaboration between them, and allows citizens to acquire more and more scientific knowledge. Citizen contribution in citizen science projects are focused but not limited in collecting data that are then used for several purposes, such as educational, people engagement in community issues, decision-making, etc. In order to accomplish that, mobile phones has been quite used in citizen science projects as a means

© Springer Nature Switzerland AG 2019
Á. Rocha et al. (Eds.): WorldCIST'19 2019, AISC 930, pp. 904–915, 2019.
https://doi.org/10.1007/978-3-030-16181-1_85

of collecting data. Mobile technologies are empowering citizen science projects enabling the general public to participate in research and contribute to scientific knowledge [7,14,21]. In the context of this ongoing study we are going to develop a mobile application for forest fire prevention and combat. This application will allows volunteers to collect environmental data to the database of the application, that will be used to verify locations with higher risk of fire through a heat-map, issue alert of fire, perform simulations in order to make predictions related to forest fire, and allocation of resources. In addition to collecting data, volunteers will also be able to analyze/explore data from the database of the application. Through our "citizen science" approach we intend to establish a network of collaborative "environmental sensors" providing weather data complemented by information on volume, nature and condition of the available fuel. The remaining of this paper is organized as follows: In second section some general concepts of citizen science are addressed; In third section eleven studies that apply mobile application in citizen science projects are addressed; In the fourth section the proposed architecture of the system is described followed by the concluding section.

2 Background

Citizen science has been contributing a lot to the development of science, fundamentally with regard to data collection; important data collected by citizens is being used in many studies in order to contribute to decision making in several areas such as management, environmental protection, natural resources management, etc. Over the last 10 years, we have seen a great growth of citizen science in many countries. Usually two paths are taken in citizen science projects: the first is concerned to building scientific knowledge; and the second is related to informing policy and encouraging public action [16].

The activity of involving the public in collect and analyze data for a particular purpose is not new. The first and largest citizen science program in the world is called Christmas Bird Count [2]. It began in 1900 with the participation of 27 individuals that spend few hours count birds in their neighborhood on Christmas afternoon. In 2004–2005, 56.623 individuals were engaged in the project, counting nearly 70 million birds at more than 2000 different locations [11]. Since then, a lot of citizen science projects have been conducted involving the public in data collection and decision-making.

2.1 Acceptance Factors in Citizen Science

Citizen science can be defined as the activity of engaging citizens in a scientific project by collecting data that are filtered and analyzed by scientists, specialists, decision-makers or the public [3,4,16].

The involvement of the public can be more than data collection. Volunteers involved in citizen science project can participate in different phases of the research such as: designing a project, collecting data, and analyzing collected data [1].

There are three important user-acceptance factors to that we should consider in citizen science project: (1) data quality; (2) privacy; and (3) motivation [8].

Data Quality. The quality of data is a big concern in citizen science because volunteers are the responsible for collecting the data. These volunteers are regular people in general amateurs. Therefore collection of data by amateurs can result in bad data submitted resulting in wrong data analysis. Otherwise volunteers can perform a good job in collecting data.

There are some ways to avoid volunteers to collect bad quality data. One way is by training users before participation, providing tutorials on how users can perform their activities [17]. Another good way to ensure data quality is applying a set of constraints in data collecting process, limiting users to follow a standard data collection procedure [19].

Privacy. Privacy is another concern to take into account. Sensor-rich mobile phones provide a set of personal information (GPS location, names, data, etc.) that users may not be comfortable to share. Citizen science projects have to ensure privacy when users do not want to share their personal information. One way to ensure privacy is to give the users the possibility to be anonymous or share information according to the users preference [8].

Motivation. Usually volunteers who participate in citizen science projects tend to be more engaged in projects in their local communities [13]; but usually the motivations for participation in citizen science projects vary. A good strategy to keep volunteers engaged is to put them part of the project from the beginning, and let them in addition to collect data, analyze/explore them [16].

3   Related Work

Several citizen science projects have been conducted and described in literature. The implementation of mobile application on activities of data collection and data visualization is more and more frequent, because of smartphones features (GPS, camera, microphone, etc.).

Kim, Mankoff and Paulos [8] developed Sensr, a software platform that enable people without technical skills to build their own mobile data collection and management tools for usage in citizen science. They found that many organizations have difficulties in implementing citizen science because of two fundamental aspects: (1) Absence of technical skills to be able to develop necessary tools for facilitate data collection and management; (2) Many of these citizen science projects do not have fund to pay the implementation of these tools. Sensr adopted the term campaign to represent a citizen science project, and author, the person who create a campaign. Anyone is able to create and manage a campaign. Sensr is constituted with two parts: (1) a web application hosted in Amazon Web Service that allows authors create and manage a campaign and where the public can access the list of active campaigns; (2) a mobile application where volunteers can subscribe and participate exploring and collecting data for the campaign, and send back to the author. Data quality and privacy were two concerns handled by Sensr. In order to guarantee data quality Sensr has a filter option that allows the author evaluate the data before be published; In case of privacy, volunteers have the option of show or not their personal information while collecting and sending data by their mobile phones.

Graham et al. [5] describe three mobile applications (What is Invasive, Project BudBurst and Picture Post) used in citizen science projects to engage citizen scientists in research. (1) What is Invasive application (mobile app) was develop in order to facilitate volunteers detect and collect data on the location of a specific invasive specie. The app provides the list of invasive species that occur in that location, allowing volunteers to select the invasive specie and take a picture of it. The app will save the data collected and the location where the data was taken.

(2) Project BudBurst app (PBB app) allows citizen scientist to capture climate change data in the stages of leafing, flowering and fruiting of plants. Using BudBurst app users upload photos, and take advantage of mobile phones to collect time, date and GPS coordinates. In order to increase the engagement of the public, BudBurst app provides a game playing (geo-caching-like game) to their data collection.

(3) Picture Post is used to take a series of photos by mobile phones digital cameras to obtain a series of digital images for observing and measuring changing vegetation over the time. Picture Post also takes advantage of mobile phones to collect personal information's such as time, date and GPS coordinates.

Jambeck and Johnsen [6] developed a multiplatform mobile application called Marine Debris Tracker (MDT). MDT was designed to allow users acting like humans "sensor" in collecting marine debris and litter data in an efficient and rapid way. MDT app uses a MySQL database hosted at University of Georgia and a PHP-based Web service that allows any internet-capable device to log marine debris data collected. MDT can be used even in that areas where there is not internet. In that cases the data are saved in memory of mobile phone and them uploaded when connected to the internet. Using MDT app user can record type, quantity, current location and description or photo of marine debris. Data is available in a public web portal that shows the five last items inserted, in a dynamic feed.

NatureNet [15] is a citizen science system developed to collect bio-diversity data in nature park settings. NatureNet let users participate directly in the model design and implementation of new features of the system. This means that users are not only engaged in data collection but also participate in model design and implementation of the system (NatureNet). NatureNet is a set of app on tabletop and mobile devices available within a nature park. When visitors come to the park they can use the mobile application to capture bio-diversity by taking photos and write a description of it. After that, these data collected by the visitor are available in the tabletop in the entrance of the park, allowing another visitors to see the photos, description and location of the biodiversity data. Using NatureNet people also participate in NatureNet initial design and NatureNet model design by submitting ideas for new features, commenting, voting and selecting ideas, and implementing new features.

SecondNose [12] is a mobile application developed with the aim of collect environmental data to monitor some air pollution indicator. SecondNose is composed with four main elements: (1) an air pollution sensor, (2) an android app, (3) a backend with collection components and (4) a web application to visualize the data. Sensordrone2, a portable device is used as air pollution sensor to collect environmental data such as temperature, light, humidity, pressure, altitude, and air pollution parameters such as Carbon Monoxide (CO) and Nitrogen Dioxide (NO2). The sensor connects to the user mobile phone by Bluetooth. The mobile android application collect data from the sensor each five minutes, and reports the temperature detected, the atmospheric pressure trend, and overall air quality. The web application provides aggregate data from all users involved in data collection and allows manipulation of these data in various granularities.

Law and Zhang [10] described a work-in-progress mobile application called BioCondition Assessment Tool (BAT), to help landholders and ordinary citizens make environmental monitoring. Monitoring the composition and condition of environment is not an easy task for citizens without experience in collect these kinds of data. Citizen must have previous knowledge to make that. In order to accomplish this issue, BAT implements augmented reality and multimedia Smartphone technology.

Concerns about and data validation in citizen science project motivated Whitsitt el al. [20] to develop the iPhen; a mobile application with a user interface properly constrained to permit collecting of valid scientific data. iPhen allows users to collect location-stamped images to complement satellite data for climate change research. Mobile phone features such as know the device location through GPS, orientation within a few degrees, image capture has been used to help defining constraints in data collection. iPhen allows user capture three images: (1) a detailed image (leaves, flowers or fruits), (2) a panoramic image of the landscape, and (3) a plant/tree image. These images are captured obeying an orientation constraint (image plain's orientation within a of 10° interval of normal to the horizon). All image are tagged with GPS location, time, device orientation and mobile device hardware.

White et al. [19] developed the Firefly Counter, a real-time spatial framework and two mobile applications. These two mobile applications (iOS and Android) collect firefly counts and measurement of ambient light, and then report these data to an online mapping application. The Firefly Counter framework is composed of four elements: (1) data collection and mapping application (an iOS app, an Android app, webform and web mapping application); (2) a web service layer that process data submission and web mapping data requests; (3) an Apache HTTP Server; and (4) PostgresSQL and PostGIS data infrastructure to support GIS raster data. The mobile app (iOS and Android) takes advantage of the Smartphone's features to collect the location of the observation. The mobile

Table 1. An overview of mobile-based citizen science projects

App Name	Paper Author	Description	Year
Sensr	Kim, Mankoff and Paulos [8]	Enable people without technical skills to build their on mobile data collection and management tools for usage in citizen science	2013
Picture Post	Graham et al. [5]	Takes a series of photos by mobile phones digital cameras to obtain a series of digital images for observing and measuring changing vegetation over the time	2011
PBB	Graham et al. [5]	Allows citizen scientist to capture climate change data in the stages of leafing, flowering and fruiting of plants	2011
MDT	Jambeck and Johnsen [6]	Allows users acting like humans "sensor" in collecting marine debris and litter data in an efficient and rapid way	2015
The What is Invasive	Graham et al. [5]	Facilitates volunteers detect and collect data on the location of a specific invasive specie	2011
NatureNet	Lou et al. [15]	Collects bio-diversity data in nature park settings. Let users participate directly in the model design and implementation of new features of the system	2014
SecondNote	Chiara et al. [12]	Collects environmental data to monitor some air pollution indicator	2014
BAT	Law and Zhang [10]	Helps landholders and ordinary citizens make environmental monitoring	2012
iPhen	Whitsitt el al. [20]	Mobile application with a user interface properly constrained to permit collecting of valid scientific data	2011
Firefly Counter	White et al. [19]	Collects firefly counts and measurement of ambient light, and then report these data to an online mapping application	2014
Leafsnap	Kress et al. [9]	Collects data about species of trees and uses visual recognition software to help identify these species by taking a photograph of theirs leaves	2018

app allows users to save firefly count records locally on the phone when Internet access is unavailable. The app also allows users to take a tutorial of the usage, reviews past submission and view a map showing where the count was taken. Kress et al. [9] described the leafsnap, a free mobile application used to collect data about species of trees and uses visual recognition software to help identify these species by taking a photograph of theirs leaves. Leafsnap let users take a photo certain specie of tree, by capturing a leaf. This Photograph is compared with over 9000 images collected and stored in a database on a remote server. Leafsnap uses software recognition to verify a set of possible matches between the photograph and the images stored, and the return the top search results. These results bring detailed images of the specie's flower, fruit, seed and bark. After that the application upload automatically the photograph and location information taken from the device, into leafsnap database on the server. Leafsnap allows analyzing the distribution of species of trees that exists in the database.

In Table 1 we show an overview of mobile-based citizen science projects.

4   Proposed System

We intend to take the advantage of mobile phone features to develop a mobile application for forest fire prevention through citizen science. This application will be used by volunteers, scientists and experts.

Volunteers will use the application to perform four main tasks (1) Notify fire forest; (2) Collect environmental data from fire-prone areas; (3) Collect data from areas where fire has already occurred; and (4) Visualize data.

Scientists and experts will use the application to analyze and explore data, and make decisions based on the available field information.

4.1   Materials and Methods

For this study we first did the state-of-the-art related to the implementation of mobile and web App in citizen science, in order to know where we are with respect to the use of mobile App in citizen science projects. We study the limitations of mobile-based citizen science mostly related to the data collecting process by volunteers, taking with account that the success of these projects depends a lot on the data collected. We started to implement the mobile-based citizen science prototype for forest fire prevention taking into account these limitations. We will go more further in our mobile-based citizen science project by implement simulation based on the data collected, using agent-based modeling, and try to make predictions based on the results of these simulation.

4.2   System Architecture

We are going to develop: (1) a mobile hybrid application (android and iOS) were volunteers collect and visualize data; and (2) a web application based on AngularJS where scientists and experts analyze and explore data, and make decisions based on data.

These two versions of the application (mobile and web) communicate to a python REST API that makes available a set of services such as (1) risk fire prediction; (2) wildfire spread prediction; (3) SOS alert; (4) heat-map display; (5) resource allocation; and (6) environment and biodiversity.

A MySQL server will be used to create and manage the database, using Amazon relational database service (RDS).

Figure 1 shows the general architecture of the system.

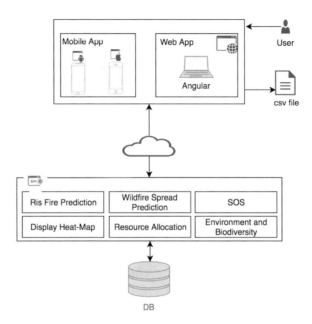

Fig. 1. System architecture

4.3   Mobile Application

Using the mobile application volunteers collect, review, send and visualize data. To be able to start using the mobile application, volunteers must create a user and login into the application. The screen of the application will have four initial tabs: Notify fire, Add record, Visualize my data, and About. In "Notify fire" tab, allows volunteers notify a fire that is currently occurring. Volunteers can take a photo of the location, set the location, date and time through mobile phone GPS, then add a description of the photo, and save the information. This information will be automatically available online.

In "Add record" tab, volunteers have two option to follow: (1) fire-prone area or (2) area where fire has occurred. In the first option, volunteers take a photo of the location, add a description of the photo, set the location, date and time through mobile GPS, then collect environmental data (temperature, relative humidity, wind speed, and rain) and save all these information. Environmental data such as temperature, relative humidity, wind speed, and rain, will be collected in order calculate Forest Fire Weather Index (FWI) in that location. We adopt the Canadian Forest Fire Weather Index (FWI) [18] an indicator of behavior and danger of fire which uses as meteorological parameters, the temperature, humidity, wind speed and rain. FWI contribute to know the risk of fire and fire behavior, and allowing decision makers take actions. In order to collect these environmental data used to calculate FWI, volunteers will be equipped with a low cost kit (sensors) that allows measuring these metereological data. Thus our mobile app will allows to calculate FWI in several points in the terrain, allowing a better resolution instead of being dependent on fixed weather stations. In the second option, volunteers collect data from an area where the fire has occurred.

In "Visualize my data" tab, volunteers have access to all data they previously added. Selecting one of these data, volunteers will be able to see the details of the data and edit some parts.

4.4   Web Application

The web application will be used by specialists in order to explore all information collected by volunteers. Only authorized people will be able to access the web application, and use all the data in decision-making.

Using the web application specialists can perform the following tasks: (1) display heat-map, (2) respond to SOS alert, (3) make predictions of risk fire and wildfire spread, (4) calculate resource allocation, and (5) save csv data.

(1) Display heat-map: based on the environmental data collect by volunteers and weather stations, specialists will be able to display a map indicating fire-prone areas.

(2) Respond to SOS alert: when a volunteer send a fire alert (notify fire) this information will be automatically available for the specialists who verify in the web application this information and allocate necessary resources to solve the situation.

(3) make predictions: based on the data the application will be able to make prediction related to risk fire and wildfire spread.

(4) calculate resource allocation: in case of eminent fire, the application has to be capable of present to the user the quantity of resources necessary to handle the fire situation.

(5) The web application will allow users to save information that they want, in the computer (Fig. 2).

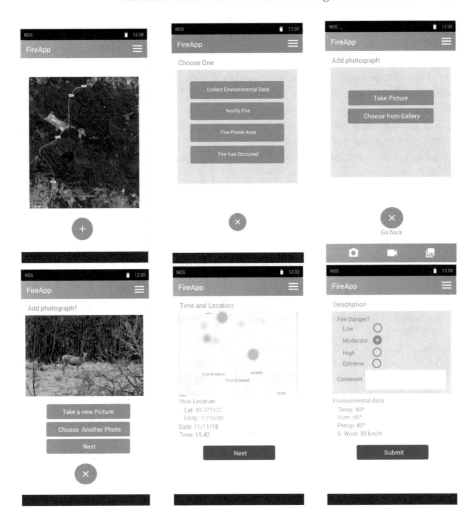

Fig. 2. Work-flow for adding record about fire-prone area.

5   Conclusion

The power of smartphones has inspired a lot citizen scientists to implement mobile application in their projects. Nowadays using mobile phones to collect citizen science data is a frequent task. The involvement of volunteers in citizen science project is becoming more and more common thanks to mobile technology features.

In this ongoing study we presented a general description related to the background of citizen science, we describe studies that apply mobile applications in citizen science projects, then we describe our proposed mobile application (proposed system, system architecture, web app experience and mobile app experience) for forest fire prevention and combat using citizen science. We already implemented a prototype (mobile and web) that collect environmental data. This prototype allows users to send these environmental data as well as georeferenced photos to the server, allowing the production of heatmaps based on calculated FWI values.

For future work we intend to implement the web and mobile hybrid application, and use the data collect to volunteers to model and simulate forest fire behavior using agent-based modelling approach and make predictions.

References

1. Bonney, R., Cooper, C.B., Dickinson, J., Kelling, S., Phillips, T., Rosenberg, K.V., Shirk, J.: Citizen science: a developing tool for expanding science knowledge and scientific literacy. BioScience **59**(11), 977–984 (2009)
2. Butcher, G.S., Niven, D.K., Sauer, J.R.: Using christmas bird count data to assess population dynamics and trends of waterbirds. Am. Birds **59**, 23–25 (2005)
3. Devictor, V., Whittaker, R.J., Beltrame, C.: Beyond scarcity: citizen science programmes as useful tools for conservation biogeography. Divers. Distrib. **16**(3), 354–362 (2010)
4. Dickinson, J.L., Shirk, J., Bonter, D., Bonney, R., Crain, R.L., Martin, J., Phillips, T., Purcell, K.: The current state of citizen science as a tool for ecological research and public engagement. Front. Ecol. Environ. **10**(6), 291–297 (2012)
5. Graham, E.A., Henderson, S., Schloss, A.: Using mobile phones to engage citizen scientists in research. Eos, Trans. Am. Geophys. Union **92**(38), 313–315 (2011)
6. Jambeck, J.R., Johnsen, K.: Citizen-based litter and marine debris data collection and mapping. Comput. Sci. Eng. **17**(4), 20–26 (2015)
7. Jennett, C., Furniss, D., Iacovides, I., Wiseman, S., Gould, S.J., Cox, A.L.: Exploring citizen psych-science and the motivations of errordiary volunteers. Hum. Comput. **1**(2), 200–218 (2014)
8. Kim, S., Mankoff, J., Paulos, E.: Sensr: evaluating a flexible framework for authoring mobile data-collection tools for citizen science. In: Proceedings of the 2013 Conference on Computer Supported Cooperative Work, pp. 1453–1462. ACM (2013)
9. Kress, W.J., Garcia-Robledo, C., Soares, J.V., Jacobs, D., Wilson, K., Lopez, I.C., Belhumeur, P.N.: Citizen science and climate change: mapping the range expansions of native and exotic plants with the mobile app leafsnap. BioScience **68**(5), 348–358 (2018)
10. Law, C.L., Roe, P., Zhang, J.: Using mobile technology and augmented reality to increase data reliability for environmental assessment. In: Proceedings of the 24th Australian Computer-Human Interaction Conference, pp. 327–330. ACM (2012)
11. LeBaron, G.S., Cannings, R.J., Niven, D.K., Sauer, J.R., Butcher, G.S., Link, W.A., Bolgiano, N.C., McKay, K.J.: The 104th christmas bird count. Am. Birds **58**, 2–7 (2004)
12. Leonardi, C., Cappelletto, A., Caraviello, M., Lepri, B., Antonelli, F.: Secondnose: an air quality mobile crowdsensing system. In: Proceedings of the 8th Nordic Conference on Human-Computer Interaction: Fun, Fast, Foundational, pp. 1051–1054. ACM (2014)

13. Lewandowski, E.J., Oberhauser, K.S.: Butterfly citizen scientists in the united states increase their engagement in conservation. Biol. Conserv. **208**, 106–112 (2017)
14. Luna, S., Gold, M., Albert, A., Ceccaroni, L., Claramunt, B., Danylo, O., Haklay, M., Kottmann, R., Kyba, C., Piera, J., et al.: Developing mobile applications for environmental and biodiversity citizen science: considerations and recommendations. In: Multimedia Tools and Applications for Environmental & Biodiversity Informatics, pp. 9–30. Springer (2018)
15. Maher, M.L., Preece, J., Yeh, T., Boston, C., Grace, K., Pasupuleti, A., Stangl, A.: Naturenet: a model for crowdsourcing the design of citizen science systems. In: Proceedings of the Companion Publication of the 17th ACM Conference on Computer Supported Cooperative Work & Social Computing, pp. 201–204. ACM (2014)
16. McKinley, D.C., Miller-Rushing, A.J., Ballard, H.L., Bonney, R., Brown, H., Cook-Patton, S.C., Evans, D.M., French, R.A., Parrish, J.K., Phillips, T.B., et al.: Citizen science can improve conservation science, natural resource management, and environmental protection. Biol. Conserv. **208**, 15–28 (2017)
17. Sheppard, S.A., Terveen, L.: Quality is a verb: the operationalization of data quality in a citizen science community. In: Proceedings of the 7th International Symposium on Wikis and Open Collaboration, pp. 29–38. ACM (2011)
18. Van Wagner, C.E., et al.: Structure of the Canadian Forest Fire Weather Index, vol. 1333. Citeseer (1974)
19. White, D.L., Pargas, R.P., Chow, A.T., Chong, J., Cook, M., Tak, I.: The vanishing firefly project: engaging citizen scientists with a mobile technology and real-time reporting framework. In: Proceedings of the 5th ACM SIGSPATIAL International Workshop on GeoStreaming, pp. 85–92. ACM (2014)
20. Whitsitt, S., Barreto, A., Hudson, M., Al-Helal, H., Chu, D., Didan, K., Jonathan, S.: Constrained data acquisition for mobile citizen science applications. In: Proceedings of the Compilation of the Co-located Workshops on DSM'11, TMC'11, AGERE! 2011, AOOPES'11, NEAT'11, & VMIL'11, pp. 267–272. ACM (2011)
21. Wiggins, A., Crowston, K.: From conservation to crowdsourcing: a typology of citizen science. In: 2011 44th Hawaii international conference on System Sciences (HICSS), pp. 1–10. IEEE (2011)

An Approach for Migrating Legacy Applications to Mobile Interfaces

Viviana Cajas[1,3(✉)], Matías Urbieta[1,2], Yves Rybarczyk[4],
Gustavo Rossi[1,2], and César Guevara[5]

[1] LIFIA, Facultad de Informática, Universidad Nacional de La Plata,
La Plata, Argentina
vivianacajas@uti.edu.ec,
{matias.urbieta,gustavo.rossi}@lifia.info.unlp.edu.ar
[2] CONICET, Buenos Aires, Argentina
[3] Facultad de Ciencias Administrativas y Económicas,
Universidad Tecnológica Indoamérica, Quito, Ecuador
[4] Intelligent and Interactive Systems Lab (SI² Lab),
Universidad de Las Américas, Quito, Ecuador
yves.rybarczyk@udla.edu.ec
[5] Centro de Investigación en Mecatrónica y Sistemas Interactivos (MIST),
Universidad Tecnológica Indoamérica, Quito, Ecuador
cesarguevara@uti.edu.ec

Abstract. Mobile applications changed unexpectedly people life and business models around the world. Nevertheless, there are old applications, called legacies, without adaptation to mobile devices, because this adaptation or migration have a considerable cost in dependence of software scope. Currently, most users bring constantly their smartphones and other devices with them, especially millennials. For that reason, some approaches try to solve this portabilization, generating certain improvements. However, in the majority of these solutions there is not a direct participation of users; do not consider their visual identity, analysis of feeling or mining of opinion. This paper proposes getting the behavior web application model with Markov heuristics from the widgets closeness matrix, prior to adaptation in order to include the user logic.

Keywords: Legacy adaptation · Markov chains · Millennials · Mobile devices

1 Introduction

By the year 2025, millennials [1] will constitute 75% of the world workforce. They are people who seek a balance between the personal and professional life, giving less importance to money, being their priority happiness in any situation; they are digital natives. For this reason, the Internet is their main tool, spending about 7 h a day, their social networks are their main way of communicating, and they carry their mobile devices - smartphones, tablets and laptops - to all parties, even at the time of sleep,

© Springer Nature Switzerland AG 2019
Á. Rocha et al. (Eds.): WorldCIST'19 2019, AISC 930, pp. 916–927, 2019.
https://doi.org/10.1007/978-3-030-16181-1_86

place them next to them [1]. Starting from this premise, the companies that have not adapted their business models to mobile devices run the risk of failing to gain the fidelity of users in the medium term. The new business models can be implemented through mobile applications [2] because, in this way, they can reach a wider range of business users and customers who mostly access Web apps from their mobile devices any moment and anywhere. Unfortunately, the applications created prior to the development of mobile devices, which are known as legacy applications [3], were not designed with the foresight to adapt automatically. The legacy system is an application that has become outdated but is still used by the users (usually an organization or company) and is not wanted to, cannot be replaced, or updated easily. Legacy application usability problems have been studied in different works [4] pointing out the usability issues arisen when the App is run in a mobile device. Some of them are: the legibility of the typography, the icons, the size, orientation of the screen, etc. Also, usability on mobile applications suggests that aesthetics graphics (balance between the colors, shapes, language, music or animation) is an important concept when evaluating the overall them [5]. In [6] the authors conducted an experiment assessing the productivity and performance of users accessing both a legacy and a mobile-friendly version of an application. The results of a controlled experiment based on two sites showed usability issues requiring more scrolling and zoom in/out events than the mobile-friendly version. Moreover, [7] collected usability experiments from 10 popular applications, where 3,575 users rated the usability using the System Usability Scale (SUS) questionnaire [8]. The average SUS rating was 77.7 out of 100, which is comparable to a C grade in the university grading scale [9]. It suggests that although mobile applications have gained a reputation as usable, they are not perfectly usable. Unfortunately, due to the lack of awareness of these issues, the users are forced to interact with applications even if they have a poor usability. Moreover, the industry tried to improve usability by increasing the width of the devices.

This work proposes to use Markov chains to analyse how the application features should be designed before a migration or portabilization from legacy and what should be omitted in the mobile application. The states of the Markov chain are formed with the screens and the controls that from here on will be generalized with the name of widgets. This model helps to understand the mechanics behind navigation. Markov chains calculation is a stochastic process where the probabilities describe the way in which the process will evolve in the future. These chains correspond to a standard mathematical technique recognized for the modeling process from the definitions of Cook and Wolf models [10], and for the development of Whittaker and Poore test cases [11]. They have also been defined as a model of the consumers that seeks to describe and predict their behaviour [12]. This model also allows being able to obtain interfaces for mobile devices. The remaining of the article is organized into eight sections. Section 2 defines the Background. Related work is described in Sect. 3. Section 4 explains the research methodology. The Markov approach for user interface design is explained in Sect. 5. An example is provided in Sect. 6. Results are shown in Sect. 7. Finally, the conclusions and future work are contemplated in Sect. 8.

2 Background

2.1 Markov Chains

They are a type of dynamic Bayesian networks [13] that predict the state of a system at a given time from the preceding state. The most important elements for the establishment of the chain are: the state space, the transition matrix, and the initial distribution [14]. The state transition diagram graphically presents the same information provided by the transition matrix, like in Fig. 1. The nodes (circles) represent the possible states, while the arrows show the transitions (the option to return to the same state is included). A transition probability defines the probability to move from one state to another (arrows in Fig. 1). Successful algorithms such as PageRank, created by Google, allow to grant a numerical value to each web page and from it, establish the order in which they appear after a search, use Markov chains. Proving that they are suitable for establishing web browsing models.

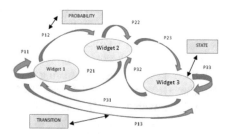

Fig. 1. Navigation Markov chain example

2.2 Systematic Layout Planning

This technique was proposed by Muther [15] whose acronym is SLP, allows the organization or distribution of the manufacturing plant improves productivity in the operations management through the reduction of movements, according to the flow of information or materials and the closeness the common areas to accomplish a process. In order to represent the relationships existing, a relational table of activities or areas is used, called closeness matrix. It is usual to express these needs through a code of letters, following a scale that decreases with the order of the five vowels, as shown in Table 1. This work makes an analogy between the software application and a manufacturing plant, which corresponds to space within the screen. Then, the distribution of elements in which users navigate must be optimal with respect to this matrix. And it must be considered according to the workflow and the distribution of the process. The efficient use of space within the screen is ensured according to the principle of optimal flow of the project.

Table 1. Closeness or proximity matrix

Key	Closeness	Value	Key	Closeness	Value
A	Absolutely necessary	5	O	Ordinary	2
E	Especially important	4	U	Without importance	1
I	Important	3	X	Undesirable	0

3   Related Work

The main related studies are summarized in Table 2. The works demonstrate that Markov chains can be a useful solution in software engineering. Nevertheless, the potentially of such a technique is still underutilized in migration processes of web to mobile applications. For this reason, we propose to use Markov to obtain in the first instance the behavior model of the web application that allows generating a more efficient adaptation based on usability and accessibility for end users, taking into account the probability determined by the most used areas, common activities, as well as, the most navigated links.

Table 2. Main characteristics of related works

Author	Evaluation/Future Work
Thimbleby et al. [14]	A tool to integrate Markov models in the design of a device with buttons (mobile, vending machines, recorders) is proposed and can be represented as finite state machines. The approach can be applied to abstract designs, prototypes, and animations, or fully operational systems
Mao et al. [16]	The authors propose an extended model (EMM) to develop plans and test methods from the components to the analysis of the system. The results of the test can gradually improve the EMM to perform regression tests and can also be used to correct programs, being a semiautomatic framework
Yanchun et al. [17]	The behavior model of a Mashup application is obtained, through a discrete Markov chain to build the model of the activities prior to building the security risk model
Chohan et al. [18]	A hierarchical Markov chain is used to perform the software product line tests. Three models are implemented: (i) captures the potential behavior of the products; (ii) keeps record the functions with the similarities or variations between the products and (iii) makes a specification of the mapping of elements in the test transitions
Nwobi-Okoye et al. [19]	The theoretical model of games and the concept of flow theory are used to obtain an optimal user interface modeled through chains of Markov. The validation was out three case studies; a web application, a desktop application and the comparison of mobile interfaces, leaving Windows Phone 7 at a disadvantage with respect to Android and iPhone

(continued)

Table 2. (*continued*)

Author	Evaluation/Future Work
Cajas et al. [20].	A systematic mapping about the portability of legacies through a migration with different strategies is developed. (i) The DOM is modified, so that the design adjusts to the size of the screens, providing an improvement in the web appearance. (ii) Model-driven development facilitates the generation of code semi-automatically from models. (iii) Mashups and (iv) middleware require high know how in order to make merge APIs and separate data sources into a single integrated interface. (v) Augmentation technique, where the user is part of the process and can select certain customization of the applications for their convenience. However, these approaches, can be used to migrate an app to a mobile version but not fully support the portabilization problem presented in this work

The hypotheses the Markovian model contemplates are the following: (i) assuming a finite number of states to describe the dynamic behavior of the widgets; (ii) assuming a known distribution of initial probabilities, which reflects either what state belongs an application widget, or the percentages of widgets in each state in the application; (iii) assuming that the transition from one current state to another in the future depends only on the current state (Markovian property); and (iv) that the probability of this transition being independent of the time stage considered (stationary property), that is, it does not change in the study time of the system.

4   Research Methodology

This study is based on the engineering method (evolutionary paradigm) [21]: observe existing solutions, propose better solutions, build or develop, measure and analyze, repeat until no further improvements are possible. For this reason, the authors began the investigation process with a Systematic Mapping [20] that studies the problem, approaches, and challenges present when migrating legacy Web applications to mobile platforms in the last decade. After examining the previously published studies, the article establishes that the Markov has not been proposed as a tool for configuring User interfaces in the migration process of legacy applications to mobile. Having said that, we present an approach based on the Markovian model. In the following section, we introduce the approach which later is instantiated in a study case [22]. Finally we conducted a simple evaluation reporting preliminary evidence about the benefit of our approach.

5   Proposed Approach

The approach aims at adapting legacy sites by modifying its structure, content, look, and feel to become a mobile friendly app. Figure 2 shows the process of adaptation. The different steps are described in the next sections.

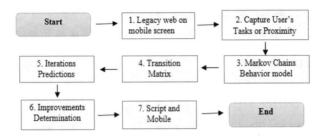

Fig. 2. Site adaptation process

5.1 Identify the Legacy Web Site

A legacy website is not properly rendered in small screens. To identify it, users can minimize or modify the size of the screen to verify if the content is rearranged in some way so that they can check whether or not the content of the layout is properly rendered. Also, they can access the site from different mobile devices, and the site is not mobile friendly when it is abusively required to perform scroll or zoom.

5.2 Define Expert Users

The selected millennial users [1] must have certain experience within the site, in order to get the simulation for expert user probabilities of navigation.

5.3 Behavior Model Development

This process must be made to each user role because the navigation cannot be general, each one has its widgets that may or may not coincide in certain cases. The first step to get the Markov model is to fill the state spaces, as in (1) defined with the widgets on the site. Then, the initial matrix (also called the initial probability vector) is filled in as in (2). Then, it is necessary to verify that P(0) adds 1. The first time, the user always access the main page of the application. For this reason, the probabilities are 1 in the widget1.

$$E = \{Widget1, Widget2, Widget3, .., Widgetn\} \tag{1}$$

$$P(0) = \{1, 0, 0, .., 0\} \tag{2}$$

5.4 Transition Matrix

The elements of the matrix of transition probabilities are non-negative, and the sum of the cells belonging to a row must result in 1. The order of transition matrix must be equals to the number of widgets (#Widget), so the algorithm can multiply the transition matrix by the initial matrix. For each widget discovered in Step 2 shown in a page, a row, and a column are placed in the matrix. Then, the matrix is filled in with the probabilities of navigation. In this case to fill in the matrix is to weigh (based on a Likert scale) the values according to the related activities, in a similar way as the closeness matrix of an SLP.

5.5 Iterations Model

Once both the initial matrix and the transition matrix are obtained, the algorithm multiplies N times until obtaining a stationary matrix that represents the behavior model. It is called stationary distribution when the initial distribution does not change when it is multiplied by the transition matrix. As long as the time goes by, it does not change with the passage of time and therefore it is called stationary or invariable distribution. It is said that a Markov Chain in discrete time admits a stationary distribution to the extent that long-term probabilities exist and is independent of the initial distribution P (0).

5.6 Improvements Determination

After calculating the stationary matrix, the probabilities for optimal site navigation should be recorded in the database. In addition, the strategies to establish the specific improvement of the site must be determined. Then, to make an initial improvement, first the algorithm must select the widgets related to 'A' and place them as close as possible on the user interface. On the contrary, the widgets related to 'X' or 'U' must be placed as far as possible or kept invisible and only displayed on request, because there is no direct relationship between them.

5.7 Evaluation of Distribution Alternatives

The Table 3 shows the scheme of the evaluation of the usability to should be implemented to proceed with the improvement of the user interface. The effect of the manipulation of the independent variable (distribution of the widgets) is reflected in the dependent variables.

Table 3. Evaluation matrix [23]

Variable	Measure	Explanation
Independent	Widgets Distribution	Screen distribution affects the distance required to perform the operations and therefore the cost and efficiency of the operation or task
Dependent	Task completion time (seconds)	The duration of tasks or parts of tasks, the time users spend in a particular mode of interaction
	Input rate	Rate of input by the user, for example using keyboard or sliders
	Use frequency	Number of keystrokes; number of mouse clicks; number of interface actions; amount of mouse activity; scroll and zoom

6 Running Example

Step 1. The Academic Management System (SGA) of the Indoamérica University [24] is used as running example. This platform allows the administration of teacher assignments, subject planning, class attendance, and register qualifications according to the regulation of the University. The cell phone version of the SGA displays unnecessary information, which can be cut off or incomplete, because there is not dynamic sizing. In consequences, the users must perform scrolling, and zoom in/zoom out to accomplish the tasks. In addition, the widgets are simply displayed in alphabetic order.

Step 2. The selected millennials teachers participate with the below proceeding:

1. List the SGA widgets of the teacher role, (Table 4).

Table 4. Tested widgets

No.	Widget name	No.	Widget name	No.	Widget name	No.	Widget name
1	Profile	5	Period	9	Attendance	13	Mail
2	Account	6	Photo	10	Teacher self-assessment	14	Teacher evaluation
3	Password	7	Search	11	Calendar	15	Schedule
4	Quit	8	Open classes	12	Grades		

2. Users in consensus must relate the widgets of Table 4 according Muther, as shown in Table 5.

Table 5. SGA closeness or proximity matrix

		1	2	3	4	5	6	7	8	9	10	11	12	13	14	15
Profile	1		5	4	3	2	1	3	4	3	1	1	2	3	1	2
Account	2			4	3	2	4	3	4	3	1	1	3	4	1	3
Password	3				3	1	1	3	2	2	1	1	2	3	1	2
Quit	4					3	3	3	3	3	3	3	3	3	3	3
Period	5						1	3	2	2	1	1	1	2	2	2
Photo	6							3	1	1	1	1	1	2	1	1
Search	7								2	2	2	2	2	2	2	2
Open Classes	8									1	1	2	1	2	1	2
Attendance	9										1	2	2	2	1	2
Teacher Self-Assessment	10											3	1	2	2	2
Calendar	11												3	2	2	2
Grades	12													2	1	2
Mail	13														3	3
Teacher Evaluation	14															2
Schedule	15															

Step 3. The space of states is establishes to determine the behavior model, such as:

   E = {Profile, Account, Password, Quit, Period, Photo, Search, OpenClasses, Attendance, TeacherSelf-Assessment, Calendar, Grades, Mail, TeacherEvaluation, Schedule}

   Next, the algorithm must establish the initial matrix distribution:

   P = {1, 0, 0, 0, 0, 0, 0, 0, 0, 0, 0, 0, 0, 0, 0}

Step 4. The transition matrix is filled in with the percentages of closeness matrix. To do so, the algorithm must add the weights of the relation scale of each pair of widgets, which corresponds to the population of each row to obtain the percentages.

Step 5. In order to make the iteration model, the algorithm must multiply the initial matrix by the transition matrix to get the stationary matrix.

Step 6. The determination of improvements is made through the analysis of screen size: (i) Reorganizing the menu according to the results obtained; (ii) Including shortcuts to the features most likely to be used; (iii) Decreasing the zoom and scroll within the pages. For an initial improvement: the algorithm groups the widgets according to nearness matrix priority as represented in Table 6. This operation consists of locating the widgets with value 'A', like Profile – My Account together; then identifying the widgets with value 'E', Profile – Opening Classes; and so on.

Table 6. Widgets profile pair priority

Key	Widgets pair	Key	Widgets pair
A	1-2;	O	1-4; 1-11; 1-14; 1-16; 1-20; 1-23; 1-24;
E	1-3; 1-7;	U	1-5; 1-9; 1-10; 1-13; 1-17; 1-19;
I	1-6; 1-8; 1-12; 1-15; 1-18; 1-21; 1-22;		

Step 7. Once raised the Markov distribution, it is necessary to determinate if the TAM of the user is improved or still require other modifications. The evaluation consists of measuring the efficiency of new widgets distribution in comparison to the old one.

7   Results

Figure 4 shows the new organization of the widgets based on the criterion of the users obtained through the closeness matrix and with mockups.

   In order to evaluate the new user interface, we conducted a preliminary evaluation that involved ten teachers who were divided into two independent groups, control and treatment, of 5 subjects each according to Nielsen recommendation. Each subject was required to complete four tasks: (i) Locate information about their schedules, (ii) Search the planning widget, (iii) Search the e-mail inbox, and (iv) Locate the grades widget. Five participants performed the tasks on the SGA application (control group) and the five others on the SGA Mockup (treatment group). The hypothesis was that the user

Fig. 3. Teacher role SGA new distribution

performance (e.g., completion time to carry out a task) should be better in the treatment group than the control group. The evaluation protocol was executed in the same way by all the subjects. They received a quick explanation of the SGA functionalities by a moderator. They were teachers who have never used the SGA application, and, moreover, they did not collaborate in design closeness matrix. In addition, two witnesses observed the behavior of subjects while performing the required tasks in their mobile device using a Chrome browser. During the assessment, the task completion time and the user interaction events count (zoom in/out, scroll, and click) metrics were evaluated. The moderator captured the time spent to complete the task as well as the required events. These values were fulfilled in a form similar to the Table 7. To evaluate the new SGA distribution a mobile application was created in Mobincube [25] that allowed to quickly replicate the new distribution obtained. The QR code of this application was shared with each participant to install and interact with the requested tasks.

Table 7. Old SGA measures vs. new SGA measures

	Old SGA Measures								New SGA Measures							
User	Task completion time seconds				UI Events				User	Task completion time/seconds				UI Events			
	T1	T2	T3	T4	T1	T2	T3	T4		T1	T2	T3	T4	T1	T2	T3	T4
U1	20	8	5	4	1	4	3	2	U6	10	2	3	2	0	0	0	0
U2	5	4	3	3	1	3	3	3	U7	1	1	1	1	0	0	0	0
U3	5	9	3	4	1	4	3	2	U8	1	3	3	1	0	0	0	0
U4	7	3	5	4	1	3	2	2	U9	1	0	0	0	0	0	0	0
U5	5	7	15	4	1	3	2	1	U10	4	3	3	1	0	0	0	0
Avg.	8,4	6,2	6,2	3,8	1	3	3	2		3	2	3	1	0	0	0	0

Table 7 shows the assessment result in an evaluation matrix of results. For each user we present the result of performing each task. The average completion time of task 1 was reduced from 8.4 s to 4 s. For task 2 it was reduced from 6.2 s to 2 s. The task 3 it was reduced from 6.2 s to 3 s. Finally, the task 4 was reduced from 3.8 s to 1 s. In the current SGA the Task 1 was the one that took the longest time in average to be completed by the users. On the other hand, the maximum optimization was reached on the task 4. The reader must note that the new SGA application mobile distribution avoided the screen glide because the widgets can be found at first sight. Also, this distribution provides a most optimal functional solution for the alphabetic previous distribution. So far so good, we have performed an evaluation presenting preliminary results supporting those benefits claimed by our approach.

8   Conclusions and Future Work

A web information system consists mainly of functional components described in terms of their behaviors and interfaces and the interconnections between components [17]. For this reason, the behavior of the user's navigation can be modeled as a Markov chain, through finite states within a transactional Web system that allow improve the software obtaining an optimized user interface promoting a more intelligent navigation. Nowadays, the user is very familiar with the closeness matrix, for this reason this approach proposes a design centered in them to interrelate software functionality in a model that they understand more easily. This methodology allows prioritizing the content of a legacy application in mobile devices, taking into account the reduced viewport with the flow of information from a user centered-design approach. Firstly, the new distribution allows the users to find faster the widgets they require. Secondly, the redistribution disposes the widgets accordingly their functional affinity. Finally, this new layout decreases the effort required to search a widget in the page, which reduces the time to complete the tasks and the size of the scrollbar. The hypotheses the Markovian model has been theoretically proved.

In addition, there is a future work to improve the development the tool that supports this Markov approach for the design of mobile interfaces, in order to measure its benefits, through a controlled experiment to evaluate its proactivity according to the use of the system to became an evolutive software. Also, the model will analyze the way in which the efficiency of the operation of a software application can be continuously improved by minimizing the operations performed (i.e. clicks, zoom, and scroll) and their cost (energy/time) according to the dynamic distribution proposals.

References

1. Abram, S.: Millennials: deal with them! Texas Libr. J. **82**(3), 96 (2006)
2. Stanley, M.: The Mobile Internet Report. Most, pp. 9–59 (2009)
3. O'Reilly, T.: What is web 2.0?: design patterns and business models for the next generation of software. O'Reilly. no. 65, pp. 17–37 (2005)

4. Zimmerman, D., Yohon, T.: Small-screen interface design: where are we? where do we go? In: IEEE International Professional Communication Conference (2009)
5. Hoehle, H., Aljafari, R., Venkatesh, V.: Leveraging Microsoft's mobile usability guidelines: conceptualizing and developing scales for mobile application usability. Int. J. Hum Comput Stud. **89**, 35–53 (2016)
6. Rivero, J.M., et al.: Improving legacy applications with client-side augmentations. In: 18th International Conference, ICWE 2018, Caceres, Spain, 5–8 June 2017, Proceedings (2018)
7. Taylor, P., Kortum, P., Sorber, M., Kortum, P., Sorber, M.: International Journal of Human-Computer Interaction Measuring the Usability of Mobile Applications for Phones and Tablets Measuring the Usability of Mobile Applications for Phones and Tablets, August 2015
8. Brooke, J.: SUS - a quick and dirty usability scale. Usability Eval. Ind. **189**, 4–7 (1996)
9. Bangor, A., Staff, T., Kortum, P., Miller, J., Staff, T.: Determining what individual SUS scores mean: adding an adjective rating scale. J. Usability Stud. **4**(3), 114–123 (2009)
10. Cook, J.E., Wolf, A.L.: Discovering models of software processes from event-based data. ACM Trans. Softw. Eng. Methodol. **7**(3), 215–249 (1998)
11. Whittaker, J.A., Poore, J.H.: Markov analysis of software specifications. ACM Trans. Softw. Eng. Methodol. **2**(1), 93–106 (1993)
12. Cheng, C.J., Chiu, S.W., Cheng, C.B., Wu, J.Y.: Customer lifetime value prediction by a Markov chain based data mining model: application to an auto repair and maintenance company in Taiwan. Sci. Iran. **19**(3), 849–855 (2012)
13. C.S.D., Murphy, K.P.: Inference and Learning in Hybrid Bayesian Networks. University of California, Berkeley (1998)
14. Thimbleby, H., Cairns, P., Jones, M.: Usability analysis with Markov models. ACM Trans. Comput. Interact. **8**(2), 99–132 (2001)
15. Muther, J.D., Wheeler, R.: Simplified Systematic Layout Planning. Management and Industrial Research Publications (1994)
16. Mao, C.-Y., Lu, Y.-S.: Testing and evaluation for web usability based on extended Markov chain model. Wuhan Univ. J. Nat. Sci. **9**(5), 687–693 (2004)
17. Yanchun, C., Xingpeng, W.: A Security Risk Evaluation Model for Mashup Application (2009)
18. Chohan, A., Bibi, A.: Optimized software product line architecture and feature modeling in improvement of SPL. In: International Conference on Frontiers of Information Technology (2017)
19. Nwobi-Okoye, C.C., Okiy, S.: Application of game theory to software user interface evaluation. Cogent Eng. **5**(1), 1–18 (2018)
20. Cajas, V., Urbieta, M., Rybarczyk, Y., Rossi, G., Guevara, C.: Portability Approaches for Business Web Applications to Mobile Devices: A Systematic Mapping, Babahoyo, 1 (2018)
21. Lázaro, M., Marcos, E.: Research in Software Engineering : Paradigms and Methods. In: CAiSE Workshops, pp. 517–522 (2005)
22. Zelkowitz, M., Wallace, D.: Experimental models for validating technology. Comput. (Long. Beach. Calif.) **31**, 23–31 (1998)
23. Hornbæk, K.: Current practice in measuring usability: challenges to usability studies and research. Int. J. Hum Comput Stud. **64**, 79–102 (2006)
24. Indoamerica, U.: Sistema de Gestión Académica (2017). https://sga.uti.edu.ec/login?ret=/. Accessed 29 Nov 2018
25. Mobincube, "mobincube" (2018). https://www.mobincube.com/es/. Accessed 01 Dec 2018

Designing IoT Infrastructure
for Neuromarketing Research

Tatjana Vasiljević[1], Zorica Bogdanović[1(✉)], Branka Rodić[2],
Tamara Naumović[1], and Aleksandra Labus[1]

[1] Faculty of Organizational Sciences,
University of Belgrade, 11000 Belgrade, Serbia
zorica@elab.rs
[2] Medical College of Applied Studies in Belgrade, 11000 Belgrade, Serbia

Abstract. The field of research is IoT infrastructure for neuromarketing. The goal is to develop a comprehensive open infrastructure based on Internet of things that can be used for conducting neuromarketing research. As a proof of concept, we have developed an infrastructure based on IoT devices and sensors for collecting data on examinee's responses to marketing content. The software part of the system includes an application for presentation of marketing contents, a non-relational database for big data analysis the collected data, and a tool for visualization of research results. A neuromarketing experiment with 30 participants has been conducted to evaluate the developed infrastructure. Results show that the developed infrastructure can be used to detect users' reactions in near real time, and to provide an adequate base for biofeedback.

Keywords: Internet of things · Neuromarketing · Biofeedback

1 Introduction

The term neuromarketing refers to the use of modern scientific methods to monitor the impact of marketing activities to consumers. Neuromarketing enables a new way of performing marketing research through the use of modern technology [1]. As such, neuromarketing has been found popular in modern companies which are oriented towards discovering, analyzing and understanding desires and needs of consumers, and which are striving to deliver the expected value to their consumers, which later leads to their loyalty. Neuromarketing helps to reveal the subconscious part of the personality of consumers. Neuromarketing tools and techniques are used with the aim to decrypt what's actually going on in the brain of consumers in time of exposure to a specific stimulus. The received biofeedback data can be used to customize marketing content to the user in near real time.

The aim of this study is to develop and evaluate the prototype of IoT infrastructure that can be used to conduct neuromarketing research. Using the concepts of wearable computing, the infrastructure is expected to provide data for examining preferences of respondents and adjusting the marketing content in real time.

© Springer Nature Switzerland AG 2019
Á. Rocha et al. (Eds.): WorldCIST'19 2019, AISC 930, pp. 928–935, 2019.
https://doi.org/10.1007/978-3-030-16181-1_87

2 Theoretical Background

2.1 Neuromarketing

Neuromarketing is an area that applies the principles of neuroscience for marketing research, studying senso-motor, cognitive and affective responses of consumers to marketing activities [2]. Neuromarketing is increasingly used in marketing because of its ability to understand the feelings of consumers [3]. As a multidisciplinary science neuromarketing relies on knowledge of psychology, neurology, neurophysiology, medicine, biology, marketing, and consumer psychology. Unlike traditional techniques and methods such as surveys, interviews and focus groups, neuromarketing has been considered a more advanced method of marketing research [4, 5].

Neuromarketing is a relatively new field of marketing that uses neuroscience methods and techniques for analyzing and understanding consumer behavior. Neuromarketing studies are based on the application of devices typically used for medical purposes (such as EEG) or specific devices for monitoring consumers' behavior (such as eye trackers) [6, 7]. The development of neuromarketing discipline is closely related to development of electroencephalography (EEG), which enables monitoring and analysis of brain activity. EEG stands for electroencephalography – a method of monitoring and recording electrical activity of the brain. The medium of this electrical activity are neurons, which communicate through electrical signals. Electrical signal is later converted into the waves using complex mathematical transformations [8].

Apart from EEG, other tools for monitoring brain activity can be used, such as functional magnetic resonance imaging (fMRI), and PET scans. Following the reactions of certain parts of the brain, and knowing their basic mental functions, researchers and marketing specialists are able to determine the type of mental processes (emotions), which takes place as a reaction to certain stimulus [4]. Further, devices for measuring different physiological parameters, such as heart rate, breathing rate, galvanic skin response, can be used for identifying the reaction of a consumer to a stimulus [9].

One of the main problems for a wider adoption an application of neuromarketing research is that the devices necessary to collect users' response data are expensive, oversized and complicated to use. A new wave of neuromarketing research has appeared with the development of internet of things and the appearance of wearable devices capable of monitoring the body functions.

2.2 IoT in Neuromarketing

Internet of Things (IoT) refers to a network of objects connected over the Internet and able to collect and exchange data using embedded sensors [10]. International telecommunication union defines IoT as "a global infrastructure for the information society, enabling advanced services by interconnecting physical and virtual things based on existing and evolving, interoperable information and communication technologies" [11]. The development of Internet of things has resulted in numerous small and portable devices that can be used to monitor emotions and users' reactions to the advertising content.

One group of such devices includes modern eye trackers that have become small and portable, so they can easily be used for field work. Eye trackers can be implemented using web cameras or dedicated devices that allow researchers to observe patterns of sight and understand what the person is looking at. It is a quick and easy way to analyze how a focus group responds to marketing tools that are used for communication. It allows researchers to find which marketing tools and techniques work well and which do not. This device helps in understanding what needs to be improved or changed in marketing in order to attract a larger number of clients [8].

Apart from eye trackers, EEG headsets are becoming a rational compromise from the standpoint of usability, amenities, prices and designs. These devices are light weight, fairly compact and easy to use. At the same time, they enable detection of a wide range of brain waves during the EEG tests products, marketing materials, and other content with high enough precision.

Ear-clip Heart Rate contains an ear-clip and a receiver module. The measured values can be displayed on a screen and can be saved for analysis. The whole system has a high sensitivity, low consumption of energy and is very portable.

Pulse sensor is essentially photoplysmograph (PPG), who is a well-known medical device used for non-invasive monitoring of heart rate. Sometimes, PPG measures oxygen levels (SpO_2). The most recent version of the hardware, the Pulse Sensor Amped, reinforces the raw signal of the previous sensor pulse and normalizes the impulse wave around medium voltage point. Pulse sensor Amped corresponds to the relative changes of intensity of light. If the amount of light that the sensor remains constant, the value of the signal will remain on (or near) 512 (middle of the Arduino range 10-bit ADC). With more light, signals are increased, with less light, the opposite. The quantity of light of the green LED diode which is reflected back to a sensor changes during each pulsation.

3 IoT Infrastructure for Neuromarketing Research

The goal of this paper is to develop an IoT infrastructure that could serve as a base for neuromarketing research. The infrastructure should be developed using the principles of wearable computing, so to enable monitoring users' reactions in real time. The final goal is to design an entirely wearable device that could measure users' reaction and provide an insight that would be used to adjust marketing content in real time.

Commercial IoT systems for neuromarketing research are usually not with open APIs, they can hardly be programmed and customized. For the purposes of integration and full customization to the specifics of each neuromarketing experiment, we have opted for an open architecture. An open architecture ensures that the developed system for neuromarketing research can be easily integrated with other systems. This approach was confirmed in previous research [12, 13].

As a proof of concept, the project enables monitoring heart rate. The system architecture is shown in Fig. 1. The system consists of physical, application and presentation layers.

The physical layer (hardware) needs to enable collection of data from sensors and send them to a database so that the data can be used in real time. The prototype was

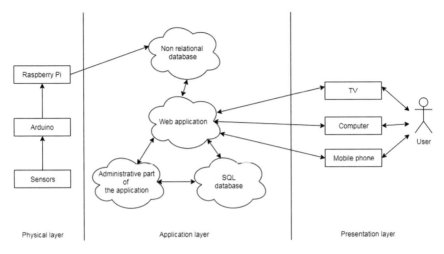

Fig. 1. System architecture

developed using available devices: Raspberry Pi microcomputer and sensors suitable for neuromarketing research [14]. Apart from Heartbeat sensor used in the prototype, any other sensor can be used in the same way.

The application layer of the system includes a database and a web application developed to manage research procedure and data. The web application was developed by modifying the Wordpress CMS. The Wordpress CMS has been chosen as an open source software that can easily be modified and customized to support the needs of our system. Wordpress has been customized so to provide the following features:

- Managing research metadata: a researcher can enter start and end date and time, and can set the research parameters. They can choose a stimulus that will be used in the research. In the developed prototype, Youtube video clips were used.
- Registering research participant: each participant can anonymously register to the system, be exposed to the stimulus, and answer survey questions.
- Analysis of research results: data from the stimulus can be paired with the data measured using sensors, so that a research can analyze the participants' reactions to the presented content.

Figure 2 shows a screenshot of the application.

Sensor data are stored in a non-relational database MongoDB, in order to create a good foundation for real-time big data analytics. MongoDB is an open-source database management system. The basic structure consists of documents that are very similar to JSON format.

The presentation layer of the system can be realized in various devices, depending on the research context (mobile phone, smart TV, computer, VR…). For the prototype, we have presented stimulus to the participants through the Wordpress web application.

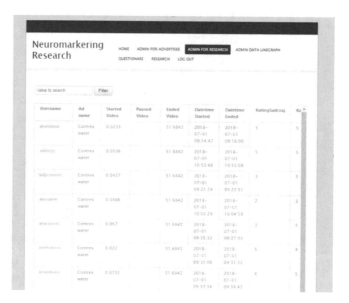

Fig. 2. Web application, page - research administration

4 Experiment

4.1 Design and Procedure

The evaluation of the developed infrastructure was done by conducting a simple neuromarketing research. Research was done in the Laboratory of e-business, Faculty of Organizational Sciences, University of Belgrade. All participants were students of the fourth year who volunteered to participate in the research. There were 30 participants: 17 male and 13 female.

At the beginning we measured pulse while a participant is at rest, to obtain a baseline. For each participant we have created an account on the described web site. Each participant filled an online questionnaire with basic information such as age, gender, marital status, etc. During this time the pulse was not measured. When respondents completed the questionnaire, they were attached to a sensor and they started watching a selected commercial. At the end, the respondents were asked to complete a second questionnaire and rate the commercial by the following criteria:

- content
- whether it carries a message
- does it hold attention
- is the duration optimal
- is it provoking an action to buy the product

The parameters that were of interest are: the pulse (heart beats per minute), the interval between heartbeats (in milliseconds), and variability of this interval associated with emotional responses.

The result of measuring where provided for every second of the commercial, in order to be able to analyze what caused the reaction in terms of attracting attention and provoking emotional reaction, which gives the possibility to customize the contents of the ad to each user individually.

As a stimulus we have selected an advertisement for the Nestle's water Contrex. It is an advertisement created for the event organized by Contrex in Paris in September 2011 under the slogan "Slimming doesn't have to be boring. Contrex, my slimming partner".

4.2   Results

Figure 3 shows the obtained data for one of the respondents. X axis shows points in time, and Y axis shows the measured value of pulse. Results show a noticeable growth of pulse at the very beginning of the advertisements which leads to the conclusion that the content of advertisements interested the subject. This is in agreement with data collected though the questionnaire, where this respondent assessed the content of the advertisement. The respondent's pulse increased in a period of the commercial when machines for exercise were shown. From this we can assume that the respondent likes to engage in physical activity. After initial growth, the pulse rate remained almost constant to the end of the commercials. We can assume that the content is interesting to the subject, which is in agreement with the given highest rating to the question "Did advertisement hold your attention?".

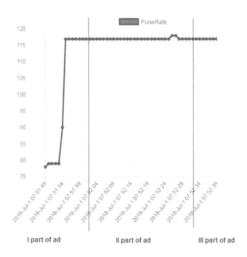

Fig. 3. Measured pulse for one respondent

Figure 4 shows correlation of scores for the contents of advertisements given by respondents and their measured pulse values. From the correlation we conclude that respondents with higher increases in heart rate in relation to the initial value have given higher scores (4 or 5) for content, for the attention that advertisement caused and for the message. We note also that the score to the question related to buying the product and measured pulse values are not correlated. This is in accordance with the campaign goal, which was more oriented towards raising awareness instead of selling products.

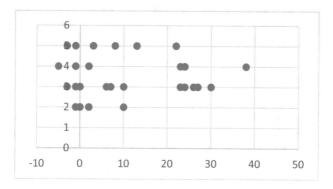

Fig. 4. Correlation of scores for the contents of advertisements and pulse values in II part of the advertisement

5 Conclusion

As the area of neuromarketing is developing, neuromarketing research methods gradually become standard tools for research in various areas. Advances in processing information, data collection and scientific theory created a good basis for improving the neuromarketing area beyond what it used to be possible [15]. Devices that enable measurement of the mentioned parameters, such as heart rate, stress, skin conductivity or eye movement, are becoming available, leading to neuromarketing research being possible in real life conditions and outside of laboratories with high-end equipment.

This research presents a prototype of an infrastructure for neuromarketing research based on the principles and technologies of internet of things and wearable computing. The infrastructure is flexible and it enables application of any suitable sensor that provides an open API.

We have described a possible research scenario, measuring pulse to detect the users' reaction to marketing content. The pulse sensor can provide only the basic information on users' reaction, and more sophisticated analyses would be possible by extending the measurements with GSR (galvanic skin response) and EEG data. Further research will be oriented to connecting more precise sensors to the developed infrastructure and obtaining more accurate data related to users' reactions to marketing content. The final goal is expected to be a full biofeedback system that would enable real-time personalization of marketing content according to users' preferences and emotional reactions.

Acknowledgement. Authors are thankful to Ministry of education, science and technological development, Republic of Serbia, grant 174031.

References

1. Genco, S., Pohlmann, A., Steidl, P.: Neuromarketing for Dummies. Wiley, Mississauga (2013)
2. Krajnović, A., Šikirić, D., Jakšić, D.: Neuromarketing and Customers Free Will. University of Zadar (2012)
3. Arthmann, C., Li, I.-P.: Neuromarketing–the art and science of marketing and neurosciences enabled by IoT technologies (2017)
4. Alčaković, S., Arežina, N.: Neuromarketing-nov način razumevanja potrošača. Osmi naučni skup sa međunarodnim učešćem, Sinergija (2011). (in Serbian)
5. Quirks.com: Find Market Research Companies, Facilities, Jobs, Articles, More|Quirks.com (2018). https://www.quirks.com/articles/what-can-measuring-brain-waves-tell-us-about-an-ad-s-effectiveness. Accessed 19 Apr 2018
6. Facial EMG: Muscles Don't Lie? - Neuromarketing, Neuromarketing (2018). https://www.neurosciencemarketing.com/blog/articles/facial-emg.htm. Accessed 4 Apr 2018
7. Farnsworth, P.B., Christopherson, R., Jensen, O.: What is ECG (electrocardiography) and how does it work? iMotions (2018). https://imotions.com/blog/what-is-ecg/. Accessed 3 Apr 2018
8. Neuromarketing Solution for Biometric Research - NeuroLab by CoolTool|CoolTool, Cooltool.com (2018). https://cooltool.com/neurolab. Accessed 19 Apr 2018
9. Yadava, M., Kumar, P., Saini, R., Roy, P.P., Dogra, D.P.: Analysis of EEG signals and its application to neuromarketing. Multimedia Tools Appl. 76(18), 19087–19111 (2017). https://doi.org/10.1007/s11042-017-4580-6
10. Atzori, L., Iera, A., Morabito, G.: The Internet of Things: a survey. Comput. Netw. 54(15), 2787–2805 (2010). https://doi.org/10.1016/j.comnet.2010.05.010
11. International telecommunication union, Common requirements of the Internet of things, ITU - T Y.4100/Y.2066 (06/2014) (2019). http://handle.itu.int/11.1002/1000/12169. Accessed 10 Jan 2019
12. Radenković, B., Despotović-Zrakić, M., Bogdanović, Z., Barać, D., Labus, A., Bojović, Ž.: Internet inteligentnih uređaja, Fakultet organizacionih nauka (2017). (in Serbian)
13. Rodić-Trmčić, B., Labus, A., Bogdanović, Z., Despotović-Zrakić, M., Radenković, B.: Development of an IoT system for students' stress management. Facta Universitatis, Series: Electronics and Energetics 31(3), 329–342 (2018)
14. Rodić-Trmčić, B., Labus, A., Barać, D., Popović, S., Radenković, B.: Designing a course for smart healthcare engineering education. Comput. Appl. Eng. Educ. (2018). https://doi.org/10.1002/cae.21901. Online ISSN: 1099-0542
15. Dooley, R.: Brainfluence: 100 Ways to Persuade and Convince Consumers with Neuromarketing. Wiley, Hoboken (2011)

Extraction of Fact Tables from a Relational Database: An Effort to Establish Rules in Denormalization

Luís Cavique[1(✉)], Mariana Cavique[2], and António Gonçalves[3]

[1] Universidade Aberta, MAS-BioISI, 1269-01 Lisbon, Portugal
Luis.Cavique@uab.pt
[2] Universidade Europeia, 1500-210 Lisbon, Portugal
MarianaCavique@hotmail.com
[3] Escola Superior de Tecnologia de Setúbal, Instituto Politécnico de Setúbal,
2914-504 Setúbal, Portugal
Antonio.Goncalves@estsetubal.ips.pt

Abstract. Relational databases are supported by very well established models. However, some neglected problems can occur with the join operator: semantic mistakes caused by the multiple access path problem and faults when connection traps arise. In this paper we intend to identify and overcome those problems and to establish rules for relational data denormalization. Two denormalization forms are proposed and a case study is presented.

Keywords: Fact tables · Join operator · Multiple access path problem ·
Fan trap · Denormalization forms

1 Introduction

The normalization forms in relational databases are a well-known subject with references in many research and teaching documents. On the other hand, the denormalization in relational databases is usually seen as the application of the join operator over a set of tables/relations Ri, R1 $|><|$ R2 $|><|$ ... Rn.

Given its simplicity, the relational model and SQL query language have been well accepted and are currently widespread in business environments. However, in some specific database schemas, the result of SQL queries does not match the expected answers. The powerful join operator can lead to undesired situations. The most common are queries with multiple paths between tables return different results (Wald and Sorenson 1984) and some relational inferences can fall into connection traps (Feng and Crowe 1999). Generally join operator problems have been neglected in the literature. The identification of SQL traps has received increasing attention in some professional literature (Business Objects 2007).

In this work we use the concept of fact table given by Kimball and Ross (2013) regarding data warehousing. Data warehouse (DW) provides a stable environment to run complex queries contrary to the use of transactional databases. So, data warehouse design is essential for sustainable Business Intelligence (BI). DW is the basis for the

© Springer Nature Switzerland AG 2019
Á. Rocha et al. (Eds.): WorldCIST'19 2019, AISC 930, pp. 936–945, 2019.
https://doi.org/10.1007/978-3-030-16181-1_88

creation of OLAP (online analytical processing) cubes and multi-dimensional analytical. In typical data warehousing, the ETL (extract, transform, load) process extracts data from operational systems to load it into the data warehouse. Beyond DW, OLAP and BI the migration of relational data to geographic information systems (Kingdon et al. 2016), object oriented tools or web environments (Karnitis and Arnicans 2015) is a relevant issue.

In this work we use the following nomenclature for the tables: lookup tables for table with only cardinality equal to one, intermediate tables for tables with cardinality 1 and N, and fact tables for tables with cardinality N, as shown in Fig. 1. We also draw all tables with relation 1:N this way: the table with a single line is drawn on top, while the table with multiple lines is drawn underneath.

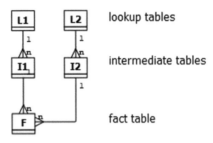

Fig. 1. lookup, intermediate and fact tables

The goal of this paper is to make a review of some neglected join operator problems and to establish two denormalization forms in order to extract fact tables from a relational database. In this work we intend using in the fact tables the same data granularity of the original relational database.

As already referred, database normalization is a well-studied subject, but the bibliography for denormalization is scarce, for this reason a related work section is not presented.

The paper is organized in 5 sections. In Sect. 2 two neglected issues in relational models are recalled: the multiple access path problem and the connection traps. In the same section poly-tree structure is presented. Section 3 presents an effort to establish two denormalization forms taking into account the previous issues. A case study is presented in Sect. 4. Finally, in Sect. 5, some conclusions are drawn.

2   Background Information

Relational model relies extensively on the join operator. However, some difficult situations can occur. This section highlights the query ambiguity, produced by the multiple access paths, which can return queries with different results, and also highlights the connection traps, in particular the fan trap, that return incorrect results.

This section presents two neglected problems with the join operator that affect the extraction of fact tables.

2.1   Multiple Access Path Problem

The Multiple Access Path Problem (MAPP) is presented in (Wald and Sorenson 1984) seeking to translate a sentence into an unambiguous database query. A detailed introduction can be found in (Wu et al. 1996). This problem does not occur in a database where there is a single path between two tables. When two or more paths occur between two tables, the corresponding queries can return different solutions.

In order to systematize the analysis of multiple-path in a schema, it is important to define and classify the paths.

Multiple-paths definition: let the database DB = $\{T_1, T_2,\ldots, T_n\}$ with T_n tables with a set of attributes. A multiple-path occurs if there is more than one path, using the foreign keys, between T_i and T_j tables.

To illustrate the Multiple Access Path Problem, a database of scientific evaluation will be used. Throughout this paper it will be used the database layout considered in Fig. 2. This database schema presents tables with relation 1:N, in which the table with a single line is drawn on top, while the table with multiple lines is drawn underneath.

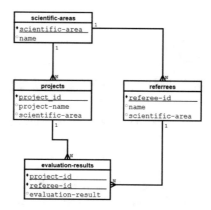

Fig. 2. Database where the Multiple Access Path Problem (MAPP) can occur

In this paper we will use relational algebra symbols: the symbol σ for selection operator, the symbol Π for projection operator, the symbol $|\!\!\times\!\!|$ for the join operator and the symbol G_{function} to identify the aggregation functions.

Choosing tables Scientific-areas and Evaluation-results, two different queries can be defined to obtain the number of evaluations by scientific area. The queries with different paths are:

Q1: G_{count} ($\Pi_{\text{scientific-area}}$ (Scientific-area $|\!\!\times\!\!|$ Projects $|\!\!\times\!\!|$ Evaluation-results))
Q2: G_{count} ($\Pi_{\text{scientific-area}}$ (Scientific-area $|\!\!\times\!\!|$ Referees $|\!\!\times\!\!|$ Evaluation-results))

Using the following data:

Scientific-areas = ((A,_), (B,_), (C,_))
Projects = ((P1,_,A), (P2,_,A), (P3,_,B))
Referres = ((R1,_,C), (R2,_,C), (R3,_C))
Evaluation-results = ((P1, R1,_), (P2, R2,_), (P3,R3,_))

The queries would return different values, Q1 = ((A,2), (B,1)) and Q2 = ((C,3)), showing that queries using different paths between two tables can return different outputs. We present poly-tree structure in order to avoid such inconsistencies.

2.1.1 Poly-tree Structure

An oriented tree is a direct acyclic graph with an in-degree node equal to one, excepting of the root, where the in-degree is equal to zero. A poly-tree is a relaxed oriented tree, with one (and only one) path between any pair of nodes, where the in-degree of a node can be greater than one. Trees and poly-trees have N nodes and N-1 arcs. The poly-tree structure was widespread in Bayesian (or belief) networks (Darwiche 2009).

The first algorithm for Bayesian networks was restricted to trees (Pearl 1982) and was followed by a generalization that became known as the poly-tree algorithm (Pearl 1986). The term poly-tree was coined by (Rebane and Pearl 1987) and the poly-tree algorithm was the very first exact inference algorithm for Bayesian networks.

The goal of a Bayesian network is a complete representation of the joint probability of a set of variables. Inference in a Bayesian network means computing the probability of a query variables set, given an evidence variables set.

The complexity of belief network inference depends on the network structure. In poly-trees (or singly-connected) structures the computational time is linear in the size of the network. When the number of arcs is equal or greater than the number of nodes, the structure is called multiply-connected, and the computational time is exponential in the worst case.

There are three basic methods to solve multiply-connected networks: clustering (or merging), loop-cutset conditioning (or split strategy) and simulation (Russel and Norvig 2003). The clustering method and the loop-cutset conditioning transform the multiply-connected network into a singly-connected network, while the simulation generates a large number of instances that give an approximated solution. In Fig. 3 an original multiply-connected belief network is shown and the poly-trees obtained by using the split method and merge method.

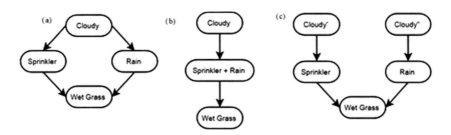

Fig. 3. (a) loopy belief network, (b) merge method, (c) split method

Both opposite strategies of splitting and merging information, using the loop-cutset conditioning/split and clustering/merging methods, contribute to the generation of poly-trees.

2.2   Fan Trap

Connection traps are well-known in database design. A detailed introduction can be found in (Feng and Crowe 1999). The term was coined by Codd when he was proposing the relational model (Codd 1970) and followed by (Date 1995).

To illustrate a connection trap, and in particular the fan trap, a database of ordered and planned sales will be used. The database with a fan schema and a data sample are shown in Fig. 4.

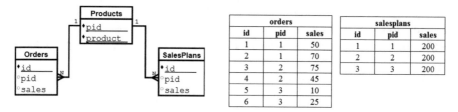

Fig. 4. Fan schema and data sample where the connection trap occurs

At the first glance, the SQL query which returns the sum of the ordered sales and planned sales can be written as follows:

```
SELECT P.product, SUM(O.sales) as ordered, SUM(SP.sales) as planned
FROM products P, orders O, salesplans SP
WHERE P.pid = O.pid
AND P.pid = SP.pid
GROUP BY P.product
```

However, the three tables' junction will inflate the value of the planned sales, returning the value of 400 for each product, instead of 200.

In Fig. 4 (on the left), the table Products (on the top) and the tables Orders and SalesPlans (on the bottom) identifies a specific pattern, called in this paper as fan schema. The fan schema in a database is synonymous of connection trap. When a fan schema is found in a database, the data must be aggregated in three steps. For the example shown, first aggregate orders, then aggregate sales and finally join orders and sales.

Connection traps, as referred by (Feng and Crowe 1999), have been defined as 'the lack of understanding of a relational composition' by (Codd 1970), 'false inference' by (Date 1995), 'represent a ternary relation as two binary relations' by (Cardenas 1985) and more explicitly as 'an intrinsic deficiency of the relational theory' by (Ter-Bekke 1992).

3   Denormalization Forms

In this section we propose two denormalization forms. The first denormalization form is when the database has a poly-tree schema, avoiding queries with multiple paths. The second denormalization form is when the poly-tree schema is divided into trees that forbid fan traps. In the second denormalization form the join operator can be applied over a set of tables/relations Ri, R1 |><| R2 |><| ... Rn.

3.1 First Denormalization Form: Poly-tree Structure

The first denormalization form (1DF) is achieved when the database acquires the shape of a poly-tree. The poly-tree database schema avoids the Multiple Access Path Problem (MAPP) already defined.

The split and merge strategies (Russel and Norvig 2003) applied in the belief networks methods (loop-cutset conditioning/split and clustering/merging) are reused for databases, in order to generate poly-trees.

In this work we adopt only the split strategy to create poly-trees. The split technique example is given with the evaluation of scientific projects schema in Fig. 5. By removing the relation of the referee scientific area, there is loss of information. Thus it is preferable to add a new table which distinguishes the scientific areas of projects from the scientific areas of the referees. The removal of multiple-path is obtained by eliminating of a relation and inserting a alias-table.

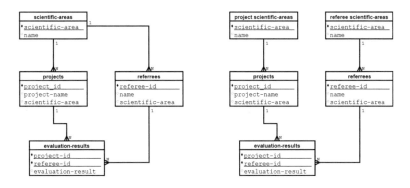

Fig. 5. Split table technique

3.2 Second Denormalization Form: Divide into Trees

The second denormalization form (2DF) is achieved when the database acquires the shape of one or many trees. In a tree database schema, with only one fact table, the fan trap does not occur.

In Fig. 6a a poly-tree, in the first denormalization form, is shown. Tables B and I are fact tables, because they only have relations with cardinality N. If we have more than one fact table the schema is not in the second denormalization form. The join of the tables C |><| D |><| G will cause a fan trap, so the poly-tree should be divided into two trees represented in Fig. 6b.

In the second normal form any join can be performed in each tree, like A |><| B |><| C |><| E |><| F or I |><| J |><| L |><| K |><| H.

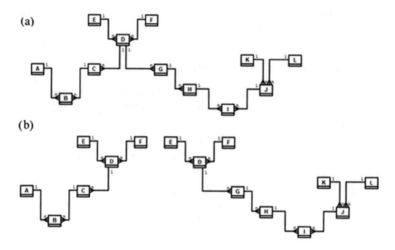

Fig. 6. Database denormalization: (a) poly-tree with two fact tables, (b) two trees

4   Case Study

To illustrate the proposed method, the MP3 database schema, with 16 tables, presented in (Ramos 2006, p. 74) will be used. The database stores information about MP3 play-lists with songs from album tracks performed by different artists. To visualize the database, in Fig. 7, we divide the tables into three levels: lookup tables level, inter-mediate tables level and fact tables level with 8, 3 and 5 tables respectively. In the same figure, multiple-path and fan trap schema can be identified.

To reach the first denormalization form (1DF) a poly-tree database schema must be found. For the MP3 database the following steps were performed:

- remove all lookup tables in order to simplify the visualization; after the denor-malization the lookup tables can be included;
- applying the split strategy to remove the multiple-path, by duplicating table Album_1 related to table Track;

Figure 8 presents the MP3 database in the 1DF with a poly-tree structure. In the 1DF the database avoids the semantic ambiguity of multiple paths, but given the number of linked fact tables fan trap can occur.

To reach the second denormalization form (2DF) a tree database schema must be found. For the MP3 poly-tree the following steps were performed:

- apply split technique by duplicating table Artist and isolating the tree (Artist_1, Composition);
- apply split technique by duplicating tables Artist, Album and Track and isolate the tree (Album_2, Track_1, Artist-Track, Artist_2);
- the previous procedures isolated tree (Album, Album-Artist, Artist) and tree (Album_1, Track, PlayList).

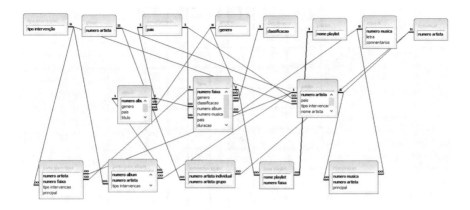

Fig. 7. MP3 database presented in three levels: lookup tables level, intermediate tables level and fact tables level

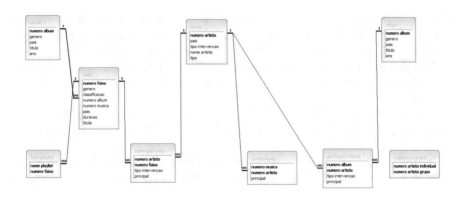

Fig. 8. MP3 database in the 1DF with a poly-tree structure

With the database in the 2DF, all the join operators can be performed without any restriction. Each fact table can be joined with the related tables. Given the 5 fact tables and the 8 lookup tables, a matrix fact tables versus dimensions (or lookup tables) can be drawn in order to give an overview of the database, as it is usual in data warehousing (Fig. 9).

fact tables/ dimensions	1	2	3	4	5
1	X		X		X
2		X		X	
3			X	X	X
4	X	X			
5	X			X	X
6		X	X	X	
7		X			
8			X		X

Fig. 9. Matrix fact table vs dimensions

5 Conclusions

The relational model is one of the most widespread ways to represent data. However, some problems can occur with the join operator: semantic mistakes caused by the multiple access path problems (MAPP) and faults when fan traps arise.

The goal of this paper is to make a review of some join operator problems and to establish two denormalization forms in order to extract fact tables from a relational database.

The first denormalization form (1DF) is achieved when the database acquires the shape of a poly-tree. The poly-tree database schema avoids the multiple access path problem.

The second denormalization form (2DF) is achieved when the database acquires the shape of one or many trees. In a tree database schema, with only one fact table, the fan trap does not occur.

This contribution is an effort to establish formal rules in denormalization and is relevant in ETL (extract, transform, load) for Data Warehousing (DW) and Business Intelligence (BI), but also for geographic information systems (GIS), object oriented tools and web environments and other systems that reuse relational data.

References

Business Objects: Designer's Guide, XI Release 2 (2007)

Cardenas, A.F.: Data Base Management Systems, 2nd edn. Allyn and Bacon, Boston (1985)

Codd, E.F.: A relational model of data for large shared data banks. Commun. ACM **13**(6), 377–387 (1970)

Darwiche, A.: Modeling and Reasoning with Bayesian Networks. Cambridge University Press, New York (2009). ISBN 0521884381

Date, C.J.: Introduction to Database Systems, 6th edn. Addison-Wesley Publishing Company, Reading (1995). ISBN 020154329X

Feng, J., Crowe, M.: The notion of 'classes of a path' in ER schemas. In: Proceedings of Third East European Conference on Advances in Databases and Information Systems, ADBIS 1999, pp. 218–231. Springer, Berlin (1999)

Karnitis G., Arnicans, G.: Migration of relational database to document-oriented database: structure denormalization and data transformation. In: 7th International Conference on Computational Intelligence, Communication Systems and Networks (CICSyN) (2015). https://doi.org/10.1109/cicsyn.2015.30

Kimball, R., Ross, M.: The Data Warehouse Toolkit: The Definitive Guide to Dimensional Modeling, 3rd edn. Wiley and Sons Inc, Hoboken (2013). ISBN 9781118530801

Kingdon, A., Nayembil, M.L., Richardson, A.E., Smith, A.G.: A geodata warehouse: using denormalisation techniques as a tool for delivering spatially enabled integrated geological information to geologists. Comput. Geosci. 96, 87–97 (2016)

Pearl, J.: Reverend Bayes on inference engines: a distributed hierarchical approach. In: Proceedings American Association of Artificial Intelligence National Conference on AI, Pittsburgh, PA, pp. 133–136 (1982)

Pearl, J.: Fusion, propagation, and structuring in belief networks. Artif. Intell. 29, 241–288 (1986)

Ramos, P.N.: Desenhar Bases de Dados com UML. Edições Sílabo, Lisbon (2006). ISBN 9789726184744

Rebane, G., Pearl, J.: The recovery of causal poly-trees from statistical data. In: Proceedings of Uncertainty in Artificial Intelligence, pp. 222–228 (1987)

Russell, S.J., Norvig, P.: Artificial Intelligence: A Modern Approach, 2nd edn. Pearson Education, London (2003). ISBN 0137903952

Ter-Bekke, J.H.: Semantic Data Modeling. Prentice Hall, New York (1992). ISBN 0138060509

Wald, J.A., Sorenson, P.G.: Resolving the query inference problem using Steiner trees. ACM Trans. Database Syst. (TODS) 9(3), 348–368 (1984)

Wu, X., Ichikawa, T., Cercone, N.: Knowledge-Base Assisted Database Retrieval Systems. World Scientific, Singapore (1996). ISBN 978-981-02-1850-8

Deep Learning in State-of-the-Art Image Classification Exceeding 99% Accuracy

Emilia Zawadzka-Gosk[1], Krzysztof Wołk[1(✉)],
and Wojciech Czarnowski[2]

[1] Polish-Japanese Academy of Information Technology, Warsaw, Poland
{ezawadzka,kwolk}@pja.edu.pl
[2] Jatar, Koszalin, Poland
wcz@jatar.com.pl

Abstract. Automatic image recognition and classification is a field of science that became popular in the recent years. Free platforms as Google Collaboratory make machine learning experiments more available to perform for everyone. The current technology enables us to use image recognition in such domains as medicine, criminology, entertainment or trading. In our research we created a state-of-the-art image classifier based on convolutional neural network model to classify ten models of old polish cars. We elaborated eight step training to fine tune the neural network. As the first step the data augmentation and precomputed activations were enabled. After that we froze all the layers but the last one, found the proper learning rate and performed interactive training. Fine-tuning, proper training and appropriate data preparation brought great results. The accuracy of cars' model recognition exceeded 99% with some room for improvement.

Keywords: Deep learning · Image classification · Interactive training ·
Data augmentation · Transfer learning

1 Introduction

Image recognition is a fast-developing part of science [1, 7]. Even few years ago, discovering specified patterns on the pictures was difficult and challenging assignment. Only projects with strong development and scientific teams were able to research this subject. Huge amount of data was also required to perform such experiments. Nowadays, thanks to the development of technology everyone can work on image recognition and classification.

Image recognition becomes an important issue in different domains like medicine [8], security [3], entertainment, trade [14] or science. For example, there are several visual diagnostic methods as MRI [5, 6], CT, USG or X-ray [9, 10]. All of them provide for the doctor valuable information about a patient, but not all present patterns are visible to human eyes. The support of automatic image recognition and classification can be a valuable help.

© Springer Nature Switzerland AG 2019
Á. Rocha et al. (Eds.): WorldCIST'19 2019, AISC 930, pp. 946–957, 2019.
https://doi.org/10.1007/978-3-030-16181-1_89

Recognizing human faces [4] can be broadly used in the criminology and security. It can be easily imagined that such technology is utilized by police or security systems. Nowadays it is commonly used for entertainment in smartphones or social media portals, as Facebook.

Object detection is a subject for different everyday activities support, from autonomous cars, assistants for blind or disabled people to detecting systems.

Online trading portals as eBay [14], have the ability to search or verify products available on their website.

Scientists and analysts, working with huge amount of visual data, use classification and recognition for the research.

Recently, the most interesting methods of image recognition are convolutional neural networks. Due to larger availability of big amount of data and great computing power, neural networks are employed more often into tasks of recognition. As the results of CNN usage are highly rewarding, we decided to use convolutional neural network in our work. In our experiment we focused on classifying old polish car brands. The results we got are very satisfying. The ratio of correctly classified images is very high.

2 Environment and Dataset

We performed our experiment in Google Collaboratory, the environment provided by Google to experiment with machine learning. It enables to use nVidia K80 graphic cards for computations. It is built on top of Jupyter Notebook, therefore work is convenient and easy to share with other researchers.

Moreover, the PyTorch machine learning library was used in our trainings. We also employed fast.ai, the library to neural network training which facilitates and accelerates the process by including best practices and recent scientific research results [12].

In our work we used dataset containing 10 models of old polish cars:

- Autobus Jelcz MZK
- Autobus Jelcz Ogorek
- Fiat 125p
- Fiat 126p Maluch
- FSC Zuk
- FSO Polonez
- FSO Warszawa
- Gazik
- Syrena 105
- ZSD Nysa.

It consists of 7300 photos of the cars with included information of their brand and 800 photos not signed, prepared for testing. Resolution of the images was reduced to only 224 × 224 px for faster computing.

3 Model and Training

For our experiments we used ResNet-34 neural network [2]. It is a convolutional deep network that won the ImageNet 2015 contest. Based on it we prepared a good quality image classifier for old Polish cars' models.

Fine tuning was performed on the ResNet. At first, the last layer of the network was replaced with the new one with 10 outputs (instead of 1000 present in original model). We decided to fit existing model to our data by finding two hyperparameters: learning rate and number of epochs. Hyperparameters describe high-level properties of our model, its complexity and learning pace. They cannot be learnt from the ordinary learning process, but by adjusting the model to the dataset. To perform it we used gradient descent method [15].

To train the network we elaborated eight step training that includes:

1. Enabling the data augmentation and precomputed activations for the training.
2. Finding the highest value of learning rate where the loss value is still decreasing.
3. Freezing all the layers except the last one and training the last layer for 1–2 epochs with precomputed activations.
4. Training the last layer for 2–3 epochs with data augmentation and cycle_len parameter equals to 1.
5. Unfreezing the frozen layers.
6. Setting 3–10 lower learning rates for the previous layers than for the next layers.
7. Finding again the highest value of learning rate where the loss value is still decreasing.
8. Training the whole CNN with cycle_mult parameter equals to 2 until network overfitting.

Figure 1 presents the schema of fine-tuning method performed on our CNN. It needs to be accentuated, that in the first steps, only the last layer is trained, whereas the previous ones are frozen.

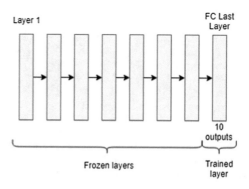

Fig. 1. Schema of CNN during fine-tuning training method

As the first step of our network training we enabled data augmentation and pre-computed activations. Next, we found learning rate (speed of our weights updates) using a technique developed in 2015 [13], based on increasing learning rate, starting from very small value and finishing when the loss stopped decreasing.

To validate learning rate of a new (not trained) model, we created new 'learner' and displayed learning rate 'training plan'. Figure 2 displays the learning rate value depending on iterations of SGD algorithm [15].

Additionally, we checked loss value to discover a point where the loss is decreasing (Fig. 3). It occurred that the decreasing tendency of the loss value is visible till learning rate value equals 0.1.

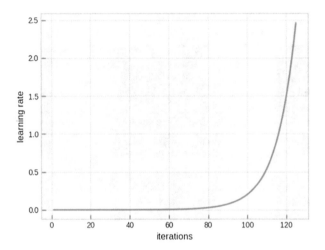

Fig. 2. Learning rate of newly created learner

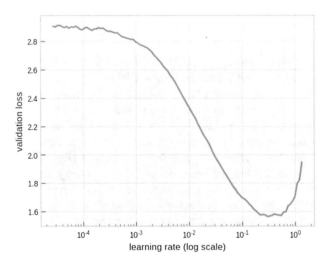

Fig. 3. Loss rate of newly created learner

Analysis of learning rate and loss values made us decide to set the learning rate at 0.08. All the layers of the CNN, except the last one, were frozen.

We trained last layer of the neural network with chosen learning rate for 1 epoch, values of loss and accuracy were as follows: loss = 0.24572 and accuracy = 0.9173.

Continuing to train the network for more epochs we discovered that the network got over-trained. It means that our model was recognizing individual images from our training set instead of generalizing. It could not be applied to our validation set with have good results. We decided to create more data by augmentation of transforms (Fig. 4). The fast.ai library enables to perform augmentation simple and fast, by setting 'aug_tfms' parameter to tfms_from_model. We decided to use transforms_side_on value and max_zoom = 1.1, to rotate, flip, change the lighting and enlarge of the pictures.

Fig. 4. Images after augmentation process

After applied data processing we could train the last layer of our CNN by 3 epochs. It resulted with loss = 0.20907 and accuracy = 0.9295.

We also set cycle_len parameter to 1. In stochastic gradient descent with restarts (learning rate annealing) which we used, it is possible to change the part of the weights' space to find more stable and accurate weights. It is called 'restart'. To force the model to move to another weights' space learning rate is increased rapidly. The cycle_len parameter describes the number of epochs between restarts (Fig. 5).

As we already trained the last layer, we would like to adapt the previous layers.

In the model we used previous layers are already trained to recognize images, so we needed to perform carefully to not destroy computed weights. As the previous layers works more generally, they need less tuning then the last ones did. Therefore, we used different learning rates for different layers. First ones: 0.0001, middle layers 0.001 and FC layers 0.01.

Additionally, we used cycle_mult parameter to multiply the length of the learning cycle. The Fig. 6 presents the learning rate changes for the last layers where cycle_mult was set to 2.

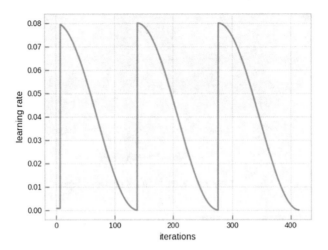

Fig. 5. Learning rate values during training with cycle_len = 1

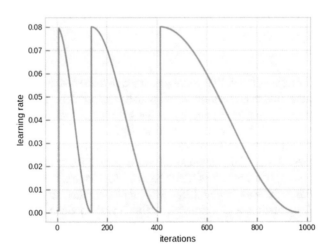

Fig. 6. Learning rate values of the last layers during training with cycle_mult = 2.

After those operations we found that our loss value decreased significantly to 0.0456 (Fig. 7).

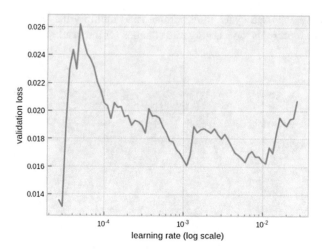

Fig. 7. Loss rate after training previous layers

We continued the training till the network was becoming over-trained. Final accuracy value we received was 0.9918.

To verify our network, we decided to perform Test Time Augmentation [16]. The TTA created predictions based on original image from our dataset and four augmentations of it. The mean prognosis from all those images was also equal to 0.99.

4 Results

The results of our tests without TTA were presented in a confusion matrix (Fig. 8). The matrix displays the final classification of old polish cars' models. The results presented on the chart are very good, vast majority of the car images was recognized correctly. Only few images occurred to be categorized wrongly.

In 98 samples of 'Autobus Jelcz MZK' 97 was classified correctly, only one was confused with 'ZSD Nysa'. Similar situation was observed for 'Autobus Jelcz Ogórek', which also was wrongly classified as 'ZSD Nysa'. 'FSO Polonez' was mistaken in one photography with 'FSC Zuk'. 'Fiat 125p' was once classified as 'Fiat 126p Maluch' and the opposite situation also occurred. 'ZSD Nysa' model was also labelled as 'Gazik'. The model with the highest number of classification errors was 'Gazik', 2 of 98 samples were classified incorrectly. Other models were categorized with the accuracy of 100%.

Having such high value of accuracy, made us more intrigued what images were problematic in classification process. We decided to check what images became recognized correctly and incorrectly in our initial model of CNN, after the initial training of the last layer. We displayed sample pictures of Syrena 105 model which were classified correctly, incorrectly and with uncertain predictions.

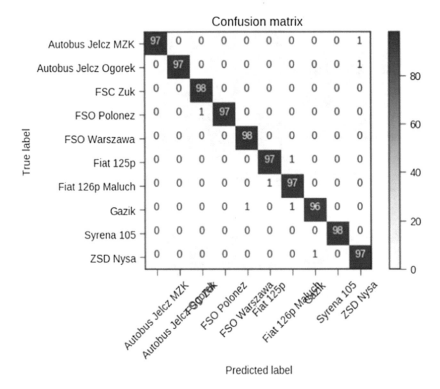

Fig. 8. Confusion matrix of old polish car model's classification system

Figure 9 presents correctly classified Syrena 105 model. According to the image quality the accuracy of the prediction is diverse. Cars on the first and third images were predicted with high accuracy, over 99%. The vehicles are in the middle of the photography, although the angle of the photoshoot is different. The second image was predicted only for 66.48% to be Syrena 105, but object on the picture is not centred and the photography is probably a part of a larger image.

Images where the photoshoot was made from the front side of the car have the highest probability values, maybe because the hood of this model is very specific. In examples presented in the Fig. 10 the system was 100% sure of the model of a car.

The black-white or negatives photos impede the recognition of the car type. Shot at an unusual angle also seems to be problematic for recognition. Examples of such incorrect classifications are present in the Fig. 11.

Figure 12 presents most incorrectly categorized cars. All the photos present Syrena 105, but all are classified as other model with quite high probability. Although, it has to be emphasized that classifying car models on these photographs is a challenging task also for human. The photo negative was a very problematic issue to be classified. The CNN classified it as a Fiat 126p with 40% accuracy. Also, the photos where the car was partially covered, or vehicle integrity was modified were categorized incorrectly.

Fig. 9. Examples of correctly classified Syrena 105 images

Fig. 10. Examples of correctly classified Syrena 105 images with 100% prediction

The example of images that received the most ambiguous results are presented in the Fig. 13. All the prediction values are close to 50%. Some of the cars are classified correctly as Syrena 105, but other two as Fiat 126p and Fiat 125p. Again, the unusual angle and the car being just a background of an image caused the problems with model recognition.

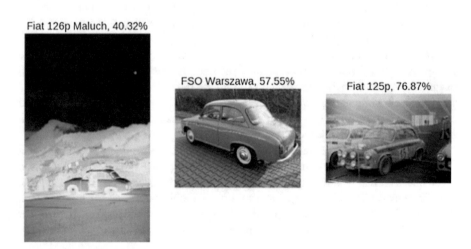

Fig. 11. Examples of incorrectly classified Syrena 105 images

Fiat 126p Maluch, 40.32%

FSO Warszawa, 41.73%

Fiat 126p Maluch, 47.77%

Fig. 12. Examples of most incorrectly classified Syrena 105 images

Syrena 105, 48.9%

Fiat 126p Maluch, 47.77%

Fiat 125p, 52.54%

Fig. 13. Examples Syrena 105 images uncertain predictions

5   Discussion and Conclusion

In this paper we introduced the state-of-the-art image classifier. Using Google Col-laboratory, PyTorch, fast.ai and the ResNet-34 neural network model sophisticated and effective image classifier was built. To create it we followed eight steps, including data augmentation, hyperparameters fitting, training of the last and previous neural network layers. We adapted learning rate and cycle parameters using stochastic gradient descent with the restarts method. Providing dataset of over 7000 images of old polish cars the good quality image recognition system was created with accuracy value about 99.18%, which means that the vast majority of the images from validation set was classified correctly. Analysing the individual cases where neural network might have difficulties with classification, we have noticed that the cars on the pictures were difficult to recognize even for humans. Different angles of the photoshoots, partially covered cars or unusual colours of the image seems to be the problem for classification task, but our final solution managed to handle with the issue. After applying our training strategy on the ResNet-34 neural network model we reached more satisfying results, where, according to the confusion matrix, all the pictures of Syrena 105 model were classified correctly.

As our solution still makes incorrect classifications, we plan to continue our research on the issue. Further experiments using ResNet-50 NN model are expected to bring better results, although we will need to use the images of higher resolution like 512 × 512 px or even greater. The accuracy of 99% is very high but still leaves room to improvement. One of the vulnerabilities of our model is poor augmentation. Only basic transformations as rotating, flipping, change of the lighting or picture enlarging were used. Applying more sophisticated graphic editor modifications for training set images can result in better network training. We are thinking about some Photoshop augmentation based on scripts and advanced filters. Finally, we would like to take more closer look at the training parameters of our convolutional neural network and the quality of the training data to improve our final accuracy. Our future work will also expand the scope of our research to the images from other domains, e.g. medical diagnostic images training datasets [11], where high accuracy of implemented solution is a crucial aspect of usefulness.

References

1. Simonyan, K., Zisserman, A.: Very deep convolutional networks for large-scale image recognition. arXiv preprint arXiv:1409.1556 (2014)
2. He, K., Zhang, X., Ren, S., Sun, J.: Deep residual learning for image recognition. In: Proceedings of the IEEE Conference on Computer Vision and Pattern Recognition, pp. 770–778 (2016)
3. Daugman, J.: Face and gesture recognition: overview. IEEE Trans. Pattern Anal. Mach. Intell. 19(7), 675–676 (1997)
4. Hong, Z.Q.: Algebraic feature extraction of image for recognition. Pattern Recogn. 24(3), 211–219 (1991)
5. Hofmann, M., Steinke, F., Scheel, V., Charpiat, G., Farquhar, J., Aschoff, P., Brady, M., Schölkopf, B., Pichler, B.J.: MRI-based attenuation correction for PET/MRI: a novel approach combining pattern recognition and atlas registration. J. Nucl. Med. 49(11), 1875 (2008)
6. Clarke, L.P., Velthuizen, R.P., Phuphanich, S., Schellenberg, J.D., Arrington, J.A., Silbiger, M.M.R.I.: MRI: stability of three supervised segmentation techniques. Magn. Reson. Imaging 11(1), 95–106 (1993)
7. Caputo, B., Dorko, G.: How to combine color and shape information for 3D object recognition: kernels do the trick. In: Advances in Neural Information Processing Systems, pp. 1399–1406 (2003)
8. Tadeusiewicz, R., Ogiela, M.R.: Automatic image understanding a new paradigm for intelligent medical image analysis. Bio-Algorithms Med-Syst. 2(3), 3–9 (2006)
9. Jain, A.K., Chen, H.: Matching of dental X-ray images for human identification. Pattern Recogn. 37(7), 1519–1532 (2004)
10. Kovalev, V., Liauchuk, V., Kalinovsky, A., Shukelovich, A.: A comparison of conventional and deep learning methods of image classification on a database of chest radiographs. Surgery 12(1), S139–S140 (2017)

11. Kumar, A., Kim, J., Lyndon, D., Fulham, M., Feng, D.: An ensemble of fine-tuned convolutional neural networks for medical image classification. IEEE J. Biomed. Health Inf. **21**(1), 31–40 (2017)
12. Howard, J., Ruder, S.: Fine-tuned Language Models for Text Classification. arXiv preprint arXiv:1801.06146 (2018)
13. Smith, L.N.: Cyclical learning rates for training neural networks. In: 2017 IEEE Winter Conference on Applications of Computer Vision (WACV), pp. 464–472. IEEE, March 2017
14. Find It On eBay: Using Pictures Instead of Words, 26 July 2017. https://www.ebayinc.com/stories/news/find-it-on-ebay-using-pictures-instead-of-words/. Accessed 14 Nov 2018
15. Johnson, R., Zhang, T.: Accelerating stochastic gradient descent using predictive variance reduction. In: Advances in Neural Information Processing Systems, pp. 315–323 (2013)
16. Howard, J.: Now anyone can train Imagenet in 18 minutes, 10 August 2018. https://www.fast.ai/2018/08/10/fastai-diu-imagenet/

Absenteeism Prediction in Call Center Using Machine Learning Algorithms

Evandro Lopes de Oliveira[1], José M. Torres[1,2], Rui S. Moreira[1,2,3(✉)],
and Rafael Alexandre França de Lima[4]

[1] ISUS Unit, University Fernando Pessoa, Porto, Portugal
{evandro.oliveira,jtorres,rmoreira}@ufp.edu.pt
http://isus.ufp.pt
[2] LIACC, University of Porto, Porto, Portugal
[3] INESC-TEC, FEUP - University of Porto, Porto, Portugal
[4] Federal University of Minas Gerais, Belo Horizonte, Brazil
rafaelfrancalima@gmail.com

Abstract. Absenteeism is a major problem faced particularly by companies with a large number of employees. Therefore, the existence of absenteeism prediction tools is essential for such companies depending on intensive human-resources. This paper focuses on using machine learning technologies for predicting the absences of employees from work. More precisely, a few prediction models were tuned and tested with 241 features extracted from a population of 13.805 employees. This target population was sampled from the help desk work force of a major Brazilian phone company. The features were extracted from the profile of the help desk agents and then filtered by processes of correlation and feature selection. The selected features were then used to compare absenteeism prediction given by different classification algorithm (cf. Random Forest, Multilayer Perceptron, Support Vector Machine, Naive Bayes, XGBoost and Long Short Term Memory). The parameterization of these ML models was also studied to reach the classifier best suited for the prediction problem. Such parameterizations were tuned through the use of evolutionary algorithms, from which considerable precision was reached, the best being 72% (XGBoost) and 71% (Random Forest).

Keywords: Absenteeism · Call center · Feature selection · Machine learning · Algorithms and classifiers

1 Introduction

The telephone service centres have a large amount of data about their employees, and this is interesting for researches in some domains, including the prediction of indicators for business management and planning. The prediction is important because the operational challenges faced by call centres managers have become more complex over time exactly by the variety and complexity of these management indicators.

© Springer Nature Switzerland AG 2019
Á. Rocha et al. (Eds.): WorldCIST'19 2019, AISC 930, pp. 958–968, 2019.
https://doi.org/10.1007/978-3-030-16181-1_90

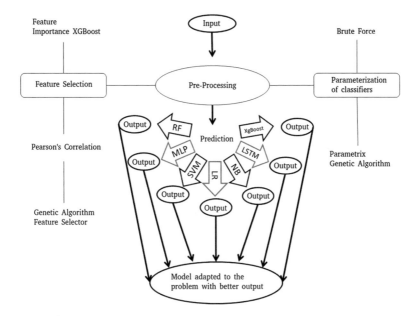

Fig. 1. Schematization to find classifier best adapted to the problem

Absenteeism is an important indicator since its high rate affect companies negatively. This indicator may represent the lack of commitment and engagement of the employees but also highlight the need for the company to promote counter measures against such problem [6].

Understanding the real causes leading professionals to become absentees is a complex task that depends on a large number of employee's variables. It is hard to find or characterize an absentee profile. In this paper a few computational techniques based on machine learning algorithms are tested to predict the absentee behavior of 13805 employees from a call center company that operates in different regions of the Brazilian territory. For that, all the available variables of the telemarketing company employees were searched for selecting 241 features. Then, these variables were pre-processed in order to select the ones that would most correlate with absenteeism and thus submit them to the models.

Besides evaluating the correlation of the variables with absenteeism, the final selection was performed in line with the evaluation of the best prediction results. This was accomplished by comparing the submission of the variables, in different combinations, to the predictive models. The tested machine learning algorithms were selected from common existing classifiers, namely: Random Forest, Multilayer Perceptron, Support Vector Machine, Naive Bayes, XGBoost, Long Short Term Memory and Logistic Regression.

The evolutionary algorithms were used together with the models to optimize the way they are parameterized. Several, randomly selected, combinations of parameters were used for comparison. As there were several combinations in the selection of variables, the same evolutionary algorithms were responsible for the selection of inputs in the models.

Finally, the conducted experiments show the best algorithmic solutions to be used with the available data. These algorithms may allow to predict the absenteeism of employees from call centres companies, and potentially apply them in other market sectors.

The scheme that represents the steps of the research can be seen in Fig. 1.

The remaining of the paper is organized as follows: Sect. 2 describes the data pre-processing mechanisms; Sect. 3 details the selection of variables; In Sect. 4, the prediction itself and in Sect. 5, results and conclusions.

2   Input and Output Data

The main goal of this research is to find out the probability of an employee being absentee at work for one or more days in a future period of seven days. This study considers an absentee employee to be someone who may have an absence greater than 50% in the planned work hours of a day.

For this, a lot of information was collected about employees, available at the target company. The study was conducted in finding correlations of the available information with absenteeism.

For this research, 241 variables (Table 1) were collected from the personal and professional profile of each employee. More precisely, the collected information focused on personal data, behavioral variables, variables directly related to absences and delays, variables related to productivity and personal variables related to the company's operational environment, as well as data obtained from some internal questionnaires.

Regarding data temporality, the set of input features from employees is formatted on a weekly-based data frame. Moreover, each data frame carries fixed and varying information about each employee (cf. Table 1). The varying features related with temporal information incorporate the absenteeism of each employee on previous weeks (cf. past absenteeism on each of the eight previous days). The fixed and varying information from the selected features is then used on a training and validation process following a nested cross-validation approach to avoid data leakage, similarly to the work described in [3].

Much of the information has been extracted from an administrative social platform (cf. Robbyson) used by employees that unites all the members of the company.

2.1   Transformation of Variables

Since these variables contain a finite number of categories, each of the categories were associated with a new variable indicating if it was *valid* or *not valid* for a particular context (cf. a value of 1 was defined as true and 0 as false).

Table 1. General features of employees

Personal Features		Work activities features		Social and admin platform features and administrative platform		Absenteeism related features		Grand total
Feature	Qty	Feature	Qty	Feature	Qty	Feature	Qty	
Individual registration	2	City of work	2	Productivity	26	Holidays	2	
Has landline	2	Distance from work	2	Mood	21	Absences at work	14	
Dependents	2	Work sector	5	System access	2	Weekly absences at work	7	
Internal question- naire	37	Work shift	4	Virtual store and character interaction	5	Absence of friends at work	1	
Individual and man- agement assessment	18	Instant manager	7	Interaction messages	9	Days to last rest	1	
Education	8	Productivity level	36	Friends on the social network	1	Last day worked	1	
Marital status	8	Worked hours	1	Login Feature	1			
Origin of person	10	Business Time	1					
Age	3	Hiring Disclosure	1					
Sex	1							
Total	**91**		**59**		**65**		**26**	**241**

In the transformation process some discrete variables were created and categorized to be submitted to the models in the same way as the binarized categorical variables previously mentioned. Some of the examples of variables created by categorization are related to productivity variables.

3 Selection of Variables

For real problems, the selection of variables is important because it speeds up the learning process and improves its quality [13].

For the present research, 3 ways of selecting variables were analyzed. These are Pearson's correlation, the genetic selector of variables algorithm and an internal variable of feature selection provided by the algorithm XGBoost that assigns importance to the features.

3.1 Pearson's Correlation

The correlation between data sets is a measure that tells how closely they are related. Among many methods of correlation analysis in practice, Pearson's product moment correlation coefficient (PPMCC) is perhaps the most prevalent and the most common measure of correlation in statistics [8].

The percentage of correlation with the output variable distributed in Table 1, from the 5 most correlated variables with absenteeism, can be seen as follows:

- 16,5% - variables related to absenteeism;
- 12% - variables related to the group of time logged;
- 10% - variables related to the percentage of bad mood;
- 9% - variables related to the virtual currency bonus;
- 8% - variables related to work break.

3.2   Genetic Selector Algorithm

The selection of variables using genetic algorithms generally assumes that there is no auto-correlation between variables [25]. This independence factor introduces a processing delay, considering the objective function used that corresponds to the value generated by the model itself.

The process followed on the experiments supporting this research is presented in Fig. 2. The result generated by one of the chosen models, is the individual's aptitude index in the population evaluation.

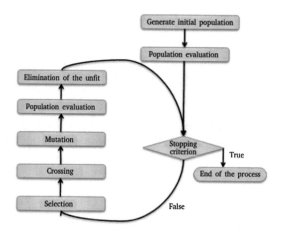

Fig. 2. Genetic algorithm

There is a set of specific variables for each model used as an algorithm that generates aptitude for the population individuals.

An individual in the population is the set of all input variables, being each of them a gene, as can be seen in Fig. 3. This figure also illustrates how cross-referencing between individuals is made. Each gene may have a value of 1 (*positive*) or 0 (*negative*), indicating if the variable will or will not fit into the classification model. At the end of the process the individual with the highest score is the one whose positively marked variables are those chosen by the selector. The experiments considered a population of 10 individuals with 241 genes each and 100 executed generations.

In most literature the parameters of the genetic algorithm found are in the range of 60 to 65% for the probability of cross-over and between 0.1 and 5%

for the probability of mutation. The size of the population and the number of generations depend on the complexity of the optimization problem and they must be determined experimentally [19], as it was done in the present work.

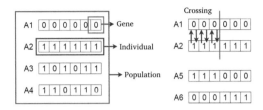

Fig. 3. Population model of the genetic algorithm

The evaluation criterion for the model applied to the present research was the area under the Receiver Operating variable Curve (ROC AUC). This is the measure that compares the precision of the classification. It is a general measure of predictability which decouples the classification assessment from the operating conditions [14].

The selection made by the genetic algorithm reached the variables that can be seen in Table 2. The results show the percentages of maintained variables after selection, considering 5 periods, and the associated ROC AUC.

Table 2. Percentage of features maintained by classification model

Models	% Features	ROC AUC
RF	25%	0,7315
MLP	15%	0,7215
SVM	19%	0,5024
NB	15%	0,6811
XGBoost	21%	0,7436
LSTM	17%	0,5801
LR	21%	0,7116

3.3 Selection of Variables in XGBoost Algorithm

One benefit of using sets of tree decision methods, such as gradient increase, is that they can automatically provide estimates of importance to the variables of a trained predictive model. It is in the construction of the trees that it recovers the scores of importance for the variables [29]. Generally, the importance of each feature provided by the score indicates how useful or valuable each of them was in building decision trees within the XGBoost model. The more an attribute is used to make key decisions, with decision trees, the greater their relative importance [29].

4   Parameters Optimization of the Models

The absenteeism prediction is performed considering, as input, the chosen variables for each of the tested models. This is done by experimenting the selectors already mentioned in the previous section.

All the 6 prediction models used in the experiment had to undergo optimization techniques of their parameters, described below, to improve the obtained results. There are several possible combinations for each of them specifically.

Optimizing the parameter values of predictive models is a complex problem that is beyond the limits of human capability [28]. By using random combinations of parameters or even the literature standards to calibrate the models it can get interesting results, but each case has its peculiarity according to the information base used.

4.1   Simple Random Sampling

In the present paper, 50 random combinations of parameter values were collected characterizing Simple Random Sampling in an empirical experiment. In a pure brute force approach the total number of combinations would be exponential to the number of parameters of each model if this limit were not established experimentally. The limit of combinations was empirically defined by making comparisons to the output values. The values which presented the best results in the execution of the models were chosen among them.

4.2   Genetic Algorithm Based Parameterization

In the present experiment, the set of parameters defines an individual of the population with each of the parameters being a gene. It can be seen in Fig. 3, but in this case the gene is not binary and may assume one of the available values defined for the corresponding parameter domain.

The aptitude function of each individual is given by the values generated when the classification models are executed. The highest value determines the best individual and consequently the best parameters. The process follows the cycle of Fig. 2.

5   Absenteeism Possibility

The expected result at the end is a list of all employees of the company with their expectation of absence. This list represents the possibility of the employee to be an absentee in one of the seven days following the current observation.

The prediction classifier provides the probability of each employee to be absent (missing more than 50% of daily working hours, according to the rules in force in the company) in at least 1 day of work over the next seven days, disregarding those who are away for any legal reason or have some specific benefit that may be entitled to be late during the seven days of prediction.

The employee absentee comes from the possibility percentage that it can happen. In order to have a greater possibility of absenteeism, a cut must be established covering a smaller number of hit employees and thus the precision will be better. Making a cut of 80%, for example as in Fig. 5, the absentees sample would be smaller than the selected sample in the cut of 50% applied to this work, but the accuracy would be better. The smaller the possibility of chosen absenteeism, the lower the accuracy and the greater the number of employees reached by the experiment. Thus, a cut of 50% is a balance between the number of reached employees and the accuracy (Fig. 4).

Fig. 4. Possibility of employee absenteeism

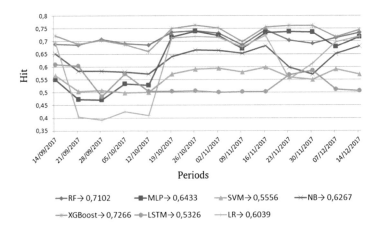

Fig. 5. AUC ROC of the classification models by period

The results of the prediction were verified considering all seven-day periods with the possibility above 50% of each employee to commit absenteeism. At the end it was demonstrated how much the results were assertive according to the output values of the experimented models.

Figure 5 represents the precision of the models tested in the seven-day periods in the three-month interval.

6   Conclusion

The reported approach for predicting absenteeism of a large number of employees, combines the parameterization of classification models and feature selection to generate the best results for this particular application scenario. The proposed work was successful in finding the more suited classifiers for the specific absenteeism problem, but also on providing a generic architecture and framework for similar prediction scenarios and problems. The created framework can easily be adapted to predict other important impact factors on enterprise business models.

From our experiences it became clear that the learning behavior of the ML models tends to be more precise when applying the genetic algorithms. The use of the evolutionary algorithms allowed to combine the best parameter values and the more suited features for obtaining the highest prediction results. This work evaluated seven ML classification models and showed that, although they are sensitive to the correlation of the input features with the response variable, many of them have a higher impact on the results when they combined strategically.

The model that obtained the best results for predicting absenteeism was the XGBoost algorithm with an AUC value greater than 72%, followed by the Random Forest model with an AUC ROC value greater than 71%. The results obtained from this experiment suggest that algorithms based on tree decision may be the best adapted to this type of prediction problem.

Even though there are other projects predicting the performance of employees with interesting results [10, 21, 30], the presented study stands out precisely for the scenario and techniques used to approach the problem. More specifically, other studies do not present a thoroughly and diverse selection of features to predict the absenteeism on an expressive amount of collaborators. In addition, this study combines the feature selection with a careful and systematic choice of algorithm parameterization, within the mentioned context, to obtain the best absenteeism prediction results.

Acknowledgements. This work was partially funded by: FCT-Fundação para a Ciência e Tecnologia in the scope of the strategic project LIACC-Artificial Intelligence and Computer Science Laboratory (PEst-UID / CEC/ 00027/ 2013); and by Fundação Ensino e Cultura Fernando Pessoa.

References

1. McCallum, A. Nigamt, K.: A comparison of event models for naive bayes text classification. In: AAAI-98 Workshop on Learning for Text (1998)
2. Carino, B.M.A., Saliby, S.E.: Prevendo a demanda de ligações em um call center por meio de um modelo de Regressão Múltipla - Gestão & Produção - SciELO Brasil (2009)
3. Courtney, C.: Time series nested cross-validation. Towards Data Science (2018). https://towardsdatascience.com/time-series-nested-cross-validation-76adba623eb9. Accessed Jan 2018
4. Hulme, C., Maughan, S., Brown, G.D.A.: Memory for familiar and unfamiliar words: evidence for a long-term memory contribution to short-term memory span. J. Mem. Lang. **30**(6), 685–701 (1991)
5. Hibbert, D.B.: Genetic algorithms in chemistry. Chemom. Intell. Lab. Syst. **19**, 277–293 (1993)
6. Cohen, A., Golan, R.: Predicting absenteeism and turnover intentions by past absenteeism and work attitudes - an empirical examination of female employees in long term nursing care facilities. Career Dev. Int. **12**(5), 416–432 (2007)
7. Cournapeau, D.: Scikit-learn. http://scikit-learn.org/stable/. Accessed Nov 2017
8. Mari, D., Kotz, S.: Correlation and Dependence. Imperial College Press, London (2001)
9. Sanchis, E., Casacuberta, F., Galiano, I., Segarra, E.: Learning structural models of subword units through grammatical inference techniques. In: ICASSP-91 Proceedings, pp. 189–192 (1991)
10. Jantan, H., Hamdan, A.R., Othman, Z.A.: Towards applying data mining techniques for talent managements. In: International Conference on Computer Engineering and Applications, IPCSIT, vol. 2, p. 2011. IACSIT Press, Singapore (2009)
11. Holland, J.: Adaptation in Natural and Artificial Systems. University of Michigan Press, Ann Arbor (1975)
12. Wainer, J.: Comparison of 14 different families of classification algorithms on 115 binary datasets. Campinas Campinas, SP, 13083-852, Brasil, 6 June 2016
13. Kira, K., Rendell, L.A.: The feature selection problem: traditional methods and a new algorithm. In: AAAI (1992)
14. Lessmann, S., Voß, S.: A reference model for customer-centric data mining with support vector machines. Eur. J. Oper. Res. **199**, 520–530 (2009)
15. Breiman, L.: Random Forest. Mach. Learn. **45**, 5–32 (2001). Kluwer Academic Publishers, Manufactured in The Netherlands
16. Dash, M., Liu, H.: Feature selection for classification. Intell. Data Anal. **1**(1–4), 131–156 (1997)
17. Mancini, L.: Call Center: estratégia para vencer. Summus editorial (2001)
18. Fernández-Delgado, M., Cernadas, E., Barro, S.: Do we need hundreds of classifiers to solve real world classification problems? J. Mach. Learn. Res. **15**, 3133–3181 (2014)
19. Miranda, M.: Algoritmos Genéticos: Fundamentos e Aplicações. Disponível em. http://www.gta.ufrj.br/~marcio/genetic.html. Acesso em: 20 outubro 2009
20. Friedman, N., Geiger, D., Goldszmidt, M.: Bayesian network classifiers. Mach. Learn. **29**, 131–163 (1997)
21. Rohit, P., Pankaj, A.: Prediction of employee turnover in organizations using machine learning algorithms a case for extreme gradient boosting. (IJARAI) Int. J. Adv. Res. Artif. Intelli. **5**(9), C5 (2016)

22. Riddle, P., Segal, R., Etzioni, O.: Representation design and brute-force induction in a Boeing manufacturing domain. Appl. Artif. Intell. Int. J. **8**(1), 125–147 (1994)
23. Guo, R., Abraham, A., Paprzycki, M.: Analyzing call center performance: a data mining approach. J. Knowl. Manag. **4**(1), 24–37 (2006)
24. Ivanir, C.. Pinto, F.R., Prado, P.K.R.M, Jose, S.R.: Um estudo sobre dashboard inteligente para apoio à tomada de decisão em uma empresa de courier. In: 13th CONTECSI International Conference on Information Systems and Technology Management (2016)
25. Leardi, R.: Application of genetic algorithm-PLS for feature selection in spectral data sets. J. Chemom. **14**, 643–655 (2000)
26. Leardi, R., Lupianez, A.: Gonzalez Genetic algorithms applied to feature selection in PLS regression. Chemom. Intell. Lab. Syst. **41**, 195–207 (1998). Lupianez Gonzalez
27. Leardi, R., Boggia, R., Terrile, M.: Genetic algorithms as a strategy for feature selection. J. Chemom. **6**(5), 267–281 (1992)
28. Smit, S.K., Eiben, A.E.: Comparing parameter tuning methods for evolutionary algorithms. In: IEEE Congress on CEC 2009 (2009)
29. Chen, T., Guestrin, C.: XGBoost: a scalable tree boosting system. In: Proceedings of the 22nd ACM SIGKDD International Conference on Knowledge Discovery and Data Mining, pp. 785–794 (2016)
30. Nagadevara, V., Srinivasan, V., Valk, R.: Establishing a link between employee turnover and withdrawal behaviours: application of data mining techniques. Res. Pract. Hum. Res. Manag. **16**(2), 81–97 (2008)
31. Vapnik, V.: The Nature of Statistical Learning. Springer, New York (1995)

Author Index

© Springer Nature Switzerland AG 2019
Á. Rocha et al. (Eds.): WorldCIST'19 2019, AISC 930, pp. 969–972, 2019.
https://doi.org/10.1007/978-3-030-16181-1

Printed in the United States
By Bookmasters